NATURE OF BIOLOGY

THIRD EDITION

Book 1

JUDITH KINNEAR MARJORY MARTIN

This edition published 2006 by
John Wiley & Sons Australia, Ltd
42 McDougall Street, Milton, Qld 4064

Offices also in Sydney and Melbourne

Typeset in 10.5/12 pt Times

First edition published 1992
Second edition published 2000

National Library of Australia
Cataloguing-in-publication data

Kinnear, J. F. (Judith F.).
Nature of biology. Book 1.

3rd ed.
Includes index.
For year 11 students in Victoria.
ISBN-13 978 0 7314 0236 6.
ISBN-10 0 7314 0236 7.

1. Biology — Textbooks. I. Martin, Marjory. II. Title.

570

Illustrated by Stephen Francis, Carolyn Gardiner, Craig Jackson, Paul Lennon, Janice McCormack, Jean Mulligan, Phil Parry, Terry St Ledger, Doone Wildin and the Wiley Art Studio
Cartography by MAPgraphics Pty Ltd, Brisbane and the Wiley Art Studio
Front cover image: Australian Picture Library/Corbis/Martin Harvey
Back cover and spine image: © PhotoDisc, Inc.
Internal design images: © Digital Vision/Martin Child, © PhotoDisc, Inc., © Viewfinder Australia Photo Library

Printed in China by
Printplus Limited

10 9 8 7 6 5 4 3 2

This book is dedicated
to friends and colleagues
who generously shared
their stories.
WATTLE
H4WKS

CONTENTS

UNIT 1

Unity and diversity

AREA OF STUDY 1 Cells in action

AREA OF STUDY 2 Functioning organisms

PREFACE

This third edition of *Nature of Biology Book 1* builds on previous editions that were positively received by teachers and students of biology. It has been thoroughly revised and updated and reflects current curriculum decisions with regard to key knowledge and skills expected of biology students.

This book continues to seek to convey a multifaceted sense of biology: as a rigorous scientific discipline with explanatory models that organise the living world for us in a meaningful way; as a dynamic science whose explanations are subject to change, rather than as a fixed and unchanging body of knowledge; as a science that impacts on everyday life, at the level of the individual where it can inform personal choices and at a societal level where it can inform community and government decisions.

We continue to emphasise recent developments in biotechnology as exemplified by the techniques of robotic surgery and the latest advances in microscopy. We have placed emphasis on case studies relevant to Australia, as well as cases of global interest. Relevant sites from the World Wide Web are identified for further investigation and research.

Included in each unit are examples to assist students to understand how biological knowledge and skills can be applied in a variety of settings. Profiles of 'Biologists at work' are also intended to increase student awareness of vocational opportunities.

We have enjoyed writing this book and we hope that our readers will also enjoy reading the text and exploring the visual images. This project was greatly enhanced by the generous cooperation of many colleagues, friends and acquaintances. In particular, we owe a special debt of gratitude to the following:

Associate Professor Leigh Ackland (Deakin University), Dr Michael Ackland (Public Health Division, Department of Health Services), Margaret Anderson, Professor Mike Archer, Ausma and Zaiga Augstsprogis, Australian Red Cross Blood Service, Dr Peter Beech (Deakin University), Dr Margaret Brumby (Manager, Walter and Eliza Hall Institute for Medical Research), Dr Lynda Campbell (St Vincent's Hospital), Nicholas Carlile (NSW National Parks and Wildlife Service), Professor David Cockayne FLS (Oxford University), Professor Suzanne Cory (Director, Walter and Eliza Hall Institute for Medical Research), Hilary Taylor Deayton, Ken and Margaret Dowling, Norah Fagan, Dr Karen Firestone (Australian Museum), Joyce Francis, Sister Jacqueline Fraser (Cabrini Hospital), Dr Christopher Gleeson, Associate Professor Dawn Gleeson (University of Melbourne), Dr Sue Hand (UNSW), Harcourt Family, Dr John Holland (Massey University), Professor Jeremy Hyams, Karen Inge (Victorian Institute of Sport), Dr Josephine Kenrick (University of Melbourne), Professor David Lambert (Massey University), Keryn Lapidge (Pest Animal Control CRC), Jean-Marc Lefebvre-Despeaux (BLUESTAR Forensic), Maurice Leslie, Ruth Leslie (Chisholm Institute of TAFE), Library Staff, Linnean Society of London UK, Jane McCooey (La Trobe University), Mark McGrouther (Australian Museum), Martin Family, Dr Bruce Maslin (CALM), Professor Julian Mercer (Deakin University), Dr Agnes Michalczyk (Deakin University), Ian Miller (AIMS), Dr Josephine Milne (Royal Botanic Gardens, Melbourne), David Noble (NSW NPWS), Julian Ophel (St Vincent's Hospital), Dr Jawahar Patil (CSIRO), Margaret Perring, Julia Quince (Massey University), Ray Riechelt (Little Desert Tours), Robotics Operating Theatre staff (Epworth Hospital), Lauren Starr (Pest Animal Control CRC), M. Suzanne Searls, Megan Short (Deakin University), Craig Sowden (The University of Sydney), Leonie Stanberg (National Herbarium of NSW), David Stuart (Carl Zeiss Pty Ltd), Dallas Sturtevant, Professor Martin Thompson (The University of Sydney), Professor A. J. Underwood (The University of Sydney), Verity Family, Graham Webb, Rob Weppler (Biological Services SA), Dr Jan West (Deakin University), Christopher Wilson (Monash University).

ABOUT THE CD-ROM

Features of the interactive *Nature of Biology Book 1* CD-ROM accompanying this textbook include:

- **Interactive text on CD-ROM**

The entire *Nature of Biology Book 1* textbook is on the CD-ROM in PDF format. There are many bookmarks provided to help navigate through the text and its accompanying resources.

- **Key term links**

Click on the bolded key words as they appear in the text to view their definitions. Click on the definition box to hide it.

- **Quick-check links**

Click on the 'Quick-check' headings to access these questions as downloadable Word files. Complete the answers on screen or print them out.

- **Chapter review questions**

Click on the 'Questions' headings on the chapter review pages to access these as downloadable Word files, which can be completed on screen or printed out.

- **Crosswords** 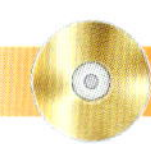 CROSSWORD

Near the 'Key words' heading on each chapter review page, a CD-ROM icon links to an interactive crossword for revision of the key terms for the chapter.

Minimum system requirements

Windows 98: Pentium II, 64 MB RAM

Windows 2K/XP: Pentium III, 128 MB RAM

Macintosh OSX: Power Macintosh G3 running OS 10.1.5, 10.2.6, 10.3, 128 MB RAM

Running the CD-ROM

Windows 98/2K/XP

Insert the CD-ROM into your CD-ROM drive.

Macintosh OSX

Insert the CD-ROM into your CD-ROM drive.

Double-click on the CD-ROM icon.

Double-click on *start.pdf*.

Troubleshooting

If you have problems with the operation of this CD-ROM:

- Check that you have the right equipment (see 'Minimum system requirements').
- Either phone, fax, email or write to John Wiley & Sons Australia.

  Phone:

  Multimedia Assistant 07 3859 9649

  Fax:

  07 3859 9755 (Attention: Multimedia Assistant)

  Email: multimedia@johnwiley.com.au

  Address:

  Multimedia Assistant
John Wiley & Sons Australia, Ltd
PO Box 1226
MILTON QLD 4064

ACKNOWLEDGEMENTS

The authors and publisher would like to thank the following copyright holders, organisations and individuals for their permission to reproduce copyright material in this book.

Images

• Associate Professor Leigh Ackland: **1** (right), **11** (bottom), **13** (bottom), **22**, **89** (photos and art) • Carl Zeiss Pty Ltd: **2**, **12** (photos), **16** (all), **17** (top) • Australian Picture Library: **3** (top) /Heritage Image Partnership Limited; **3** (centre) / Science and Society Picture Library; **3** (bottom) /Robert Visser; **6** (left) /Corbis/Bettmann; **6** (right), **7** (top left) / Bettmann; **102** /Leo Meier; **228** (centre) /Corbis/Stephen Frink; **257** (top left) /Corbis; **388** (right) /Chris Mattison; **477** (lower right) /Photo Index/Evan Collins • photolibrary.com: **7** (top right), **282** (top), **434** (bottom right) /Science Photo Library/Dr Jeremy Burgess; **11** (top left) /Oxford Scientific Films/Peter Parks; **11** (centre left) /Oxford Scientific Films/Alastair MacEwen; **23** (left) /Science Photo Library/Jan Hinsch; **23** (centre and right) /Science Photo Library/Steve Gschmeissner; **36** (photos), **358**, **368** (centre right) /Photo Researchers, Inc.; **37** (left), **47** (bottom right), **67** (right) /Photo Researchers, Inc./Biophoto Associates; **82** (top) /Science Photo Library/Andrew Syred; **87** (top) / Oxford Scientific Films/Harold Taylor ABIPP; **96** /Dr Kari Lounatmaa; **101** (right) /Gary Lewis; **132** (bottom), **144** (top), **438** (left) /Science Photo Library; **138** (centre) / Science Photo Library/James King Holmes; **180** (top) / Oxford Scientific Films/John Cheverton; **244** (top left) / Oxford Scientific Films/Lloyd Nielsen; **256** (forest) /Geoff Higgins; **355** (both) /Photo Researchers, Inc./A Mckeone Carolyn; **373** /Oxford Scientific/Tobias Bernhard; **375** (bottom) /Science Photo Library/Science Pictures Ltd; **378** (top left) /Peter Arnold Images/Seitre Roland; **379** (top) / Oxford Scientific/Robin Bush; **380** (centre) /Oxford Scientific/Clive Bromhall; **400** (lower left) /Peter Arnold Inc.; **401** (right) /Oxford Scientific Films/Stan Osolinski; **418** (right) /Science Photo Library/Astrid & Hanns-Frieder Michler; **419** (right) /Ted Mead; **420** /Oxford Scientific Films/David Fox; **421** /Oxford Scientific Films/R Jackman; **423** (bottom left) /Oxford Scientific Films/Michael Fogden; **434** (top) /Science Photo Library/British Technical Films; **437** /Manfred Thonig; **439** (top right) /Oxford Scientific/ Doug Allan; **471** /Pacific Stock/Doug Perrine; **495** (top left) /Oxford Scientific Films/David B Fleetham; **495** (top right) / Oxford Scientific Films/Javed Jafferji; **506** (centre left) / Science Photo Library/Ted Mead • © The Royal Society: **7** (centre right) • Courtesy of Royal Botanic Gardens, Kew: **7** (bottom left) • © The Brian J. Ford Science Website: **7** (bottom right) • The authors: **13** (top), **19** (centre and bottom left), **24** (right), **37** (lower right) /Photo by Teresa Dibbayawan, **43** (all), **45** (photos), **46** (tissues, left and right), **53** (bottom), **62**, **67** (bottom left), **70** (photos), **85** (left), **86** (both), **87** (bottom), **90** (photos), **98** (photos), **99** (both), **103** (left), **105**, **111** (left), **113**, **124** (photos), **128**, **129** (both), **145** (top), **149** (top right), **153** (both), **158** (centre), **164** (photos), **166** (top left), **167** (top), **175** (left) / Courtesy: Daisy Day, **181** (strawberry plant), **182** (bottom), **183**, **184** (twins), **187** (photos), **188** (top right), **210** (top and bottom right), **216** (bottom), **218** (bottom) illustrations by French artist Charles Le Sueur (out of copyright), **221** (top), **223**, **225** (centre left) Reproduced by permission of the Linnean Society of London, **226** (left), **231** (both), **237** (flowering plant and moss), **253** (all), **271** (photos), **272** (right), **303** (bottom), **331** (left), **332** (both), **333** (all), **334** (bottom), **389** (top), **390** (top), **397** (photos), **399** (both), **400** (lower right), **400** (top left and right), **412** (top left and right), **413** (bottom left, centre and right), **414** (top left), **415** (both), **422** (top), **430** (lower left), **432** (left), **434** (bottom left), **442** (right), **445**, **483** (photo), **505** (photos), **514**, **528** (bottom) /Photo by Margaret Dowling, **529** (bottom), **529** (top), **530** (top left and right, and lower left), **533**, **534** • Cuong Huynh and Megan Short, Deakin University: **15** (both), **19** (centre right) • Vincent A. Fischetti, PH.D, Rockefeller University, New York: **24** (left), **34** (top), **47** (top left) • © Viewfinder Australia Photo Library: **29** (frog) • Anne Simpson, NANO Major National Research Facility at the Electron Microscope Unit, The University of Sydney: **33** (lower right) • Julian Ophel, St Vincent's Hospital: **33** (top right), **34** (left), **35** (left), **47** (bottom left) • Courtesy Dr Peter Beech: **38** (both), **39**, **80** (photos) • John Wiley & Sons, Inc.: **50**, **71**, **109**, **135** (bottom), **138** (left), **146** (top), **188** (bottom), **245**, **350**, **364** (top), **464** and **488** (left) all from G Brum et al. *Biology: Exploring Life*, 2nd edition, John Wiley & Sons Inc., 1994, pp. 101, 12, 571, 589, 597, 626, 215, 13, 1020, 430, 951 and 991 respectively • BlueStar Forensic: **51**, **52** courtesy Ph Esperanca, MS • Professor Jeff Gawthorne: **63** • © PhotoDisc, Inc.: **66** (maple leaf), **93** (right), **104** (background), **175** (right), **210** (centre right), **310** (left), **351** (left), **365**, **416** (upper centre), **419** (left), **425** (top), **439** (top left) • © Graham Edgar: **68** (photos), **273** (centre), **412** (centre left and right) • Conly L. Rieder: **75** micrograph by Dr Conly L. Rieder, Wadsworth Centre, N.Y.S. Dept of Health, Albany, New York 12201-0509 • Newspix: **76** (top) / Ross Swanborough; **217** /Troy Bendeich; **258** (top) /Nicole Emanuel; **280** (top) /Steve Strike; **352** (centre) /Jeff Henderson; **482** /Colin Murty; **503** /Matthew Munro; **524** (bottom) /David Crosling • Clinical Cell Culture: **77** • Dr Andrew S. Bajer: **78** (right column images), **91** (top left) • Elsevier: **79** reprinted from *Trends in Cell Biology*, Vol. 15, Issue 1 front cover, © 2005, with permission from Elsevier • Michael Marko: **82** (bottom) reproduced with permission from Michael Marko, NIH/NCRR Resource for Visualization of Biological Complexity, Wadsworth Center, Albany, NY USA • Associate Professor Lynda Camp: **83** (both) • Ocean Earth Images: **94**, **192** (both) /Kevin Deacon • Lochman Transparencies: **95** (right) /Geoff Taylor; **291** (lower left) /Hans & Judy Beste; **398** (bottom) /Dennis Sarson; **416** (centre right) /Marie Lochman • Fre Farm: **103** (right) • ANTPhoto.com.au: **120** (left) and **360** /Fredy Mercay; **194** (top) and **292** (top right) /Klaus Uhlenhut;

210 (centre left) and **431** (bottom) /© Fenton Walsh; **226** (centre), **256** (desert), **297** and **341** /Frank Park; **227** (top) /G B Baker; **237** (fern) /Karin Clanelli; **244** (top right), **348** (top), **414** (bottom right) and **480** (left) /Ralph & Daphne Keller; **245** (top) and **426** (centre) /Paddy Ryan; **248** (top right) and **261** (top) /BG Thomson; **258** (right) and **273** (right) /Pete Atkinson; **260** (left) /Martin Harvey; **261** (centre) and **316** /Dick Whitford; **261** (bottom) and **405** (top right) /Pavel German; **263** (top left) and **285** (right) /Rob Blakers; **268** /David Whelan; **285** (top left) /Otto Rogge; **285** (centre left), **289**, **414** (bottom centre), **468** (right) and **476** (left) /© Dave Watts; **287** /J Burt; **288** (bottom) and **530** (lower right) /G Cheers; **290** (left), **413** (upper centre) and **416** (bottom) /Ken Griffiths; **292** (top left) and **489** (right) / Ted Mead; **292** (lower left) and **414** (centre) /Allan Burbidge & Julie Raines; **292** (lower right) and **384** (top) /Kelvin Aitken; **313** and **342** /Cyril Webster; **320** (left) ANTPhoto. com.au; **343** (centre) /Densey Clyne; **364** (bottom) /T J Hawkeswood; **378** (centre) /Jack Cameron; **385** (top right) / John Weigel; **388** (left) /Rudie Kuiter; **389** (bottom) /Murray Price; **395** and **417** (right) /G E Schmida; **398** (centre left) / JP & ES Baker; **404** /N.H.P.A.; **414** (bottom left) /Michael Cermak; **423** (top right) and **478** (top left) /Ron and Valerie Taylor; **436** /Gunther Schmida; **458** (right) /Norbert Wu; **474** (lower right) /© Rik Thwaites; **480** (right) /Denis O'Byrne; **489** (top left) /Bill Bachman; **531** (left) /Jutta Hosel; **531** (right) /Cliff & Dawn Frith • Courtesy of Dr Ken Richardson: **120** (right) • Karen Inge: **122** • © Cabrini Hospital: **130** • Courtesy of Valda Leslie: **139** (bottom left and right) • The Picture Source: **144** (bottom) • Coo-ee Historical Picture Library: **146** (right) • CSIRO Land and Water — SA: **159** (left) reproduced with permission from Tom Hatton, CSIRO Land and Water; **500** (right), **524** (top) and **525** (centre right) © Copyright CSIRO Land and Water. Photos by Willem van Aken • Visuals Unlimited: **161** (top left) © Dr George J. Wilder • © Bryan G. Bowes: **161** (bottom left and right) • National Library of Australia: **162** reproduced by permission of the National Library of Australia, Tyrell Collection 53/10 • Dennis Kunkel Microscopy, Inc.: **174**, **178**, **496** (bottom right) • © Zefa Images/Bill Bachman: **177** • Micrographia: **179** © micrographia.com • Mark Simmons: **180** (bottom), **210** (top left) • Science Photo Library: **181** (top left) /photolibrary. com/Science Photo Library/Dr Jeremy Burgess • Getty Images: **181** (bottom) /Neil Fletcher & Matthew Ward; **382** (right) / Photographer's Choice/Tim Davis; **422** (bottom right) /Taxi/David Maitland; **424** (bottom) /National Geographic/Mattias Klum; **496** (bottom left) Stone/Bob Thomas • Royal Botanic Gardens Sydney: **184** (top left and right) © Botanic Gardens Trust, Sydney • AAP Images: **185** (top) /AAP Images/AP Photo/APTV; **185** (lower) /AP via AAP/Tony Gutierrez • HortResearch: **190** /M Heffer, HortResearch • Roy Caldwell: **194** (centre) • Stephen Nebauer — a Bear Image: **196** • Oceanwide Images Picture Library: **197** (bottom) © Rudie Kuiter/oceanwideimages. com • Dr Jan West: **209** (both) • Carol Grabham: **210** (bottom left), **225** (top), **255** (right) • Australia Post: **214** original stamps held in the National Philatelic Collection, Australia Post • Botanic Gardens Trust: **215** (left), **216** (right) /Ken Hill, © Botanic Gardens Trust, Sydney • Densey Clyne: **215** (centre) • Australian National Botanic: **216** (top left) /G. McEwin © Australian National Botanic Gardens; **228** (right) © M. Fagg, Australian National Botanic Gardens • C. Freebairn: **218** (top left) photo by C. Freebairn, DPI & F, Queensland, courtesy *The Good Bug Book*, 2nd edition • Rob Weppler: **218** (top right) Biological Services, Loxton, SA • © Hans Brunner: **220** (cross-sections and patterns) from 'Hair ID' CD-ROM by Hans Brunner and Barbara Triggs, Ecobyte Pty Ltd, published by CSIRO Publishing • Austral International: **226** (right) /First Light • Bruce Maslin: **228** (left), **232** (all) Department of Conservation and Land Management, Western Australia • Monterey Bay Aquarium Research: **233** /Greg Pio for MBARI © 1997 • Christopher J. Earle: **234** (top) www. conifers.org • Dave Noble: **234** (lower) • PLoS Biology: **236** /The Public Library of Science Biology www. plosbiology.org • © Peter Storer: **237** (pine), **398** (top left) • University of Wisconsin: **248** (top left) © Mike Clayton/ University of Wisconsin Plant Teaching Collection • © Brand X Pictures: **248** (bottom) • Jane Scott: **260** (right) To find out more about this and all the other species of Western Australian flora, visit FloraBase at http://florabase.calm. wa.gov.au • Geoscience Australia: **264** © Commonwealth of Australia, Geoscience Australia (2005) • David Riggs: **265** (all) David Riggs and Jennene Paris/Blue Office Productions • CSIRO Marine Research: **266** © CSIRO; **477** (top left) © CSIRO Marine and Atmospheric Research • Dr John Holland: **267** (all) • Courtesy of Craig Sowden & the Centre for Research on Ecological Impacts of Coastal Cities, University of Sydney: **272** (centre) • Image Quest Marine: **272** (left) /Peter Parks; **273** (left) /Roger Steene; **431** (centre) /Masa Ushioda; **494** /Peter Batson • Australian Bureau Meteorology: **274** (top and centre) © Commonwealth of Australia. Mark Jenkin, Bureau of Meteorology; **274** (bottom) © Commonwealth of Australia. Ian Forrest, Bureau of Meteorology; **279** (screenshots) © Commonwealth of Australia/ Bureau of Meteorology • CSIRO Publishing: **279** (lower right) reproduced from *Ecos*, a CSIRO magazine reporting on scientific research related to the environment (Spring 1992), with permission of CSIRO Publishing. www. publish.csiro.au/ecos • MAPgraphics Pty Ltd, Brisbane: **280** (bottom), **284**, **472** (right), **473** (both maps), **523** (right) • Commonwealth Copyright Administration: **281** (graph) based on data from the Bureau of Meteorology © Commonwealth of Australia. Reproduced by permission • Macmillan & Co Ltd UK: **282** (centre left and right) from *The Botany of the Living Plant* by Frederick Orpen Bower 1923, London, Macmillan • Martin Thompson: **315** (both) • Fairfax Photo Library: **319** (top) /Craig Sillitoe; **442** (centre) /Rick Stevens • © CFA Victoria: **319** (centre and right) • M. Suzanne Searls: **324** (top), **334** (top), **368** (top right and bottom left) • © Auscape: **345** (bottom) /Jim

Frazier; **375** (top) /Kevin Deacon; **378** (right), **380** (top) /Glen Threlfo; **382** (left), **391** (all), **506** (bottom left) /D Parer & E Parer-Cook; **390** (bottom), **423** (bottom right), **506** (centre right) /Jean-Paul Ferrero; **412** (bottom) /John McCammon; **422** (bottom left) /Densey Clyne; **521** /Wayne Lawler; **523** (left) /Jean-Marc La Roque • SeaPics.com: **346** /Brandon D. Cole; **347** (bottom left) /Doug Perrine; **439** (bottom right) /Robin W Baird • Dr Chris Gleeson: **361** • © Digital Vision: **368** (centre left and bottom right), **387** /Stephen Frink • Out of copyright: **368** (top left) George Graves (d. 1834), 'Le Regne Animal', Paris, Fortin, Masson et Cie, Libraires, 1826 • Dr J Floor Anthoni: **374** (left) Seafriends Marine Conservation and Education Centre • Carol Buchanan: **374** (right) • Ross Armstrong: **376** Ocean Wildlife • © Banana Stock: **389** (centre) • Phillip Colla Photography: **406** • Woods Hole Oceanographic Institution (WHOI): **407** (top) /photo by Chris Knight; **408** (bottom centre) /photo by Carl Wirson; **408** (bottom left) /photo by Robert Hessler; **408** (top left and bottom right) /photo by Robert Ballard • Stacey Tighe: **407** (bottom) reproduced with permission of Stacey Tighe, University of Rhode Island Graduate School of Oceanography, funded by the Joint Oceanographic Institute, Washington DC • © University of Georgia: **410** • © Mary Malloy: **416** (upper left) / Nitrographics • © EyeWire Images: **417** (left) • UNSW Press: **418** (left) reproduced from A. M. Young, *A Field Guide to the Fungi of Australia* with permission of UNSW Press • Professor David M Lambert: **424** (top), **472** (left) • © Dr Klaus Hellrigl: **427** (top) • © Missouri Botanical Garden: **427** (left) • © Dr Daniel L. Nickrent: **429** (all) • ASGAP — Association of Societies: **430** (top) © Owen Roberts, reproduced with permission from his family and the Association of Societies for Growing Australian Plants (ASGAP) • © Nitragin, Inc.: **431** (top) • © Steve Nicol: **438** (right) • © Digital Stock/ Corbis Corporation: **442** (left) • Orbimage: **443** © ORBIMAGE/NASA, from the SeaWiFS Project • Australian Antarctic Division: **450** photo by Wayne Papps © Australian Antarctic Division, Commonwealth of Australia • © IT Stock: **456** • Dr Josephine Milne: **467** • Australian Antarctic Division: **472** Courtesy AAD, © Commonwealth of Australia • © Rod Seppelt: **474** (upper right) • Craig Johnson: **477** (top right) reproduced with permission from Craig Johnson and the Department of Sustainability and Environment Victoria • Jonathan Majer: **477** (lower left) data courtesy of H.F. Recher and J.D. Majer • © Australian Institute of Marine Science: **478** (right), **485** (right) • Ian Miller: **479** • U.S. Census Bureau: **481** (lower left and right), **496** (top left) data from the U.S. Census Bureau International Data Base • Louis Gross: **491** (graphs) /M. Beals, L. Gross & S. Harrell • CSIRO Science Image Online: **493** © CSIRO Sustainable Ecosystems Division. Image supplied by Science Image Online • © Image Source: **495** (lower right) • West Australian Newspapers: **500** (main) courtesy *The West Australian* • Nicholas Carlile: **501** (right) • © Colin G. Wilson: **504** (left and right), **513** (upper left) • NASA: **507** /Reto Stockli, NASA GSFC, http://visibleearth.nasa.gov; **508** (bottom) /Feldman, G. C., C. R. McClain, Ocean Color Web, MODIS Reprocessing, NASA Goddard Space Flight Center. Eds. Kuring, N., Bailey, S. W. 24 August 2005. http://oceancolor.gsfc.nasa.gov/; **532** (right) /Jeff Schmaltz, MODIS Rapid Response Team, NASA/GSFC. http://visibleearth.nasa.gov • University of Wisconsin Space: **508** (top) courtesy of the Space Science and Engineering Center, University of Wisconsin-Madison • © Grahame Webb: **510** • John Thorp: **513** (bottom left and right) material sourced from www.weeds.org.au and reproduced with permission from John Thorp • Dr Jawahar Patil, Senior Research Scientist at CSIRO Marine Laboratories: **516** (bottom) • CSIRO Mathematical: **520** © CSIRO, Land Monitor • Victorian National Parks Association: **522** graph based on data from 'Floods and the Barmah-Millewa Forest' by Tony Ladson. Reproduced with permission from the Victorian National Parks Association • Mark Schneegurt: **525** (bottom) /Roger Burks (University of California at Riverside), Mark Schneegurt (Wichita State University) and Cyanosite (www.cyanosite.bi.purdue.edu). Reproduced with permission • NASA Earth Observatory: **532** (left) /Jeff Schmaltz, MODIS Rapid Response Team, NASA-GSFC • Courtesy BHP Billiton: **535**, **536** (all) • NASA — Ozone Processing Team: **538** (all) • © River Murray Urban Users Committee Inc. and the Murraycare Program: **539** (top) • © The Murray–Darling Basin Commission: **539** (bottom).

Text

• Jennifer Isaacs: **102** table summary extracted from Isaacs, J. 'Bush Food', Weldon, 1989. Reproduced with permission from Jennifer Isaacs • Out of copyright: **257** Alfred Lord Tennyson • Curtis Brown Australia: **278** extract from 'My Country' © Dorothea Mackellar. Reproduced by permission of Curtis Brown Australia.

Every effort has been made to trace the ownership of copyright material. Information that will enable the publisher to rectify any error or omission in subsequent editions will be welcome. In such cases, please contact the Permissions Section of John Wiley & Sons Australia, Ltd who will arrange for the payment of the usual fee.

UNIT 1

UNITY AND DIVERSITY

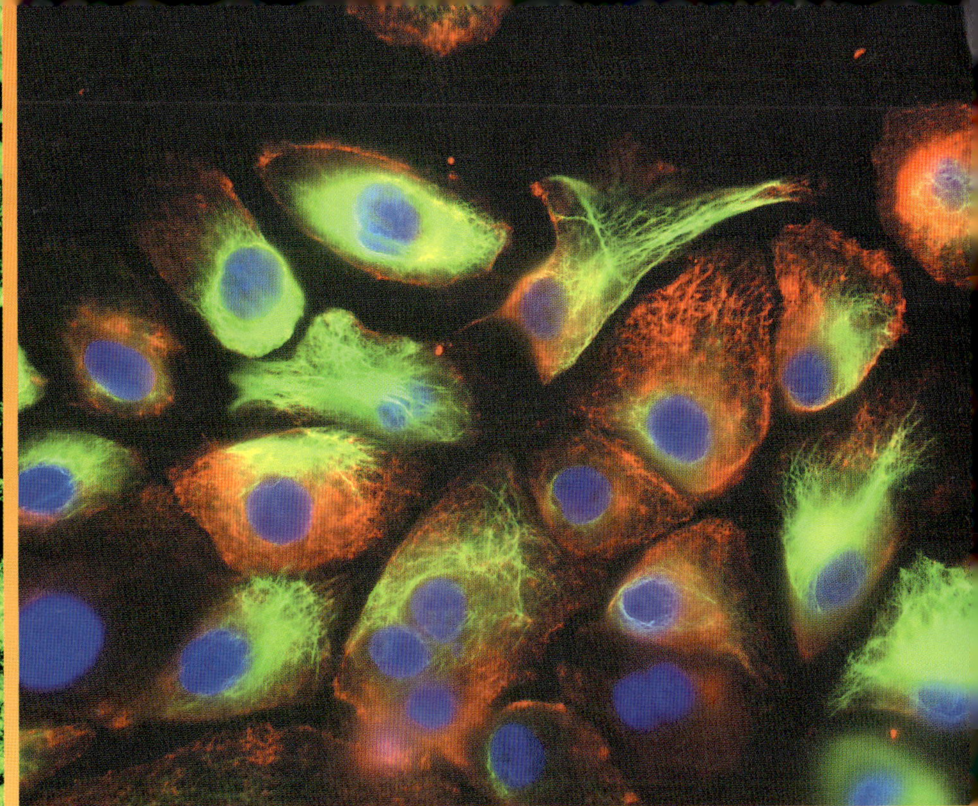

AREA OF STUDY 1

Cells in action

1 Cells: discovery and exploration

KEY KNOWLEDGE

This chapter is designed to enable students to:

- appreciate the historical development of microscopy techniques
- investigate current and emerging technologies in light and electron microscopy
- understand the importance of technological advances to our knowledge of life forms and cells.

Figure 1.1 Examination of high-resolution three-dimensional brilliant fluorescence images is now possible with current stereomicroscopes such as this SteREO Lumar.V12 manufactured by Carl Zeiss Pty Ltd. The stereomicroscope has lenses specially developed for use with fluorescence and its operation is completely motorised. The focus of an object can be rapidly set and precisely reproduced with the use of human interface panels (HIP). A system control panel (SyCoP) is designed for use by either a right- or left-handed person and combines joystick, buttons and a touch screen — a design similar to a computer mouse so that the operator can control the microscope while still viewing through the eyepiece. In this chapter, we will discuss significant historical developments in microscopy techniques and the latest advancements in microscope technologies.

Life on Earth ... and beyond?

Is there (or was there ever) life on Mars?

At the turn of the twentieth century, an American astronomer, Percival Lowell (1855–1916), drew maps of the surface of planet Mars that showed intricate patterns of linear structures that he called canals. He argued that these canals were not natural features but were artificial constructions produced by intelligent life. Figure 1.2a shows Lowell's drawings of canals on Mars. Figure 1.2b shows a typical area on the surface of Mars as revealed by the *Viking Lander* in 1976. Definitely no canals! Definitely no evidence of life, intelligent or otherwise!

(a)

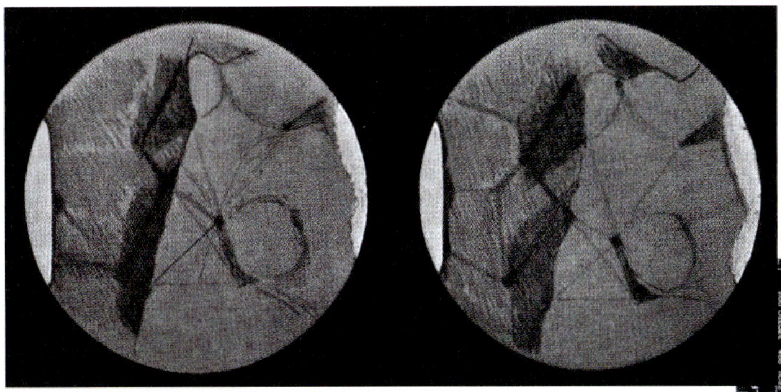

(b)

Figure 1.2 **(a)** Canals on Mars based on observations made from Earth by Lowell in the early 1900s, and **(b)** the surface of Mars as revealed by the *Viking Lander* in 1976. Can you suggest a possible reason for Lowell's observations being flawed?

The *Viking Lander* carried instruments to test for the existence of living organisms on Mars (note the trenches dug by the soil retrieval scoop in figure 1.2b), but the results of the tests were inconclusive.

ODD FACT

In July 1976, when the *Viking Lander* reached the surface of Mars, it became the first spacecraft to land on the surface of another planet.

Then, in 1996, sensational headlines worldwide publicised the claim by NASA scientists that life once existed on Mars. This claim was based on studies of a meteorite that originated from that planet. The evidence included the presence of tiny structures within the meteorite (see figure 1.3) that were said to be fossilised microbes (tiny living organisms). However, other scientists disputed this claim and argued that these microbe-like structures could be produced by chemical reactions. Again, the evidence for life on Mars was inconclusive.

Another development occurred in January 2004 when two Rovers landed on the surface of Mars to study its rocks and minerals. Data from these Rovers provided evidence that liquid water once existed on Mars. We know that liquid water is essential for life. We also know that microbes can survive in extreme environments on Earth, such as in rocks deep below ground, in ice-sealed lakes, in glaciers high on mountains and in cold dry valleys of Antarctica. Based on these facts, it remains possible that life does or did exist on Mars.

There are plans to launch a Mars Science Laboratory from Earth to Mars in December 2009. This mobile laboratory will search for evidence of life — past or present — on Mars using instruments that can detect organic compounds, such as proteins and amino acids, that are made only by living organisms.

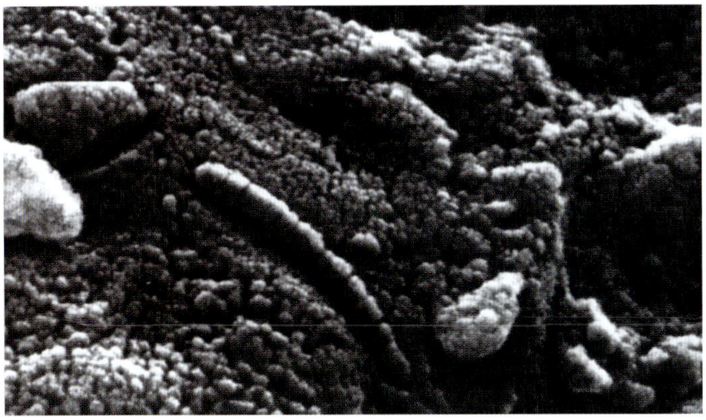

Figure 1.3 Scanning electron micrograph image of part of a meteorite (known as ALH84001) from the surface of Mars that landed in the Antarctic. While the elongated structures look like microbes (tiny living organisms) they are *not* universally accepted as being fossilised microbes.

Scientists expect that, if life exists now or existed in the past on Mars, this extraterrestrial life will be like the microbes that live today in extreme environments on Earth. Microbes, like all living things, are organised into microscopic 'compartments' known as **cells**. Each microbe typically consists of just one cell and the internal contents of each cell are separated from the external environment by a membrane boundary. The strongest direct evidence for past or present extraterrestrial microbial life on Mars would be the discovery of structures that can without any doubt be identified as cells.

Let us now look in more detail at the historical development of ideas and technological advances that have contributed to our knowledge and understanding of life forms, and their living compartments or cells.

Cells and microscopes: an introduction

Cells are the basic structural and functional units of all living things (figure 1.4). Although most cells are too small to be seen with the unaided eye, **microscopes** give enlarged images of cells and the structures they contain, and make it possible for us to examine cells with great detail.

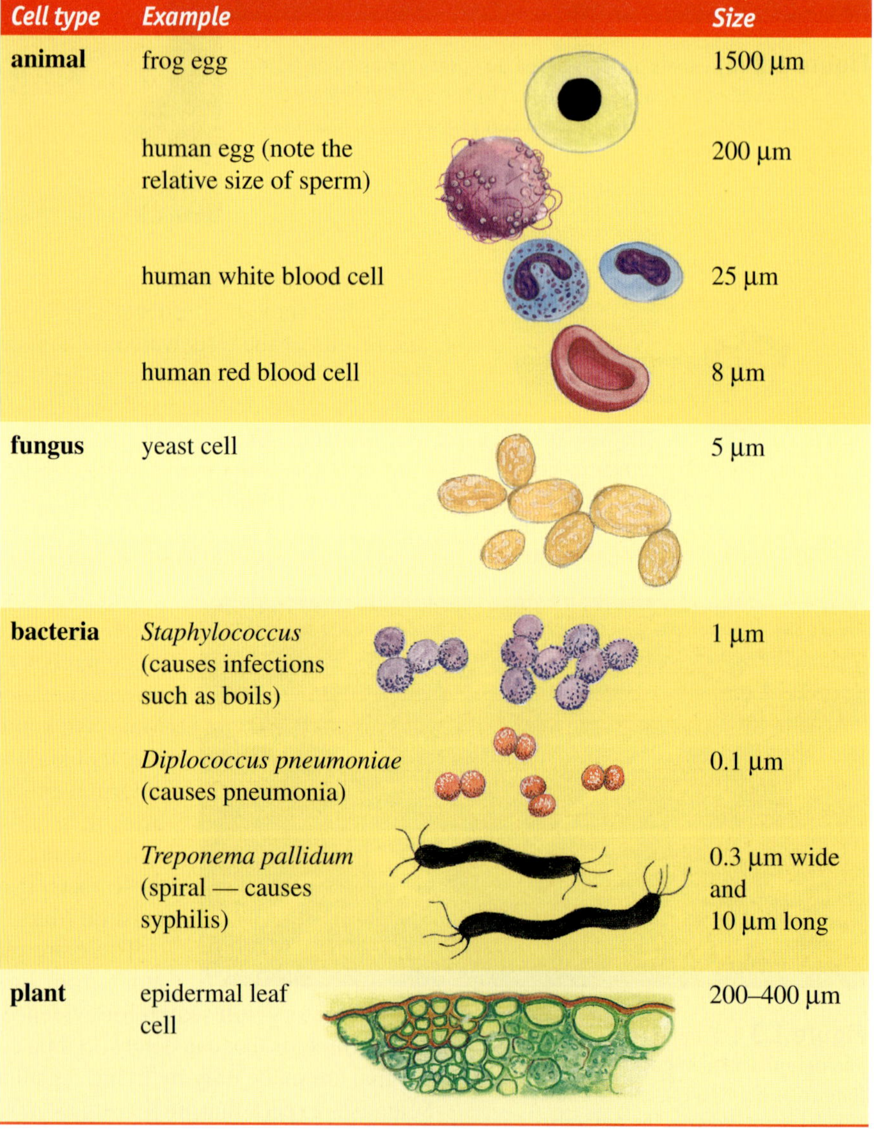

Cell type	*Example*	*Size*
animal	frog egg	1500 μm
	human egg (note the relative size of sperm)	200 μm
	human white blood cell	25 μm
	human red blood cell	8 μm
fungus	yeast cell	5 μm
bacteria	*Staphylococcus* (causes infections such as boils)	1 μm
	Diplococcus pneumoniae (causes pneumonia)	0.1 μm
	Treponema pallidum (spiral — causes syphilis)	0.3 μm wide and 10 μm long
plant	epidermal leaf cell	200–400 μm

Figure 1.4 Most human cells typically range in diameter from about 8 to 25 micrometres (μm) or 0.008 to 0.025 mm. In comparison, a hair from a man's beard is about 200 μm (0.2 mm) wide. Typical bacterial cells range from 0.1–1.5 μm and giant *amoeba* are about 1000 μm wide. Note the sizes of various kinds of cells (1 mm = 1000 μm). Cells are not drawn to scale.

The development of microscopes over the centuries has depended on the development of glass, then on glass being made into lenses, the development of different kinds of lenses and their assembly to form microscopes.

The increase in our understanding of cells has paralleled:

- the improvements in and development of new kinds of microscopes
- the variety of different techniques available, including stains, sectioning and using different kinds of light.

The advanced microscopes of today have a history dating back more than three thousand years to when the first glass was made by Phoenician sailors. A summary of some of the important steps in the development of the microscope and our understanding of cells is presented in table 1.1.

Table 1.1 Some important milestones in the development of microscopes

Date or period	Person and development
> 3000 years ago	Glass beads first made by Phoenician sailors, who were from an area now known as Lebanon
250 BC–AD 100	In China, the first recorded uses of optical lenses
AD 79 (excavated 1748)	People of Pompeii used glass-crystal lenses
1200–1250	Robert Grosseteste, Bishop of Lincoln, UK, made a primitive but functional magnifying glass.
1590s	Dutch lens-makers, Hans Janssen and his son Zacharias, used two lenses to develop the first compound microscope, called 'telescope' by some writers.
1605–1619	Cornelius Drebbel (1572–1633), a Dutch/English inventor of many scientific instruments, developed a machine for grinding lenses and improved the quality of compound microscopes.
1605–1614	Galileo Galilei (1564–1642), an Italian, refined the Janssen microscope into a high-quality astronomical telescope. He also further developed the microscope and may have been the first to examine and describe living tissue. He described the cuticle of a fly as being covered in fur.
between 1605–1610	Galileo was a prominent member of the Accademia dei Lincei (Academy of the Lynx) that introduced the term '*microscopio*' — a lens for the examination of very small objects.
1665	Englishman Robert Hooke (1635–1703) published *Micrographia.* He describes 'cells' in a piece of cork (page 6 and figure 1.5) and draws many cell types. Hooke's microscope could magnify 14–42 times.
1674	Antony van Leeuwenhoek (1632–1723), a Dutch cloth merchant, built a microscope with a magnifying range from 50 to 300 times. He was the first to make descriptive drawings of protozoa, bacteria, spermatozoa and red blood cells (page 6 and figure 1.6, page 7).
1733	Englishman Chester Moor Hall used lenses made of different kinds of glass to invent the achromatic lens that removed many of the optical distortions of previous lenses.
1738	German Johann Lieberkuhn added a metal reflector to the microscope to increase light falling on a specimen.
1831	Robert Brown (1773–1858), a Scottish botanist and naturalist, described the nucleus in orchid cells (figure 1.7, page 7 and page 8).
1838	Two Germans, botanist Matthias Schleiden (1804–1881) and zoologist Theodor Schwann (1810–1882), suggested that cells are the basic structural units of all plant and animal matter.
1851	Binocular microscope (viewing with two eyes) constructed by Professor Riddell
1878	Germans Ernst Abby and Carl Zeiss produced improved oil-immersion microscope lenses that significantly increased the ability to magnify cells (figure 1.10, page 11).
1931–1933	German Ernst Ruska developed the electron lens and used several to make the first electron microscope.
1936	Swedish Torbjorn Oskar Caspersson used an ultraviolet microscope to study cells.
1938	Dutch Fritz Zernike built the first phase contrast microscope enabling examination of transparent cells and micro-organisms without the need to stain or kill.
1955	Marvin Minsky of the USA invented the confocal scanning microscope.
1969	Scientists in Holland, Britain and America developed confocal laser scanning microscopy. Americans Paul Davidovits and David Egger announced they were able to 'optically section' thin slices of three-dimensional specimens such as a cell (as in figure 1.15, page 12).
2003	PlasDIC is a special form of a differential interference contrast microscope in which special prisms are used to reveal high-resolution, three-dimensional details of a specimen illuminated with non-polarised light (figures 1.21 and 1.22 on page 16).
2004	Laser scanning microscopy LSM 5 *LIVE* scans living cells at speeds of up to 1010 frames per second. Allows a better understanding of cellular processes and study into cellular interaction mechanisms.

In this chapter, we will consider some of the people and technologies, and their contribution to our understanding of cells. We will also consider the characteristics of the following tools used for viewing cells.

- Light microscopes:
  - Simple light microscope
  - Compound light microscope
  - Phase-contrast microscope
  - Fluorescence microscope
  - Scanning confocal microscope
  - PlasDIC microscope
- Electron microscopes:
  - Transmission electron microscope
  - Scanning electron microscope

Cells: an historical overview

ODD FACT

Robert Hooke was a scientist, inventor and architect who drew up plans for the rebuilding of London after the Great Fire of 1666. Hooke built the vacuum pump that Boyle used in his experiments on the pressure and volume of gases. Hooke also formulated the law that describes the behaviour of springs when stretched.

In his book, *Micrographia*, published in 1665, English scientist Robert Hooke (1635–1703) describes how he used a microscope to examine thin slices of cork from a tree and saw small box-like compartments that he called 'cells'. Robert Hooke is credited as the person who discovered cells. In fact, Hooke was not looking at living cells. What he saw were the remains of dead and empty plant cells (see figure 1.5). However, Hooke's observations were important because he was the first to realise that this plant material had an organised structure at the microscopic level.

(a)

(b)

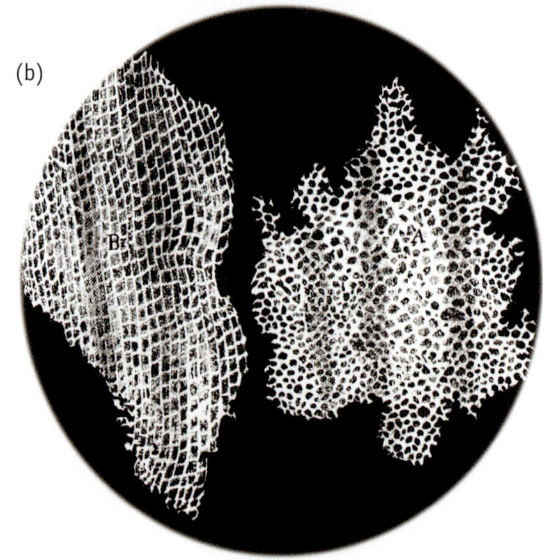

Figure 1.5 **(a)** The microscope that Robert Hooke built and used to examine thin slices of plant material. What name did he give to the minute building blocks that he saw? What light source might Hooke have used for this microscope? The specimen for examination was placed on a specimen holder. **(b)** First drawings made in 1665 of 'cells' from a thin piece of cork. Were these living or dead cells?

In 1674, a few years after Hooke discovered cells, a Dutch cloth merchant, Anton van Leeuwenhoek (1632–1723), used a simple microscope (see figure 1.6a, page 7) to observe material that he scraped from between his teeth. After examining this material, he wrote:

> … in the said matter there were very many little living animacules, very prettily a-moving.

What Leeuwenhoek saw were probably the first bacterial cells to be viewed (see figure 1.6b). Although Leeuwenhoek's original interest with microscopes was to examine fibres in the cloth he traded, he was inspired by the publication of Hooke's *Micrographia*.

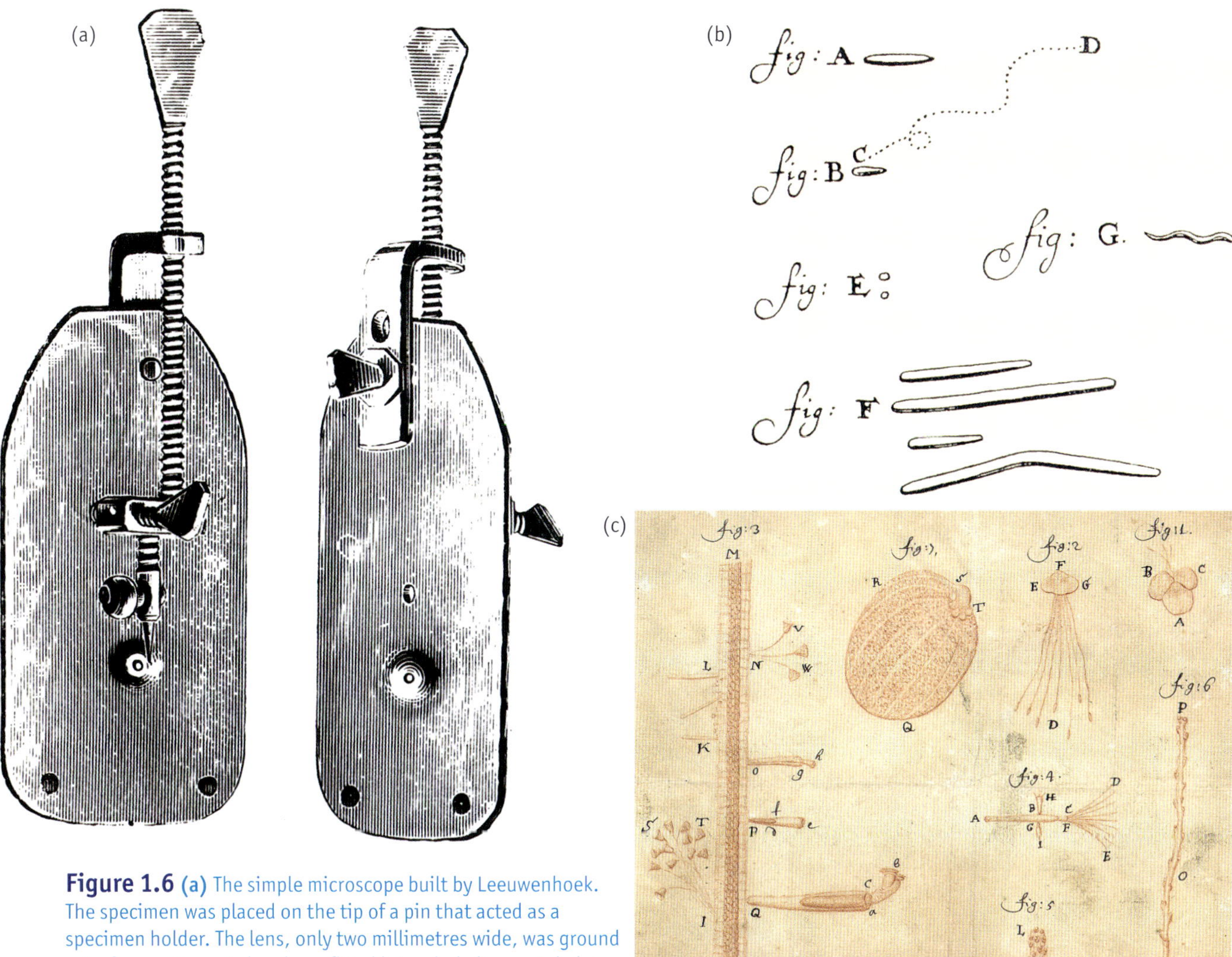

Figure 1.6 **(a)** The simple microscope built by Leeuwenhoek. The specimen was placed on the tip of a pin that acted as a specimen holder. The lens, only two millimetres wide, was ground out of a quartz crystal and was fitted into a hole in a metal plate. The instrument was held up to the eye and the specimen viewed through the lens. **(b)** Some of the 'little animacules' seen by Leeuwenhoek. These were various bacteria. **(c)** Algal and other cells from pond water as drawn by Leeuwenhoek

(a) (b)

Figure 1.7 **(a)** Wax medallion of Robert Brown, made in 1852 **(b)** One of the microscopes used by Brown in his observations of pollen and other plant cells. Note the mirror (closest to the base) and the fine-adjustment knob for movement of the specimen platform above it.

ODD FACT

Robert Brown was the botanist who accompanied Sir Joseph Banks on the *Investigator*, when Captain Matthew Flinders charted the southern coast of Australia from 1801 to 1805. Brown and Banks collected nearly 4000 specimens of different species of plants, most of them unknown to Western science at the time. Brown also discovered molecular movement, now called 'Brownian movement'.

Leeuwenhoek built over 50 simple microscopes to examine material from different sources. Figure 1.6c shows Leeuwenhoek's drawings of some of the algal and other cells he observed in pond water.

More than 150 years later, in 1831, Robert Brown (1773–1858), a Scottish botanist (see figure 1.7, page 7), was involved in a dispute about how pollination and fertilisation occurred in plants. During his studies with orchids, on 13 June 1831 he made a note that:

> It appears that each cell has … on its inner side a spherule or at least orbicular corpuscle …

Brown called this structure the **nucleus** of a cell. Others, including Leeuwenhoek, had observed nuclei but Brown was the first to introduce the concept of a nucleated cell as the unit of structure in plants. Brown had no idea about the importance of the nucleus and had some doubt about whether each cell needed one.

Recognising the pattern: the Cell Theory

By the early 1800s, the accepted idea was that plants and animals were composed of globules, called cells, and formless material. Brown had enhanced this idea by describing nuclei in cells of orchid plants. These views were to be extended by two German biologists.

ODD FACT

Schleiden was initially educated as a barrister. Because of his lack of success in this profession, he attempted suicide, shooting himself in the forehead, but recovered. Schleiden then turned to the study of natural science and medicine and became a professor of botany.

In 1838, a German botanist, Matthias Schleiden (1804–1881), suggested that cells were the basic structural unit of all plant matter. A German zoologist, Theodor Schwann (1810–1882), independently proposed that animals were aggregates of cells arranged according to a definite law. In sharing their ideas over dinner in October 1838, the two biologists came to recognise that both plant and animal tissues have a cellular organisation. Nearly 200 years after Hooke first described cells, the basic structural pattern of living things was finally recognised.

The recognition that all kinds of living things share a common structural unit — the cell — provided the foundation of one of the major unifying themes of biology. All living things are composed of cells and substances produced by cells or developed out of cells. Because of this unity of structure, results of studies of cells from one type of organism can be used to make predictions about cells from other kinds of organisms. Schwann wrote in 1839:

> The elementary parts of all tissues are formed of cells in an analogous, though very diversified manner, so that it may be asserted, that there is one universal principle of development for the elementary parts of organisms, however different, and that this principle is the formation of cells.

This basic idea arising from the work of Schwann and Schleiden, published in 1839, is known as the **Cell Theory**:

> All living things consist of one or more organised structures that are called cells or of products of cells.
>
> Cells are the basic functional unit of life.

Figure 1.8 According to one theory, living things could arise from dead matter; for example, leaves could become animals or fish, depending on where they fell.

A German doctor, Rudolf Virchow (1821–1902) added to the understanding of cells by providing a new answer to the question: How are new living things produced? Past answers to this question included **spontaneous generation**, the idea that living things could arise from nonliving matter or dead matter. Another idea was that living things developed from globules that gathered to form a compact mass and then became organised into cells.

In 1858, Virchow challenged these old ideas with his concept of **biogenesis** (from bio = life; genesis = origin, creation). He proposed that new cells come from existing cells, and in one of his famous lectures said:

> … no development of any kind begins *de novo* [from new] … Where a cell arises, there a cell must have previously existed just as an animal can spring only from an animal, a plant only from a plant … No developed tissue can be traced either to any large or small simple element, unless it be a cell.

Virchow's contribution extended the Cell Theory to include the basic concept:

> New cells are produced from existing cells.

In 1862, the French biologist, Louis Pasteur (1822–1895) carried out experiments that conclusively disproved the old idea of spontaneous generation, and supported the view that new cells are produced by existing cells.

Life span of cells

The terms unicellular and multicellular refer to organisms built of one or more building blocks respectively.

Cells of a multicellular organism do not necessarily live as long as the organism itself. Some cells have a relatively short life and are constantly being replaced.

The average life spans of some human cells are as follows:

- stomach cells 2 days
- mature sperm cells 2–3 days
- skin cells 20–35 days
- red blood cells about 120 days.

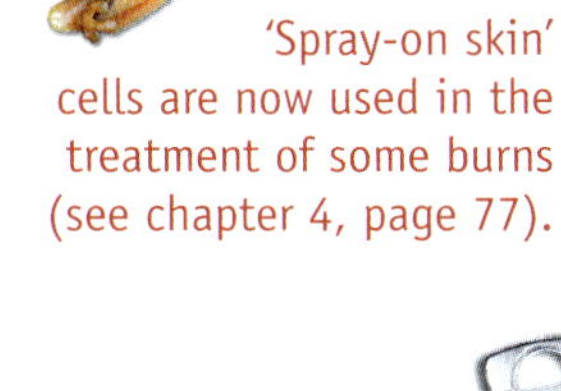

'Spray-on skin' cells are now used in the treatment of some burns (see chapter 4, page 77).

A person can make a blood donation because the cells removed can be replaced by new cells. Skin can be taken from one part of a person's body and grafted onto another area where the skin tissue has been completely destroyed. Skin cells from the undamaged area will reproduce to replace the cells removed.

In contrast, other types of cell, such as brain cells, have long life spans and are not replaced during a person's lifetime. If brain tissue is damaged, most cell types in the brain cannot reproduce to replace the damaged cells.

KEY IDEAS

- Cells were first identified and named by Hooke in 1665.
- The nucleus in a cell was identified and named by Brown in 1831.
- The Cell Theory arose in the mid-1800s.
- The Cell Theory recognises that all living things are composed of one or more cells and that new cells are produced by existing cells.
- The life span of cells in a multicellular organism varies.
- The unit of measure often used in relation to cell size is the micrometre (µm).

QUICK-CHECK

1 Suggest why cells were not discovered by the Greek physician Hippocrates (died 357 BC).
2 Who is credited with the discovery of the basic building block of living organisms?
3 Who is credited with the discovery of the cell nucleus?
4 What was the important contribution by Schleiden and Schwann to biology?
5 Identify one commonplace idea about the origin of living things before Virchow.
6 Which person is more likely to have permanent damage after an accident: person A who survives after blood loss or person B who survives after some loss of brain tissue? Explain.
7 How many micrometres (µm) are there in a millimetre (mm)?

Tools for viewing cells

With few exceptions, individual cells typically are too small to be seen with an unaided eye. Because of this, the study of cells has depended on the use of instruments, such as microscopes. There are many different kinds of microscopes but they can be broadly divided into two groups, light and electron.

Light microscopes

Some microscopes used for viewing a dissection or a small organism use light being reflected from the surface of the organism.

Hooke needed a **light microscope** to see the dead cells present in cork. Light microscopes (LMs) increase the ability of the human eye to see tiny objects. LMs can reveal objects such as the unicellular organism in figure 1.11a that are too small to be seen, or details that are too minute to be resolved with an unaided human eye. LMs use visible light that illuminates and passes through a specimen. When tissues are examined using an LM, a slice of tissue just a few cells thick is viewed. The tissue is usually stained with a dye (see page 11) to make it more visible.

In the 1990s, a new technique, known as near-field scanning optical microscopy (NSOM), was developed for viewing cells and other objects. The technique allows organelles that are too small to be resolved with a normal light microscope to be seen.

Simple light microscope

Light microscopes (LM) use glass lenses. Early LMs with only one lens, like the kind used by Hooke and van Leeuwenhoek, are called **simple light microscopes**. They are similar to a magnifying glass.

Compound light microscope

Microscopes with at least two sets of lenses are called **compound light microscopes** (CLM). Most compound light microscopes have several **objective lenses**, each of a different **magnification** (see figure 1.9). The amount of magnification you obtain when using a light microscope, that is, how large the object appears, depends on the magnification powers of both the objective lenses and **eyepiece (ocular) lenses** you use. The magnification you obtain of an object is calculated by multiplying the magnification (power) of the objective lens by the magnification of the ocular lens you use.

The highest magnifications are obtained with the use of an **oil immersion objective lens**. Light usually travels in a straight line through a particular medium. As light passes from one medium to a different medium, the rays change direction — they are refracted. Hence, as light passes through a glass slide holding a specimen and into air above, rays are refracted and there is a reduction of light entering the objective lens (figure 1.10). A reduction of light reduces the clarity of an image. With an oil immersion objective lens, oil of the same refractive index as glass is placed between the glass slide and the objective lens. This reduces the loss of light due to refraction and higher magnifications are possible. Oil immersion lenses usually have a magnification of 100×, compared with the usual maximum of 40× with a dry lens.

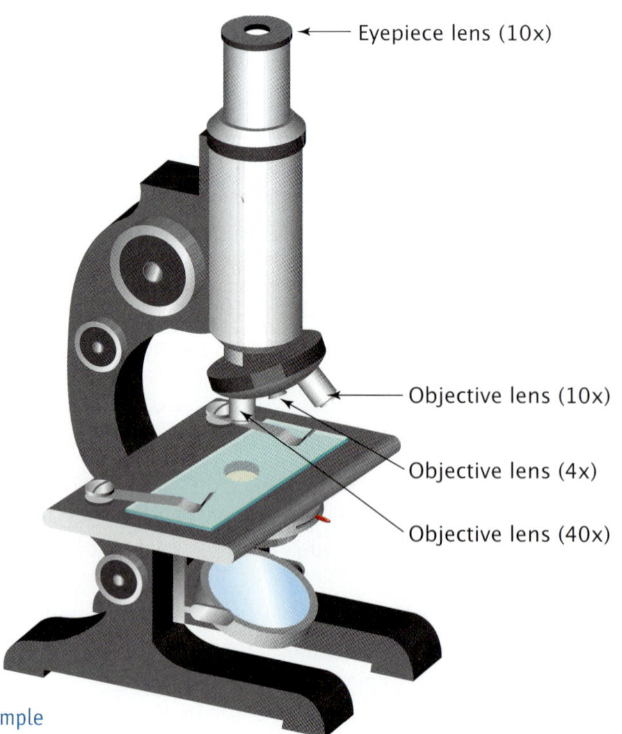

Figure 1.9 An example of a compound light microscope

Figure 1.10 Note that the use of oil (with the same refractive index as glass) between the specimen and the objective lens increases the amount of light passing through the optical system of the microscope by reducing refraction. Higher magnification objective lenses can be used and a clearer image of the specimen is obtained.

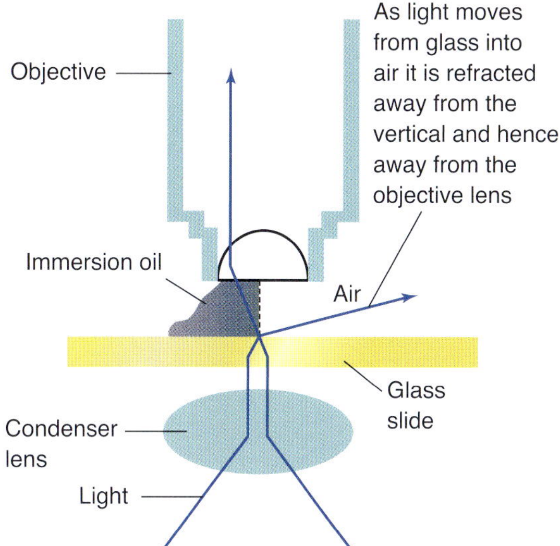

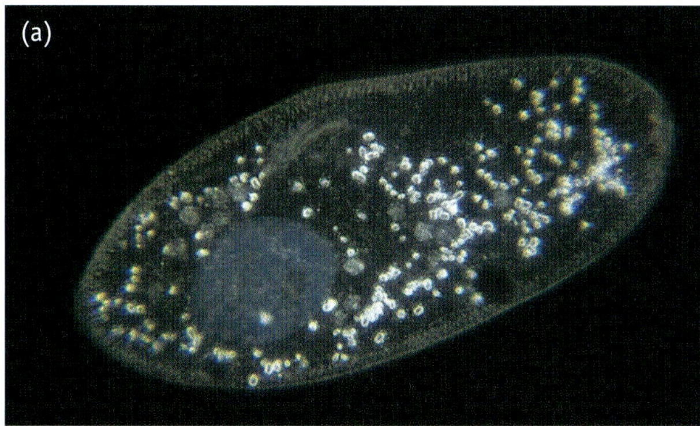

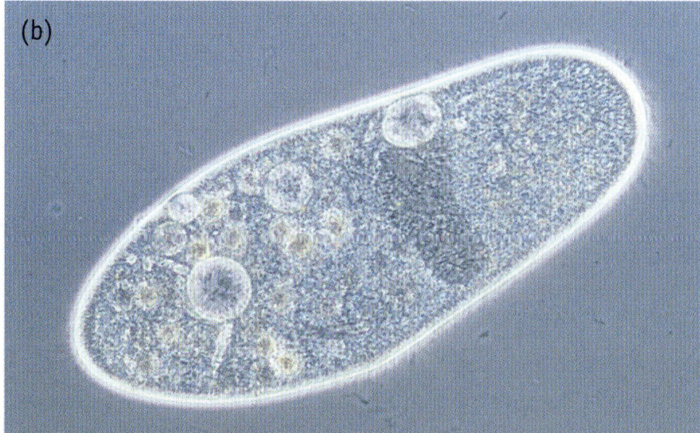

Figure 1.11 (a) Image of *Paramecium*, a unicellular organism, as seen with a light microscope (b) Same type of cell as seen with a phase contrast microscope

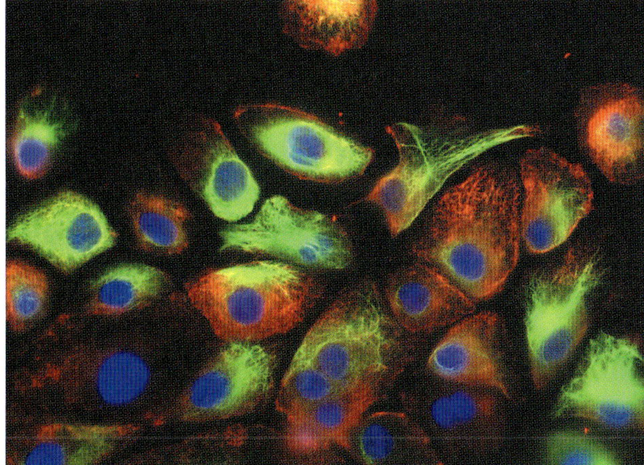

Figure 1.12 Cancerous breast cells viewed with a fluorescence microscope after staining for the presence of vimentin (green) and keratin (red)

Characteristics of the lenses also influence a microscope's **resolution**. Resolution is the ability to see two points that are close together as two separate points. Our eyes have limited resolving power: they may interpret two small spots that are close together as a single blurred spot. We use microscopes to resolve things that our eyes are unable to see; with an appropriate microscope we can distinguish the two small spots. But microscopes also have a limit to their resolving power. A poor quality microscope might simply magnify the blur we see into a larger blur. The wavelength of light used, as well as the characteristics of the lenses, influence the size of an object that can be resolved with a microscope. The smaller the wavelength of light used, the smaller the size discernible. Standard light microscopes use visible light.

Cells are virtually colourless and hence are difficult to see under a standard LM. **Staining** is required. Groups of cells are also cut into thin slices before staining. These treatments necessarily kill cells and often distort cell features. During the twentieth century, other kinds of microscopes and techniques as described below were developed for viewing and analysing cells.

Phase contrast microscope

The **phase contrast microscope** is a modified compound light microscope (CLM) that was developed to observe unstained, intact living cells (figure 1.11). These microscopes use the fact that different parts of a cell transmit and change the direction of light to varying degrees and enhance that difference. The image produced has highly contrasting bright and dark areas.

Fluorescence microscope

Another kind of CLM is the **fluorescence microscope**, which uses ultraviolet (UV) light to reveal compounds that have been stained with fluorescent dyes that bind to particular compounds in a cell. The colour of fluorescence depends on the particular fluorescent stain being used and the nature of the compound to which it is attached (see figure 1.12).

Scanning confocal microscope

Another development has been the **scanning confocal microscope** (figure 1.14).

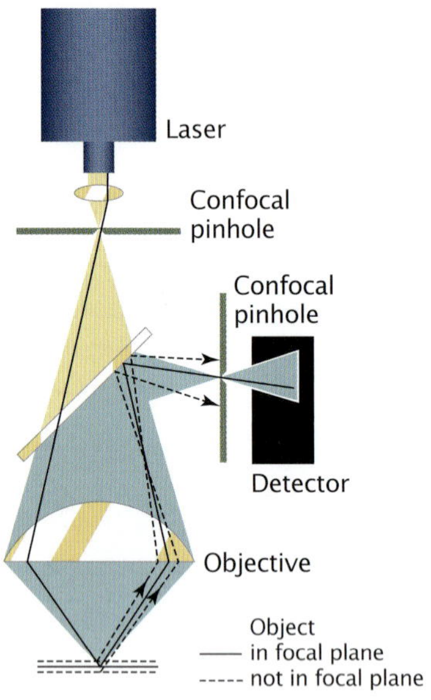

Figure 1.13 In a confocal microscope, light outside the focal plane is excluded from the detector by a pinhole aperture.

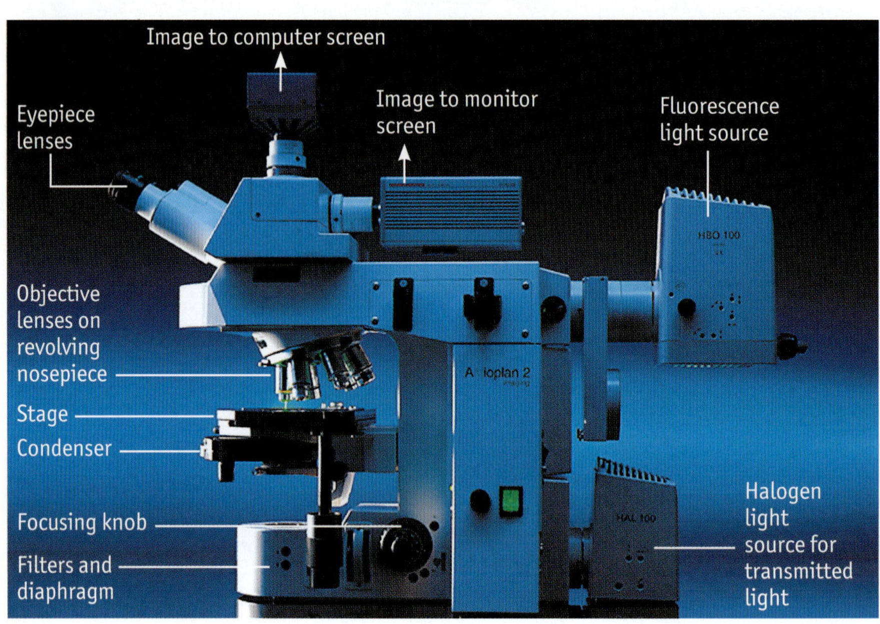

Figure 1.14 Confocal microscope — note the parts that you recognise from the microscopes you use in practical classes. Other features allow the microscope to be connected to computer and video systems. Because of the detailed work possible, settings used often by an operator can be stored in a computer and fed back to the microscope as required.

Figure 1.15 A *Drosophila* embryo that has been laser scanned. The left-hand side shows selected optical slices of the embryo. The right-hand side is a projection of the entire 47 optical slices.

Confocal microscopy uses laser light and special optics to allow a viewer to look at successively deeper layers of an object, such as a cell or micro-organism, without having to cut it into the many thin sections required by traditional light microscopy. Fluorescent stains are also used. Fluorescence coming from the specimen is focused by an objective lens through a pinhole aperture to a detector (figure 1.13). Fluorescence from out-of-focus planes above and below the in-focus plane is not transmitted. The fact that out-of-focus images are not transmitted means that the only image received by the detector is a sharply in-focus image of a thin slice of specimen (see figure 1.12 on page 11). The blurriness of out-of-focus parts is no longer observed.

Another advantage of confocal microscopy is that it can be combined with scanning microscopy, in which a user can perform 3-D microscopy of fluorescently labelled specimens or reflective surfaces. A computer is used to digitise the image of each section of the specimen obtained from confocal microscopy and the results are combined to give a 3-D image of the specimen that can be rotated and viewed from various aspects (see figure 1.15).

Microscopes are now commonly used in combination with computers and automated cameras. Computers can analyse the shape, colour and density of images seen under a microscope and enable biologists to easily carry out tasks that would otherwise be too difficult or too time consuming (figure 1.16).

Biologists such as Associate Professor Leigh Ackland use microscopes and techniques such as those described above and later in the chapter. Read what Leigh writes about her work on page 13.

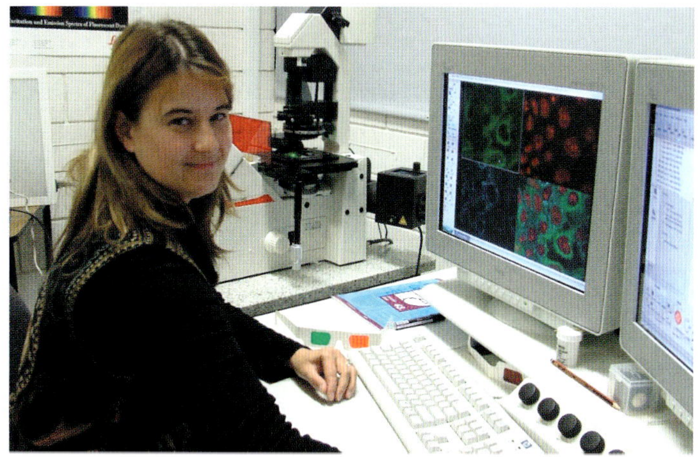

Figure 1.16 Confocal microscope and associated equipment. Note the glasses on the bench that are used for 3-D microscopy

BIOLOGIST AT WORK

Associate Professor Leigh Ackland — Molecular Biologist

Associate Professor Leigh Ackland is a Research Scientist and Senior Lecturer at the Centre for Cellular and Molecular Biology, School of Biological and Chemical Sciences, at Deakin University. Leigh writes:

'Since my early days as a research biologist, I have been very interested in studying the biology of cells by getting them to grow outside the body, using tissue culture techniques. Recent knowledge of the nutritional requirements of different types of cells has enabled biologists to grow cells taken from parts of the body such as skin, gut, breast and placenta. Using this approach, we have learned much about the life of a cell. In culture, cells grow, become specialised to carry out different functions, communicate with each other and their environment and eventually die.

'The study of cells in tissue culture has provided a wealth of information about the behaviour of normal cells and diseased cells, such as cancer cells. Cancer cells start their life as normal cells but undergo changes causing them to grow without control, to lose contact with each other and with the substrate to which they are attached. These changes can lead cells to spread around the body.

'I became interested in finding out about how normal breast cells turn into cancerous cells when I started working with a human breast cancer cell line which had the unusual capacity to develop into different subtypes of cells. This line provided an opportunity for us to develop a model of the human breast for studying cancer. Breast cancer cells arise from the glandular part of the breast which consists of epithelia (refer to figure 4.20, page 89). Different epithelial cells can be identified by the types of structural proteins (cytoskeleton) they contain. Together with members of my laboratory, I developed a tissue culture model to represent the glandular structures of the normal breast.

'Using this model, we have shown that the normal behaviour of the cells could be converted to the abnormal behaviour characteristic of cancer cells. The converted cells were not able to form proper contacts with each other and with the extracellular environment. They showed other changes, including alteration in expression of different intracellular markers, in particular one cytoskeleton marker called vimentin. Vimentin protein has been correlated with the degree of invasiveness of the cancer.

Figure 1.17 Associate Professor Leigh Ackland. When cells are not being used for experiments they can be frozen in liquid nitrogen. First, an anti-freeze agent is added to the culture to prevent damage to cell membranes.

'Cancer is a very complex disease. Many research scientists are working on different aspects of it, ranging from the role of the immune system and the role of cellular microenvironment to the epidemiology of cancer. These studies will all contribute to understanding how cancer cells arise and what factors are important in the progression of the disease.

'My interest in science was kindled by my maternal grandfather, a chemistry teacher, who took me on expeditions to places like museums and questioned the science behind what we saw. His inspiration stimulated me to take science at school and then at university.'

Electron microscopes

Transmission electron microscope

In the 1930s, the **transmission electron microscope** (TEM) was developed. Instead of light, a beam of electrons with a much shorter wavelength passes through and is used to illuminate specimens. Instead of glass lenses that control the passage of light rays in LMs, a TEM has a series of electromagnets that each create an electromagnetic field to control the path of the electron beam. Figure 1.18 shows a comparison between the internal structure of a light microscope and a transmission electron microscope. Note the similarities; note the differences.

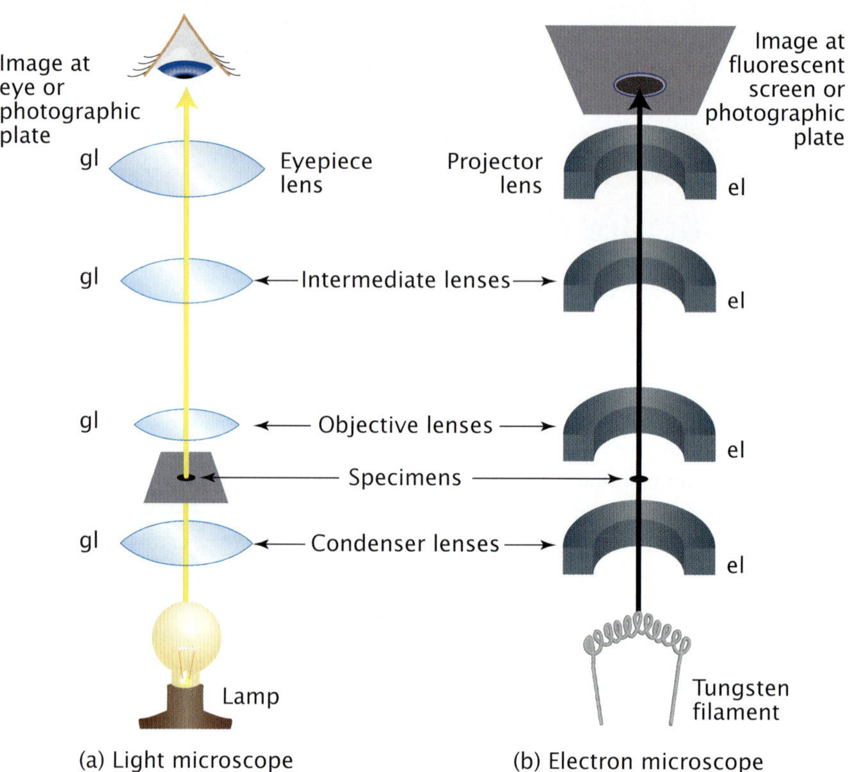

Figure 1.18 **(a)** Optical system of a light microscope (LM). The light source is visible light and glass lenses (gl) produce an image that can be detected by an eye or other appropriate receptor such as a camera. **(b)** Optical system of a transmission electron microscope (TEM). A tungsten filament emits a beam of electrons which is controlled by a series of electromagnetic lenses (el). In this figure, the orientation of the TEM system has been reversed to allow direct comparison of its components with those of a light microscope. In reality, the filament is at the top and the viewing screen at the bottom so a TEM resembles an inverted light microscope.

TEMs have a much greater resolving power than light microscopes (see table 1.2) because of the short wavelengths of electron beams. TEMs have revealed the presence of many kinds of cell organelles and have shown the complex internal structure that exists within cells. (See pages 33 and 34, figures 2.14b and 2.16a, which show part of the internal structure of a cell as seen with a TEM.)

Table 1.2 Comparison of the human eye, light microscope (LM) and transmission electron microscope (TEM). The smaller the separation distance between two objects, the larger the resolving power of the instrument being used.

	Human eye	*LM*	*TEM*
smallest resolvable separation distance	0.1 mm (100 µm)	0.000 2 mm (0.2 µm)	0.000 000 5 mm (0.0005 µm)
source of illumination	light rays	light rays	electron beam

Scanning electron microscope

The **scanning electron microscope** was released in 1965. This instrument is able to provide detailed images of surfaces (see figure 1.19). An electron gun produces an electron beam that is focused onto one spot on the surface of a specimen and is then scanned back and forth along the specimen's surface. The surface releases another set of electrons from the specimen and these form an image on a small fluorescent screen. Depending on their size, whole organisms can be scanned (figure 1.19).

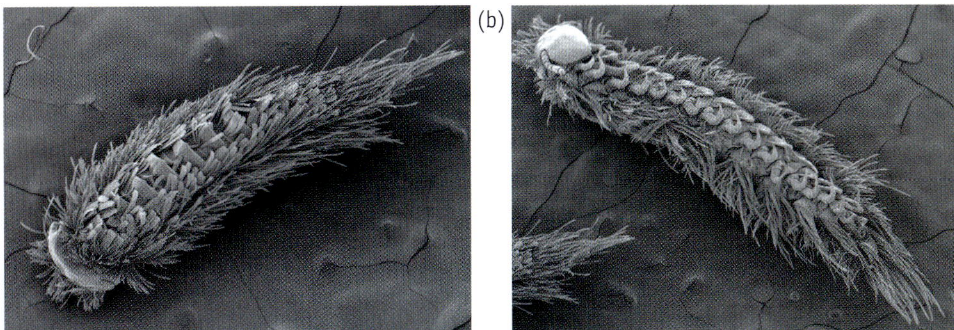

Figure 1.19 Scanning electron micrograph of the pin cushion millipede, *Phryssonatus novaehollandiae*. Note **(a)** the plates and hairs along the dorsal surface and **(b)** the pairs of legs and hairs visible on the ventral surface of the same organism. The adults of this species grow to about 4 mm in length and are abundant in the sand and soils of Victoria.

Although electron microscopes have greater resolving power than light microscopes, they can be used only with dead cells or organisms. The current ability of modern light microscopes, such as the confocal microscope, to allow the detailed study of living cells and identification and location of specific molecules in a cell make light microscopes more appropriate for some settings in spite of their reduced resolution. The size of the cell or organism under examination is also important in the choice of instrument. Figure 1.20 outlines the limits of use of the unaided eye, light microscopes and electron microscopes.

mm = millimetre
µm = micrometre
nm = nanometre

A logarithmic scale is one in which each marked unit moving up the scale is 10 times larger than the next. This contrasts with a linear scale in which each marked unit is the same size as the next.

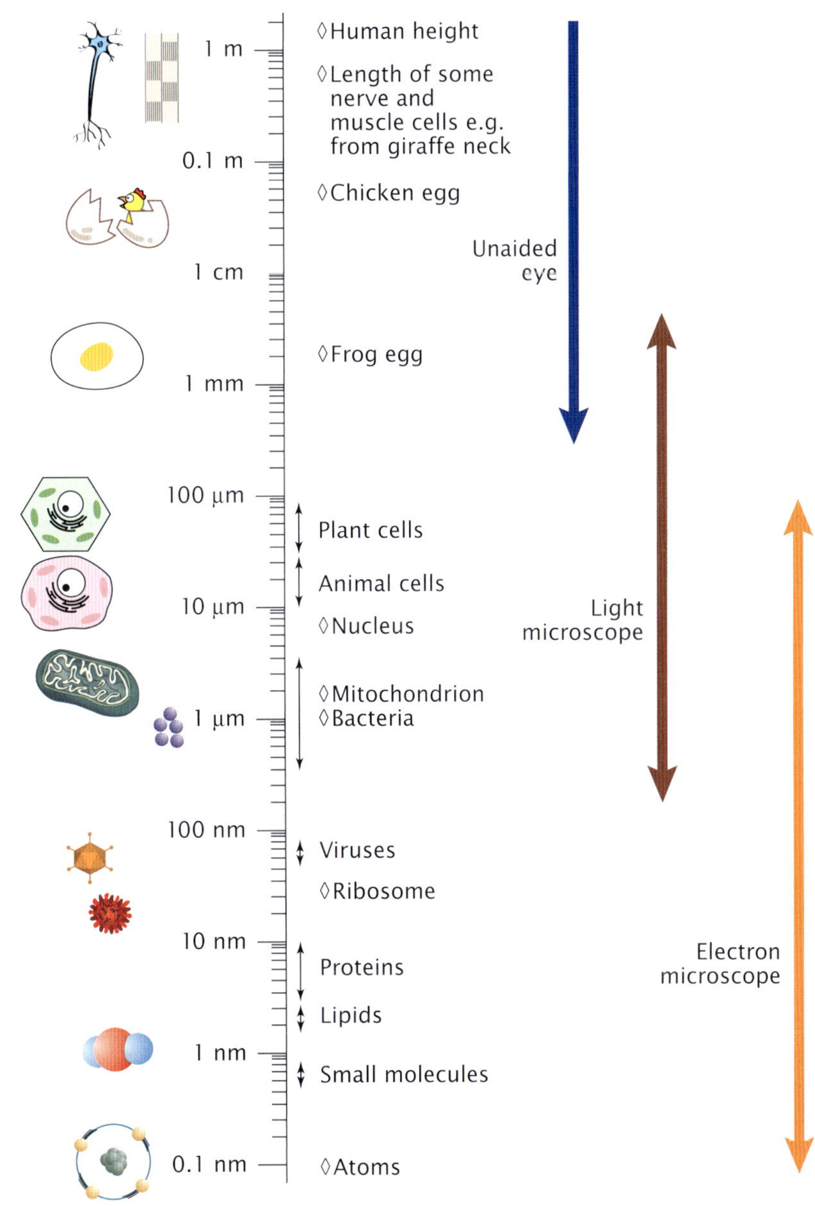

Figure 1.20 The arrows on the right-hand side indicate the ranges over which viewing is possible with an unaided eye, light microscope and electron microscope. The logarithmic scale indicates the size of organism, cell or cell part visible by the particular tools. 1 mm = 1000 µm; 1 µm = 1000 nm.

Recent developments in current systems

Light microscopes

Light microscopes have now been in use for centuries. As new technologies and materials became available, the power and capabilities of microscopes have changed significantly. Development continues. Computers have facilitated the use of many microscopes. Other advancements include modification of existing lens systems and automation in cells examination. We will consider two of these recent technological advances.

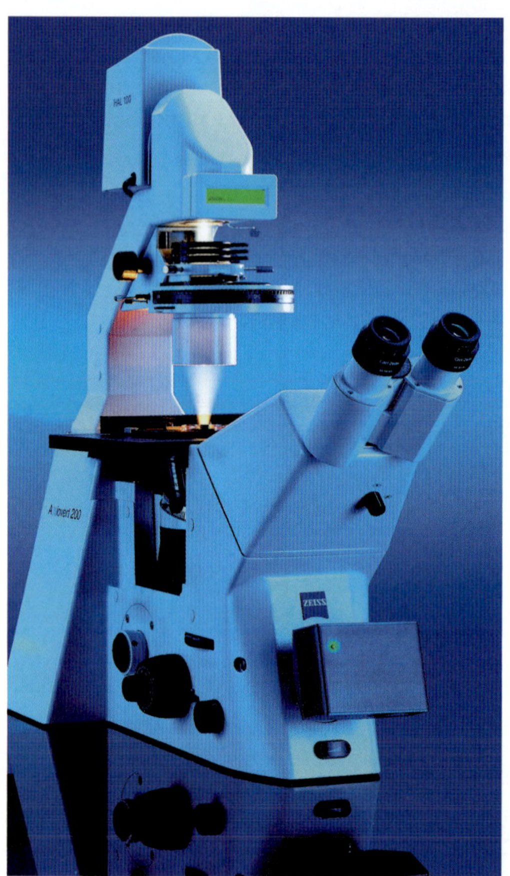

Figure 1.21 Differential interference contrast microscope. Note that the object being examined is illuminated from above. This is called an inverted system and is important because it provides a more stable system when manual manipulation of an object is required.

Differential interference contrast (DIC) microscopes

Differential interference contrast (DIC) microscopes (figure 1.21) are used to obtain and examine three-dimensional impressions of an object. The images are achieved using specially designed prisms to split and then recombine the light. A recent modification involves the way in which the light is split (the PlasDIC technique). This has significantly improved the optical resolution and use of the microscope.

The PlasDIC is a system that gives first-class, three-dimensional views of an object. This is particularly important when fine-detailed manipulation of living cells is required. One example of this is the need to inject a sperm into the cytoplasm of an egg (oocyte) in some cases of *in vitro* fertilisation (figure 1.22).

When a woman fails to conceive a child, it is sometimes due to a low sperm count in the semen of the male. In such cases, fertilisation can sometimes be achieved by manually injecting a single sperm into a mature egg. A good three-dimensional view of the egg is essential to check that all parts of the egg are in good condition, hence ensuring the highest possible chance of success of the injection.

Also in this modified optical system, plastic dishes can be used instead of glass to hold any specimen. This is important, not only because cells grow better in plastic receptacles than in glass, but also because plastic equipment is far cheaper to use.

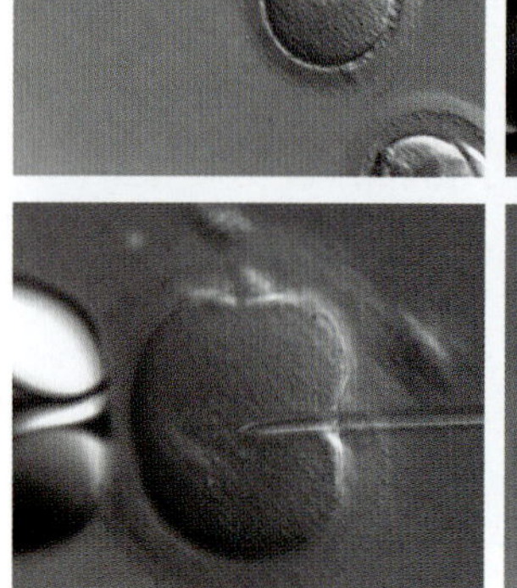
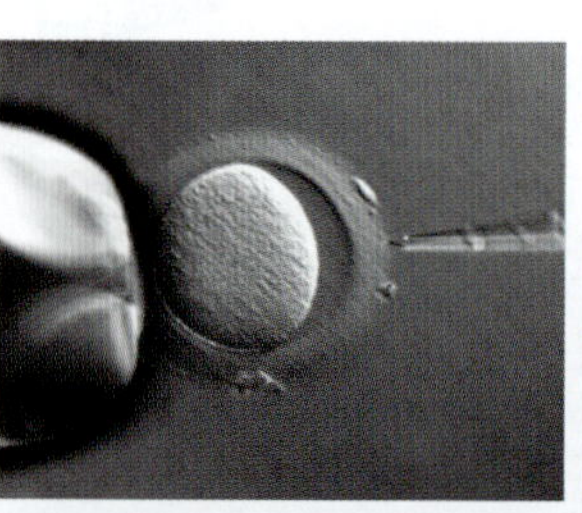
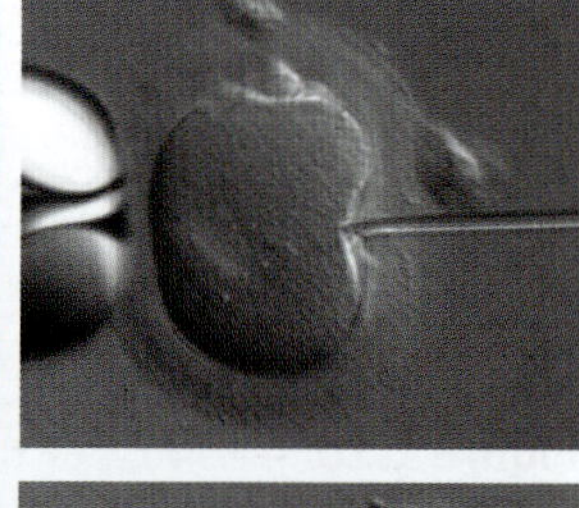
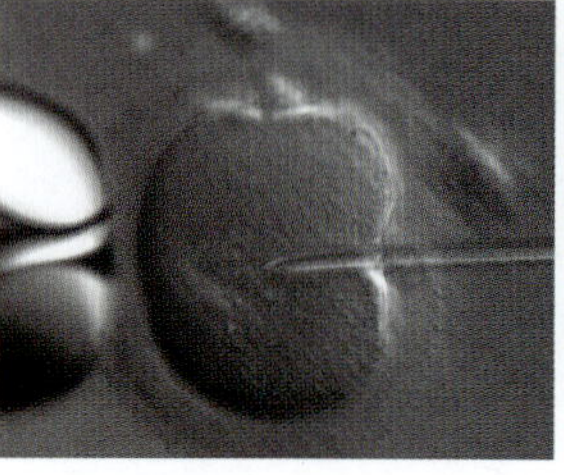
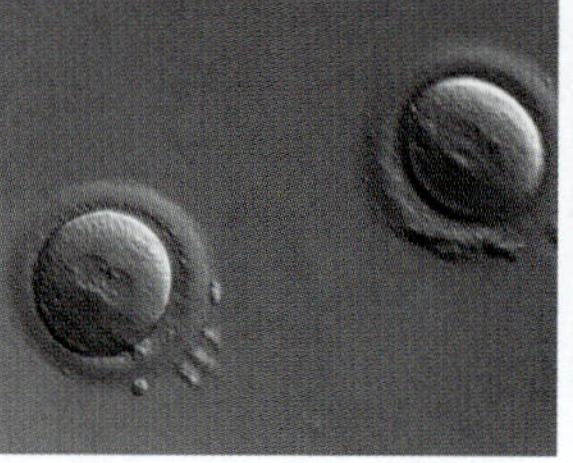
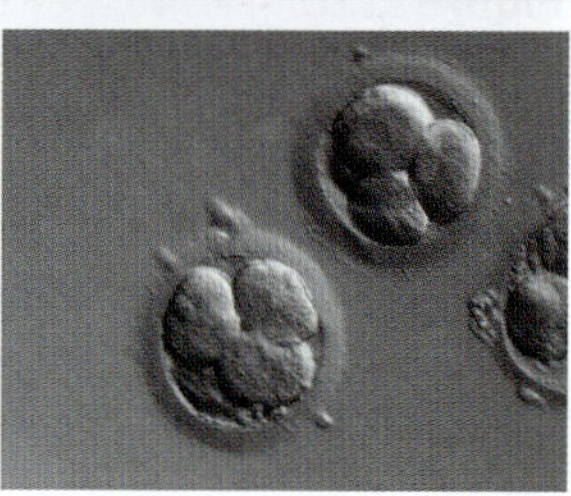

Figure 1.22 Injection of a sperm into the cytoplasm of an oocyte as viewed with a Zeiss PlasDIC system

Automatic scanning of cells

In medical diagnosis, there are situations in which many thousands of cells must be examined in a search for rare defective cells. Automatic scanning is now possible using motorised scanning systems (figure 1.23).

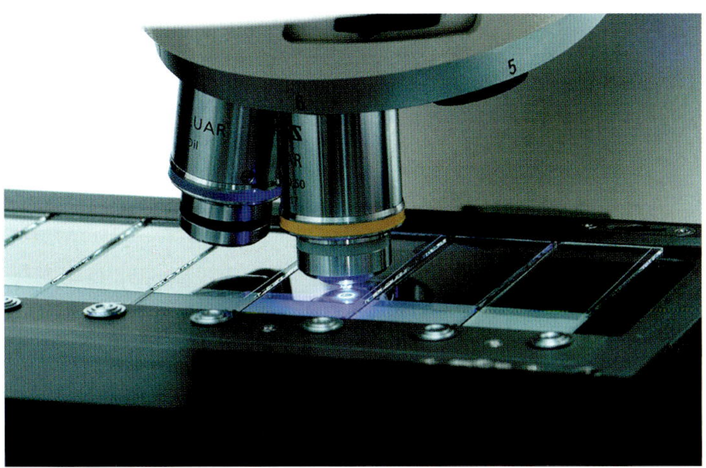

Figure 1.23 Automatic scanning of up to 7000 cells per second is possible with motorised scanning systems. Rare defective cells are identified. A computer records details about each defective cell and its position on the particular slide. A simple click of a computer mouse allows re-examination of any defective cell.

The technique combines two powerful components. The first is a mechanised evaluation platform holding the specimen slides. The platform searches through and analyses the cells, up to 7000 per second, and singles out those that are defective. The second powerful component is a computer that stores the results of any highlighted defective cell, as well as data related to the particular slide and the position of the cell. This means that defective cells can be readily found at a later time by simply clicking the computer mouse.

The system operates with both single cells, such as in a blood culture, and with groups of cells, for example, cells in a section of breast tissue. Fluorescent stains are often used in such systems.

Electron microscopes

Two other techniques are important for electron microscopy — **freeze fracture** and **shadowing**.

Freeze fracture

In freeze fracture, a small block of living or dead tissue is rapidly frozen in liquid nitrogen. Virtually no change occurs to the molecules of the specimen involved. The frozen tissue is placed into a vacuum chamber and broken with the sharp edge of a knife. The fracturing of the specimen exposes internal structures and their surfaces that can be examined after further treatment (see figure 2.20c, page 37).

Shadowing

In the shadowing technique, fractured pieces of specimen are exposed to, and partly covered by, heavy metal such as platinum or gold that is evaporated from a heated wire to one side of the vacuum chamber. Because the metal atoms come from one side of the chamber (see figure 1.24), the thickness of the metal film reflects the contours of the parts that are covered. The parts are then covered with a layer of carbon atoms that is transparent to electrons. This layer strengthens the metal replica. The specimen fragments are dissolved away leaving metal replicas that can be examined with an electron microscope.

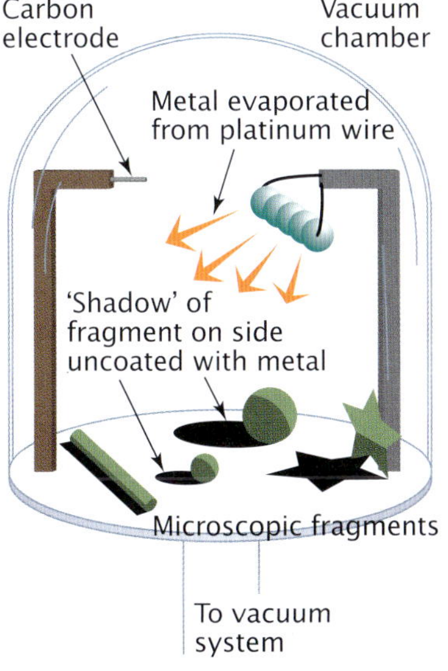

Figure 1.24 The technique of shadowing occurs in a vacuum chamber.

In this chapter we have examined a range of tools and techniques important in the study of cells, the basic structure of all living things. Although there are basic features that all cells share, there are also significant distinguishing features. We will explore the structure and function of cells of different kinds of organisms in the next chapter.

KEY IDEAS

- Various types of microscopes can be used to examine cells.
- Light microscopes (LMs) reveal details about the arrangement of cells and the internal structure of cells.
- Compound light microscopes (CLMs) have at least two sets of lenses: objective lenses and ocular (or eyepiece) lenses.
- Cells are often stained with one or more dyes to make their various components easier to see.
- Phase contrast microscopes allow the study of unstained living cells.
- Fluorescent microscopes reveal details of chemical substances present.
- Confocal microscopes use lasers to produce a sharply in-focus image of a thin layer of a specimen.
- Electron microscopes use beams of electrons instead of beams of light.
- Transmission electron microscopes (TEMs) reveal fine detail of the internal structure of cells.
- Scanning electron microscopes (SEMs) reveal details of cell surfaces.
- Technological advances involving equipment and stains associated with microscopy continue to be developed.

QUICK-CHECK

8 If you had a choice of any kind of light microscope, identify, giving a reason, the most appropriate one for viewing the following:
  a a living *amoeba*
  b a section of stained plant tissue
  c the transfer of a nucleus from one cell into another.

9 True or false? Briefly explain your choice.
  a All kinds of light microscopes use visible light to illuminate objects.
  b If the objective lens of a light microscope has a 5× magnification and its ocular lens is 10×, then the magnification obtained of an object being viewed is 15×.
  c The use of an oil immersion lens increases the magnification capability of a microscope.

10 If you had a choice of any kind of electron microscope, identify, giving a reason, the most appropriate one for viewing the following:
  a the surface of a layer of cells
  b a section of brain tissue
  c a small insect about 1 mm long.

11 True or false? Briefly explain your choice.
  a Electron microscopes can be used to view living and non-living tissues.
  b The resolving power of a TEM is greater than that of a LM.
  c TEMs and SEMs are equally appropriate to use for viewing minute organisms.

BIOCHALLENGE

1

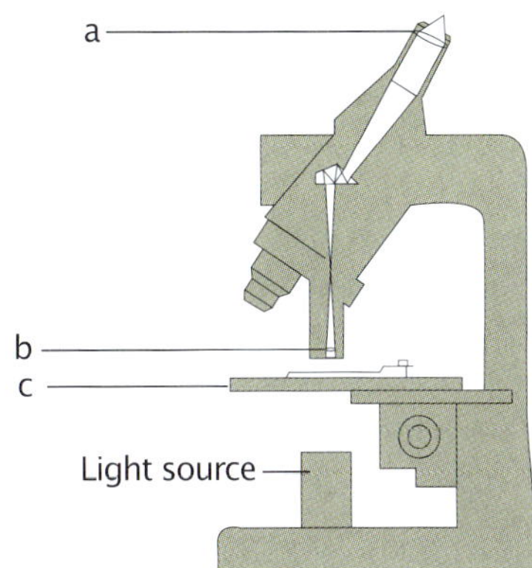

Match these labels to the parts indicated on the diagram of a compound light microscope:

- objective lens
- eyepiece
- stage.

2

5X
10X
Eyepieces

10X
40X
100X oil
Objective lenses

The magnifications are shown on these objective lenses and eyepieces from a light microscope. What are the minimum and maximum magnifications possible with these lenses?

3

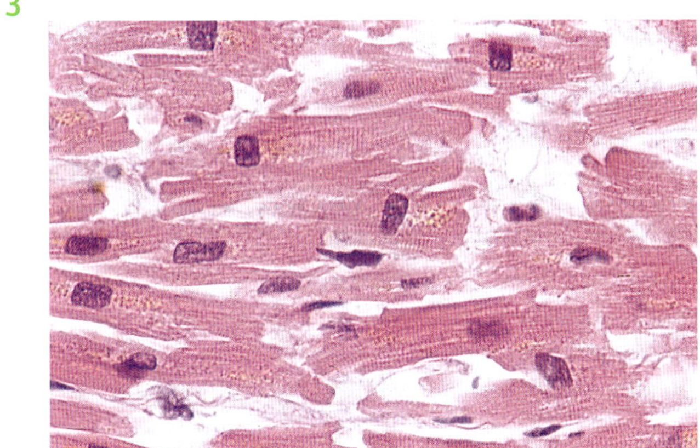

What kind of light was used to obtain this image of cells from an animal?

4

Given the detail in this image of part of a small animal, what kind of microscope was used?

5

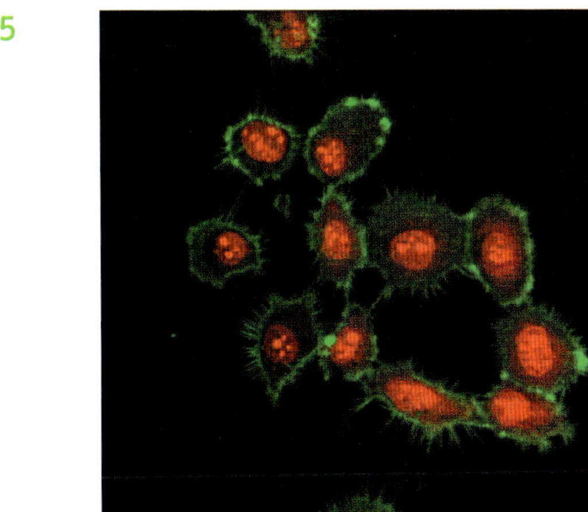

What kind of light was used to obtain this image of cells from an animal?

6 A small microbe was placed on a microscope slide that had a scale grid with lines at regular intervals of 10 µm etched into its surface. The microbe was examined with a light microscope. The result was:

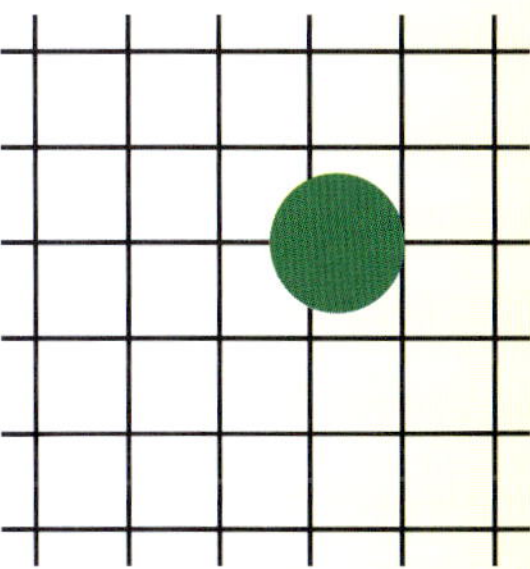

Given the dimensions of the grid, what is the diameter of the microbe?

CHAPTER REVIEW

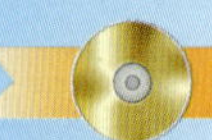

CROSSWORD

Key words

biogenesis
Cell Theory
cells
compound light microscopes
eyepiece (ocular) lenses
fluorescence microscope
freeze fracture
light microscope
magnification
microscopes
nucleus
objective lenses
oil immersion objective lens
phase contrast microscope
resolution
scanning confocal microscope
scanning electron microscope
shadowing
simple light microscopes
spontaneous generation
staining
transmission electron microscope

Questions

1 ***Making connections*** ➧ The key words listed above can also be called concepts. Concepts can be related to one another by using linking words or phrases to form propositions. For example, the concept 'compound light microscope' can be linked to the concept 'lenses' by the linking phrase 'contains at least two' to form a proposition. An arrow shows the sense of the relationship:

compound light microscope —contains at least two→ lenses

When several concepts are related in a meaningful way, a concept map is formed. Because concepts can be related in many different ways, there is no single, correct concept map. Figure 1.25 shows one concept map containing some of the key words and other terms from this chapter.

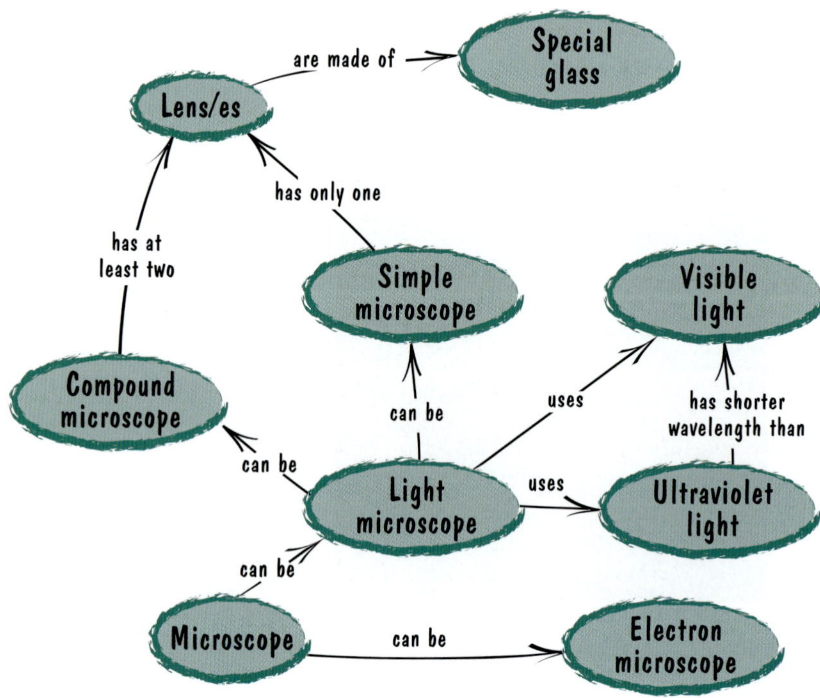

Figure 1.25 Example of a concept map

Use at least eight of the key words, and other words of your choice, to make a concept map relating to microscopes and the contribution they have made to the development of the Cell Theory and our ability to examine cells.

2 *Apply your understanding* ▸ You wish to examine a number of specimens. Refer to figures 1.4 (page 4) and 1.20 (page 15). Which microscope would you use to examine each of the following?

a the surface of a cell membrane

b a frog egg

c a clear view of cytoskeletal fibrils in a cell

d a general overall view of a plant cell

e very small structures in cell cytoplasm

Using scientific terminology/conventions ▸
When cells are being examined, dimensions are generally given in terms of micrometre (μm). It is important that you try to gain some understanding of how this measure relates to measurements of length that you are already familiar with, measurements such as metre (m), centimetre (cm) and millimetre (mm). Questions 3 and 4 are designed to give you practice at understanding these comparisons.

3 a How many millimetres (mm) are there in a metre (m)?

b How many times larger than a millimetre (mm) is a metre (m)?

c How many micrometres (μm) are there in a millimetre (mm)?

d How many times larger than a micrometre (μm) is a millimetre (mm)?

4 Fill in the following blanks.

a 1 μm = ________ nm

b 1 ________ = 10^{-9} m

c 1 μm = ________ m

5 *Communicating ideas* ▸ Explain why using an oil immersion objective lens has advantages over objective lenses that are used without the application of oil.

6 *Interpreting and communicating information using the Web* ▸ Go to www.jaconline.com.au/natureofbiology/natbiol1-3e and click on the 'Microscopy' weblink for this chapter. This information on microscopy consists of four pages. After reading page 1, go to page 2.

a Compare the image obtained with a confocal microscope with a wide-field image obtained with a standard microscope. Describe the difference(s) between the two images shown on page 2.

b What is the key feature of confocal microscopy that results in a sharp image being observed?

Now go to page 3 on fluorescence microscopy.

c What are the three components necessary for successful fluorescence microscopy?

d What were the colours resulting from the use of the triple stain?

e What components did the triple stain reveal as being present in the cell?

Now go to page 4.

f Indicate one situation in which you might choose to use differential interference contrast rather than phase contrast microscopy.

You may wish to browse through other relevant sections of this website.

2 Structure and function of cells

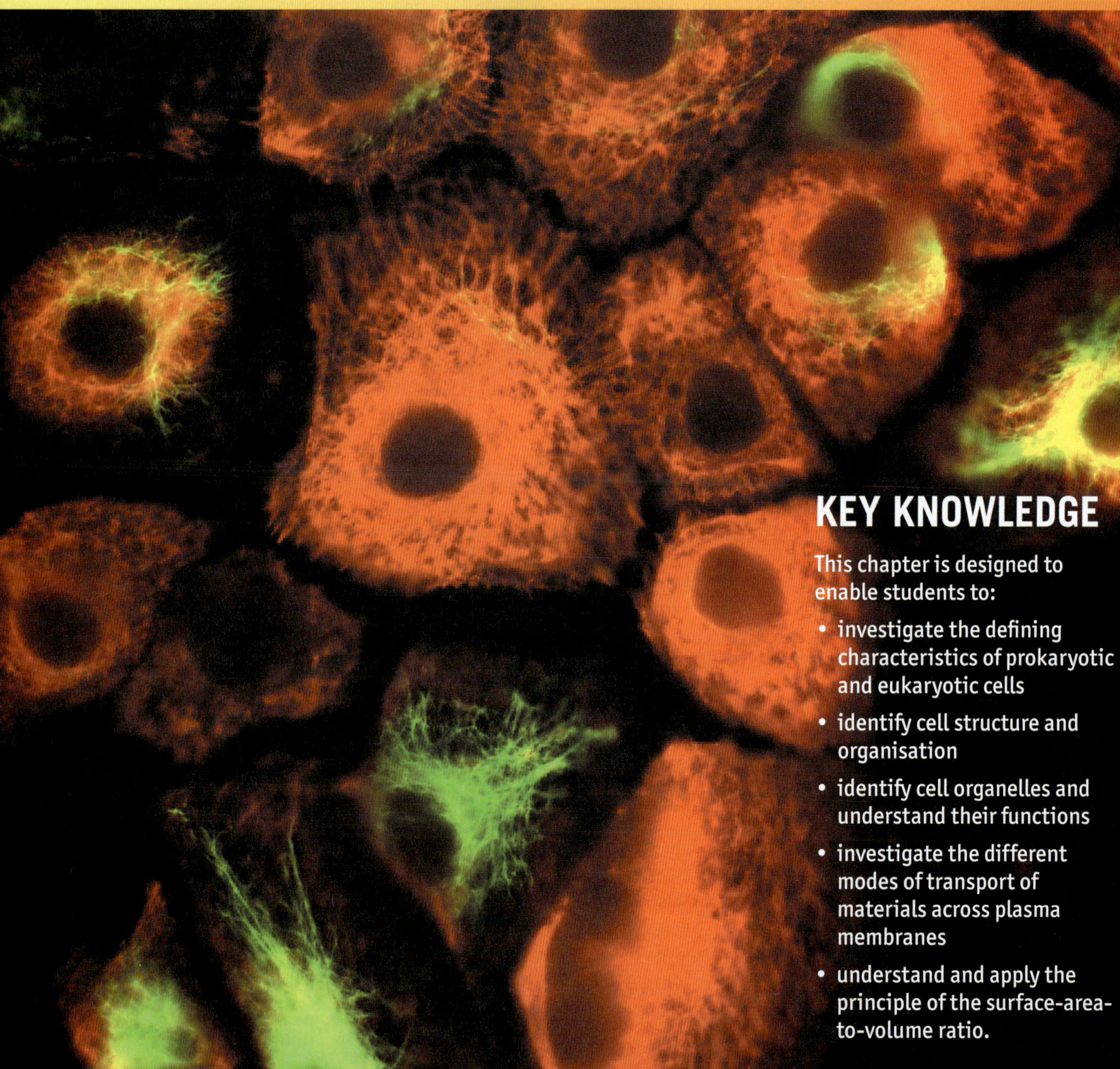

KEY KNOWLEDGE

This chapter is designed to enable students to:

- investigate the defining characteristics of prokaryotic and eukaryotic cells
- identify cell structure and organisation
- identify cell organelles and understand their functions
- investigate the different modes of transport of materials across plasma membranes
- understand and apply the principle of the surface-area-to-volume ratio.

Figure 2.1 Breast cells viewed with a confocal fluorescence microscope. A confocal microscope only collects light that is in focus so that only in-focus images are seen. The cells have been treated with special stains to highlight two different cytoskeleton proteins, vimentin (green), found in cells within cancerous tissue, and keratin (red). The cell membranes such as the nuclear envelope and the plasma membrane are not visible in this image because they have not received special staining. The position of each nucleus is made obvious by the black spherical area within each cell. Cell boundaries are located in the black area surrounding the stained part of each cell. In this chapter, we will consider the basic structure and organisation of prokaryotic and eukaryotic cells and how those structures relate to processes that are essential for the maintenance of life.

Clues from a pond

In July 1991, in Connecticut, USA, two young boys were fishing at a local pond. They were attacked by three older boys and severely beaten until almost unconscious. The older boys then threw the younger boys into the pond where they were in danger of drowning. Fortunately one of the young boys was sufficiently conscious to save himself and his friend.

Three suspects were apprehended, but what material evidence was available to place them at the crime scene?

Diatoms are single-celled golden-brown algae that are common in both sea water and fresh water. Whatever the shape of a diatom cell (figure 2.2), its plasma membrane is surrounded by a wall made of silica (glass) and pectin. The wall is in two parts that fit together like a lid on a box with the cell inside the box. Contact between a diatom cell and its environment takes place through thousands of tiny holes in its 'glass' coat.

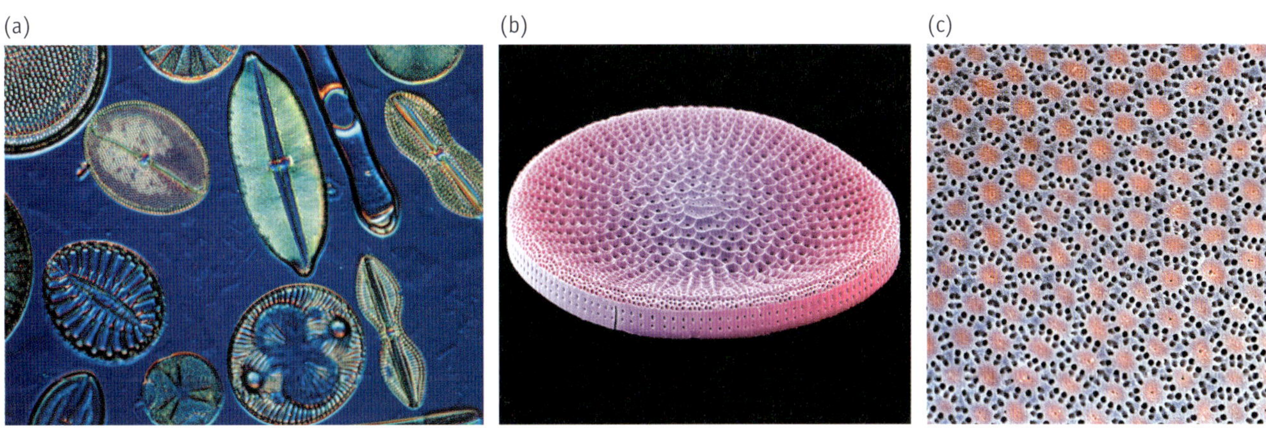

Figure 2.2 **(a)** Each species of diatom has a distinctive shape. **(b)** A close-up of the silica shell of a diatom (580×). The silica and pectin walls of diatoms survive long after the cells that made them have died. **(c)** An even closer view of the surface of the silica shell (2340×) shows the detailed patterns and perforations.

There are innumerable different species of diatoms commonly found, but research has shown that the ratio of numbers of each of the species is different from location to location. The ratio at a particular site is characteristic of that site, even if it changes from season to season.

Let's return to the attacks at the pond and the need for material evidence. Samples were collected from the pond and compared with diatoms found in the mud on the shoes of both the victims and the suspects. The same 25 different freshwater species were isolated from each of the samples. Statistical testing indicated that there was no difference in the population ratios in each of the samples. So the police had material evidence that the suspects were at the scene of the crime. They were, in fact, guilty of the attack.

Diatoms are used extensively in forensic cases where it is important to establish whether death occurred in water and, if so, what kind of water and at which location. Each pond and stream has its own populations — and diatoms that live in still water do not generally populate running waters. It is the presence of a rigid wall that survives after the death of the diatom cell that leaves a trace and this can be followed long after the death of a person.

The popular face of forensic science, as it is portrayed in many television programs, tends to rely heavily on evidence gained from the DNA and other analyses of animal tissues, such as blood, skin, hair and semen. However, forensic botany also has an important role. In addition to diatom studies, the type of plant material found at a crime scene and knowledge about where it grows may be an important clue. Did a person die where the body was found or was it transported from elsewhere?

Analysis of food in the stomach can indicate the time of death after a meal. A plant cell has a wall of **cellulose** surrounding its plasma membrane.

Deposits of diatom coats have accumulated for millions of years to form thick layers on ocean floors that are now part of geological structures. The coats crumble into a fine white powder that is mined and used in a variety of cleaning applications, including toothpaste.

ODD FACT

Archaeologists and palaeontologists examine fossilised faeces to study any bones and undigested parts of fruit and vegetables. This can help to establish the diets of prehistoric humans and other animals.

As cellulose is not digested, information about whether the person's last meal was high vegetable or low vegetable may be a clue as to where the person last ate and may lead investigators to people who saw the person there.

In this chapter we will consider the specialised structures that are found in different cells and how those structures relate to processes that are vital for the maintenance of life.

Looking at cells

Examination of cells using various microscopes reveals much about their internal organisation. Each living cell is a small compartment with an outer boundary known as the **cell membrane** or **plasma membrane**. Inside each living cell is a fluid, known as **cytosol**, that consists mainly of water containing many dissolved substances.

Another feature shared by all living cells is DNA, the genetic material that controls all the metabolic activities of a cell.

In contrast to these shared features, living cells can be classified into two different kinds on the basis of their internal structure:

- **Prokaryotic cells**. These have little defined internal structure and, in particular, lack a clearly defined structure to house their DNA. Organisms that are made of prokaryotic cells are called **prokaryotes** and include all **bacteria** (figure 2.3a) and all **archaeans**, another group of microbes (refer to chapter 8).
- **Eukaryotic cells**. These have a much more complex structure (see figure 2.3b) than prokaryotic cells. All eukaryotic cells contain many different kinds of membrane-bound structures called **organelles** suspended in the cytosol. These organelles include a nucleus with a clearly defined membrane called a **nuclear envelope**. The DNA of a eukaryotic cell is located in the nucleus. Organisms that are made of eukaryotic cells are called **eukaryotes** and include all animals, plants, **fungi** and **protists**, the single-celled organisms. Although a nucleus is usually visible with a light microscope, many organelles are visible only with electron microscopes.

Organelles are held in place by a network of fine **protein filaments** and **microtubules** within the cell, collectively known as the **cytoskeleton**. The filaments of the cytoskeleton are visible with an electron microscope, but require special staining to be seen with a confocal microscope (figure 2.1, page 22).

(a)

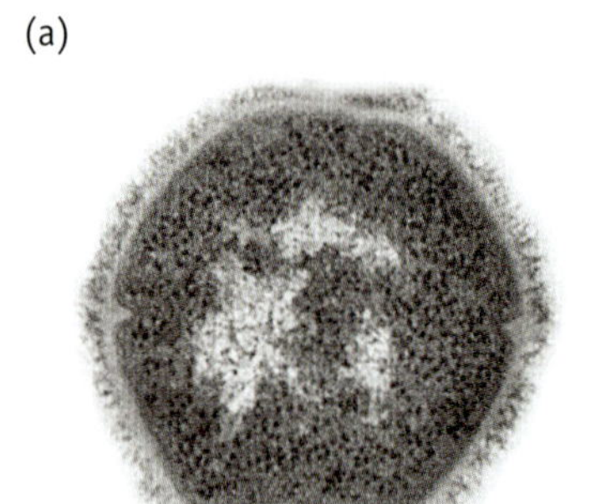

(b)

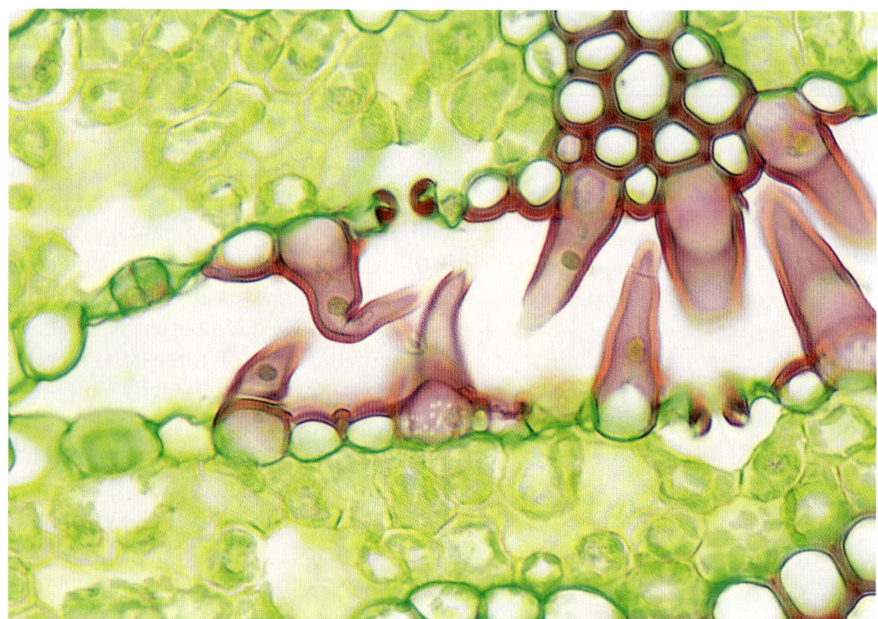

Figure 2.3 **(a)** Magnified image of a bacterial cell, *Streptococcus pyogenes*. Note the lack of a distinct, membrane-bound nucleus. **(b)** Transverse section of marrum grass, the eukaryote *Ammophila arenaria*, magnified 400×. The cells stained pink are hair cells. Note the circular structure, a nucleus, visible in some of the hair cells.

The plasma membrane boundary

The boundary of all living cells is a plasma membrane which controls entry of dissolved substances into and out of the cell. A plasma membrane is an ultra thin and pliable layer with an average thickness of less than 0.01 µm (0.000 01 mm). A plasma membrane can be seen using an electron microscope.

	Prokaryotes	*Eukaryotes*
Plasma membrane	present	present
Function	boundary of a cell; maintains the internal environment of a cell by controlling the movement of substances into and out of the cell	

A plasma membrane contains both lipid and protein. A more recent model of the plasma membrane is shown in figure 2.4. This model suggests that a plasma membrane consists of a double layer of lipid, and that proteins are embedded in this layer forming channels that allow certain substances to pass across the membrane in either direction. This model is known as the fluid mosaic model.

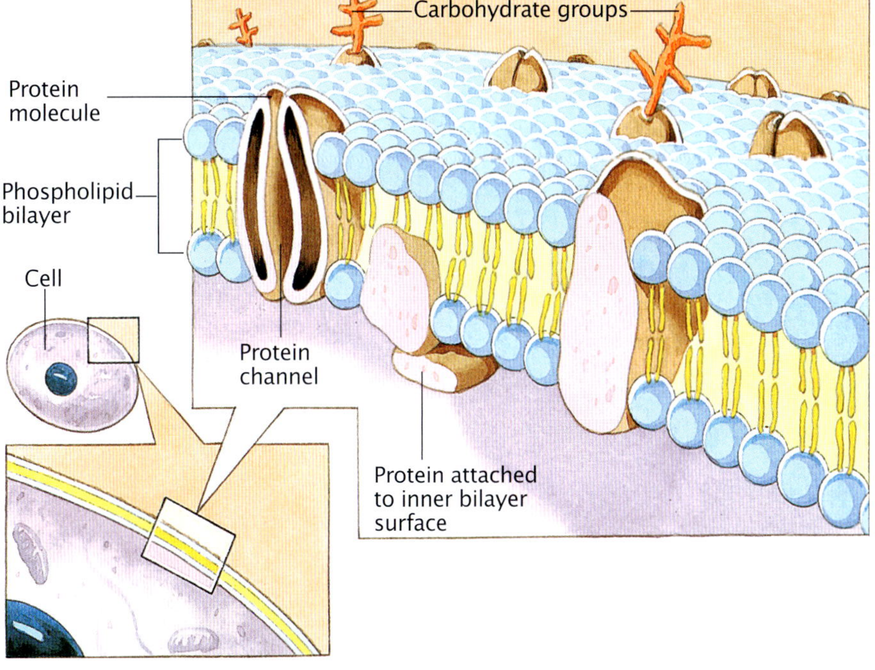

Figure 2.4 The fluid mosaic model proposed by Singer and Nicholson for the structure of the plasma membrane of a red blood cell. The lipid in the membrane is a complex lipid known as phospholipid. What does this name suggest about this lipid? The carbohydrate groups in the outer surface form part of antigens (see chapter 3).

Movement in and out of cells

All cells must be able to take in and expel various substances across their membranes in order to survive, grow and reproduce. Generally, these substances are in solution, but in some cases, may be tiny solid particles.

Because a plasma membrane allows only some dissolved materials to cross it, the membrane is said to be a **partially permeable** boundary (see figure 2.5). ('Partially permeable' is also known as selectively or differentially or semi-permeable.) Dissolved substances that are able to cross a plasma membrane — from outside a cell to the inside or from the inside to the outside — do so by various processes, including diffusion and active transport.

Figure 2.5 A partially permeable barrier allows only some substances to cross it.

SURFACE-AREA-TO-VOLUME RATIO

- Why are cells small?
- The surface of some cells is elaborately folded. What is the importance of these outfoldings?
- Some animals have a greatly flattened shape. How might this affect their survival?

Consider the surface area of cells compared with their volumes. This value is sometimes called the **surface-area-to-volume ratio (SA:V ratio)**. The SA:V ratio of any object is obtained by dividing its area by its volume.

'Area' refers to the coverage of a surface. One unit of measurement of area is a square centimetre (cm^2). 'Volume' refers to the amount of space taken up by an object. One unit of measurement of volume is the 'litre' (L), but the volume of solid matter, such as a brain, is sometimes expressed in units such as 'cubic centimetres' (cm^3). (Note: For a sphere, SA = $3\pi r^2$ and V = $\frac{4}{3}\pi r^3$, where r = radius.)

Looking at SA:V ratio

Examine the following data. Notice that as a sphere increases in size, its surface-area-to-volume ratio decreases.

Radius of sphere	SA:V
1 unit	3.0
2	1.5
3	1.0
6	0.5
10	0.3

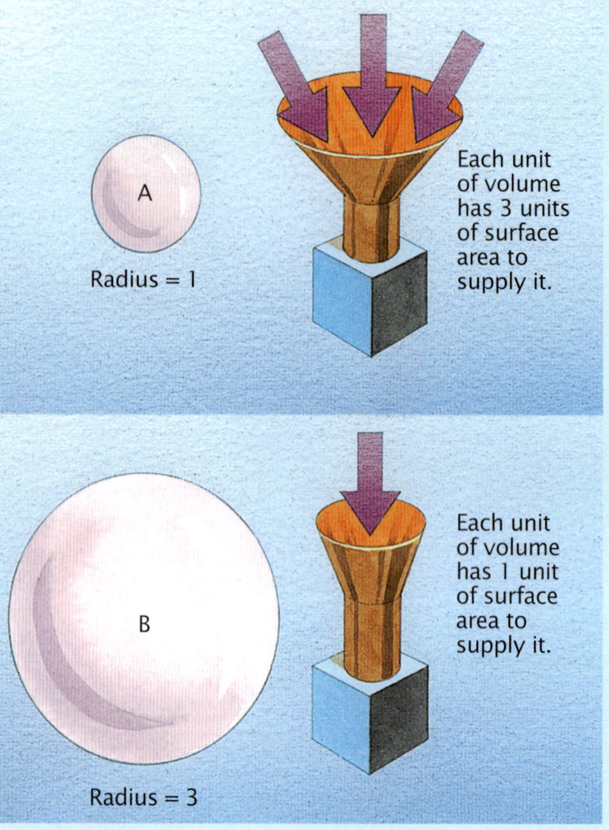

Figure 2.6 As a sphere increases in size, the amount of surface area for each unit of volume decreases.

What does this change in surface-area-to-volume ratio values mean? The SA:V ratio of a shape identifies how many units of external surface area are available to 'supply' each unit of internal volume. So, for a sphere with a radius of one unit, each unit of volume has three units of surface area to supply it. In contrast, for a sphere with a radius of three units, each unit of volume has only one unit of surface area to supply it. This interpretation is shown diagrammatically in figure 2.6. In general, as a particular shape increases in size, the SA:V ratio of the shape decreases.

As part of staying alive, cells must take in supplies of essential material from outside to meet their energy needs. Cells have to move wastes from inside to the outside. Efficient uptake and output of material is favoured by a higher surface-area-to-volume ratio. It is reasonable to suggest that cells are limited to small sizes so that their surface areas are large enough to let in essential material fast enough to meet their needs, and to allow waste materials to diffuse out fast enough to avoid the cells being poisoned by their own wastes.

Same volumes, different shapes

Cells differ in shape. How does the shape of a cell affect its SA:V ratio?

Look at the SA:V ratios of some different shapes in table 2.1 below. Although they differ in shape, they have the same volume of one litre (1000 cm^3).

The SA:V ratio varies according to shape. The flat sheet has about 10 units of surface area for each unit of volume. In contrast, the sphere has just half a unit of surface area for each volume unit. What are the biological consequences of the conclusion? Cells with outfoldings can exchange matter with their surroundings more rapidly than cells lacking this feature.

A cell has an area of 10 units and a volume of two units. What is its SA:V ratio?

Two cells (*P* and *Q*) have the same volume, but cell *P* has a surface area that is ten times greater than that of cell *Q*. Which cell would be expected to take up matter from its surroundings at the greater rate? Why? What can be reasonably inferred about the shapes of the two cells?

A cell has many outfoldings on its surface. How would these outfoldings affect its surface area as compared with a cell with a 'smooth' surface?

Summary

As a structure increases in size, its surface-area-to-volume ratio (SA:V) decreases.

Various shapes differ in their SA:V ratios, with this ratio being highest in flattened shapes and lowest in spheres.

The size of cells is limited by SA:V ratios since these ratios influence the rate of entry and exit of substances into and out of cells.

Table 2.1 The SA:V ratios for different shapes with the same volume vary depending on the shape. Which shape has the highest SA:V ratio?

Shape of container	*Surface area (cm^2)*	*Volume (cm^3)*	*Approx. SA:V ratio*
flat sheet (100 × 100 × 0.1 cm)	10 040	1000	10.0
cube (10 × 10 × 10 cm)	600	1000	0.6
flat pancake (height: 1 cm; radius: 17.8 cm)	2 112	1000	2.1
sphere (6.2 cm radius)	483	1000	0.5

Free passage: diffusion

Diffusion is the net movement of a substance, typically in solution, from a region of high concentration of the substance to a region of low concentration. The process of diffusion does not require energy.

Figure 2.7 shows a representation of this process for dissolved substance X. At all times, molecules of X are in random movement. At first, some molecules collide with and cross the plasma membrane into the cell (see figure 2.7a). As long as substance X is more concentrated outside the cell than inside, more collisions causing molecules of X to move from outside to inside occur than collisions from the opposite direction. As a result, a net movement of molecules of substance X occurs from outside to inside and the concentration of X inside the cell rises (figure 2.7b). Eventually, the numbers of collisions occurring on both sides of the membrane become equal. At that time (figure 2.7c), the number of molecules of X passing into the cell is equal to the number passing out. Diffusion stops at the stage when the concentration of substance X is equal on the two sides of the membrane.

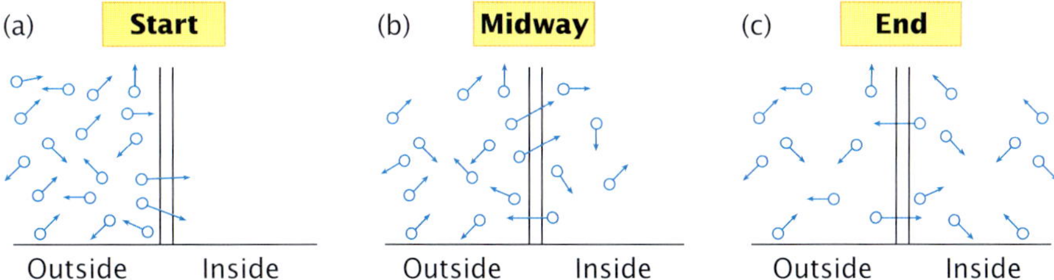

Figure 2.7 Diffusion in action **(a)** At the start, substance X starts to move into the cell because of random movement that results in some collisions with the membrane. **(b)** Midway, molecules of substance X are moving both into and out of the cell, but the net movement is from outside to inside. **(c)** When the concentration of X is equal on each side of the membrane, the number of collisions on either side of the membrane is equal and the net movement of molecules of substance X stops. Does this mean that collisions of molecules of substance X with the membrane stop?

One special case of diffusion is known as **osmosis**. The process of osmosis occurs when a net movement of water molecules occurs by diffusion across a cell membrane either into or out of a cell. Read the box on page 28 which outlines the movement of water into and out of a cell when it is placed in a strong sugar solution (figure 2.8a) and pure water (figure 2.8b) respectively.

Substances that can dissolve readily in water are termed **hydrophilic**, or 'water-loving'. Some substances that have a low water solubility or do not dissolve in water are able to dissolve in or mix uniformly with lipid. These substances are termed **lipophilic** (sometimes called hydrophobic). Examples of lipophilic substances include alcohol and ether. Lipophilic substances can cross plasma membranes readily. This observation provides indirect evidence for the presence of lipid in the structure of the plasma membrane. The rapid absorption of substances, such as alcohol across plasma membranes, appears to be related to the ability of alcohol to mix with lipid.

OSMOSIS — A SPECIAL DIFFUSION CASE

Consider red blood cells suspended in a strong sugar solution, as shown in figure 2.8a. Water molecules can pass through the plasma membrane in either direction, but the sugar molecules cannot cross the membrane. What will happen? The cell will shrink due to water loss. A solution that has a higher concentration of dissolved substances than a solution or a cell to which it is being compared is called a **hypertonic** solution.

When a plant cell is placed in a strong sugar solution, the plasma membrane shrinks away from the cell wall which retains its shape. Now consider cells suspended in pure water, as shown in figure 2.8b. In this case, there will be a net inflow of water molecules into the cell. Why? The cell will swell and then burst. A solution that has a lower concentration of dissolved substances than a solution or cell to which it is being compared is called a **hypotonic** solution.

If a cell is placed in a solution and there is no net gain or loss by the cell, the solution must have the same solute concentration as the cell and is called an **isotonic** solution.

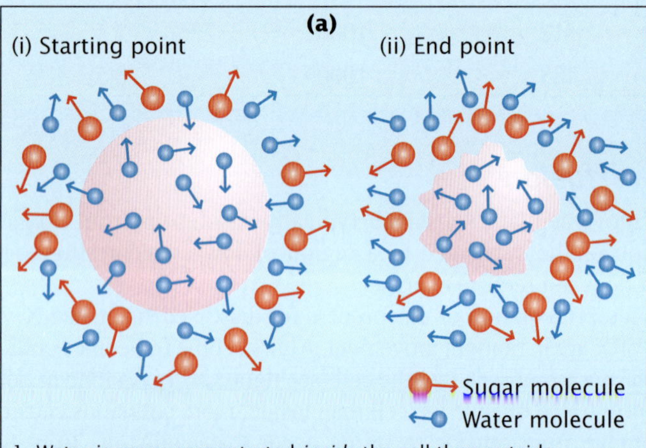

1. Water is more concentrated *inside* the cell than outside.
2. Water molecules move in random directions and some collide with the plasma membrane.
3. Initially, the number of water molecules inside the cell colliding with the plasma membrane and moving *out* is greater than the number outside moving *in*.
4. These differential rates of random collisions with the plasma membrane produce a net outward flow of water molecules from the cell.
5. The cell shrinks because of this water loss.

(b)

(i) Starting point

(ii) End point

Organic molecule

Water molecule

1. Water is more concentrated *outside* the cell than inside.
2. Water molecules move in random directions and some collide with the plasma membrane.
3. Initially, the number of water molecules outside the cell colliding with the plasma membrane and moving *in* is greater than the number inside moving *out*.
4. These differential rates of random collisions with the plasma membrane produce a net inward flow of water molecules into the cell.
5. The cell swells because of this water gain and then bursts.

Figure 2.8 **(a)** Cell in strong sugar solution **(b)** Cell in pure water

The movement of some substances across the plasma membrane is assisted or facilitated by carrier protein molecules. This form of diffusion, involving a specific carrier molecule, is known as **facilitated diffusion** (see figure 2.9a). The net direction of movement is from a region of higher concentration of a substance to a region of lower concentration, and so the process does not require energy. Movement of substances by facilitated diffusion mainly involves substances that cannot diffuse across the plasma membrane by dissolving in the lipid layer of the membrane. For example, the movement of glucose molecules across the plasma membrane of red blood cells involves a specific carrier molecule.

Paid passage: active transport

Active transport is the net movement of dissolved substances into or out of cells against a concentration gradient (see figure 2.9b). Because the net movement is against a concentration gradient, active transport is an energy-requiring (endergonic) process. Active transport enables cells to maintain stable internal conditions in spite of extreme variation in the external surroundings.

In some situations, active transport of salts occurs. Animals that live in fresh water, for example frogs (figure 2.10), tend to lose salts by diffusion across their

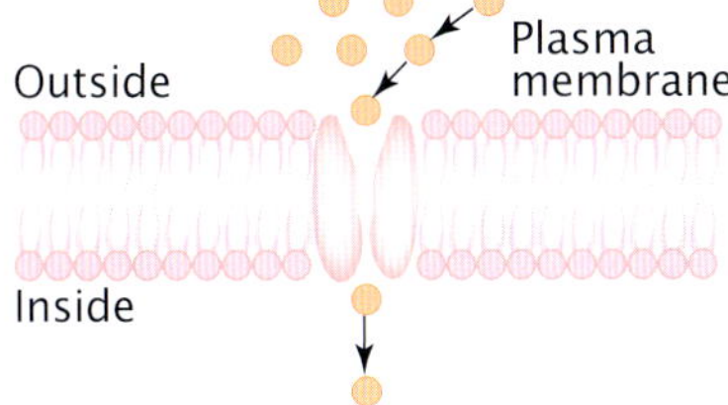

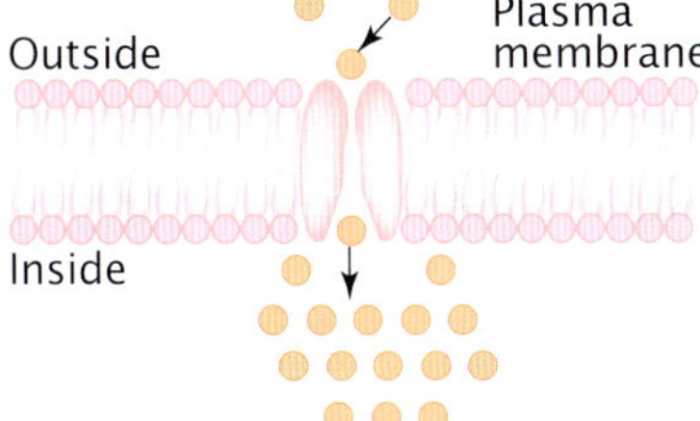

Figure 2.9 **(a)** Facilitated diffusion occurs with substances that cannot dissolve in the lipid layers of the plasma membrane. **(b)** Active transport. Does this process require an input of energy?

skin-cell plasma membranes into the surrounding fresh water. Energy in the form of **adenosine triphosphate (ATP)** is used to transport salt molecules against a concentration, from the surrounding water where salt concentration is low, across plasma membranes into cells where the salt concentration is very high.

This process involves a carrier protein for each substance that is actively transported. If the carrier protein for a particular substance is defective, the organism may show a disorder. In human beings, a defect in the carrier protein involved in the active transport of chloride ions (Cl^-) has been found to be the cause of the inherited disorder, cystic fibrosis.

Figure 2.10 To balance the loss of salts that occurs from frog skin cells by diffusion, energy is used to drive active transport of salts from a region of low concentration in the surrounding water, across plasma membranes, into the frog skin cells that have a high concentration of salts.

Some bacteria thrive in highly salty water where other organisms cannot survive (see table 2.2). How do these halophytic ('salt-loving') bacteria maintain a stable internal environment?

Table 2.2 Conditions inside and outside halophytic bacterial cell

	Water concentration	*Salt concentration*
outside the bacterial cell	low	HIGH
inside the bacterial cell	HIGH	low

Salt molecules do not readily cross the plasma membrane. A net movement of water molecules occurs down the concentration gradient from inside the cell to outside. However, the bacteria have an efficient mechanism for active transport of water. Water molecules are actively transported into the cell at a rate that compensates for the loss of water by osmosis, so that the internal conditions in the bacterial cell remain stable. Energy is needed to power this 'water pump'. Placed in the same very salty conditions, cells of other organisms would shrivel and dehydrate.

Bulk transport

Solid particles can be taken into a cell. For example, one kind of white blood cell is able to engulf a disease-causing bacterial cell and enclose it within a **lysosome** sac where it is destroyed. Unicellular protists, such as *Amoeba* and *Paramecium*, obtain their energy for living in the form of relatively large 'food' particles that they engulf and enclose within a sac where the food is digested (see figure 2.11a).

Refer to page 36 for more information on lysosomes.

Note how part of the plasma membrane encloses the material to be transported and then pinches off to form a membranous **vesicle** that moves into the cytosol (figure 2.11b). This process of bulk transport of material into a cell is called **endocytosis**. When the material being transported is a solid food particle, the type of endocytosis is called **phagocytosis**.

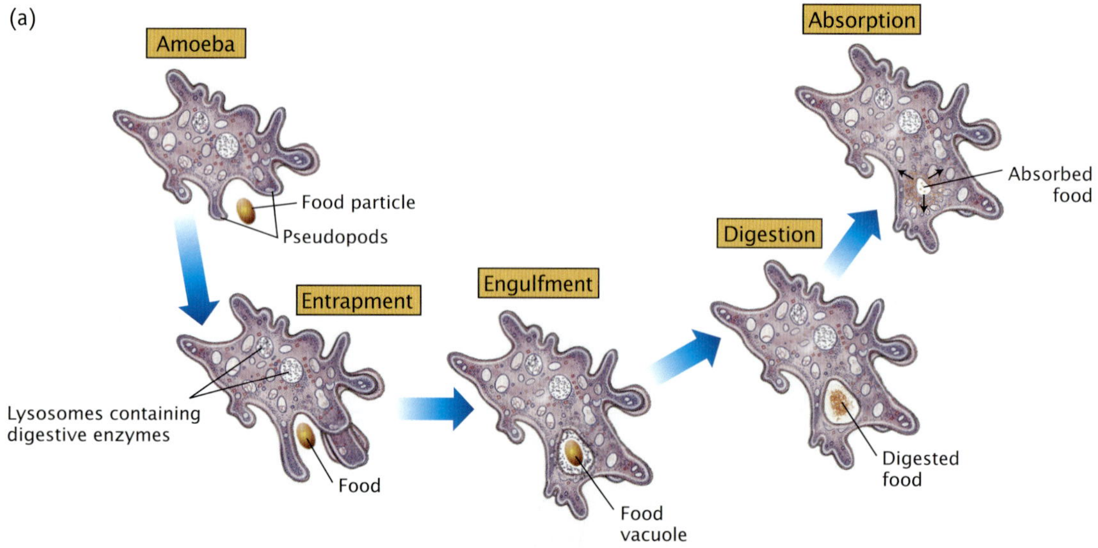

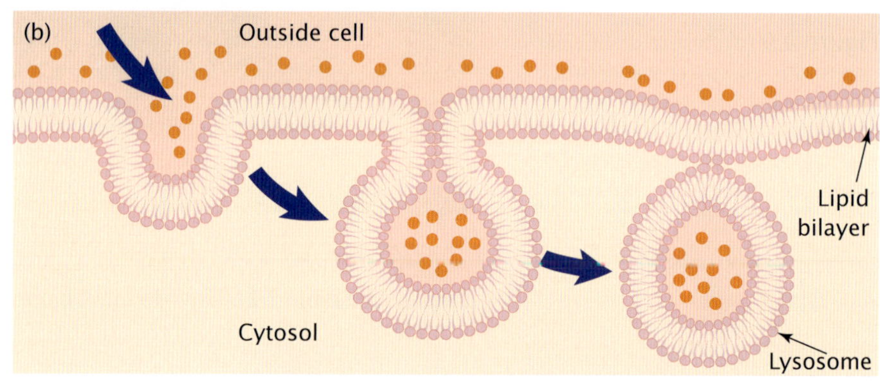

Figure 2.11 **(a)** Transport of a solid food particle across the membrane of an *Amoeba*. **(b)** Endocytosis occurs when part of the plasma membrane forms around a particle to form a vesicle, which moves into the cytosol.

Endocytosis
bulk transport of material into a cell

If material is solid → the process is called **phagocytosis** from the Greek *phagos* = 'eating' and *cyto* = 'cell'

If material is fluid → the process is called **pinocytosis** from the Greek *pinus* = 'drinking' and *cyto* = 'cell'

Figure 2.12 Endocytosis — a summary

Although some cells are capable of phagocytosis, most cells are not. Most eukaryotic cells rely on **pinocytosis**, a form of endocytosis that involves material that is in solution being transported into cells. Note the summary in figure 2.12.

Bulk transport out of cells (for example, the export of material from the Golgi complex, discussed on pages 34–5) is called **exocytosis**. In exocytosis, vesicles formed within a cell fuse with the plasma membrane before the contents of the vesicles are released from the cell (see figure 2.13). If the released material is a product of the cell (for example, the contents of a Golgi vesicle), then 'secreted from the cell' is the phrase generally used. If the released material is a waste product after digestion of some matter taken into the cell, 'voided from the cell' is generally more appropriate.

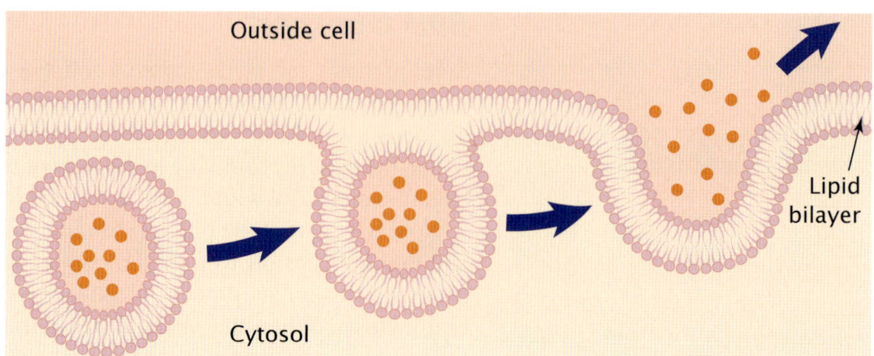

Figure 2.13 Exocytosis (bulk transport out of cells) occurs when vesicles within the cytosol fuse with the plasma membrane and vesicle contents are released from the cell.

Cell walls

	Prokaryotes	Eukaryotes		
		Fungi	Plant	Animal
Cell wall	present	present	present	absent
Function	semi-rigid, protective structure deposited by the cell outside the cell membrane			

The plasma membrane forms the exterior of animal cells. However, in plants, fungi and bacteria, a rigid **cell wall** lies outside the plasma membrane. The absence of a cell wall is characteristic of organisms in Kingdom Animalia.

Composition of cell wall

The cell wall varies in composition between plants, fungi and bacteria (see table 2.3).

Table 2.3 Composition of cell wall in various types of organisms. Why are animals excluded?

Type of organism	*Compounds present in cell wall*
plants	include cellulose
fungi	include chitin
bacteria	include complex polysaccharides

In some flowering plants, the original or **primary cell wall** in certain tissues becomes thickened and strengthened by the addition of lignin to form **secondary cell walls**. This process provides great elastic strength and support, allowing certain plants to develop as woody shrubs or trees.

KEY IDEAS

- The plasma membrane forms the boundary of each living cell.
- Several different processes exist whereby substances may cross plasma membranes.
- Cell walls lie outside the plasma membrane of plant, fungal and prokaryotic cells.

QUICK-CHECK

1 What is meant by the label 'partially permeable' in reference to the plasma membrane?
2 Which of the following is an energy-requiring process?
  a osmosis
  b diffusion
  c active transport
  d facilitated diffusion
3 What is the function of a cell wall?

Cell organelles

The nucleus: control centre

	Prokaryotes	*Eukaryotes*
Nucleus	absent	present
	DNA is dispersed in cell	encloses the DNA

Cells have a complex internal organisation and are able to carry out many functions. The control centre of the cells of animals, plants, algae and fungi is the nucleus (see figure 2.24, page 41). The nucleus in these cells forms a distinct spherical structure that is enclosed within a double membrane, known as the nuclear envelope. Cells that have a membrane-bound nucleus are called eukaryotic cells. The regular presence of a nucleus in living cells was first identified in 1831 by a Scottish botanist, Robert Brown (1773–1858) (see pages 7–8).

The term 'chromosome' means 'coloured body'. The fact that the cell of each species contains a definite number of chromosomes was first recognised in 1883.

Cells of organisms from Kingdom Monera, such as bacteria, contain the genetic material (DNA), but it is not enclosed within a distinct nucleus. Cells that lack a nuclear envelope are called prokaryotic cells.

A light microscope view reveals that the nucleus of a eukaryotic cell contains stained material called **chromatin** that is made of the genetic material **deoxyribonucleic acid (DNA)**. The DNA is usually dispersed within the nucleus. During the process of cell reproduction, however, the DNA becomes organised into a number of rod-shaped chromosomes (refer to chapter 4, pages 83–4). The nucleus also contains one or more large inclusions known as **nucleoli** which are composed of **ribonucleic acid (RNA)**.

Textbook diagrams often show a cell as having a single nucleus. This is the usual situation, but it is not always the case. Your bloodstream contains very large numbers of mature red blood cells, each with no nucleus. However, at an earlier stage, as immature cells located in your bone marrow, each of these cells did have a nucleus. Some liver cells have two nuclei.

Mitochondrion: energy-supplying organelle

	Prokaryotes	*Eukaryotes*
Mitochondria	absent	present
Function	site of production of much of the ATP required by a cell	

ODD FACT

Skeletal (voluntary) muscles are ones that you can move at will and that you use when you stand up or throw a ball. Skeletal muscle consists of long fibres formed from the fusion of many cells. As a result, these muscle fibres contain many nuclei, and are said to be multi-nucleate. Is a muscle fibre an example of one cell with many nuclei?

Living cells use energy all the time. The useable energy supply for cells is chemical energy present in a compound known as ATP (adenosine triphosphate) (see figure 2.14). The ATP supplies in living cells are continually being used up and must be replaced.

ATP is produced during **cellular respiration** (or just simply respiration). In eukaryotic cells, most of this process occurs in organelles known as **mitochondria** (singular = mitochondrion) which form part of the cytoplasm. Mitochondria cannot be resolved using an LM, but can be seen with an electron microscope. Each mitochondrion has an outer membrane and a highly folded inner membrane. Mitochondria are not present in prokaryote cells.

The role of mitochondria in respiration is discussed further in chapter 3.

Prokaryotes obtain their energy from a range of sources. This will be explored in your later studies of biology.

(a)

HO–P–O–P–O–P–O–CH₂ structure:

Triphosphate

Adenine

D-ribose

Adenosine

(b)

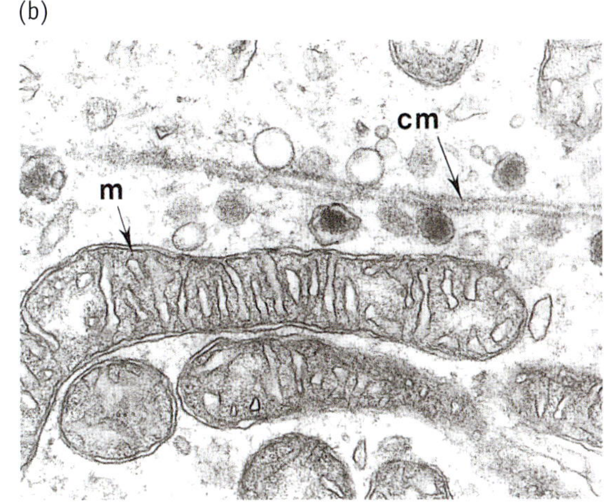

(c)

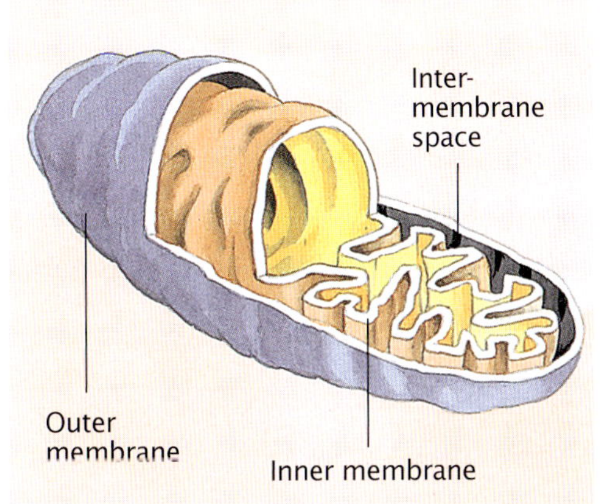

(d)

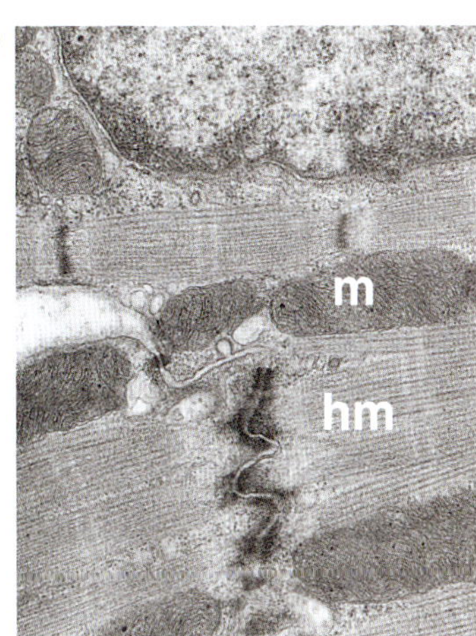

Figure 2.14 **(a)** Chemical structure of adenosine triphosphate (ATP), which has three phosphate groups, and so, adenosine tri(= 3) phosphate **(b)** Electron micrograph of mitochondrion (× 78 000) (from the Greek 'mitos' = thread, and 'chondrion' = small grain). Which is more highly folded — the outer membrane or the inner membrane? m = mitochondrion, cm = cell membrane. **(c)** 3-D representation of a mitochondrion **(d)** Mitochondria in heart muscle. Suggest why heart muscle (hm) contains large numbers of mitochondria.

Ribosomes: protein factories

	Prokaryotes	*Eukaryotes*
Ribosomes	present	present
Function	site of protein synthesis	

Living cells make proteins by linking amino acid building blocks into long chains. Human red blood cells manufacture haemoglobin, an oxygen-transporting protein; pancreas cells manufacture insulin, a small protein which is an important hormone; liver cells manufacture many protein enzymes, such as catalase; stomach cells produce digestive enzymes, such as pepsin; muscle cells manufacture the contractile proteins, actin and myosin.

Ribosomes are the organelles where production of **proteins** occurs. These organelles, which are part of the cytoplasm, can only be seen through a TEM (see figures 2.15 and 2.16, page 34).

Ribosomes are not enclosed by a membrane. The structures of prokaryotic and eukaryotic ribosomes are almost identical and function in a similar way. Although ribosomes are free within prokaryotic cells, in eukaryotes many are attached to membranous internal channels, called endoplasmic reticulum, within the cell. Chemical testing shows that ribosomes are composed of protein and ribonucleic acid (RNA).

ODD FACT

Many biologists agree with the hypothesis that, thousands of millions of years ago, mitochondria were free-living organisms, like bacteria. This hypothesis suggests that these organisms became associated with larger cells to form a mutually beneficial arrangement. This idea is supported by the fact that mitochondria contain small amounts of the genetic material DNA. The size of a mitochondrion is about 1.5 μm by 0.5 μm. This is similar to the dimensions of a typical bacterial cell.

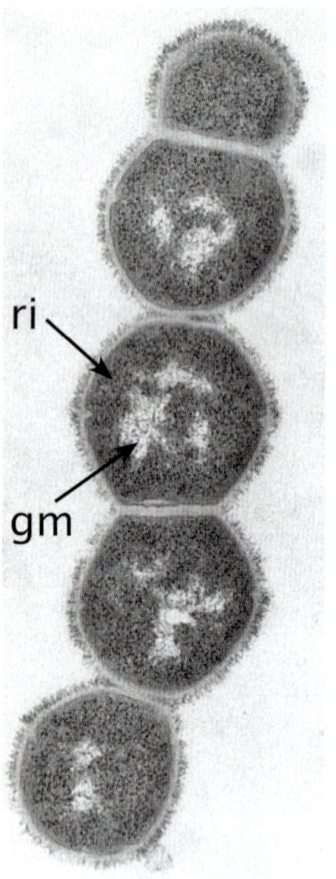

Figure 2.15 Prokaryotic cells. Note the many ribosomes (ri) in each cell, the lack of any internal membranous structure and the dispersed genetic material (gm).

Endoplasmic reticulum and Golgi complex: transport, storage and export

	Prokaryotes	*Eukaryotes*
Endoplasmic reticulum	absent	present
Function	series of membranous channels for transport	
Golgi complex	absent	present
Function	stacks of membranous sacs that package materials for transport	

The proteins made by some cells are kept inside those cells. Examples are contractile proteins made by muscle cells and the haemoglobins made by red blood cells. Other cells, however, produce proteins that are released for use outside the cells. The digestive enzyme, pepsin, is produced by cells lining the stomach and released into the stomach cavity; the protein hormone, insulin, is made by pancreatic cells and released into the bloodstream.

Transport of substances *within* cells occurs through a system of channels known as the **endoplasmic reticulum (ER)**. Figure 2.16 shows part of this system of channels in a cell. The channel walls are formed by membranes.

Endoplasmic reticulum with ribosomes attached is known as **rough endoplasmic reticulum**. Without ribosomes, the term smooth endoplasmic is used.

A structure known as the **Golgi complex** (also called Golgi apparatus or Golgi bodies) is prominent in cells that shift proteins *out of* cells.

This structure consists of several layers of membranes (see figure 2.17). The Golgi complex packages material into membrane-bound bags or vesicles for export. These vesicles carry the material out of the cell.

(a)

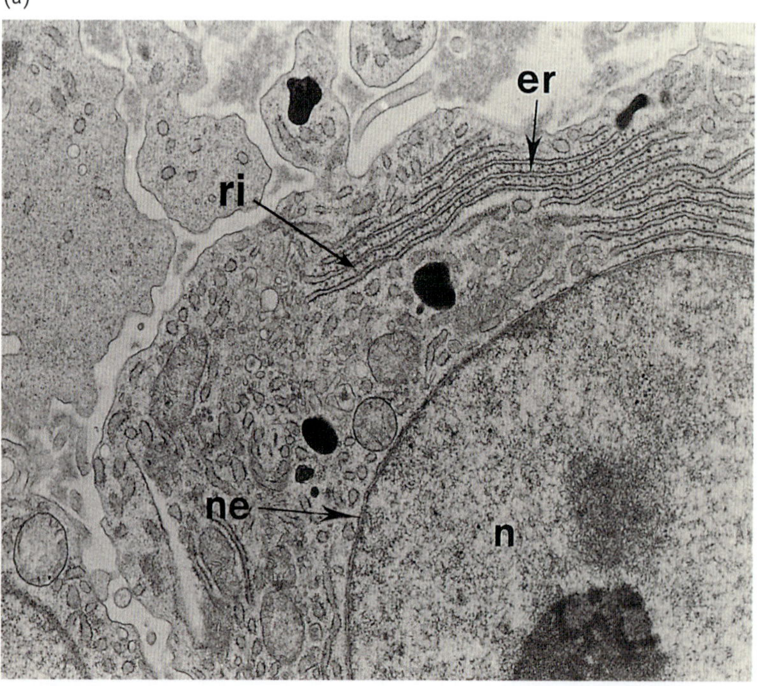

(b)

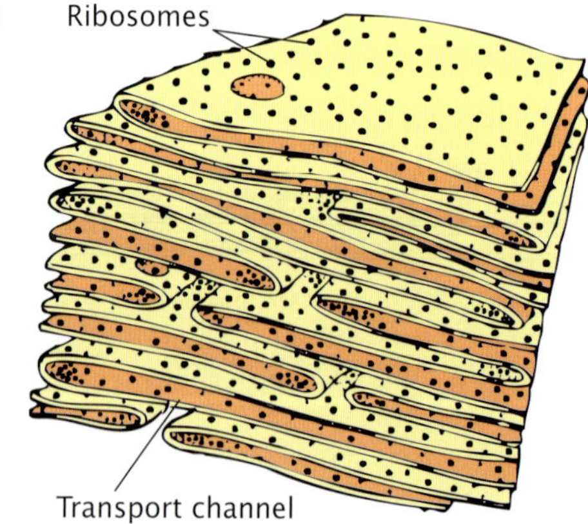

Figure 2.16 **(a)** Electronphotomicrograph showing channels of the endoplasmic reticulum (ER) (x 45 000) (er = endoplasmic reticulum with ribosomes, ri = ribosomes, ne = nuclear envelope, n = nucleus) **(b)** 3-D representation of endoplasmic reticulum with ribosomes

Both the endoplasmic reticulum and Golgi complex also synthesise some materials. You will study this aspect of their function in *Nature of Biology Book 2, Third edition*.

(a)

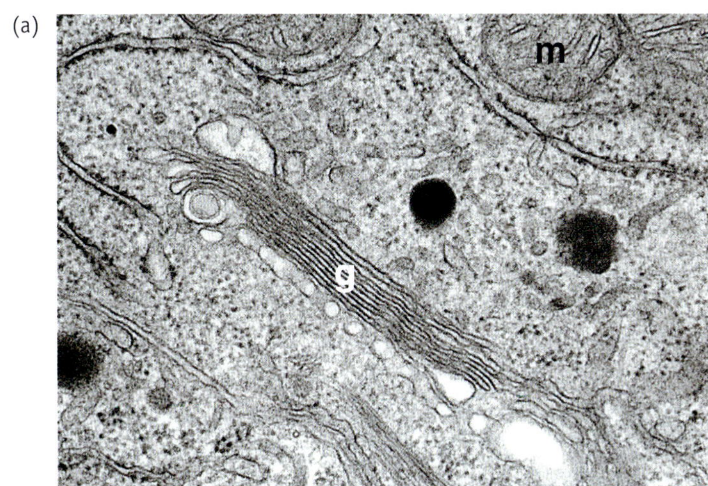

(b)

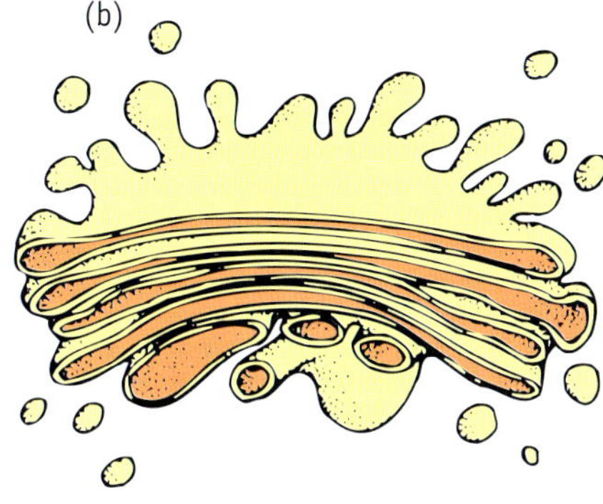

Figure 2.17 **(a)** Electronmicrograph (× 60 000) showing a Golgi complex in a cell (g = Golgi complex, m = mitochondrion) **(b)** 3-D representation of a Golgi complex

KEY IDEAS

- Prokaryotic cells lack any internal membrane-bound organelles.
- In eukaryotic cells, the nucleic acid DNA is enclosed within the nucleus, a double-membrane-bound organelle.
- Living cells use energy all the time, principally as chemical energy present in ATP.
- Mitochondria are the major sites of ATP production in eukaryotic cells.
- Ribosomes are tiny organelles where proteins are produced.
- The endoplasmic reticulum (ER) is a series of membrane-bound channels, continuous with the membrane of the outer nuclear envelope, that transport substances within a cell.
- The Golgi complex packages substances into vesicles for export.

QUICK-CHECK

4 True or false? Briefly explain your choice.
  a A nucleus from a plant cell would be expected to have a nuclear envelope.
  b Bacterial cells do not have DNA.
  c A mature red blood cell is an example of a prokaryotic cell.
5 Suggest why the nucleus is called 'the control centre' of a cell.
6 Is the major site of ATP production the same in a plant cell as in an animal cell?
7 A scientist wishes to examine ribosomes in pancreatic cells.
  a Where should the scientist look — in the nucleus or in the cytoplasm?
  b What kind of microscope should the scientist use?
8 A substance made in a cell is moved outside the cell. Outline a possible pathway for this substance.

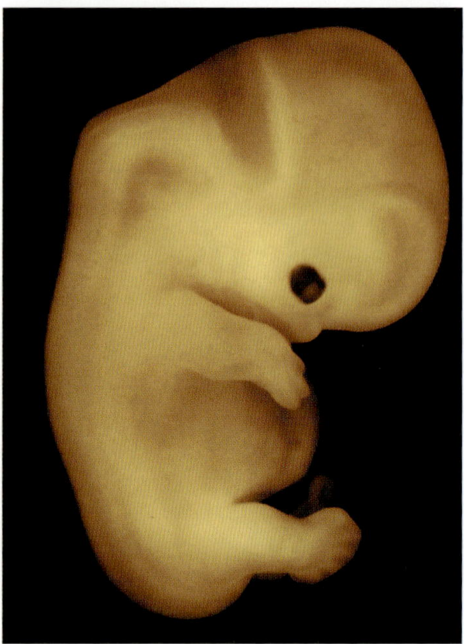

Figure 2.18 Note the webbing between the fingers in an early embryo at 6 weeks development.

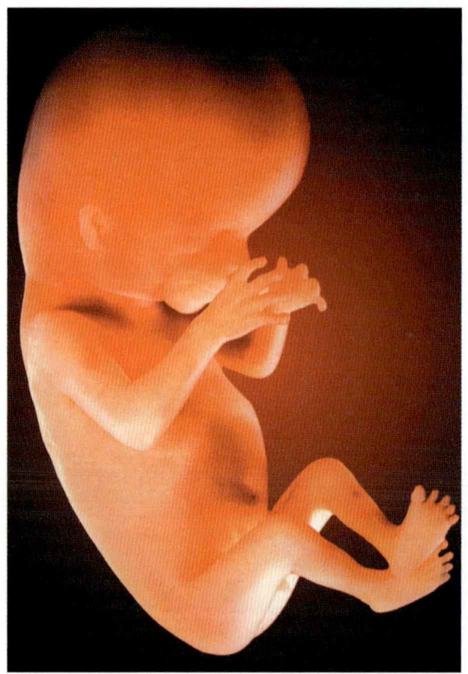

Figure 2.19 Later in embryonic development as in this fetus at 12 weeks, individual fingers are visible because the cells of the webbing have been destroyed by enzymes secreted by lysosomes.

Lysosomes: controlled destruction

	Prokaryotes	*Eukaryotes*
Lysosomes	absent	present
Function		principal site of digestion within a cell

The human hand is a marvellous living tool that allows a person to grasp objects, manipulate and investigate them. Typically, a human hand has five digits that are separated from each other along their length. This is not always the case — a rare condition, known as syndactyly (pronounced sin-*dack*-till-ee), in which the fingers are fused, can occur. How does this happen?

During human embryonic development, the hands appear first as tiny buds with no separate digits (see figure 2.18). The separation of the fingers normally occurs on about the 52nd day of development (see figure 2.19). This separation involves the 'programmed death' of groups of cells between the fingers. The process of programmed cell death is called **apoptosis**. If this programmed cell death does not occur, the fingers and toes form but they remain fused.

Animal cells have sac-like structures surrounded by a membrane and filled with a fluid containing dissolved digestive enzymes. These fluid-filled sacs are known as lysosomes. Lysosomes can release their enzymes within the cell, causing the death of the cell. This process of controlled 'self-destruction' of cells is important in development: lysosomes appear to play a role in the controlled death of zones of cells in the embryonic human hand so that the fingers become separated.

Lysosomes contain digestive enzymes and are the principal sites for digestion of large molecules and unwanted structures within a cell.

Chloroplasts: sunlight trappers

	Prokaryotes	*Eukaryotes*		
		Fungi	*Plant*	*Animal*
Chloroplasts	absent	absent	present	absent
Function		site of photosynthesis and storage of starch		

Solar-powered cars have travelled across Australia. The power source for these cars is not the chemical energy present in petrol, but the radiant energy of sunlight trapped and converted to electrical energy by solar cells. Use of solar cells is becoming more common in Australian households and it is not unusual to see solar cells on a roof.

Solar cells are a relatively new technology. However, hundreds of millions of years ago, some bacteria and all algae and then land plants developed the ability to capture the radiant energy of sunlight and to transform it to chemical energy present in organic molecules, such as sugars. The remarkable organelles present in some cells of plants and algae that carry out this function are known as **chloroplasts** (see figure 2.20a). The complex process of converting sunlight energy to chemical energy present in sugar is known as **photosynthesis**.

Photosynthesis is discussed further in chapter 3, pages 66–68.

Chloroplasts can be easily seen through a LM. They are green in colour owing to the presence of light-trapping pigments known as **chlorophylls**. Each chloroplast has an outer membrane and also has an intricate internal structure consisting of many folded membrane layers, called **grana**, that provide a large surface area where chlorophylls are located. **Stroma** is fluid between the grana.

Prokaryotic cells do *not* have chloroplasts. Some kinds of bacteria, however, possess pigments that enable them to capture the radiant energy of sunlight and use that energy to make sugars from simple inorganic material. These are known as photosynthetic bacteria.

The length of a typical chloroplast is 5 to 10 μm. In comparison, the length of a mitochondrion is about 1.5 μm. In 1908, the Russian scientist, Mereschkowsky, suggested that chloroplasts were once free-living bacteria that later 'took up residence' in eukaryotic cells. Some evidence in support of this suggestion comes from the fact that a single chloroplast is very similar to a photosynthetic bacterial cell.

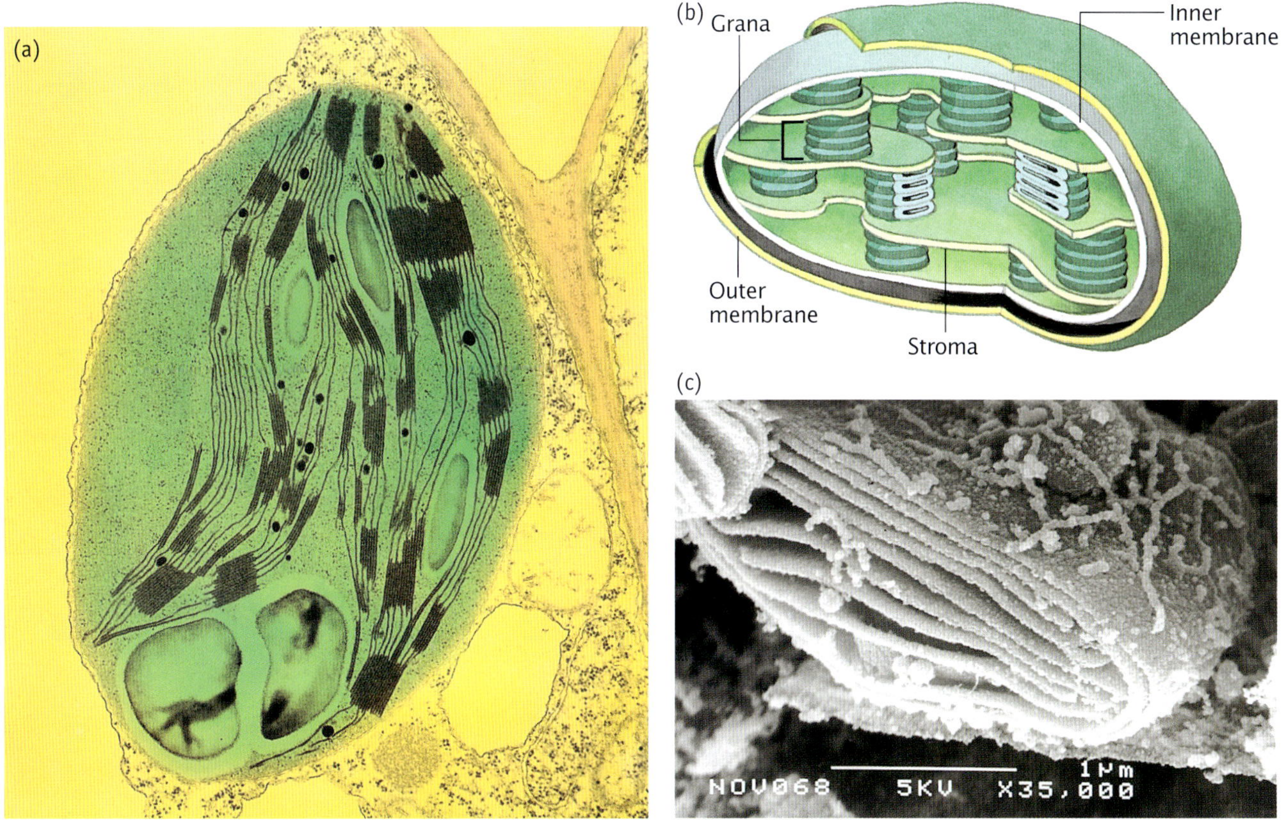

Figure 2.20 **(a)** Internal structure of chloroplast showing many layers of membranes **(b)** 3-D representation. Where are chlorophylls located? **(c)** Scanning electronmicrograph (× 78 000) of fractured red algae chloroplast. Note fine tubular endoplasmic reticulum on outer surface of chloroplast envelope (scale bar = 1 μm).

Other membrane-bound structures

Other small membranous structures found in the cytosol of eukaryotic cells include the **endosomes** (animal cells only) and **peroxisomes**. (These are dealt with in more detail in *Nature of Biology Book 2, Third edition*.) Many plant cells also contain **vacuoles**, some very large that almost fill a cell. Vacuoles are filled with a fluid, mostly water, containing a number of different materials in solution, including plant pigments.

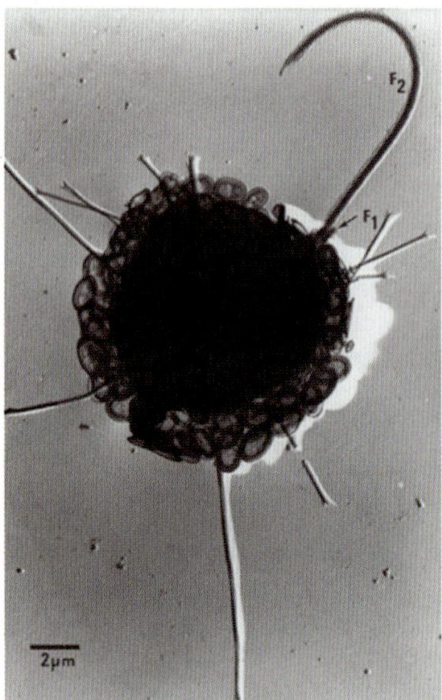

Figure 2.21 *Thaumatomastix*, a colourless marine protist. Note the two flagella F1 and F2. The scales and spines that cover the entire cell are made of silica.

Flagella and cilia: whipping around

Some bacterial cells and other single-celled organisms have a whip-like structure that is attached to the plasma membrane and protrudes through the cell wall (see figure 2.21). This structure is usually known as a **flagellum** (plural = flagella, from the Latin word meaning 'whip'). What role might this structure serve?

The rotation of a flagellum results in the movement of the organism. Some bacteria have many flagella, such as the bacteria that cause typhoid (*Salmonella typhosa*). Other bacteria, such as species of *Pseudomonas,* have one flagellum or a cluster of several flagella at one end.

Many eukaryotic cells have one or many whip-like structures on their cell surfaces. When many such structures are present, they are termed **cilia** (singular = cilium, from the Latin word meaning 'eyelash'); when only one or two are present, they are termed flagella (figure 2.21).

In eukaryotes, each cilium and flagellum is enclosed in a thin extension of the plasma membrane. Inside this extension of the membrane are fine protein filaments known as microtubules. In the human body, the cells lining the trachea or air passage have cilia that project into the cavity of the trachea. The synchronised movement of these cilia assists mucus to travel up the trachea to an opening at the back of the throat. Other human cells that have flagella include sperm cells.

Dr Peter Beech, a cell biologist, carries out research on the replication of cells and their organelles. Figures 2.21 and 2.22 and figure 4.8a (page 80) show some of his results. Read what he has to say about his work.

BIOLOGIST AT WORK

Dr Peter Beech — Cell biologist

Dr Peter Beech is a Research Scientist and Senior Lecturer in the School of Biological and Chemical Sciences at Deakin University in Melbourne. Peter writes:

'Like many kids who watched Jacques Cousteau on television exploring the world's oceans, I wanted to be a marine scientist. I spent summers at the beach wondering about how I could get a job working with the sea. I was told "go to uni, study science and then see what grabs you". It was good advice, and I quickly discovered that biology was indeed for me.

'My first lab project was on identifying algal scales, the beautifully intricate cell coverings of many **phytoplankton** (figure 2.22). This work required an electron microscope, and I was thus irreversibly led into the world of the subcellular, where I could see scales being made, as well as the other cellular organelles — many of which are also found in our own cells.

'My PhD was on how certain phytoplankton made their scales and deposited them on the cell surface, as well as how they made their flagella. Flagella are the whip-like appendages that beat to propel cells through the water — sperm tails are flagella. I was not the first to realise that by looking at protists (as algae and many other mostly unicellular eukaryotes are known), we could learn a lot about cells. Many protists are ideally suited to laboratory culture and experimentation. Phytoplankton, for example, are unicells that have all they need to get by in life on their own. Often all that is needed to grow them in the lab is light and clean sea water or pond water.

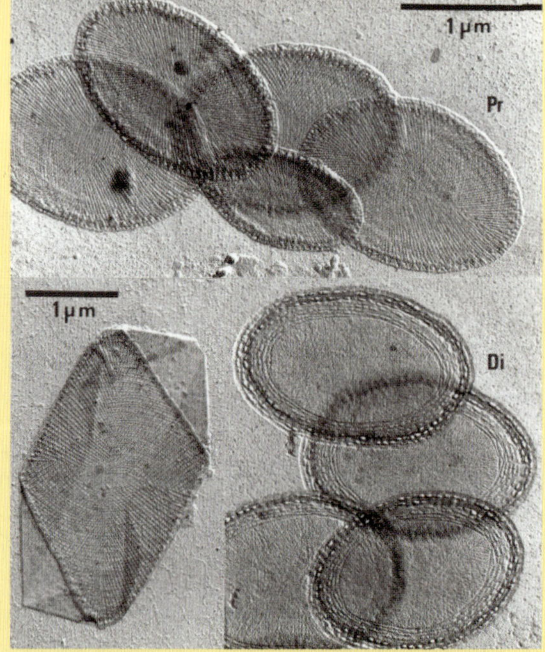

Figure 2.22 A transmission electron micrograph of body scales made by an algal cell, of *Chrysochromulina pringsheimii*. The scales and their intricate patterns are constructed of polysaccharide fibrils and are made inside the cell. Pr = proximal side of the scale, Di = distal side.

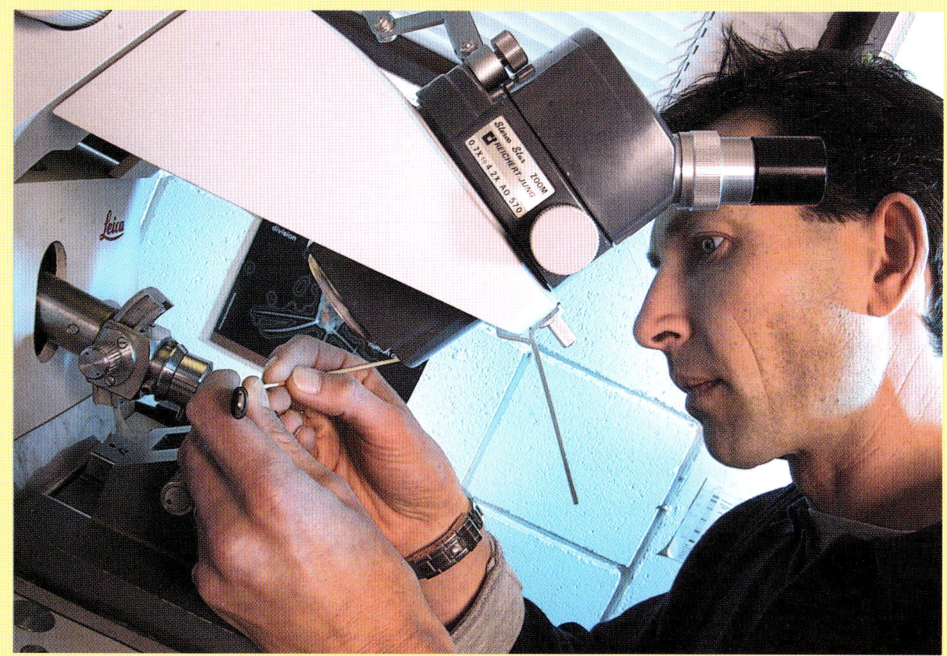

Figure 2.23 Dr Peter Beech using an ultramicrotome to cut very thin (70 μm) sections of plastic-embedded cells for the transmission electron microscope. The dark line projecting down from the ultramicrotome is a side view of a thin, clear screen that protects the thin sections from the breath of an operator.

'From wanting to be a marine scientist, I thus became a cell biologist. I had post-doctoral research jobs in algal cell biology in Germany and the USA. The latter, at Yale University, was as part of a team investigating a newly discovered phenomenon called intraflagellar transport. IFT, as it became known, is a great example of how protists, in this case the unicellular, green, soil alga *Chlamydomonas*, can open our eyes to principles that are important for all cells. In 1993, a PhD student at Yale, named Keith Kozminski, showed that the two flagella of *Chlamydomonas* exhibited a novel movement that shuttled "rafts" of particles up and down the flagellum — like express lifts between the penthouse and lobby of a building. The movement was unrelated to flagellar beating and probably evolved to deliver building materials to the growing flagellar tip.

'We now know that IFT works in our eyes too. All vertebrates have modified flagella (cilia) in their retinas. Even though these cilia do not beat, they are an intricate part of the rod and cone cells in which they are found: they are the transport tunnels through which newly-made photosensory pigments (rhodopsins) pass before they are assembled into light-detecting discs. We now know that the rod and cone cells use IFT to transport the rhodopsins to the photo-receptive discs. Thus, without IFT, we'd be blind. In fact, we'd have all sorts of problems. Recent work indicates that IFT is important for the very existence of all cilia, from those in our sperm or oviducts, to those in the kidney. Thus, thanks to a dirt dweller with two bold flagella, we can now begin to understand the fundamentals of numerous diseases involving cilia.

'In my own lab, we continue to use protists to learn about all cells. We study how the two main energy-producing organelles of eukaryotes split into two to reproduce; mitochondria perform cellular respiration, and chloroplasts are the sites of photosynthesis in plants in algae. Though these two organelles do very different jobs (mitochondria make ATP from sugars, and chloroplasts make sugars using light energy), they have similar evolutionary histories. Mitochondria and chloroplasts arose separately a billion or so years ago through the capture of bacteria by early cells. The bacterium that gave rise to the chloroplasts already had the capacity for photosynthesis, and was probably similar to present-day blue-green "algae" (cyanobacteria). But how do mitochondria and chloroplasts now divide? We know that new mitochondria and chloroplasts, like bacteria, can only arise from the division of pre-existing individuals. So perhaps organelle division molecules are the same as those used by the bacteria? It turns out that many mitochondria, such as those of the alga *Mallomonas* (see figure 4.8a on page 80), appear to divide using a protein called FtsZ — and yes, FtsZ is used by bacteria to divide — nicely re-confirming that mitochondria really are bacteria that now specialise in power production for larger cells. Furthermore, we also know that chloroplasts use FtsZ to divide. Interestingly though, the mitochondria of lots of different organisms, including those of animals, fungi and land plants, have independently dumped the bacterial division mechanism, and developed their own. Why? — we now have the fun job of finding out.

'One of my joys as a university lecturer is, of course, to teach. In my cell biology classes, protists rule!'

Putting it all together

The cell is both a unit of structure and a unit of function. Organelles within one cell do not act in isolation, but interact with each other. The normal functioning of each kind of cell depends on the combined actions of its various organelles, including plasma membrane, nucleus, mitochondria, ribosomes, endoplasmic reticulum and Golgi complex.

In some cells, the plasma membrane is very highly folded. This folding expands the surface area across which materials move into or out of cells while the internal volume remains unchanged. This produces an increase in the surface-area-to-volume ratio (SA:V) of cells.

Consider a cell that produces a specific protein for use outside the cell. Table 2.4 identifies the parts of a cell involved in this process.

Table 2.4 Parts of a cell involved in producing a specific protein

Structure	*Function*
plasma membrane	structure that controls the entry of raw materials, such as amino acids, into the cell
nucleus	organelle that has coded instructions for making the protein
ribosomes	organelles where amino acids are linked, according to instructions, to build the protein
mitochondrion	organelle where ATP is formed; provides an energy source for the protein-manufacturing activity
endoplasmic reticulum	channels through which the newly made protein is moved within the cell
Golgi complex	organelle which packages the protein into vesicles for transport across the plasma membrane and out of the cell

Figure 2.24 shows the typical structures of an animal and a plant cell, including the organelles involved in the processes outlined in table 2.4. Examine the two cells. Note the presence of protein filaments in each cell. These give a cell shape; they form a kind of 'internal skeleton' for the cell and also provide a system for movement during, for example, mitosis (see chapter 4, page 77 onwards).

KEY IDEAS

- Lysosomes can digest material brought into their sacs. Lysosomes play a role in organised cell death.
- Chloroplasts are relatively large organelles found in photosynthetic cells of plants and algae.
- Chloroplasts have an external membrane and layers of folded internal membranes and contain pigments called chlorophylls.
- Chloroplasts can capture the radiant energy of sunlight and convert it to chemical energy in sugars.
- Structures known as flagella are present on many prokaryotic cells.
- Cilia or flagella are present on many eukaryotic cells.
- Flagella and cilia are cell organelles associated with movement.

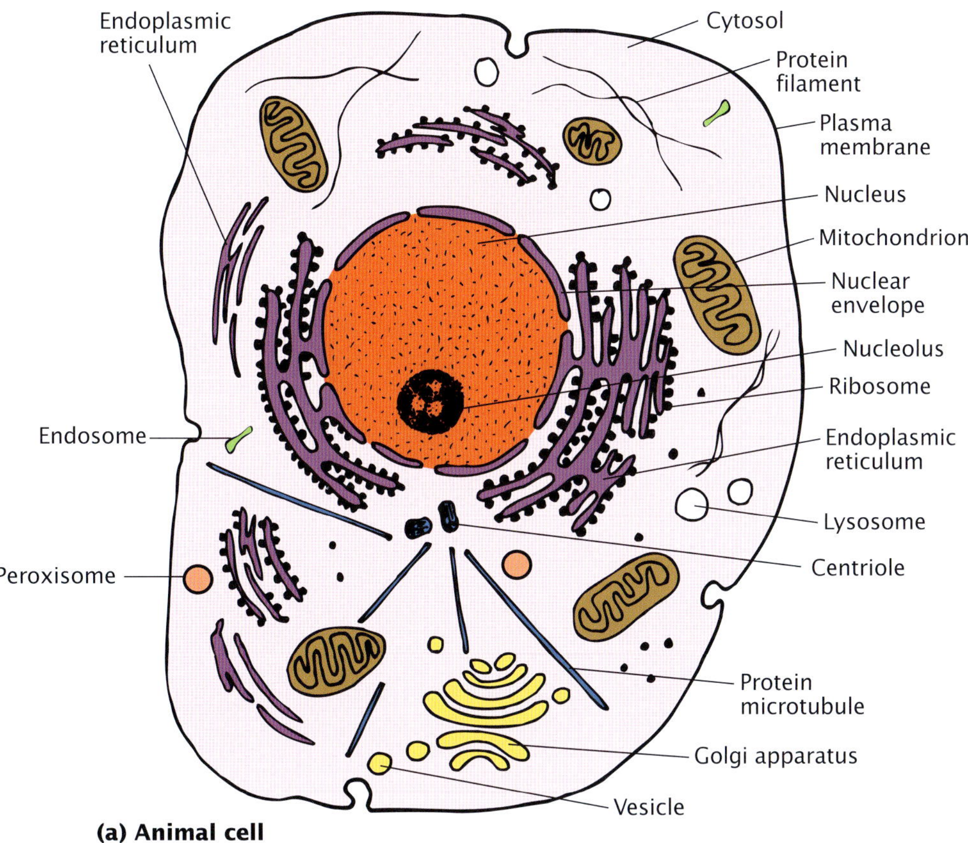

(a) Animal cell

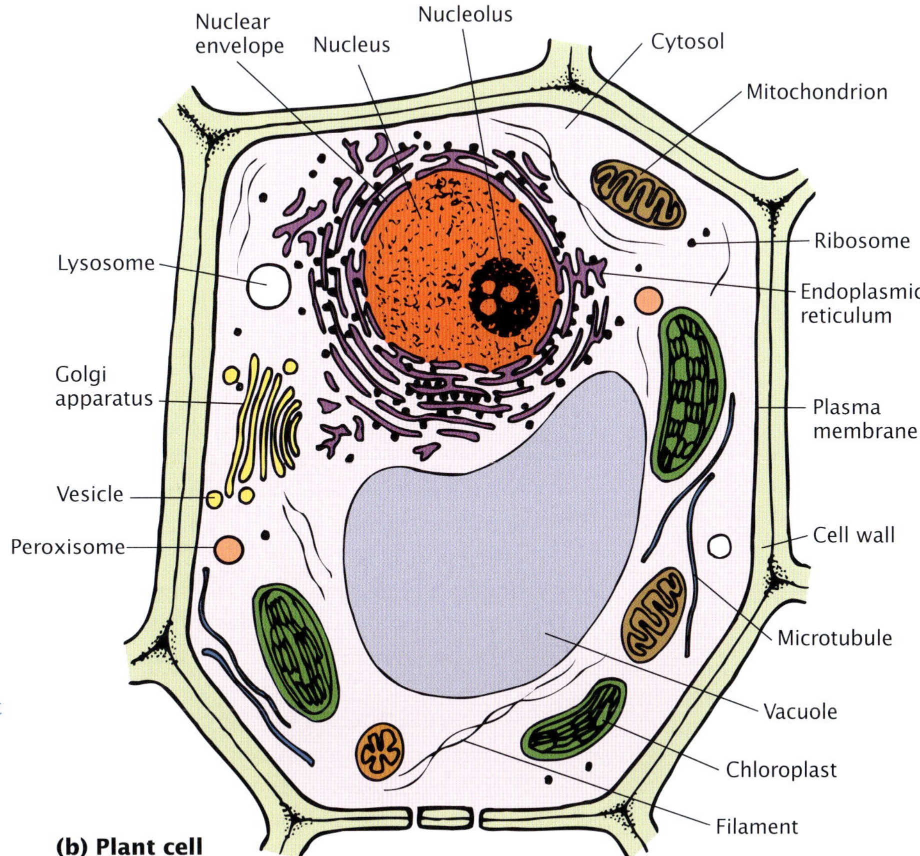

(b) Plant cell

Figure 2.24 Compare **(a)** an animal cell with **(b)** a plant cell. What organelles are found in both of the cells? What organelles are unique to either plant or animal cells? What other differences in structure are there between the two cells?

QUICK-CHECK

9 Lysosomes are sometimes called 'suicide bags'. Suggest why this name is given.
10 Identify the following as true or false and briefly justify your answers.
  a Plant cells without chloroplasts can capture the energy of sunlight.
  b Chloroplasts can be seen through an LM.
11 List one location in the human body where cells with cilia are found.
12 Consider a cell with cilia beating on its surface. Identify one other organelle that would be expected to assist in the action of these cilia.
13 List four cell organelles that are involved in the process of making protein. What is the contribution of each organelle to this process?
14 Does an amoeba have organs? Explain.

Cells in multicellular organisms: levels of organisation

Unicellular organisms must carry out all the metabolic processes necessary for life. They are complex cells capable of independent existence. In contrast, multicellular organisms have millions of cells that depend on each other for survival. During development of a multicellular organism, groups of cells become specialised to perform particular functions that serve the whole organism. Specialised cells have fewer functions than those found in a unicellular organism but the functions they have are very highly developed. In addition, each group of specialised cells must coordinate with other specialised cells. We will consider the different levels of organisation that interact to ensure proper functioning for the whole organism.

Tissues

When cells that are specialised in an identical way aggregate to perform a common function, they are called a **tissue**. Different kinds of tissue (see figure 2.25) serve different functions in an organism. For example, cardiac muscle is a particular kind of muscle tissue found only in the heart. Epidermal tissue is a general name for any tissue that forms a discrete layer around a structure. It may be a layer of plant cells forming the outermost cellular layer of leaves or it may be the outer layers of human skin.

You will recall from pages 26–7 that the surface-area-to-volume-ratio (SA:V) of a cell is important in determining the cell's efficiency to move materials across its membrane and that the higher the SA:V ratio of a cell, the more efficient it is in carrying out those functions. The need for small cells can be graphically demonstrated with regard to groups of cells (figure 2.26, page 44). Exchange of materials between tissues and their environments has the potential to be far more efficient if the tissue is made up of many small cells rather than fewer larger cells.

This potential for efficiency of small cells becomes a reality only if each of the cells in a group of cells is close to a delivery mechanism, capable of providing material to and removing material from the cells (figure 2.27, page 44). A mass of small cells without a delivery system has no advantage over a single large cell.

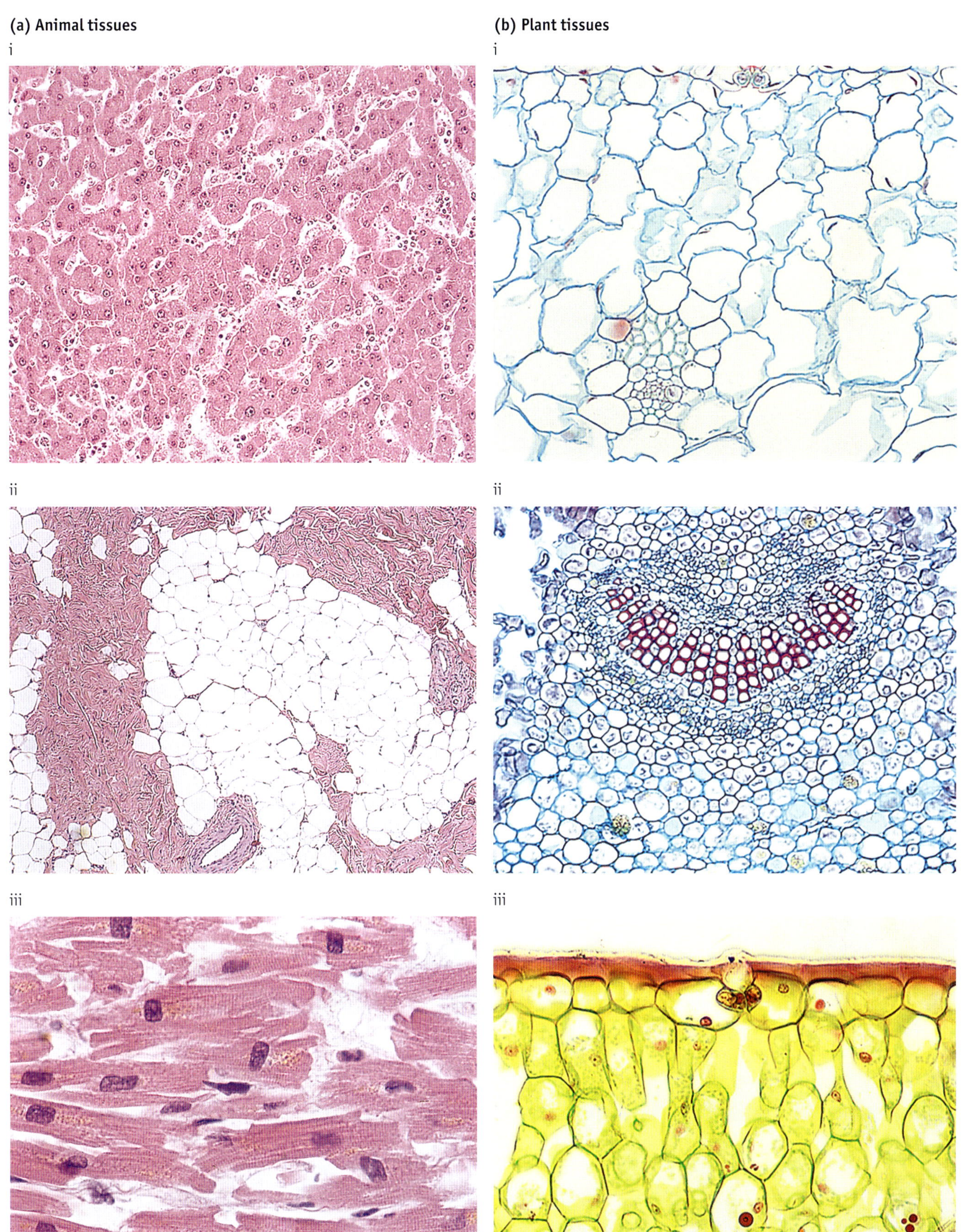

Figure 2.25 Various tissues **(a)** Different animal tissues: (i) Liver (ii) Fat or adipose (iii) Cardiac muscle **(b)** Different kinds of plant tissue in a leaf: (i) Parenchyma (ii) Vascular — transporting tissue, and (iii) Epidermal with cuticle

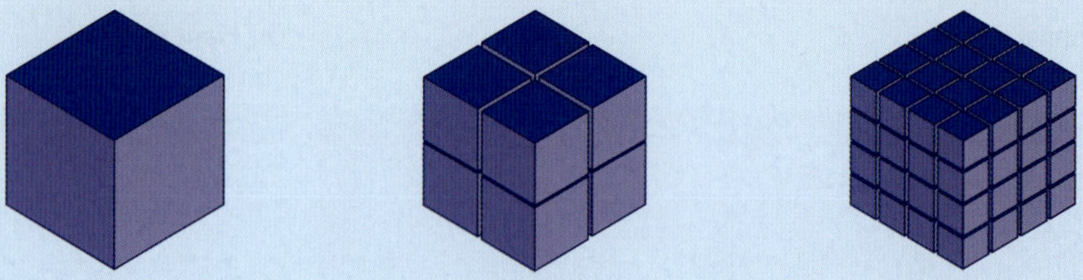

Number of cells filling a particular space	one	eight	sixty-four
Total surface area (height × width × number of sides × number of cells)	$= 1 \times 1 \times 6 \times 1$ $= 6$	$= 0.5 \times 0.5 \times 6 \times 8$ $= 12$	$= 0.25 \times 0.25 \times 6 \times 64$ $= 24$
Total volume (height × width × length × number of cells)	$= 1 \times 1 \times 1 \times 1$ $= 1$	$= 0.5 \times 0.5 \times 0.5 \times 8$ $= 1$	$= 0.25 \times 0.25 \times 0.25 \times 64$ $= 1$
Surface-area-to-volume ratio (area ÷ volume)	$6 \div 1$ $= 6$	$12 \div 1$ $= 12$	$24 \div 1$ $= 24$

Figure 2.26 The number of cells occupying a particular space influences the rate of movement of materials into and out of the mass occupying the space. The greater the overall surface-area-to-volume ratio, the greater the efficiency of movement of materials. Arbitrary units have been used in this example.

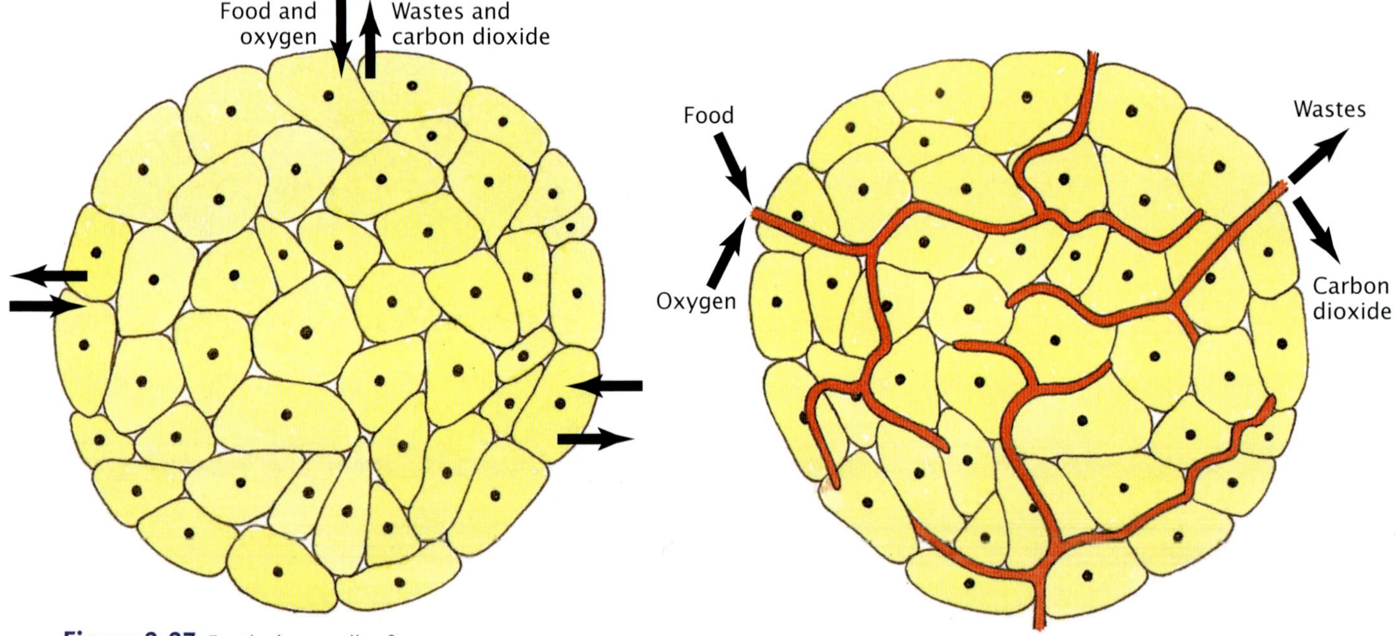

Figure 2.27 For the inner cells of a tissue to operate as efficiently as the outer cells, they must have a delivery system that transports food and gas to them and takes away wastes. In many animals, the delivery system is the blood circulatory system.

Organs

In multicellular organisms, groups of different tissues often work together to ensure a particular function is successfully performed (figure 2.28). A collection of such tissues is called an **organ**. Your stomach is an organ. Tissues of the stomach include an epithelium, smooth muscle cells and blood (see figure 2.28a). Other organs include your heart, brain and kidneys. A plant leaf is an organ. Tissues of a leaf include an epithelium, vascular tissue and parenchyma tissue (see figure 2.28b). Other plant organs include its root, stem and flower.

Organ systems

Your digestive system comprises various organs that work together to ensure that the food you eat is digested and that the nutrients it contains are absorbed and

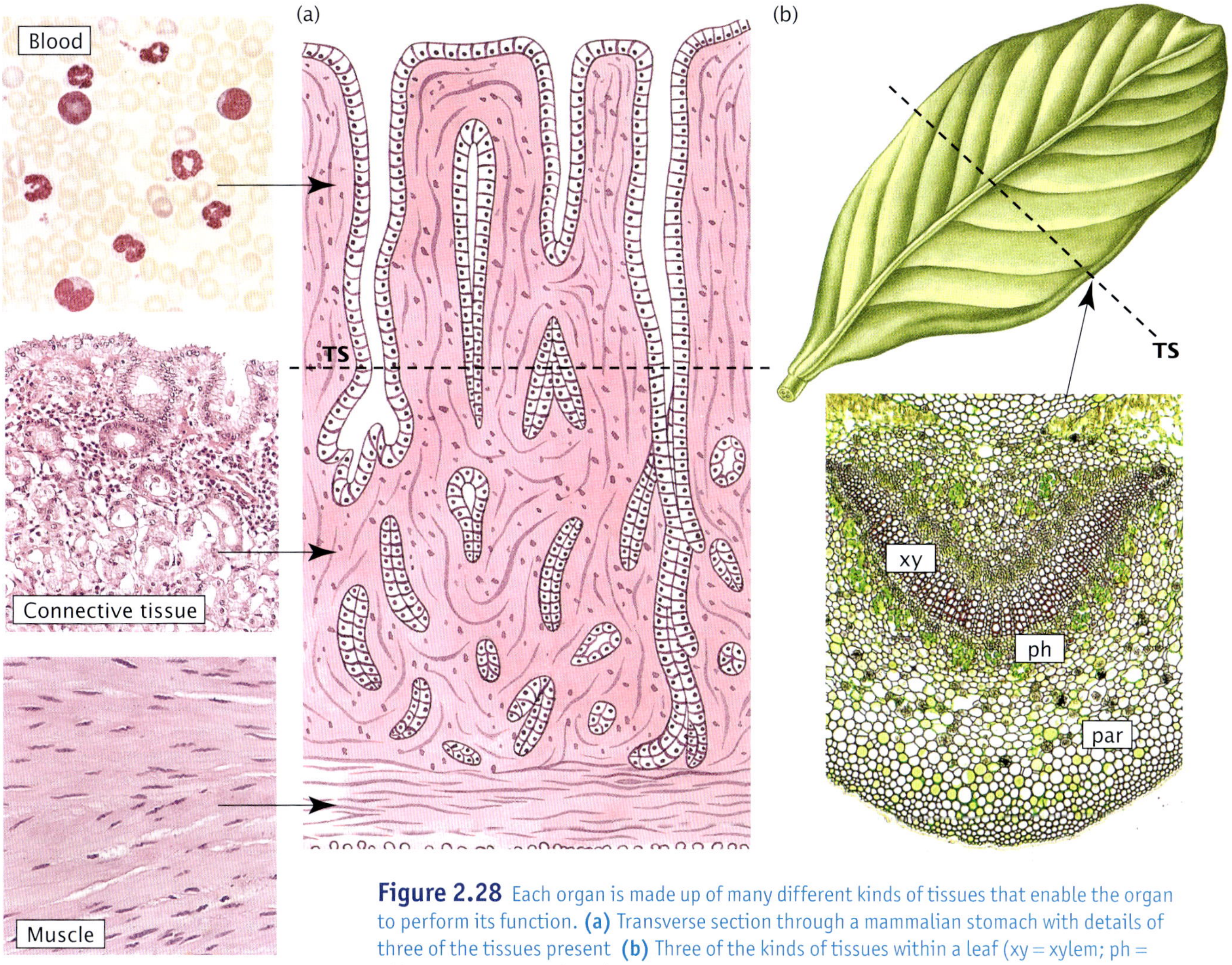

Figure 2.28 Each organ is made up of many different kinds of tissues that enable the organ to perform its function. (a) Transverse section through a mammalian stomach with details of three of the tissues present (b) Three of the kinds of tissues within a leaf (xy = xylem; ph = phloem; par = parenchyma)

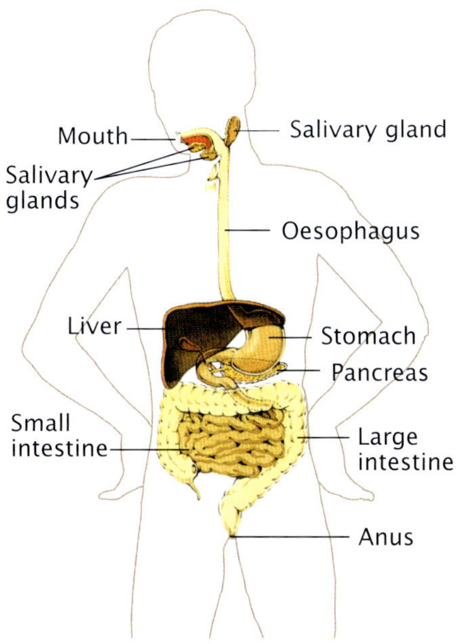

Figure 2.29 The main organs of a human digestive system

transported to all cells of your body. This organisation is called an **organ system**. Your digestive system commences with your mouth and includes organs such as your teeth, oesophagus, stomach, intestines and liver (figure 2.29). Once digested food has been absorbed by cells lining the intestine, it is transported by the blood circulatory system throughout the body. This system links with the respiratory system where it picks up oxygen, also for delivery.

As blood delivers nutrients and oxygen to all tissues, it collects nitrogenous and gaseous wastes for delivery to the excretory systems of the body.

Because plants do not move from place to place, their energy needs are far less than mobile animals. Hence, plants lack the equivalent of complex organ systems such as the respiratory and digestive systems of animals. Green plants produce their own food through photosynthesis and this process also delivers oxygen directly to some cells. Other cells rely on diffusion to receive oxygen. The extensive root system of a plant ensures that it absorbs sufficient water to meet the plant's requirements. An extensive vascular system delivers that water throughout the plant; however, there is relatively little difference in the structure of the various parts of a plant vascular system compared with differences found in systems of an animal.

We will consider some of the organ systems of animals and plants in greater detail in later chapters. A summary of the levels of organisation in multicellular organisms is shown in figure 2.30, page 46.

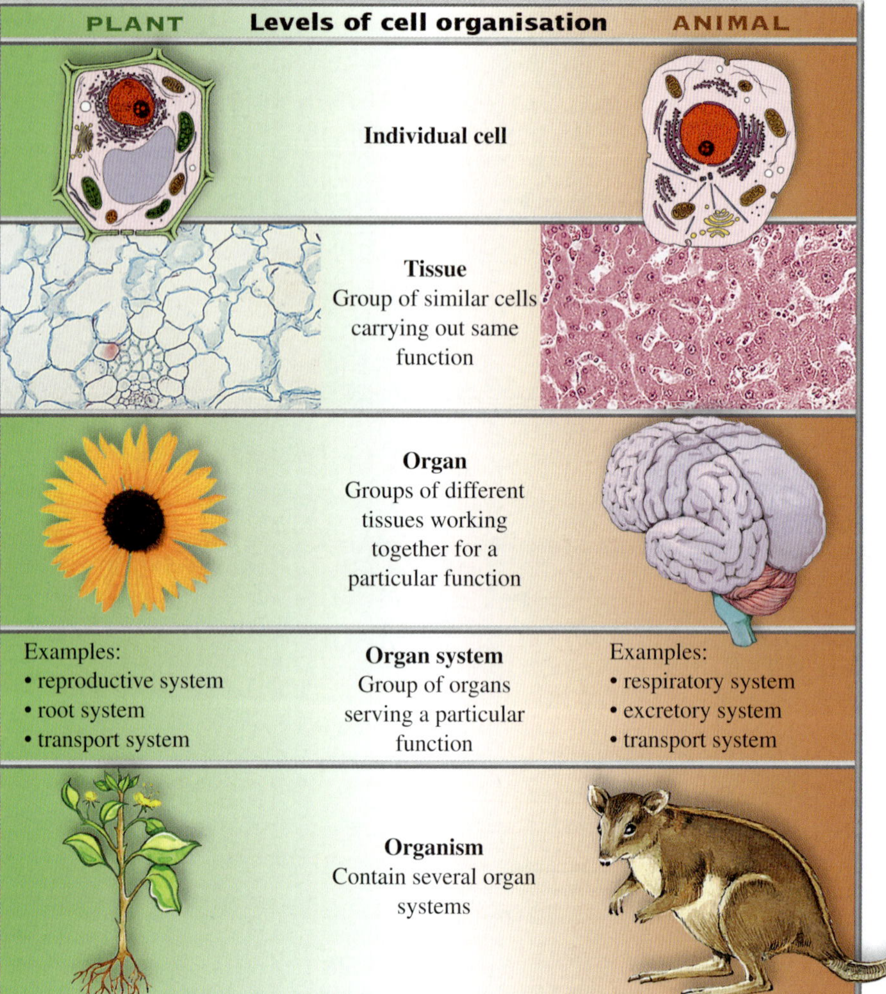

Figure 2.30 Similar cells group together to form a tissue. Different groups of tissues combine to form an organ. Different organs work together to form an organ system that has a particular function. The systems of a multicellular organism include the hormonal and nervous systems that coordinate and control the whole organism.

KEY IDEAS

- Single-celled organisms are able to carry out all the metabolic processes necessary for life.
- In multicellular organisms, cells become differentiated to perform specialised functions.
- The different levels of organisation of cells in multicellular organisms are single cell, tissues, organs, systems and the whole organism.
- Individual cells in a group of cells must be able to receive an adequate supply of materials and get rid of wastes.
- Each system serves the needs of other systems.

QUICK-CHECK

15 What characterises a tissue, an organ and an organ system?

16 Classify each of the following as tissue, organ or system.

- nerve cells in the tip of a finger
- a flower
- a human liver
- fleshy part of an apple
- nose, trachea and lungs
- layer of fat around a kidney

BIOCHALLENGE

1

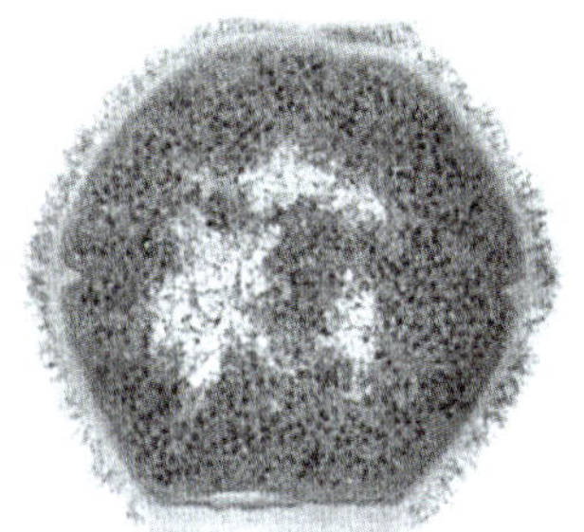

Explain whether this cell is prokaryotic or eukaryotic.

2

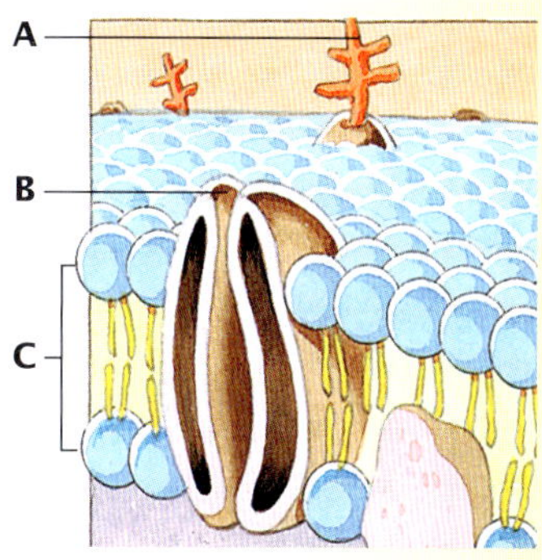

Name the parts of the cell membrane that are labelled A, B and C.

3

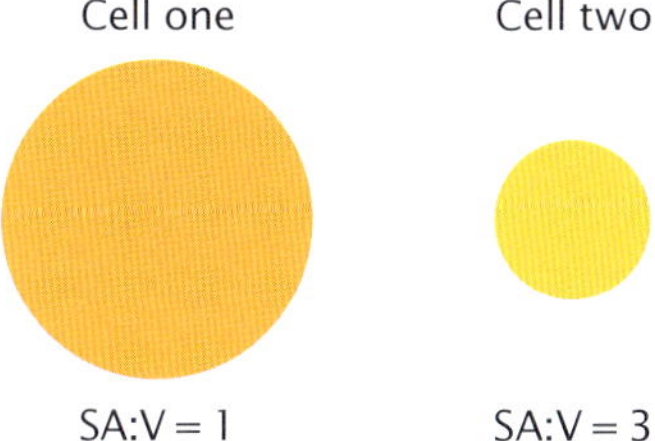

If both of these cells are in the same environment, which has the capacity to absorb more nutrients per unit volume, per unit time?

4

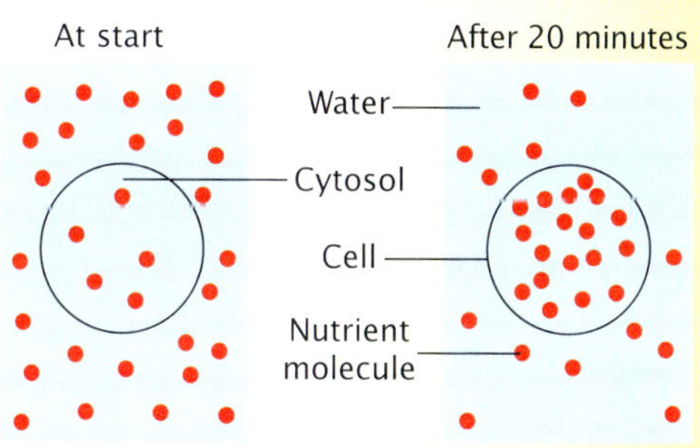

Explain whether the process occurring in this diagram is active transport or diffusion.

5

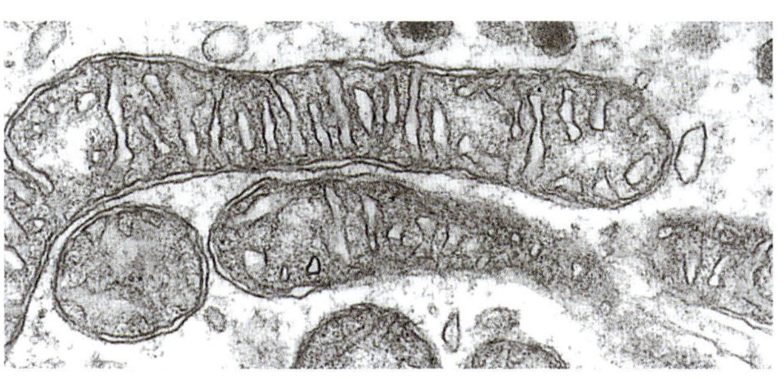

Name the organelle and describe its function.

6

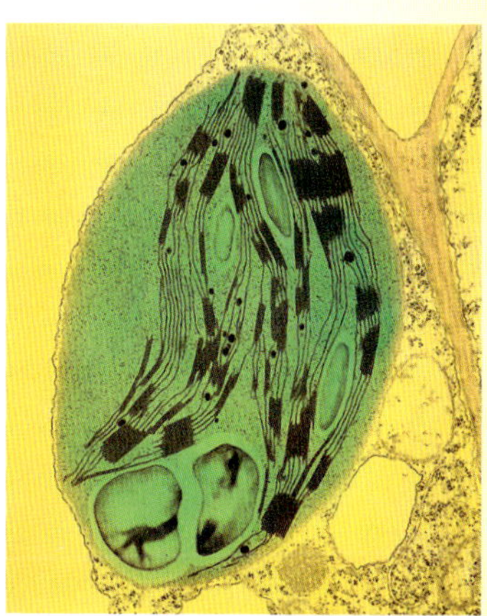

Where in a plant cell would you find this structure? What is its function?

CHAPTER REVIEW

Key words

CROSSWORD

active transport
adenosine triphosphate (ATP)
apoptosis
archaeans
bacteria
cell membrane
cell wall
cellular respiration
cellulose
chlorophylls
chloroplasts
chromatin
cilia
cytoskeleton
cytosol
deoxyribonucleic acid (DNA)
diatoms
diffusion
endocytosis
endoplasmic reticulum (ER)
endosomes
eukaryotes
eukaryotic cells
exocytosis
facilitated diffusion
flagellum
fungi
Golgi complex
grana
hydrophilic
hypertonic
hypotonic
isotonic
lipophilic
lysosome
microtubules
mitochondria
nuclear envelope
nucleoli
organ
organ system
organelles
osmosis
partially permeable
peroxisomes
phagocytosis
photosynthesis
phytoplankton
pinocytosis
plasma membrane
primary cell wall
prokaryotes
prokaryotic cells
protein filaments
proteins
protists
ribonucleic acid (RNA)
ribosomes
rough endoplasmic reticulum
secondary cell walls
stroma
surface-area-to-volume ratio (SA:V ratio)
tissue
vacuoles
vesicle

Questions

1 ***Making connections***

a Use at least eight of the key words above to make a concept map relating to the organelles observed in the cytosol of a plant cell. You may use other words in drawing up your map.

b Use at least six of the key words above to make a concept map relating to the movement of substances across a cell membrane. You may use other words in drawing up your map.

2 ***Applying your understanding*** Identify five locations in a typical cell where membranes are found. Describe how membranes in these various locations assist in the function of cells.

3 ***Communicating understanding*** Substances can enter or exit a cell through various processes.

a Prepare a table with the following headings:

Name of process *Energy cost*

Identify the processes by which material crosses the cell membrane and complete the table.

b Identify one other useful heading and add it and the relevant information to your table.

(a) Before experiments

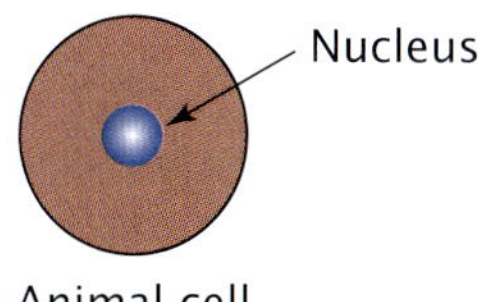

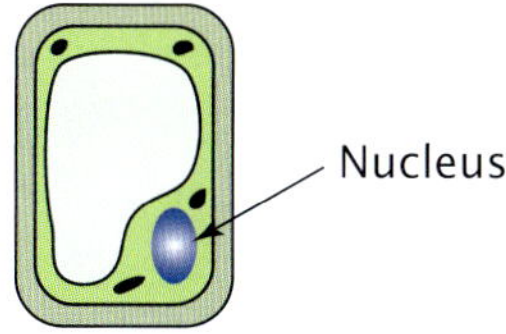

(b) After experiments

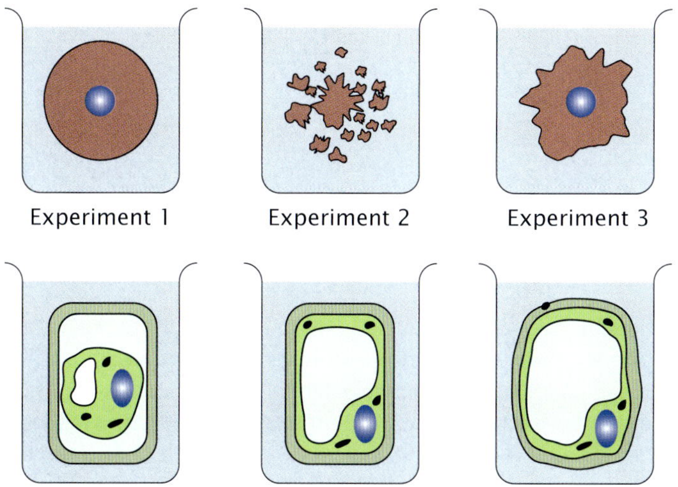

Figure 2.31
(a) Cells before experiment
(b) After experiment

4 ***Analysing data and drawing conclusions*** ▸ In a series of six experiments, animal cells and plant cells were placed in solutions of different concentrations.

Solution 1: distilled water

Solution 2: same concentration as the cytosol of the cells

Solution 3: higher concentration than the cytosol of the cells

The initial appearance of the cells was as shown in figure 2.31a. After several minutes in the solutions the cells appeared as shown in figure 2.31b.

Which solution had been used in each of the experiments? Explain what has happened to the cell in each experiment.

5 ***Communicating understanding*** ▸ Where are the following in a eukaryotic cell?

a control centre of a cell
b site of control of entry or exit of substances to or from a cell
c energy source for cell
d internal transport system
e site of packaging for export from cell
f 'self-destruct button' for cell

6 ***Applying your understanding*** ▸

a List the following in order of decreasing size from largest to smallest:

i cell　ii tissue　iii mitochondrion
iv nucleus　v nucleolus　vi ribosome.

b List the following in order from outside to inside a leaf cell:

i nuclear envelope　ii cell wall　iii plasma membrane
iv cytosol　v nucleolus.

7 ***Analysing information and drawing conclusions*** ▸ Suggest possible explanations for each of the following observations:

a Flight muscle fibres of bats contain very large numbers of mitochondria.
b One kind of cell has a very prominent Golgi complex, while another kind of cell appears to lack this organelle.
c Chromosomes were seen in many cells of the root tip tissue of a flowering plant.
d After being soaked in water, a limp lettuce leaf becomes crisp.

8 ***Communicating ideas*** ▸ Discuss the validity of each of the following statements.

a A tissue contains groups of cells where each group has quite a different function.
b Delivery mechanisms are important if a group of small cells is to operate more effectively than one large cell.
c The surface-area-to-volume ratio of a cell influences the rate at which substances can enter or exit the cell.

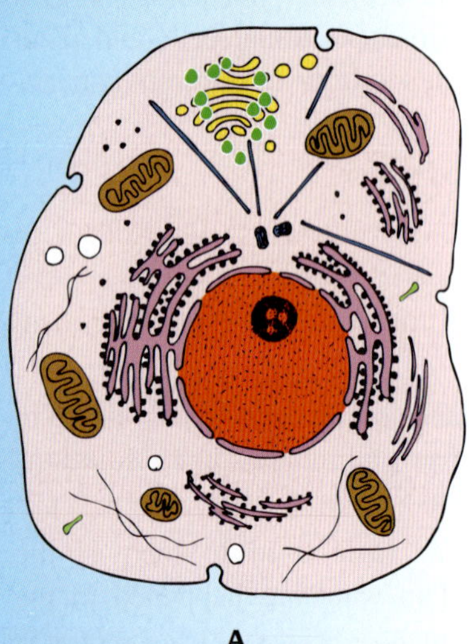

A

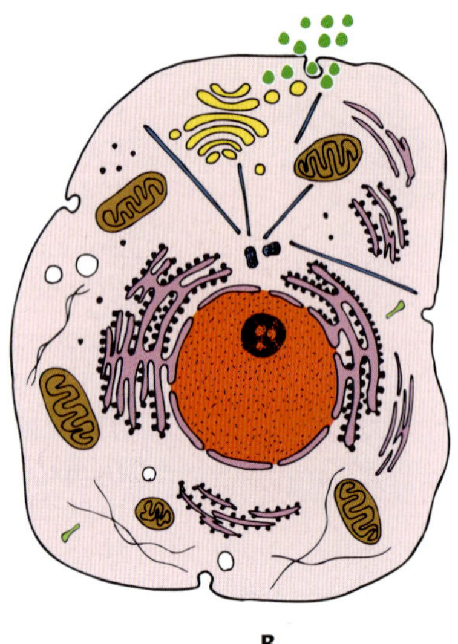

B

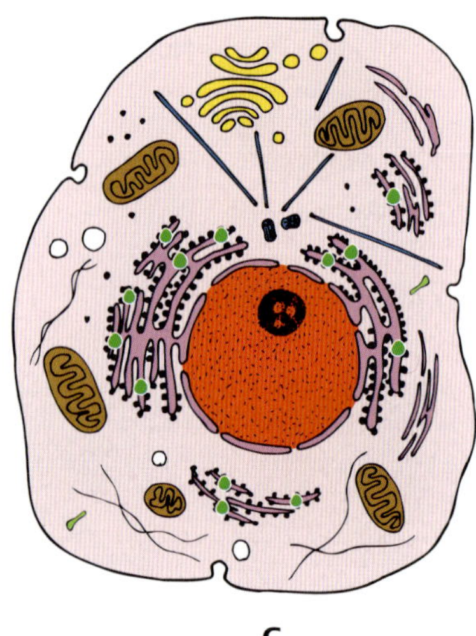

C

Figure 2.32

9 ***Analysing information*** ▸ A scientist carried out an experiment to determine the time it took for a cell to manufacture proteins from amino acids. The scientist provided the cell with radioactively labelled amino acids and then tracked them through the cell to establish the time at which protein synthesis commenced. He monitored the cell 5 minutes, 20 minutes and 40 minutes after production started in order to track the proteins from the site of synthesis to a point in the cell from which they were discharged from the cell.

The scientist made an image of the cell at each of these times but forgot to mark each image with its correct time. The images are given in figure 2.32. Radioactivity is indicated by the green spots.

a Which cell corresponds to each of the particular times of viewing? List the correct order according to time of viewing.

b On what grounds did you make your decision?

10 ***Using the web*** ▸ Go to www.jaconline.com.au/natureofbiology/natbiol1-3e and click on the 'Cell organelles' weblink for this chapter.

a Locate the definition given for the term 'lysosome'. Do you agree or disagree with the definition? Explain your answer. (Check descriptions or definitions given in other resources or at other sites if you are unsure.)

b The website provides a number of ways in which you can test your knowledge of cell organelles. Try them out. Which way works best for you?

11 ***Using the web*** ▸ Go to www.jaconline.com.au/natureofbiology/natbiol1-3e and click on the 'Biology Project' weblink for this chapter. Scroll down and click on 'Prokaryotes, Eukaryotes and Viruses'. Click on 'Prokaryotes', read the information on the page then answer the following questions.

a What is the 'simple statement' used to summarise prokaryotes?

b What are three of the possible shapes found within prokaryotic cells?

c Compared with a typical eukaryotic cell, how much DNA is found in a prokaryotic cell?

d Explain what you think is meant by the statement: 'Eukaryotes have enslaved some of your "brethren" to use as energy generating mitochondria and chloroplasts'.

3 Composition of cells

KEY KNOWLEDGE

This chapter is designed to enable students to:

- develop a knowledge and understanding of the composition of cells
- understand the relationship between the nature of various substances found in cells and the functions they perform in those cells
- understand the general role that enzymes play in cellular biochemical processes
- understand the inputs and outputs of the significant stages in the biochemical processes of photosynthesis and cellular respiration
- analyse and evaluate unfamiliar problem situations related to cells.

Figure 3.1 Footprints at a crime scene. No blood was visible to the naked eye but murder had occurred. The luminescence you see here is due to the reaction of minute traces of haemoglobin, a protein in blood, with a chemical called BLUESTAR ® FORENSIC. The chemical was sprayed because detectives suspected this was the scene of the crime. BLUESTAR ® FORENSIC is a recently developed product and gives a better result than those obtained with Luminol, a chemical that has starred in many television shows about forensic investigations. In this chapter, we explore the major compounds of cells and the functions of those compounds, including their links with the biochemical reactions of photosynthesis and cellular respiration.

Cells reveal clues to a crime

Figure 3.2 The luminescence indicates that an 'invisible' compound has reacted with BLUESTAR ® FORENSIC that was sprayed in this area. Further testing proved the compound was haemoglobin.

Date: 5 May 2000. Place: Palo Alto, California, USA. A woman's body lies at the foot of the stairs leading from the kitchen to the basement of her home. Her grief-stricken husband tells police that he arrived home to find his dead wife in the basement near the foot of the stairs. The only blood visible is on the basement floor, around the woman's head. Upstairs, the kitchen is spotlessly clean. Apparently, a tragic accident has caused the fatal head injuries to Kristine Fitzhugh, wife of Kenneth Fitzhugh.

A post-mortem examination shows that the injuries Kristine suffered are not those that would result from falling downstairs. Five days later, police return to the Fitzhugh house. In the dark, they spray a chemical solution around the kitchen. They take photographs which show glowing patches of bluish-green luminescence on areas of the kitchen wallpaper, on a chair and a cabinet and on the floor leading from the kitchen to the top of the basement stairs. The solution they sprayed is known as Luminol (BLUESTAR® FORENSIC is a new product).

Luminol is a chemical that produces a short-lived bluish glow when it reacts with the oxygen-carrying protein (haemoglobin) found in red blood cells. Luminol can detect minute traces of blood, fresh or old, even at a dilution of one part per million. Because of this extreme sensitivity, Luminol can reveal blood traces even in a crime scene that has been wiped clean of any visible traces of blood.

Luminol can also react with substances other than blood, for example, bleach. Consequently, further tests are carried out in the Fitzhugh home which show the substance in question is indeed blood. Samples taken from clothing in the laundry basket and from between the kitchen floorboards are identified as blood, and then more specifically as Kristine's blood.

Kenneth Fitzhugh is charged with his wife's murder. The prosecution asserts that, during an argument in the kitchen, Kenneth Fitzhugh beat his wife, inflicting fatal head injuries. He moved her body to the foot of the basement stairs, arranging it to appear as if she had died as a result of a fall. He then cleaned the blood from the kitchen and left the house. Kenneth Fitzhugh was found guilty of his wife's murder, due in part to his wife's red blood cells and the haemoglobin they contained.

In this chapter, we investigate the major groups of compounds that make up cells and explorc how some of these compounds are associated with processes such as photosynthesis and cellular respiration.

Materials to build and fuel cells

Symbol	*Element*
C	*Carbon*
H	*Hydrogen*
N	*Nitrogen*
O	*Oxygen*
P	*Phosphorus*
S	*Sulfur*

We have already mentioned some of the kinds of materials found in living cells in chapter 2. The main groups of compounds found in cells are carbohydrates, proteins, lipids and nucleic acids. These carbon-based compounds are called **organic compounds** and all contain the elements carbon, hydrogen and oxygen. Some also contain the elements nitrogen, phosphorus and sulfur. These elements comprise the main elements in the human body. In addition to these compounds we will also consider water, minerals and vitamins.

Water — essential for life

Water is the most abundant compound in our bodies. About 50 per cent of an average adult female is water. Males have less fat than females and have an average water content of 60 per cent. Newborn babies are about 75 per cent water.

Figure 3.3 The average distribution of water for an average adult female in three body regions

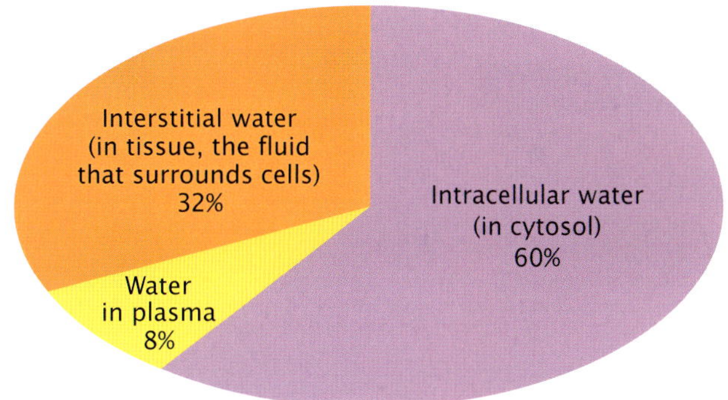

The average adult human body contains about 40 to 42 litres of water distributed across three interconnected compartments. The distribution of water in these broad locations of the body, for an adult female, is shown in figure 3.3. As indicated in table 3.1, a high water content is typical of living organisms.

Table 3.1 Typical water content of some organisms

Organism	Average percentage water by mass
human	
adult male	60
adult female	50
newborn baby	75
bacterium	80
jellyfish	95
rat	67
mushroom	88
typical green plant	75

Figure 3.4 Water content of the raw vegetables shown here ranges between 82 and 96 per cent. The value will change for some on cooking.

The water content in plants also varies. The raw vegetables shown in figure 3.4 vary between 82 per cent and 96 per cent water.

Metabolism goes well in water

Metabolism includes all the chemical reactions that occur in an organism. It involves **catabolism**, the breakdown of compounds to release energy and other compounds or atoms, as well as **anabolism**, the synthesis of new compounds from simpler ones. These reactions occur most readily in solution. Water is the predominant solvent in the body and, as many organic compounds dissolve in water, metabolism occurs in a watery solution. Water facilitates metabolism.

Water molecules stick together

Each water molecule consists of a combination of a single oxygen atom with two hydrogen atoms (see figure 3.5a). Each hydrogen atom is linked to the oxygen atom by a strong **covalent bond**. Although a water molecule overall has a neutral charge, the oxygen at the end of a covalent bond is slightly negative and the

Table 3.2 Water content in tissues

Tissue	*Percentage*
blood (red cell)	60
blood (plasma)	92
muscle	75
cartilage	70

hydrogen atoms are slightly positive areas. Individual molecules of water are highly attracted to each other such that the negative oxygen of one molecule of water is attracted to the positive hydrogen of another water molecule. In fact, the oxygen of a water molecule attracts one hydrogen in each of two other water molecules. Water molecules are said to be highly **cohesive**. They stick together, held by so-called **hydrogen bonds**, which are weaker than covalent bonds (see figure 3.5).

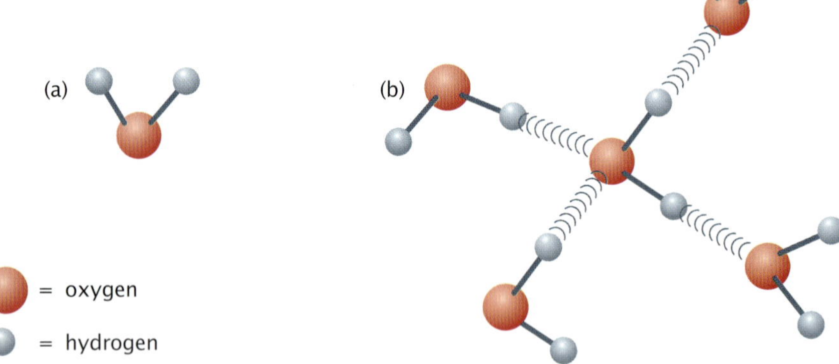

Figure 3.5 Each water molecule is made of a single oxygen atom combined with two hydrogen atoms. **(a)** The hydrogen atoms are joined to the oxygen with covalent bonds. **(b)** Water molecules are attracted to each other and hydrogen bonds form between them. Hydrogen bonds are shown as broken lines in the diagram.

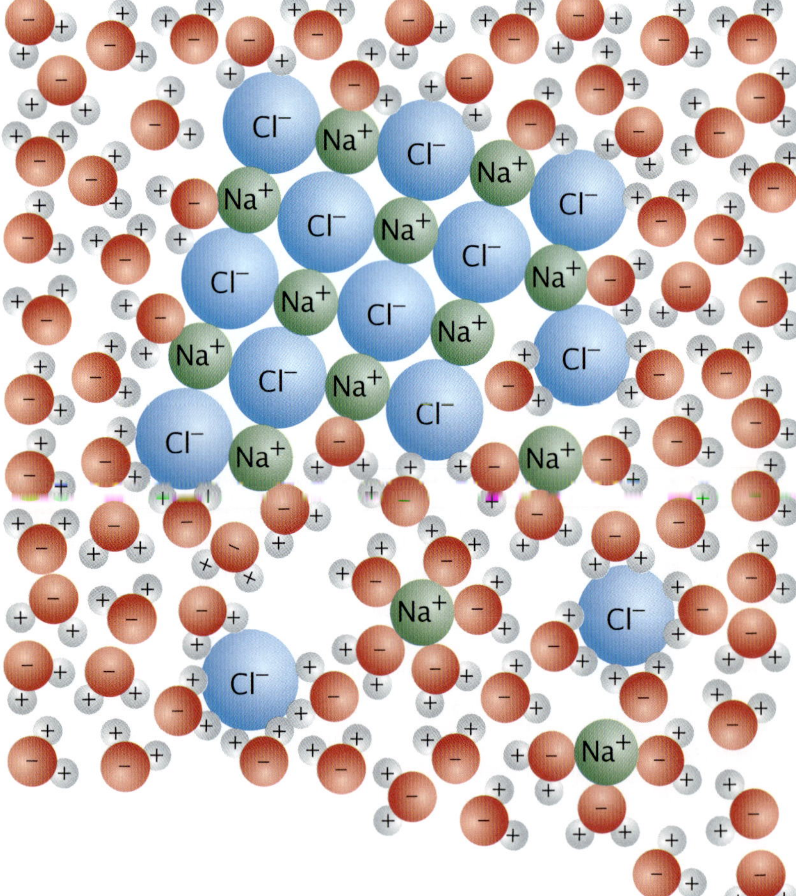

Figure 3.6 A grain of salt contains many sodium chloride molecules. When salt is placed in water, salt molecules dissociate because of the attraction of sodium and chloride to different parts of the water molecules. A ring of water molecules surrounds each sodium and chloride atom and they remain in solution.

Water is a versatile solvent

Water is the predominant solvent in living organisms. Its versatility as a solvent is due to the cohesive nature of the molecule that we examined in the previous section. Substances that dissolve readily in water are called **hydrophilic** or **polar**. Substances that tend to be insoluble in water are called **hydrophobic** or **non-polar**. How would you classify fats?

How do substances dissolve in water? A substance such as sodium chloride (salt, NaCl) dissociates into its parts — positively charged sodium ions and negatively charged chloride ions — when it comes into contact with water molecules. This occurs because the positive sodium ions are attracted by the negative oxygen and the negative chloride ions are attracted by the positive hydrogen of water. These attractions are sufficient to break the **bonds** between sodium and chloride. A ring of water molecules surrounds each sodium and chloride atom and they remain in solution (see figure 3.6).

Water is the major component of cells

Water is by far the main component of living human cells. Although the percentage of water in different cells may vary (see table 3.2, page 54), 70 per cent is a reasonable average for us to consider. In such a human cell, the percentage of major chemicals are as outlined in figure 3.7.

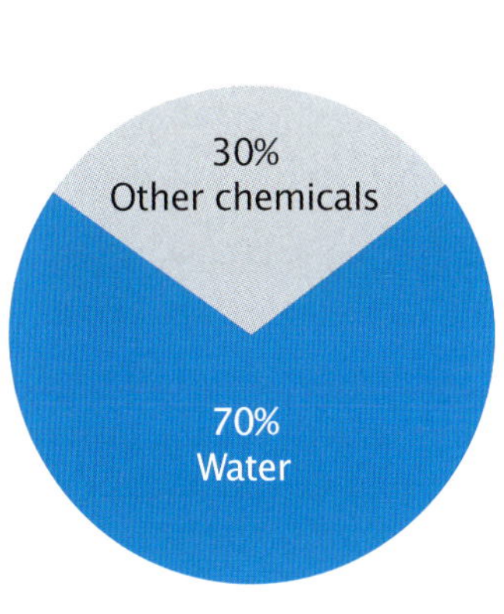

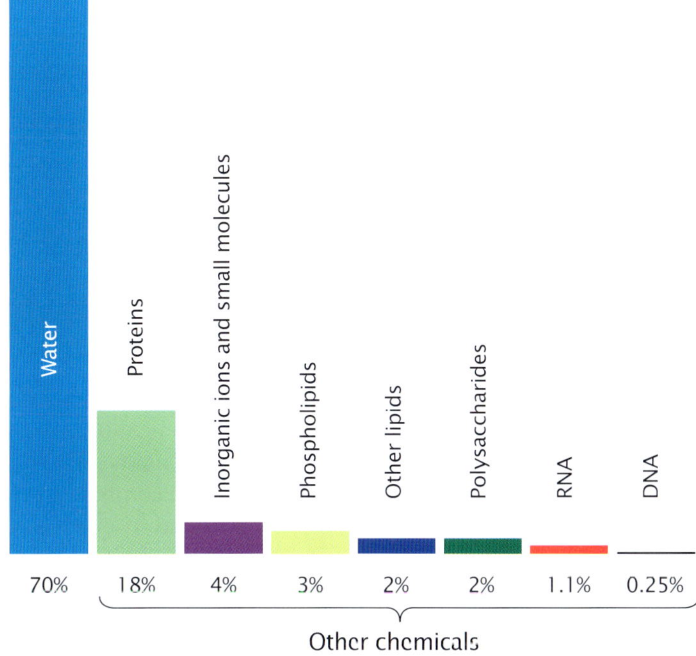

Figure 3.7 The graphs show the percentage of total cell weight of major chemicals in a typical human cell.

KEY IDEAS

- Carbohydrates, proteins, lipids and nucleic acids are organic compounds.
- Metabolism refers to all the chemical reactions that occur in a living organism.
- Water is the predominant solvent in living organisms.
- Water molecules are highly cohesive because of the attraction between hydrogen and oxygen atoms.
- Substances vary in their ability to dissolve in water.

QUICK-CHECK

1 Name the three elements found in all organic compounds.
2 What is the difference between:
  a catabolism and anabolism?
  b hydrophilic and hydrophobic substances?
3 Which is the stronger: a covalent bond or a hydrogen bond?

Organic compounds

Organic molecules are often large molecules made of smaller sub-units that are bonded together in various ways. Compounds formed in this way are called **polymers**. The sub-units are called **monomers**. We can classify molecules on the basis of the kind of sub-unit they contain. Some examples are shown in figure 3.8.

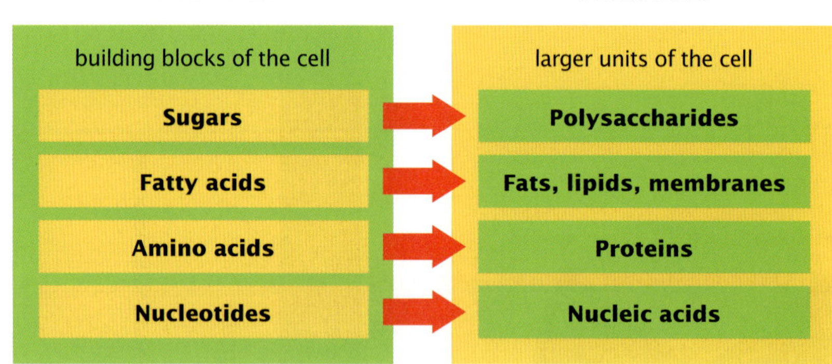

Figure 3.8 The four main groups of organic molecules in cells. Note the monomers that make up the polymers of each group.

The kinds of organic molecules we will consider are **carbohydrates**, **proteins**, **lipids**, and **nucleic acids**. We will examine the basic unit of structure of each, how the basic units combine to form the complex molecules, where each kind of molecule is found in a cell and the function of the molecules. We will also examine **minerals** and **vitamins** and their importance in cells.

Carbohydrates — energy-rich

The basic unit of carbohydrates is a sugar molecule also called a **monosaccharide**. The most common monosaccharide is **glucose** and units of glucose combine in different ways to form different kinds of **polysaccharides**. Carbohydrates containing one or two sugar units are sometimes called simple; those containing many sugar units are called **complex carbohydrates** or polysaccharides. **Starch**, **cellulose** and **glycogen** are all polysaccharides composed entirely of glucose molecules and yet their properties are very different. The difference in property relates to the way in which the glucose molecules are linked together (see table 3.3).

Table 3.3 Basic structure and function of various groups of carbohydrates

Type	*Example*	*Function*	*Structure*
monosaccharides	glucose	energy source in plant and animal cell	
disaccharides	sucrose	form of carbohydrate for transport in plants	
polysaccharides	starch	form of energy storage in plants	
	glycogen	form of energy storage in animals	
	cellulose	structural carbohydrate in plant cell walls	

Carbohydrates are important in plants as structural material. The firm outer covering of insects and spiders is also a polysaccharide — **chitin**. Cellulose is the most abundant organic compound on Earth. Carbohydrates are also stored and used as a source of energy for plant and animal cells (see figure 3.9).

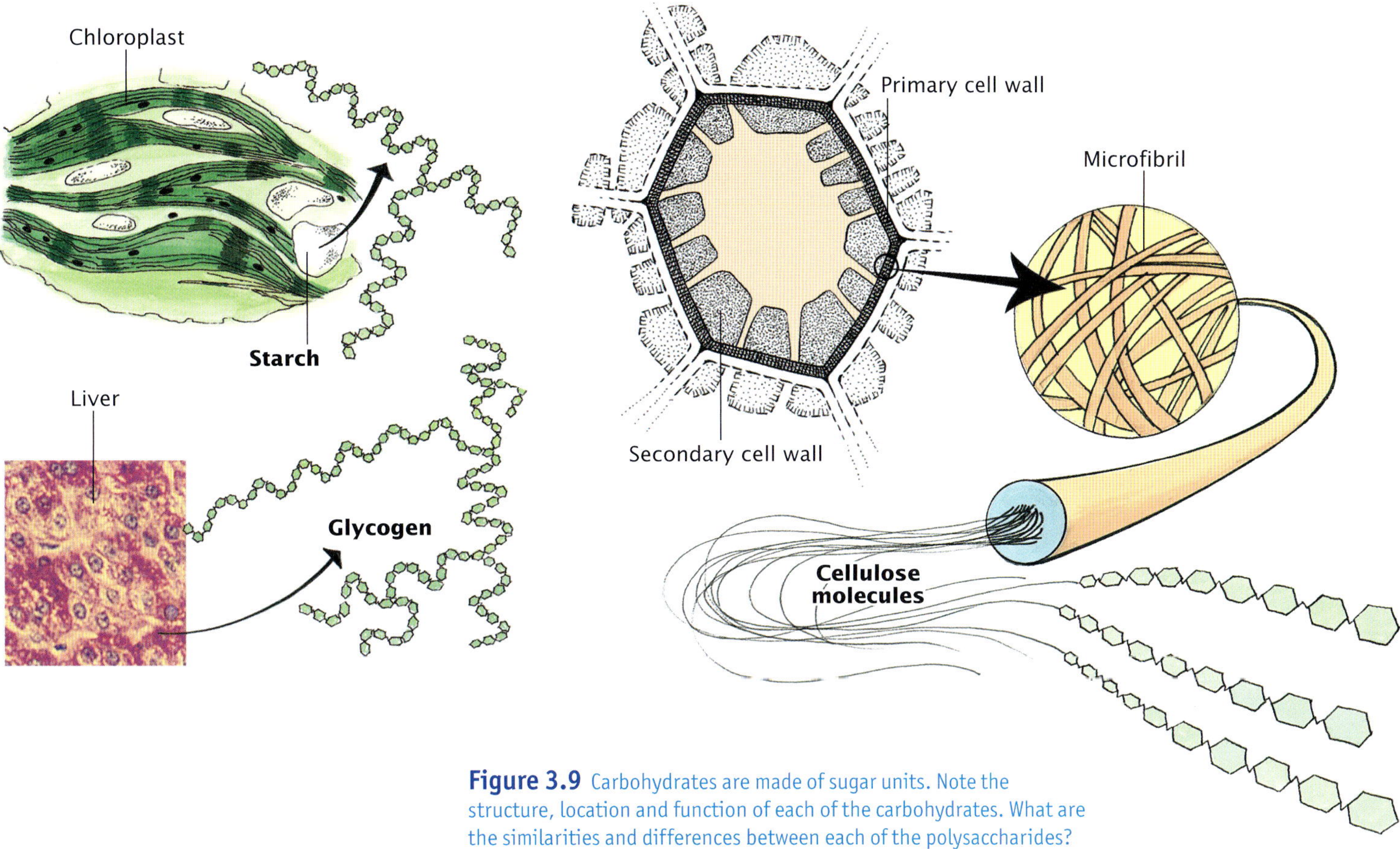

Figure 3.9 Carbohydrates are made of sugar units. Note the structure, location and function of each of the carbohydrates. What are the similarities and differences between each of the polysaccharides?

Cells require energy all the time. Energy is used for movement, for making new material for growth and repair, for maintaining internal conditions within a narrow range and for detecting and responding to environmental changes. When glucose is broken down in cells to smaller products during cellular respiration, the chemical energy present in the glucose is released. A portion of this energy is harvested and captured in ATP molecules (discussed further on pages 69–70).

Proteins

Proteins are large molecules built of sub-units called **amino acids** (figure 3.10a). Note that all amino acids and hence proteins contain nitrogen as well as carbon, hydrogen and oxygen. Some proteins also contain sulfur.

There are 20 naturally occurring amino acids (refer to appendix B) and each amino acid has one part of its molecule different from other amino acids (figure 3.10a and b). Two amino acids join together when a peptide bond forms between them (figure 3.10c). When a number of amino acids join in this way, a polypeptide is formed (figure 3.10d). Each type of protein has its own particular sequence of amino acids. Polypeptide chains become folded in different ways depending on their function.

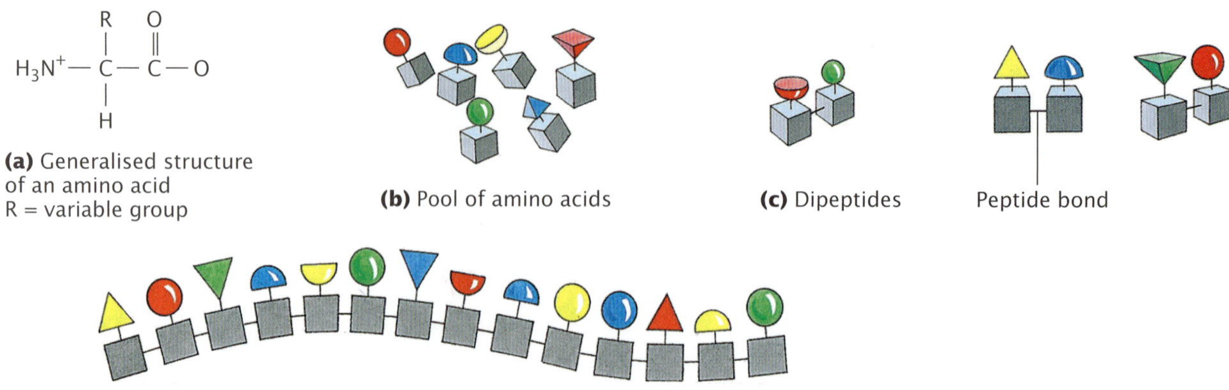

Figure 3.10 Proteins are large molecules made from amino acid sub-units.

Proteins may be structural, for example in cell membranes (see figure 3.11) or the protein filaments within a cell (refer to figure 2.24, page 41). The haemoglobin of red blood cells is protein. The **enzymes** that take part in all the metabolic processes of every living cell are also proteins. Some hormones are proteins.

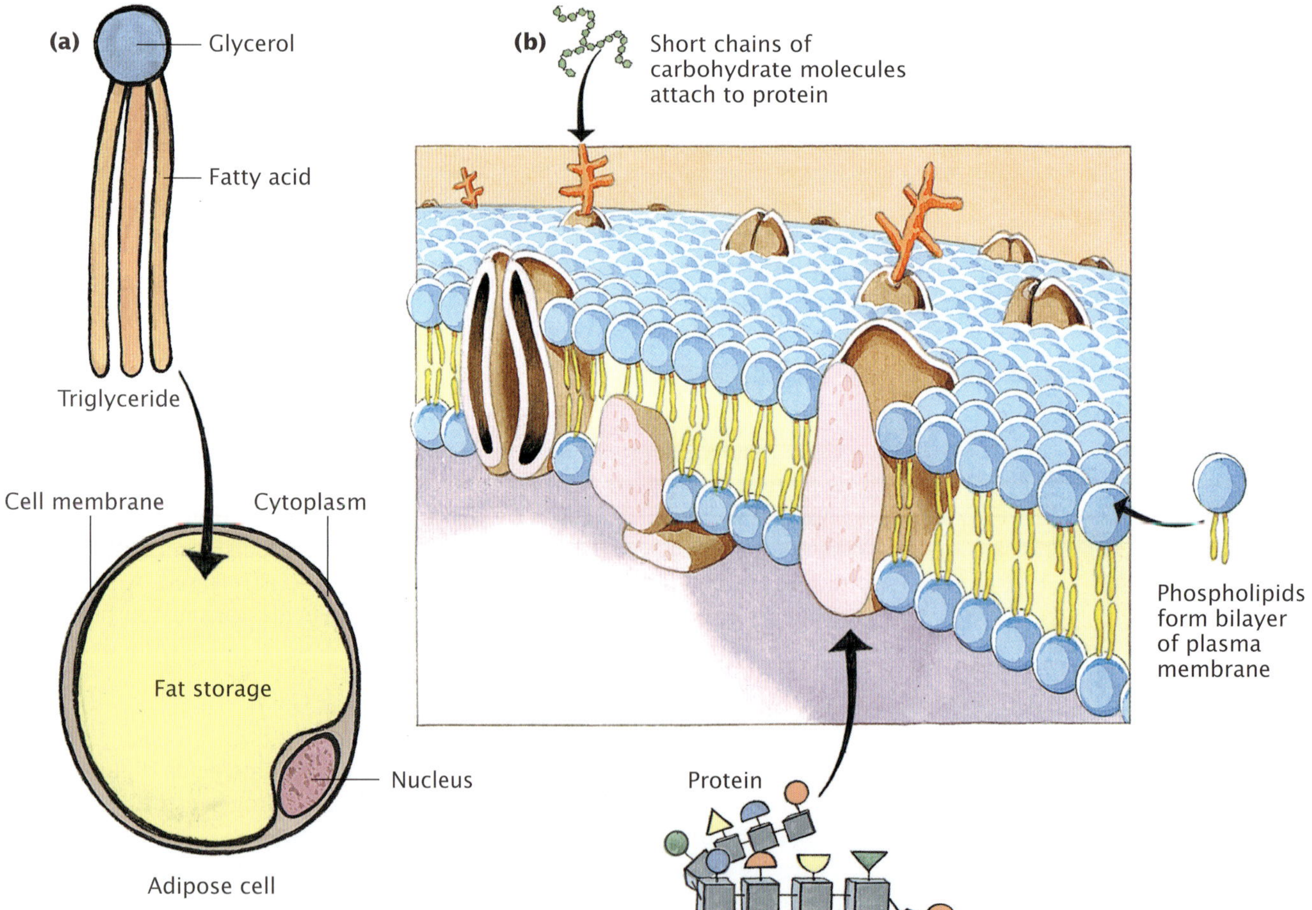

(a) Glycerol
Fatty acid
Triglyceride
Cell membrane
Cytoplasm
Fat storage
Nucleus
Adipose cell

(b) Short chains of carbohydrate molecules attach to protein
Phospholipids form bilayer of plasma membrane
Protein

Figure 3.11 (a) Fats are important energy stores, particularly in adipose cells. (b) Phospholipids are an important component of cell membranes. Note the carbohydrate molecule attached to the protein in the cell membrane.

Animals would have to carry twice as much carbohydrate, compared with a weight of fat, to have the same energy store.

Lipids

Lipid is the general term for fats, oils and waxes. They have little affinity for water. Fats and waxes are generally solid at room temperature and oils are liquid. A fat molecule is made of two kinds of molecules, **fatty acids** and **glycerol**. **Triglycerides** (figure 3.11a, page 58) are a common form of fats. These fats have a single glycerol molecule to which three fatty acid molecules are attached. The fatty acid molecules may be the same or different and lack affinity for water. Hence, fats also have little or no attraction for water and are insoluble in it. Fats and other compounds insoluble in water are called hydrophobic.

Phospholipids, another kind of fat, have two fatty acids attached to a glycerol. They also have a phosphate group attached to the glycerol and other small groups attached to the phosphate to make different kinds of phospholipids. Phospholipids are a major component of cell membranes (see figure 3.11b).

On a weight basis, fat stores twice as much energy as the same weight of polysaccharide so fats are important energy stores in cells, particularly animal cells because animals carry their energy store around with them. Plants are able to rely more heavily on polysaccharide energy storage.

Nucleic acids

There are two kinds of nucleic acid. One is **deoxyribonucleic acid (DNA)** that is located in chromosomes in the nucleus of eukaryotic cells. It is the genetic material that contains hereditary information and is transmitted from generation to generation. The second kind is **ribonucleic acid (RNA)** that is formed against DNA which acts as the template.

Deoxyribonucleic acid

The genetic material deoxyribonucleic acid (DNA) is a polymer of nucleotides. Each nucleotide unit has a sugar (deoxyribose) part, a phosphate part and an N-containing base. The sugar and phosphate parts are the same in each nucleotide. There are four different kinds of nucleotides because four different kinds of N-containing bases are involved. The four different N-containing bases are **adenine**, **thymine**, **cytosine** and **guanine** and the four different nucleotides are denoted by the letters A, T, C and G because of the kind of base each contains (figure 3.12).

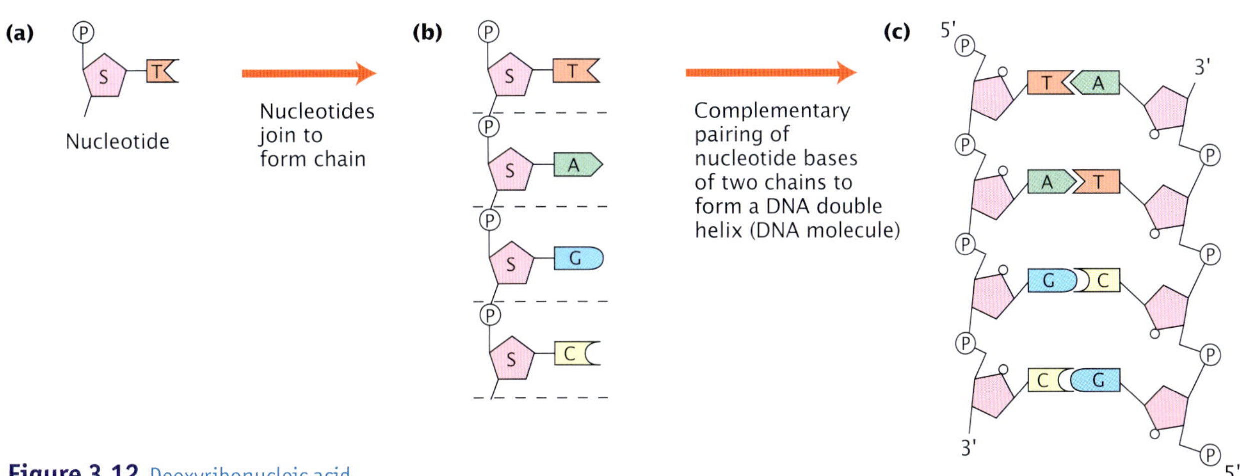

Figure 3.12 Deoxyribonucleic acid is made from nucleotide sub-units. Each DNA molecule is made of two complementary chains of nucleotides.

Examine figure 3.12 (page 59). The nucleotide sub-units (a) are assembled together to form a chain (b) in which the sugar of one nucleotide is bonded with the phosphate of the next nucleotide in the chain. Each DNA molecule contains two chains (c) that bond with each other because the bases in one chain pair with the bases in another. The base pairs between the two strands, namely A with T, and C with G, are said to be complementary base pairs.

Now examine figure 3.13. The two chains form a double-helical molecule of DNA (a) that combines with certain proteins to form a chromosome (b). These chromosomes reside in the nucleus of a cell (c) and the DNA they contain carries genetic instructions that control all functions of the cell.

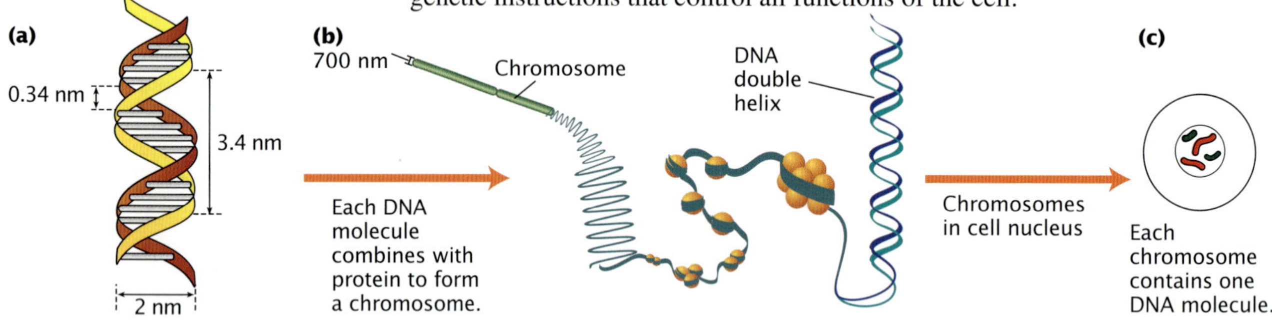

Figure 3.13 In eukaryotic cells, each DNA molecule combines with proteins to form a chromosome.

Ribonucleic acid

Ribonucleic acid (RNA) is also a polymer of nucleotides. It differs from DNA in that it is an unpaired chain of nucleotide bases and exists in three different forms. RNA is constructed from four different bases, three of which — adenine, guanine and cytosine — are identical to those in DNA. The fourth nucleotide is uracil that is capable of pairing with A (figure 3.14).

The three different forms of RNA are:

- messenger RNA (mRNA), formed against DNA as a template. mRNA carries the genetic message to the ribosomes where the message is translated into a particular protein.
- ribosomal RNA (rRNA) which, together with particular proteins, makes the ribosomes found in cytosol
- transfer RNA (tRNA), molecules that carry amino acids to ribosomes where they are used to construct proteins.

The strand of nucleotides in each of the RNAs is folded in a different way.

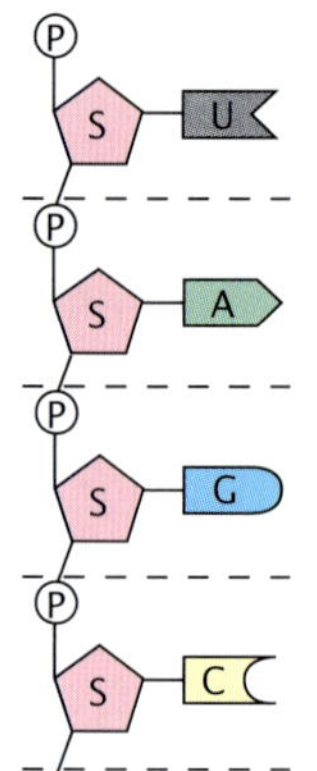

Figure 3.14 The four nucleotide sub-units, uracil, adenine, guanine and cytosine, from which the three RNAs are constructed.

KEY IDEAS

- The major compounds that make up living cells are different kinds of carbohydrates, lipids, proteins and nucleic acids.
- Each major compound has a specific role. Some exist only in either plant or animal cells; others play an important role in both.

QUICK-CHECK

4 Two proteins in a cell each contain the same number of amino acids and yet have quite different functions. Explain.

5 Consider the carbohydrates glucose, starch, cellulose and glycogen. In what ways are these compounds similar and in what ways are they different?

6 Explain how the hydrophobic and hydrophilic parts of phospholipid molecules influence the way they orientate in cell membranes.

7 In addition to C, H and O, which element is found in all proteins?

Minerals

Minerals are inorganic ions required by both animal and plant cells. In humans, minerals make up about six per cent of the body (see table 3.4). Some, for example calcium and phosphorus, are present in relatively large amounts. Others, for example cobalt and molybdenum, are present only in trace amounts.

Table 3.4 Percentages of minerals in the body mass of humans

Mineral	Percentage	Mineral	Percentage
calcium	2.0	copper	0.000 15
potassium	1.0	iodine	0.000 04
phosphorus	1.0	manganese	0.000 03
sulfur	0.25	cobalt	trace
sodium	0.15	molybdenum	trace
chloride	0.15	fluoride	trace
magnesium	0.05	selenium	trace
iron	0.005	chromium	trace
zinc	0.002		

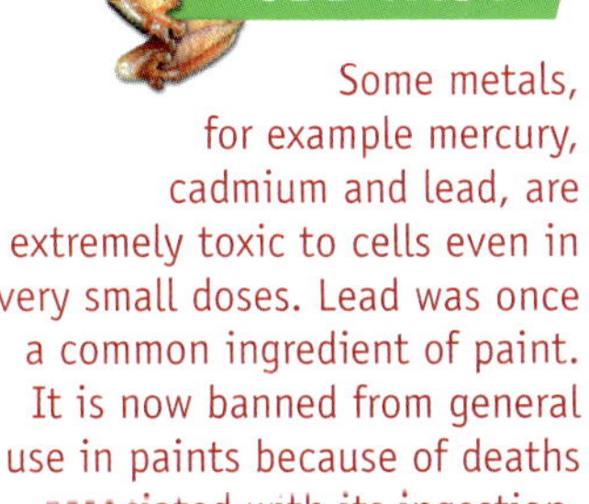

Some metals, for example mercury, cadmium and lead, are extremely toxic to cells even in very small doses. Lead was once a common ingredient of paint. It is now banned from general use in paints because of deaths associated with its ingestion.

Minerals play a role in metabolic processes of cells and are incorporated into many structures produced by cells. For example, animal body structures such as bone and teeth have significant mineral content in the form of calcium. The cell walls of plants contain minerals such as silicon and boron and rely on calcium for their middle lamella. Minerals also contribute to cell manufacture of many hormones, enzymes and vitamins.

Plants and animals cannot manufacture minerals. Animals must obtain their minerals from their diet; plants generally obtain theirs from the soil. Table 3.5 shows a sample of minerals important for plants and animals.

Table 3.5 A sample of minerals important for animal and plant processes

Mineral nutrient	FUNCTIONS	
	Animals	Plants
Calcium	Bones and teeth, transmission of nerve impulses, enzyme reactions	Middle lamella of cell walls, enzyme reactions
Potassium	Osmotic balance of cells, transmission of nerve impulses	Establishing cell turgor, co-factor for many enzyme reactions
Sodium	Osmotic balance of cells, transmission of nerve impulses	Substitutes for potassium in some functions
Phosphorus	Metabolism of fat, protein and carbohydrate, cellular respiration	Cellular respiration
Iron	Cytochromes — needed by all cells for respiration	Cytochromes — involved in photosynthesis, nitrogen fixation and cellular respiration

Although some minerals are required in relatively small amounts, they are still vital for normal healthy functioning of cells. For example, copper is an essential mineral for cellular respiration. A deficiency of copper supply to cells results in a significant deficiency of production of energy-rich ATP. Some babies are unable to supply sufficient copper to their cells and this has a fatal effect. Read about Professor Julian Mercer and his work on copper deficiency in babies.

BIOLOGIST AT WORK

Professor Julian Mercer — Molecular geneticist

Professor Julian Mercer is a Research Scientist and Director of the Centre for Cellular and Molecular Biology, School of Biological and Chemical Sciences, Deakin University in Melbourne. Julian writes:

From my school days I was always interested in chemistry. After school I completed a BSc majoring in biology and chemistry at the Australian National University. After completing my honours year in organic chemistry, I decided that the biological sciences were more interesting to me than pure chemistry, and the developing field of molecular biology was particularly exciting. At the time, the mechanisms of protein synthesis on the ribosome were still being worked out, and I undertook a PhD to work in this area in the Biochemistry department at the University of Adelaide. My PhD involved a lot of organic synthesis; I was making the end of transfer RNA to see if it would work in peptide bond formation.

After completing a PhD, a young scientist needs to find a laboratory to continue postdoctoral research training. I chose to return to the Australian National University because I had an offer of a Fellowship to work on mRNA isolation, which moved me closer to molecular biology than chemistry.

After three years in Canberra, I moved to the Laboratory of Molecular Biology in Cambridge, UK, for another two years postdoctoral work. What an exciting place to work in! The Laboratory of Molecular Biology was, and still is, an amazing place. At the time I was there, five Nobel Prize winners were working in the same building. These included Francis Crick who, with James Watson, successfully worked out the structure of DNA. It was truly awe-inspiring to meet and talk with these famous scientists — people who had laid the foundations of molecular biology. One of the wonderful aspects of science is the opportunity to meet and work with extraordinary people from all cultures, and also to live and work in other countries. You realise that science is an international culture, one that overcomes many of the cultural barriers that may arise in other aspects of life.

I completed my conversion into a molecular geneticist when I joined Professor David Danks in the Genetics Research Unit (which later became the Murdoch Institute) at the Children's Hospital in Melbourne. The field was just entering the era of cloning disease genes and I undertook projects aimed at cloning the gene affected in PKU and the gene affected in the X-linked copper disorder, Menkes disease. I did not know then, but my investigation of Menkes disease would occupy me for more than twenty years.

Figure 3.15 Professor Julian Mercer in his laboratory

It is not widely known that copper is an essential element, and an adequate supply of this metal is vital for normal development. Copper forms part of many important enzymes. Perhaps the most important is cytochrome c oxidase which is involved in aerobic respiration in mitochondria. Babies with Menkes disease cannot supply enough copper to cytochrome c oxidase and this impairs their ability to form ATP. The deficient ATP production has a catastrophic effect on brain development (the brain is a big user of ATP) and boys with Menkes disease usually die within 3 years. Fortunately the disease is rare.

Menkes disease is an X-linked condition found in about 1/100 000 boys born; very few girls are known to have had the condition. Affected children have a range of other abnormalities that can be related to copper deficiency:

- reduced hair pigmentation because tyrosinase, which is part of the pathway that makes the dark pigment melanin, needs copper
- weak artery walls leading to aneurisms and weak bones, because copper is needed for an enzyme that helps in connective tissue formation
- unusual hair, which tends to be a bit like steel wool in texture and is an important diagnostic feature.

Figure 3.16 Copper deficiency in sheep causes 'steely wool', as in this sample. The white banding in the black wool of this sheep is due to the lack of copper that results in reduced tyrosinase and hence reduced pigment.

Interestingly, Professor Danks discovered that Menkes disease was caused by copper deficiency because he was aware of Australian research that showed copper deficiency in sheep causes 'steely wool' (see figure 3.16), very like the unusual hair in boys with Menkes disease.

We worked for many years to try to find the gene affected in this disease, but had success only when a rare female was diagnosed with Menkes disease. Her condition arose as a result of a genetic accident at a very early stage of her development. A portion of one chromosome broke away and became inserted into the Menkes gene, disrupting the normal function of that gene. If we could find the location at which the piece had been inserted — that is, the breakpoint in the X-chromosome — we could locate the position of the Menkes gene. After much work we were successful. The rare female was a chance event that we were able to use to our advantage.

Chance events over the years have provided scientists with opportunities for insights about situations that have long puzzled them. The isolation of the Menkes gene was an international race between four laboratories. Eventually three succeeded and the three papers were published together, so each laboratory received credit for their discovery. The fourth laboratory isolated the wrong gene.

Vitamins

Vitamins are a group of organic compounds that occur in minute quantities in food. Although the requirement for vitamins may vary from animal to animal, they are unable to make them and hence require vitamins in their diets. Although it was known for centuries that certain diseases could be 'cured' by the addition of a range of fresh foods to the diet, vitamins were not discovered and extracted from food until early in the twentieth century.

The name of the water-soluble vitamin folic acid is derived from *folium*, Latin for 'leaf', and is found in green plants, fresh fruit, yeast and liver. The active co-enzyme form of folic acid is tetrahydrofolate.

When vitamins were first discovered, their chemical structures were unknown. This resulted in the use of a series of letters — A, B, and so on — to designate different vitamins. Now, the chemical structures are known and vitamins tend to be called by their chemical names.

Vitamins can be divided on the basis of their chemical nature into two groups: fat-soluble and water-soluble vitamins. Refer to table 3.6 (page 64) for more information about vitamins.

Vitamins are essential for many of the chemical reactions that occur in cells. The water-soluble vitamins, except for vitamin C, are components of co-enzymes. Co-enzymes are small molecules necessary for the normal functioning of many of the enzyme-controlled reactions within cells. The function of fat-soluble enzymes is not as well understood as that of water-soluble ones but it is known that they have an important role in blood coagulation, bone structure and vision (see the information on 'Enzyme function and specificity' on page 65).

Table 3.6 Vitamins — their source and functions

Vitamin	*Source*	*Required for*	*Recommended daily intake*
Fat-soluble vitamins			
A retinol	fish-liver oils, butter and margarine, green and yellow vegetables, carrots, yellow-fleshed fruits, tomatoes, egg yolk, liver and kidneys, whole milk	epithelial tissues — skin, linings of nose, mouth, digestive and urinary tracts; vision in dim light — forms visual purple in retina of eye	infants and children 250–750 μg; adults 750 μg; lactating women 1200 μg
D calciferol	liver, eggs, and can be made by the body	stimulates absorption of calcium and phosphorus in bone and teeth formation	5–10 μg; children and pregnant women have greatest need because of bone growth
E tocopherols	wheatgerm, butter and margarine, bread, green leafy vegetables, whole-grain products	prevents damage to cell membranes; protects fats and vitamin A from destruction by oxidation	10–15 mg
K phytomenadione	green vegetables, tomatoes, cabbage, cauliflower, potatoes, cereal, eggs	formation of prothrombin, essential for blood clotting	children 15–100 mg; adults 70–100 mg
Water-soluble vitamins			
B_1 thiamine	seafood, meat, whole-grain products	the release of chemical energy from carbohydrate	children 0.5–1.2 mg; women 0.6–0.8 mg; men 0.8–1.1 mg
B_2 riboflavin	meat, eggs, green vegetables, mushrooms, whole-grain products, pasta	protein metabolism; helps maintain healthy skin and eyes; growth of new body tissue	children 0.5–1.2 mg; women 0.8–1.0 mg; men 1.0–1.4 mg
B_3 niacin	leafy vegetables, whole-grain products, peanut butter, potatoes, tuna, eggs	enzyme systems that convert carbohydrates, proteins and fats into energy; aids synthesis of hormones	children 9–20 mg; women 10–13 mg; men 14–18 mg
B_6 pyridoxine	yeast, wheat germ, cereals, liver, meat, soya beans, peanuts, egg yolk	many enzyme reactions and for development of red blood cells	adolescents 1–2.2 mg; adults 1–1.5 mg; lactating women 1.6–2.2 mg
B_{12} cyanocobalamin	liver, meat, eggs, oysters, sardines	development of red and white blood cells; also involved in metabolism	adults 2 μg; pregnant and lactating women 3–3.5 μg
folate	liver, kidneys, yeast, mushrooms, leafy green vegetables	reduction of neural tube defects; development of red blood cells and metabolism of protein; linked with vitamin B_{12}	adults 200 μg; pregnant women 400 μg; lactating women 300 μg
biotin	liver, meat, fish, yeast, egg yolk, milk, smaller amounts in grains and vegetables	metabolism of fat and protein; growth and function of nerve cells	children 65–200 μg; adults 100–200 μg
pantothenic acid	liver, meat, fish, egg yolk, yeast, cereals, peanuts and vegetables; also found in 'royal jelly' (fed to the queen bee) but this is most expensive and not in any way a source superior to others	metabolism of carbohydrate, fat and protein	estimated 5–10 mg
C ascorbic acid	citrus and other fruits, tomatoes, leafy vegetables, potatoes, rosehips	connective tissues, bones, teeth; promotes wound healing and absorption of iron	children 30–50 mg; adults 30 mg; pregnant and lactating women 60 mg

Note: In an attempt to set a standard for an adequate diet for groups of the population, health authorities introduced a scheme of recommended dietary intakes (RDI). These recommendations are often quoted now in labelling of nutrients on foods, but it is important to realise that an RDI does not necessarily equal the requirement of a particular individual.

Variation exists in the amount of nutrients required by different individuals. Variation is due to a number of factors including genetic factors, body size, activity level and state of health. An RDI value should satisfy the requirements of 95 per cent of the population but some people require less than the RDI and others require more. A person with a varied and well-balanced diet will receive adequate amounts of vitamins and minerals.

ENZYME FUNCTION AND SPECIFICITY

Enzymes are protein molecules that increase the rate of the reactions that occur in living organisms. Without enzymes, metabolism would be so slow at body temperature and pH that insufficient energy would be available to maintain life. Many enzymes are intracellular; they are used within cells that produce them. Others (for example, the digestive enzymes) are extracellular. They are secreted by cells and act outside those cells.

The compound being acted on by an enzyme is called a **substrate**. The compounds obtained as a result of the enzyme action are called the products. Enzymes are highly specific in their action; each enzyme acts on only one kind of substrate. This is because the shape of one enzyme matches that of one substrate at a particular region known as the **active site** of the enzyme, as shown in figure 3.17.

An enzyme is named for the substrate it associates with. The enzyme maltase acts on the disaccharide maltose. Lipase acts on lipid, and amylase on amylose. Note the -ase endings in the name of most enzymes. The few exceptions include trypsin and pepsin.

Although enzymes participate in reactions, they are not used up in the reactions and are available for reuse. The rate of an enzyme reaction is influenced by pH, temperature, enzyme concentration and substrate concentration. Why do you think the activity of an enzyme is destroyed at high temperatures?

Many enzymes require the presence of other factors before they act. If the factor is an organic molecule (for example a vitamin), it is called a **coenzyme**. Some poisons, for example cyanide and arsenic, block the active sites of enzymes and hence interfere with cell metabolism.

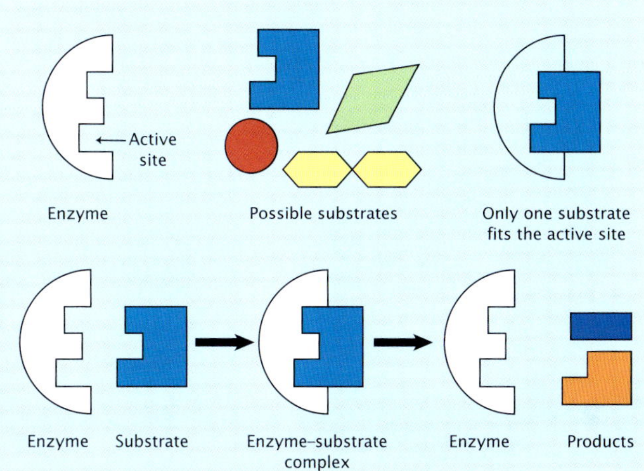

Figure 3.17 Representation showing that each enzyme reacts with only one kind of substrate. This is because the active site of an enzyme matches the shape of a particular substrate and the two are able to come closer together. This matching of shapes is often called the 'lock and key' theory of enzyme action.

KEY IDEAS

- The body requires a number of minerals, some in relatively large amounts, others in trace amounts.
- Minerals play a role in metabolic processes of cells and are incorporated into many structures produced by cells.
- Copper is an essential mineral for cellular respiration.
- Vitamins are a group of organic compounds and are essential for many of the chemical reactions that occur in cells.
- Vitamins can be divided on the basis of their chemical nature into two groups: fat-soluble and water-soluble vitamins.
- Each enzyme acts on only one kind of substrate.

QUICK-CHECK

8 Name, and give the function of, two minerals that the body requires in relatively large amounts and two that it requires in trace amounts.

9 a Name four foods that are a source of fat-soluble vitamins.
  b Name four foods that are rich in water-soluble vitamins.

10 What feature of enzymes controls which substrate they act on?

Producers at work: photosynthesis

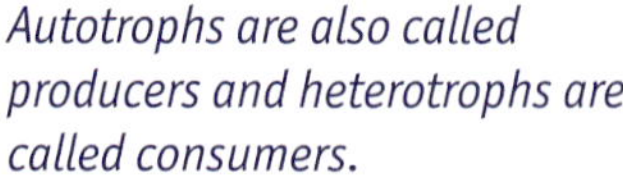

Autotrophs are also called producers and heterotrophs are called consumers.

How do carbohydrates originate? Using the energy of sunlight, plants, algae and some protists (such as phytoplankton) can make organic molecules, such as sugars, by **photosynthesis**. Organisms with this ability are termed **autotrophic**. Other organisms, such as animals and fungi, that depend, directly or indirectly, on the organic compounds produced by producers are called **heterotrophic**.

Photosynthesis is the process in which **light energy** is transformed into **chemical energy** stored in sugars. In a typical producer, such as a terrestrial flowering plant, the complex series of reactions in photosynthesis can be summarised as follows:

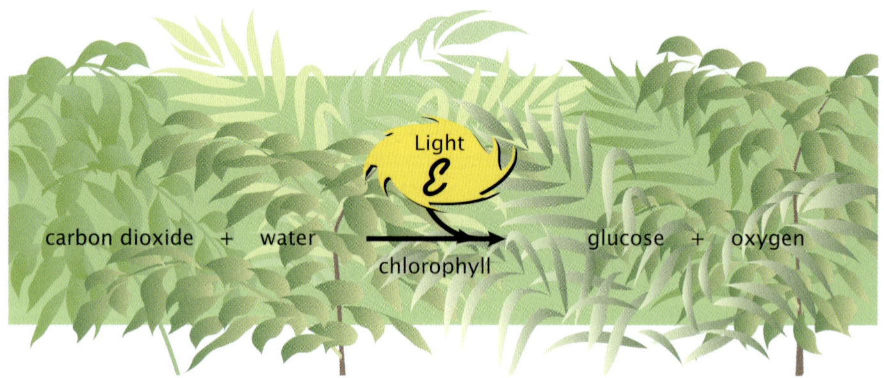

In fact, the complete balanced equation for photosynthesis is

$$6CO_2 + 12H_2O \xrightarrow[\text{light}]{} C_6H_{12}O_6 + 6O_2 + 6H_2O$$

This is often simplified to:

$$6CO_2 + 6H_2O \xrightarrow[\text{light}]{} C_6H_{12}O_6 + 6O_2$$

showing net consumption of water only.

The general word equation for photosynthesis is the summary for a chain of biochemical reactions, as we will see later. One way of summarising the inputs and outputs of photosynthesis is shown in figure 3.18.

Figure 3.18 The inputs and outputs of the process of photosynthesis

In fact, the details of the chemical reactions that make up photosynthesis are more complex than the equation shown here. You will be introduced to more information about the chemistry of photosynthesis as you progress through your biology studies. For now, we will explore where photosynthesis occurs and how different wavelengths of light can influence the process.

$-CH_3$ in chlorophyll a

$H-CHO$ in chlorophyll b

Porphyrin ring

Mg

Figure 3.20 Molecular structure of chlorophyll a

Figure 3.21 Deciduous leaves in early autumn

Chloroplasts: where the action is!

The cytoplasm of cells in the leaf tissue where photosynthesis occurs contains specialised organelles, known as **chloroplasts**. Each chloroplast has an outer membrane and folded inner membranes joined to form stacks of flattened discs, know as **grana** (singular = granum) (see figures 2.20 (page 37) and 3.19). The enzymes needed for the reactions in photosynthesis are located inside the chloroplasts, with some located in the grana membranes and some in solution in the **stroma**.

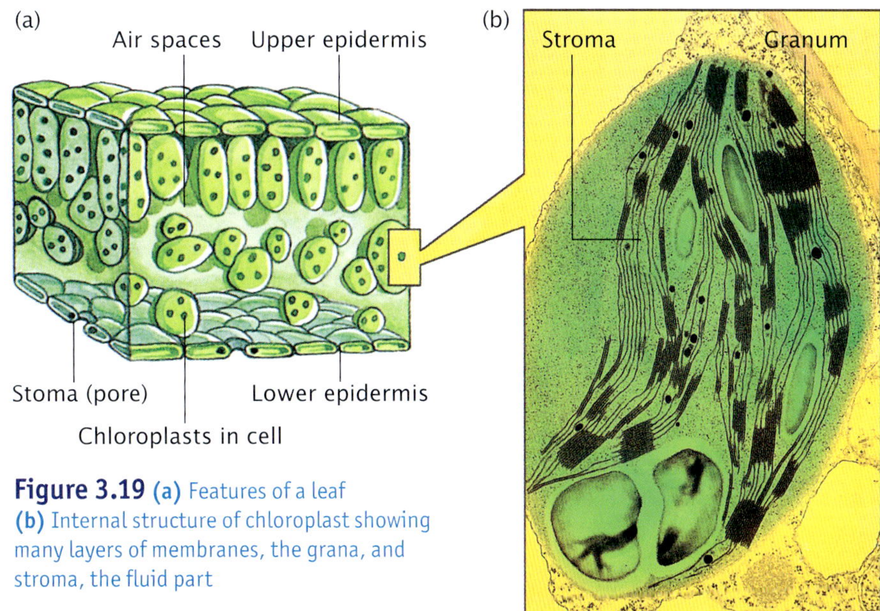

Figure 3.19 (a) Features of a leaf (b) Internal structure of chloroplast showing many layers of membranes, the grana, and stroma, the fluid part

Light-trapping pigments

Radiant energy from the sun includes a range of wavelengths:

- visible light, in the range from about 0.4 to 0.7 μm
- infra-red (IR), with wavelengths greater than 0.7 μm
- ultraviolet light (UV), with wavelengths shorter than 0.4 μm.

Various pigments can trap light energy. The major light-trapping pigments are green **chlorophylls** (see figure 3.20) located on the grana membranes. Other kinds of light-trapping pigments, known as **accessory pigments**, are also found in the chloroplasts of various organisms (see table 3.7).

Table 3.7 Occurrence of various light-trapping pigments in plants and algae

Pigment	Organisms where found
chlorophyll a	all plants and all algae
chlorophyll b	all plants and green algae
chlorophyll c	brown algae and phytoplankton
chlorophyll d	red algae
carotenoids	all plants and green algae
phycobilins	red algae

Carotenoids are red, orange and yellow pigments that are normally masked by the green chlorophylls. In temperate climates in early autumn, deciduous trees begin to withdraw the chlorophyll from their leaves, exposing the carotenoids and other pigments that are normally hidden (see figure 3.21). Phycobilins are blue-green (phycocyanin) and red (phycoerythrin) water-soluble pigments.

Trees that lose their leaves during one short period of the year are said to be deciduous. In contrast, Australian native trees drop leaves in small numbers over the entire year. Because these trees do not become leafless, they are termed evergreens.

The various pigments trap light energy of different wavelengths (see figure 3.22). Consequently, the presence of accessory pigments extends the range of light wavelengths that can be absorbed by a plant and converted to chemical energy. Light energy absorbed by accessory pigments must be transferred to chlorophyll a before it can be converted into chemical energy. If the accessory pigments were removed from a plant cell, what would you predict would happen to the rate of photosynthesis?

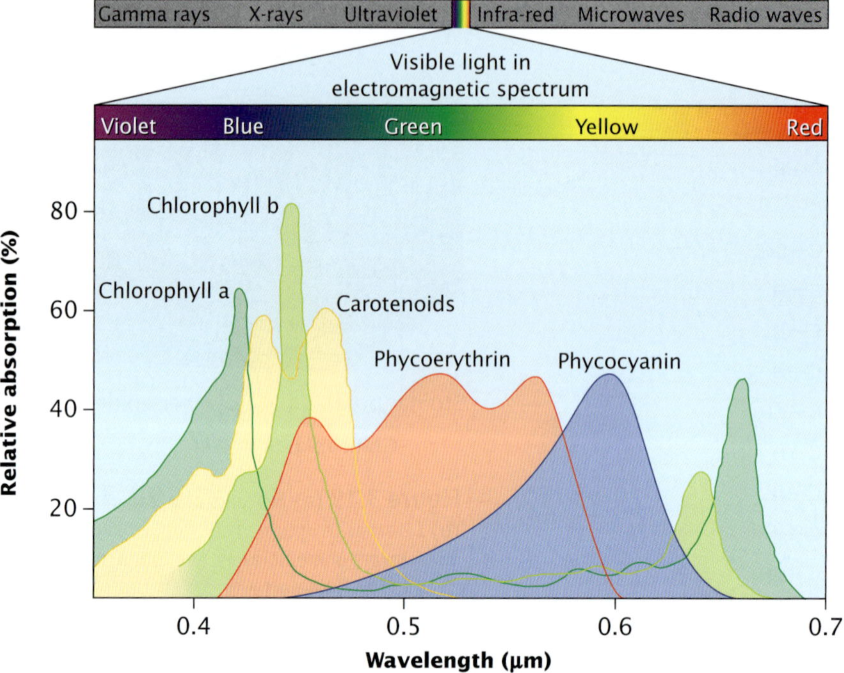

Figure 3.22 Absorption of light of various wavelengths for different plant pigments. Which pigment absorbs yellow light best? What wavelengths of light (colours) are best absorbed by chlorophylls?

Photosynthesis is most efficient in light of red and blue wavelengths. These wavelengths do not penetrate very far below the surface of water. In deeper waters most of the light wavelengths available are blue-green, which accessory pigments can absorb. The phycobilins, in particular, contribute to photosynthesis in deeper waters and these are found in seaweeds (see figure 3.23). Seaweeds that have colour when brought to the surface of the water often look black when looked at in water. Can you determine why this is so? Think about absorption and reflection of different wavelengths of light.

Figure 3.23 Different algae: **(a)** green seaweed, *Caulerpa remotifolia*, found at depths of up to 10 metres **(b)** brown seaweed, *Macrocyctis pyrifera*, also found at depths of up to 10 metres **(c)** red seaweed, *Callophyllis lambertii*, found at depths of up to 35 metres

Accessing energy: cellular respiration

All living organisms require energy to maintain life. The energy required to maintain cellular functions is in the form of adenosine triphosphate (ATP) (see figure 3.24). Where does this ATP come from? As we saw on pages 29 and 32, ATP is formed when energy is released during **cellular respiration** of glucose, a carbohydrate.

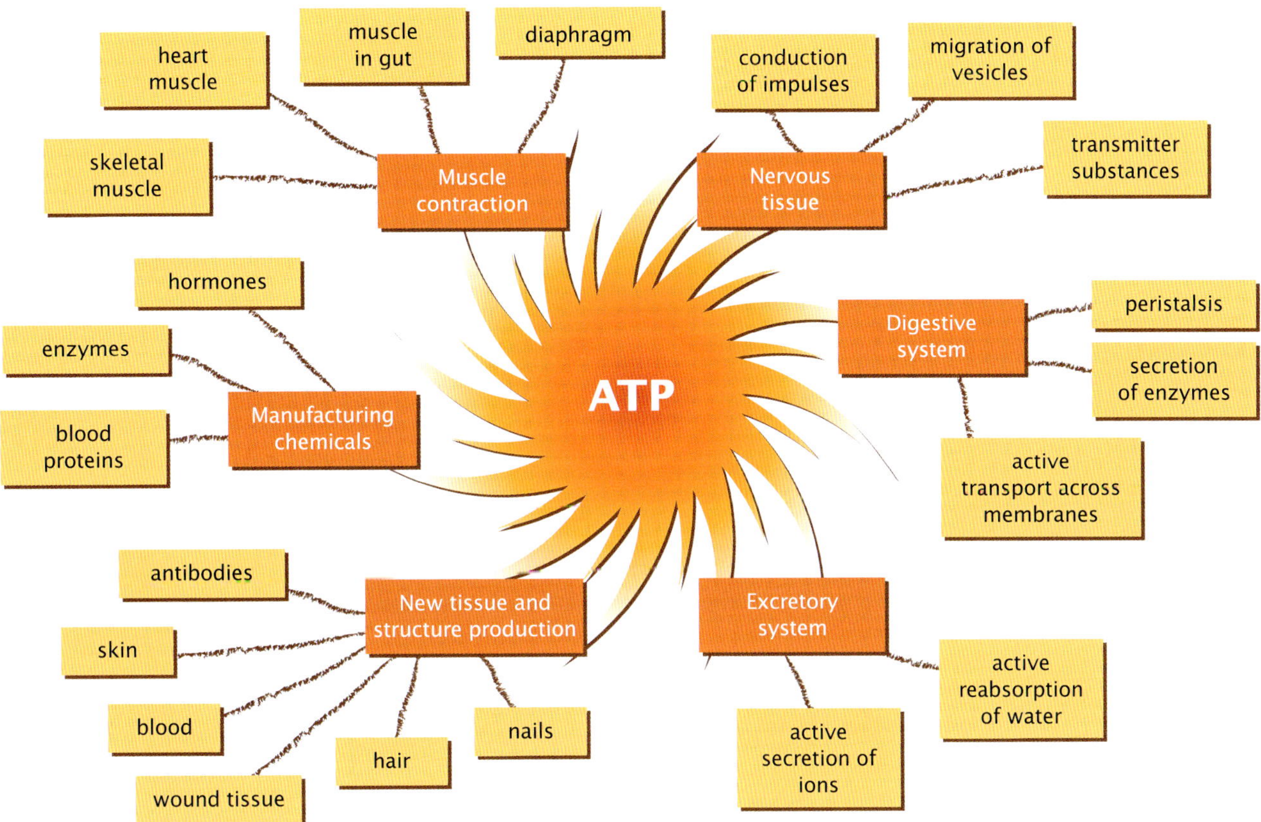

Figure 3.24 Energy is necessary for life. ATP, a high-energy compound, is a major source of energy for many functions in the human body, some of which are outlined here.

The transfer of chemical energy from glucose to ATP occurs through a coupling of chemical reactions (see figure 3.25).

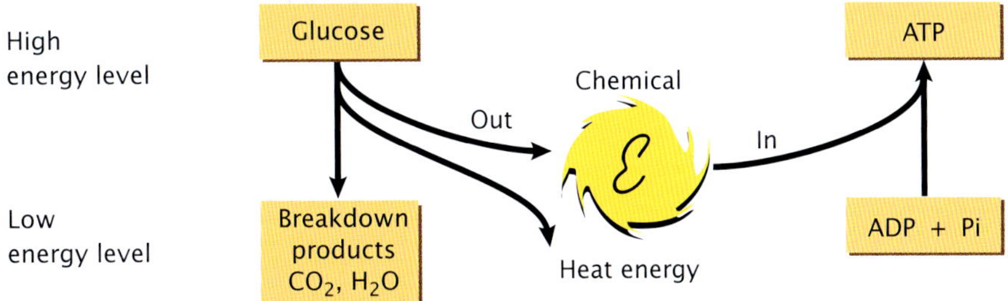

Figure 3.25 Energy released from reactions involving glucose is transferred to the energy-requiring process of ATP production. Is the process 100% efficient? (Pi = inorganic phosphate; ADP = adenosine diphosphate)

The process of energy transfer from glucose to ATP is not 100 per cent efficient. Some of the chemical energy is converted to and 'lost' as heat energy. In general, the energy transfers in cellular respiration are 40 per cent efficient — about 40 per cent of the chemical energy present in glucose is transferred

ODD FACT

Built-in electric blanket! Brown fat tissue is found in young mammals and in adult mammals of species that hibernate. Brown fat is metabolised and the energy released appears as heat energy, not as chemical energy in the form of ATP.

to ATP and the remaining 60 per cent appears as heat energy. So, living cells produce heat. The box below describes how heat production from respiring fruit and vegetables contributes to their deterioration.

Cells cannot use heat energy to drive energy-requiring activities, such as muscle contraction or transport against a concentration gradient. However, in mammals and birds, the heat energy released from cellular respiration is trapped by insulating layers of fat, fur or feathers and is the internal source of the heat needed to maintain their core body temperatures within narrow ranges.

When cellular respiration involves the use of oxygen, the term **aerobic respiration** or **aerobic cellular respiration** is used and the overall equation is:

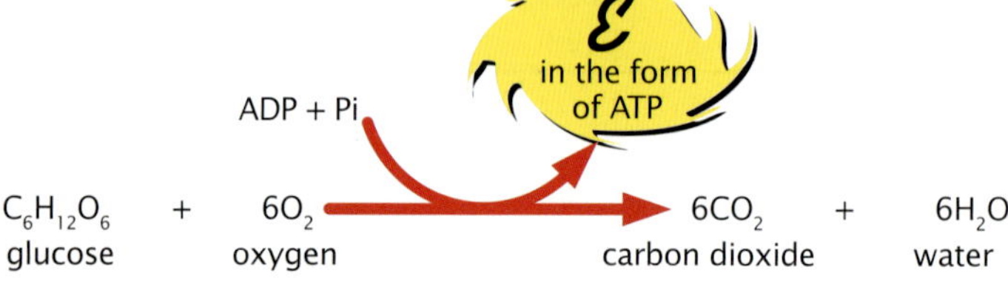

ODD FACT

The end products of anaerobic respiration in yeast are the alcohol ethanol and carbon dioxide.

Aerobic respiration of glucose generally yields 36 molecules of ATP per molecule of glucose.

In some tissues, respiration occurs without the involvement of oxygen and is referred to as **anaerobic respiration**. Anaerobic respiration is far less efficient than aerobic respiration in converting the energy in glucose into energy in ATP yielding just 2 ATP molecules per molecule of glucose. The end products of anaerobic respiration in human muscle are lactic acid and carbon dioxide.

CELLULAR RESPIRATION IN FRUIT AND VEGETABLES

Like all living tissue, fruit and vegetables carry out cellular respiration and produce some heat energy. The higher the rate of respiration, the more heat energy is produced and the greater the chance that deterioration of the plant material will occur. The rate of respiration varies in different produce (see table 3.8).

Respiration rates depend on the storage temperature and the stage of ripeness of the plant material. For example, broccoli at 20°C respires nearly 15 times faster than broccoli at 0°C. Produce that is highly perishable generally has a high respiration rate and a high rate of heat evolution. Unless this heat is removed, the temperature of the produce increases resulting in an even higher rate of respiration, greater release of heat energy and more rapid deterioration.

Several techniques are used to slow respiration rate and prevent deterioration:

- cooling of fruit and vegetables after harvesting and transport to markets in refrigerated trucks
- keeping the produce in a confined space in an atmosphere with reduced oxygen and increased carbon dioxide. The reduced oxygen results in a slower rate of aerobic respiration and lower heat production. However, the level of gases must be carefully controlled because too little oxygen can cause anaerobic respiration or fermentation which can also ruin produce.
- coating the surface of produce, such as apples and bananas, with wax. This coating lengthens shelf life by slowing the intake of oxygen and the escape of carbon dioxide.

Table 3.8 Examples of approximate respiration rates (watts/tonne)

Fruit or vegetable	*Respiration rate*
storage at 0°C	
broccoli	212
lettuce	72
celery	61
peaches	23
cabbage	15
plums	6
storage at 2°C	
asparagus	155
storage at 14°C	
bananas (ripening)	111
bananas (green)	40

Figure 3.26 The living cells in fruit and vegetables respire.

Levels of biological organisation

In chapter 2 (pages 42–46), we introduced you to the various levels of organisation of cells. Figure 3.27 shows how the molecules and compounds we have discussed in this chapter fit in the organisation of cells and of the living world. The progression of complexity from subatomic to atomic extends into molecules and compounds, then to cells that become functional entities, capable of sustaining life and being the basic building block of all living organisms.

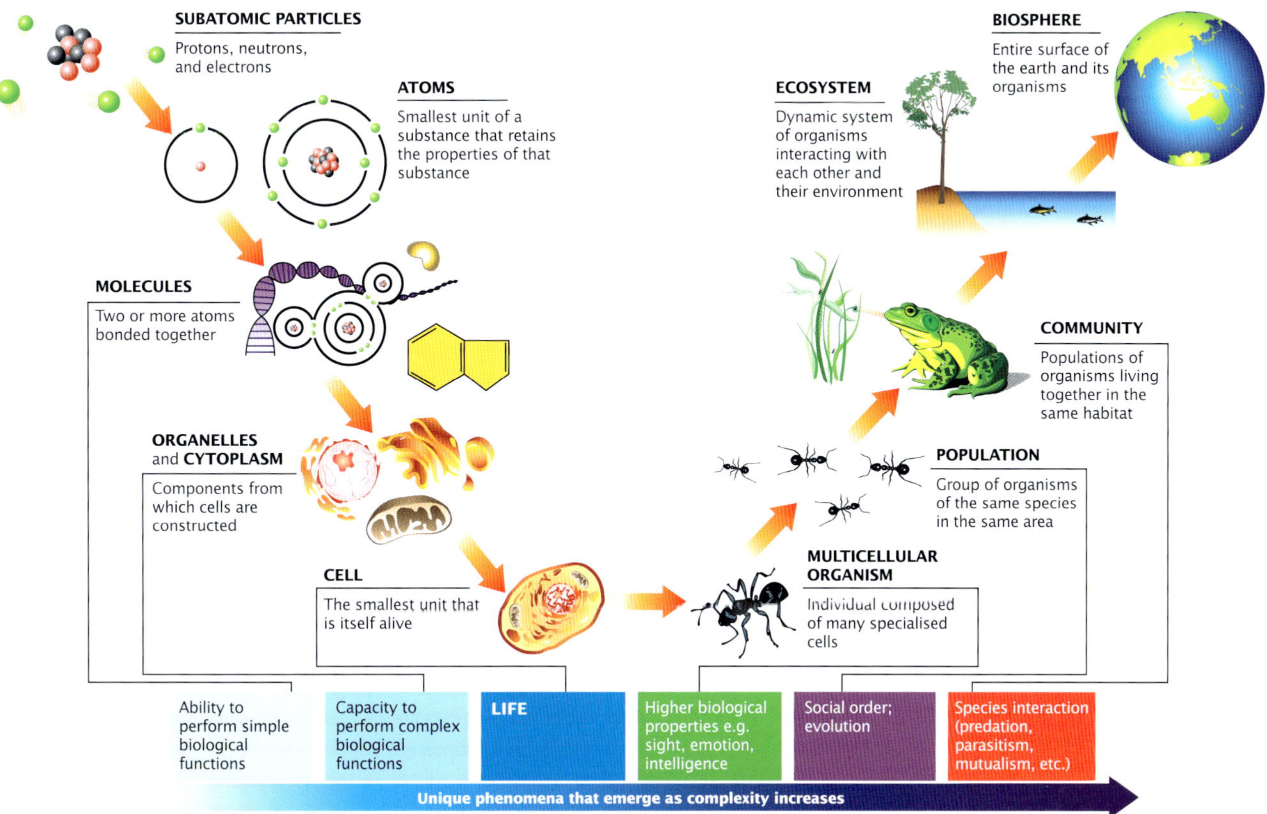

Figure 3.27 Levels of biological organisation. As each level increases, structural complexity increases and unique phenomena may emerge.

KEY IDEAS

- Photosynthesis is the process of converting light energy to chemical energy stored in sugars.
- Organisms that can make organic molecules by photosynthesis are called autotrophs and include all plants, algae and some bacteria.
- Adenosine triphosphate (ATP) is formed when energy is released during cellular respiration of glucose, a carbohydrate.

QUICK-CHECK

11 What is the difference between an autotroph and a heterotroph?
12 Where in a cell is light energy transformed to chemical energy?
13 Name the inputs and outputs in photosynthesis.
14 Where in a cell does aerobic respiration occur?
15 Name the inputs and outputs of (a) aerobic respiration; (b) anaerobic respiration.

BIOCHALLENGE

1

Many compounds contain the element carbon, symbol C. For example, the formula for the mineral calcite is $CaCO_3$. Explain whether you would classify calcite as an organic compound.

2

Mammalian tissue cells live in a moist environment of *tissue fluid*. This fluid contains many different materials in solution. Would you expect these materials to be monomers or polymers?

3

A 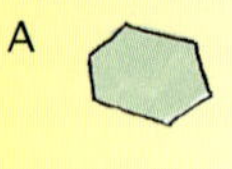

B 

C 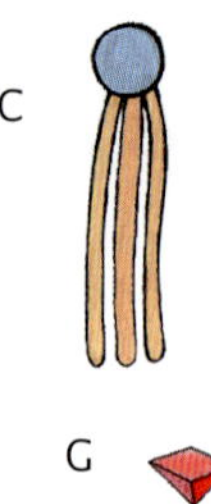

D 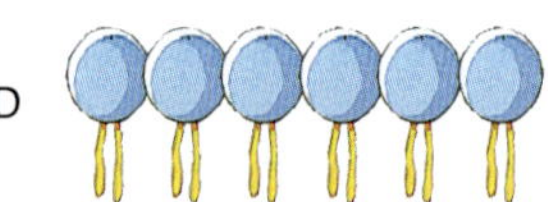

E 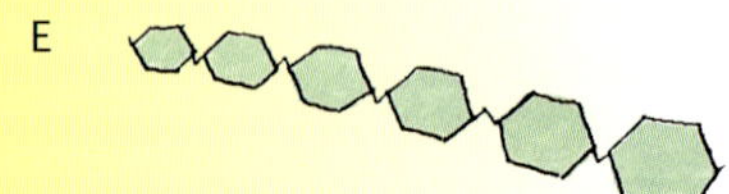

F 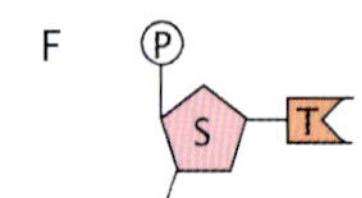

G H 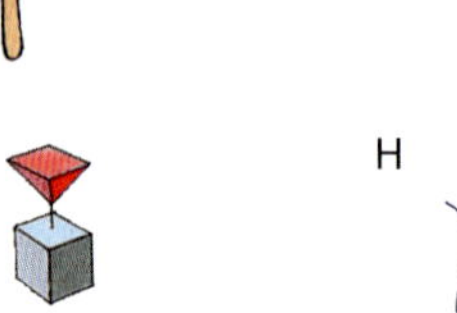

Examine the illustrations above. Which of the structures is:

a an amino acid?
b a lipid polymer?
c a carbohydrate monomer?
d an amino acid polymer?
e a nucleic acid polymer?
f a carbohydrate polymer?

4 The temperature at which an enzyme is most efficient is the temperature at which the rate of a reaction involving the enzyme is at its highest point. It is reasonable to assume that this temperature is close to the body temperature of an organism.

Scientists investigated the rates of reaction of three enzymes over a range of temperatures. The enzymes were taken from the following warm-body organisms:

- Arctic gull, body temperature 34°C
- human, body temperature 37°C
- western pewee bird, body temperature 44.8°C.

The results for two of the enzymes investigated are given in the following graph.

a From which organism is enzyme D likely to have come?
b If the results from the third enzyme were plotted, where would you expect them to be in relation to the two results already on the graph?

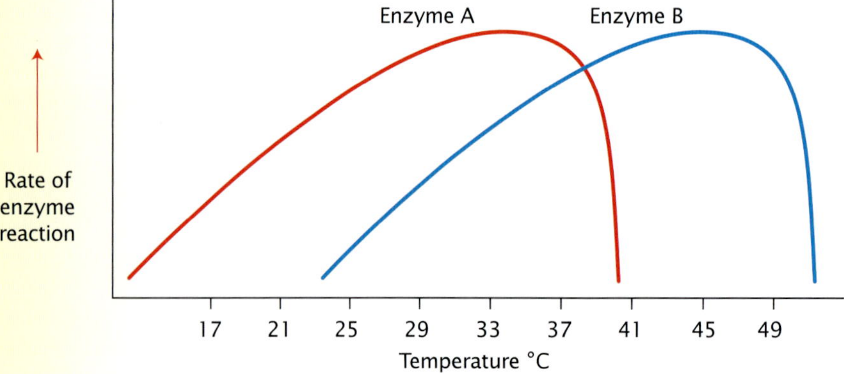

CHAPTER REVIEW

Key words

CROSSWORD

accessory pigments
active site
adenine
aerobic cellular respiration
aerobic respiration
amino acids
anabolism
anaerobic respiration
autotrophic
bonds
carbohydrates
catabolism
cellular respiration
cellulose
chemical energy
chitin
chlorophylls
chloroplasts
coenzyme
cohesive
complex carbohydrates
covalent bond
cytosine
deoxyribonucleic acid (DNA)
enzymes
fatty acids
glucose
glycerol
glycogen
grana
guanine
heterotrophic
hydrogen bonds
hydrophilic
hydrophobic
light energy
lipids
metabolism
minerals
monomers
monosaccharide
non-polar
nucleic acids
organic compounds
phospholipids
photosynthesis
polar
polymers
polysaccharides
proteins
ribonucleic acid (RNA)
starch
stroma
substrate
thymine
triglycerides
vitamins

Questions

1 *Making connections between concepts* ▸ Prepare a concept map, using key words and phrases from the list above. You may use any additional concepts that you wish.

2 *Understanding scientific terminology* ▸ Many vitamins are called coenzymes. What is the function of such vitamins?

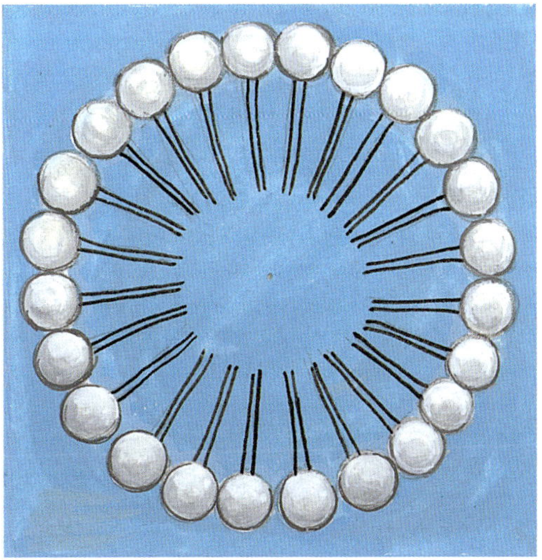

Figure 3.28

3 *Applying understanding to new contexts* ▸ When phospholipids are added to water, they aggregate (come together) as shown in figure 3.28. Explain why the molecules come together in this way.

4 *Interpreting and communicating ideas* ▸ Discuss the validity of each of the following statements.

a The source of oxygen used in aerobic respiration is the same for plants and animals.

b All nucleic acids are identical.

c All enzymes act in the same way because they are all made of proteins.

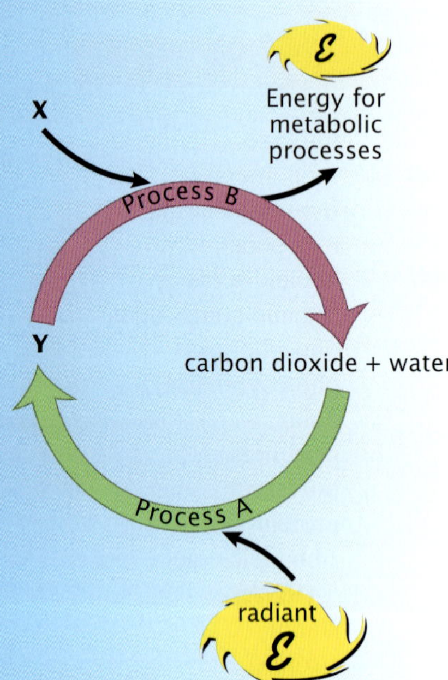

Figure 3.29

5 ***Making connections between concepts*** The diagram in figure 3.29 is sometimes called the 'biology energy wheel'. Process A is photosynthesis.

a Name process B.

b Name the input element X.

c Name the output product at Y.

d In what form must energy be before it can be used for metabolic processes in a cell?

6 ***Analysing information and making connections*** All living cells in humans are close to a blood supply that delivers oxygen that is then available for aerobic cellular respiration. Skeletal muscle is also capable of carrying out anaerobic respiration and does so at certain times.

a What is the advantage of skeletal muscle being able to carry out anaerobic respiration as well as aerobic respiration?

b Describe the conditions that individuals might be under when their skeletal muscles engage in anaerobic respiration.

7 ***Application of concepts*** Fruit and vegetables are sometimes transported over very long distances between grower and consumer. For many products, refrigerated transport is essential. Why is such transport needed to ensure that the produce is in good order on its arrival at markets?

8 ***Analysing and evaluating information*** Suggest explanations for the following observations.

a It is cheaper to keep a large pot plant alive than a small dog.

b A greengrocer found that his refrigerator costs were greater when he stored broccoli in his cold room than when he stored the same amount of cabbages.

9 ***Analysing and evaluating information*** Discuss the validity of each of the following statements.

a A tissue contains groups of cells where each group has quite a different function.

b Delivery mechanisms are important if a group of small cells is to operate more effectively than one large cell.

c The surface-area-to-volume ratio of a cell influences the rate at which substances can enter or exit the cell.

10 ***Using the web*** Go to www.jaconline.com.au/natureofbiology/natbiol1-3e and click on the 'Photosynthesis' weblink for this chapter. Select 'Center for Study of Early Events in Photosynthesis' from the given list. Read 'The Power of Green' and examine the drawing on the page.

a The drawing on the page implies some similarity between photosynthesis in a plant and the water wheel of a mill. Explain what similarity you think the artist was indicating in the drawing.

b Select 'Educational Resources' from the index bar. Then select 'What is photosynthesis?' Select 'Introduction to Photosynthesis and its Applications' and read the section headed 'Basics'.

i Why do we see most leaves as the colour green?

ii Suggest why wavelengths of light below 330 nm are likely to damage cells.

4 Cell replication

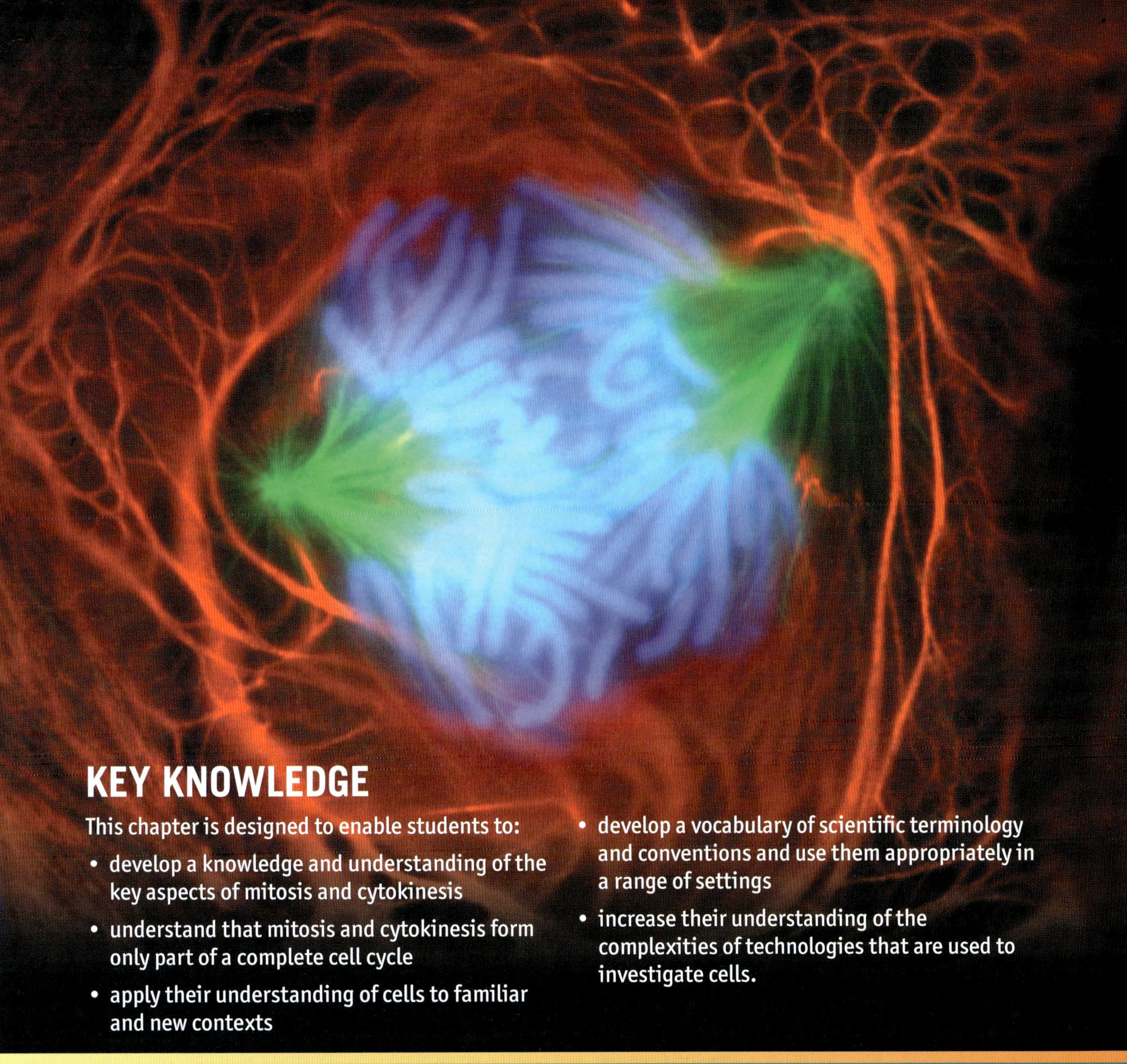

KEY KNOWLEDGE

This chapter is designed to enable students to:

- develop a knowledge and understanding of the key aspects of mitosis and cytokinesis
- understand that mitosis and cytokinesis form only part of a complete cell cycle
- apply their understanding of cells to familiar and new contexts
- develop a vocabulary of scientific terminology and conventions and use them appropriately in a range of settings
- increase their understanding of the complexities of technologies that are used to investigate cells.

Figure 4.1 Genetic material, carried in chromosomes, must be duplicated and separated when cells reproduce. The process by which cells carry out this duplication and separation is called mitosis. In this image of a cell from a newt (*Notophthalmus*, $2n = 22$), the chromosomes, stained blue, have already replicated and are attached to protein microtubules of a spindle, stained yellow-green, all surrounded by a keratin cage, stained red. Chromosomes are drawn to each of the poles of the spindle as the microtubules contract. After mitosis, the cytoplasm is separated by cytokinesis and two identical daughter cells are formed.

In this chapter, we investigate the various stages of the cell cycle, the characteristics of each stage and the biological significance of the process.

'Spray-on skin'

Figure 4.2 Dr Fiona Wood was awarded Australian of the Year for 2005 for her work on developing an improved method of skin-cell regeneration leading to improved and more rapid treatment for people with skin burns.

When Dr Fiona Wood (figure 4.2) of the Royal Perth Hospital was made Australian of the Year for 2005, it was in recognition of her work related to the treatment of severely burnt people. For about ten years prior to March 2003, Dr Wood had been developing improved methods for growing replacement skin. When 28 Australians were badly wounded and burnt in an explosion in Bali, Indonesia, it was decided that they should be returned to Australia as soon as possible for treatment. They were sent to Dr Wood and Australians followed their progress through the daily press. 'Spray-on skin', known commercially as CellSpray, and Dr Wood became famous.

Normal intact skin (figure 4.3) provides a covering for the body. The outermost part of the outer layer, the epidermis, comprises dead cells. Beneath this dead outer layer is a layer of living epidermal cells that can regenerate and repair when damage occurs.

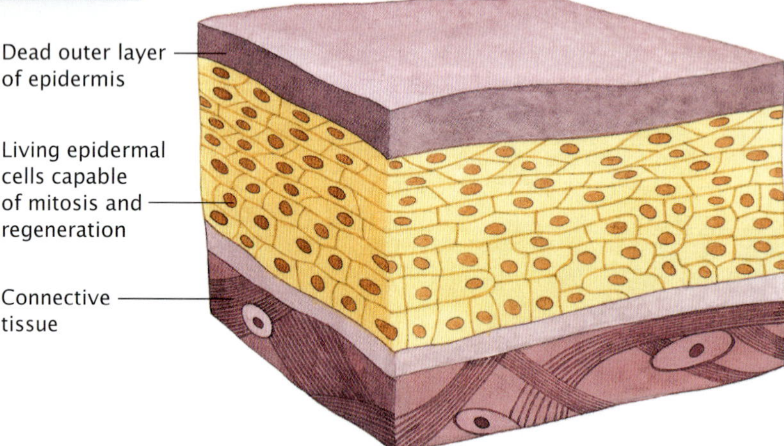

Figure 4.3 Section showing the outermost layer of skin. Note the different parts of the epidermis.

Treatment of an area where skin cells have been severely damaged through burning or some other trauma involves trying to get new skin to grow over the damaged area. The first step is to remove epidermal cells from an uninjured part of the skin of the patient. In older and more traditional methods for replacing burnt skin, these collected cells were grown in plastic dishes until they formed sheets of cells that could then be transplanted over the burnt area. There can be problems with this technique.

One problem was that it took considerable time — up to 21 days — to grow the sheets of cells that were sufficiently large to cover extensively burnt areas. Also, the sheets began to act like skin and the surface cells formed keratin and died so that they were less active growers when the transplant was carried out. Scarring tended to be more severe the longer the patient waited to be treated and the longer wait also increased the chance of infection and other complications with the wounds.

Dr Wood's research has concentrated on finding a way of shortening the time between the burn and the application of replacement skin and, out of that research, CellSpray has been developed.

Uninjured skin cells from the patient are the starting point. These skin cells are incubated with special nutrients that stimulate the cell cycle and are grown in this way for about five days. A suspension containing these actively replicating cells is then sprayed over the burnt areas. The cells continue to replicate and migrate

It has been estimated that each person replaces, on average, about 18 kg of skin cells during a lifetime. Dandruff, skin cells from our scalps, represents just a fraction of the skin cells we must replace.

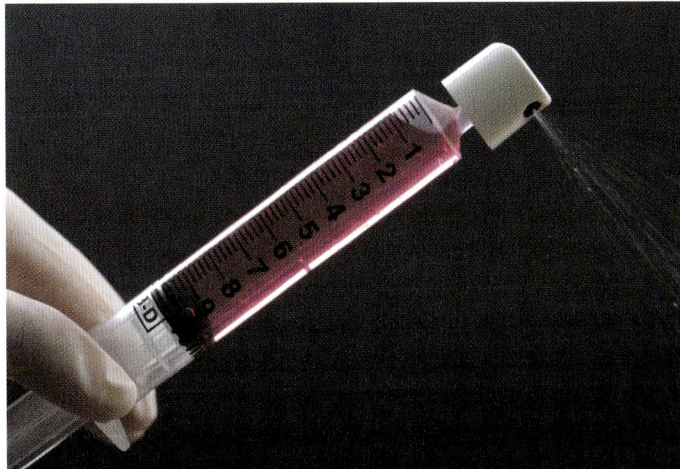

Figure 4.4 'Spray-on skin' or CellSpray, developed by Dr Fiona Wood. Skin cells taken from a patient are cultured and allowed to replicate. A suspension of these cells is sprayed onto burnt areas where they continue to grow and form a new skin.

so that they spread and grow over the damaged area. Spraying also means that larger areas can be treated at any one time. Some scarring may occur but it appears to be less than that occurring with traditional methods.

The science involved in growing new skin cells is possible because living skin cells are able to regenerate. We continually shed our old skin cells and so we continually need to replace them. Skin cells are continually being replaced by the cell cycle, a process that results in the production of two new cells, each identical to the parent cell that gave rise to them. **Mitosis** is an important part of that cycle and involves the replication of the genetic material in the cell. The cytoplasm of a cell is shared between the two new cells at **cytokinesis**.

In this chapter, we consider in some depth the importance of mitosis and cytokinesis. We also explore where these processes occur in a range of animals and plants.

Nuclear division leads to reproduction of cells

New cells are constantly being produced in multicellular organisms. We have already mentioned cell reproduction playing a role in the **regeneration** of skin cells. In mammals, red blood cells, skin cells and gut cells are constantly being produced to replace cells that have died. Replacement cells are produced only by reproduction of existing cells.

As we have seen in chapter 2, the cells of eukaryotes typically have a nucleus, which contains the genetic material **deoxyribonucleic acid (DNA)**. DNA is found in thread-like structures called **chromosomes** and influences the characteristics and controls all the functions that go on within an individual. As cells reproduce, it is critical that the genetic material is also reproduced so that any new cells produced have the same amount and kind of genetic material as the parent cell. The correct distribution is vital because any error may result in serious defects in a cell and ultimately in an organism.

The process that ensures the same amount and kind of genetic material is transmitted from one generation to the next as cells reproduce is called mitosis.

Mitosis

Mitosis is a process of nuclear division in which the replicated genetic material is separated and two new nuclei are formed (see figure 4.5, page 78). Replication of the whole cell is completed only after the cytosol and organelles in the cytosol separate around the two new nuclei that are formed during mitosis. The separation of cytosol and the organelles it contains is called cytokinesis (see figure 4.5 and pages 79–80).

Before mitosis begins, chromosomes are too slender to be visible in a cell. As replication of the genetic material begins, the chromosomes become shorter and thicker and are more easily seen (see figure 4.5). From that point, their behaviour can be studied using a light microscope.

Generally each chromosome is single-stranded and consists of one molecule of DNA. However, at certain times during the reproduction of a cell, a chromosome is double-stranded and consists of two molecules of DNA.

STARTING POINT: One cell containing four single-stranded chromosomes

i. Nucleus well defined at late interphase. Animal cells have a pair of centrioles in an aster of microtubules close to nuclear envelope. Chromosomes not visible but their DNA has already duplicated.

Interphase

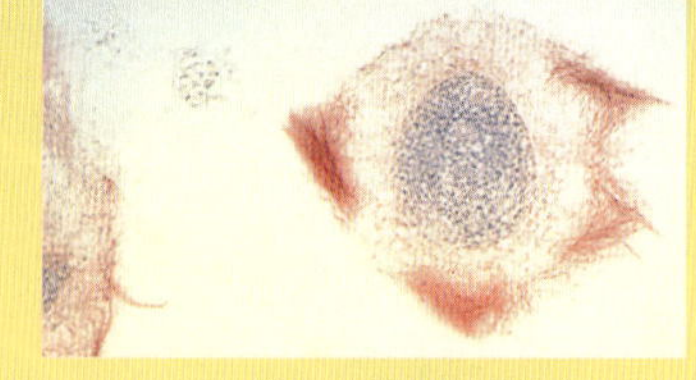

Interphase

MITOSIS

ii. Chromosomes become visible early in mitosis. At first they appear thin and long but gradually become thicker and shorter. Later, the chromosomes can be seen to be double-stranded, held together at the centromere. The replicated centrioles move apart; microtubules of the mitotic spindle continue to extend from the centrioles.

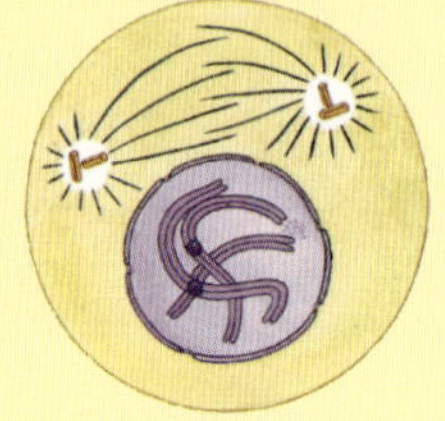

Prophase

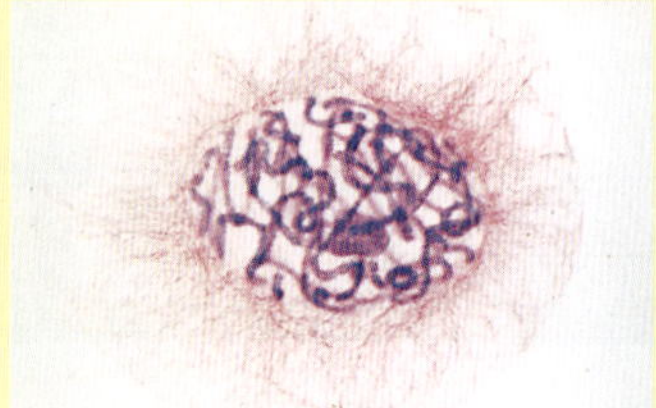

Prophase

iii. Mitotic spindle fully formed between the pairs of centrioles at the two poles of the spindle. The double-stranded chromosomes (each strand is called a chromatid) line up around the equator of the cell. From the side, they form a line across the middle of the cell. How would they appear if viewed from above?

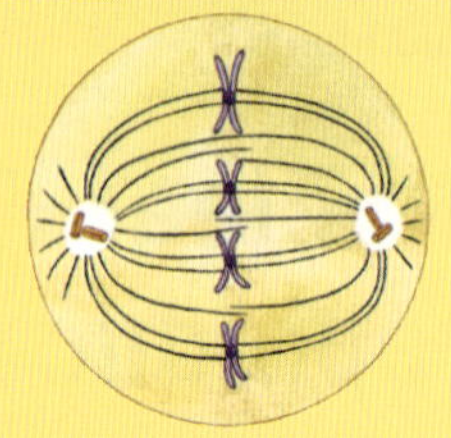

Metaphase

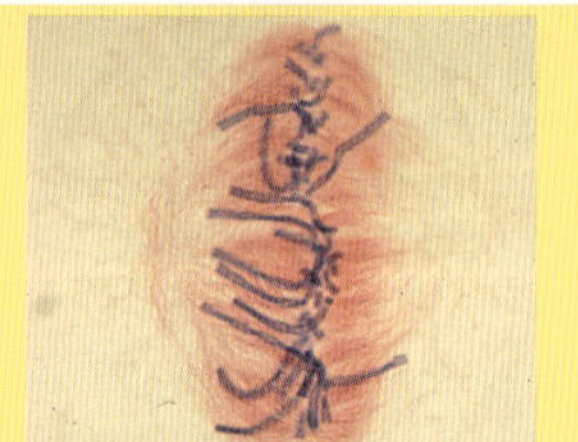

Metaphase

iv. Each centromere divides, so that the single-stranded copies of each chromosome move to opposite ends of the cell as the tubules shorten. This migration is orderly and results in one copy of each chromosome moving toward each end of the spindle.

Anaphase

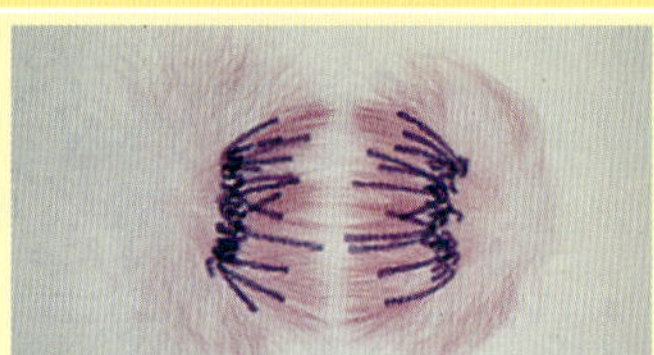

Anaphase

v. The chromosomes become thinner and less obvious. A new nuclear membrane begins to form around each group of chromosomes. This completes the process of mitosis.

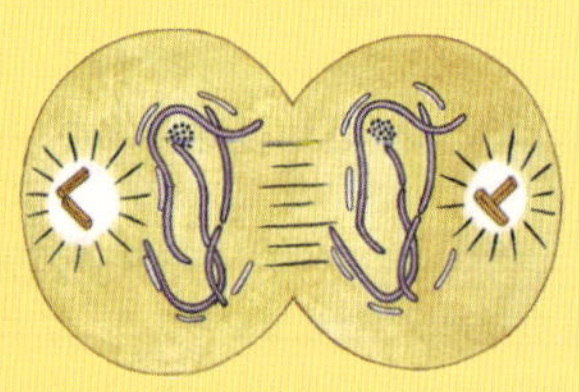

Telophase

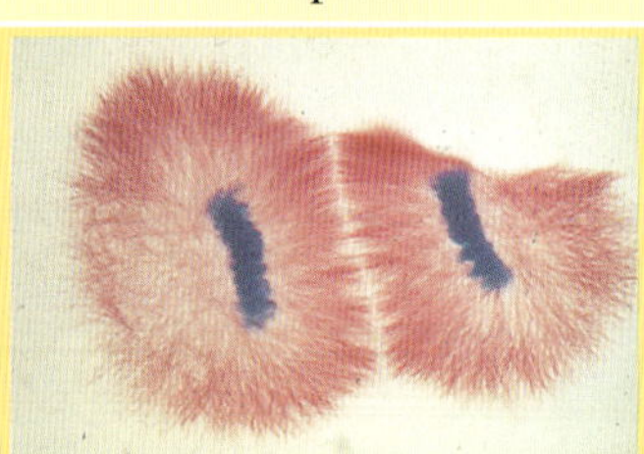

Telophase

vi. Division of the cytoplasm by a process called cytokinesis is completed, new membranes form enclosing each of the two new cells (and cell walls in the case of plants) which become interphase cells.

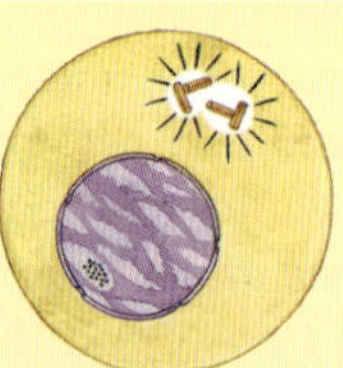

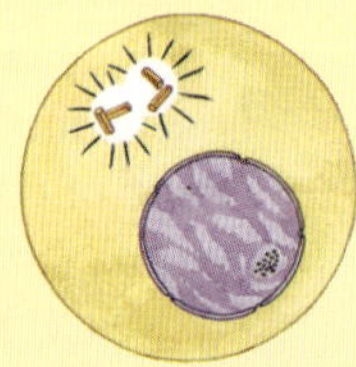

Interphase

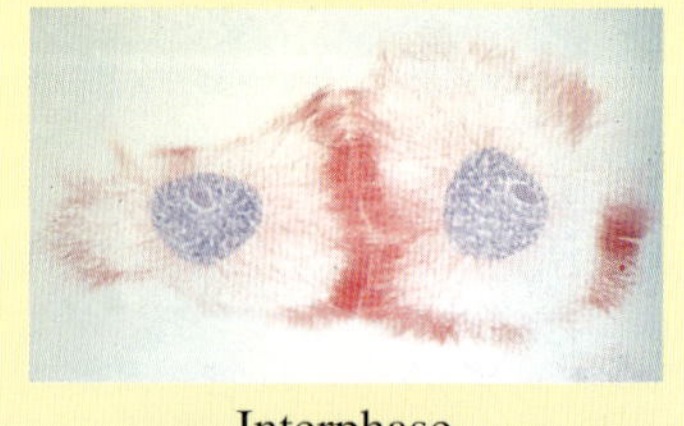

Interphase

END POINT: Two cells each containing four single-stranded chromosomes

Figure 4.5 Summary of mitosis and cytokinesis. The drawings (middle column) show a stylised version in an animal cell containing four chromosomes. The light micrographs (third column) show mitosis in the endosperm of the seed of an African blood lily, *Scadoxus katherinae Bak* (18 chromosomes in each cell). Chromosomes are stained purple and microtubules are stained pink. Note the changes in chromosomes and the formation and distribution of microtubules and fibres as the cell moves through the cell cycle. Two daughter cells form from each cell by the completion of the cell cycle.

Individual chromosomes first become visible as double, thread-like structures held together in a constricted region. Each of these threads is called a **chromatid** and the position where they are held together is called a **centromere**. The fact that the chromosomes are double-stranded and therefore contain two molecules of DNA indicates that the genetic material in the parent cell has already been replicated (see figure 4.10).

The chromosomes continue to shorten and thicken and the nuclear membrane disintegrates. At the same time, the very fine protein fibres or microtubules in the cytosol move towards the nucleus. The function of the fibres is to guide the movement of the chromosomes in the cell. The fibres become arranged in the cell rather like the lines of longitude on a globe to form a structure called a **spindle**. The chromosomes become attached by their centromeres around the 'equator' of the spindle.

Two things then happen. The centromeres split so that there are pairs of chromosomes, and the spindle fibres contract. The contraction of the spindle fibres is responsible for the movement of the chromosomes towards the poles of the spindle. The movement of the new chromosomes is very ordered. One of the new chromosomes from each pair moves to one end of the spindle; its identical pair moves towards the opposite pole. The end result is a set of chromosomes at each end of the spindle. Because the new chromosomes behave in an orderly way, the set of chromosomes at one end of the spindle is identical with the set of chromosomes at the other end of the spindle.

ODD FACT

If a chromosome fails to attach to spindle fibres, its two chromatids separate to become chromosomes, but move at random in the cell.

The chromosomes at each end of the spindle begin to lengthen and become less visible as distinct structures. At the same time, the protein fibres disperse back into the cytosol and a nuclear membrane develops around each group. Generally, the separation of the genetic material is followed by another significant event, cytokinesis.

Cytokinesis

In January 2005, the journal *Trends in Cell Biology* (figure 4.6) announced a series of special articles on research into cytokinesis under the title 'Cytokinesis: the great divide'. In the first of these articles, Professor Jeremy Hyams of Massey University wrote:

> Cytokinesis brings the curtain down on the cell cycle; it is the final dramatic act in which one cell becomes two.

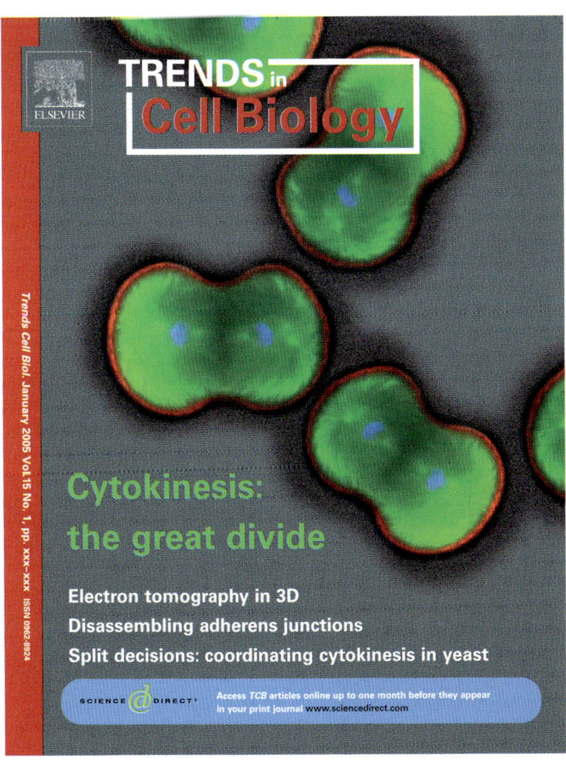

Figure 4.6 The front cover of the journal in which research into cytokinesis is discussed

As the two new nuclei form at the end of mitosis, the cytosol and organelles, such as mitochondria and chloroplasts, surround each nucleus and cytokinesis occurs. Minor differences occur during cytokinesis in different organisms. Generally in animals, the bridge of cytoplasm between the two new nuclei narrows as the plasma membrane pinches in to separate the nuclei and cytoplasm into two new cells (figure 4.7a, page 80). In plant cells, a cell plate forms between the two groups of chromosomes and develops into a new cell wall for each of the newly produced cells (figure 4.7b).

Mitosis is essentially the same in plant and animal cells. The small differences that do exist are not related to the genetic material, nor do they impact on the biological significance of the process. The biological significance is that, through mitosis, a cell is able to reproduce and give rise to two new cells identical to each other, and identical to the original cell. The two new cells contain exactly the same number of chromosomes as the original cell and the same kind of genetic material as the original cell. The outcome of mitosis is summarised in figure 4.5 (page 78).

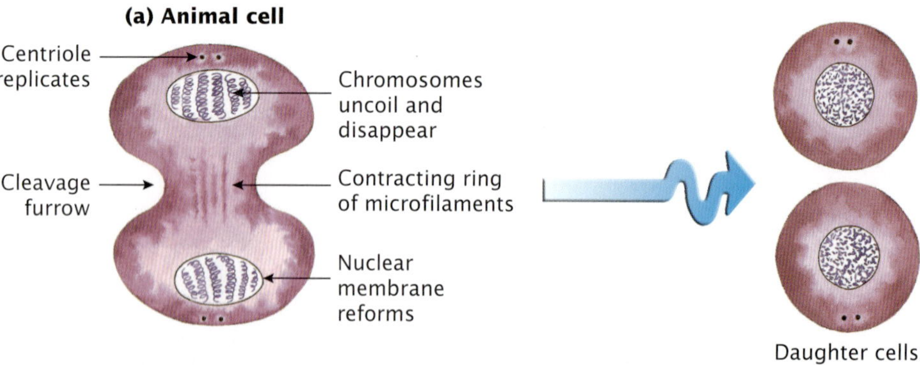

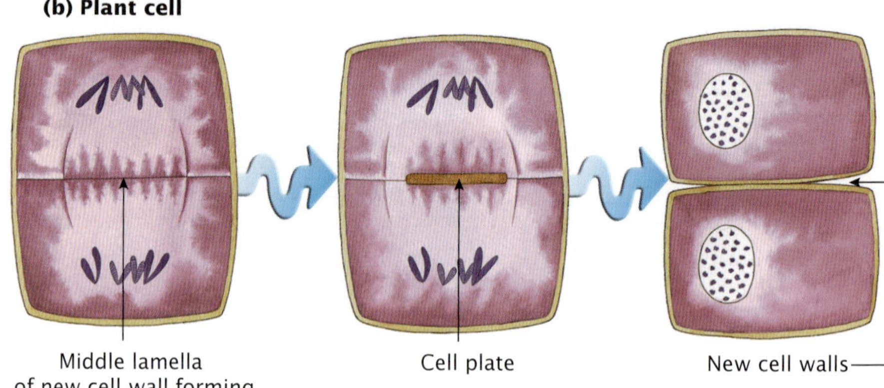

Figure 4.7 Minor differences are visible in plant and animal cells during mitosis and cytokinesis.
(a) An animal cell has a pair of centrioles at each pole of the spindle and a ring of contracting filaments that separates the cytosol and organelles during cytokinesis.
(b) In a newly replicating plant cell, a cell plate forms between the two groups of chromosomes and gives rise to a new cell wall for each new cell.

Organelles such as mitochondria and chloroplasts also replicate

We have seen that mitosis is followed by cytokinesis. This is essential so that the two new nuclei formed can each be combined with cytosol to give two new cells. Obviously the organelles such as mitochondria and chloroplasts within the cytosol must also be replicated during the cell cycle, otherwise cells would contain an ever decreasing number of these structures.

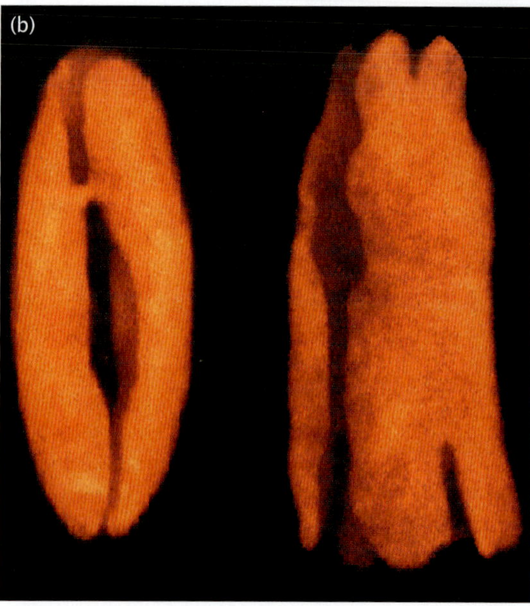

Figure 4.8 **(a)** *Mallomonas splendens*, a golden-brown, single-celled alga **(b)** Chloroplast autofluorescence in two cells of *M. splendens* taken (with a confocal microscope) at the same magnification as (a). On the left, a cell at interphase shows the two lobes of a single chloroplast joined by a narrow connection. On the right, a replicating cell in which the chloroplast is also replicating. Note the connection has broken and the two lobes are each now single chloroplasts that are beginning to constrict. The mitochondria, shown as superimposed red images in (a), would also replicate.

Just as a nucleus contains DNA that must replicate before two new nuclei are formed, so do mitochondria and chloroplasts. These two organelles contain DNA that must replicate before the organelles divide. The alga *Mallomonas splendens* (see figure 4.8a) has a single chloroplast composed of two lobes joined by a narrow connection. As a cell of *M. splendens* replicates, its chloroplast must also replicate. During replication of the chloroplast, the narrow connection breaks and each of the two lobes grows and constricts to give two, two-lobed chloroplasts (see figure 4.8b). Organelles such as chloroplasts and mitochondria can arise only from pre-existing organelles. Cells can arise only from pre-existing cells.

Dr Peter Beech, a cell biologist, carries out research on the replication of cells and their organelles. Figure 4.8, page 80, shows some of his results. Read what he has to say about his work in chapter 2, on pages 38–39.

How long is a cell cycle?

The time taken for a newly formed cell to mature and then give rise to two new cells is called the **cell cycle** (see figure 4.9). The total time taken for one cycle can vary greatly, from as short as 20 minutes to as long as several weeks, but it usually lasts about 10 to 30 hours in plants and 18 to 24 hours in animals.

At what stage of the cell cycle is the genetic material actually replicated? As we have mentioned, it must be before the chromosomes first become visible during mitosis. Figure 4.9 shows the various phases in a complete cell cycle. The phase between successive mitoses is called **interphase** and it is during a restricted period of interphase, termed the S (for synthesis) period, that DNA is replicated in preparation for reproduction of the cell.

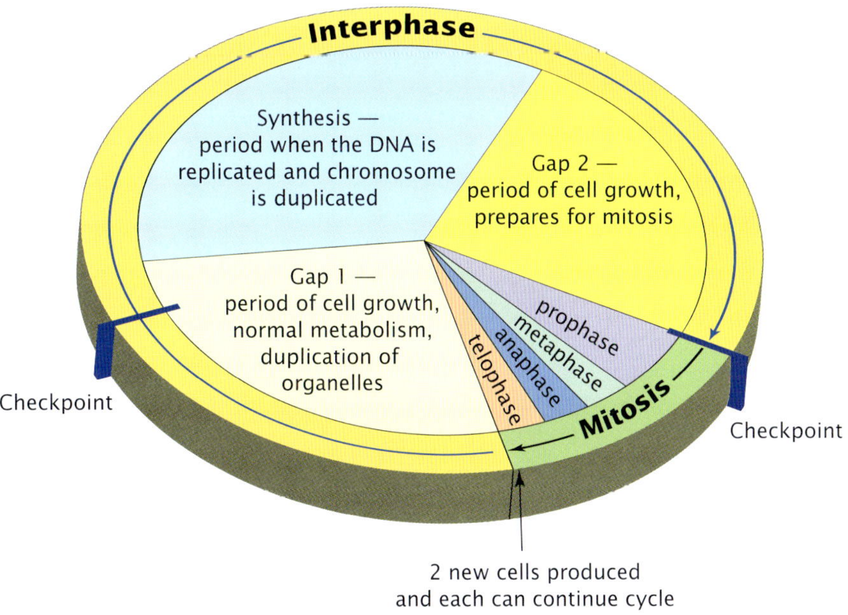

Figure 4.9 The total time taken for one mitotic cell cycle can vary greatly from organism to organism. Note the checkpoints at which there appears to be self-checking to ensure that mistakes have not occurred during the synthesis of DNA or replication of the cell.

The time of replication can be easily identified. Since one of the building blocks found in DNA is thymidine, the time of DNA replication can be identified as corresponding to that time when the cells are actively taking up and incorporating radioactive thymidine into DNA.

Choosing a cell at the right stage of the cycle may impact on the success or otherwise of cloning experiments where a nucleus from one cell is inserted into another cell for development.

The S period is flanked by G (or gap) phases during which cell growth takes place. The G phases also seem to be times at which the cell checks its DNA for mistakes (as shown by 'checkpoints' in figure 4.9). Gap 1 seems to include an examination for mistakes in DNA that may have arisen during the replication of the cell. In Gap 2, cells check for mistakes that may have occurred during the synthesis of DNA in the S phase. If a cell does not receive a go-ahead signal at a checkpoint, it exits the cycle.

ODD FACT

Spindle microtubules attach to a centromere at two specialised regions called kinetochores.

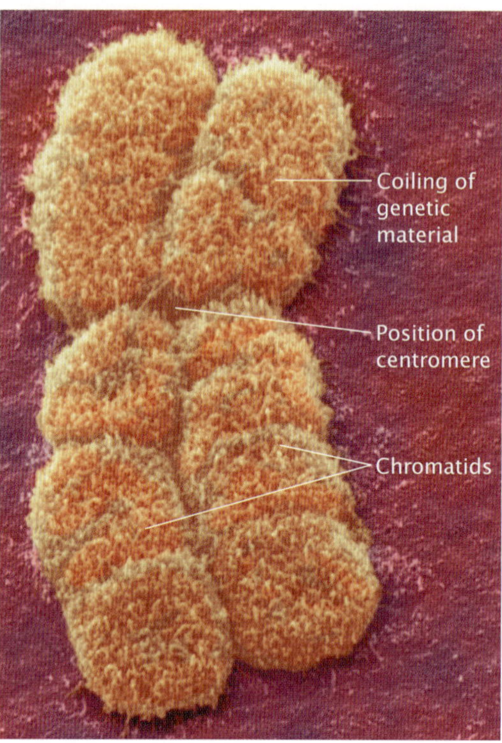

Figure 4.10 Coloured electron photomicrograph (EM) of a human chromosome, showing its two chromatids, during mitosis

During the G1 stage of interphase, each chromosome contains a single molecule of DNA. After replication of DNA in the S phase, the chromosomes duplicate but the DNA remains in its extended state and so chromosomes are not readily seen. After a cell enters mitosis from the S phase, the DNA and proteins in the newly formed chromosomes become coiled and condensed so that the chromosomes become increasingly visible.

Refer to figure 4.10. It is clear that each chromosome has two distinct strands or chromatids that are still connected to each other at the centromere region. This region is indicated by an indent in the chromosome. The centromere is a region of highly condensed DNA and protein.

BIOTECH — ELECTRON TOMOGRAPHY

In the past, electron microscope (EM) images often resulted in only two-dimensional pictures. Using equipment in special ways now enables researchers to obtain three-dimensional pictures of very small, highly specialised areas, such as the centromere and kinetochores.

Lasers for microsurgery are now used to slice chromosomes in living cells. Using a technique called **electron tomography**, three-dimensional images of the small pieces obtained are reconstructed from a large number of photographs taken at different angles by the type of electron microscope shown in figure 4.11. This EM is located at the Resource for the Visualization of Biological Complexity at the Wadsworth Center (Albany, New York) and is used by Professor Conly Reider and his co-researchers (see also figure 4.1, page 75).

Figure 4.11 Modern 400 kV JEOL JEM4000FX analytical, energy-filtered cryo-electron microscope. Three-dimensional pictures are constructed using information obtained from many two-dimensional images taken from different angles with the EM. This technique is called electron tomography.

For convenience, mitosis is divided into the arbitrary stages: **prophase**, **metaphase**, **anaphase**, **telophase** (see table 4.1). The duration of each stage varies from species to species.

Table 4.1 A summary of the stages of mitosis

Stage of mitosis	*Observation*	*Inference*
prophase	Chromosomes are double-stranded.	There are two molecules of DNA in each chromosome.
metaphase	Chromosomes are attached by the centromere around the equator of the spindle.	Centromeres are about to divide.
anaphase	New chromosomes move from the equator to the poles of the spindle.	Centromeres have divided and the spindle fibres are contracting.
telophase	Two groups of identical chromosomes exist.	The two groups are genetically identical.

How many chromosomes?

ODD FACT

In the echidna (*Tachyglossus aculeatus*), there are 63 chromosomes in somatic cells of males and 64 in females. Male echidna have three sex chromosomes, denoted X_1, X_2 and Y; females have an $X_1X_1X_1X_2$ sex chromosome complement.

Each species has a characteristic number of chromosomes in each of its body cells. Human body cells contain 46 chromosomes (see figure 4.12). In most species of mammal, the males and females have the same number of chromosomes. The collection of chromosomes includes a pair of **sex chromosomes**. The sex chromosomes in male mammals comprise one X and one Y chromosome, and the sex chromosomes of females comprise two X chromosomes.

The remaining 44 chromosomes in human body cells are called **autosomes**. These comprise 22 pairs of chromosomes and each pair is identified by a number from one to 22. Each pair of autosomes makes up a **homologous** pair. The pair of X chromosomes in a female are also homologous; that is, they are alike in size and shape and carry genetic material that influences the same characteristics. When the chromosomes of a cell are paired in this way the cell is said to be **diploid**. The same term, diploid, is used for the organism from which the cell is taken.

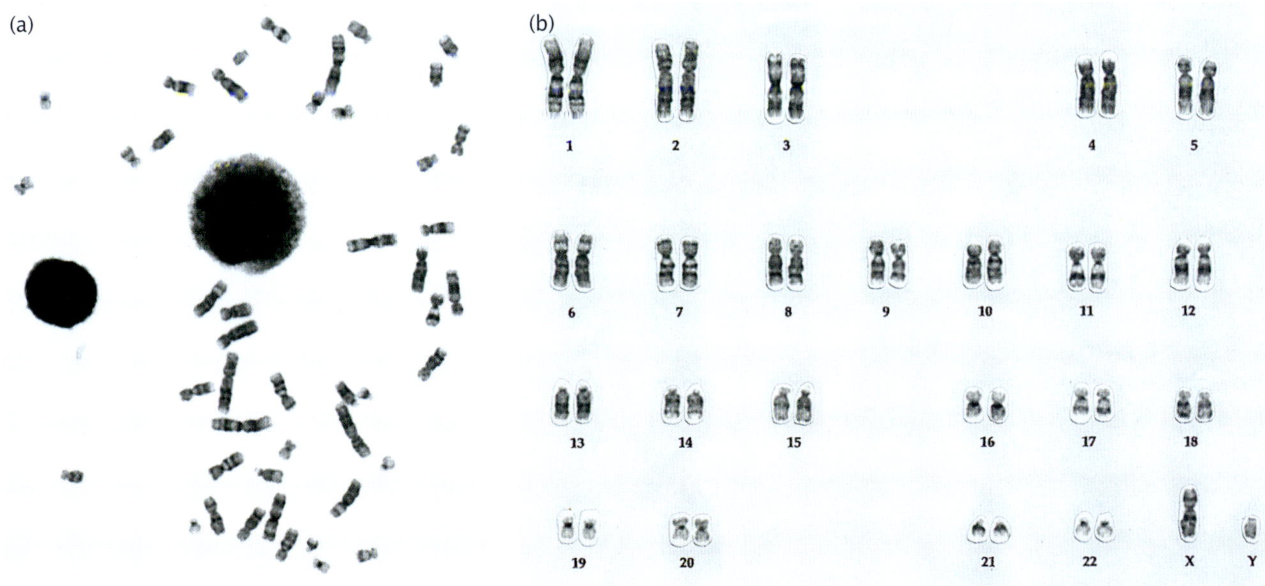

Figure 4.12 **(a)** The chromosomes from a somatic cell of a human male. This is called the 'metaphase spread'. **(b)** Chromosomes from (a) arranged into a karyotype. Note that the 46 chromosomes are arranged in 23 pairs. When chromosomes can be paired this way, the organism is said to be diploid and the number of chromosomes is called the diploid number.

There is no relationship between the size of an organism and the chromosome number in its somatic cells. The largest mammal, the blue whale (*Balaenoptera musculus*), has a chromosome number of 44 (see table 4.2). In contrast, small mammals such as the dog (*Canis familiaris*) and the mouse (*Mus musculus*) have chromosome numbers of 78 and 40 respectively.

Figure 4.13 The size of an animal and its chromosome number are not related.

Table 4.2 Chromosome numbers of somatic cells of some plants and animals

Animals	*Diploid number*	*Plants*	*Diploid number*
black-tailed wallaby (*Wallabia bicolor*)	10 (females); 11 (males)	mulga (*Acacia aneuran*)	26
blue whale (*Balaenoptera musculus*)	44	*Banksia* spp.	28
brush-tailed possum (*Trichosurus vulpecula*)	20	bread wheat (*Triticum aestivum*)	42
common wombat (*Vombatus ursinus*)	14	*Eucalyptus* spp.	22
echidna (*Tachyglossus aculeatus*)	64 (females); 63 (males)	*Grevillea* spp.	20
cat (*Felis catus*)	38	*Hakea* spp.	20
Indian elephant (*Elephas maximus*)	56	*Leptospermum* sp.	22
koala (*Phascolarctos cinereus*)	16	lettuce (*Lactuca sativa*)	18
Pacific dolphin (*Delphinus bairdii*)	44	maize (*Zea mays*)	20
platypus (*Ornithorhynchus anatinus*)	52	pineapple (*Ananas comosus*)	50
red kangaroo (*Macropus rufus*)	20	she-oak (*Casuarina torulosa*)	26
dog (*Canis familiaris*)	78	strawberry (*Fragaria ananassa*)	56

KEY IDEAS

- Cells reproduce during the cell cycle.
- Cells can reproduce only if the genetic material is replicated.
- The duplication of cells involves mitosis and cytokinesis.
- The two newly formed cells each have the same kind and amount of genetic material as the parent cell.
- Each species has a characteristic chromosome number.

QUICK-CHECK

1 What is the genetic material of eukaryotes?
2 What are the phases of the cell cycle and what event/s occur at each phase?
3 At what stages of mitosis are the chromosomes double-stranded?
4 What is the chromosome number of the human species?
5 How many chromosomes are there in one of your bone marrow cells? Each of your skin cells? Each of your white blood cells?

Where does mitosis occur?

We saw at the start of this chapter that skin cells regenerate. In fact, this regeneration is usually the normal process of replacement that occurs throughout our lives and special techniques are used to enhance that replacement in times of accident.

Mitosis occurs in different tissues in different animals and plants.

Mitosis in mammals

Mammalian embryos arise from a single living cell that has been formed when a sperm fertilises an egg. This single diploid cell divides by mitosis, followed by cytokinesis, time and time again, to give a multicellular structure. Eventually these cells begin to undergo specialisation and different tissues form — heart tissue, brain tissue, bone tissue, cartilage, skin and many other different kinds.

Human skin is made of many layers of cells and the outer layers are continually being worn away. Cells in deeper layers under the skin continually divide by mitosis and replace the cells lost from outer layers.

Cells in bone marrow continually divide to provide an ongoing supply of red and white blood cells as older ones wear out and are removed from the bloodstream.

Most highly specialised cells are unable to divide by mitosis and, if they are damaged, a person may be seriously impaired. For example, nerve cells cannot divide so that an accident involving a head wound in which a significant portion of the central nervous system is damaged can lead to paralysis or other permanent damage to some part of the body. One organ that is able to regenerate to some extent is the liver.

Spore formation by fungus

The fungus or mould you see on bread or fruit grows by mitosis. A single cell, a fungal spore, lands on food and grows into a mass of thread-like hyphae. Specialised stalks, each with a spore case at its tip, grow up from the mass of hyphae (see figure 4.14). Mitosis occurs within the spore case and thousands of black spores are formed. On maturing, the spore case splits open and the tiny, light spores are scattered. When conditions are favourable, each spore germinates and grows into a new hyphal mass.

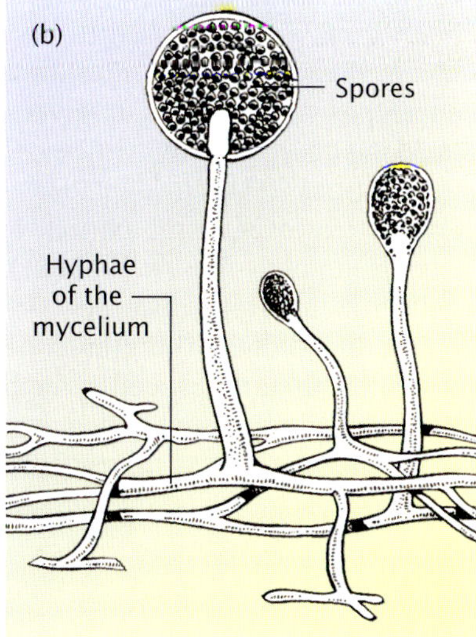

Figure 4.14 The fungus on a rotting tomato **(a)** comprises a mass of white threads or hyphae. Asexual reproduction occurs at the tips of some hyphae and large numbers of black spores are formed **(b)**, each genetically identical with the parent.

New plants from leaves

Some plants, for example, *Bryophyllum* sp., have meristematic-type tissue at notches along the edges of their leaves. This tissue is able to reproduce to give rise to new cells. Rounded structures grow out from the notches (see figure 4.15) and develop into small plants that drop to the soil and take root. What process is responsible for this growth?

Figure 4.15 Asexual development of new plants from the leaf margin of *Bryophyllum* sp. Note the rounded structures and the small plants developing from these.

AFTER THE BUSHFIRE — PRODUCING NEW PLANT CELLS

Bushfires are common in many areas of Australia. Although trees may appear to be burnt to a point that one might think they are dead, a picture such as the one in figure 4.16 (taken just six weeks after the area was devastated by bushfire) clearly shows this is not the case. It is clear from the photograph that the fire has completely destroyed the undergrowth of grasses, shrubs and herbs. Fire-blackened trees with their scorched dead canopy of leaves are in the back, while, in the foreground, the burnt trunks of rough-barked eucalypt trees are visible. One tree is already showing signs of regrowth; it is a thick-barked eucalypt whose thick outer layer of protective bark has insulated the underlying living tissues from the effects of the fire.

The trunk of a eucalypt does not usually show growing shoots. However, if the normal leaf canopy is destroyed, as happened in this fire, buds which are present beneath the bark will grow and reproduce new green leafy shoots, known as **epicormic shoots**. The growth of epicormic shoots involves the production of new cells. The buds below the bark contain tissue called **meristem** which is made of cells that are able to reproduce to give rise to new cells. These new cells are identical with each other and identical to the parent cell.

Figure 4.16 The new shoots from the trunk of a burnt eucalypt tree develop as a result of mitosis in buds present beneath the bark. The buds do not develop unless the canopy is destroyed, as has happened in this case.

New liverworts from cells in a cup

Liverworts, class Hepatica, are small plants that have a flat, fleshy, leaflike structure from which **rhizoids** extend into the soil. The name 'liverwort' is derived from the shape of the organism — rather like that of a liver — and the Anglo-Saxon word for herb — wort. As you might predict from the name, it was once thought that this plant might be useful in the treatment of liver diseases.

In addition to reproducing sexually, liverworts reproduce asexually by means of fragmentation of parts of the plant. Also, liverworts produce gemmae, small multicellular bodies produced in special cuplike structures called gemma cups (see figure 4.17). When rain falls, the gemmae are splashed out of the cup. Gemmae are produced from cells of the parent plant by mitosis. When they grow into new plants they do so by mitosis. The new liverwort plants produced by growth of the gemmae are genetically identical to the parent plant from which they were derived.

Figure 4.17 A new plant develops from each of the small bodies that splash out of the gemma cups on a liverwort plant. The new plants are genetically identical to the parent plant.

Mitosis in insects and other invertebrates

Planaria, phylum Platyhelminthes, are flatworms that live in water. They are one of the few animals that can reproduce asexually by regeneration. The parent breaks into two or more pieces and each piece grows into a new planarian. The new parts are produced by mitosis of cells and each new planarian is an exact copy of the parent.

If a starfish loses some of its 'arms', new ones are regenerated by mitosis (see figure 4.19).

Figure 4.18 If a starfish is cut into two, each half can regenerate into a whole.

Figure 4.19 If a starfish loses some of its 'arms', they regrow. Here you can see six new 'arms' on a damaged starfish.

Like all animals, a developing insect embryo grows by mitosis within its egg. Once hatched, an insect may go through several forms before it reaches adulthood. It may go through a number of **moults** as a caterpillar, during which time the structure discarded during a moult must also be replaced by new cells formed by mitosis.

Some caterpillars pupate, during which stage a firm casing is formed around the body. A **pupa** does not eat and yet the body of the insect goes through a major reorganisation. The cells of the caterpillar break down within the pupal case. This 'soup' provides the raw material for embryonic type cells that have been dormant within the caterpillar but now become active within the pupa. These cells undergo mitosis and give rise to the tissues of the adult insect. Hence the caterpillar develops into an adult fly.

Other insects may not pupate, but grow through a series of moults and then grow wings from small pads of embryonic cells that carry out mitosis followed by specialisation.

Control mechanisms can fail

We have examined how mitosis is essential for the production of new cells. For example, skin and blood cells are continually dying and must be replaced. The death of cells is a natural feature of healthy tissue. This 'programmed' cell death is called **apoptosis** and in healthy tissues is balanced by the production of new cells by mitosis. A breakdown in this balance can occur. If too much apoptosis or too little mitosis occurs, there will be a deficiency of the particular kind of cell and a degenerative disease such as Alzheimer's disease can develop. If there is an excess of cells a tumour develops. If a tumour continues to grow and invades healthy groups of cells it is said to be malignant.

You will learn more about apoptosis in your Biology studies next year.

Breast cancer is the most common cancer in Australian females and accounts for about 26 per cent of all cancers in women. Figure 4.20 (page 89) shows how cancer spreads in breast tissue (a and b) and demonstrates the spread of cancer cells (c). The usual control mechanisms of cells fail to operate in cancer cells. Currently, not a lot is known about the mechanisms of cancer and current research is concentrating on analysing the cellular and molecular changes, as well as changes in the micro-environment of cells, that occur during the growth of cancer cells (refer to the box on Associate Professor Leigh Ackland, chapter 1, page 13). A better understanding of these aspects and the interaction between the various parts within cancer cells increases the chance that improved treatments and cure rates may be found for those with cancer.

KEY IDEAS

- Mitosis occurs in a range of different tissues in different plants and animals.
- Some eukaryotes reproduce asexually from a single cell.
- Uncontrolled cell replication can lead to cancers.

QUICK-CHECK

6 What are two examples of mitosis in plants?
7 What is one example of mitosis in an animal?
8 What is the relationship between apoptosis and cancer?

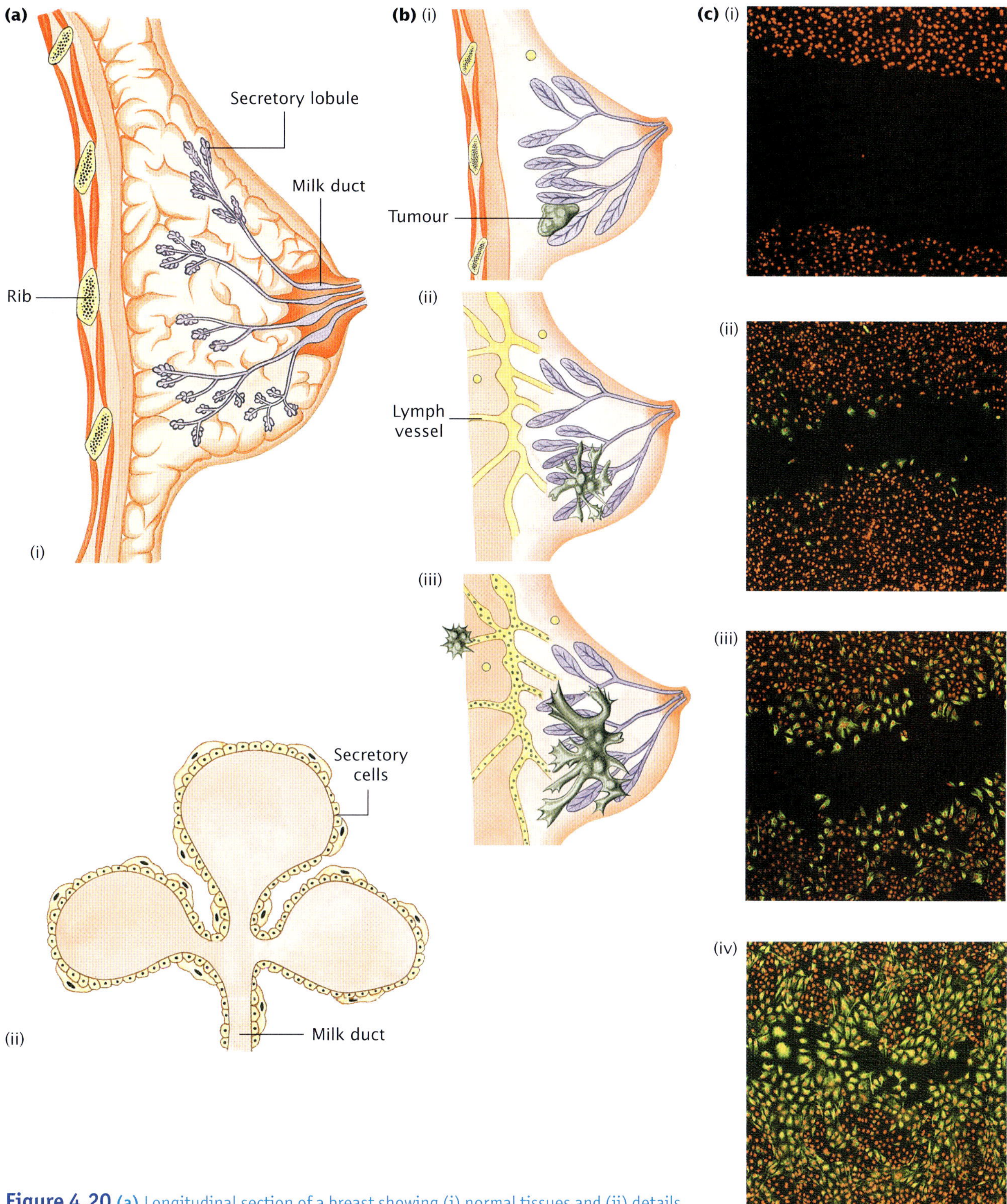

Figure 4.20 **(a)** Longitudinal section of a breast showing (i) normal tissues and (ii) details of a secretory lobule and duct **(b)** Development of a tumour, then cancer from a single cell. As a cancer progresses, epithelial cancer cells leave the primary tumour, invade surrounding tissue and enter the blood and lymph vessels which carry the cancer cells to the other organs. **(c)** Breast cancer cells. When a gap, simulating a duct, is made *in vivo* in a culture of breast cancer cells (i), the cancer cells (stained green) migrate to fill the space (ii) to (iv). This is a model of what happens *in vivo* where cancer cells are motile and produce secondary cancers away from their initial source. Migrating cancer cells derived from breast epithelium express the protein vimentin which stains green. The function of vimentin is unknown but is not expressed by normal epithelial cells except during development.

BIOCHALLENGE

1

A number of cells were monitored as they completed one cell cycle. The average amount of DNA per cell was measured and graphed over the time it took for the completion of one cycle. The graph obtained is shown at right.

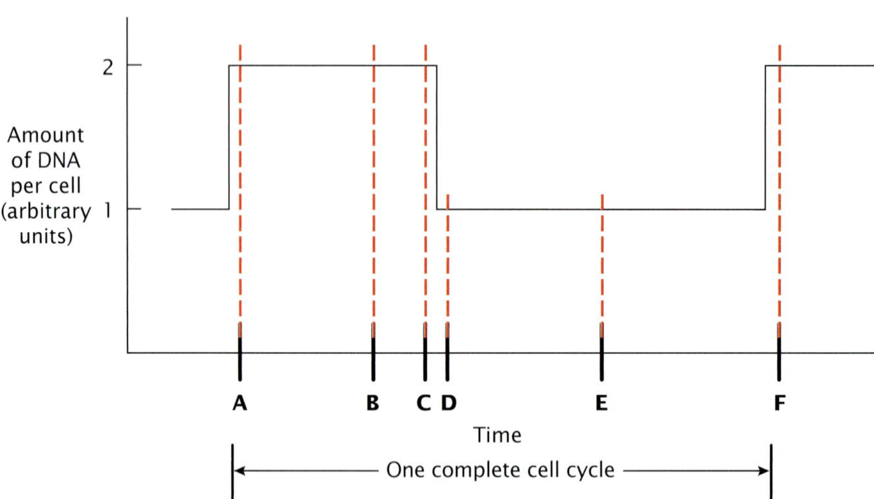

a The letters ABCDEF in the graph represent different times in a cell cycle. What are the stages indicated?

b At the same time, sample cells were examined. The cells examined were as follows:

1

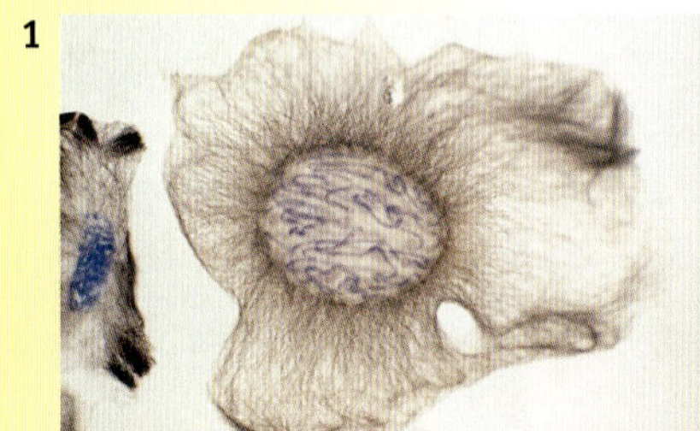

2

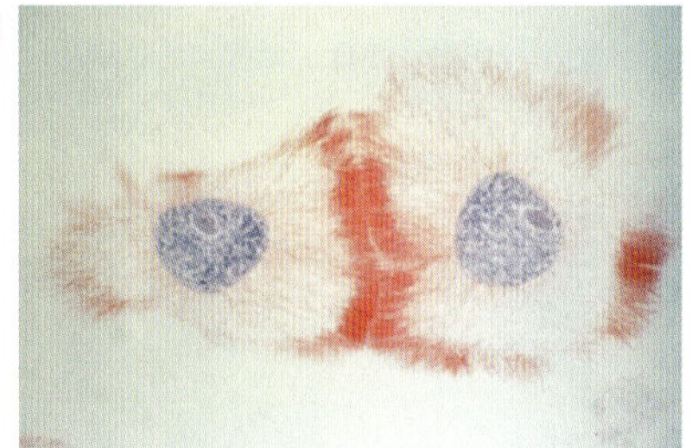

3

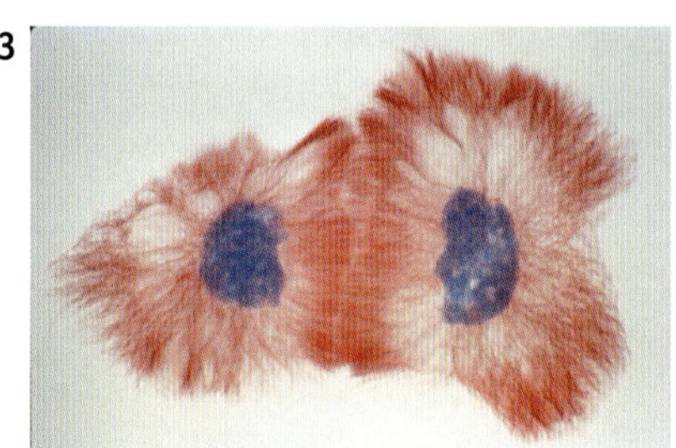

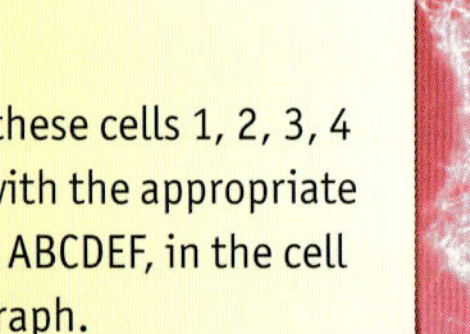

Match these cells 1, 2, 3, 4 and 5 with the appropriate points, ABCDEF, in the cell cycle graph.

4

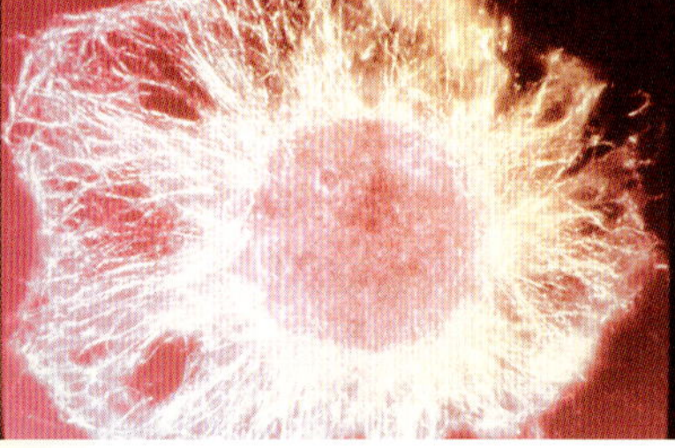

5

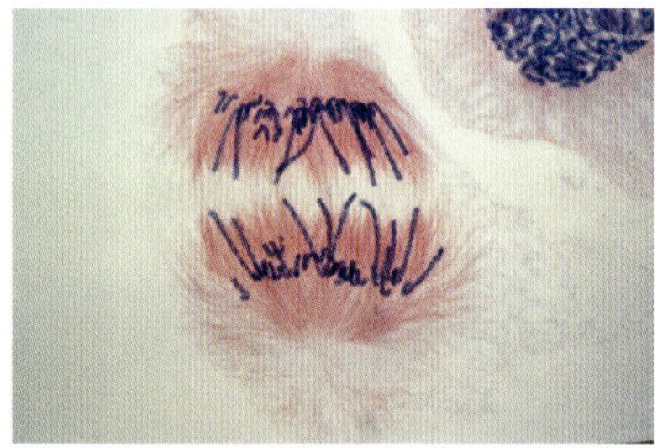

2

Order the following events in animal cell replication.

a Alignment of chromosomes on the spindle equator
b Attachment of microtubules to centromere region
c Breakdown of nuclear envelope
d Condensation of chromosomes
e Decondensation of chromosomes
f Duplication of centromere
g Elongation of the spindle
h Pinching of cell into two
i Re-formation of nuclear envelope
j Separation of centromeres
k Separation of sister chromatids

3

You are examining a cell undergoing mitosis. You are asked whether the cell is from a plant or an animal.

a What three features would you look for in terms of their presence or absence in order to determine the answer to the question you have been asked?
b Explain what you would expect in each case.

CHAPTER REVIEW

Key words

CROSSWORD

anaphase
apoptosis
autosomes
cell cycle
centromere
chromatid
chromosomes
cytokinesis
deoxyribonucleic acid (DNA)
diploid
electron tomography
epicormic shoots
homologous
interphase
meristem
metaphase
mitosis
moults
prophase
pupa
regeneration
rhizoids
sex chromosomes
spindle
telophase

Questions

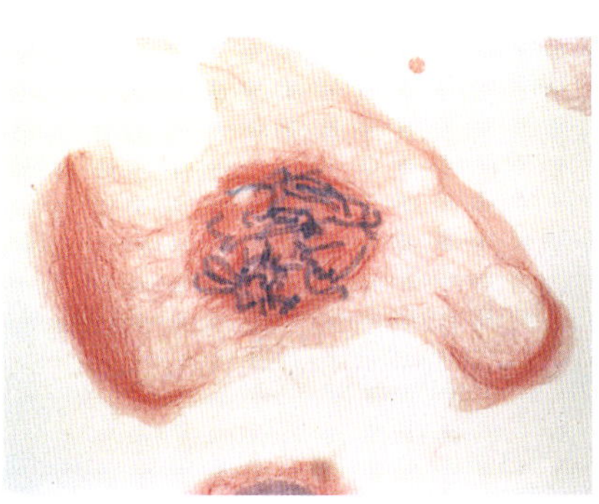

Figure 4.21 Cell of an African blood lily undergoing mitosis

1 *Making connections* ▸ Use as many as possible of the chapter key words to construct a concept map.

2 *Applying understanding* ▸ The image at left (figure 4.21) shows a cell of the African blood lily, *Scadoxus katherinae* Bak ($2n = 18$), undergoing mitosis.

a Describe the appearance of the chromosomes as you would see them under a powerful light microscope.

b What difference, if any, would you see in the chromosomes if you examined them after the spindle had been formed and its contraction commenced?

c How many chromosomes would you expect to see in a leaf cell?

d How many chromosomes would you expect to see in a root cell?

3 *Applying understanding* ▸ A cell containing 24 chromosomes reproduced by mitosis. A genetic accident occurred and one of the resulting cells had only 23 chromosomes.

a How many chromosomes would you expect in the other cell produced? Explain why.

b At what stage of cell reproduction do you think the genetic accident occurred?

4 *Interpreting and applying understanding of a new concept* ▸ Grafting is a technique used with some plants. In grafting, two pieces of living plant tissue are connected in such a way that they will unite and subsequently behave as one plant. For example, the shoot of one kind of plant can be grafted onto the root of another kind of plant (see figure 4.22).

The shoot of a pear tree, *Pyrus communis*, was grafted onto the root of a quince tree, *Cydonia oblonga*, and then allowed to grow. The chromosome number of pear is 68 and the chromosome number of quince is 34.

a After several years' growth, how many chromosomes would you expect in the leaves of the tree?

b How many chromosomes would you expect in cells of a newly grown root? Explain.

c You will note that the chromosome number of pear is twice the chromosome number of quince. Does this mean that a pear cell will contain twice the amount of genetic material as a quince cell? Explain.

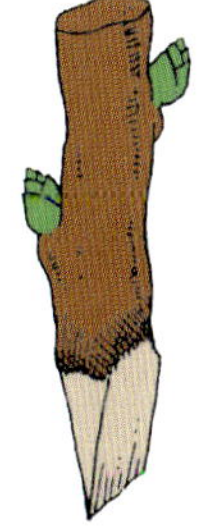

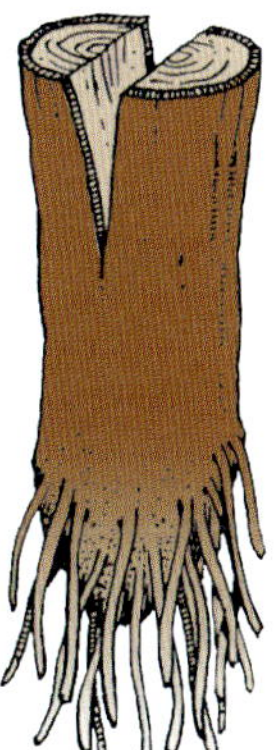

Figure 4.22 A slit is made in the bark of the stock and the bud graft with its own piece of bark is slipped inside. The graft is held in place with tape or twine and the wound covered with grease to exclude fungi and reduce evaporation.

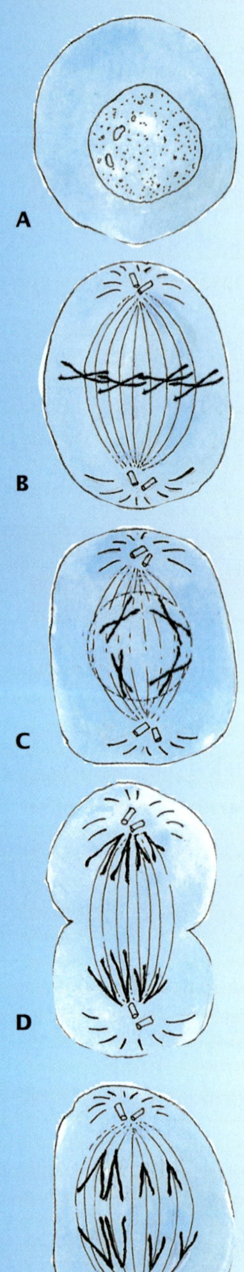

Figure 4.23

5 *Analysing and evaluating information* ▸ Do you agree or disagree with each of the following claims about mitosis?

a The nuclear envelope is visible throughout the process.

b Mitosis would occur in the developing limb of a larval frog.

c Mitosis in plants is significantly different from mitosis in animals.

d Mitosis is accompanied by replication of cell organelles such as mitochondria and ribosomes.

6 *Analysing and interpreting information* ▸ The illustration at left (figure 4.23) shows a series of drawings, all of the same cell at some stage during mitosis.

a Starting with cell A, place the drawings in the sequence that the stages would occur during mitosis.

b Draw what you would expect to see next in the sequence.

7 *Making connections between concepts* ▸ The length of the cell cycle can vary greatly from one kind of cell to another.

a Suggest how this may relate to the length of life of a cell in a particular site in the body of an organism.

b In which part of the human body would you expect to find cells with the shortest life span?

8 *Making connections between concepts* ▸ During mitosis, chromosomes become attached to microtubules that vary in length during the mitotic process. Explain when microtubules would be at their greatest length and when they would be at their shortest.

9 *Applying understanding to new concepts* ▸ Some drugs used in the treatment of some cancers act on microtubules. They act by interfering with the normal contraction and extension capabilities of microtubules.

a Explain the effect you would expect such drugs to have on mitosis and cell replication.

b Why would such drugs be useful in cancer treatment?

10 *Using the web* ▸ Go to www.jaconline.com.au/natureofbiology/natbiol1-3e and access the 'Cells alive' weblink for this chapter.

a Below the heading 'Interactive' on the left-hand side, choose 'Cell Cycle' from the menu. Run the 'Cell Cycle' under animations and ask for checkpoints. This model presents three checkpoints.

i Where in the cycle are each of the three checkpoints and what is being checked at each of the points?

ii Why is 'resting' a misleading word to use with respect to a cell at any phase during the cell cycle?

iii Is there any aspect of the animation that you would suggest changing? Explain your answer.

b Still using the weblink you accessed for part (a), select the option 'Mitosis' from the left side menu under 'Interactive'. Study the animated cycle. What is the diploid number of the cell shown?

c Now select 'Take a quiz' and choose the quiz on 'Cell Biology'. Take the quiz with a partner so that you can discuss your answers to each question. How many questions did you answer correctly? How many times did you take before you gave the correct answer? Visit the site more than once if necessary to check your learning.

UNIT 1

UNITY AND DIVERSITY

AREA OF STUDY 2

Functioning organisms

5 Obtaining energy and nutrients for life

KEY KNOWLEDGE

This chapter is designed to enable students to:

- distinguish between autotrophs and heterotrophs in terms of nutrient requirements for life
- identify the structural features of photosynthetic organisms that facilitate their ability to photosynthesise
- understand the systems for processing nutrients in a range of different heterotrophic organisms.

Figure 5.1 Autotrophs, such as plants and algae, soak up sunlight to gain the energy they need for living and for building complex organic matter from simple organic matter. In contrast, heterotrophs, such as animals, obtain their energy from their food. No food means no energy and no organic matter — and so, no life! This whale shark, a heterotroph, feeds on small marine organisms that are the source of both energy and organic matter for this magnificent animal. In this chapter we consider and contrast the ways in which autotrophic and heterotrophic multicellular organisms obtain their nutrients and energy.

Heterotrophs and autotrophs

The full moons of March and April each year signal the start of the **spawning** season of the corals of Ningaloo Reef in Western Australia. Soon after full moon, the corals release enormous numbers of eggs into the warm waters of the reef so that the surrounding waters are like a nutrient soup. This abundance of food attracts other small marine animals — krill, crabs, small fish — which gather there both to feed on the released eggs and to spawn themselves. Into the seas abutting the Ningaloo Reef also come larger fish that are attracted by the concentrated richness of marine life on which they can feed. The most remarkable of these annual autumn visitors to the reef are whale sharks (*Rhincodon typus*).

ODD FACT

Whale sharks are the largest fish found in the seas and can reach a maximum length of about 18 metres. (Some whales are larger but they are mammals.) Whale sharks are docile and are tolerant of human divers who swim with them on the seaward side of Ningaloo Reef during autumn.

Whale sharks are the largest of the sharks. Unlike smaller shark species that feed on large fish and seals, whale sharks feed on tiny marine animals, such as krill, plankton and small fish.

Whale sharks have two feeding techniques. One is the *passive* technique. In this case, a whale shark simply swims for long periods through waters rich in krill with its wide mouth open, engulfing food along with enormous quantities of water (see figure 5.1). The water and krill reach a filtering system formed by comb-like gill rakers deep in the whale shark's throat (figure 5.2). Here, the tiny animals are trapped and later swallowed while the water exits via the gill slits. The other feeding technique of the whale shark is an *active* one. In this case, the shark pursues **prey**, such as a dense school of tiny fish, and actively sucks them into its mouth.

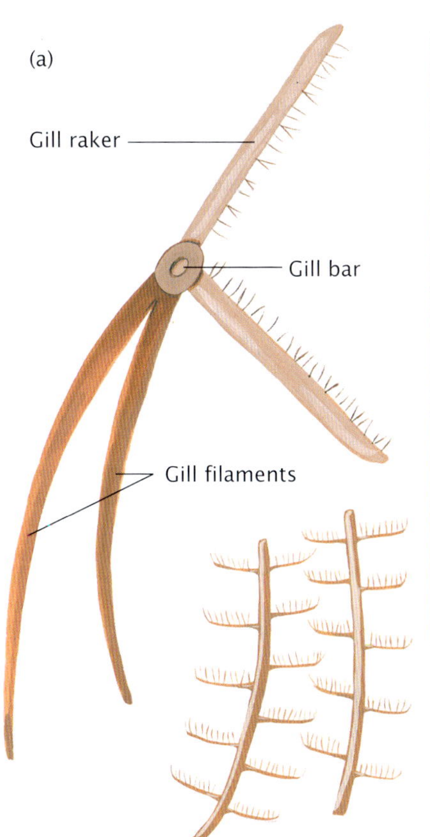

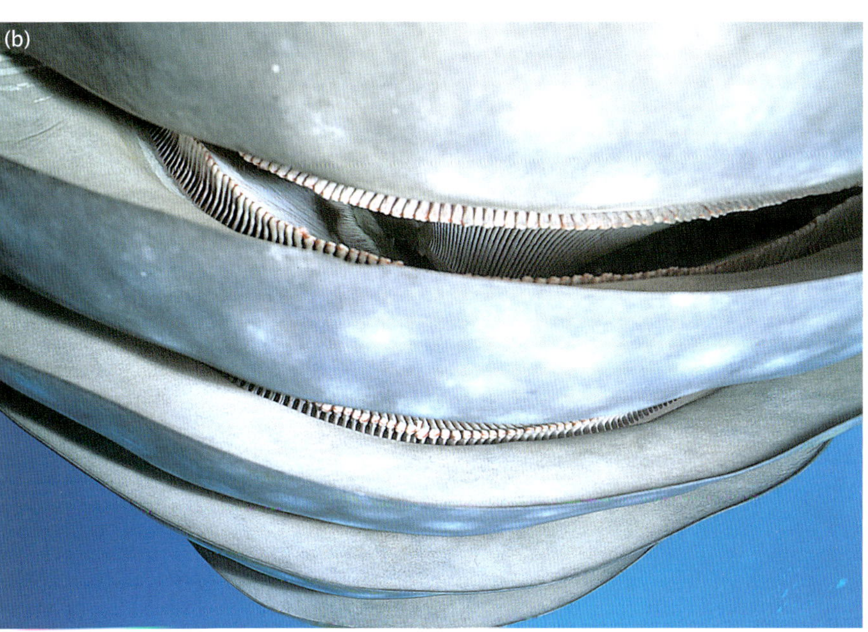

Figure 5.2 Gill rakers, shown in the photograph above and in diagrammatic form in **(a)**, are arranged like the teeth of a comb. They form the filtering system that enables a whale shark to retain small food items and simultaneously release water back to the sea. The gill rakers can be seen in **(b)**, packed together deep inside the open gill slits.

Whale sharks, along with all other animals, all fungi and some bacteria are **heterotrophs**. Heterotrophs obtain the energy for living and the material for building and repairing their structure from organic matter in their surroundings. The organic matter used by each heterotroph is its food. In contrast, **autotrophs**, such as plants and algae, just soak up sunlight to gain the energy they need for living and build their organic matter from simple inorganic matter taken up from their surroundings (for terrestrial autotrophs, from air and soil; for aquatic autotrophs, from the water).

Each heterotroph has structural and physiological features and displays behaviours that assist it to obtain its food. Some of the features that equip a whale shark to obtain its food include its enormous mouth gape and the gill rakers that enable it to filter out tiny marine animals from water. Behaviours that contribute to its success in feeding are its active and passive feeding techniques.

In this chapter we will consider in some detail the different nutrient requirements of multicellular **autotrophic** and **heterotrophic** organisms. We will also outline how their different structures and functions relate to their nutritional needs.

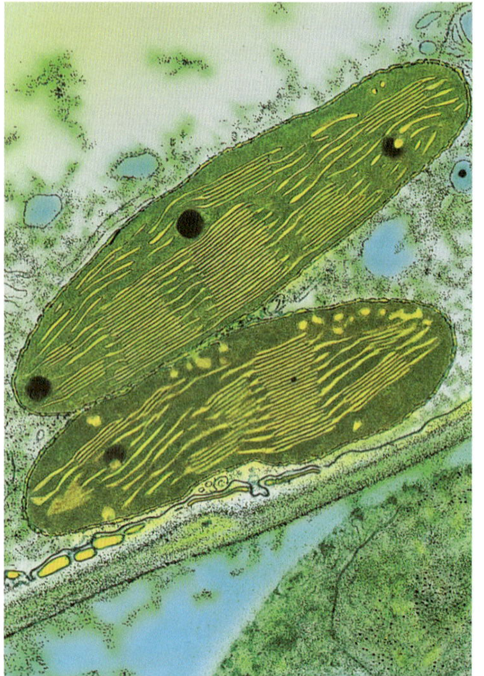

Figure 5.3 Transmission electron microscope view of two chloroplasts with a plant cell. What is the source of their green colour?

Autotrophs and photosynthesis

We saw in chapter 3 (page 66) that plants, algae and some protists (such as phytoplankton) can make organic molecules, such as sugars, by **photosynthesis**. They do this using the energy of sunlight. Organisms with this ability are called *autotrophic*. (Other organisms, such as animals and fungi, that depend, directly or indirectly, on the organic compounds produced by producers are called *heterotrophic*.)

Photosynthesis occurs in the **chloroplasts**, found in the cytosol of some cells (see figure 5.3).

Ins and outs of photosynthesis

Photosynthesis is a series of complex biochemical reactions in which sunlight energy is converted to chemical energy in sugars. As we saw in chapter 3, the process can be summarised in a relatively straightforward equation (see page 66) but essentially involves two sets of reactions:

- The first set of reactions depends on the availability of light and the presence of chlorophyll.
- The second set of reactions is independent of light but depends on the products of the first set.

You will study these two sets in more detail in your future studies of biology. For now, let us think of the overall result of the two sets of reactions put together as an 'assembly line'. Examine the assembly line of photosynthesis in figure 5.4 then consider the questions listed in the table below. Check the answers given for each question by following what happens on the assembly line during each 'shift'.

Table 5.1 An analysis of the photosynthesis assembly line

Shift 1 of the assembly line	
Question 1 What ingredients are required by the first shift?	*Answer 1* Sunlight, water (H_2O) and adenosine diphosphate (ADP). Note that chlorophyll already exists in the equipment.
Question 2 What is produced during the first shift?	*Answer 2* Oxygen, hydrogen ions (H^+) and adenosine triphosphate (ATP)
Shift 2 of the assembly line	
Question 3 What ingredients are required for the second shift?	*Answer 3* Carbon dioxide (CO_2), hydrogen ions (H^+) and adenosine triphosphate (ATP)
Question 4 What is produced during the second shift?	*Answer 4* Glucose ($C_6H_{12}O_6$) and adenosine diphosphate (ADP)
Question 5 In what ways do the shifts rely on each other?	*Answer 5* Hydrogen ions (H^+) and adenosine triphosphate (ATP) produced during the first shift are used in the second. Adenosine diphosphate (ADP) produced by the second shift is used by the first shift.

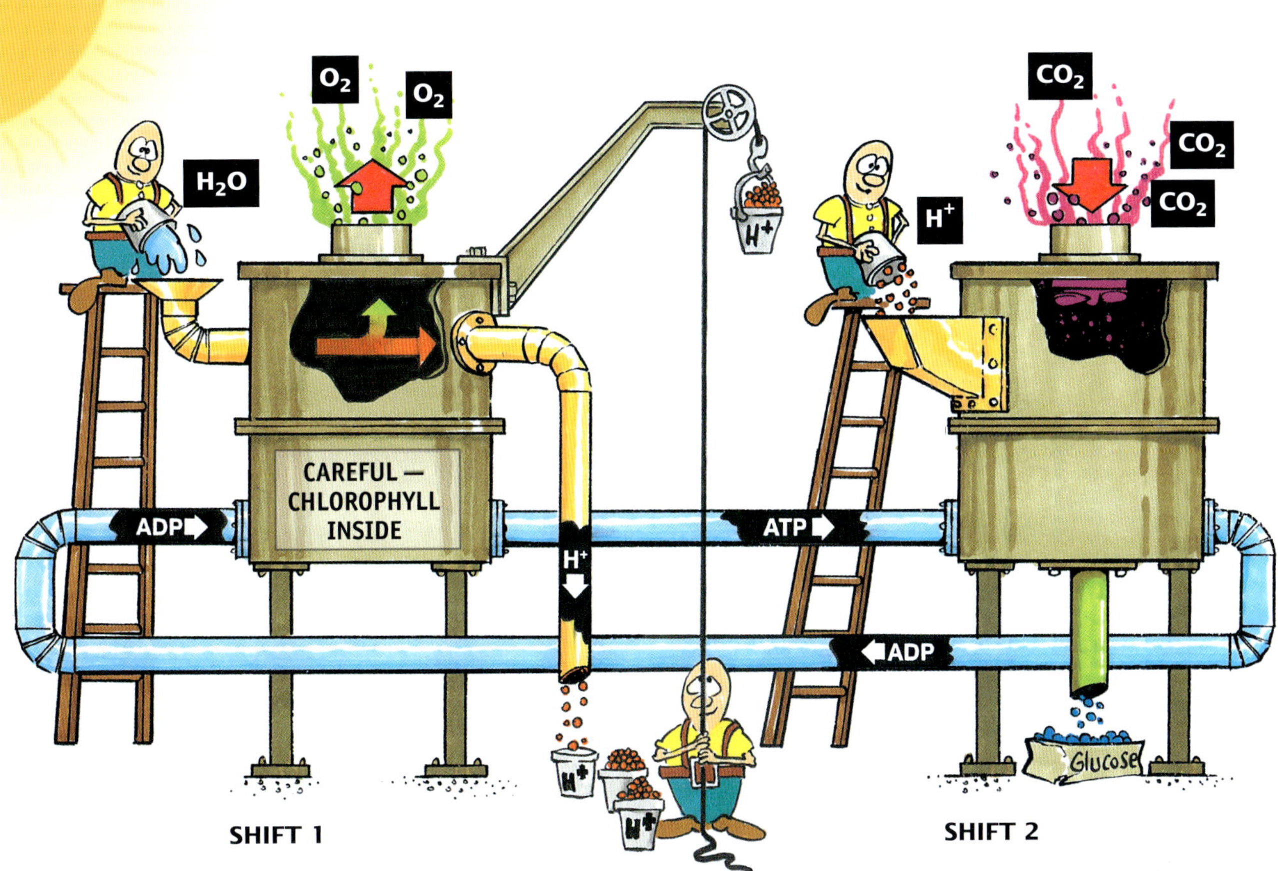

Figure 5.4 The photosynthesis assembly line. Note the requirements (inputs) and the products (outputs) of each 'shift'.

If we consider each assembly line we have:

Shift 1:

water + ADP —(sunlight / chlorophyll)→ oxygen + hydrogen ions + ATP

Shift 2:

carbon dioxide + hydrogen ions + ATP → glucose + ADP

Combining the shifts and noting the use of material between shifts, the photosynthesis equation becomes:

carbon dioxide + water —(sunlight / chlorophyll)→ glucose + oxygen

The summary of figure 5.4 takes us back to the simplified equation of photosynthesis that was introduced in chapter 3 (page 66):

$$6CO_2 + 6H_2O \xrightarrow[\text{light}]{} C_6H_{12}O_6 + 6O_2$$

In the assembly line, sunlight provides the energy required to drive the first shift. The ATP produced by shift one is a high-energy compound that provides the energy to drive the set of reactions that build glucose from carbon dioxide and hydrogen ions.

Of what significance are the outputs, glucose and oxygen, to living organisms? We will come back to this question later in the chapter when we consider cellular respiration.

Plant structures in relation to photosynthesis

Photosynthesis requires the presence of pigments capable of absorbing light energy. These light-trapping pigments were discussed in chapter 3 (pages 67–8). What about requirements such as water and carbon dioxide? What structures and characteristics of leaves and other plant parts ensure a ready supply of the materials required for photosynthesis? How are photosynthetic products transported from leaves to all non-photosynthetic cells throughout a plant?

Consider figure 5.5. The three broad regions of a plant we must consider are leaves, stem or trunk, and roots.

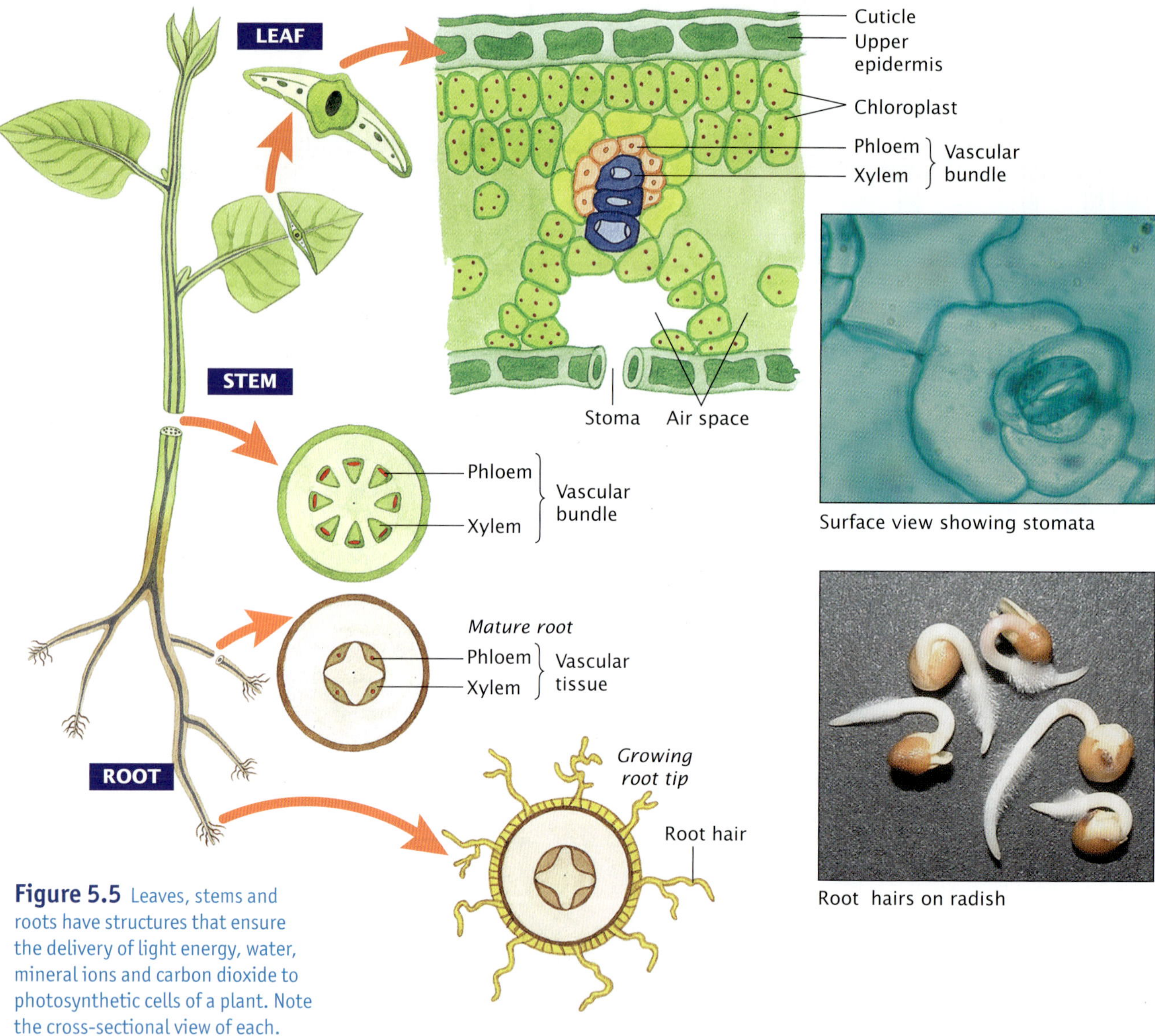

Figure 5.5 Leaves, stems and roots have structures that ensure the delivery of light energy, water, mineral ions and carbon dioxide to photosynthetic cells of a plant. Note the cross-sectional view of each.

Leaves

- Their flat shape provides a large surface area exposed to sunlight.
- Each photosynthetic cell contains many chloroplasts enabling it to trap the energy of sunlight.
- The many stomata (pores) on one or both leaf surfaces provide access into the leaf for carbon dioxide.
- The thinness and the presence of internal air spaces in leaves enables the ready diffusion of carbon dioxide to photosynthetic cells in the leaf tissue.

Refer to page 163 in chapter 6 for additional information about gaseous exchange in plants.

- The extensive network of vascular tissue contains xylem vessels to transport water to photosynthetic cells and phloem tissue to transport the products of photosynthesis from these cells to non-photosynthetic cells throughout a plant.

Stems
- Xylem vessels and fibres give rigidity to a stem and assist in its upright stance.
- Branching of stems allows layers of leaves at different levels of a plant, hence increasing the total area available for sunlight.
- Xylem vessels transport water and minerals from roots to all aerial parts of a plant.
- Phloem transports products of photosynthesis from photosynthetic cells to non-photosynthetic tissue throughout a plant.

Roots
- Extensive root system taps a significant volume of soil for water and mineral salts (see figure 5.6).
- Just behind the tip of each young root, thousands of root hairs grow. Each is a single cell and has a shape that gives a high surface-area-to-volume ratio. Hence, overall, there is a significant increase in the area available for absorption of water and mineral salts compared with a root of similar size that lacks root hairs.
- Oxygen from air in soil diffuses through root hairs into plants.

The characteristics of leaves, stems and roots ensure that these structures combine to provide the sunlight energy, carbon dioxide and water that a plant requires for photosynthesis.

(a)

Figure 5.6 **(a)** The oldest grapevine in the United Kingdom has been at Hampton Court Palace for 234 years (as at 2005). **(b)** The important role of the roots in the ongoing health of this vine is emphasised by this sign on a vacant allotment next to the glasshouse built around the vine.

KEY IDEAS

- Heterotrophs include all animals, all fungi and some bacteria.
- Autotrophs — all plants, algae and some bacteria — differ from heterotrophs in being able to make their own organic material from simple inorganic substances, typically using the energy of sunlight.
- Food for a heterotroph is any organic matter that it can obtain, break down and use as a source of energy and organic matter.
- Photosynthesis is the process of converting the energy of sunlight to chemical energy in sugars.
- The raw materials of photosynthesis are carbon dioxide and water, and the products are sugar and oxygen.
- In plants and algae, photosynthesis occurs in chloroplasts.
- The structures of the various parts of a plant maximise the plant's ability to obtain the raw materials necessary for photosynthesis.

QUICK-CHECK

1 What is the essential difference between an autotroph and a heterotroph?
2 Write a definition of 'food' for:
   a a heterotroph
   b an autotroph.
3 Do all cells of a plant carry out photosynthesis? Explain.
4 Where in a cell is light energy transformed to chemical energy?
5 What structures of a plant facilitate delivery of sugars to non-photosynthetic cells throughout the plant?
6 Considering the overall reaction of photosynthesis:
   a what are the 'inputs' for photosynthesis?
   b what are the products of photosynthesis?
7 In what form is carbohydrate typically stored in a plant?

ODD FACT

Listen for your meal! A five-kilometre stretch of motorway in Lancashire, UK, has been given a special silent surface so that birds can get their next meal. Owls use both their hearing and sight to hunt in an adjacent nature reserve and the traffic noise has made it very difficult for them to hear the rustlings of the mice and other small animals that are their prey. The Highway Agency laid the silent road surface at a cost of nearly $A7 million in response to the concern of residents, farmers and the Council for the Protection of Rural England.

Food for heterotrophs

Consider paper, rotting logs, wool, nectar, grass, a dead wallaby, tadpoles, old boots, dung, the skin between human toes, Christmas beetles, rope, eucalyptus leaves … what do these have in common?

None of these items would appear on a restaurant menu, but they are all sources of energy and organic matter for one or more species of heterotroph. Along with bread, moussaka, fettuccine and fish and chips, which are human foods, the listed items are foods for other kinds of heterotroph.

What eats what? Some silverfish eat paper, some toadstools consume rotting logs, larvae of the clothes moth feed on wool, honey possums feed on nectar, eastern grey kangaroos eat grass, Tasmanian devils (see figure 5.7) eat dead wallabies, some freshwater fish eat tadpoles, various fungi thrive on the leather in old boots, dung beetles eat (guess what!) dung, a *Trichophyton* fungus finds moist human skin a delicacy (see figure 5.8), sugar gliders consume Christmas beetles and a marine fungus consumes rope. Can you identify a heterotroph that feeds on eucalyptus leaves?

Figure 5.7 The Tasmanian devil (*Sarcophilus harrisii*), the largest of the flesh-eating marsupials, is found only in Tasmania.

Figure 5.8 The fungus *Trichophyton rubrum* causes athlete's foot as it 'feeds' on moist human skin between the toes.

Heterotrophs as a group exploit a remarkable range of foods — any source of organic matter can be food for one or more different kinds of heterotroph.

To be food for a heterotrophic organism, a substance must contain organic matter that can be broken down and used by the heterotroph to supply it with the chemical energy for living and the organic matter to build and repair its own structure.

Changes through time

The production and use of food is basic to human survival and forms an integral part of human culture. Throughout history, certain events have had an impact on what food has been eaten.

The discovery of fire, the growing of crops and the domestication of animals led to changes in the types of food available. Exploration in the late fifteenth century led to an exchange of food resources between different cultures which radically affected diets and nutritional standards at that time.

Some of the exchanges were advantageous. The widespread distribution of tea and coffee by the early 1800s led to a decline in certain diseases which were caused by micro-organisms in water. Because people boiled water to make the new beverages, they killed disease-causing organisms in the water. This was an incidental (or unplanned) outcome, since no-one knew about micro-organisms, and their disease-producing effects were recognised only in the late 1800s.

On the other hand, new foods can lead to new diseases. When sugar was first introduced to Europe, it was very expensive. Dental caries (tooth decay) became a new disease of the rich. A visitor to Queen Elizabeth I of England (1533–1603) commented on her 'bad black teeth' and attributed them to her 'too great use of sugar'.

Initially, poorer people could not afford sugar and their teeth remained relatively healthy. Dental caries became more widespread much later, particularly during the nineteenth century when the use of sugar increased rapidly. Other changes in diet at about the same time also contributed to a decline in dental health; for example, white flour began to replace the more nutritious wholemeal, stone-ground wheat flour. It is only in recent years that there has been a reduction in dental caries owing to the increase in knowledge about dental care.

ODD FACT

George Washington (1732–1799), the first President of the United States, was one of the first people to wear false teeth.

As elsewhere in the world, the kind of food consumed by Australians has also changed over time. The early inhabitants of Australia, the Aborigines, hunted animals and collected plant life from the surrounding bush. During good seasons, they ate well. In bad seasons, because there were no domesticated animals or cultivated crops, food was often difficult to find. During such periods, the diet was supplemented by dried fruits that had been preserved in times of plenty.

After European settlement of Australia in 1788, the Aboriginal diet became increasingly European as the source of their traditional foods declined.

FOOD FROM THE BUSH

Some Aboriginal communities in parts of Australia still obtain a balanced diet from indigenous plants and animals. This food is referred to as 'bush tucker' or 'bush food'. Many Aborigines are experts on bush tucker. They know about species that may be toxic and how to treat these so that they are safe to eat. The Aborigines know when best to harvest each species, and how to read the 'tell-tale' signs that indicate when foods are just right to collect.

The traditional diet includes animals such as kangaroo and large fish which the men hunt. Other foods such as shellfish, crabs and various kinds of plant food are gathered by the women. Plant food includes a rich variety of fruits, nuts, and plant parts such as roots and stems. A sample of bush tucker found in various regions is shown in table 5.2. European settlers used traditional Aboriginal foods such as fruits from the lilly-pilly tree (*Acmema smithii*) to make jam.

Figure 5.9 A selection of some of the nutritious bush food eaten by Aboriginal people

Table 5.2 Aboriginal people collected and ate a wide range of plant food.

Region	*Plant source*	*Food and its value*
Arnhem Land	river wattle, *Acacia difficilis*	Gum from the bark — source of carbohydrate and fibre
Central desert	mulga, *Acacia aneura*	Seeds used for damper, high in protein and carbohydrate
	bush tomato, *Solanum ellipticum*	Fruit has high water and carbohydrate content, some protein, fat and vitamins
Northern Australia	boab, *Adansonia gregorii*	Fruit is eaten, trunk and roots tapped for water
all states	misletoe, *Amyema linophyllum*	Berries are high in vitamin C
Cape York, Queensland	Queensland almond, *Elaeocarpus bancroftii*	Nuts — rich in carbohydrate, protein and fat

From settlement until the mid-twentieth century, the Australian food pattern was essentially constant and based on that of the British. Since the late 1940s, the range of foods available in Australia has been influenced by the large number of different groups that have migrated to Australia. Can you think of some examples of foods of Greek origin? Of Italian origin? Of Lebanese origin? Of Vietnamese origin?

A recent change to diet has been the inclusion of organic foods by some consumers for whom this is a priority. Organic foods are considered by some to be more natural products as they are produced without the application of many of the fertilisers and sprays now used. Animals may be raised and kept under different conditions. Two examples of organic produce are shown in figure 5.10.

Figure 5.10 Organic food is now available for consumers for whom this is a priority. **(a)** Clem Kandiliotis, a fruiterer, stocks organic products. **(b)** Shane and Jayne Hill at FRE FARM in Victoria produce organic free-range eggs and chicken meat. Chickens are raised in an environment free of herbicides, hormones and antibiotics. They also have loyal guard dogs that protect the chickens from wild dogs, foxes, dingoes, cats, eagles and hawks.

What animals need for digestion

How do heterotrophs access the nutrients and energy in the food they eat? For vertebrates, **digestion** of food is the first important step that must occur after the food has been eaten.

Some of the nutrients in food (such as minerals and some vitamins) eaten by animals dissolve in water and can readily enter cells. Other nutrients (such as polysaccharides, fats and proteins) are too large to pass through cell membranes. These large molecules must be broken down into smaller units to allow **absorption** by cells. Digestion is the chemical process of breaking down large organic molecules to a size that can be absorbed.

Generally, four steps occur before a substance becomes available to the cells of an animal. These are:

1. *Ingestion of food:* Ingestion occurs when a food source in the environment is captured and taken into an animal's mouth — this is called eating.
2. *Mechanical breakdown of the ingested food:* This results in large pieces of food being broken down into many smaller pieces of food. Why is this process important? If we consider the surface area (SA) of a piece of food compared with its volume (V), we obtain the **SA:V ratio**. This SA:V ratio of a shape identifies how many units of external surface area are available to 'supply' each unit of internal volume.

   In general, as a particular shape decreases in size, the SA:V ratio of the shape increases (see figure 5.11, page 104). As the food is broken down into smaller and smaller pieces, although the total volume of food remains the same, the total area of food exposed to the action of enzymes increases. The greater the surface area over which the enzymes act, the faster the digestion will take place.
3. *Secretion of various digestive enzymes onto the food:* These enzymes catalyse the breakdown of organic molecules into smaller units that can cross cell membranes. This chemical breakdown of complex organic molecules is called digestion. The energy and matter contained in the organic molecules become available for use by an animal.

Different structures are used to trap and ingest food. Humans use a hand-to-mouth process. Different animals use a wide range of strategies to obtain the different kinds of food they eat.

Refer back to chapter 2, figures 2.6 and 2.26, where surface-area-to-volume (SA:V) ratio was introduced.

4. *Absorption of digested food:* The organic molecules pass through membranes of cells lining the digestive tract and then pass into body fluids.

Different animals eat different kinds of food. The way in which an animal carries out the steps necessary to obtain the energy and matter in the food depends on the kind of food eaten.

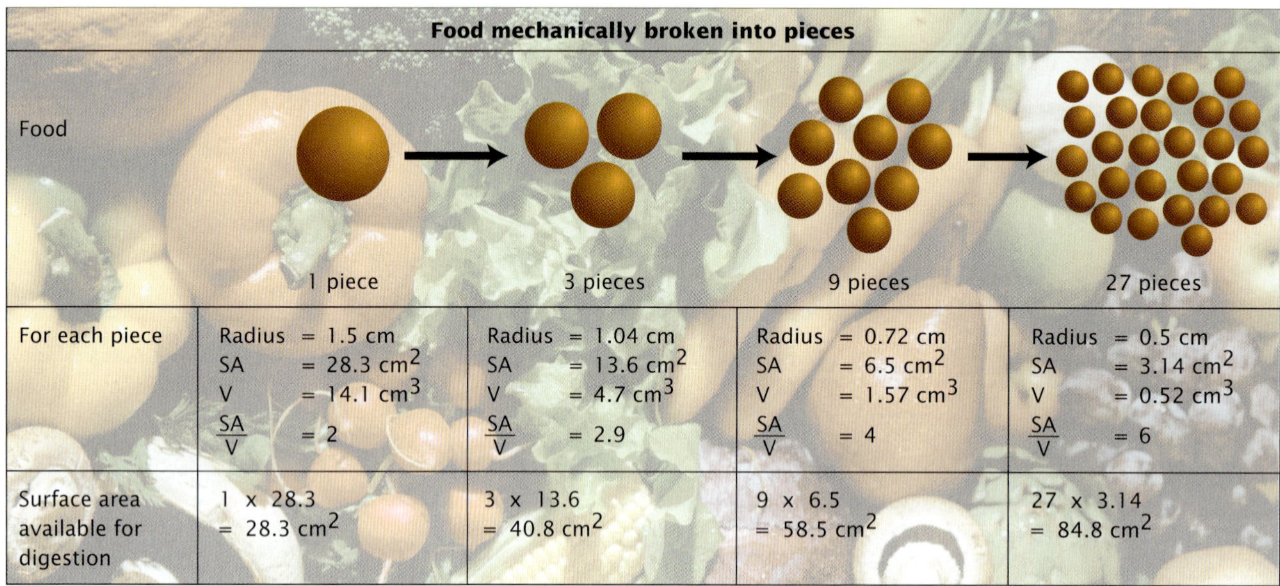

Food mechanically broken into pieces				
Food	1 piece	3 pieces	9 pieces	27 pieces
For each piece	Radius = 1.5 cm SA = 28.3 cm^2 V = 14.1 cm^3 $\frac{SA}{V}$ = 2	Radius = 1.04 cm SA = 13.6 cm^2 V = 4.7 cm^3 $\frac{SA}{V}$ = 2.9	Radius = 0.72 cm SA = 6.5 cm^2 V = 1.57 cm^3 $\frac{SA}{V}$ = 4	Radius = 0.5 cm SA = 3.14 cm^2 V = 0.52 cm^3 $\frac{SA}{V}$ = 6
Surface area available for digestion	1 x 28.3 = 28.3 cm^2	3 x 13.6 = 40.8 cm^2	9 x 6.5 = 58.5 cm^2	27 x 3.14 = 84.8 cm^2

Note: SA (sphere) = $4\pi r^2$ Vol. (sphere) = $\frac{4}{3}\pi r^3$

Figure 5.11 As food is broken down into smaller pieces, the total area available for enzyme action increases. The time taken for digestion decreases.

Mechanical breakdown of food

In order to provide as large an area as possible for enzyme action, pieces of ingested food are broken down into smaller pieces that are mixed with digestive enzymes. In many animals, jaws and teeth play an important role in this mechanical breakdown of food.

Jaws and teeth

The jaws surround the mouth and are opened and closed by muscles. Teeth are hard bony appendages found on the jaws of many animals. Movement of toothed jaws breaks food into pieces.

Most living vertebrates, except birds, have teeth or some kind of teeth-like structures. One exception is the short-beaked echidna (*Tachyglossus aculeatus*). It has a short snout that it uses to break into nests of ants or termites that it then catches with its sticky tongue. The echidna chews its food between two horny plates at the back of the mouth. Other animals, such as the platypus (*Ornithorhynchus anatinus*), may have teeth for only part of their life. Infant platypuses have teeth that eventually fall out but are not replaced. Adult platypuses grind their food between hard plates on the upper and lower jaws.

Four kinds of teeth

There are four different kinds of teeth found in most mammals — incisors, canines, premolars and molars. When they are present, these four kinds of teeth occur in the jaws in the same order from front to back. About two-thirds of each tooth is embedded firmly in the jawbone (see figure 5.12). The teeth at the front of the mouth, the incisors, are generally used to get food into the mouth. They have sharp edges and cut chunks of food from larger pieces. The pointed canine teeth pierce and tear the food. The premolars and molars have ridged surfaces to grind the food so that it is easy to swallow.

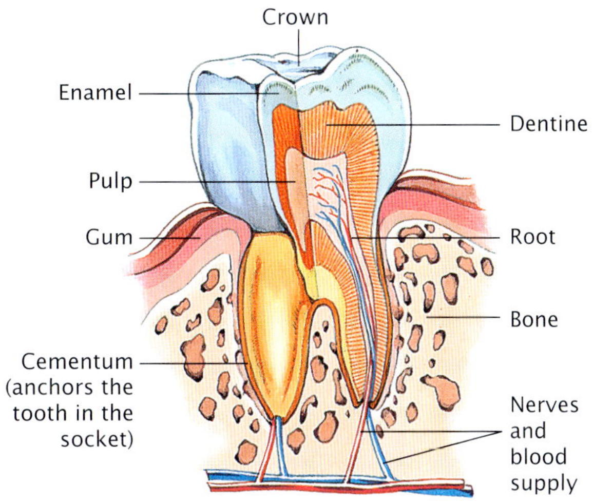

Unlike most mammals which have different kinds of teeth, some mammals such as common dolphins (*Delphinus delphis*) have teeth that all have the same structure (see figure 5.13). Their teeth have a conical shape and are used to grasp the fishes on which they feed. Dolphins do not use their teeth for chewing.

Figure 5.12 Longitudinal section through a tooth to show how it is bedded in the gum and bone of the jaw

Figure 5.13 Dolphins grasp fish with their regular conical teeth.

carrion = decaying flesh of dead animals

The Tasmanian devil (*Sarcophilus harrisii*) is the largest surviving Australian carnivorous marsupial. Although it has a preference for carrion, a Tasmanian devil will eat a range of live animals from grubs to mammals larger than itself. When it catches large animals, the Tasmanian devil, like many carnivores, is unable to swallow its prey whole. When the jaws of the Tasmanian devil are closed, the canine teeth break off pieces of the prey. Carnivorous mammals use their jaws and teeth to chew their food (see figure 5.14a).

Many animals live only on plant material. Some herbivores eat fruit; others live mainly on leaves or grass. Many herbivorous mammals do not have canine teeth. They have gaps between the incisors and the premolars and this makes it easier for the animal to move food around the mouth with the tongue (see figure 5.14b). Many herbivores grind their food for most of the day and so there is great wear on their molars. Some herbivores such as rabbits have large, chisel-like incisors that grow continuously (see figure 5.14c). Because the teeth wear away continuously they remain a reasonably constant size. Most herbivores have two sets of teeth. One set, the milk teeth, is present in infancy. These are replaced by a permanent adult set.

Some animals eat both plant and animal material. These animals include humans and are called **omnivores**.

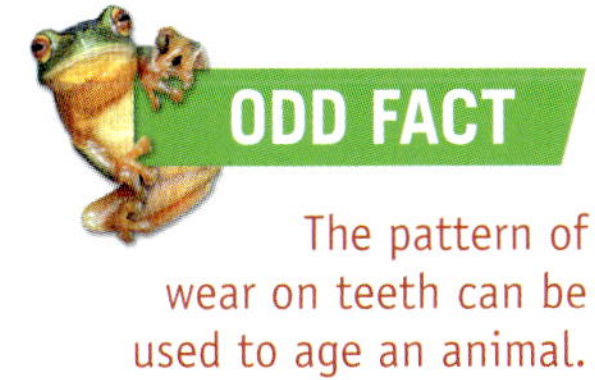

The pattern of wear on teeth can be used to age an animal.

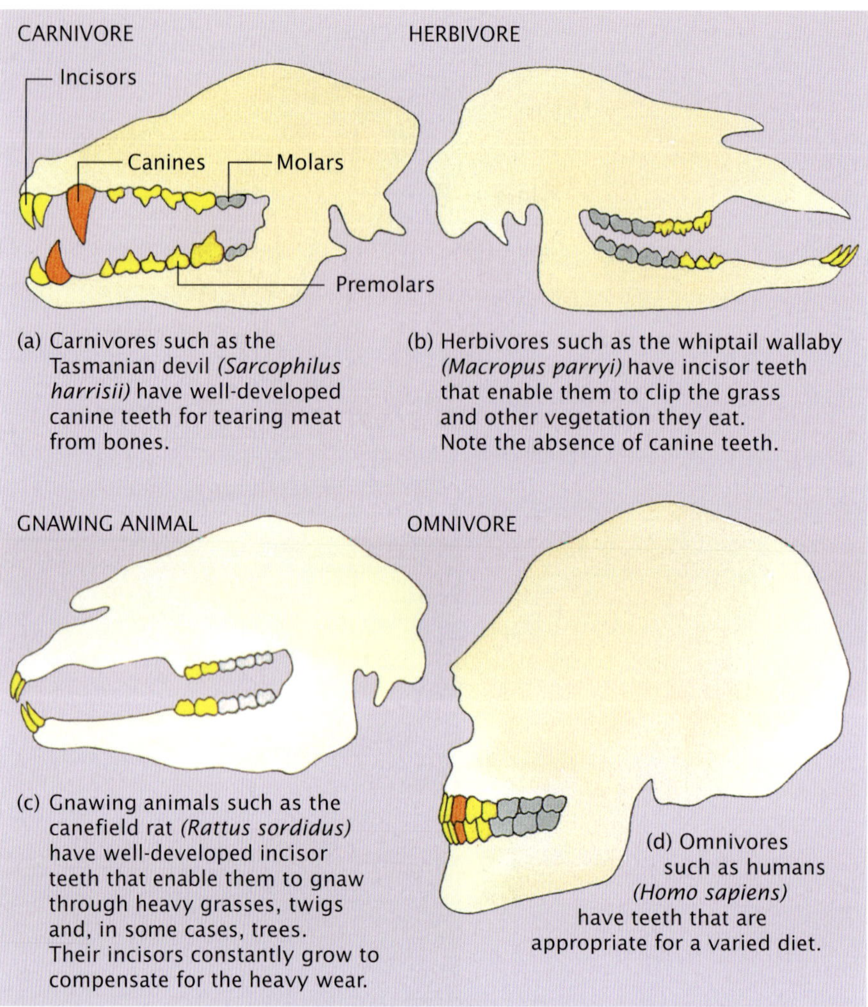

Figure 5.14 The differing teeth arrangements of some animals

KEY IDEAS

- In Australia over the last 60 years, the consumption of different foods has changed.
- Food contains large molecules that must be broken down into smaller parts before it can be used by animals.
- The digestion of food involves mechanical breakdown by structures such as teeth into smaller pieces.
- The digestion of food also involves secretion of enzymes that break down molecules into smaller molecules for absorption.

QUICK-CHECK

8 Why was the introduction of sugar to Europe a significant event in nutritional history?
9 How does the SA:V ratio of food after mechanical breakdown compare with the SA:V ratio of unchewed pieces of the same food?
10 In what ways do the teeth of carnivores, herbivores and omnivores differ?
11 What is digestion?
12 Why is digestion necessary in most animals?

The digestive system

The digestive system of almost all vertebrates is similar in basic structure with the same essential parts (see figure 5.15).

Note the function outlined for each part. Because different vertebrates have different diets, the kind and amounts of different complex organic molecules they consume may vary. For example, some grasses may be more fibrous than others and require different treatment for digestion to be completed. These differences can be reflected in the structures of digestive systems of different vertebrates.

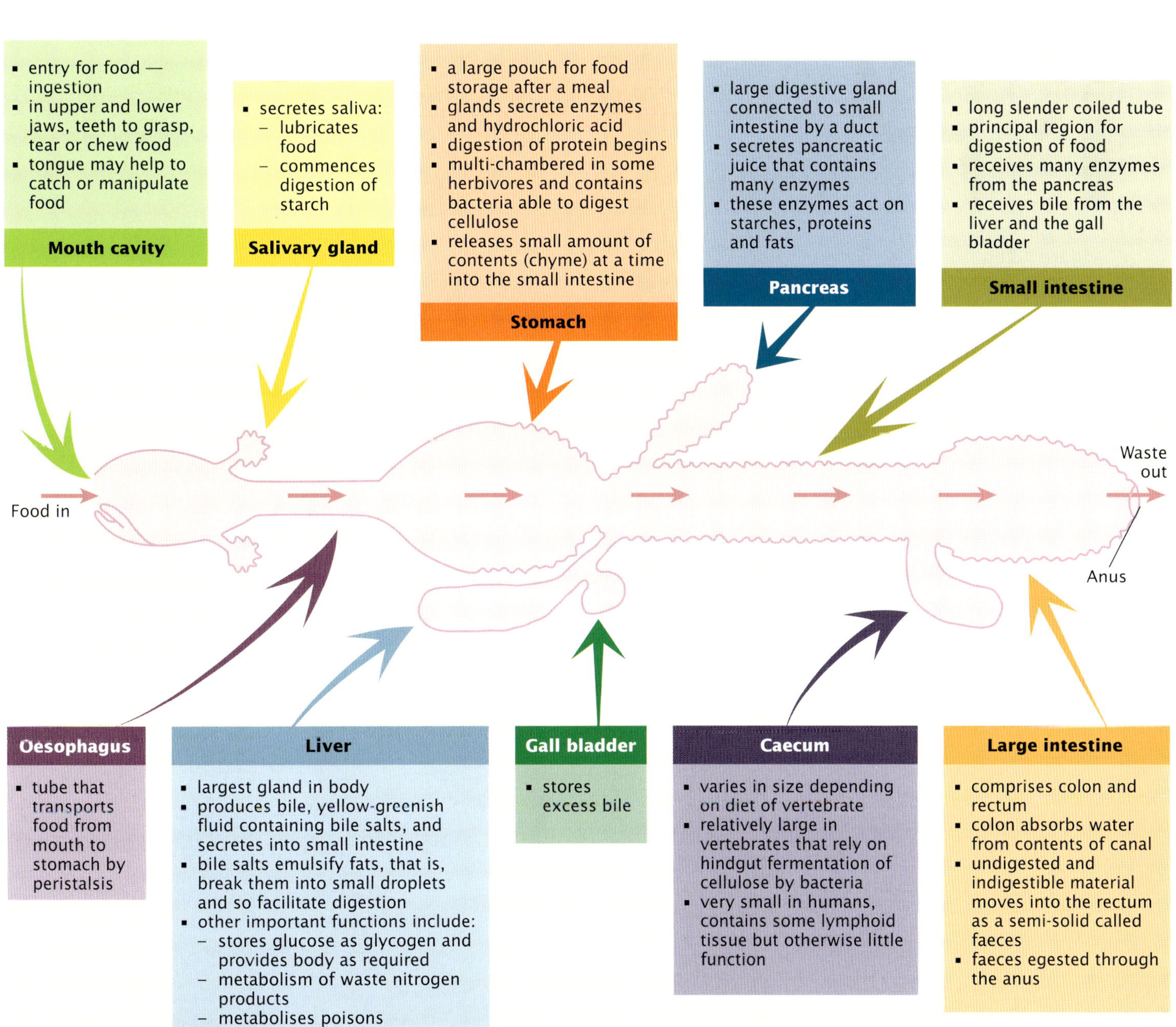

Figure 5.15 A generalised diagram of the basic structure of the digestive system of vertebrates. A mucous membrane lines the entire length and may contain glands that secrete mucus and other materials. Wavy lines indicate areas in which glands may be found.

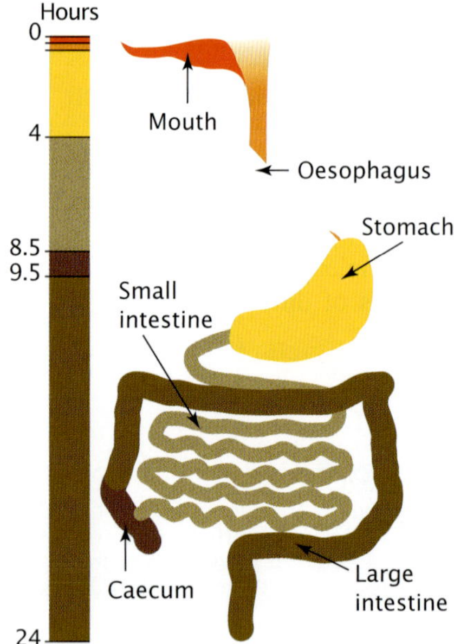

Figure 5.16 Food takes, on average, 24 hours to pass through the digestive system. It stays in the stomach for between one and four hours, then spends about four hours in the small intestine and 10 to 15 hours in the large intestine.

We will consider the digestive system in different vertebrates in relation to the kind of food they eat. First we will examine the digestive tract of a human to gain a better understanding of the function of each part of the tract. Humans are omnivores — they eat both plant and animal material. This contrasts with **herbivores**, that eat only plant material, and **carnivores**, that kill and eat other living animals.

Digestion in people

The **alimentary canal** is also called the gastro-intestinal tract, the digestive canal, or simply the gut. It is a muscular tube, approximately eight metres long, and extends from the mouth to the anus. It includes the mouth, oesophagus, stomach, duodenum, small intestine, and large intestine (colon and rectum). Muscular contractions of the wall of the gut move food along the tract. These contraction movements are called **peristalsis**. The total time for food to pass through the gut is about 24 hours. Figure 5.16 shows the approximate time food stays in each section of the intestine.

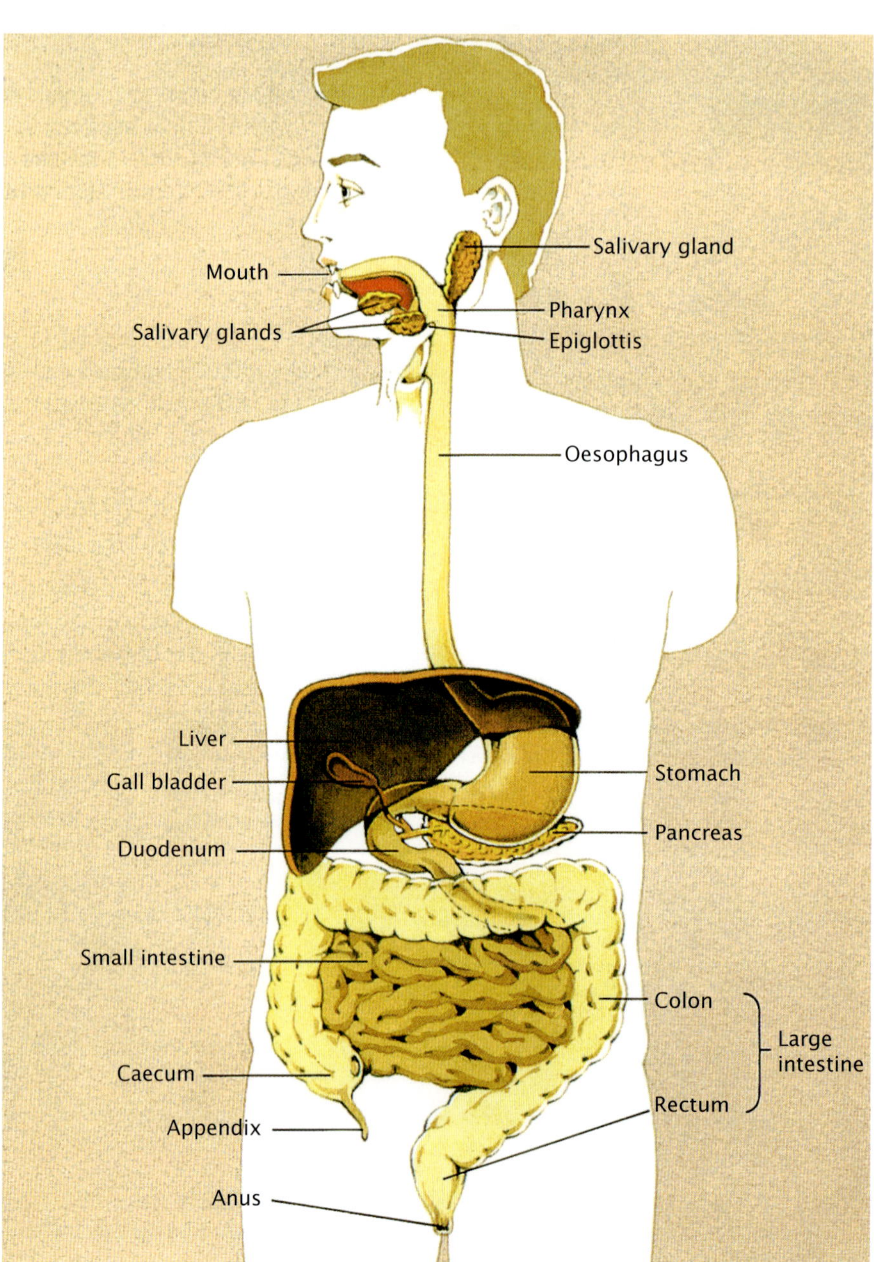

Figure 5.17 Human digestive system. The liver has been moved to one side to show the gall bladder, stomach and pancreas. In reality the liver is much larger and covers these organs.

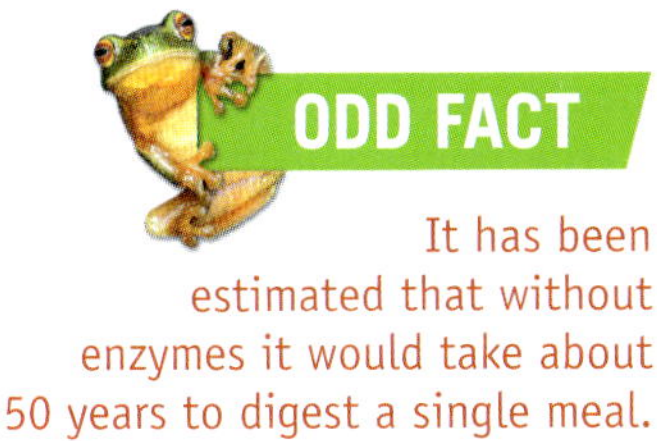

It has been estimated that without enzymes it would take about 50 years to digest a single meal.

Ducts drain from the liver, the gall bladder and the pancreas into the duodenum. Digestive **enzymes** and other substances enter the canal via these ducts. Walls of the gut also secrete digestive enzymes. The alimentary canal, together with the organs that secrete chemicals into the canal, is called the **digestive system**.

Figure 5.17 shows the relative positions of the major parts of a human digestive system. These are the parts of the system that will process food so that the energy and nutrients it contains become available for use by cells of the person.

Let us follow the processes as a roast-beef sandwich moves along the digestive canal. Before it even enters the mouth, the roast-beef sandwich has an effect on the digestive system. Saliva has been secreted into the mouth in anticipation of the food. More saliva is secreted as the person ingests the sandwich and subjects it to mechanical digestion in the mouth.

The mouth

When the sandwich is chewed in the mouth, both mechanical and chemical changes take place. The bread and roast beef are ground into smaller pieces by the teeth. As these actions proceed, the saliva released from salivary glands mixes with the food. The food becomes a soft, flexible mass that is easy to swallow.

Saliva contains the enzyme **amylase**, which acts on the starch molecules and begins to break them down into shorter chains such as dextrins and disaccharides. The saliva also dissolves some of the food, which stimulates the tastebuds on the tongue. This enables a person to detect and recognise the flavours of the food.

A person decides when to swallow the food in the mouth. A rounded mass of food, a **bolus**, is forced to the back of the mouth at the start of swallowing. Once the swallowing process has begun, it then becomes automatic.

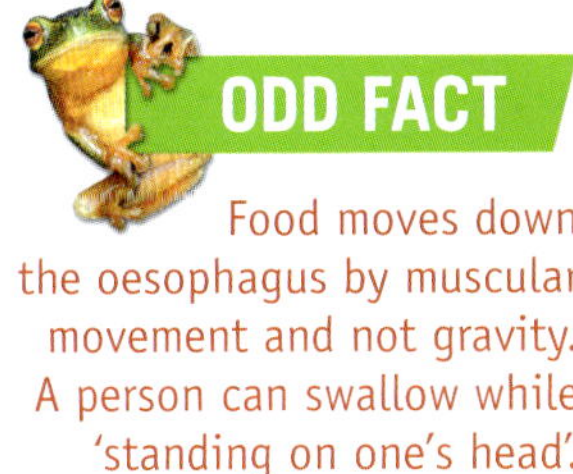

Food moves down the oesophagus by muscular movement and not gravity. A person can swallow while 'standing on one's head'.

The oesophagus

The **oesophagus** is a soft, muscular tube, about 20 to 23 centimetres long in an adult. It secretes **mucus** and transports food from the mouth to the stomach.

Contraction of muscles in the walls of the oesophagus moves the food from the mouth to the stomach by peristalsis. It takes food up to five seconds to move from the mouth to the stomach (see figure 5.18).

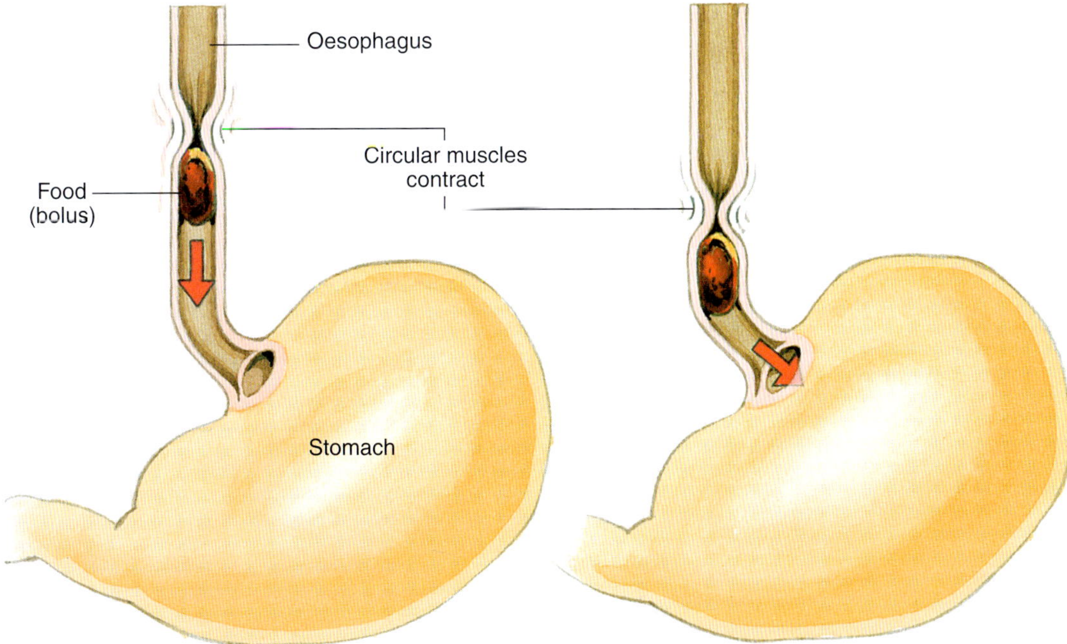

Figure 5.18 In peristalsis, food is pushed along the oesophagus when muscles above the food contract. The contraction moves like a wave along the oesophagus.

The openings to the oesophagus and the trachea are very close to each other. When food is swallowed, the epiglottis 'closes' the trachea so that food goes down the oesophagus. If food blocks the trachea, choking can occur.

The stomach

Helicobacter pylori bacteria cause stomach ulcers. In 2005, two Australian scientists, Barry J. Marshall and Robin Warren, received a Nobel prize for this discovery made in the 1980s.

Mechanical and chemical processes continue to act on the sandwich in the stomach (see figure 5.19). The **stomach** is an expanded part of the gut that is just below the diaphragm. It is about the same size as a large banana when it is empty with a volume of about 50 millilitres, but readily stretches (owing to muscles in its wall) up to 1.5 litres when food is swallowed. As the stomach is stretched, nerve endings in the wall of the stomach are activated, and cause a reflex action that leads to the secretion of gastric juice. **Gastric juice** is a mixture of mucus, enzymes, hydrochloric acid and water. These various components of gastric juice are secreted by different kinds of cells in the stomach wall.

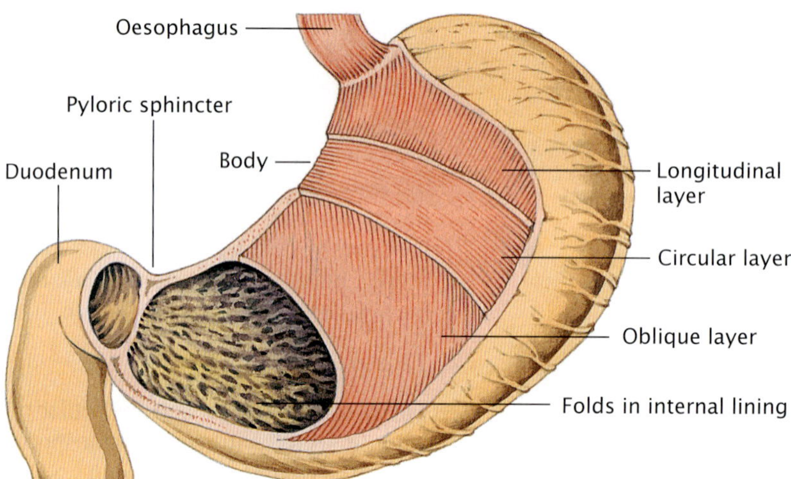

Figure 5.19 Food in the stomach is churned around by the continual contraction and relaxation of the muscle layers in the wall. Note the folds in the internal lining of the stomach. What function might these serve?

The small pieces of bread, margarine and roast beef are churned around so that they are thoroughly mixed with the gastric juices. The thin, soupy mixture containing swallowed food and saliva and gastric juice is called **chyme**.

Enzymes produced in the digestive system act in the same way as those that were described in chapter 3 (see the box on page 65).

The gastric juices contain hydrochloric acid and pepsinogen, which is the inactive form of the enzyme **pepsin**. Pepsinogen is converted into the active enzyme, pepsin, in the presence of the acid conditions (pH of 2) provided by the hydrochloric acid. The acid also helps to break the food into even smaller particles. Why do you think pepsin is secreted in an inactive form?

The enzyme pepsin acts on peptide bonds between amino acids. This action begins digestion of proteins. So digestion of protein in the bread and roast beef begins in the stomach. Pepsin acts on the long chains of amino acids of the protein, and breaks them down into shorter chains called peptides. Although peptides are much shorter chains of amino acids than proteins, they are still too large to be absorbed by the cells of the intestine. Further breakdown is necessary.

On average, food stays in the stomach between one to four hours. The more carbohydrate in a meal, the sooner it leaves the stomach. The more fat a meal contains, the longer it stays in the stomach. Meals rich in protein stay longer than carbohydrate but less than fatty meals.

During digestion in the stomach, hormones are also produced in the stomach wall. These are transported to body parts such as the pancreas and gall bladder to signal that food is in the stomach and will shortly enter the small intestine.

The small intestine

The peristaltic waves in the stomach that mix the food also force small amounts of food out of the stomach through a muscular opening called the **pyloric sphincter** into the small intestine. The food, or chyme, that moves from the stomach into the small intestine is like a thick soupy mixture. The **small intestine** is about 6.2 metres long, 2.5 centimetres in diameter, and has three main regions — the **duodenum**, **jejunum**, and **ileum**. Food remains in the small intestine for about three to six hours.

Digestion in the small intestine relies on secretions from the intestine as well as secretions from three organs that lead into the small intestine — the **pancreas**, **liver** and **gall bladder**. Look back at figure 5.17 (page 108) and note the position of these organs in relation to the human alimentary canal. Note also how the ducts from the pancreas and the gall bladder combine to form the common bile duct.

In people with **cystic fibrosis**, abnormally thick mucus can block this bile duct and prevent pancreatic enzymes from entering the duodenum. Cystic fibrosis sufferers must obtain their pancreatic enzymes in capsule form to aid digestion and prevent malnutrition (see figure 5.20).

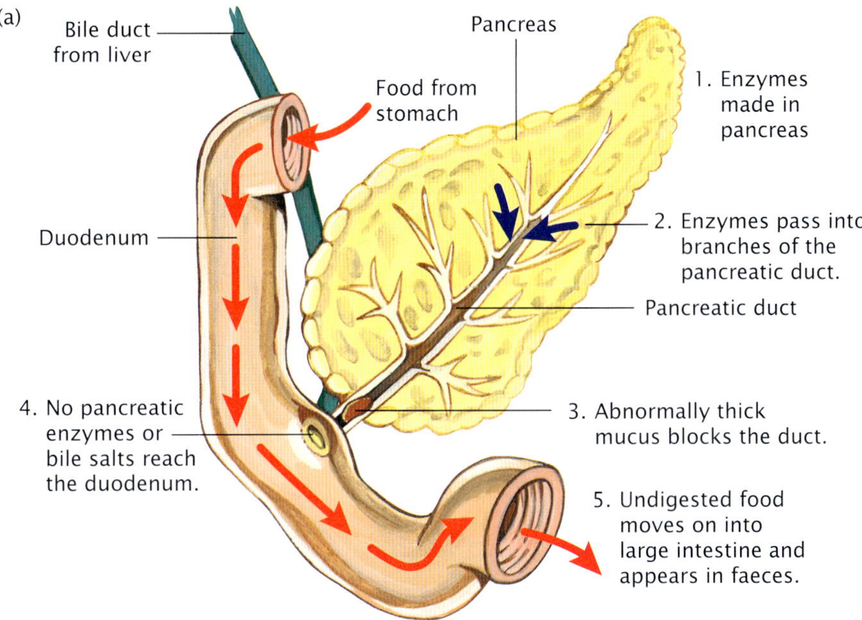

Figure 5.20 **(a)** In a person suffering from cystic fibrosis, pancreatic enzymes are unable to enter the duodenum because abnormal mucus blocks the pancreatic duct. Without these enzymes, food in the duodenum is not fully digested. **(b)** One day's supply of tablets for a person with cystic fibrosis. Note the enzyme granules.

(b)

The duodenum

The duodenum is about 25 centimetres long and is the first part of the small intestine. By the time the churned-up sandwich leaves the stomach, enzymes and other substances have already been secreted into the duodenum.

Secretions from the pancreas

Secretions from the pancreas enter the duodenum via the pancreatic duct. Pancreatic secretions include:

- bicarbonate that neutralises the acid food from the stomach. This is necessary because the optimum pH for enzyme action in the duodenum is 7.6.
- amylases, which are enzymes that digest carbohydrates. The digestion of carbohydrate that began in the mouth continues in the duodenum.
- protein-digesting enzymes that continue the digestion of proteins in the sandwich. Protein-digesting enzymes from the pancreas are secreted in an inactive form. For example, the pancreas secretes the enzyme trypsinogen. This is converted into the active form, trypsin, inside the gut. The trypsin then activates the other protein-digesting enzymes secreted by the pancreas.

- lipases, which are enzymes that digest fats and phospholipids into glycerol and fatty acids
- nucleases, which are enzymes that digest ribonucleic and deoxyribonucleic acids. These compounds are components of cells, particularly of cell nuclei. They provide no energy for the body and are not considered as nutrients.

The role of the liver and the gall bladder

Cells of the liver produce bile. The bile passes out of the liver via the **bile duct** into the duodenum. If there is no food in the intestine, bile moves from the bile duct into the gall bladder where it is stored and concentrated. When food is in the duodenum, bile is secreted from the gall bladder via the bile duct into the duodenum.

ODD FACT

Bilirubin is the pigment released when red blood cells are broken down. Iron and globin from red blood cells are recycled and used by the body. Bilirubin is broken down in the intestine.

Bile is a greenish fluid which contains bile salts, cholesterol and pigments such as bilirubin. In the intestine, the bile salts break up large fat droplets into smaller droplets of fat, a physical process called **emulsification**. It occurs as bile is mixed with the soupy mixture that was once the sandwich. Bile emulsifies the fats in the margarine and in the beef of the sandwich. The formation of smaller droplets of fat results in a much larger surface area for action by lipase enzymes. Lipase from the pancreas acts on the small droplets of fat and breaks them down into fatty acids and glycerol. These are small enough to be absorbed by the cells lining the gut. About 10 per cent of fat in the sandwich remains undigested by the time the food leaves the gut.

The liver is involved in many metabolic functions, including the synthesis of plasma and other proteins.

As well as digestion, the liver has other important functions. If liver function is significantly impaired, a liver transplant may be required.

Why do fatty foods satisfy our appetites?

Fats take longer to digest than other foods, so they need to spend a greater length of time in the small intestine. If the duodenum detects fatty food leaving the stomach, it allows only a limited amount through. Only when this is digested and moved along the intestine is more fatty food allowed into the duodenum from the stomach.

When our stomach holds food for a longer period it will be some time before we feel hungry again. If we don't feel hungry we say we are satisfied. Fatty foods tend to satisfy our appetites.

The jejunum and ileum

Food in the duodenum moves along the gut into the jejunum (about 2.5 metres long) and then into the ileum (about 3.6 metres long), the next parts of the small intestine. The pancreatic enzymes continue to act on any undigested food. In addition, cells of the jejunum and ileum secrete aminopeptidases — enzymes that complete the digestion of proteins.

$$\text{peptides} \xrightarrow{\text{aminopeptidases}} \text{amino acids}$$

Enzymes that act on specific carbohydrates are also secreted by cells of the small intestine. They complete the digestion of disaccharides, the double sugars. For example, maltase acts on maltose to break it down into two molecules of glucose.

Other enzymes act on the other disaccharides. Food has to be digested down to the basic units such as amino acids and glucose before it is of use to the body. These molecules are small enough to pass across the cell membranes of the intestine and are then transported through the body.

The amylase enzymes secreted by the pancreas and the enzymes secreted by cells in the small intestine complete the digestion of the sugars and starches in the bread of the sandwich. The resultant single sugar molecules can be absorbed by the intestine cells.

Protein-digesting enzymes from the pancreas and the small intestine digest protein in the sandwich's bread and meat that was not digested in the stomach. The amino acids from the proteins are small enough to be absorbed by the cells lining the gut. About 15 per cent of the protein in a meal remains undigested. Thus the digestion of the sandwich is completed as it moves along the small intestine. A summary of digestion is given in figure 5.24 (page 115).

The structure of the different parts of the alimentary canal relate to the functions of the part (see figure 5.21).

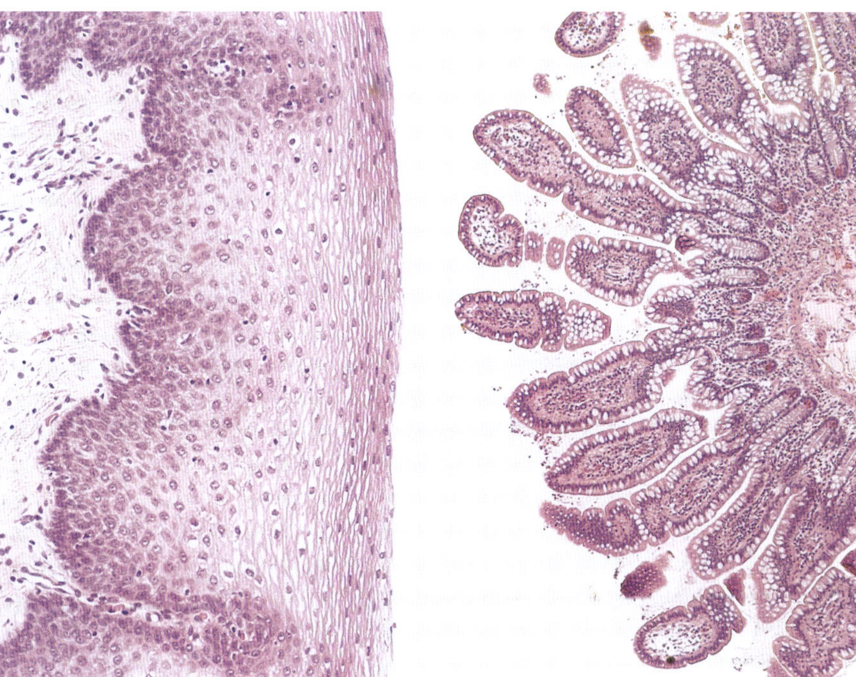

Figure 5.21 Two sections from different parts of the alimentary canal. Which is from the oesophagus and which is from the small intestine? What are the reasons for your decision?

Absorption in the small intestine

The small intestine has a structure that is especially designed for absorption of the products of digestion. Look at figures 5.22 and 5.23 (page 114). The lining of the gut has folds so that there is a greater surface area than if the inner surface were cylindrical. The surface area is further increased because the undulating surface is covered with the minute fingerlike projections called **villi**. The surface of each villus is covered with thousands of microvilli. The increase in surface area provided by the folds increases the rate at which absorption occurs.

Amino acids, glucose, fructose, galactose, glycerol and fatty acids from the sandwich will be absorbed across the cells of each villus. There will be a high concentration of the products of digestion in the gut and so, initially, much of the absorption of these materials into the cells lining the gut is by diffusion. That is, they will move from a region of high concentration to a region of lower concentration. These compounds will be absorbed into the blood and lymph vessels below the epithelial cells of the intestine.

In addition, active transport occurs. The cells lining the gut expend energy to absorb material against a concentration gradient into an area that already has a high concentration of that material.

Most of the glycerol and fatty acids are absorbed into the **lacteals**. These are blind extensions of the lymphatic system. The lymph vessels transport glycerol and fatty acids to the veins near the neck where they enter the bloodstream. Some glycerol and small fatty acids are absorbed into the blood capillaries.

About 90 per cent of absorption of nutrients takes place in the small intestine. The remaining 10 per cent takes place in the stomach and large intestine.

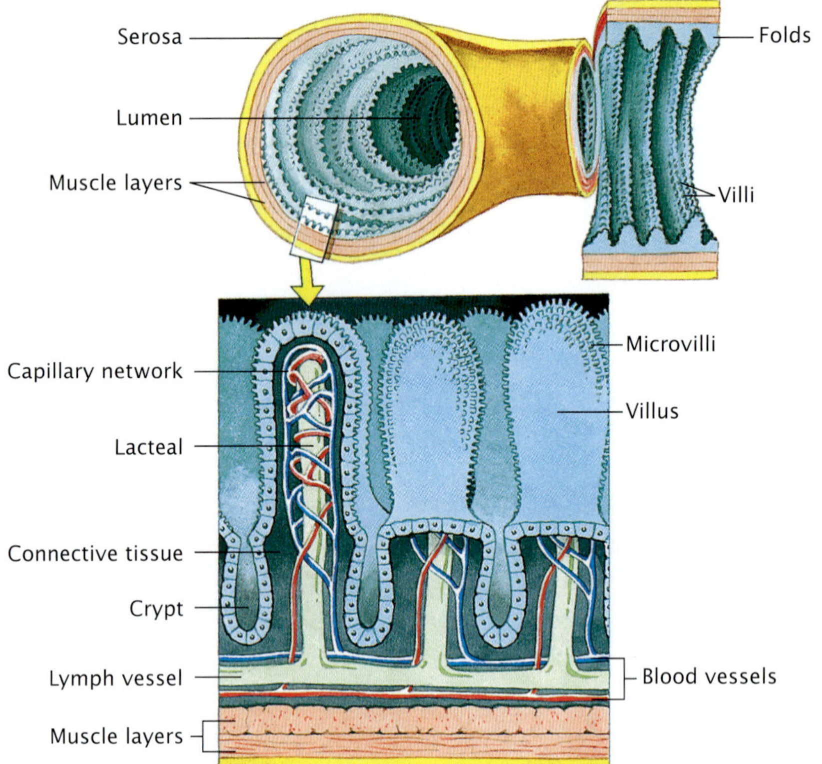

Figure 5.22 The intestine (*top*) is a muscular tube with many internal folds that greatly increase the surface area. Capillaries and lacteals (*bottom*) that absorb digested materials

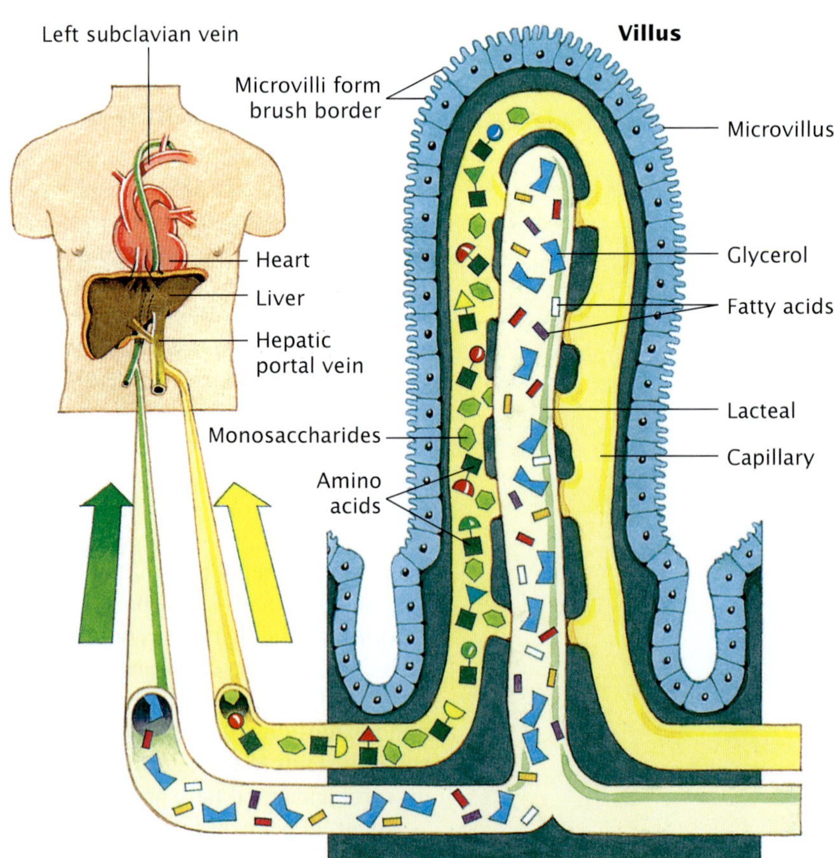

Figure 5.23 Amino acids and monosaccharides are absorbed into the capillaries in the villi lining the wall of the intestine and travel to the liver via the hepatic portal vein. Most glycerol and fatty acids are absorbed into lacteals and enter the bloodstream via veins near the neck.

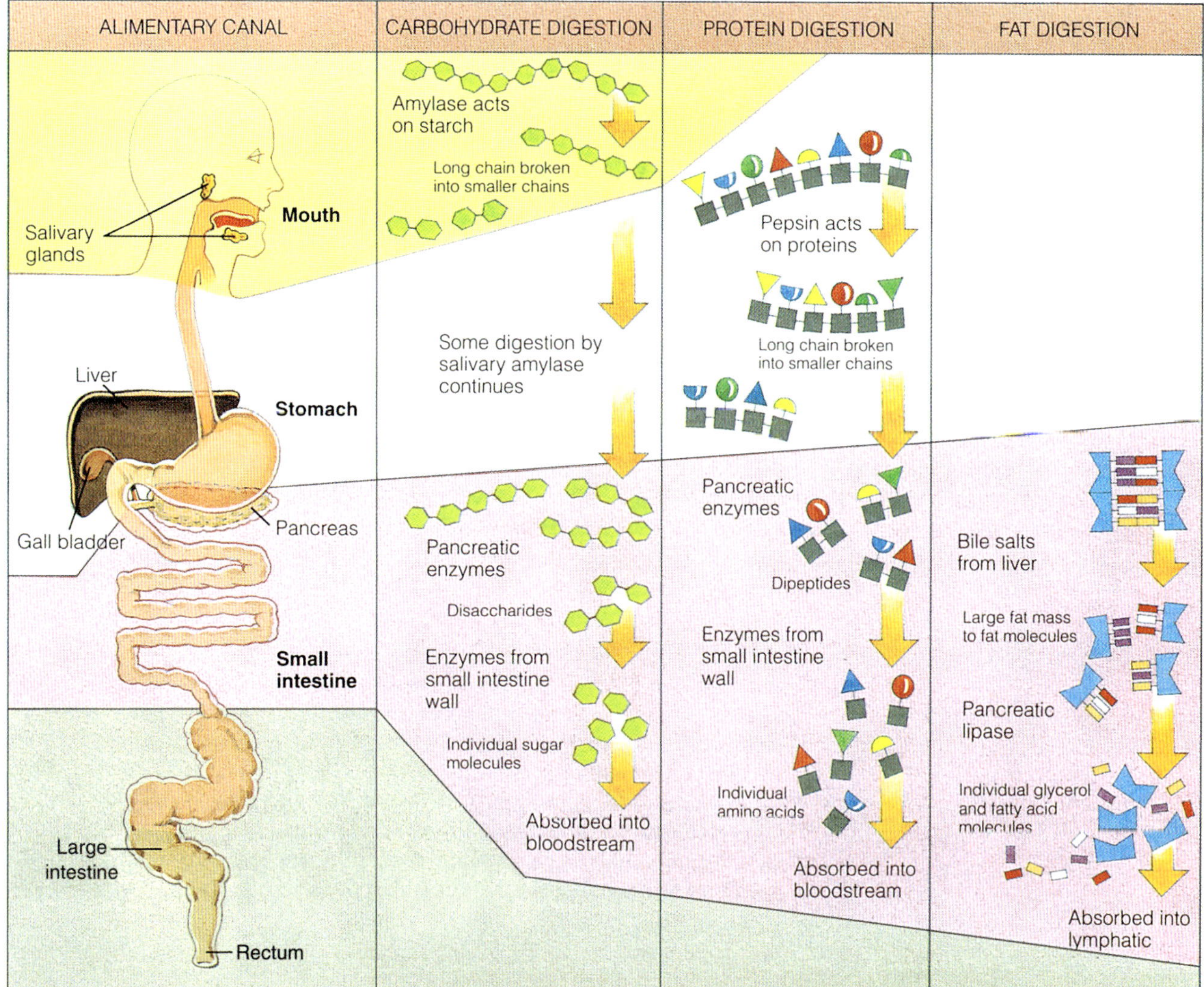

Figure 5.24 Complex food compounds must be broken down into small units before they can be absorbed by cells lining the alimentary canal. In which region is each kind of compound digested? What are the end products in each case? The products of digestion can all be absorbed by cells lining the alimentary canal.

About 2300 millilitres of water is obtained from food and fluids each day and enters the alimentary canal. In addition, up to 7000 millilitres of water enters the canal as enzymes and mucus are secreted by cells in the canal wall. Hence, over 9000 millilitres of fluid enter the intestines each day. About 95 per cent, over 8500 millilitres, of this is absorbed in the small intestine along with the products of digestion mentioned above. If this does not occur, as in diarrhoea, the body quickly becomes dehydrated because of the continued loss of water from the cells lining the alimentary canal.

Water is absorbed by the cells of the intestine by osmosis. Only about 500 millilitres of water are passed on to the large intestine daily.

The monosaccharides, amino acids, glycerol and fatty acids that have been digested from the roast-beef sandwich are initially transported to the liver for processing. Later, they are transported to other cells and provide energy and matter.

The large intestine

After the absorption of nutrients in the small intestine, the remainder of the gut contents move into the large intestine. The **large intestine** is about 1.5 metres long and consists of two main parts, the **colon** and the **rectum**. The name 'large intestine' really comes from its diameter — it averages about 6.5 centimetres.

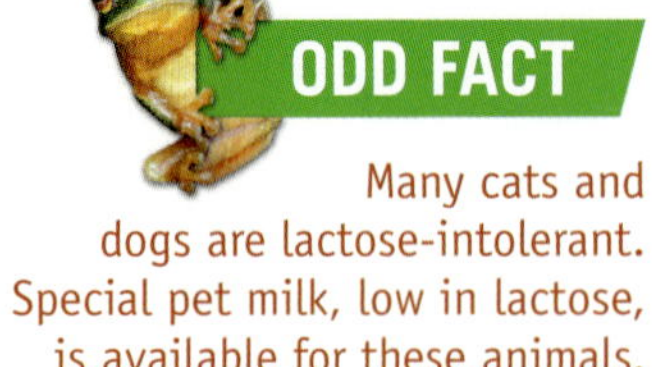

Many cats and dogs are lactose-intolerant. Special pet milk, low in lactose, is available for these animals.

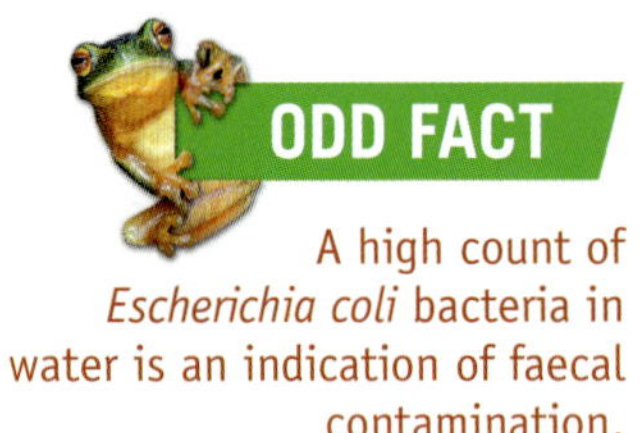

A high count of *Escherichia coli* bacteria in water is an indication of faecal contamination.

Considering the sandwich that was eaten, the residue that moves into the colon contains five grams of undigested fibre from the bread in the sandwich. It also contains undigested food (about 15 per cent of eaten protein and 10 per cent of eaten fat). Bacteria in the colon digest about 30–40 per cent of the fibre into fatty acids and also produce small amounts of some vitamins, particularly vitamin K. These products are absorbed by the walls of the colon along with water and electrolytes such as sodium or chloride.

Water moves freely, in either direction, across the membranes of the intestine at any point along its length. The amount of water absorbed in the large intestine depends on how long the remains of the food stay in the region. The longer they stay, the more water is removed. Generally, up to about 400 millilitres of water are absorbed in the large intestine each day.

Peristalsis moves the contents of the colon into the rectum which is the last 20 centimetres of the large intestine. The contents of the rectum are called **faeces**. In addition to food remains and water, faeces also contain bacteria, dead cells shed by the gut, the dried remainder of enzymes, undigested protein, fat and fibre, and inorganic matter. Faeces generally comprise about two-thirds water. The colour and texture of faeces is influenced by the kind of food eaten, the amount of fluid intake and breakdown products excreted by the liver into the intestine.

Faeces leave the gut by the **anus**. On average, it takes food about 24 hours to move through the gut and faeces leave the gut about once per day. A number of factors influence the movement of faeces. Particular diets can affect this — a diet rich in carbohydrate is likely to result in more frequent elimination of faeces.

Excessive amounts of air or gas sometimes accumulate in the stomach or intestine. This condition is called flatus. As much as 500 millilitres of air may be swallowed with a meal or with carbonated drinks. The gas in the stomach is often expelled by belching (burping) but some of it may remain and move around as the food mixes and give the gurgling sound we sometimes hear. Flatus also occurs if large amounts of gas are released during the breakdown of foods in the gut. Bacterial fermentation in the gut can produce up to 700 millilitres of gas. Gas in the intestine is expelled through the anus.

Energy absorption

Remember that energy and nutrients were trapped in the sandwich. How much of this becomes available for the cells of the body? Table 5.3 gives some measures of energy absorbed from the sandwich. The sugars and starches are completely digested and absorbed. About 10 per cent of fat and 15 per cent of protein remain undigested, so the energy in the undigested food is not available for use by the body. About 35 per cent (1.8 grams) of the fibre from the sandwich will be digested in the colon. Of the energy in the sandwich, about 90 per cent becomes available for use by the body.

Table 5.3 Energy asorbed from a sandwich after digestion

Energy source	*g*	*kj*
sugar and starch	26	416
protein	27.2	462
fat	19.8	733
fibre	1.8	29
		1640

ALCOHOL

About 6 per cent of the average adult Australian diet consists of alcoholic beverages. For the 1996–1997 year, for Australian adults 18 years and over, the average consumption of beer and wines was 127 litres and 25.3 litres respectively, and has remained reasonably constant since that time. More recent statistics use the consumption of pure alcohol, that is, the amount of alcohol contained in the beverages, rather than the total volume of the beverage consumed. Although spirits and liqueurs have the highest energy content per millilitre of the alcoholic beverages (see table 5.4), it is important to remember that the volume of each of the beverages consumed may be very different. This influences the amount of alcohol consumed. The idea of standard drinks (see figure 5.25) is used to assist people to monitor what they drink.

Table 5.4 Energy content of a range of alcoholic beverages

Beverage	*Alcohol (g) per 100 mL*	*Energy (kj) per 100 mL*
special light beer	0.9–0.95	61–70
light beer	2.0–2.6	84–120
medium light beer	3.0–3.5	100–125
regular beer	4.5–5	145–170
diabetic ale	4.5–4.6	119–126
stout	5–6	188–200
spirits	up to 40	920
dry sherry	16	481
sweet shery	16	568
red and dry white wine	10	284
champagne	10	315
liqueurs	23	1303
port	18	660

Note: *Diabetic ale has a low sugar content (0.8–1.0 g per 100 mL) compared with regular beer (2.6–3.2 g per 100 mL).

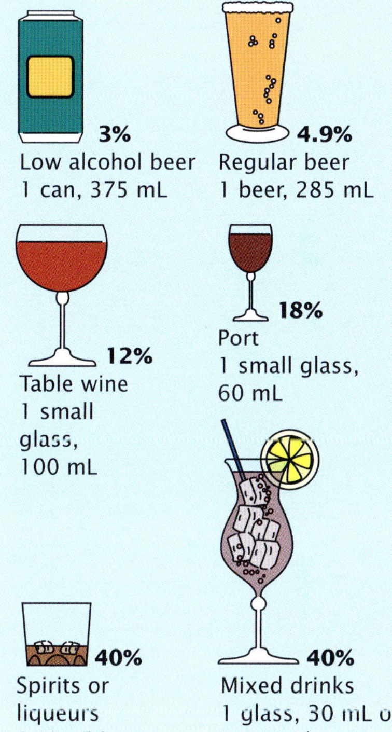

Figure 5.25 Different types of alcoholic drink contain different amounts of pure alcohol. A standard drink is defined as one that contains 10 g of pure alcohol.

Alcohol in the blood

As soon as a person has a drink of alcohol, the alcohol is absorbed into the body. However, two people who drink the same amount of alcohol over the same period of time are unlikely to have exactly the same blood alcohol level. Factors such as the rate of absorption of alcohol in a person, the amount of food eaten, and body size may be important. A female will generally have a blood alcohol level higher than that of a male of the same size who has consumed the same amount of alcohol. The reasons for this are not clear. One factor is that a greater proportion of a female's body is made up of fat and fat does not absorb alcohol; thus a higher proportion of the alcohol is retained in the blood.

(continued)

The rate at which alcohol is metabolised also varies. About 90–95 per cent of alcohol absorbed is metabolised by the liver. If excessive alcohol is consumed over a long period, the liver is likely to be seriously affected, and cirrhosis of the liver may develop. The rate of removal of alcohol from the blood is shown in figure 5.26.

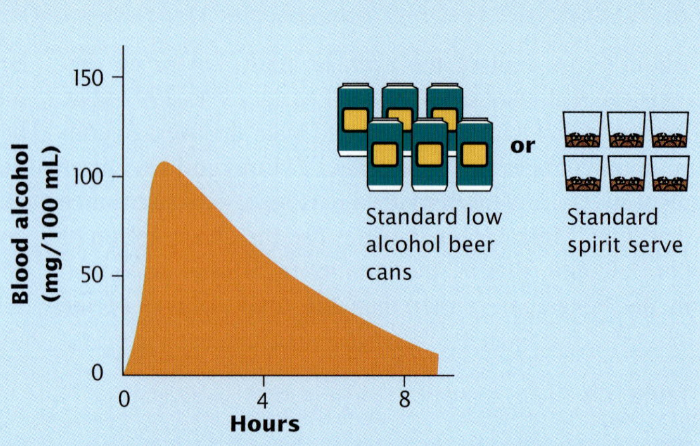

Figure 5.26 Although there is individual difference, the rate of removal of alcohol from the blood is about 15 mg/100 mL/hour. The graph shows the average rate of loss of blood alcohol after consumption of 60 g of pure alcohol. Does the source of the alcohol have an effect on rate of loss?

Measurement of blood alcohol concentration

There is a direct relationship between the amount of alcohol in the blood and the amount in the deep parts of the lungs. The higher the blood alcohol level of a person, the greater the amount of alcohol that diffuses into the deep parts of the lungs. Blood alcohol testing is generally carried out indirectly with a breath test. The instrument used measures the amount of alcohol in the breath. This is then translated against a scale of blood alcohol content. In some cases, an analysis is carried out on the blood of a person. The effect that alcohol has on driving behaviour is summarised in figure 5.27.

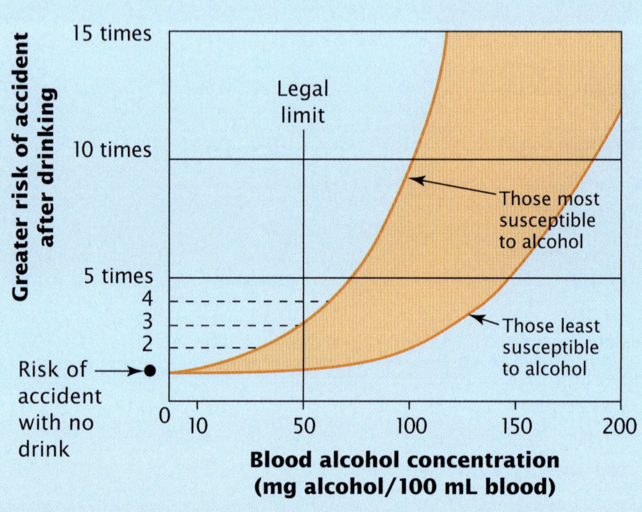

Figure 5.27 Although alcohol may help some people to relax, even at the lowest levels it also increases the risk of having an accident when driving. Concentration and judgement abilities are impaired. What is the legal blood alcohol limit for an 'L' driver, a 'P' driver and a fully licenced driver in your state?

Digestion in herbivores

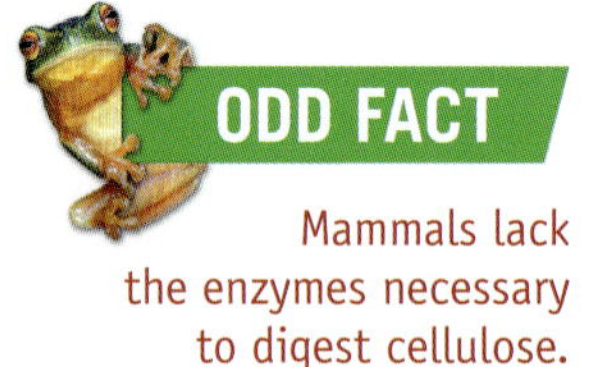

Mammals lack the enzymes necessary to digest cellulose.

Most large animals, including humans, have some bacterial fermentation in the digestive tract. In carnivores and omnivores this plays a minor role in the digestion and supply of nutrients for the animal. This contrasts with the situation in some herbivores.

Herbivores live on plant material. A significant proportion of carbohydrate in plant material is cellulose. The energy in cellulose is not available to a herbivorous mammal unless the cellulose can be processed in some way to a form that can be absorbed by the body. To do this, these animals make significant use of bacteria to ferment the grasses they eat. The broken-down products from the bacteria are absorbed by the gut, and energy from the cellulose becomes available for use by the herbivore.

In order to accommodate the large numbers of bacteria that are needed to ferment the grasses, the intestine is expanded at some part along its length. The size and location of these expansions indicate the extent to which the process is important for the mammal. The greater the need for energy from cellulose,

In one year, fermentation in the rumen of an average dairy cow produces about 90 kilograms of methane, a flammable gas. The energy contained in this gas is equivalent to the energy contained in 120 litres of petrol.

the greater will be the modification to the gut. Some mammals have modified regions of the colon and caecum for bacterial digestion. These are the hindgut fermenters. Dugongs, for example, are **hindgut fermenters**. In other mammals, the stomach or part of the oesophagus is vastly enlarged to accommodate bacterial digestion. These are the **foregut fermenters**.

Foregut fermenters

Foregut fermenters have vastly expanded sections of the digestive system, particularly the stomach. For example, in cattle, an expanded stomach can have a volume of up to 300 litres and constitute 15 per cent of the body mass. Very large volumes of a mixture of bacteria and fibrous material are retained for extended periods in the expanded sections to allow the breakdown of the plant material. The total digestive system can account for 40 per cent of the body mass.

Some foregut fermenters, the ruminants, have several compartments to their stomachs.

Ruminants — one stomach, many compartments

Cattle, sheep and goats are **ruminants**. They graze and chew grass which is then swallowed for the first time. The food passes into the first compartment of the stomach, the rumen (see figure 5.28). This is a huge chamber in which bacteria begin the fermentation of the food. The fermentation breaks down the cellulose in the grasses. The food then passes into the second compartment of the stomach, the reticulum, which is also called the honeycomb stomach. Look at tripe in a butcher's shop and you will see why this name is used.

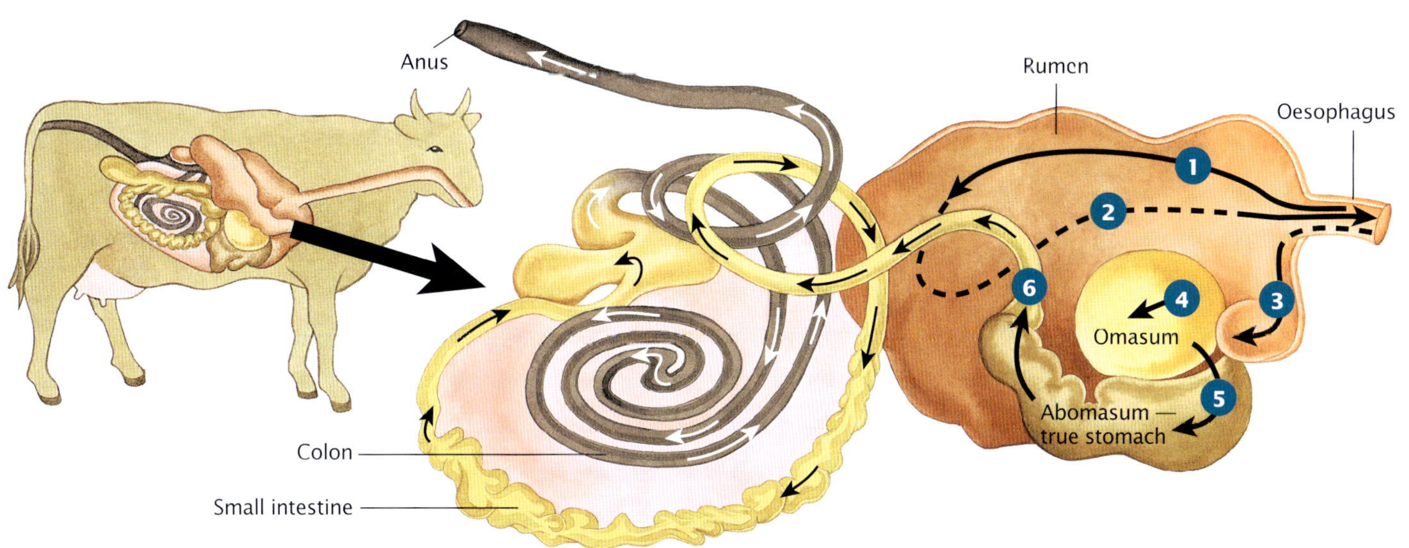

Figure 5.28 (1) Food eaten by the cow goes into the rumen. After fermentation, it moves into the reticulum. (2) From there the cud moves back into the mouth for further chewing before it is swallowed into the omasum (3 and 4). Food then goes into the abomasum (5), the true stomach, before it passes into the duodenum (6).

ODD FACT

Cellulose-digesting bacteria require neutral conditions so cows must keep their acid and bacteria apart: bacteria in the rumen and acid in the caecum.

The reticulum prepares the food for regurgitation. Wads of partly digested food, called cud, move from the reticulum back into the mouth. When an animal 'chews its cud', it is chewing its food for the second time and grinding it into a paste with its molars. It also adds saliva to the food — in sheep and goats, this is 10–15 litres per day; in cows, 200 litres per day.

When food is swallowed for the second time it goes into the third compartment of the stomach, the omasum, that has two parts. From there it passes into the

The koala, *Phascolarctus cinereus*, has a caecum about 2 metres long that acts as a fermentation chamber for *Eucalyptus* leaves eaten. This is longer than the caecum of any other animal of equivalent size (see figure 5.30, page 121)

fourth stomach, the abomasum. This is also called the rennet stomach and it is the equivalent of the human stomach. Gastric juices are secreted and some food is digested by the animal's own enzymes. When milk is swallowed by a young calf, into what part of the stomach would you expect it to go?

Although some sugar is produced by the fermentation of the fibre in the stomach, this is used by the fermenting bacteria and not by the ruminant host. The host absorbs fatty acids produced by the bacteria, which are then transformed into glucose by the liver.

Protein in the food is digested by enzymes of the bacteria. Some of the amino acids are absorbed through the wall of the reticulum but most are absorbed by the bacteria for growth and reproduction. The bacteria eventually die and move into the abomasum, where they are digested and become an important source of amino acids for the host. Bacteria also digest some food in the colon of ruminants.

The elaborate process of ruminant digestion allows an animal to use a wide-spread but generally indigestible source of food: fibrous plant material, particularly the grasses. In addition, bacteria can make many vitamins, particularly those in the B group. These are available for absorption by the host animal. Ruminants require vitamins A and D in their diets.

Digestion in a nectar and pollen feeder

Some heterotrophs obtain their food from just one or a few restricted sources and so have a highly specialised diet. The honey possum, *Tarsipes rostratus* (figure 5.29a), is an example of such specialisation.

Adult honey possums are tiny with a mass ranging from 7 to 12 grams. They feed exclusively on the pollen and nectar of native plants growing in heathlands in the south-west of Western Australia. Honey possums have very long brush-tipped tongues with which they lick up pollen and probe deeply into flowers to mop up nectar (see figure 5.29b).

The nectar requires no digestion and provides immediate sugar and water for the honey possum. Combs on the roof of the mouth remove the pollen grains from the brush-tipped tongue. These pollen grains are digested within six hours in a relatively short, simple digestive system. Short digestive systems (see figure 5.30) are typical in mammals that live on plant juices, nectar and pollen.

The energy requirements of a 7-gram honey possum have been estimated at 47 kilojoules per day. If the pollen and nectar from one *Banksia* can provide about two kilojoules of energy daily, about how many such banksias would the honey possum need to visit to meet its daily energy needs?

Figure 5.29 (a) A tiny honey possum on a *Banksia* flower head **(b)** A magnified image of the brush-like tip of the honey possum's tongue

Digestion in carnivores

The digestive system of almost all vertebrates is similar in basic structure with the same essential parts (figure 5.15, page 107) and is the link between an animal's diet and its nutritional needs. We have seen that some herbivores are unable to readily access the nutrients and energy in the plant material they eat and have special features of their digestive systems to provide accommodation for the bacteria that break down the cellulose for them. Carnivores have no such problem.

Carnivorous mammals feed mainly on herbivorous animals. The Tasmanian devil (*Sarcophilus harrisii*, figure 5.7, page 101), the largest of the flesh-eating marsupials, eats any material of animal origin but is primarily a carrion eater. It has powerful jaws and sharp teeth and is able to eat every part of a dead animal, including the skull.

Protein is much easier to digest than cellulose. Hence, carnivores can digest their protein diets relatively easily compared with the digestion of plant material by many herbivores. Carnivores have a relatively shorter digestive system than many herbivores and have a very reduced or small caecum. Figure 5.30 shows the digestive system of a carnivore compared with those from foregut and hindgut-fermenter herbivores.

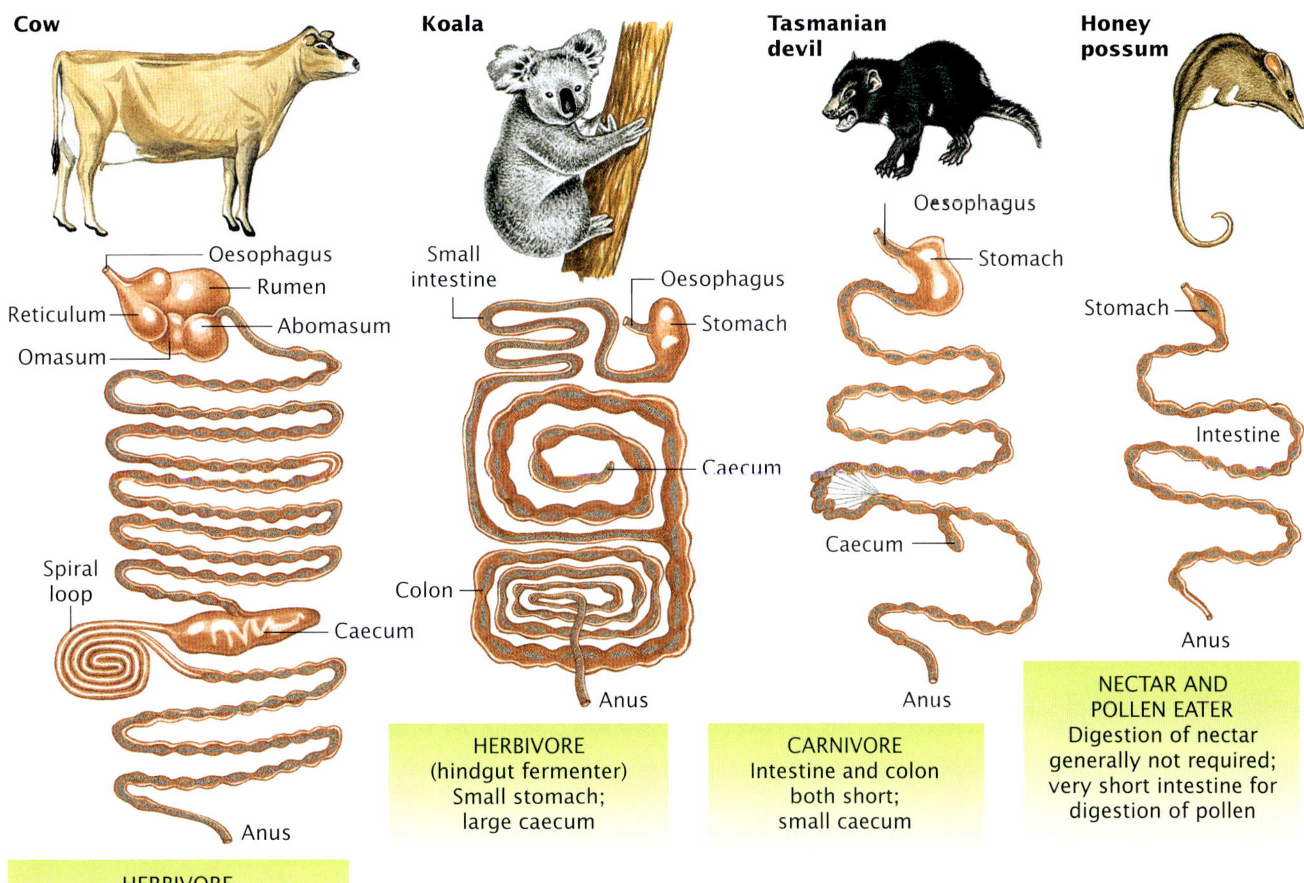

Figure 5.30 Compare the length, and the relative size of the different parts, of the alimentary canal of a carnivore with those of three different herbivores. Each system is designed to cater for the different diets to ensure that each animal receives sufficient energy and nutrients from the food eaten. Note that organs such as the liver and pancreas have been omitted for simplicity.

BIOLOGY IN THE WORKPLACE

Karen Inge — sports dietitian

Figure 5.31 Karen Inge

I work in the area of sports nutrition. I was always interested in working in the health area and with my love of people, fascination with science, passion for food and the fact I was always on a diet, I was destined to be a dietitian.

I started my career as a dietitian at Peter McCallum Cancer Clinic then moved to the Caulfield Community Care Centre where I conducted weight-loss groups and nutrition-education programs for people with diabetes, planned individual diets for people with medical problems and worked with high-risk groups, such as alcoholics, drug addicts and the homeless.

I gave up the work I loved in order to do further study in nutrition at Deakin University and this really changed the direction of my career. I was required to do a research project and by chance was given the opportunity to work with the Collingwood Football Club. They had just lost another Grand Final and were looking at ways to improve their performance.

In 1980, there was little understanding about the role of nutrition in performance. Football players were still eating high-protein, high-fat meals such as steak and fried eggs two hours before the game. People only ate high carbohydrate foods like pasta if they were Italian. Players were thought to be 'tough' if they could get through training or a game without needing a drink. There was no restriction on alcohol after a game and the concept of nutrition recovery was unknown.

My project was assessing the cardiovascular risk profile of football players. In those days we compared the weight of players with the Metropolitan Life Insurance tables to assess whether they were overweight. Because athletes have a higher muscle mass, which weighs more than fat, they were often in the overweight category while being relatively lean. It became important to investigate other methods of assessing body composition to more accurately understand their percentage of body fat and lean body mass. Today we refer to body composition assessment as kinanthropometry.

My first paper presented at the International Nutrition Conference in the United States was on the body composition of Australian Football players. At the conference I met with dietitian colleagues, some of whom were working with Olympic athletes, and gained insights into the fascinating area of sports nutrition.

When I returned to Australia, I set up a private practice at McKinnon Sports Medicine and continued to work at the Collingwood Football Club for seven years. During that time I co-authored *Food for Sport* which was the first Australian book about sports nutrition. Exercise scientists were researching this new and exciting area but dietitians were able to introduce the practical component. Dietitians could help athletes increase their endurance, power and strength as well enhance their overall performance.

Following the success of *Food for Sport* and my work with elite athletes around the country, I left Collingwood to work with the Hawthorn Football Club at the end of the 1987 season. The eight years with Hawthorn were great. In the first two years, Hawthorn won back-to-back premierships. The players were fit, highly skilled and extremely talented. My role was to ensure that their body fat levels were low to help improve their endurance and speed. Players were encouraged to eat appropriate high carbohydrate, low fat meals and to monitor their blood levels for specific nutrients, such as iron. I introduced pre-game eating and hydration strategies as well as recovery regimes for after games and training. I spent three nights a week at the club and attended every game. Even for interstate trips, plane and hotel meals were planned.

The success of the Hawthorn Football Club enhanced my career as a dietitian. I received job offers from food companies. I worked for the Ricegrowers Cooperative to develop a nutrition program for children to complement a swimming program that the Cooperative had sponsored. I have worked with a range of food companies, for example Uncle Tobys, Jalna Dairy Foods, SPC Ardmona and the Kiwifruit industry, as their media spokesperson and helping them develop healthier food products.

I continue to be involved with elite athletes – those from the Victorian Institute of Sport for 15 years, the women's artistic gymnastics team for 10 years, the Opals basketball team in preparation for the 1996 and 2000 Olympics as well as the Australian Ballet Company. Individuals and different activities have specific nutritional requirements.

My interest in children's nutrition is reflected in my latest book, *Let's Eat Right for Kids!*. It's all about translating the science of nutrition into the foods we need to eat that keeps me stimulated. That is the nature of science — it is continually challenging the thinking of the day and that's why I love what I do.

Accessing energy in organic compounds

When you see the term 'respiration', remember that it may refer to the process of cellular respiration, but it may also refer to the process of breathing by a terrestrial vertebrate.

Digestion of plant and animal materials results in simple molecules, including glucose, being absorbed into the bloodstream and transported to all cells.

How do cells access the chemical energy that is stored in these molecules? The transfer of chemical energy from organic compounds such as glucose to ATP occurs through a series of chemical reactions, most of which occur in mitochondria (pages 32–3). These reactions are referred to as **cellular respiration.**

The energy required to drive the reactions in cells comes from ATP.

Cellular respiration occurs *all the time* in the cells of all living things — plants, animals, fungi, protists and bacteria. Although cellular respiration comprises a number of biochemical reactions, when it takes place in the presence of oxygen it can be summarised by the equation:

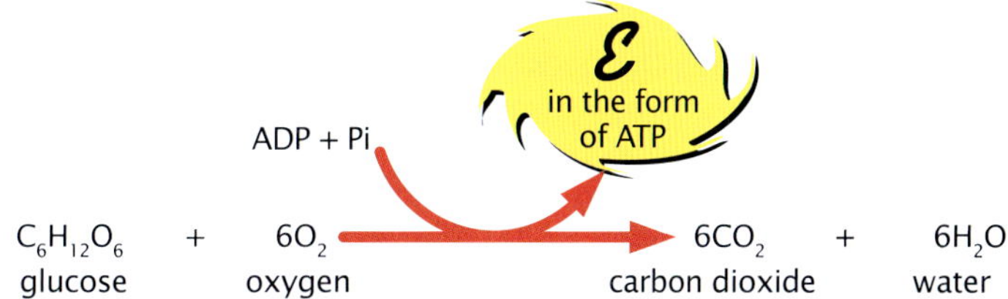

Figure 5.32

What are the 'inputs' of cellular respiration? What are the 'outputs'?

The process of energy transfer from glucose to ATP is not 100 per cent efficient. In general, about 40 per cent of the chemical energy present in glucose is transferred to ATP and the remaining 60 per cent appears as heat energy.

KEY IDEAS

- Specialised parts of the digestive system vary in different animals.
- Foregut-fermenters (herbivores) have very large and complex stomachs.
- Herbivores that are hindgut-fermenters have large caecums.
- Nectar eaters have relatively short, simple digestive systems.
- Carnivores have relatively short alimentary canals compared with those of herbivores.
- Chemical energy stored in organic matter is converted to ATP by cellular respiration.

QUICK-CHECK

13 Humans and cows are both mammals and yet they have quite differently structured stomachs. Explain the importance of these differences.

14 Why do some animals such as honey possums have relatively short digestive tracts?

15 Why are carnivores able to gain sufficient nutrients and energy from the food they eat even though their alimentary canals are relatively shorter than those of many herbivores?

16 Explain why photosynthesis and cellular respiration are essential for all ecosystems.

BIOCHALLENGE

1 Examine the following images. Identify any autotrophic and any heterotrophic organisms.

a

b

c

d

e

2 Name the inputs and outputs labelled A-E below.

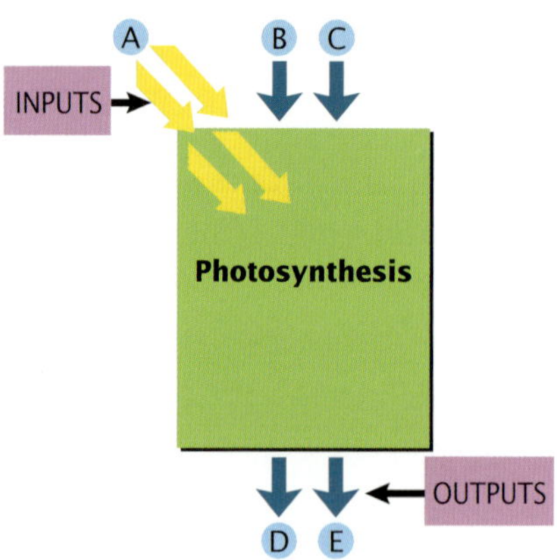

3 The diagram below represents a mixture of two different enzymes and four different kinds of complex compound.

a Explain which compound enzyme 1 can associate with.

b Explain which compound enzyme 2 can associate with.

c Select one of the compounds not acted on by either of the enzymes shown. Draw an enzyme that you think would catalyse its breakdown. Is it possible to have a different drawing from another student and still be correct? Explain your answer.

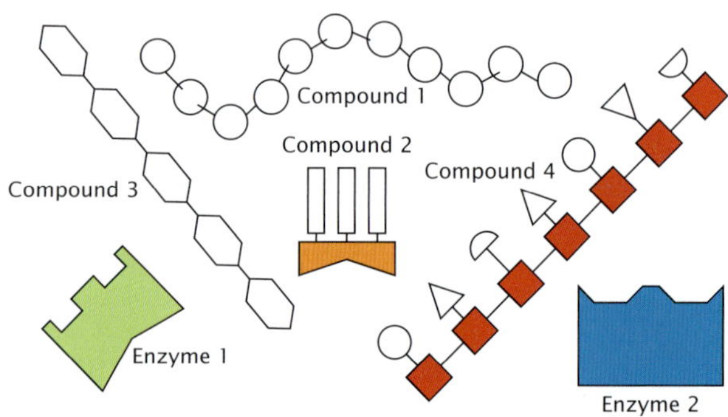

4 a What is the summary equation for aerobic cellular respiration?

b What does the following statement mean? 'Energy transfer in cellular respiration is less than 100% efficient.'

c Outline what happens to heat energy that is released during cellular respiration.

CHAPTER REVIEW

Key words

CROSSWORD

absorption
alimentary canal
amylase
anus
autotrophic
autotrophs
bile
bile duct
bolus
carnivores
cellular respiration
chloroplasts
chyme
colon
cystic fibrosis
digestion
digestive system
duodenum
emulsification
enzymes
faeces
foregut fermenters
gall bladder
gastric juice
herbivores
heterotrophic
heterotrophs
hindgut fermenters
ileum
jejunum
lacteals
large intestine
liver
mucus
oesophagus
omnivores
pancreas
pepsin
peristalsis
photosynthesis
prey
pyloric sphincter
rectum
ruminants
SA:V ratio
small intestine
spawning
stomach
villi

Questions

1 *Making connections* ▸ Use at least six of the key words from this chapter to construct a concept map.

2 *Applying your understanding* ▸ Indicate whether each of the following organisms is autotrophic or heterotrophic.

a a lemon tree

b the ringworm fungus of a dog

c a breastfed baby

d a red seaweed

3 *Communicating your understanding* ▸

Refer to figures 3.19 (page 67) and 5.5 (page 98).

a List three characteristics of leaves that facilitate their ability to photosynthesise.

b The stomata of some plants close for an hour or more at about midday. Explain whether there is likely to be any advantage for the plant.

c When plants are removed from soil and transplanted elsewhere, they often wilt for a few days even when a ready supply of water is available. They generally recover after a few days. Explain this observation.

4 *Communicating your understanding* ▸ Photosynthesis occurs in two stages, a light-dependent stage and a light-independent stage.

a What is the equation for each stage?

b Combine the equations of part a to demonstrate how the simplified overall equation of photosynthesis is obtained.

c In what way do the products of each stage serve the other stage?

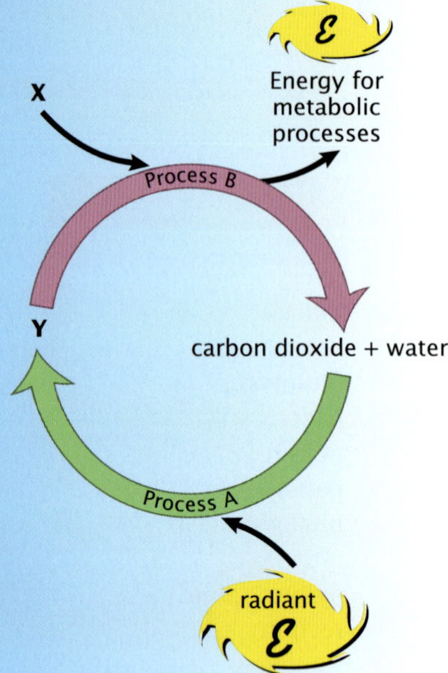

Figure 5.33

5 ***Analysing information and drawing conclusions*** ▸ The diagram in figure 5.33 is sometimes called the 'biology energy wheel'. Process A is photosynthesis.

a Name process B.

b Name the input element X.

c Name the output product at Y.

d In what form must energy be before it can be used for metabolic processes of a cell?

e With reference to ecosystems, explain how what is represented in this diagram serves as a 'biology energy wheel'.

6 ***Communicating biological information*** ▸

a What is the role of mechanical digestion in vertebrates?

b What is the role of chemical digestion in vertebrates?

7 ***Applying understanding to new contexts*** ▸ The diagram in figure 5.34 shows two structures, one found in animals and the other in plants.

a A student claimed that both structures must have a role in absorption. Explain what characteristic of the structures would lead to a student making the claim for the particular function.

b Outline two ways in which the role of each of the structures differ from each other.

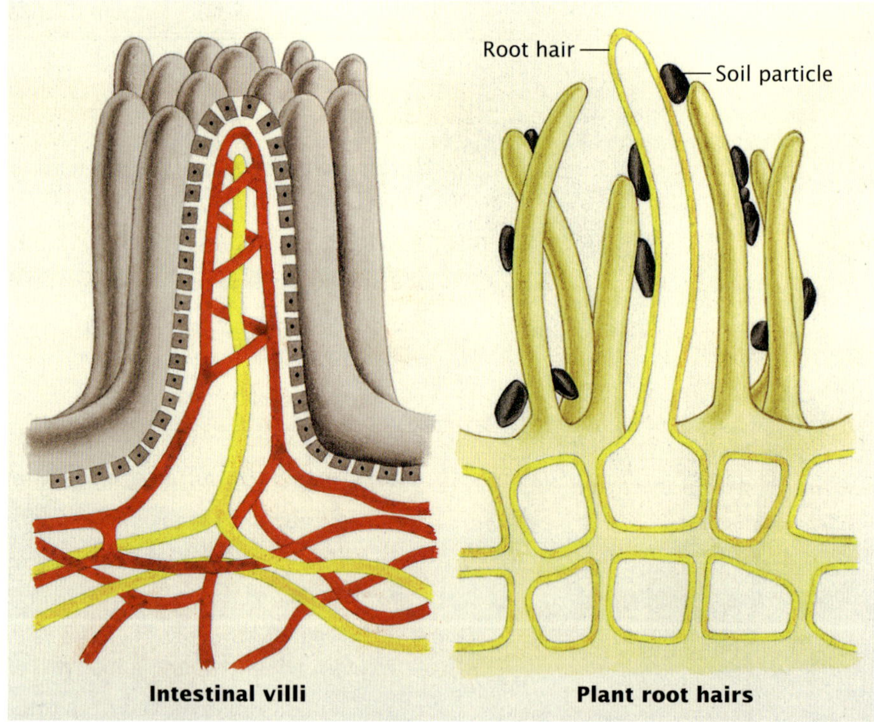

Figure 5.34

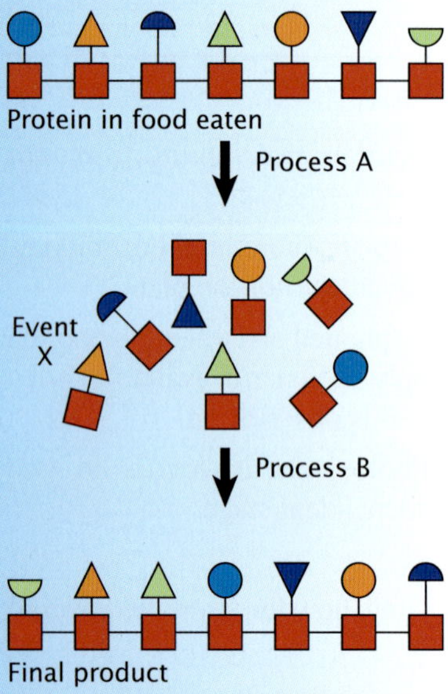

Protein in food eaten

Process A

Event X

Process B

Final product

Figure 5.35

8 ***Applying understanding to familiar and new contexts*** ▸ Look at figure 5.35.

a What class of compounds do the symbols in the protein represent?

b Where does process A occur in the body?

c What kinds of compound assist in this process?

d What happens to the compounds represented at event X?

e Where in the body does process B occur?

f How would you account for the difference in the sequence of the symbols in the final product compared with the sequence in the protein in the food eaten?

g Would other sequences be possible? Explain.

9 *Analysing information and communicating ideas* ▸ Some plant cells store starch that is later broken down into glucose as required by the plant. Explain whether or not you think this breakdown is digestion.

10 *Communicating ideas* ▸ Is bile an enzyme? Explain.

11 *Applying your understanding* ▸ Sometimes a person's gall bladder is surgically removed. What impact would this have on the person's digestion and how might the person have to modify eating habits?

12 *Applying understanding to new concepts* ▸

a A new animal is found and it has a relatively short alimentary canal. Discuss what kinds of food the animal is most likely to eat.

b Spiders and insects are sometimes called miniature meat-eaters. What predictions would you make with regard to the kinds of digestive enzyme they produce and the relative need for each kind?

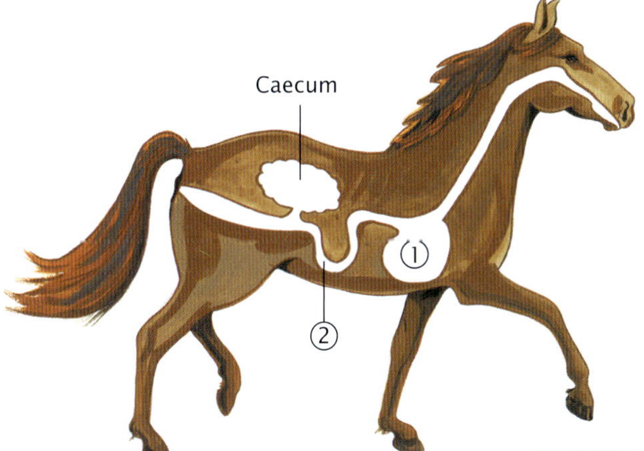

Figure 5.36

13 *Analysing information and drawing conclusions* ▸ Figure 5.36 shows the outline of the digestive system of a horse.

a What are structures 1 and 2?

b Note the size of the caecum. What predictions would you make about differences and similarities in the digestive processes of a horse and a cow, both grass-eating animals?

c Refer to figure 5.30 (page 121). What is one structural difference between the alimentary canal of a horse and that of a carnivore? What is the significance of that difference?

14 *Interpreting data* ▸ The changes in percentage composition of the solid fractions of the milk of the tammar wallaby, *Macropus eugenii*, are shown in figure 5.37.

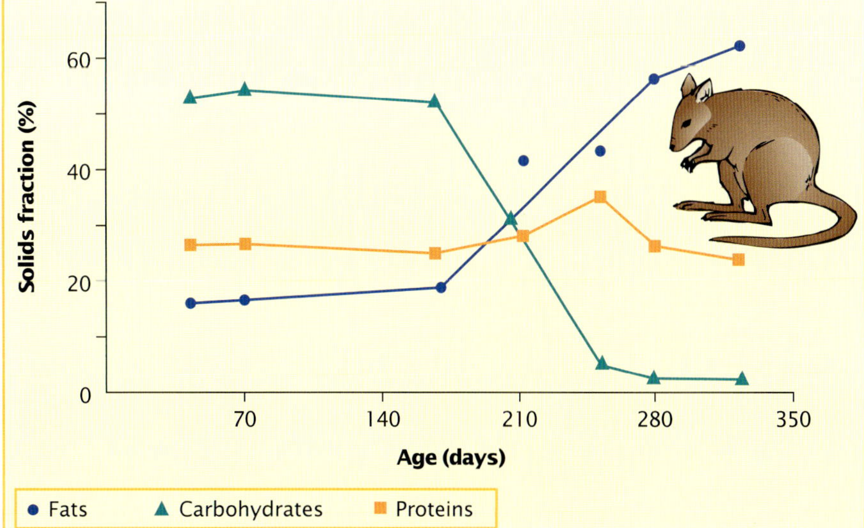

Figure 5.37 Changes in milk of the tammar wallaby during development of an offspring, showing relative proportions of fats, proteins and carbohydrates

a What is the most common solid component in milk 70 days after birth of the young wallaby?

b What is the most common solid component in milk 280 days after birth of the young wallaby?

6 Distribution of materials

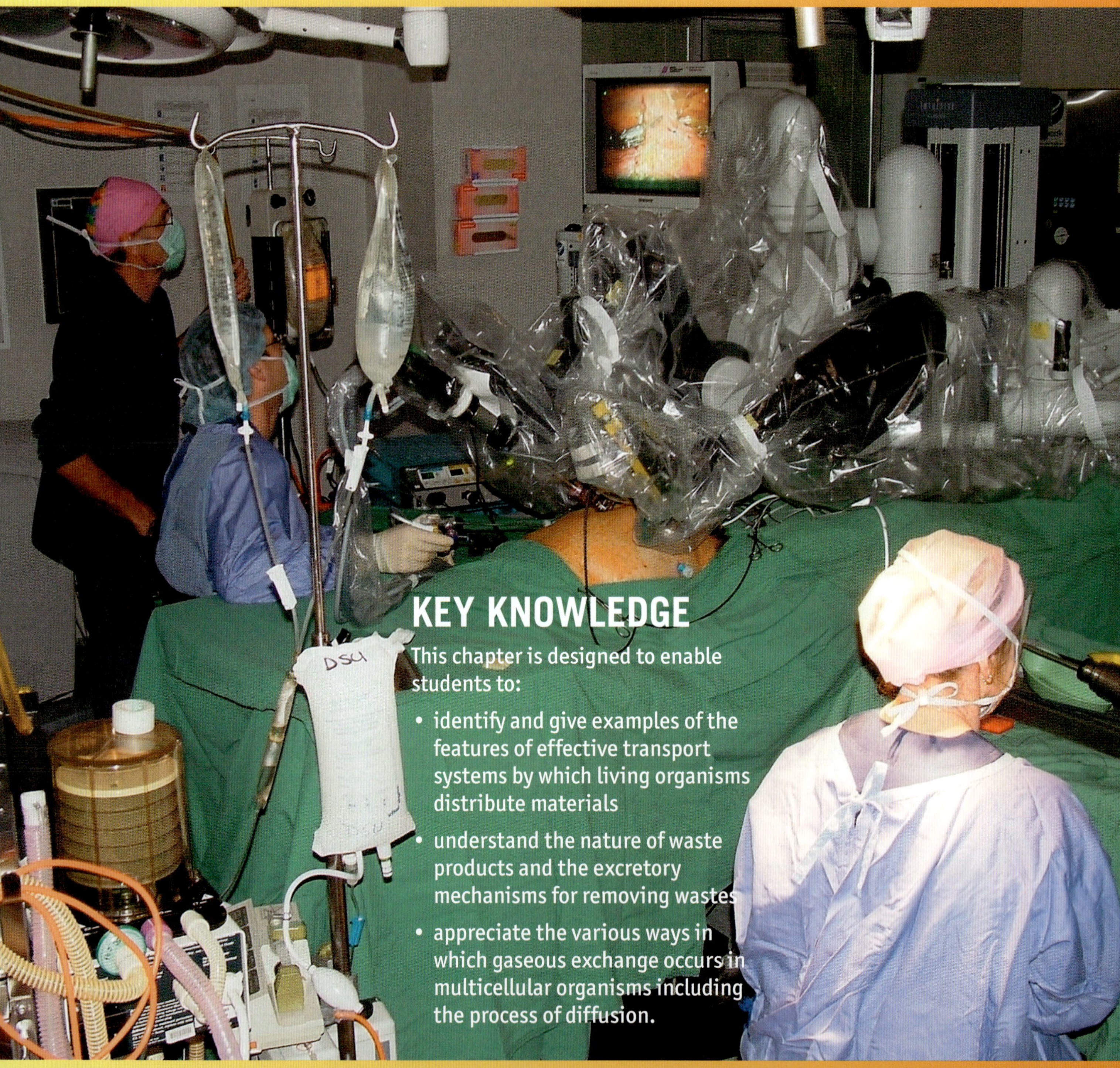

KEY KNOWLEDGE

This chapter is designed to enable students to:

- identify and give examples of the features of effective transport systems by which living organisms distribute materials
- understand the nature of waste products and the excretory mechanisms for removing wastes
- appreciate the various ways in which gaseous exchange occurs in multicellular organisms including the process of diffusion.

Figure 6.1 In 2004, Epworth Hospital, Melbourne, was the first Australian hospital to install a da Vinci Surgical System that allows robotic assistance to perform invasive surgery through small incisions. Note the 'robot' wrapped in a variety of sterilised plastic bags, the patient's abdomen into which three small tubes, or ports, have been inserted and, from left, the technician, assistant surgeon and scrub nurse. The surgeon is in a console out of sight in this picture. Initial use of the Epworth system is for heart and urogenital surgery. In this chapter, we consider transport systems in organisms, including the role of the heart in animals. We also discuss the role of those systems in the transport of nutrients, gases and wastes.

Robotic surgery: risk reduction

When a surgeon performs surgery as if looking into a mirror, it is called 'counter intuitive'.

All surgery involves some risk. This is particularly true of traditional open surgery because of the intrusive nature of the procedure. Recent techniques such as laparoscopy have reduced those risks. Laparoscopy is a technique in which a fibre optic instrument is inserted through a small hole in the abdominal wall, or into other structures such as knee joints, to allow small-scale surgery. However, it has two drawbacks: a surgeon has to perform the surgery as if looking into a mirror and receives a two-dimensional image only.

The da Vinci Surgical System cost $3 million. View more photographs of the system and the instruments at www.intuitivesurgical.com (select 'Products').

The most recent developments in surgery (figure 6.1) involve the latest advances in robotics and computer-enhanced technology. The surgeon, who sits in a console (figure 6.2a) and receives a three-dimensional view because a stereoscopic camera has been inserted through one of the ports, can operate with hands moving as they would in open surgery. The assistant surgeon monitors the position of the camera to ensure that the surgeon obtains the best view of the surgery site. The scrub nurse also monitors the surgery (figure 6.2b) because she must manage the sterilised instruments and have appropriate ones ready to make changes as the operation proceeds. The instruments are reduced in size and have been modelled on movements of a human wrist to ensure maximum flexibility. They are inserted through two ports close to the surgery site and the third port carrying the camera.

(a)

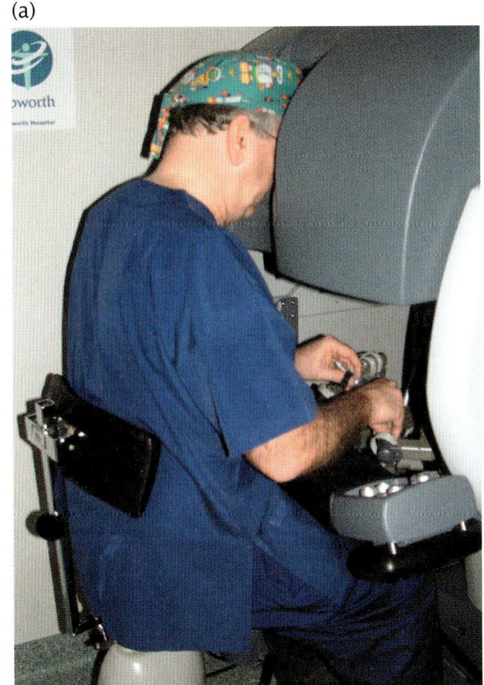

(b)

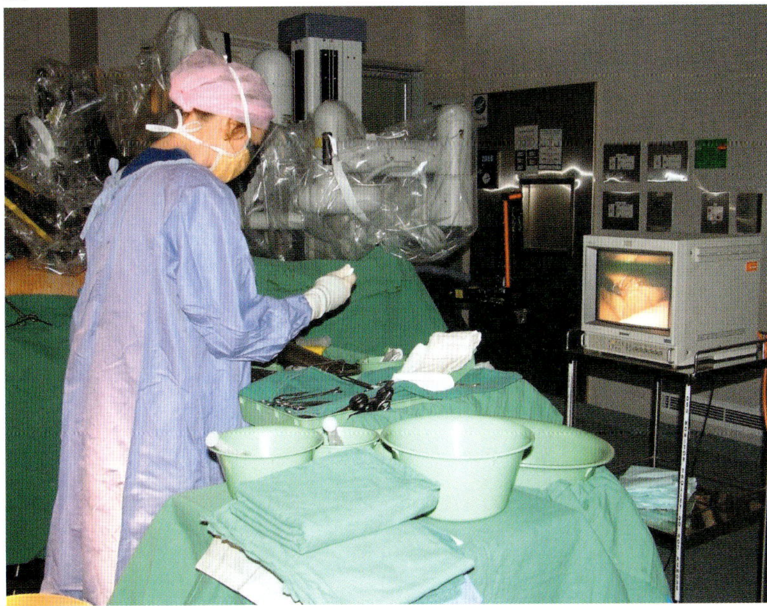

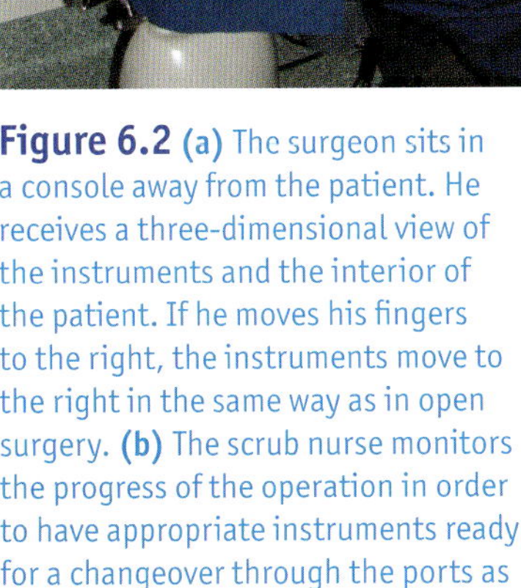

Figure 6.2 **(a)** The surgeon sits in a console away from the patient. He receives a three-dimensional view of the instruments and the interior of the patient. If he moves his fingers to the right, the instruments move to the right in the same way as in open surgery. **(b)** The scrub nurse monitors the progress of the operation in order to have appropriate instruments ready for a changeover through the ports as required.

The nature of the surgery is such that there is great benefit for a patient. There is some variation depending on the type of operation but benefits include:

- less time in hospital; 2–3 days instead of 8–12 days
- faster recovery; return to normal activity in 10–14 days instead of 4–6 weeks
- far less blood loss and less need for transfusions
- less post-operative pain, post-operative complications and scarring.

Because a surgeon is seated with head resting against a pad in the console, long operations are less exhausting. In addition, the system is constructed so that any slight tremor occurring in the surgeon's hands is virtually removed.

Although the surgeon sits at a console in a corner of the robotic operating theatre at Epworth Hospital, surgery could be performed from remote locations. The original concept for the system developed from ideas about remote surgery for wounded soldiers and surgery that may be required by travellers in space.

BIOTECH

A modern operating theatre

A non-robotic modern operating theatre is also a complex biotech facility. Figure 6.3 shows an operating theatre with heart surgery in progress.

1. **Hewlett Packard monitor** supplies details to the anaesthetist and displays ECG and invasive monitoring pressures
2. **anaesthetist**
3. **syringe pumps** add heparin or other chemicals to the blood line
4. **myocardial temperature probe** measures temperature of myocardium muscle
5. **surgical draping**
6. **surgeon**
7. **headlight** ensures optimal light on target organ for surgeon
8. **instrument nurse**
9. **circulating nurse**
10. **assistant surgeon**
11. **cathode lab monitor** provides visual data of heart rhythms of patient undergoing angioplasty
12. **preparation room** contains a trolley prepared for the operative procedure
13. **telephone**
14. **whiteboard** where patient details are recorded throughout the procedure (e.g. age, weight and allergies)
15. (a) **clock**
15. (b) **emergency clock** measures tourniquet or clamp time accurately
16. **light switches** for overhead surgical lights
17. **X-ray display**
18. **writing desk**
19. **linen and rubbish trolleys**
20. **headlight power source**
21. **monitor to bypass pump** which is a computer control providing visual display of pressures
22. **saturation monitor** displays venous saturation and haematocrit of blood
23. **vaporiser** adds anaesthetic agent as required
24. **CO_2** used to blow-dry grafts
25. **perfusionist** monitors and regulates fluid movement between patient and machinery
26. **heart–lung bypass machine** replaces function of heart and/or lungs during operation
27. **blood warmer**
28. **surgical lights**
29. **pendent** provides access to gas and electrical power supplies
30. **slave monitor**
31. **labels**
32. **syringes** for blood gases and heparinisation
33. **cell saver** spins out blood, separating fat and blood for reinfusion
34. **cardioplegia solution** to achieve systemic hypothermia

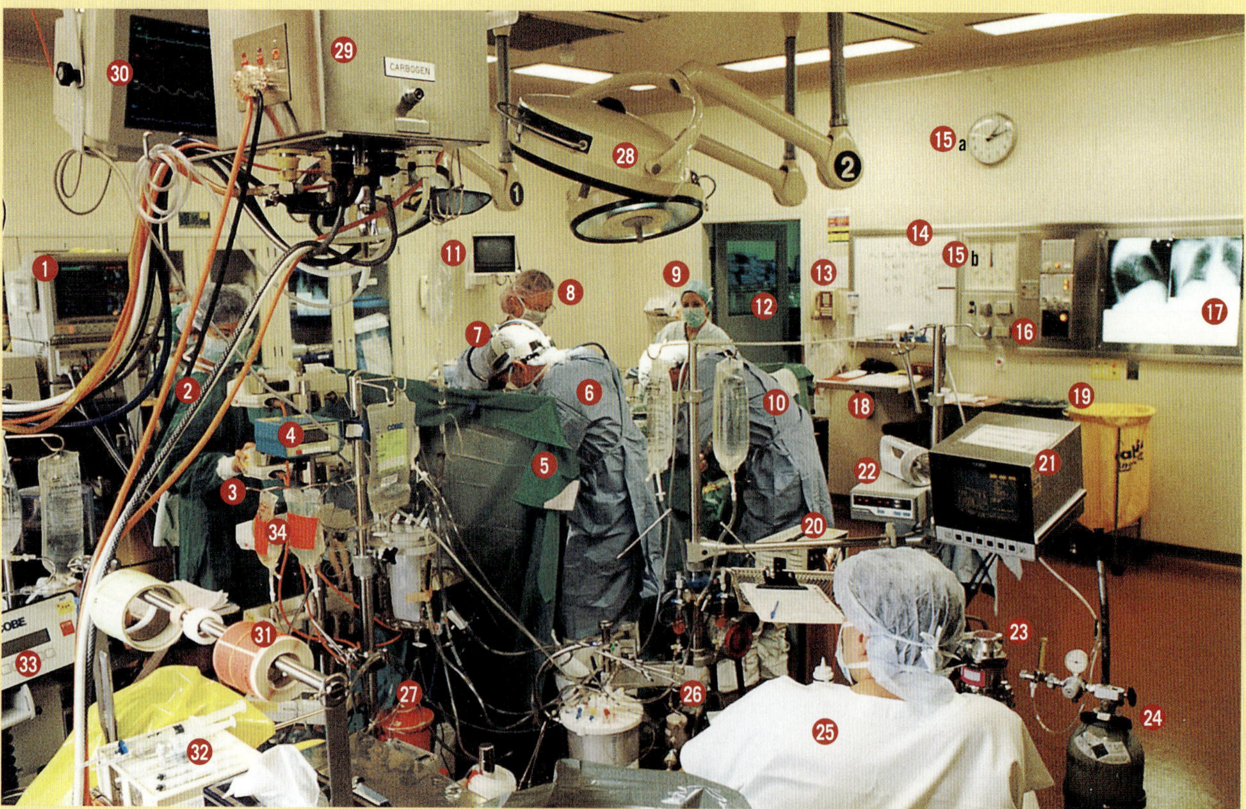

Figure 6.3 Equipment in a modern non-robotic operating theatre. Why do you think a blood warmer is required?

Transport systems

Michael Servetus (1511–1553), a Spaniard who postulated the existence of a pulmonary circulation in 1553, was burnt to death at the stake in Geneva for his heresy.

In chapter 5, we considered the formation and preparation of nutrients for plants and animals. The oxygen that moves from the external environment across membranes within the respiratory systems must be transported throughout an organism and then cross membranes into living cells. Also, the carbon dioxide produced by those cells must be transported to respiratory membranes that act as excretory membranes in releasing the gas into the external environment.

In this chapter, we consider the systems responsible for the transport of oxygen and carbon dioxide in animals and plants as well as the transport of water, nutrients and a range of other materials. We also consider the origins and transport of nitrogenous metabolic wastes from cells to the excretory systems that expel them from an animal's body. Excretion in plants is contrasted with that of animals.

Blood circulatory system of mammals

In higher animals, transport of materials both to and from the cells of tissues is generally carried out by a **blood circulatory system**. What are the characteristics of this system that facilitate its functions?

The blood circulatory system is the mechanism that delivers nutrients and oxygen to all cells of a multicellular organism. The circulatory system consists of:

- blood vessels through which blood is transported
- the heart, a muscular pump that keeps the blood moving.

We will consider the circulatory system of humans which is a system typical of all mammals.

Prior to 1616, the popular belief was that blood moved forwards and backwards along the same blood vessels within the body. In 1616, Dr William Harvey of St Bartholomew's Hospital in London demonstrated that blood leaves the left side of the heart and is pumped to all parts of the body, excluding lungs, before it returns to the right side of the heart. From there it goes to the lungs and back to the left side of the heart and from there it is once again pumped around the body.

Blood travels through the heart twice during each circulation of the body: once after circulating through the pulmonary system (the lungs) and once after circulating through the remaining parts of the body (figure 6.4).

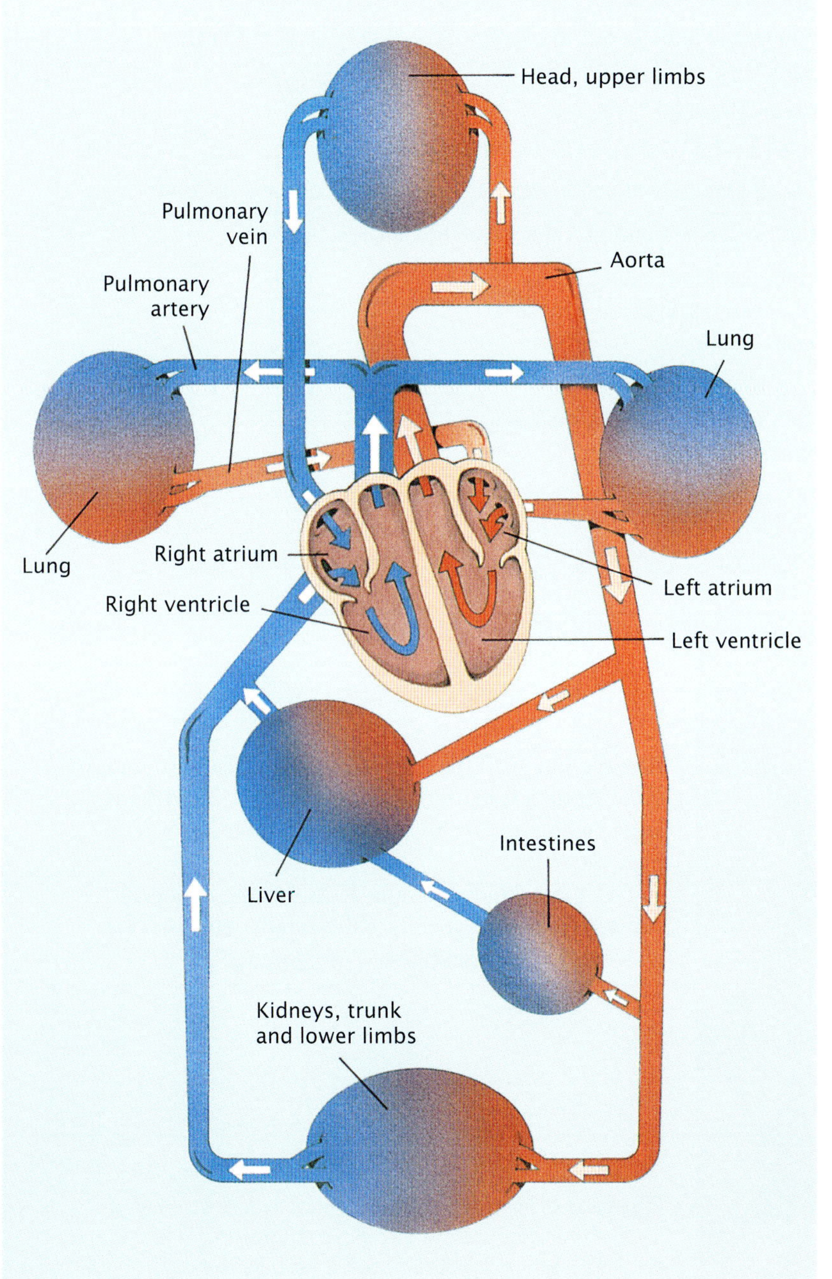

Figure 6.4 Main routes for blood circulation. Arrows within vessels indicate the direction of blood flow. Arteries branch into smaller vessels called arterioles, then capillaries. From capillaries, blood flows through venules into veins and is transported back to the heart.

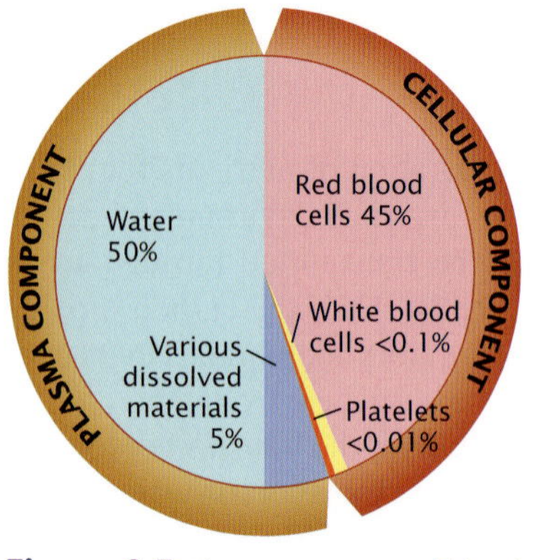

Figure 6.5 The components of blood

Components of blood

Blood is made up of a fluid part, called plasma, and cells suspended within this fluid (figure 6.5).

Plasma

Blood is about 55 per cent plasma, which is itself about 90 per cent water. The balance of 10 per cent of the plasma is made up of the various compounds and molecules dissolved in the water part. The materials transported by plasma and the significance of each are summarised in table 6.1.

Table 6.1 Materials transported by plasma and the significance of each

Component of plasma	*Significance*
water	solvent for non-cellular materials in blood suspends cellular components
nutrients: • glucose • amino acids • fatty acids • vitamins	supply: • source of energy for cell • building material for renewal of cell components
nitrogenous waste and carbon dioxide	produced during metabolism and must be removed from cells and the body
oxygen	small amount carried in solution (refer to red blood cells, page 133)
plasma proteins such as: • albumin • globulins • fibrinogen	• bind hormones and fatty acids • antibodies — react with foreign material • acts in blood clotting
ions such as: • sodium • calcium • chloride • magnesium • potassium • bicarbonate	contribute to a variety of actions such as: • stabilising pH • osmotic balance • regulation of permeability of plasma membranes
hormones	transported from sites of manufacture to particular target cells or organs around the body

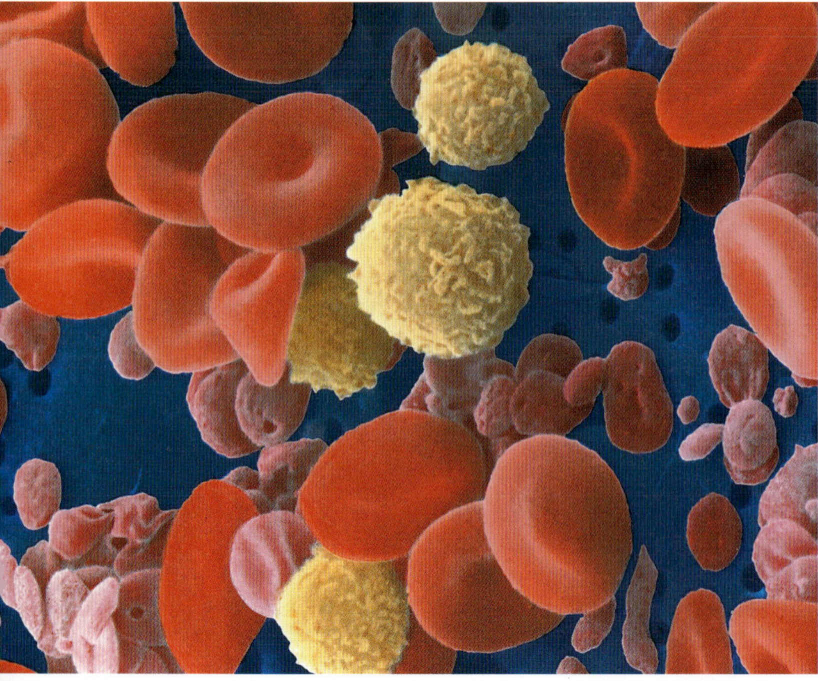

Blood cells

Blood contains three different kinds of cells (figure 6.6). These are red blood cells, also called **erythrocytes**, white blood cells, also called **leucocytes**, and platelets.

The three kinds of cells found in the blood are derived from special cells, called **stem cells**, in bone marrow. This is summarised in figure 6.7.

Figure 6.6 Photomicrograph of human blood. Note the many red blood cells, white blood cells (yellow) and platelets (pink). Platelets are specialised fragments of larger cells.

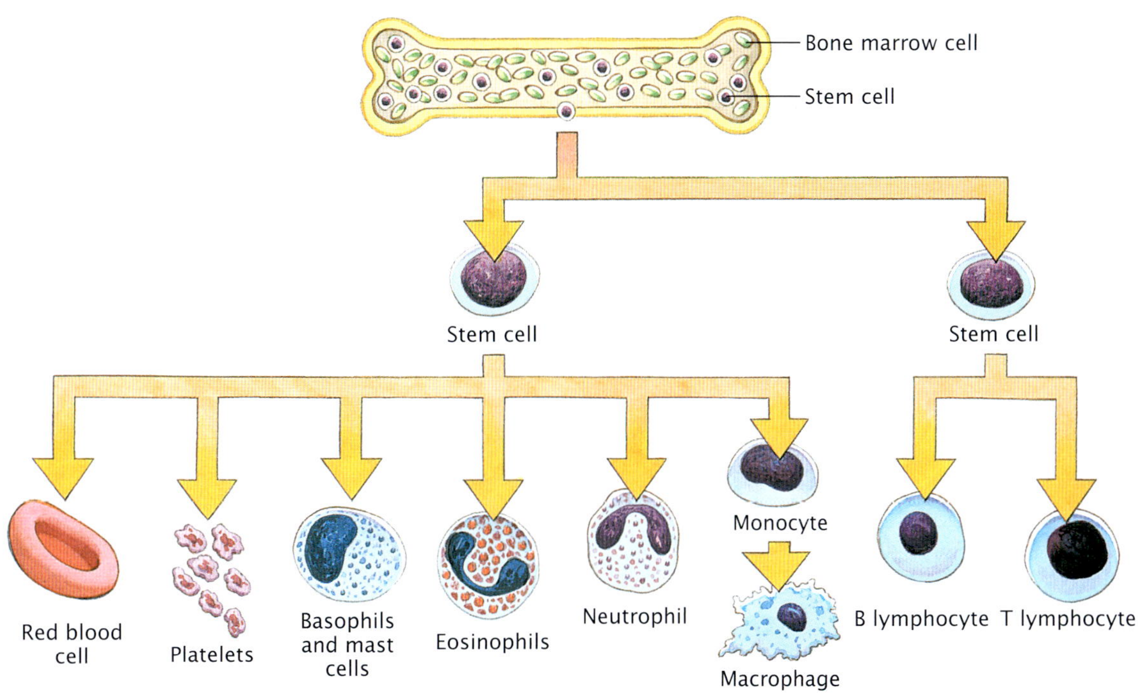

CELL TYPE	Red blood cell or erythrocytes	Platelets	White blood cells or leucocytes
FUNCTION	Transport oxygen and to a lesser extent carbon dioxide	Important in blood clotting	Active in defence against infection and in the immune system
NUMBER (per mm^3 of blood) in healthy person	Typical male: 5.4×10^6 Typical female: 4.8×10^6	250 000 to 500 000	5000 to 7000
SURVIVAL TIME in healthy person	About 120 days	About one week	May survive for several days. If person has an infection may survive for only a few hours.

Figure 6.7 Blood cells all develop from special cells, called stem cells, in bone marrow. Stem cells continually reproduce by mitosis and then differentiate.

Red blood cells

Red blood cells contain haemoglobin, a red iron-containing protein that readily combines with oxygen to form oxy-haemoglobin. This is the form in which virtually all oxygen is transported to all cells of the body.

Males generally have a higher metabolic rate than females and hence have a greater need for energy and hence oxygen. The blood of a male carries about 10 per cent more red blood cells than the same volume of blood in a female.

ODD FACT

The blood volume of an average-sized adult male is from five to six litres compared to the four to five litres of an average-sized adult female.

White blood cells

The general function of white blood cells is to combat infection. They can move out of the bloodstream, in much the same way as amoebas move and crawl, through connective and epithelial tissues. Many white blood cells are phagocytic, which means that they ingest bacteria and other foreign material.

ODD FACT

People who require a bone marrow transplant (for example, people with types of cancer or leukaemia) can be given cord blood (i.e. blood drained from an umbilical cord after the birth of a baby) as it it rich in bone marrow cells. About 100 mL of this blood is drained from each cord.

Platelets

Platelets are specialised fragments of larger cells and play an important role in the clotting of blood when damage occurs to blood vessels.

Vessels to transport blood

Blood is transported around the body in three different kinds of blood vessels. These are the larger vessels (arteries and veins) which are linked by the smallest of blood vessels — capillaries.

Arteries

Arteries are thick-walled vessels that carry blood away from the heart. Arteries branch into smaller and smaller arteries. The smallest arteries are called arterioles. Arteries have thick walls to withstand the pressure of the blood as it is pumped from the heart but their walls are far too thick to allow the ready movement of materials and gases between blood and tissue cells. A much thinner barrier is required if diffusion is to occur. As arterioles enter tissues, they branch into microscopic vessels called capillaries (see figure 6.8).

Although the flow of blood into capillary networks is controlled by arterioles, additional control occurs through the contraction of smooth muscle cells that act as **sphincters** in the networks (see figure 6.8).

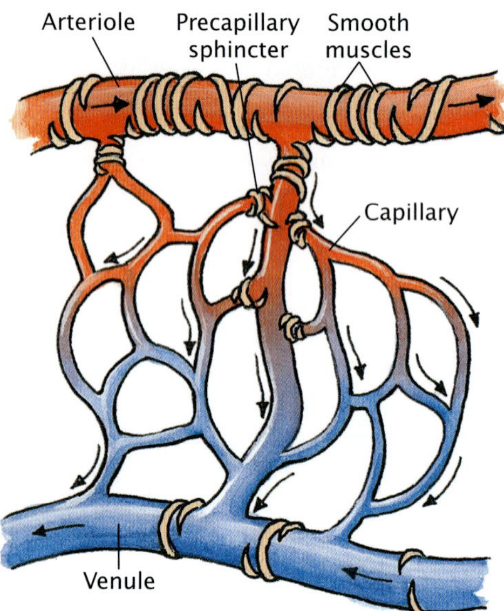

Figure 6.8 A capillary bed branching from an arteriole and ultimately feeding into a venule. Note the smooth muscle cells, called precapillary sphincters, that regulate blood flow through the capillary network.

Capillaries

Capillaries are thin-walled vessels of between 5 and 8 μm diameter. Their walls are one cell thick and materials either pass through these cells or pass between the cells as they leave or enter the blood (see figure 6.9).

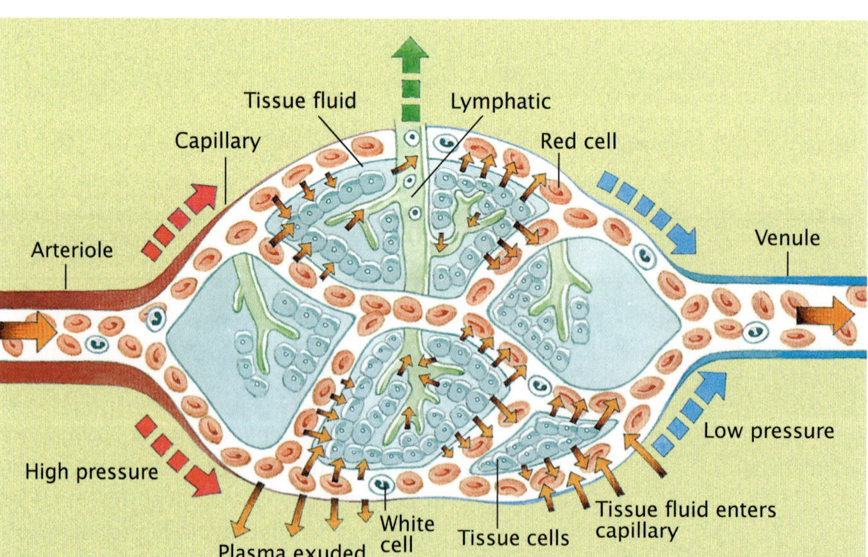

Figure 6.9 The relationship between blood capillaries, tissues and lymphatics. The diagram also indicates the movement of material between capillaries and tissue cells.

Refer to the drawing and cross-section of a capillary in figures 6.10 and 6.11. Oxygen diffuses through the endothelial cells into surrounding **tissue fluid** and cells. Oxygen moves down a concentration gradient. Carbon dioxide makes the reverse journey, also down a concentration gradient. Water and small water-soluble molecules, such as glucose and inorganic ions, diffuse through gaps between the endothelial cells. Some proteins leave the capillaries through endothelial spaces or across the cells in vesicles. Most protein stays in the blood vessels. **Phagocytic** white blood cells can squeeze between endothelial cells.

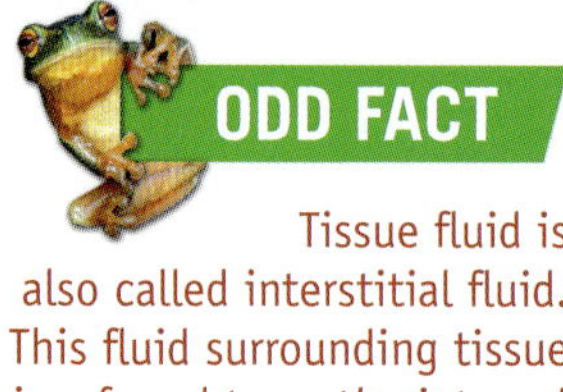

Tissue fluid is also called interstitial fluid. This fluid surrounding tissue is referred to as the internal environment.

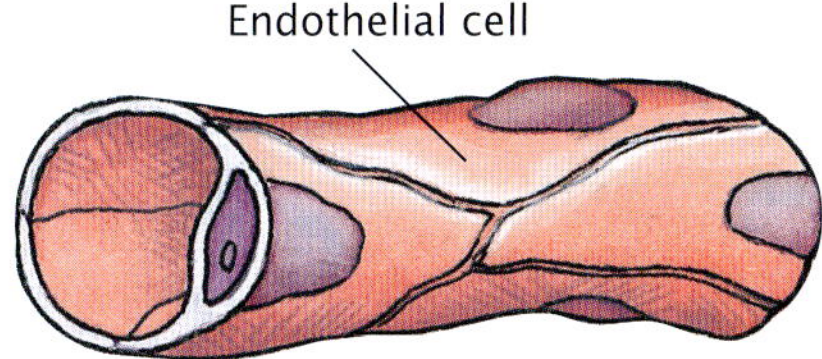

Figure 6.10 The wall of a capillary is made of a single layer of flattened endothelial cells.

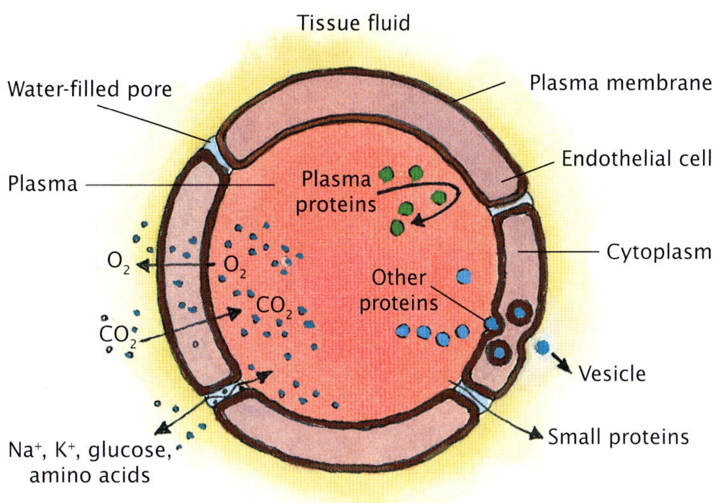

Figure 6.11 Cross-section through a capillary. Note the different ways in which different materials leave and enter a capillary.

Veins

Blood moves from the capillaries into **venules**, which combine to form larger vessels called **veins**. These are relatively thin-walled vessels that transport blood back to the heart (see figure 6.12).

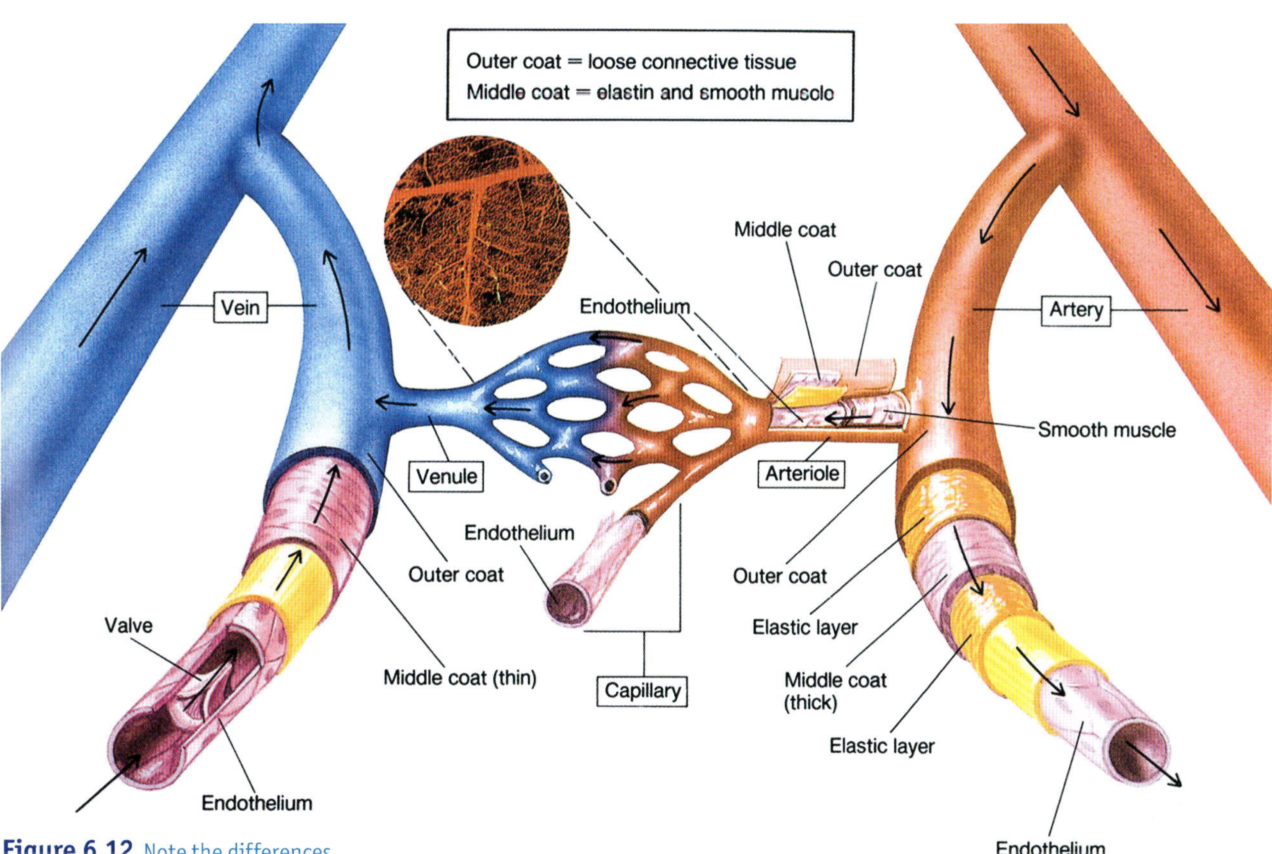

Figure 6.12 Note the differences in structure of the walls of the five different kinds of blood vessels: artery, arteriole, capillary, venule and vein. In what way is the structure of each kind of vessel related to its function?

Blood in veins is under lower pressure than blood in arteries because it has travelled further from the heart. Veins have valves that prevent backflow of blood (see figure 6.13, page 136). Contraction of muscles near veins also helps to squeeze the veins and move the blood along.

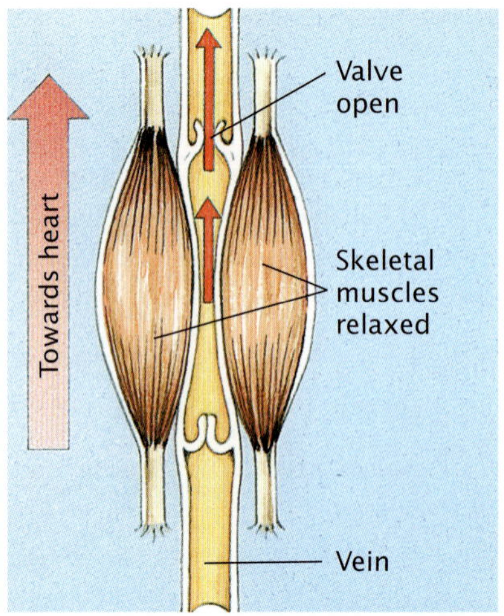

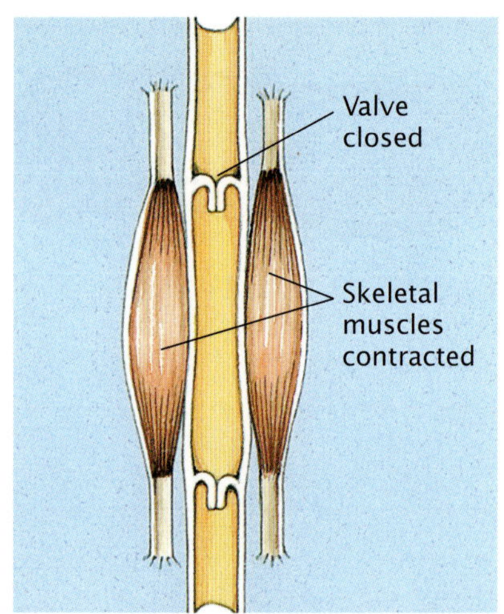

Figure 6.13 Valves in veins prevent a backflow of blood in those vessels. The contraction of muscle alongside the vessels helps to move the blood towards the heart.

KEY IDEAS

- Blood is an important transporting tissue.
- The blood of mammals is made up of several different components.
- Arteries are thick-walled vessels that transport blood away from the heart.
- Veins are relatively thin-walled vessels that transport blood to the heart.
- Arteries and veins are connected by networks of arterioles, very thin-walled vessels called capillaries and venules.
- Nutrients and oxygen diffuse from capillaries into tissue fluid then into surrounding tissue cells.

QUICK-CHECK

1 What is the function of red blood cells, white blood cells and platelets?
2 Name two components that are transported in solution by the plasma.
3 Why do arteries have thicker walls than veins?

The heart

Blood must be continually on the move — to collect oxygen from the lungs, to collect nutrients from the intestines, to transport the nutrients and oxygen to all the tissues of the body, and to carry carbon dioxide and other wastes away from cells to those specialised regions where they are disposed of by the body. The **heart** is the pump that keeps the fluid moving.

The heart can be thought of as two pumps joined together (see figure 6.14). The right side of the heart receives blood from the head and other parts of the body and pumps it to the lungs. Because the blood has come from tissues, it will be relatively low in oxygen and high in carbon dioxide.

In the lungs, carbon dioxide is released from the blood. Oxygen is absorbed from the air in the lungs and combines with the haemoglobin of the red blood cells.

The left side of the heart receives blood from the lungs and pumps it to all other tissues of the body via the main artery known as the **aorta**.

A force from behind is the main factor responsible for the continual movement of blood. A beating heart provides this force. If a heart stops beating, blood stops flowing.

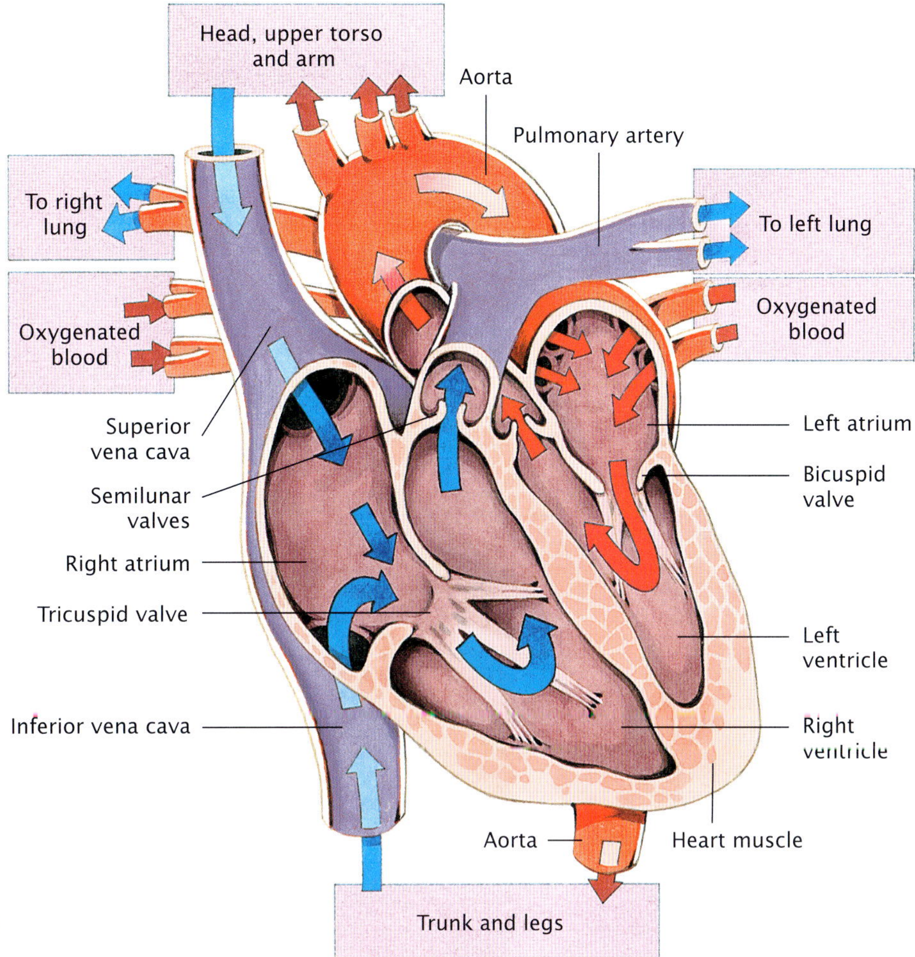

Figure 6.14 The human heart is typical for mammals. The pressure generated when the muscle of the thick wall of the ventricles contracts forces blood from the heart to the lungs and other parts of the body. Valves in the heart prevent backflow of blood into the atria when the heart contracts.

Inside the heart

The right and left sides of the heart are separated by the septum, a muscular wall. Each side of the heart has two chambers, an **atrium** and a **ventricle**. After blood moves into the atrium on the right side of the heart, it is squeezed into the right ventricle. The walls of the atria are relatively thin because not much cardiac muscle is required to move blood from an atrium into a ventricle.

The right ventricle pumps blood into the **pulmonary artery**. The muscular wall of the ventricle is quite thick because it must pump the blood with sufficient force for it to get to the lungs and back to the heart again. Blood is prevented from moving back into the atrium when the ventricle contracts by valves that separate the two chambers. The valves are flaps of muscle tissue that close over when blood moves out of the ventricle. As the heart relaxes between beats, a valve in the pulmonary artery prevents backflow of blood into the right ventricle.

Prior to birth, there is a hole in the septum between the right and left atria. If this hole does not close at birth, a baby is said to have a hole in the heart and may require surgery to close it.

Blood travels from the lungs via the **pulmonary veins** to the heart and enters the left atrium. It is squeezed into the left ventricle from where it is pumped out through the aorta to other parts of the body. Valves preventing the backflow of blood are located between the left atrium and ventricle and the left ventricle and aorta.

Blood vessels and the heart: what goes wrong?

Blood vessels, or the valves in them, can become blocked or damaged. The box on page 139 outlines the treatment called coronary angioplasty that is available for a blocked coronary artery.

If a vital blood vessel is severely damaged, it may be replaced by the transfer of a vessel from within the person. For example, the blocked parts of a coronary artery may be bypassed by grafting another vessel onto the coronary artery (see figures 6.15 and 6.16). The number of bypass grafts depends on the severity of the blockages.

Heartbeat

In infants, the normal heart rate varies between 120 to 140 beats per minute. The rate gradually slows to about 80 beats per minute at puberty. The normal rate in adults varies between 60 and 90 beats per minute with an average of about 72.

The rhythmic beating of the heart is controlled by the pacemaker, a small clump of specialised cells in the wall of the right atrium of the heart. An electrical impulse spreads from the pacemaker through the cells of the walls of the atria, causing the muscle cells of the atria to contract, which forces blood into the ventricles. The electrical impulse continues through another clump of special cells and spreads through muscle cells of the ventricles, which contract. This forces blood from the heart.

The electrical impulse that passes through heart muscle tissue generates currents that can be detected on the body's surface. These are used to examine the pattern of changes that take place as a heart beats. A record of the electrical changes that take place as a heart beats is called an electrocardiogram (ECG) (see figure 6.17). The instrument used to make the record is an electrocardiograph. A study of the electrocardiogram of an individual helps a doctor to diagnose an abnormality in the person's heartbeat.

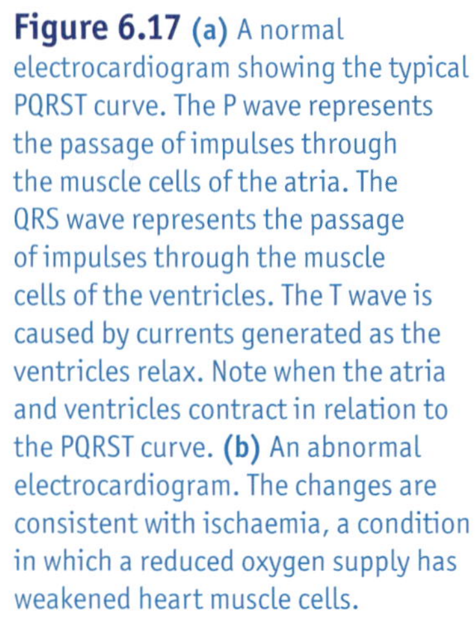

Figure 6.15 Heart showing a triple bypass required because of three blocked arteries

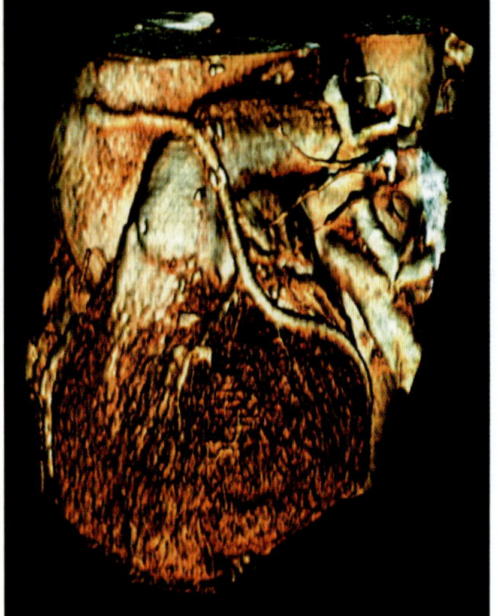

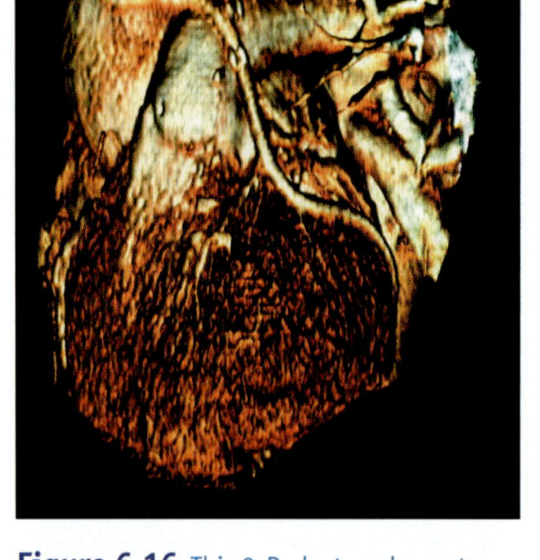

Figure 6.16 This 3-D electron beam tomography (EBT) scan shows a heart with a coronary artery bypass graft — the narrow vein running from lower right to top left. The vein, usually taken from the patient's leg, bypasses obstructions in the arteries to enable blood to flow freely again to the aorta.

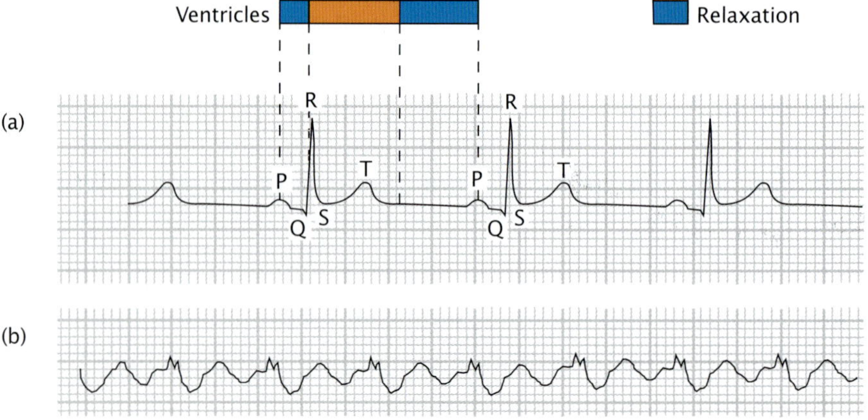

Figure 6.17 **(a)** A normal electrocardiogram showing the typical PQRST curve. The P wave represents the passage of impulses through the muscle cells of the atria. The QRS wave represents the passage of impulses through the muscle cells of the ventricles. The T wave is caused by currents generated as the ventricles relax. Note when the atria and ventricles contract in relation to the PQRST curve. **(b)** An abnormal electrocardiogram. The changes are consistent with ischaemia, a condition in which a reduced oxygen supply has weakened heart muscle cells.

BIOTECH

Coronary angioplasty — widening a blocked coronary artery

Coronary **angioplasty** is a non-surgical technique, used to remove an obstruction in a coronary artery (see figure 6.18). It is performed under local anaesthetic and generally requires only a short stay in hospital. A long, narrow, hollow tube (catheter) is inserted into an artery through a small incision, generally in the groin. X-ray images on a screen are used to guide the catheter up through the aorta to the heart arteries until the blockage is reached. A thinner catheter, tipped with a miniature deflated balloon, is guided through the first catheter until the balloon is in the blockage area. The balloon is inflated and deflated several times. The plaque is pushed against the artery wall and the artery is widened slightly. A fine metallic mesh tube, called a stent, is then inserted into the artery to prevent the artery wall from collapsing. The stent remains inside the artery and within a few weeks the natural lining of the artery grows over it.

Note the difference in diameter of the coronary artery before and after angioplasty as shown in figure 6.19.

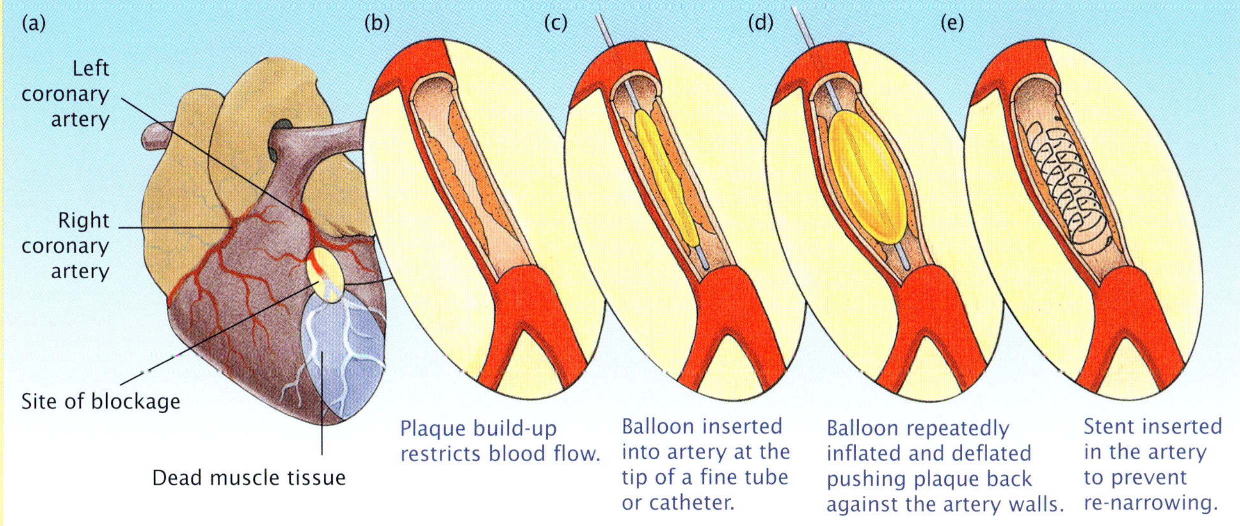

Figure 6.18 If a coronary artery becomes blocked **(a)** heart muscle close to the blocked vessel may die. **(b)** The blockage is due to the build-up of fatty deposits (atherosclerosis or plaques). **(c)** A balloon at the tip of a fine tube or catheter is inserted into the artery and when it reaches the blocked area it is repeatedly inflated **(d)** and deflated. The artery widens and the plaque is pushed against the artery wall. **(e)** A fine metallic mesh tube, called a stent, prevents a re-narrowing of the artery at the previously restricted area.

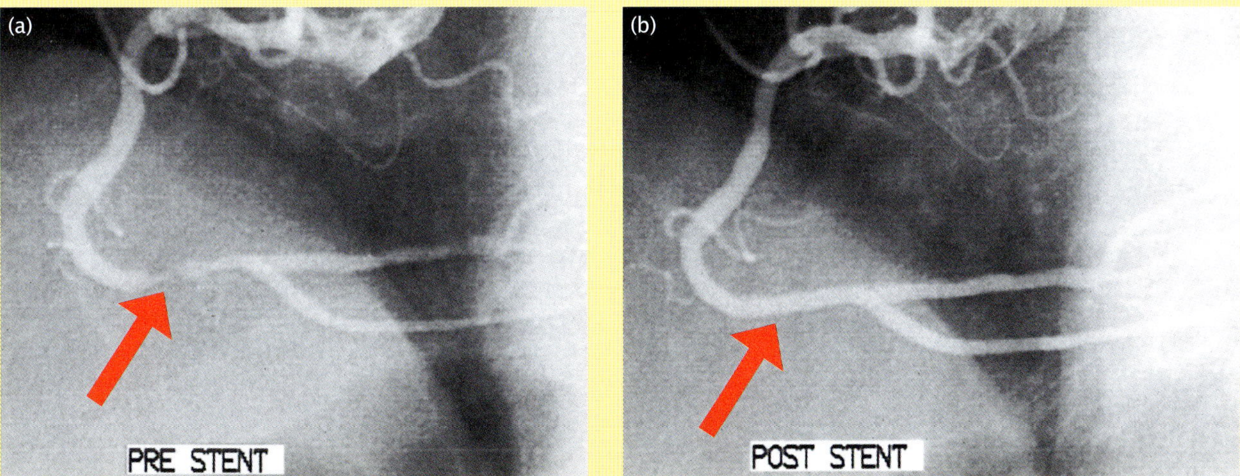

Figure 6.19 Image of a coronary artery **(a)** before and **(b)** after angioplasty. Special dyes are used to observe the flow of blood. This patient had a heart attack 10 days before having angioplasty. He was discharged from hospital two days after the procedure. Arrows show affected segments.

Different circulatory systems

Closed circulatory system

Birds have a connection between the two jugular veins in the neck so that blood can be shunted from one side to the other as the head is turned around.

Although material carried by mammalian blood leaves the circulatory system, blood itself remains in closed vessels as it circulates the body. Hence the system is called a **closed circulatory system**. In fact all vertebrates have a closed circulatory system although the nature of the heart and the circulation route may vary between different groups of vertebrates (figure 6.20). In each of these three different kinds of closed circulatory system, the system comprises three parts:

- blood, the fluid that moves around a series of vessels and carries nutrients and oxygen to all parts of the body
- a pumping organ, the heart, that forces blood through the system
- vessels, that carry blood to the tissues and then return blood to the heart.

Because blood does not come into direct contact with the organs and tissues of the body, nutrients and gases must diffuse through membranes of thin-walled vessels to get to cells.

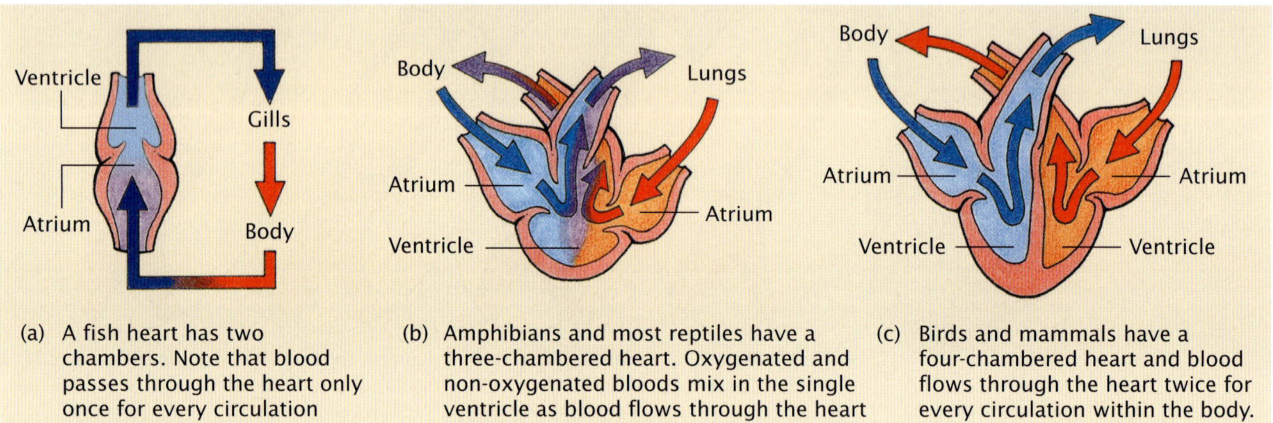

Figure 6.20 The heart and circulation route of blood in various vertebrates shows differences.

Open circulatory system

In an **open circulatory system** the blood is contained in vessels for some of the time, leaves the vessels to bathe tissues, then returns to the vessels.

Insects (phylum Uniramia, class Insecta) have open circulatory systems, but the blood contains no pigment. It looks a little like raw eggwhite. Look at figure 6.21. The blood, or haemolymph, of an insect fills the spaces between the internal organs. The blood moves from around the tissue into the heart through pairs of openings. The wall of the heart contracts and blood is forced through the aorta, from where it is forced out to bathe the internal organs and tissues.

Insect blood contains a number of ions and organic molecules in solution that are available for the cells it surrounds. Disaccharide sugars are carried in the blood. Insects have a high level of amino acids in their blood, and fats are often carried in association with a blood protein.

Insect blood does not transport oxygen. Insects have a series of air-filled tubes, or trachea, that provide oxygen to living cells (see page 148).

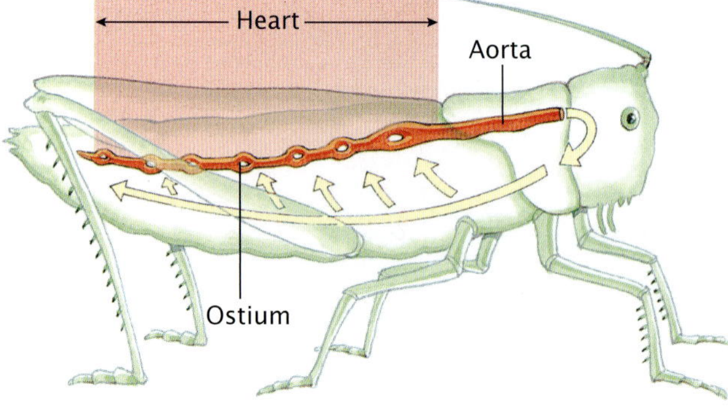

Figure 6.21 An insect has an open circulatory system. Blood bathes the internal organs, enters the heart, is pumped into the aorta, then out of the aorta to surround the tissues again.

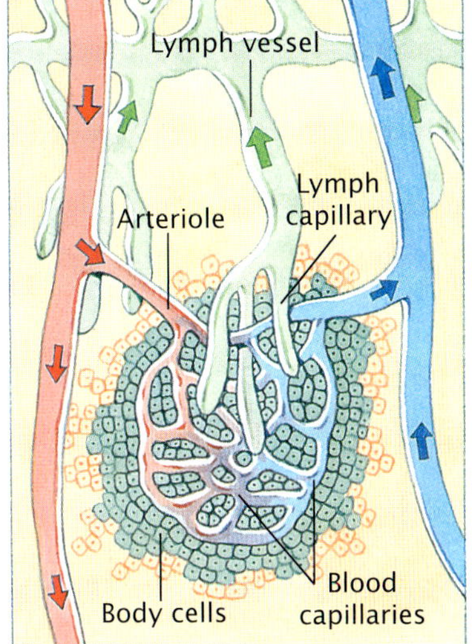

Figure 6.22 Lymphatic system (shown in green) and its relationship to the blood circulatory system. Note the blind endings of the lymphatic capillaries.

The lymphatic system

We saw earlier that capillaries have very thin walls, just one cell thick. We have also seen that oxygen and nutrients diffuse from blood into the tissue fluid surrounding cells. Fluid, some protein, and white blood cells also escape from the capillaries and move into tissue fluid, about three litres of fluid daily.

Although some of the fluid and protein that leaks out moves back into the capillaries, most of the excess tissue fluid around cells is collected by a special series of vessels that make up the **lymphatic system**. Lymph capillaries are blind-ending, thin-walled and begin in spaces between tissue cells. Small lymph capillaries combine to form larger and larger vessels. When tissue fluid enters a lymphatic, it is called lymph. You can see the relationship between a blood and a lymph capillary in figure 6.22.

Figure 6.23 shows that the lymphatic system is a one-way system. It transports lymph from the tissues to the subclavian veins where it is returned to the bloodstream.

The lymphatic system has no heart to act as a pump but it does have valves in the vessels that prevent a backflow of the lymph. In addition, the breathing action, muscular movements and pressure from adjacent blood vessels help move lymph through the lymphatic system.

The thicker areas in the lymphatic system are lymph nodes. Lymph is strained through the nodes and any micro-organisms or other foreign material are usually retained and destroyed there.

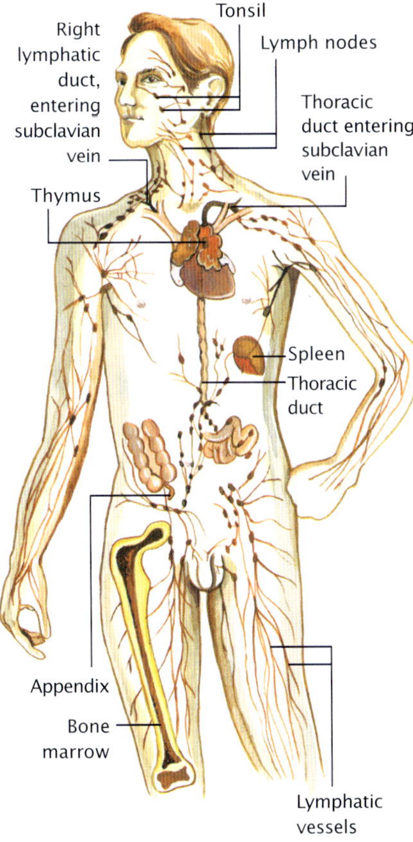

Figure 6.23 The lymphatic system is widely distributed through the body.

Transport of nutrients

Plasma of the blood transports nutrients to all living cells. These nutrients are the products of digestion in which enzymes break down large organic molecules, such as polysaccharides, fats and proteins, to compounds that are small enough to cross cell membranes. Other organic compounds, such as glucose, that are already small enough to cross cell membranes do not need to be digested.

Figure 6.24 summarises the fate of organic compounds during digestion.

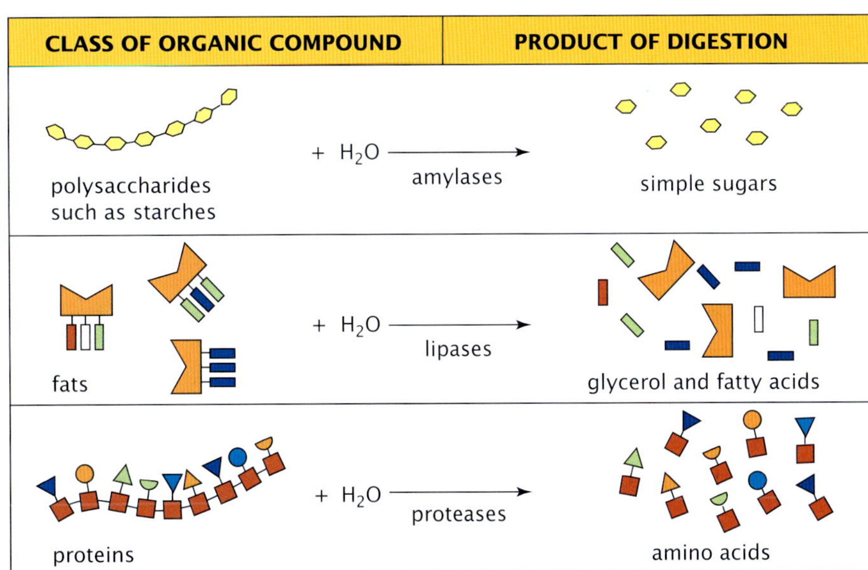

Figure 6.24 During digestion, large organic compounds are hydrolysed (broken down) to smaller compounds that are small enough to pass into cells.

ODD FACT

A portal vein is any vein that starts and ends with capillaries. The hepatic portal vein has one capillary bed in the gut wall and another in the liver. All vertebrates have a hepatic portal system. The hepatic portal vein (from the gut to the liver) should not be confused with the hepatic vein (from the liver to the major vein (vena cava) that returns blood to the heart).

Simple sugars

Simple sugars (monosaccharides), such as glucose and fructose, cross the membranes of cells lining the gut and are absorbed into blood in capillaries in the gut villi. From there, sugars are carried via the hepatic portal vein to capillaries in the liver. The hepatic portal arrangement enables the liver to act on compounds absorbed from the gut before they enter the general circulation. The liver may remove some compounds for storage or may alter other compounds. Simple sugars other than glucose are converted into glucose in the liver.

Glucose transported to the liver after a meal may be in excess of the body's needs. Some glucose is removed from the bloodstream and converted in the liver to the high molecular weight storage carbohydrate, **glycogen**, which is deposited in liver cells. The human liver can store up to about 100 grams of glycogen.

Some of the excess glucose leaves the liver via the hepatic vein and is transported through the circulatory system to the skeletal muscle tissue. Here, excess glucose is taken up by muscle tissue and converted to glycogen for storage. Human skeletal muscle tissue can store over 300 grams of glycogen.

ODD FACT

Blood must transport the products of digestion to all cells where they are used to provide energy for cellular processes and structural units for building and repair.

Amino acids

The digestion of proteins produces amino acids. Amino acids absorbed in the gut enter the hepatic portal vein and are transported to the liver. From the liver, amino acids are carried by the blood to all parts of the body where they are used.

Lipids

When fats are digested, fatty acids and glycerol are produced. Some of these diffuse through cells lining the intestine into nearby blood capillaries and are transported in blood, via the hepatic portal vein, to the liver.

KEY IDEAS

- The muscular heart contracts rhythmically and acts as a pump to maintain circulation of the blood.
- Various kinds of biotechnology are used to assist diseased hearts.
- Some animals have a closed circulatory system, and some have an open circulatory system.
- Excess fluid is returned from the tissues to the bloodstream via lymphatic vessels in the lymphatic system.
- Lymphatic vessels contain one-way valves.
- The composition of blood varies as it circulates the body.
- The liver has a major role in storing and converting organic compounds.

QUICK-CHECK

4 In which chambers of a mammalian heart is oxygenated blood found?
5 Is blood in each of the following vessels oxygenated or deoxygenated?
 a aorta
 b pulmonary artery
 c superior vena cava
6 What does tissue fluid contain?
7 What are the three main parts of a closed circulatory system?
8 What does lymph contain?
9 What is the role of one-way valves in lymphatic vessels?
10 In what form is carbohydrate transported (a) in blood and (b) in the liver?

Transport of gases

Blood transports oxygen to all cells within a vertebrate and transports carbon dioxide away from those cells. Special organs exist to facilitate the movement of oxygen into the blood and the exit of carbon dioxide from the blood. Different groups of animals have different organs for this process of **gaseous exchange**. Mammals, reptiles, birds and adult amphibians have **lungs**, fish have gills and insects have trachea. Let us look at the organs that, along with the lungs, make up the **respiratory system** in humans and their role in the transport of gases.

Gills and trachea are discussed on pages 146–9.

Pharynx and larynx

When we breathe in, air enters the nose and moves into the nasal cavity. Coarse hairs filter out any large particles and the air becomes moist and is warmed by the blood in the capillaries. The air passes from the nasal cavity into the **pharynx**, or throat, a tube about 13 centimetres long (see figure 6.25a). Part of the pharynx transports food as well as air and is lined with mucous membrane.

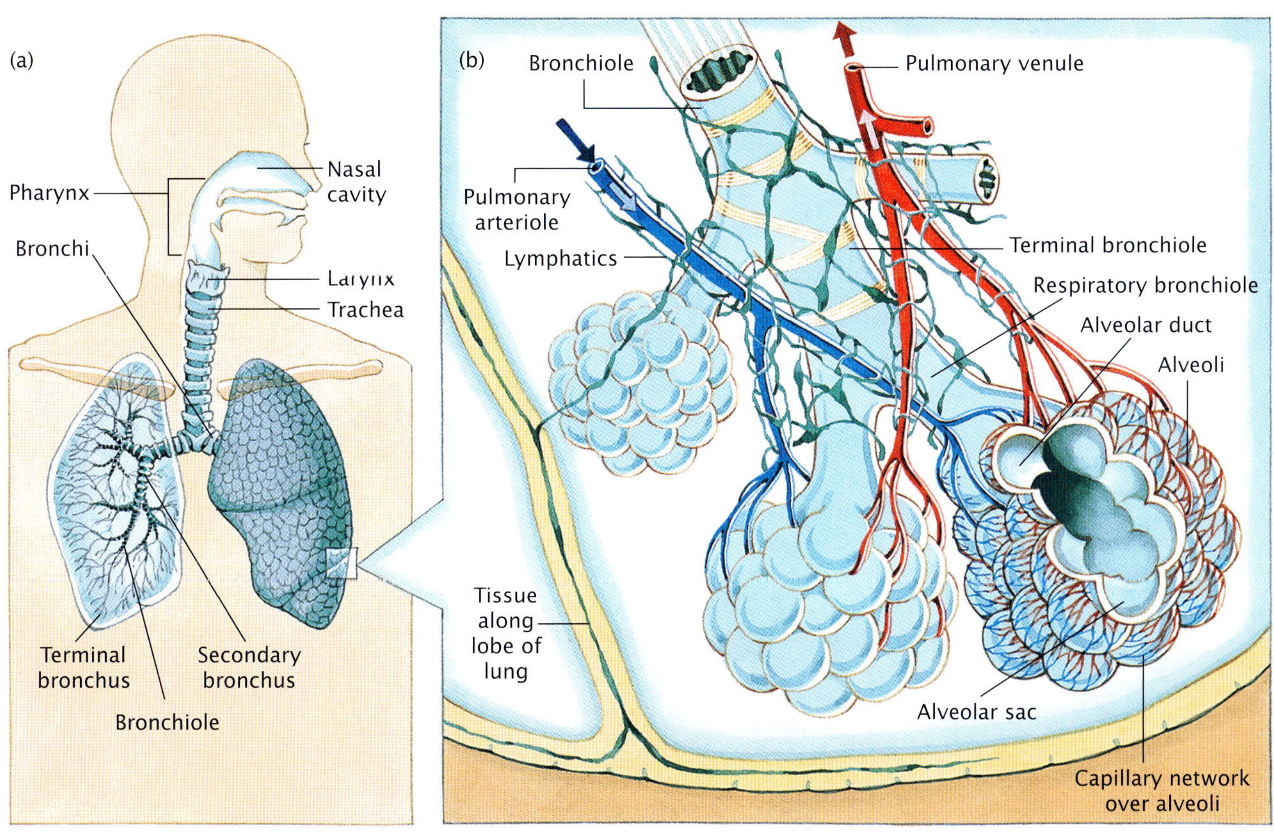

Figure 6.25 **(a)** Organs of the respiratory system **(b)** A portion of the lung expanded to show details

Air in the pharynx enters the **larynx**, or voice box, which is anterior (in front of) the oesophagus. The larynx is supported by cartilage pieces, the largest one being the Adam's apple which is particularly noticeable in males. When food is being swallowed, the opening into the larynx, the glottis, is covered by a flap of tissue called the epiglottis.

The lining of the larynx helps trap dust and other particles not removed in the nasal cavities.

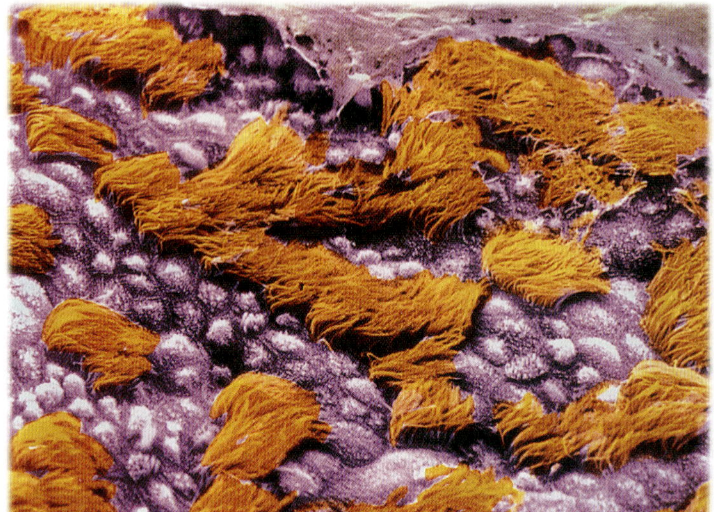

Figure 6.26 Scanning electron micrograph of the lining of the trachea showing cilia (orange) and goblet cells (purple)

Trachea

From the larynx, air moves into the **trachea**, or windpipe. This is about 12 centimetres long and about 2.5 centimetres in diameter. It is reinforced with incomplete rings of cartilage to ensure that the walls cannot collapse and block off the air supply. The lining has goblet cells and cilia which help trap any dust that has passed through the larynx (see figure 6.26).

Bronchi and bronchioles

Air moves from the trachea into two **bronchi** (singular = bronchus). Each bronchus goes into a lung and branches into smaller and smaller tubes called bronchioles (see figure 6.25b). The sequence of branching is:

Because the branching of the bronchus and bronchioles in the lung looks like the branching of the limbs of a tree, this structure is sometimes called the bronchial tree.

The structure of the tubes changes as they branch. The amount of cartilage in the wall is reduced until there is no cartilage in the bronchioles. At the same time, smooth muscle in the walls of the tubes increases. The cells lining the terminal bronchioles are cuboid and have no cilia.

ODD FACT

The vocal cords are elastic ligaments that surround the glottis. The ligaments vibrate if air is directed against them and sounds are made. Other parts of the respiratory system, such as the pharynx, the mouth and the nasal cavities, are used to convert the sounds into meaningful speech.

The lungs: two-way gaseous exchange

The lungs are paired, lobed organs lying in the thoracic cavity (the chest). They are generally shaped like a cavity or sac, with many pouches or lobes and air-filled spaces that increase the surface area (see figure 6.27). When air is taken into these structures, oxygen is absorbed into the body and carbon dioxide, a waste product of cells, is released.

The lungs are made up of a series of tubes (as we described above) and the tissue at the end of the respiratory bronchioles. Each respiratory bronchiole

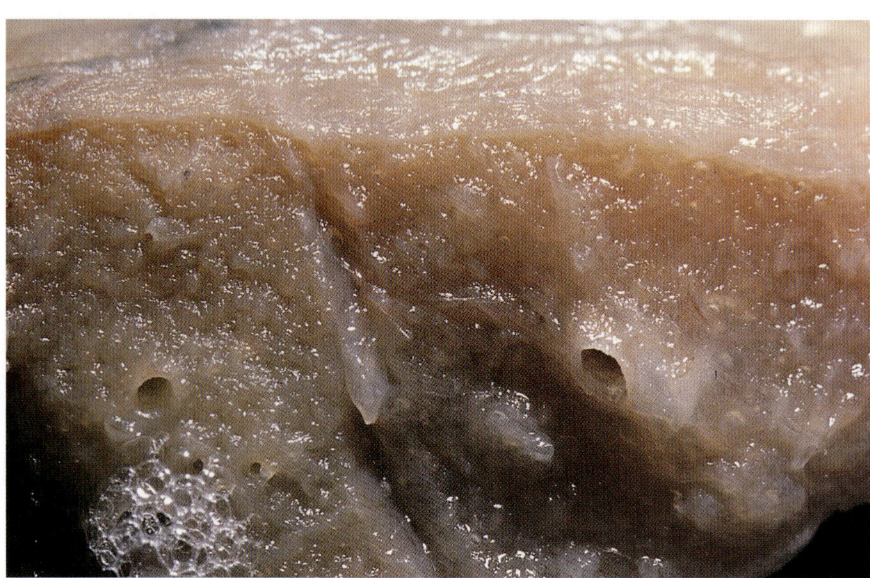

Figure 6.27 Healthy lung tissue is moist, soft and spongy because of the many air-filled spaces.

leads into an **alveolar duct**. Each alveolar duct leads into an alveolar sac. Small pouches, or **alveoli**, surround each alveolar sac (see figure 6.28). The linings of the alveolar duct and sac and of the alveoli are made of a single layer of thin flattened cells. Lung tissue also contains connective tissue and blood and lymph vessels. Each alveolus is surrounded by a capillary network.

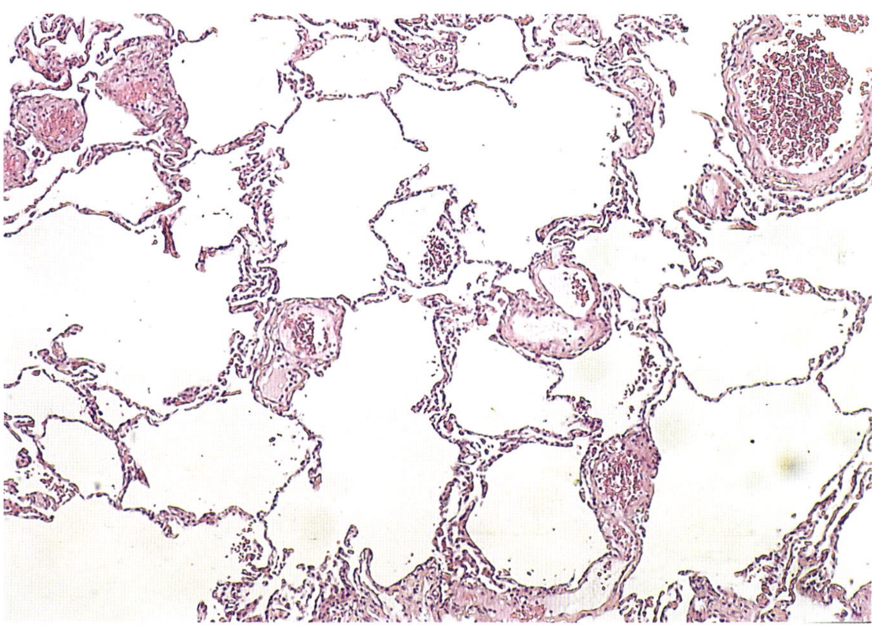

Figure 6.28 Photomicrograph showing a section of alveolus tissue. Note the very thin walls marking the alveoli. These walls contain the thin membranous lung tissue and small capillaries. Capillaries absorb oxygen from air in the alveoli and release carbon dioxide.

The passage of air from the outside into a lung follows the sequence:

nose → nasal cavity → pharynx → larynx → trachea → bronchus → secondary bronchus → bronchiole → terminal bronchiole → respiratory bronchiole → alveolar duct → alveolar sac → alveoli.

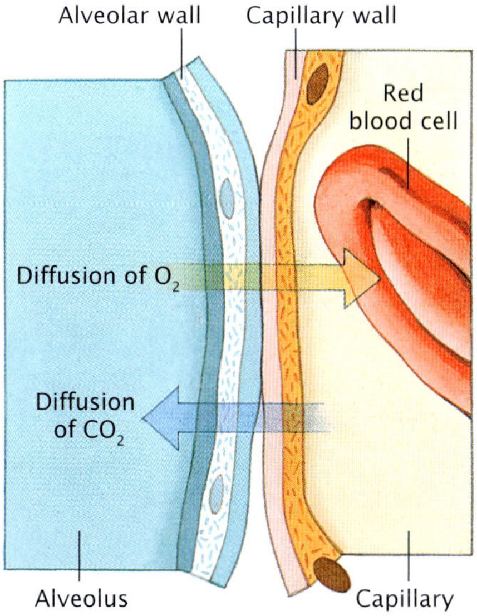

This branching of tubes, which ends in about 300 million alveoli, provides an extremely large surface area across which gases are exchanged. It has been estimated that the surface area of the alveoli is about 70 square metres. Compare this with the surface area of the skin of an average adult — about two square metres!

The alveolar–capillary membrane is very thin (only about 0.004 millimetres thick). Oxygen diffuses from the air in the alveoli across the alveolar–capillary membrane into the blood where it combines with haemoglobin in the red blood cells. At the same time, **carbon dioxide** diffuses from the blood, mainly from the plasma, into the alveoli (see figure 6.29).

Figure 6.29 The alveolar membrane and capillary wall form a very thin membrane across which oxygen and carbon dioxide readily diffuse.

Although most carbon dioxide is carried in the blood in the form of bicarbonate ions (70 per cent) in plasma, about 23 per cent combines with haemoglobin and the balance is dissolved in plasma.

When a person breathes in and out, air moves in and out of the lungs. Air going into the lungs contains more oxygen than air leaving the lungs, and air leaving the lungs contains more carbon dioxide than air entering the lungs. There is never a complete changeover of air in the lungs. When we breathe out, there is always some air left in the lungs. This mixes with air that enters the lungs. We can reduce the amount of air that is left in the lungs by breathing more deeply; however, there is always a mixture of new and used air in the lungs. Read more about the breathing process on page 146.

BREATHING IN AND OUT — WHAT HAPPENS?

What happens in the chest cavity when a person breathes? The events in this process are outlined in figure 6.30.

The diaphragm is a layer of muscle that separates the thorax (the chest cavity) from the abdomen. Carbon dioxide levels in the blood control breathing. When the concentration of carbon dioxide reaches a certain level, the breathing centre of the brain is stimulated and a signal is transmitted to the diaphragm, which contracts. The rib cage moves upward and outward. These movements increase the volume of the chest cavity and the lungs are able to expand. Air rushes in through the nose or mouth to fill the extra volume available in the lungs. This whole process is called breathing in, inspiration or inhalation (see figure 6.30a).

Even a small increase in the amount of carbon dioxide in the blood will result in more rapid and deeper breathing. When the signal from the breathing centre ceases, the diaphragm relaxes and the rib cage moves downward and inward. The volume of the chest cavity is reduced and air is forced out of the lungs. This is called breathing out, exhalation or expiration.

It is impossible for you to hold your breath indefinitely. The build-up of carbon dioxide in the blood acts on the breathing centre in the brain, which eventually forces you to recommence breathing.

What happens when oxygen is restricted?

Generally, humans drown if they are submerged in water for more than about four minutes. The tissues are deprived of oxygen, the brain is irreversibly damaged and the person dies. There have been reports, however, of people who have been revived and suffered no permanent brain damage even though they have been submerged in water for relatively long periods, such as 20 or 30 minutes. This may occur if the person has been submerged in very cold water. If this happens, the body temperature drops, the metabolic rate of the tissue falls, and the oxygen requirement is also reduced. The production of carbon dioxide is reduced. The heartbeat is reduced in much the same way as that of a seal or whale during a dive.

Fashion can have adverse effects on the body. When tiny waists became fashionable in nineteenth-century Europe, women wore very tight corsets which forced the organs into abnormal positions (see figure 6.31). Lungs were restricted, the air available was reduced and women often fainted, or 'swooned'. Smelling salts were often used to revive a fainting woman — the pungent odour of ammonia in the salts would cause her to gasp in air, and so regain consciousness. Sometimes corset laces were cut to allow the woman more air, and she would recover. In about 1881, Sir Hamo and Lady Thornycroft founded the Healthy and Artistic Dress Union — devised to relieve ladies from heavy and restrictive clothing.

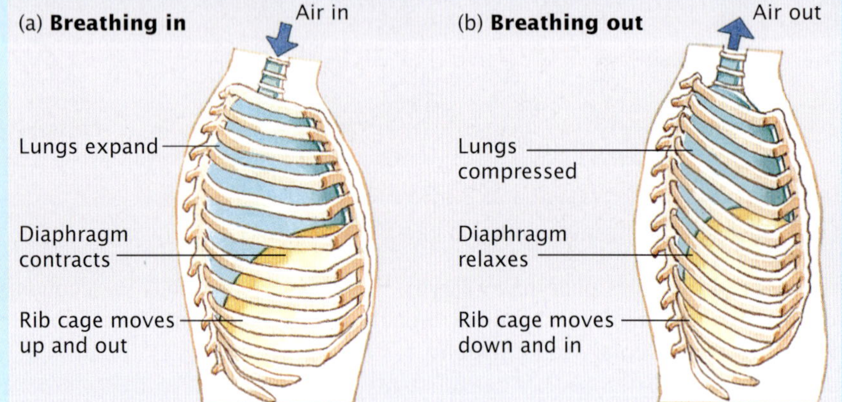

Figure 6.30 **(a)** When the muscular diaphragm contracts, the rib cage moves up and out, the chest cavity expands, the lungs expand to fill the cavity and air is drawn in. **(b)** When the diaphragm relaxes, the rib cage lowers, the lungs are compressed and air is forced out.

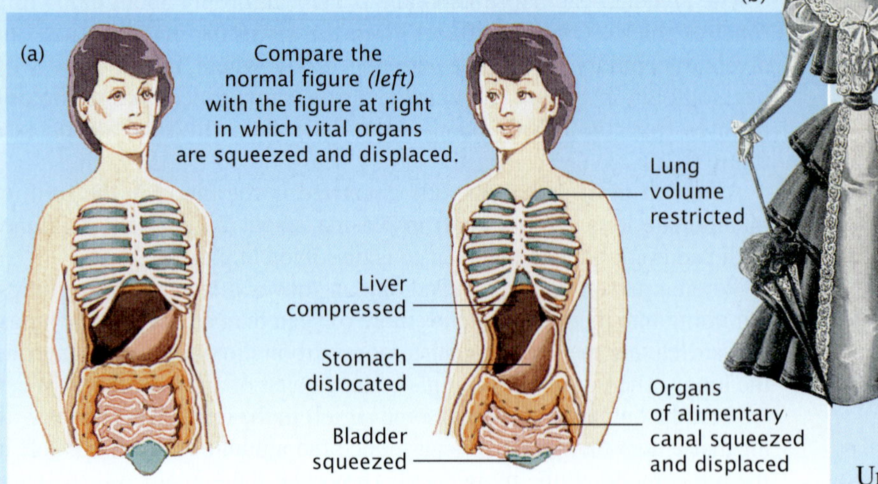

Figure 6.31 **(a)** The impact of corsets on women **(b)** Nineteenth-century fashion

Gills: gaseous exchange in fish

Fish live in water. Depending on the temperature and salinity of water, the amount of oxygen available is only about three to five per cent of what it is in air. A fish is able to live in such an environment partly because it is buoyed up by water. A fish requires less energy to maintain its balance than does a land animal. This reduced energy requirement means a reduced need for oxygen. Also, because of the watery environment, no energy is required to keep respiratory surfaces moist. The gaseous exchange surfaces of fish are in **gills** which are highly efficient gas-exchange organs.

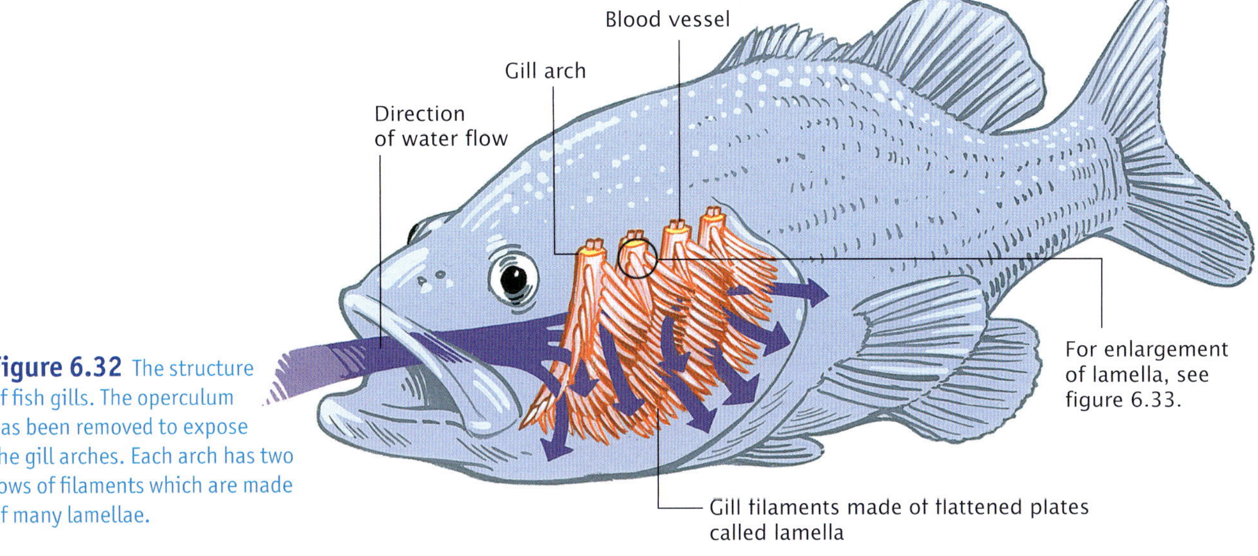

Figure 6.32 The structure of fish gills. The operculum has been removed to expose the gill arches. Each arch has two rows of filaments which are made of many lamellae.

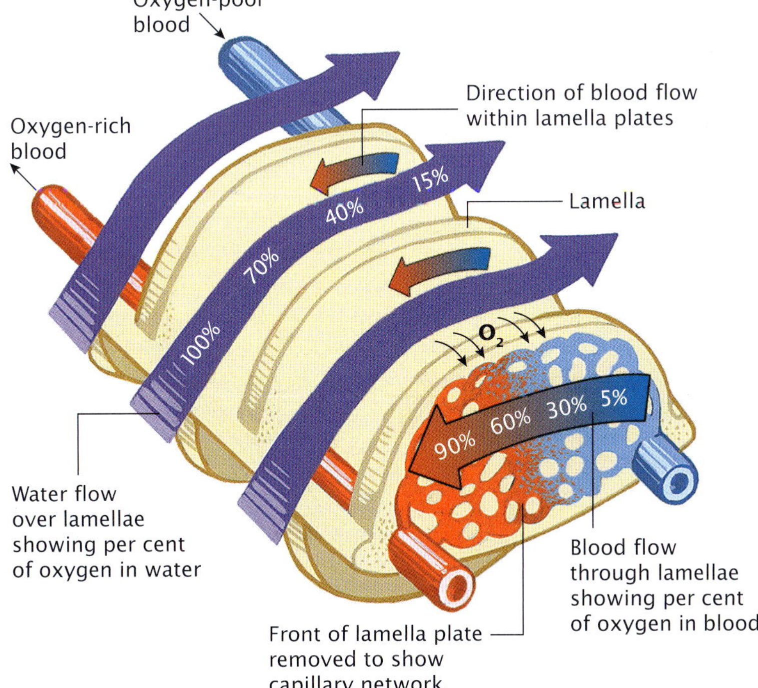

Figure 6.33 Lamella detail. Compare the direction of water flow over plates with that of blood flow in capillaries. Countercurrent flow ensures maximum transfer of oxygen and carbon dioxide between blood and water.

On each side of a fish, close to the head, is a flap called an **operculum**. Under each operculum is a series of four gill arches. Each gill arch extends into two filaments that are made of many **lamellae** which are the gas exchange surfaces. Note the structure of the gills and their filaments in figure 6.32 above.

The lamellae are a series of flat plates that are the gas-exchange surfaces of the gill filaments. Three lamella plates are shown in figure 6.33. The front of one of the plates has been removed to expose the capillary network in the plate. Note the direction of blood flow in the capillaries compared with the direction of water flow over the plates. Oxygen in the water diffuses across a very short distance to reach the red blood cells that pass, virtually single file, through the capillaries in the lamellae.

When two fluids are moving in opposite directions and some compound, or heat, is transferred from one to the other fluid, the system is called a **counter-current**. In the fish gill system, oxygen diffuses from water to red blood cells in the capillaries and carbon dioxide diffuses from blood in lamellae capillaries to the surrounding water. Note that the opposing directions of flow mean that a concentration gradient is extended along the width of the lamella. If movement was in the same direction, equilibrium with regard to concentration would soon be reached and once that occurred there would be no further transfer.

When a fish's mouth is open and the lower part of the mouth relaxed, water enters and accumulates in the mouth. Then the mouth closes and the base of the mouth is raised. This exerts a force on the water, the opercula open and water flows over the gills and exits the fish. If a fish is moving forward, the rate of movement over the gills generally increases.

Gaseous exchange in insects: trachea

Insects have a series of air-filled tubes, or trachea, that provide oxygen to living cells. These trachea make up the respiratory system (see figure 6.34).

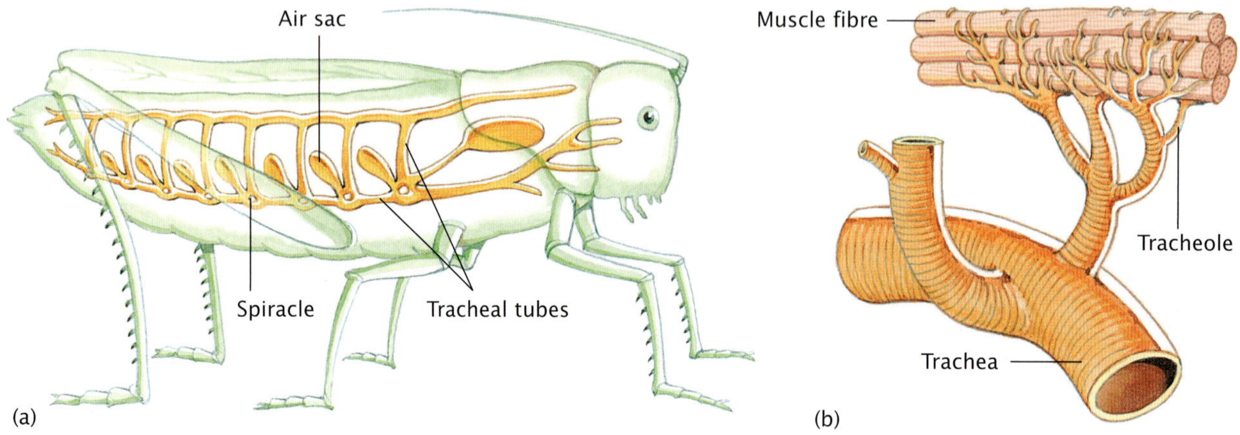

Figure 6.34 The respiratory system of an insect **(a)** comprises a number of fine tubes or trachea. **(b)** Tracheoles carry oxygen to and carbon dioxide away from cells.

Air enters the system via special inlets called **spiracles** and moves into the main trachea that continues to branch, getting smaller each time. Eventually, branches end with a group of fine **tracheoles** that are in close contact with cells as shown in figure 6.34b. Most insects are small and so diffusion provides sufficient oxygen for their needs. Very active flying insects have an increased need for energy, and so have an increased need for oxygen. In these insects, air sacs expand and contract with the up and down of wing beats. This bellow-like action forces air from large air-filled trachea into fine tracheoles. During flying, a two-gram locust moves 1.5 litres of air in and out of the body each hour.

Tracheole ends make direct contact with body cells (see figure 6.34) and oxygen diffuses from tracheoles across cell membranes into the insect cells. Some tracheoles contain fluid at the ends that make contact with cells. When insect muscle becomes active, the concentration of material inside the cells increases and fluid is withdrawn from the tracheoles. In this way, air is drawn closer to the tissue and oxygen diffuses rapidly into the tissue. Carbon dioxide diffuses out of the tissue into tracheoles.

The openings of the trachea, the spiracles, vary considerably in structure and often have muscular flaps that can be closed. Some spiracles have small chambers beneath them from which the trachea branch out. Chambers and trachea all have a cuticular lining continuous with the cuticle of the insect. This cuticle is made of chitin and protein. The cuticular linings of trachea also have spiral threads that add strength and prevent the tubes from collapse. Chitin is absent from the very fine tracheole endings that make contact with cells and diffusion of gases occurs.

ODD FACT

Trachea are very small. They can start as small as 1μ in diameter at their origin and reduce to 0.2μ at their endings. When Mt St Helens, in Washington State, USA, erupted in 1980, significant numbers of insects suffocated and died. Fine ash from the eruption blocked their tracheal systems.

Some spiracles are covered by a cuticular meshwork that retards the entry of foreign material. Figure 6.35 shows the detail of spiracles on the posterior end of a brown blowfly (*Calliphora stygia*) larva. Note that the spiracles lead into two main trachea from which other trachea branch out. Some insects have aquatic larvae. Some of these have trachea and must come to the surface of the water to obtain air but others develop anal gills that absorb sufficient oxygen from water.

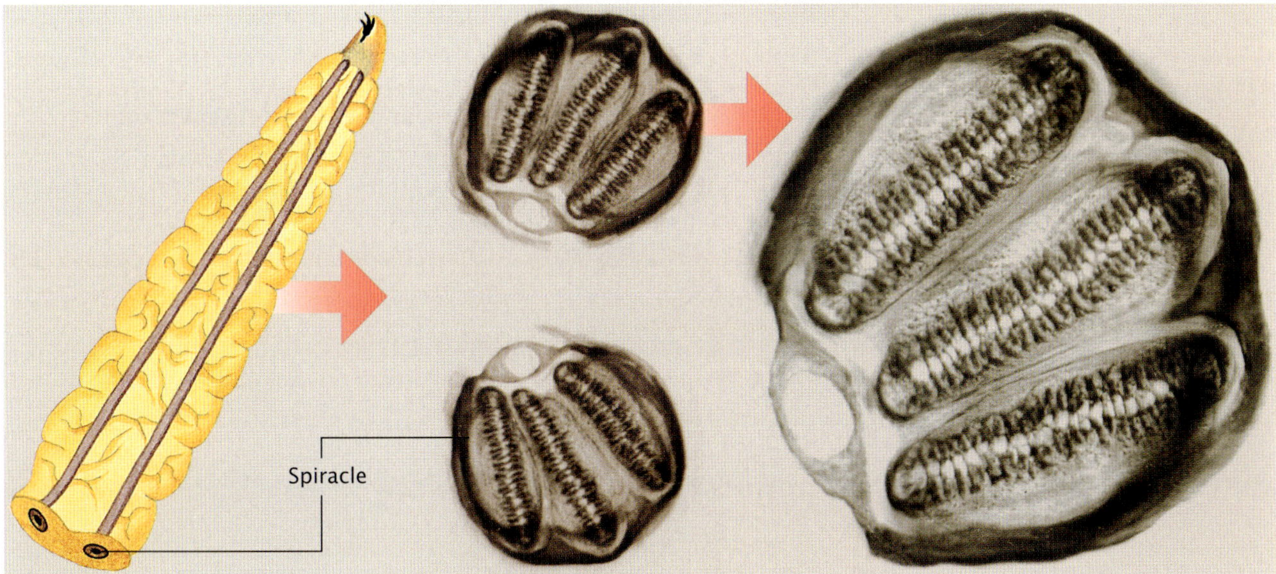

Figure 6.35 A fully grown larva of the brown blowfly, *Calliphora stygia*, has two multilobed spiracles at its posterior end. Each spiracle leads into a main trachea which gives rise to many smaller trachea.

Comparing positions of respiratory surfaces

In addition to acknowledging the role of the skin as a gaseous exchange surface in some smaller organisms, we have discussed three main types of respiratory organs — lungs, gills and trachea. The position of each of these organs in relation to the remainder of an animal's body is summarised in figure 6.36.

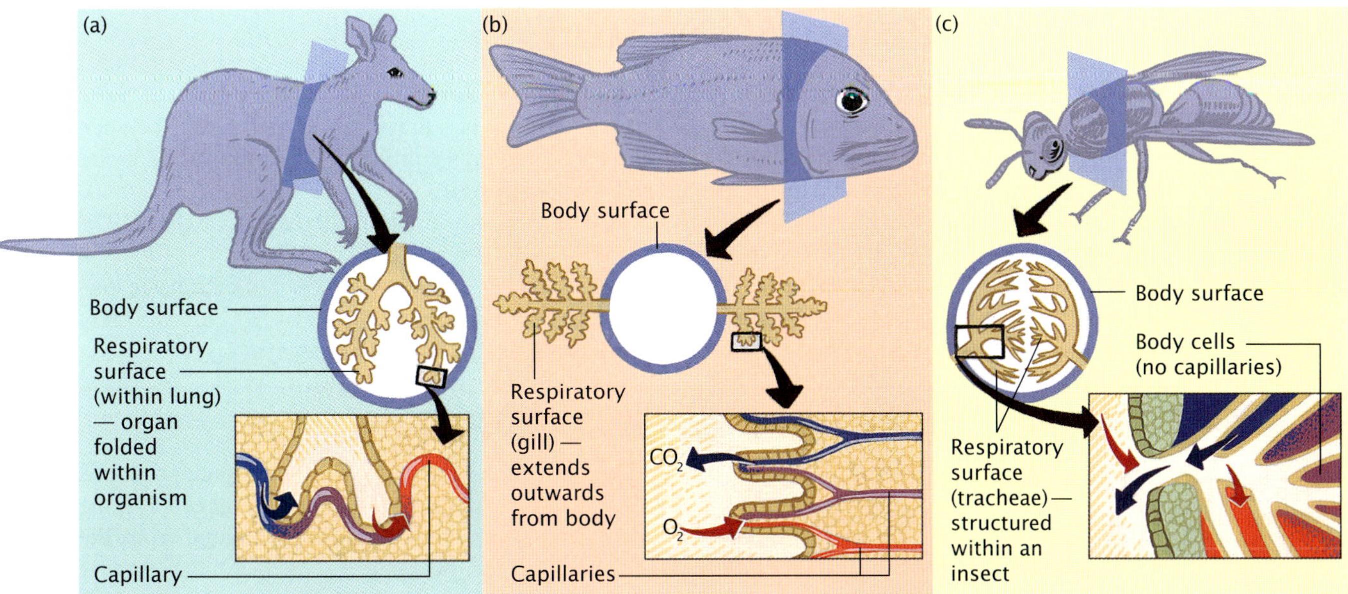

Figure 6.36 The three main kinds of organs that carry the gaseous exchange surfaces in animals are: **(a)** lungs, found in mammals and amphibians **(b)** the gills of fish and **(c)** the tracheae of insects.

KEY IDEAS

- Animals exchange gases with their external environment through special surfaces in their respiratory system.
- Mammals exchange gases with their external environment via respiratory surfaces in their lungs.
- The gas exchange surfaces of fish are located in gills.
- In insects, air-filled tubes called trachea transport oxygen to the respiratory surfaces.
- Trachea branch into very fine tracheoles that come into direct contact with cells.

QUICK-CHECK

11 Name one gas that animals require from their external environment. What parts of the respiratory system would this gas pass as it travels from the external environment to the respiratory surface of a human lung?
12 Name one gas that crosses the respiratory surfaces of lungs to enter the external environment of mammals.
13 Give two reasons why a fish can survive in water even though the oxygen in water is only about four per cent of what it is in air.
14 Name one significant difference between lungs and gills with regard to structure and their positions in relation to the remainder of the body of the animal that has them.
15 What are insect spiracles?
16 What is the pathway for oxygen as it moves from the external environment of an insect to its respiratory exchange surface?

Transport of wastes

Metabolic activities produce wastes that cannot be utilised by the body. If these wastes accumulate in the tissues that produce them, conditions would be such that those metabolic activities could no longer proceed. The removal of metabolic wastes from the body is called **excretion**.

What are the main wastes and how does the vertebrate body remove them? The main wastes are:

- carbon dioxide — excreted via the lungs as a waste product of cellular respiration (as discussed earlier in this chapter)
- **nitrogenous wastes** such as **urea** — excreted via the kidneys.

As well as urea, humans also excrete relatively small amounts of uric acid, ammonia and creatine. Creatine is produced when creatine phosphate is metabolised in muscles.

Nitrogenous wastes in animals

When animals metabolise protein, nitrogen-containing compounds are produced as waste. If these are left to accumulate, vital tissues become damaged and the animal will die. Nitrogen-containing compounds that are eliminated from the body are called nitrogenous wastes.

Different animals excrete different nitrogenous wastes because they have different amounts of water available to them. When protein is metabolised in a cell, **ammonia** is formed. Ammonia is toxic to cells and must be excreted immediately in lots of water or converted to some other non-toxic compound (see figure 6.37).

In organisms without a ready supply of water, ammonia is converted to urea, which is less toxic and requires less water for removal. **Uric acid**, a compound formed when nucleotides are metabolised, is also produced from ammonia and is the least toxic of these nitrogenous wastes. It requires little water for excretion.

The water required to remove these wastes is an obligatory part of the water balance equation for animals.

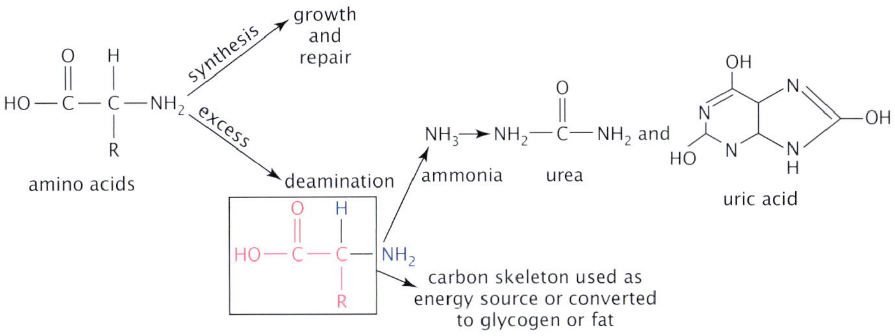

Figure 6.37 The ammonia formed when protein is metabolised is toxic to cells. If copious water is available to an animal, ammonia will be excreted immediately; if not, it must be converted into a less toxic form, either urea or uric acid.

Most nitrogenous wastes are removed from the mammalian body by the **urinary system**. We will consider the urinary system of humans as an example of a mammalian nitrogenous waste excretory system and demonstrate its relationship with the transport system.

The urinary system

Water is the most common compound in the body and its concentration is regulated to ensure that water gains and losses are balanced. The urinary system plays a significant role in maintaining the balance of water in the body. For example, if you drink copious amounts of water, more water is absorbed from the gut and your **kidneys** will produce a greater volume of watery waste.

The kidneys do much more than maintain the water balance of the body. The kidneys are the organs responsible for extracting nitrogenous waste from the blood and facilitating its removal from the body. The liquid waste that is filtered from the blood by the kidneys is called **urine**.

The urinary system consists of two kidneys, two **ureters**, a **bladder**, and a **urethra**. The function of each of these parts is summarised in figure 6.38.

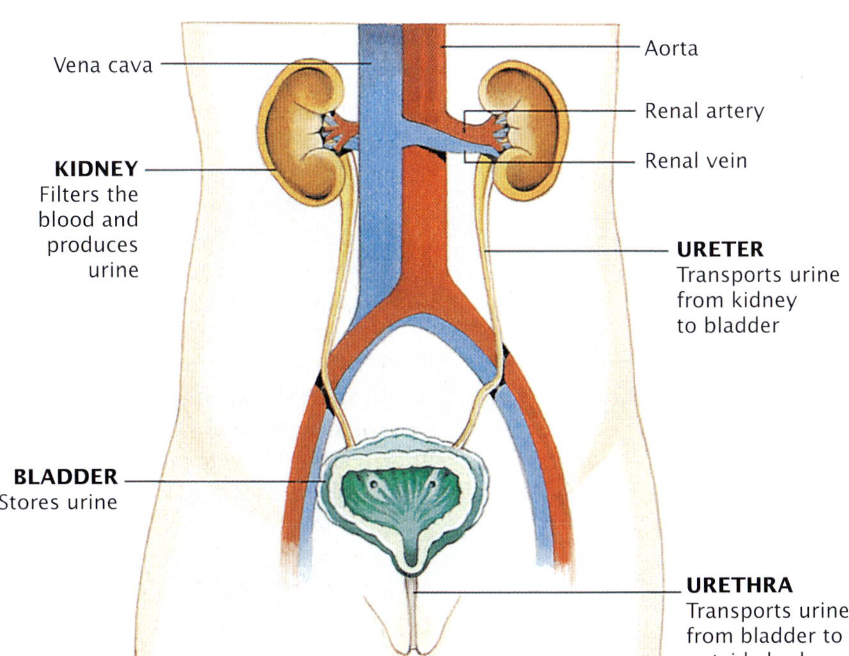

Figure 6.38 Components of the human urinary system

Kidneys

You have two kidneys. They are flattened bean-shaped organs, about 11 centimetres long, five to 7.5 centimetres wide and about 2.5 centimetres thick. Each kidney weighs about 150 grams, and they are located on each side of the spine, partly protected by the rib cage and embedded in a mass of fat for protection. Kidneys are vital organs; you would survive for only a few days without them.

Kidneys are adapted for filtering wastes from the blood. They also excrete hormones and other substances such as vitamins that would otherwise build up in the body. Kidneys also maintain a correct balance of ions in the blood by excreting those that are in excess. This helps the pH (or degree of acidity) of the blood.

A thin layer of cells, the capsule, surrounds each kidney. If you have ever had lamb's kidney at home you have probably peeled off this layer.

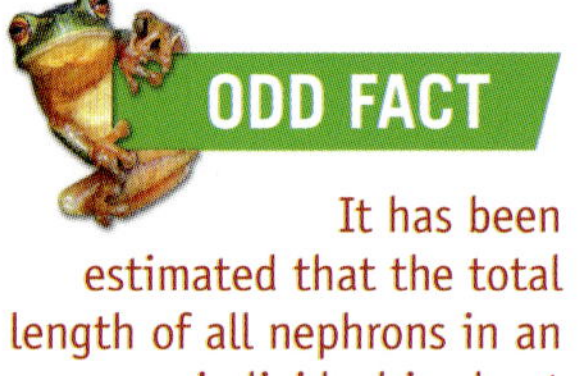

It has been estimated that the total length of all nephrons in an individual is about 85 kilometres.

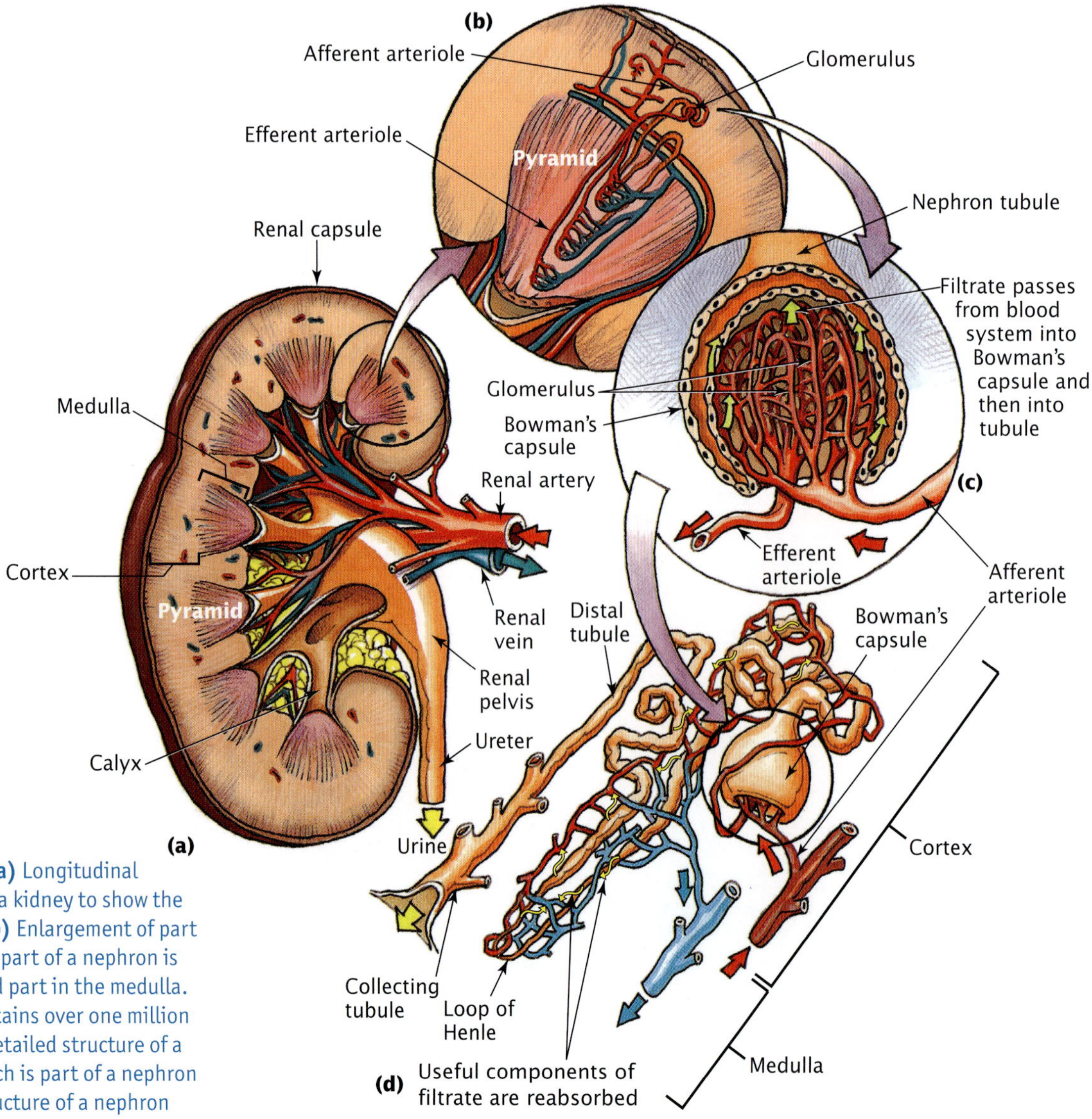

If you slice longitudinally through a kidney you can distinguish its different parts (figure 6.39a). The 'dent' in one side of the kidney is called the hilum. This region is where blood vessels enter and leave the kidney. The tube, the ureter, which transports urine away from the kidney, also leaves through the hilum.

The outermost layer of kidney tissue is the cortex, a granular-looking layer that extends into the second, striated layer, the medulla (see figures 6.39a and b). The functional units of the kidney, **nephrons**, are found in these regions.

Figure 6.39 **(a)** Longitudinal section through a kidney to show the internal areas **(b)** Enlargement of part of (a). Note how part of a nephron is in the cortex and part in the medulla. Each kidney contains over one million nephrons. **(c)** Detailed structure of a glomerulus, which is part of a nephron **(d)** Detailed structure of a nephron

(a)

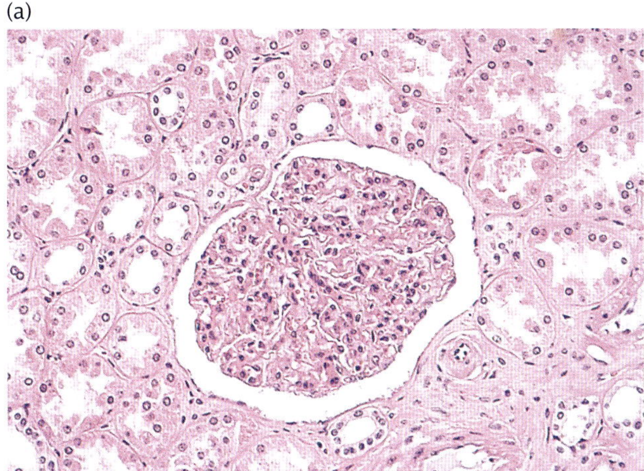

(b)

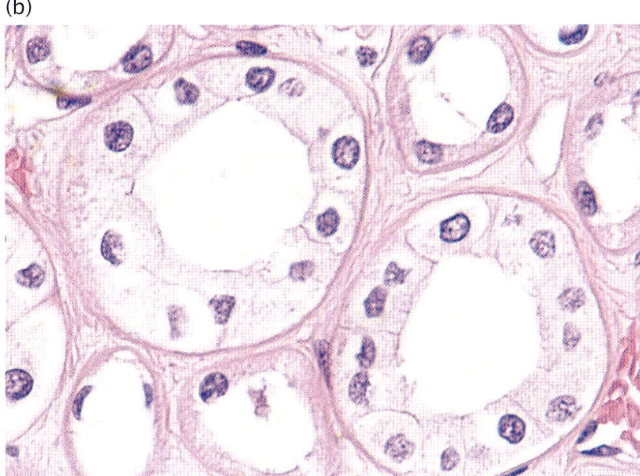

Figure 6.40 Photomicrographs of sections through a kidney showing: **(a)** glomerulus in Bowman's capsule **(b)** transverse section through kidney tubules

Each kidney contains about one million functional units or nephrons. It is the process of filtration and reabsorption in the nephrons that regulates body fluid composition. The artery that enters the kidney, the renal artery, branches in much the same way as any other artery into arterioles and finally branches into capillaries. These capillaries form small clusters, each called a **glomerulus** (see figures 6.39c and 6.40a). The glomeruli give the cortex its granulated appearance.

Each glomerulus is surrounded by a double-walled cup, formed by the 'pushed-in' end of a **tubule**. The double-walled cup is called the **Bowman's capsule** (see figures 6.39c and d). The remainder of the tubule coils away from the glomerulus and eventually joins up with a collecting tubule, also called a duct. Note that part of the tubule, the **loop of Henle**, is U-shaped and extends into the medulla of the kidney. The loops of Henle (see figure 6.39d) give the medulla a striated appearance.

The arteriole that leaves the capillary network in the glomerulus winds around the tubule before it joins with other vessels to form a vein that leaves the kidney.

Formation of urine

Waste materials are filtered from the blood in the glomerulus. Note that the arteriole leaving a glomerulus has a smaller diameter than the arteriole entering a glomerulus. This means that the blood in the glomerular capillaries is under pressure. Because blood is under pressure, substances will be forced through the 'holes' in the capillary walls and the inner wall of the Bowman's capsule. About one-fifth of all plasma that passes through a glomerulus is filtered through the capsule into the tubule. Only large compounds such as proteins and the blood cells do not pass through. The filtrate that moves into the tubule from the glomerulus contains water, nitrogenous wastes, nutrients and salts. Obviously some of these can be used by the body and need to be retained in some way, otherwise dehydration and starvation would quickly occur.

As fluid moves along the nephron tubule, some things that have been filtered out in the glomerulus are reabsorbed into the capillaries that leave the glomerulus and wind around the tubule (see figures 6.39d and 6.40b). Some material diffuses from the tubule back into the bloodstream. Other materials are actively transported against a concentration gradient (see figure 6.41, page 154).

Transport against a concentration gradient requires the expenditure of energy. Hence the kidney tubule relies on both diffusion, also called passive transport, and active transport for its function.

Some materials are added to the filtrate in the tubules by the cells of the tubule. Some drugs such as penicillin are secreted. Some ions are also excreted. The secretion of ions helps the body control the pH or acidity of its blood and other tissues. As hydrogen ions are secreted into the tubule, sodium ions are displaced from the fluid and reabsorbed by the body.

ODD FACT

Cats excrete more concentrated urine than humans. If cats are fed food with high levels of magnesium, and drink little water, they may develop bladder stones from crystals of magnesium and the urethra may become blocked. This is called feline urologic syndrome (FUS) and can be fatal if it is not treated. Both sexes suffer from the condition but males are more likely to have a blocked urethra. This condition can be minimised by ensuring that cats have access to fresh water.

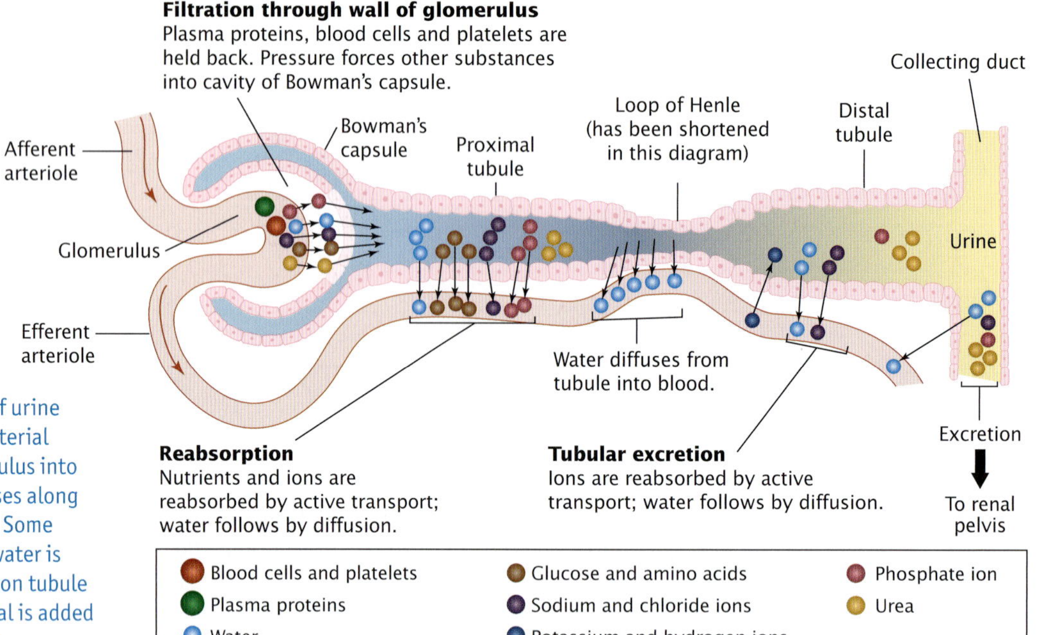

Figure 6.41 Summary of urine formation in mammals. Material forced through the glomerulus into the Bowman's capsule passes along the tubule of the nephron. Some material and much of the water is reabsorbed from the nephron tubule by the blood; other material is added to the filtrate in the tubule.

The fluid that reaches the end of the nephron tubule is urine. Because of the reabsorption that has taken place along the tubules, only about one per cent of the fluid that is filtered by the glomeruli actually leaves the kidney.

The composition of a person's urine, how much is produced and its pH depend on the body needs of that person. For example, a diet high in protein leads to higher concentrations of nitrogenous wastes in the urine. If a person is on medication of some kind, that material may be found in urine. High doses of vitamin C will result in that vitamin being found in the urine. A person of average health on a mixed diet will produce 1200–1500 millilitres of urine daily. The composition of this is shown in table 6.2.

Table 6.2 The composition of urine from a person of average health on a mixed diet

Component	Percentage
water	95–97
solids, including:	3–5
urea 30 g	
creatinine 1–2 g	
ammonia 0.5 g	
uric acid 1 g	
ions (salts) 25 g	

Mammals produce **hypertonic** urine. Hypertonic urine contains a higher concentration of solutes than do the body fluids of the same animal. Other animals, for example freshwater bony fish and amphibians, produce **hypotonic** urine. Hypotonic urine contains a lower concentration of salts than do the body fluids of the same animal.

Transport of urine

The urine is carried away from each kidney by the ureters. Each of these tubes is about 25–30 centimetres long and has a diameter of just over one centimetre at its widest part.

Two ureters, one from each kidney, transport the urine to the bladder, a muscular bag which stores the urine until it leaves the body. The urine is transported from the bladder to outside the body by the urethra. The act of getting rid of urine from the body is called micturition, more generally known as urinating or voiding.

Excretion in fish

Excretion of nitrogenous waste presents no problem for fish. They have a ready supply of water and are able to excrete the nitrogenous waste they produce without damage to their tissues, although freshwater and marine (also called saltwater) fish deal with the matter in different ways (see figure 6.42).

Because the tissues of a freshwater fish are hypertonic to its surroundings, water enters the gills by osmosis and ammonia is excreted in large volumes of dilute or hypotonic urine. This removal of large quantities of water by the kidneys is important so that freshwater fish can maintain an appropriate water balance in their bodies. Marine fish tend to lose water across their gills because the salt concentration inside their body is less than that outside. To counter this loss, a marine fish drinks sea water. The salt that is absorbed in the intestine with the water is carried by the blood to the gills where special salt-secreting cells transport it across gill membranes into the sea. Any ammonia leaves via the gills. The small amount of **isotonic** urine produced by marine fish carries magnesium sulfate salts and urea out of the body.

Isotonic urine contains the same concentration of salts as the body fluids of the same animal.

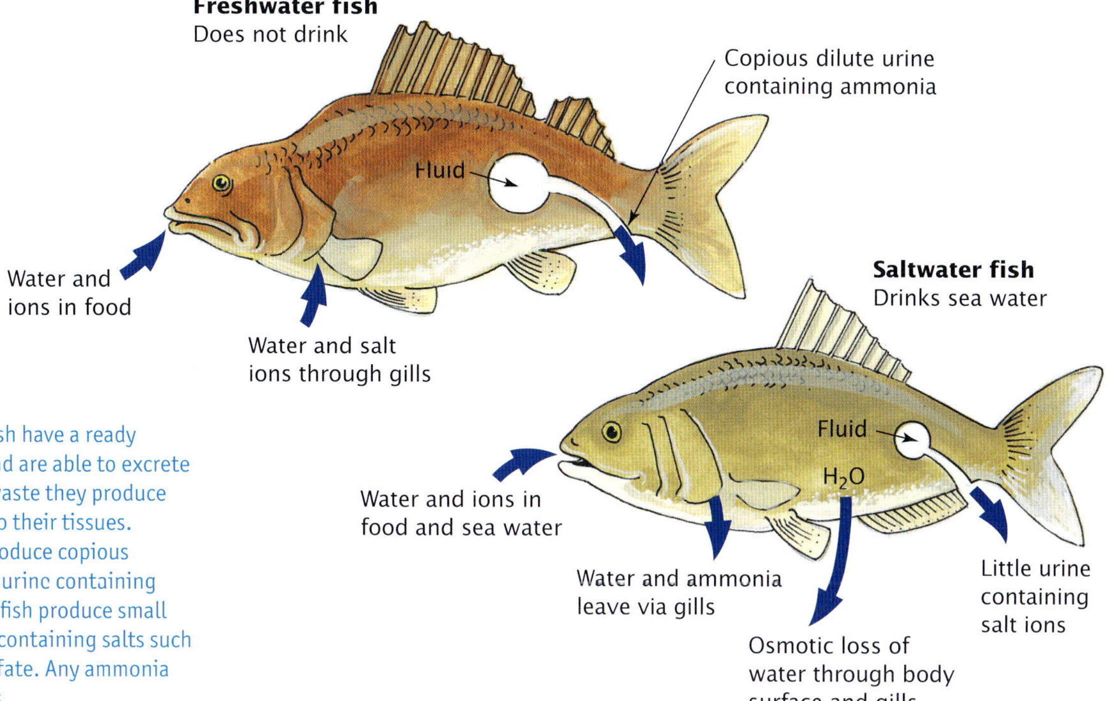

Figure 6.42 Fish have a ready supply of water and are able to excrete the nitrogenous waste they produce without damage to their tissues. Freshwater fish produce copious amounts of dilute urine containing ammonia. Marine fish produce small amounts of urine containing salts such as magnesium sulfate. Any ammonia leaves via the gills.

Excretion in insects

Insects have specialised excretory tubes called **Malpighian tubules**. They extend into the body cavity from the intestine and float freely in the blood of the open circulatory system (see figure 6.43). Wastes and fluid pass from the blood into the tubules where nitrogenous wastes are converted into uric acid crystals. Material in the tubules is then emptied into the intestine. Fluids are reabsorbed by the intestine and uric acid crystals are excreted through the anus.

Spiders also have Malpighian tubules.

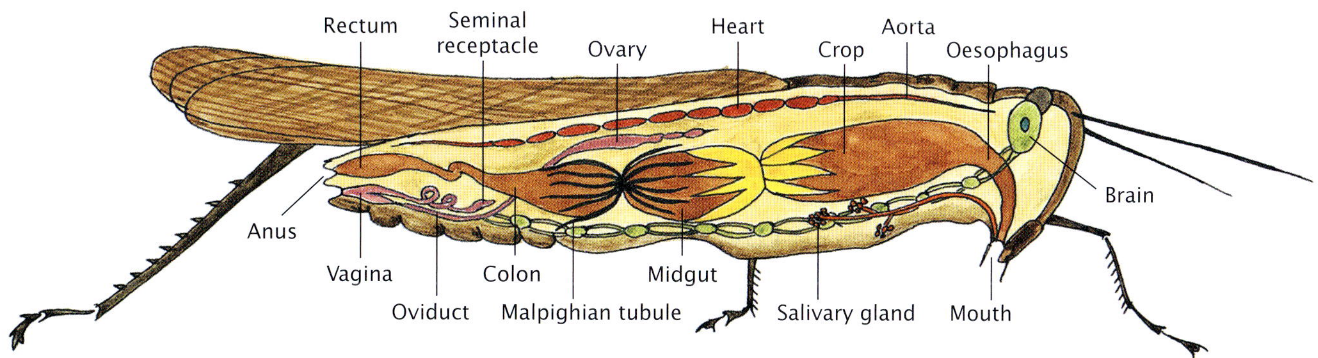

Figure 6.43 Wastes and fluid pass from the blood of a grasshopper into the Malpighian tubules. In what form is nitrogenous waste excreted in insects?

Comparing transport systems in animals

We have considered the transport systems in a range of vertebrate animals and in an insect. A comparison of the general features of each is given in table 6.3.

Table 6.3 Comparison of the general features of the transport systems in three different kinds of animals.

Feature	*Mammal*	*Fish*	*Insect*
nature of system	blood enclosed within vessels and has no direct contact with tissue cells	blood enclosed within vessels and has no direct contact with tissue cells	open system in which blood directly bathes tissue cells
oxygen carrier	contains pigment in red blood cells; high oxygen-carrying capacity	contains pigment in red blood cells; high oxygen-carrying capacity	has no function in carrying oxygen
carbon dioxide carrier	in solution in plasma as bicarbonate ions; some combines with haemoglobin	in solution in plasma	no function in carrying carbon dioxide
nutrient carrier	in solution in plasma	in solution in plasma	in solution in colourless blood
pumping mechanism	four-chambered heart	two-chambered heart	single tube heart
number of times blood travels through heart	twice per circulation within body	once per circulation within body	once per circulation within body

Insects have no oxygen-carrying pigment in their transport system. Although the insect transport system enables all cells to obtain sufficient nutrients, the amount of oxygen in solution is minimal. Insects require an additional system, the tracheal system, to ensure sufficient oxygen to meet the cellular respiration needs of their tissues.

Mammals have a far more efficient transport system than insects. Not only does it supply sufficient nutrients to cells, it also has a pigment, haemoglobin, in the red blood cells that confers a very high oxygen-carrying capacity on that system. In addition, the 'double-pump' heart provides a pressure that ensures very rapid circulation of the blood around the body. This provides a continuous supply of nutrients and oxygen to all cells.

Comparing excretory systems in animals

Different animals excrete their nitrogenous wastes in different forms. A summary of the kinds of excretory organs and the wastes they produce for humans, fish and insects is shown in table 6.4.

Table 6.4 Comparison of the nitrogenous excretory systems of humans, fish and insects

Characteristic	*Human*	*Freshwater fish*	*Saltwater fish*	*Insect*
structure that filters waste from blood	nephrons in kidneys	nephrons in kidneys	nephrons in kidneys, also gills	Malpighian tubules
length of loop of Henle	long	virtually none	long	not applicable
type of urine with respect to body fluids	hypertonic	hypotonic	isotonic	hypertonic (crystals)
main nitrogenous waste excreted	urea	ammonia	ammonia	uric acid crystals

The most concentrated urine that humans can produce is five times the concentration of dissolved material in their plasma.

Note that the basic structure that filters nitrogenous waste from blood in humans and fish is the nephron. The differences in the length of the loop of Henle are related to the differences in the need to conserve water. The loop of Henle is that region of the nephron where much of the water filtered by the nephron is reabsorbed into the bloodstream. Humans have a need to conserve water whereas freshwater fish have an ongoing supply from their environment.

Although saltwater fish do have a ready supply of water, their body fluids have a lower ion concentration than that of the sea water in which they live. Therefore they lose water across their body surfaces by osmosis. Saltwater fish conserve water by reabsorption of water in the kidney and hence produce only small amounts of urine. Most of their ammonia is removed across their gills. This contrasts with freshwater fish which have an unlimited supply of water to make copious volumes of urine to remove all the toxic ammonia they produce.

Insects lack the transport system of humans and fish that collects nitrogenous wastes from tissue cells and delivers them to kidneys. In insects, blood bathes the tissues and the free ends of Malpighian tubules float in the blood. Nitrogenous wastes diffuse from cells into blood, then into Malpighian tubules which convert the waste into uric acid crystals. These crystals pass into the gut and are discharged from the body by the digestive system. Hence, insects have no specific need for water for excretion from the body to occur, unlike humans and fish that require water to produce urine for discharge of nitrogenous waste.

KEY IDEAS

- Metabolism of protein results in the production of nitrogenous wastes.
- Different animals excrete their nitrogenous waste in different ways and through different compounds.
- In vertebrates, the kidney of the urinary system is the organ responsible for removing nitrogenous waste from the blood.
- The functional unit of the kidney is the nephron.
- In insects, Malpighian tubules remove nitrogenous waste from the blood.
- Different animals excrete different nitrogenous wastes because they have different amounts of water available to them.

QUICK-CHECK

17 List the materials that pass into Bowman's capsule when blood enters the glomerulus of a nephron.

18 What is the main excretory product in human urine?

19 In what kind of environment would you expect to find an animal that excretes ammonia?

20 What would you infer about an animal that had either no, or a very short, loop of Henle in its nephrons?

21 What is the main excretory product in insects?

22 Name one organism that produces:

a hypertonic urine

b hypotonic urine

c isotonic urine.

Transport in plants

Plants transport materials in **vascular tissue**, which is made of **xylem** and **phloem**.

- Xylem transports water and dissolved minerals. Some minerals, the macronutrients, are required in relatively large amounts for normal growth. Others, the micronutrients (also called trace elements), are needed in relatively small amounts.
- Phloem transports sucrose, produced in photosynthetic tissue, to other regions of a plant, as well as hormones and any other organic material made by the plant. Such a system is necessary because only cells containing chloroplasts can photosynthesise and all other living cells in a plant rely on photosynthetic cells for their carbohydrate nutrients.

The structure of these two tissues is outlined in figure 6.44.

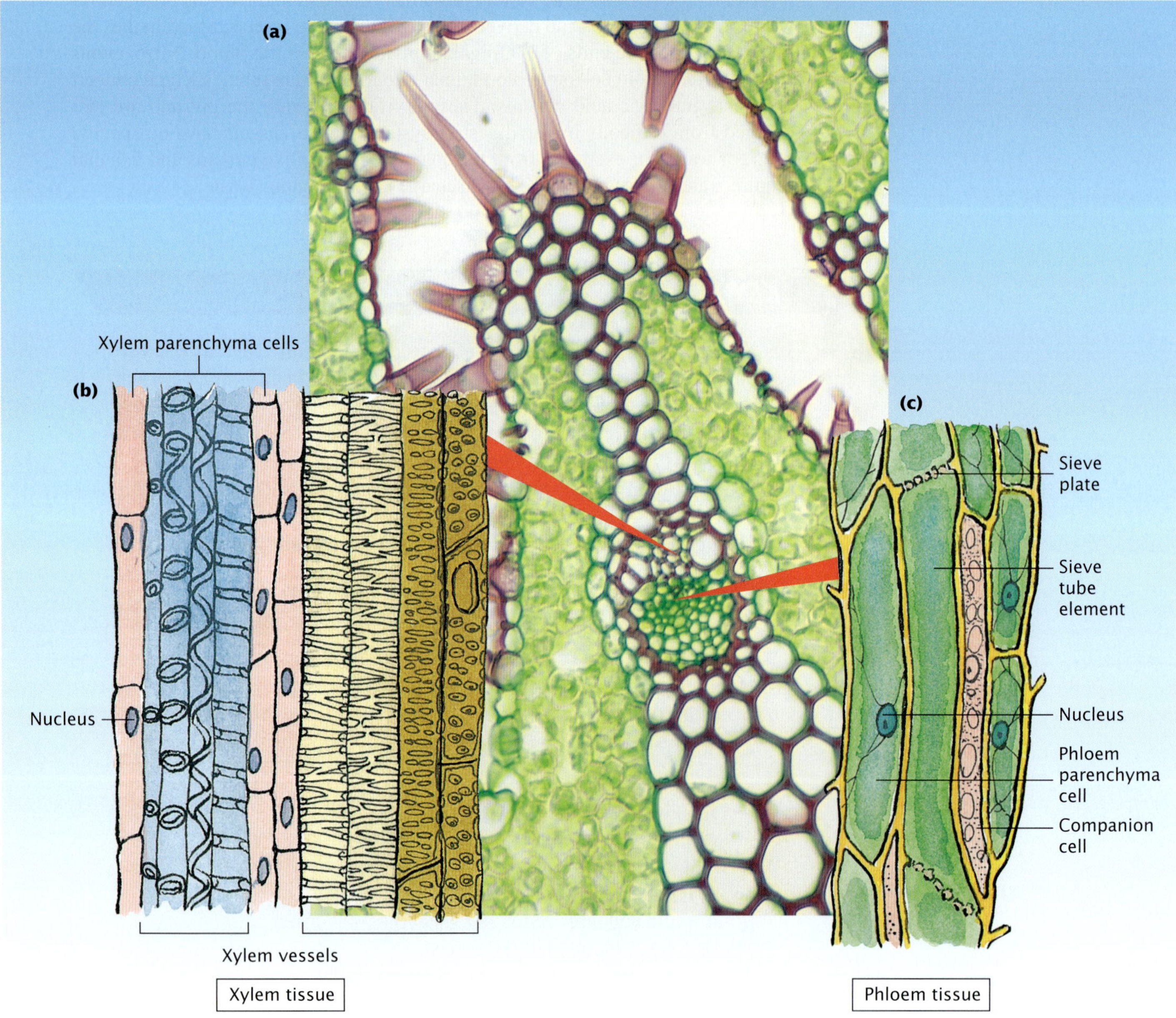

Figure 6.44 **(a)** Transverse section through portion of a marrum grass, *Ammonophila arenaria*, showing a vascular bundle. **(b)** Longitudinal section through xylem tissue. Note the xylem parenchyma cells and xylem vessels with their different kinds of lignified thickening. **(c)** Longitudinal section through phloem tissue. Note the phloem parenchyma, companion cells and the lack of nuclei in sieve tube elements.

BIOTECH

Measuring water flow in trees

(a)

Scientists can insert a measuring device called a heat-pulse meter into the trunk of a tree (see figure 6.45a) and calculate the water flow in the stem. The graph in figure 6.45b shows the results from the heat-pulse meter and indicates the volume of water used in early October by a large poplar box tree growing in Queensland.

(b)

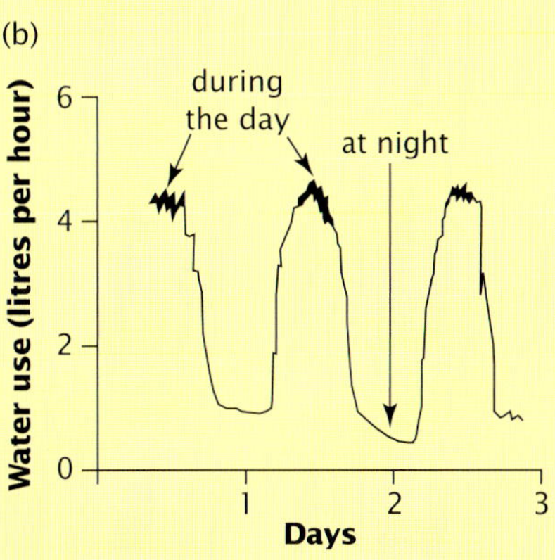

Figure 6.45 **(a)** Heat-pulse meter in the trunk of a tree **(b)** Graphical results from a heat-pulse meter in early October for a large poplar box tree growing in Queensland

How material moves in vascular tissue

Movement of water through xylem

Water is absorbed by root hairs, moves through the cortex into the xylem in the vascular tissue, and is transported throughout a plant, often to great heights. If a plant relied on air pressure alone to force water in the soil into the roots and up the stem, the maximum height of any plant would be about 10 metres. Some trees are up to 100 metres tall and so there must be some special features that make it possible for water to get to the top of such trees. Energy provided by the sun plays a major role in the process.

You will know from your study of photosynthesis that gases enter and leave the leaf through the stomata. Most of the water lost by the plant in transpiration moves out through stomata. What is the chain of events when transpiration occurs?

The walls of the mesophyll cells are moist and the air spaces around them contain water vapour. As the sun shines, stomata open and gases are able to diffuse in and out of the leaf. Water vapour moves through the stomata into the air surrounding the leaf. Water evaporates from the wall of the mesophyll cells to replace that lost from the air spaces.

As water evaporates from the surface of the mesophyll cells, water moves out of the cells to ensure that the walls are kept moist. In turn, water moves from the small xylem vessels into the mesophyll cells.

When water moves out of the xylem in the leaf, it is replaced by water that is sucked in from the xylem leading into the leaf. In effect, water vapour moving out through the stomata sets up a chain reaction in which water in xylem is moving through the vessels by the pulling or sucking movement of water ahead of it. Because of the pulling action, water in xylem vessels is under tension that is transmitted along the whole water column (see figure 6.46).

What keeps water molecules together as they are pulled through a plant? Why does the column not break?

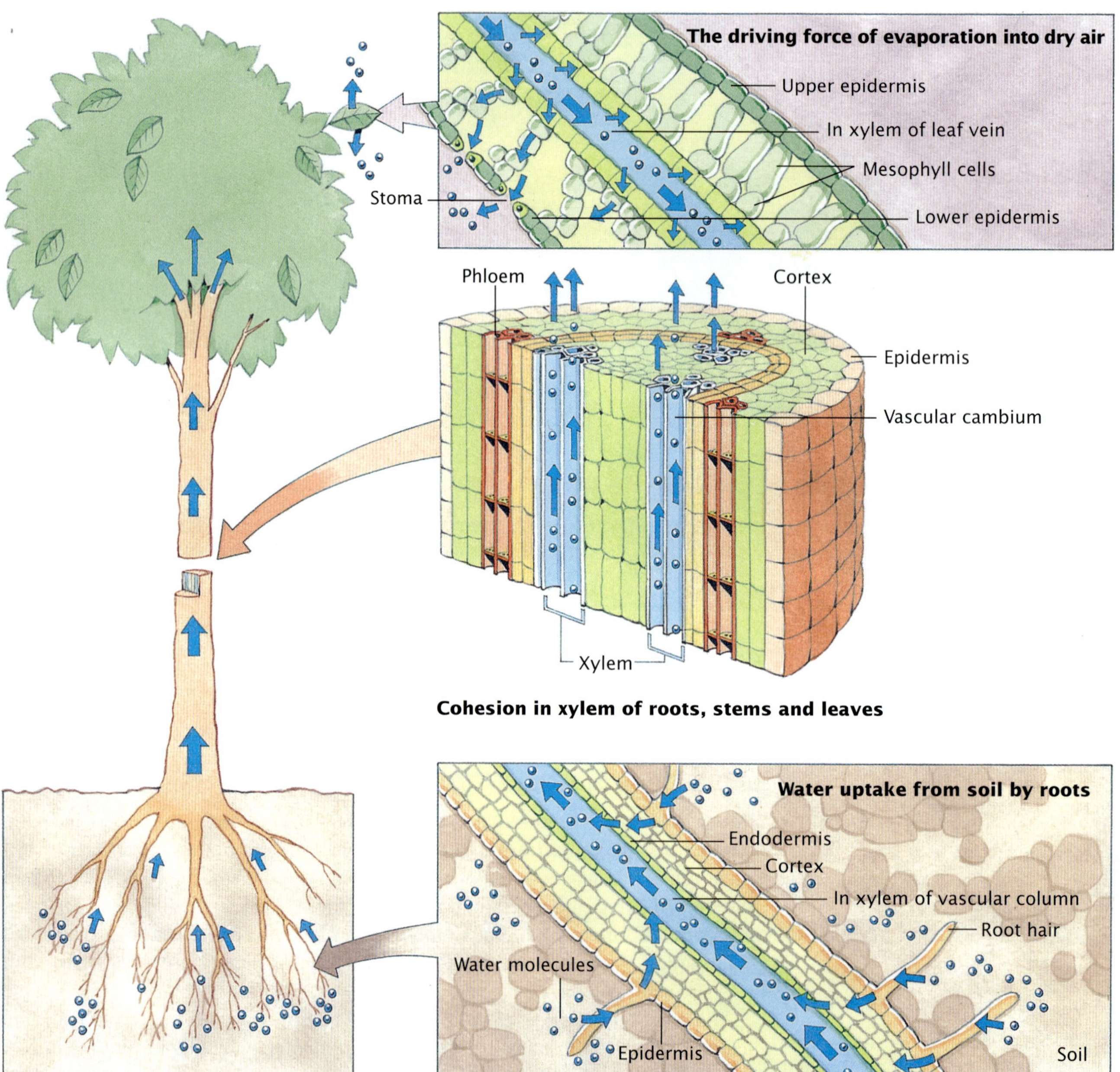

Figure 6.46 Water is absorbed from the soil by roots and passes into xylem tissue that carries it to all parts of the plant. Some of the water is used by cells but much passes through stomatal pores in leaves and is lost by transpiration.

Water has a tensile strength because of the **cohesion** of the molecules. They tend to stick together and the smaller the diameter of the tube the molecules are in, the greater the tensile strength. Water molecules stick to the walls of xylem vessels (see figure 6.47). The cohesive property of water molecules prevents their pulling apart or pulling away from the walls of the xylem vessels as they are sucked up through the xylem (see figure 6.48 on page 161).

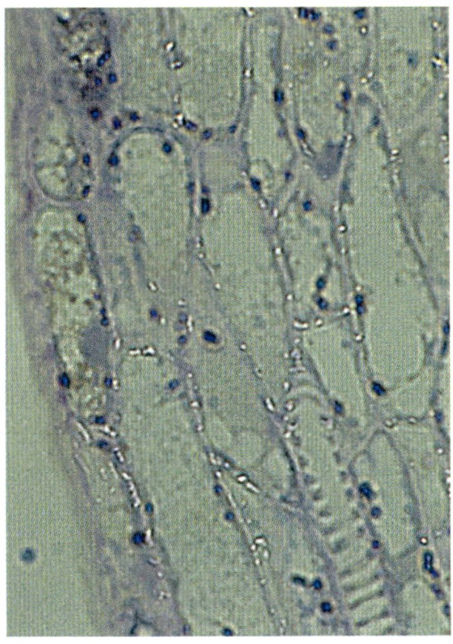

Figure 6.47 Longitudinal section of xylem tissue from a dicotyledon *Acacia* sp. Note the elongated xylem vessels, one with spiral thickening.

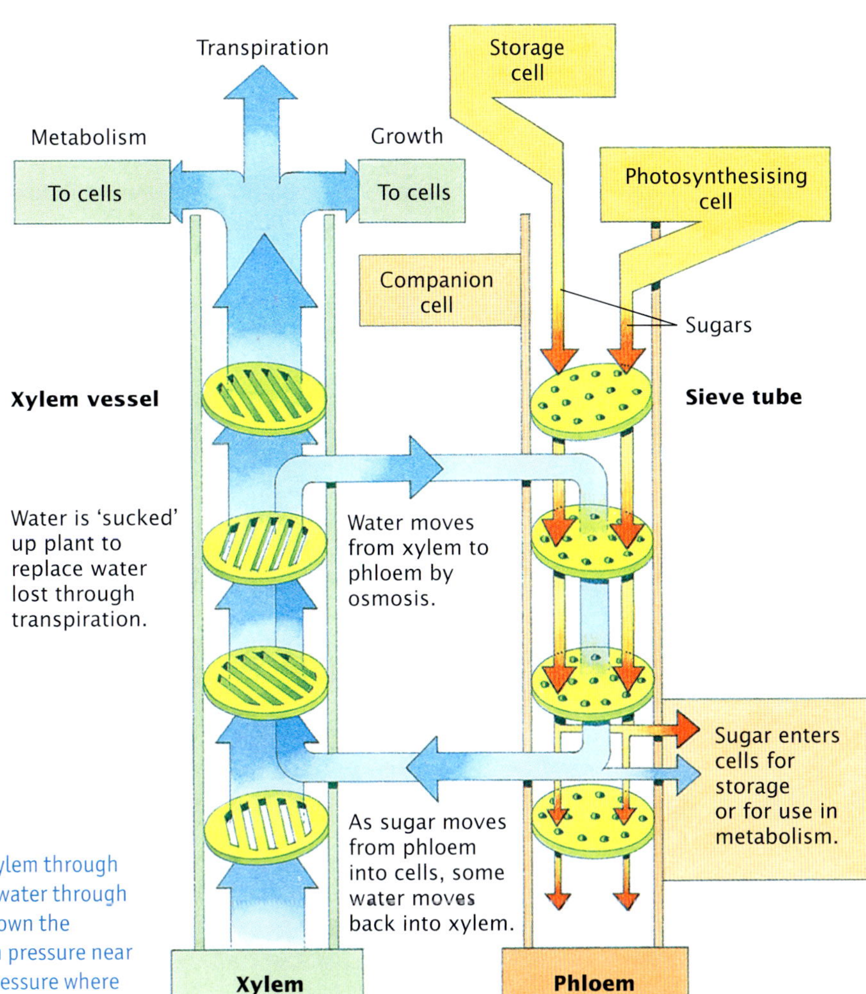

Figure 6.48 (left) Loss of water from xylem through transpiration and use in metabolism pulls water through xylem vessels. (right) Sugars are pushed down the phloem. Sugar moves from a region of high pressure near its site of production to regions of lower pressure where sugars are used.

(a)

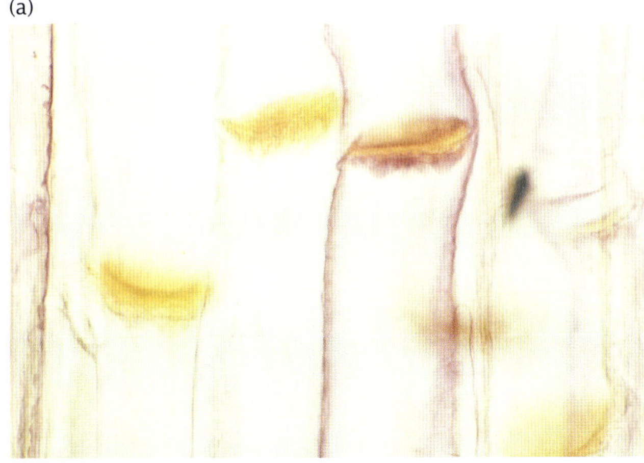

(b)

Figure 6.49 Sections of the phloem of the dicotyledon *Cucurbita* **(a)** Longitudinal section showing sieve tubes in the stem. Note the transverse sieve plates. **(b)** Transverse section showing sieve plates with sieve pores

Movement of organic substances through phloem

Soluble organic substances are transported by phloem tissue (see figure 6.48). These substances travel from the leaves, where they are synthesised, to other parts of the plant where they are used or stored. Phloem also transports organic substances from storage sites to other regions of the plant for use.

The transport of organic material through a plant is called **translocation**.

Sucrose: from leaves to phloem

Photosynthesising leaves produce sugars so are the source of organic matter for all parts of the plant. Sugars are actively transported, mainly as a sucrose, from leaves to other parts of the plant via sieve tubes. The energy required for the active transport is provided by companion cells. As the concentration of sugar in phloem increases, water moves from the xylem by osmosis into the sieve tubes in that region. This increased volume of water increases the pressure in sieve tubes (see figure 6.48, page 161). Note the sieve tube elements in figure 6.49.

Glucose is produced during photosynthesis. It is either stored locally as starch or transported by phloem in the form of the disaccharide sucrose to other parts of the plant.

Sucrose exits the phloem

Sucrose is actively transported from the phloem into cells (see figure 6.48) where it is either stored as starch or broken down into the monosaccharides used in metabolism. Parts of a plant that use organic material made elsewhere in the plant are called sinks; for example, storage tissues such as potatoes and in actively growing tissues.

When sugar leaves phloem to enter a sink, the effect is to increase the concentration of water in the sieve tube in that region. Water moves out of the sieve tube. The loss of water from the sieve tube close to the sinks results in a decreased pressure in the sieve tubes in that region.

In summary, a region of higher pressure exists in sieve tubes near the sugar source and a region of low pressure exists near the sugar sink. This difference in pressure pushes sugar-rich sap through sieve tube columns from the sources (leaves) to the sinks (for example, roots). When the sugar arrives at sinks, it is used in different ways:

- It is stored as starch.
- It is used as subunits for building structural components of cells.
- It is used as an energy source in cellular respiration (see figure 6.50).

The sap of *Sarcostemma australe* is one of many Aboriginal traditional medicines and is used on small skin lesions.

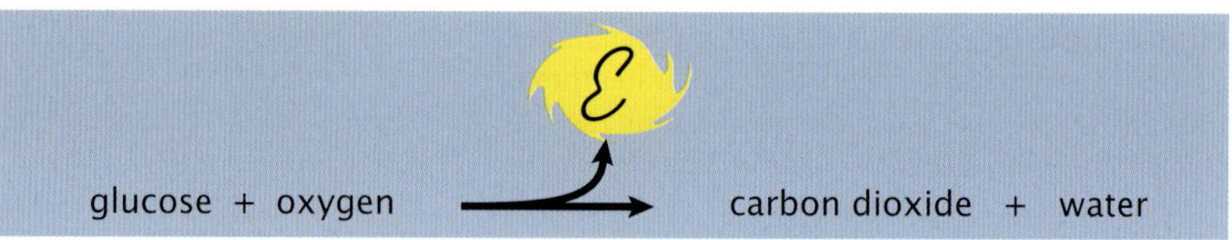

Figure 6.50 Sucrose may be converted to glucose and fructose which are metabolised to provide energy for the plant.

RINGBARKING

In the early days of farming in Australia, settlers used the technique of ringbarking to kill trees. In ringbarking, a ring about 30 centimetres wide is cut around the trunk of a tree. The bark and the outermost layer of the vascular tissue, the phloem, are removed. Carbohydrate produced in leaves can no longer be transported to the roots. The roots die and can no longer absorb water. Without water, the remainder of the tree, including the leaves, dies.

Why was ringbarking used by farmers? They required grazing land for sheep and cattle. Killing a tree by ringbarking meant that once the leaves on the tree died, there was plenty of light to allow growth of pasture up to the base of the tree. In addition, the tree no longer competed with grass for nutrients and water.

The alternative of chopping trees down and removing them was not really an option. Chainsaws and bulldozers were not available and ringbarking provided a much quicker way than chopping down and removing the trees immediately. A man with a sharp axe could ringbark dozens of trees in a day. Dead trees could be removed at a later stage and burnt or used for firewood as required. However, the excessive removal of trees has damaged ecosystems.

Figure 6.51 These trees have been killed by ringbarking. Once the land had been 'claimed' by ringbarking, settlers often took many years to clear the land completely.

KEY IDEAS

- Vascular tissue in plants comprises xylem and phloem.
- Water and mineral ions enter a plant through the roots and are transported throughout a plant in the xylem.
- Sugars are made by chloroplasts in chlorophyll-containing tissues and are conducted to other parts of the plant through phloem.

QUICK-CHECK

23 Why are root hairs so important to the water transport system of a vascular plant?

24 Comment on the validity of the following statement: Xylem tissue and phloem tissue each contain different kinds of cells.

25 By what mechanism does water reach the tops of tall trees?

26 In which part of the vascular tissue would you expect to find a higher concentration of:

a water

b sucrose

c plant hormone

d mineral ions?

Gaseous exchange in plants

Many plants have specialised structures for exchanging gases.

Green plants carry out photosynthesis, a process in which **light energy** is transformed into **chemical energy** stored in sugars. Such plants are called **producers**. In a typical producer, such as a terrestrial flowering plant, the complex series of reactions in photosynthesis can be summarised as in figure 6.52.

For such a process to occur in a cell, carbon dioxide must be available. You will also note that oxygen is a product of photosynthesis. Where is the photosynthetic tissue and how does it obtain the carbon dioxide it needs? What happens to the oxygen produced? Although some carbon dioxide is produced by plant cells as they carry out respiration, the amount produced in this way is far too little to provide sufficient for photosynthesis and the subsequent growth that does occur. Carbon dioxide must come from the air surrounding a plant.

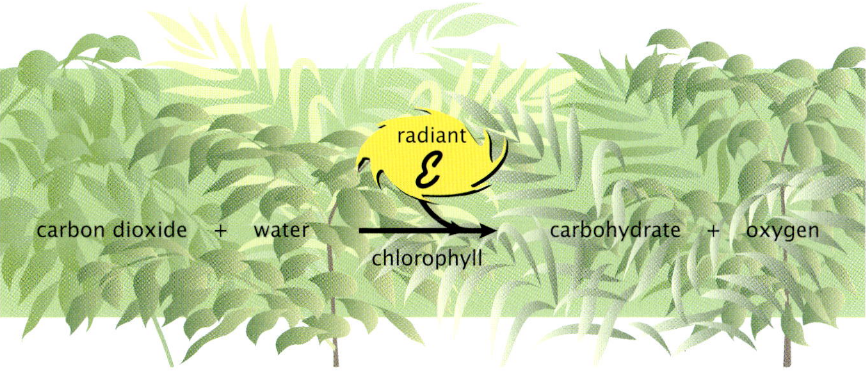

Figure 6.52 The reactions of photosynthesis

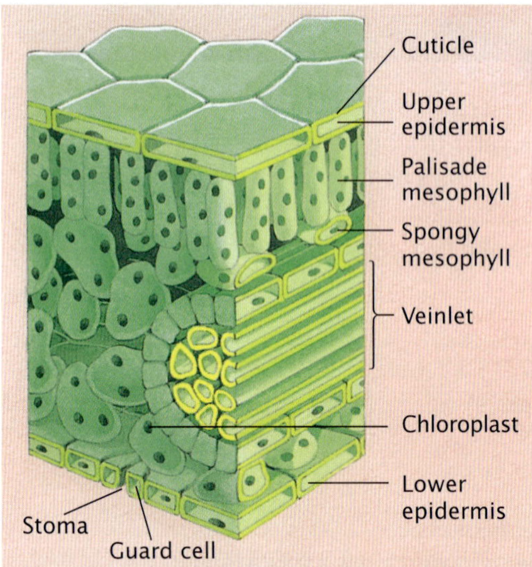

Figure 6.53 The arrangement of tissues in a dicotyledonous leaf. Note the chloroplast-containing cells and a stoma in the lower epidermis.

In a terrestrial flowering plant, photosynthetic cells are principally located in green leaves. A cross-section of a typical green leaf is shown in figure 6.53. Note the chloroplast-containing cells. Also note the **stoma**, a small opening, situated in the lower epidermis of the leaf.

Stomata: gateways in leaves for gaseous exchange

Leaves contain many pores, known as stomata (singular: stoma) (see figure 6.54). Each stoma is surrounded by two **guard cells**. The stoma opens or closes depending on the turgor of the guard cells. If the guard cells have a high water content then they are **turgid** and the pore is open. As the guard cells lose water, they become more **flaccid** and the pore closes.

Two special features of guard cells cause them to change shape in a particular way as they take up water. In many, the wall adjacent to the pore is thicker than the wall away from the pore. In addition, there are a number of cellulose micro-fibrils in guard cells and these are oriented in a way that makes the wall adjacent to the pore less elastic than the wall away from the pore. When water enters guard cells, the wall away from the pore stretches more than the inner wall and this results in an oval pore appearing in the centre as the two guard cells expand (see figure 6.55).

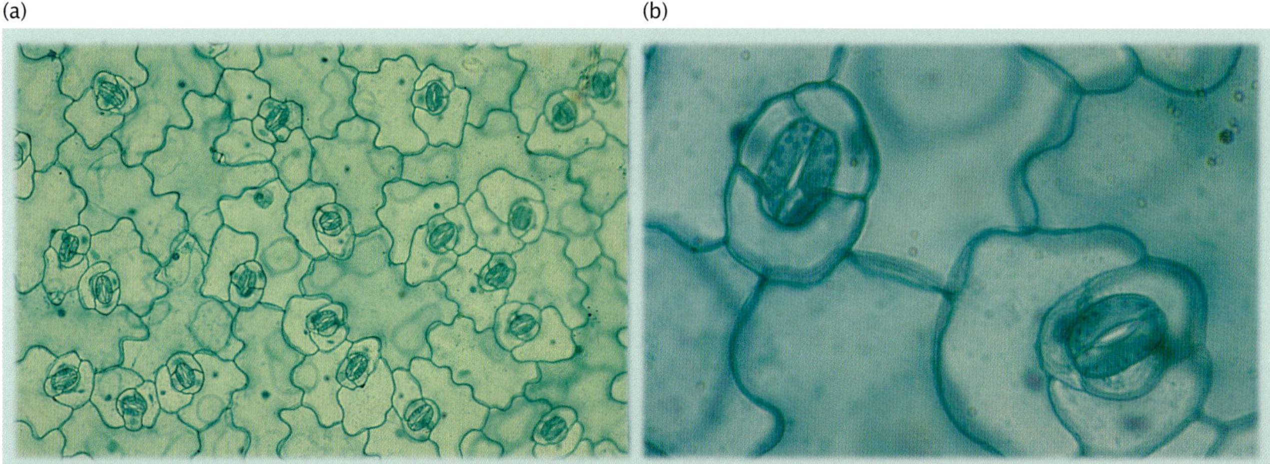

Figure 6.54 Stomata in dicotyledon leaf **(a)** Surface view of leaf (× 100) showing distribution of stomata **(b)** Surface view of leaf (× 400). Note epidermal cells, guard cells and their chloroplasts, and stomal pores.

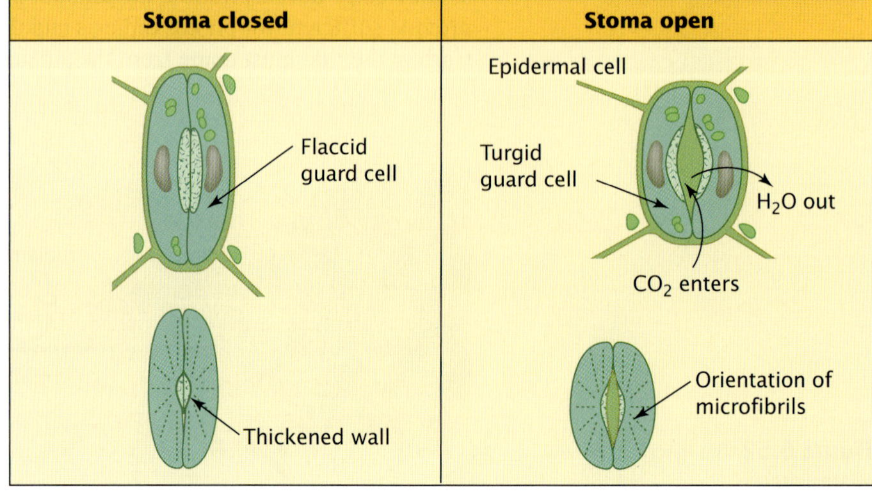

Figure 6.55 Inner and outer walls of guard cells may be of different thicknesses. An outer thinner wall can stretch more than a thicker inner wall. Microfibrils in guard cells also influence the extent to which walls of guard cells can stretch. As the outer walls of guard cells stretch, the stoma (pore) opens.

The 'dew drops' at the tip of some leaves in early morning in fact come from within the plant. The drops are excess water absorbed by roots when stomata are closed. It is pushed through the plant and out through hydathodes, permanently open pores.

What causes the change in guard cell turgor? Guard cells contain chloroplasts, whereas the surrounding epidermal cells do not. It was once argued that photosynthesis in guard cells was responsible for the change in water uptake and hence turgor. As light increases at the start of a day, photosynthesis increases and sugar is produced. The sugar content is higher than that in surrounding cells and this causes water to move from surrounding cells into the guard cells which swell and the stomatal pores open.

It is now accepted that although this argument seems to make sense it has some weaknesses. Stomata can open in very dim light which is insufficient for the production of much sugar. At other times, the opening occurs more quickly than would be possible if production of sugar was the main factor. Although the full story is not understood, it is known that carbon dioxide concentration declines because photosynthesis commences as light becomes available. This change results in a significant number of ions, particularly potassium ions, moving into guard cells. Water follows the ions into the cells.

Stomata in the epidermis, particularly the lower epidermis (refer table 6.5), of leaves provide the doorways through which carbon dioxide enters the air spaces below the epidermis. From there, it diffuses across cell membranes and is utilised in photosynthesis. Some of the oxygen produced in photosynthesis will be used in cellular respiration by the cells but the excess diffuses from the cells, into air spaces within the leaf and then from the leaf, via stomata, into the atmosphere.

Table 6.5 Average number of stomata per square centimetre on different leaves. Note the difference between the numbers on upper and lower surfaces on most plants. Monocotyledons have relatively even numbers on both surfaces.

Plants	*Upper epidermis*	*Lower epidermis*
dicotyledons		
apple (*Malus sylvestris*)	0	29 400
bean (*Phaseolus vulgaris*)	4 000	28 100
cabbage (*Brassica oleracea*)	14 100	22 600
nasturtium (*Tropaeolum majus*)	0	13 000
potato (*Solanum tuberosum*)	5 100	16 100
tomato (*Lycopersicon esculentum*)	1 200	13 000
monocotyledons		
corn (*Zea mays*)	5 200	6 800
oat (*Avena fatua*)	2 500	2 300

Open stomata: potential loss of water vapour

Under conditions of an adequate water supply, the air spaces within a leaf are saturated with water vapour. Generally the air surrounding a leaf has a much lower concentration of water vapour than the air within leaf spaces. Under these conditions, inevitably, open stomata allow water vapour to diffuse from within a leaf to the surrounding atmosphere. Under dry conditions, there is a danger that this loss of water vapour may be excessive, causing leaves and ultimately a whole plant to wilt.

The loss of water vapour from the surfaces of a plant is called transpiration *and occurs mainly through the stomata.*

Some plants have special characteristics to reduce the loss of water vapour from leaves during times of stomatal opening. One example is *Hakea*, an Australian plant whose leaves have sunken stomata (figure 6.56, page 166).

Hakea stomata are located well below the leaf surface. The upper surface of the leaf epidermis is thickened (figure 6.56b) and the whole epidermis is propped up by thick-walled sclerotic cells. Note the shape of the 'pit' of still air above the stomal opening. This pit becomes saturated with water vapour and forms an effective 'barrier' that reduces the rate of loss of water vapour from within the leaf when the stoma is open.

You will consider additional features of plants that assist in the reduction of water loss in your further studies of biology.

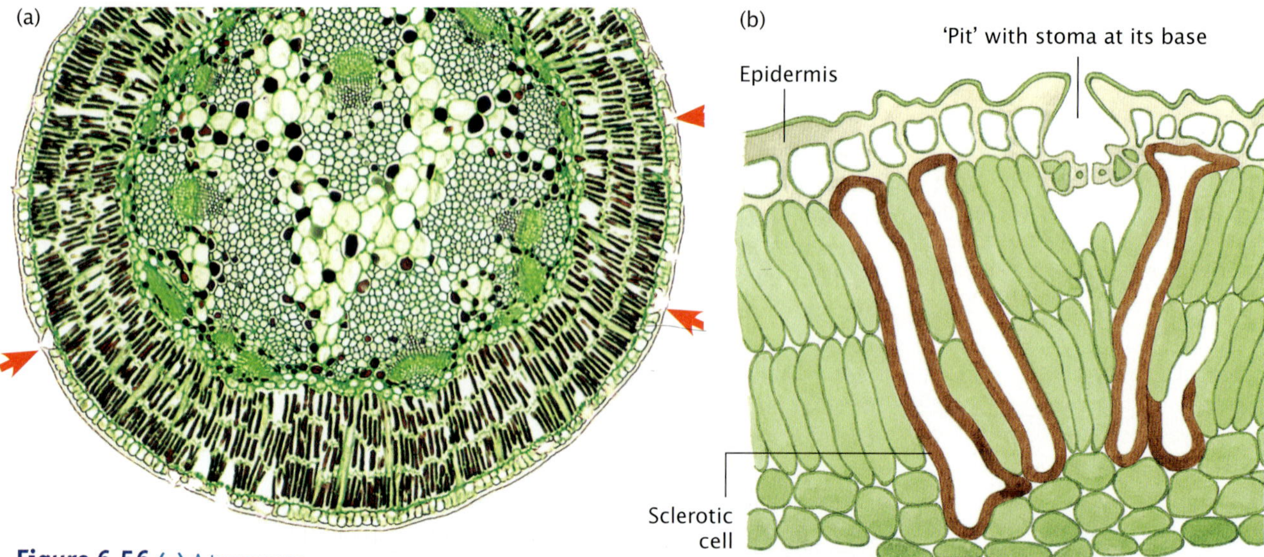

Figure 6.56 **(a)** A transverse section through a *Hakea* leaf with indications (arrows) of positions of some of its stomata **(b)** Details of a sunken stoma of *Hakea*. Note the position of the stoma below the surface of the leaf and the 'pit' above the stomal opening.

Lenticels: gateways in stems for gaseous exchange

Stems and roots generally have a woody or corky outer layer that is impervious to the movement of water and gases and yet the living tissues beneath these layers must still obtain oxygen if they are to survive. **Diffusion** over the long distances between stomata in leaves and living tissue in woody stems is just not possible. The living tissues in woody stems obtain their oxygen through special openings called **lenticels**. Lenticels develop when parenchyma cells below the bark or cork reproduce and erupt through the outer layers of the stem. Oxygen is able to diffuse into these loosely packed parenchyma cells and carbon dioxide to diffuse out (figure 6.57).

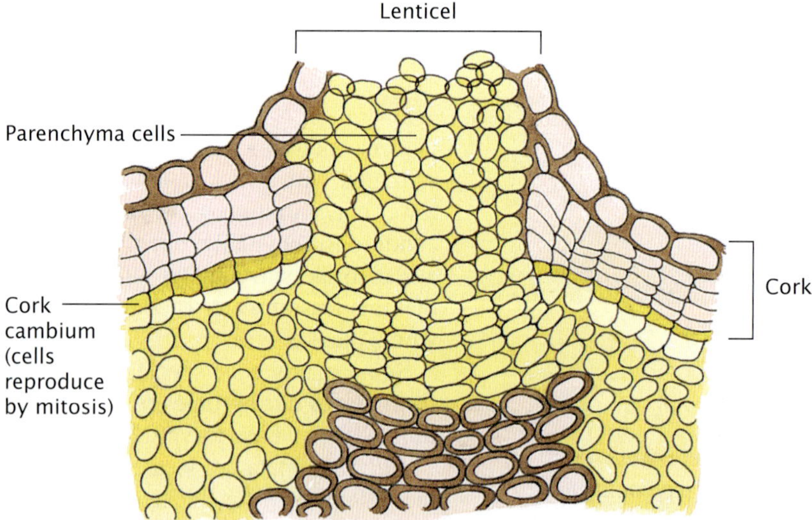

Figure 6.57 Transverse section through the lenticel in a woody stem

Lenticels have no opening or closing mechanism in the same way as stomata. Nor do parenchyma cells below lenticels contain chlorophyll.

Fruit also has living tissue below tough outer skins. Consider an apple that contains living cells that we eat. The tough outer skin of an apple contains many lenticels that allow movement of gases into and out of this living tissue. Examine an apple. The tiny dots you see are the lenticels.

Lenticel openings are vulnerable to infection. Bacteria, viruses and fungi can readily invade the soft tissue exposed by the formation of a lenticel. Refer to figure 6.58.

Figure 6.58 A Granny Smith apple. Note the many tiny dots — lenticels. A fungal infection, *Pezicula alba*, has occurred through at least three of the lenticels and penetrated into the soft flesh under the skin.

Gaseous exchange across root hair surfaces

Living cells in roots also require oxygen for respiration. The outer surfaces of fine root hairs are kept moist by water in the soil. Porous soil also contains small cavities filled with air (figure 6.59). Oxygen diffuses into root hairs and carbon dioxide diffuses out.

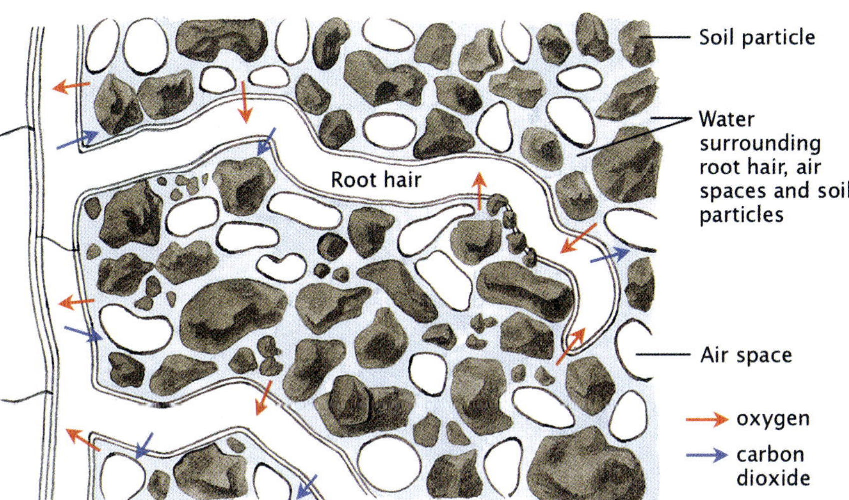

Figure 6.59 Root hairs in the soil. Note the direction of gaseous exchange.

KEY IDEAS

- Plants have a variety of surfaces through which gases are exchanged with the external environment.
- Photosynthetic plants require carbon dioxide from their external environment.
- Photosynthetic plants release oxygen into their external environment.

QUICK-CHECK

27 What are the gateways in leaves that allow gaseous exchange between a plant and its external environment?

28 By what means does a plant control the aperture of these gateways?

29 What are plant lenticels?

30 How does gaseous exchange take place between the roots of a plant and the external environment of those roots?

Excretion in plants

Plant cells tolerate far greater fluctuations in their environment than do animals. Animals have highly developed excretory systems to ensure that waste materials are removed from cells as soon as they are produced. Plants have no equivalent system because they either use the waste or store the waste in situations that pose no problem for the continued function of the cell.

In plants, the term **ergastic substances** is used instead of excretory products. These substances include stored food materials such as starch grains and other cell inclusions, often insoluble compounds, some of which are waste products.

Carbon dioxide produced by respiration of cells can be used in photosynthesis during the hours of light. Excretory products such as resins and oils accumulate in vacuoles and sometimes in cell walls. Wastes may be stored in solution or in crystal form in vacuoles (see figure 6.60). For example, spinach stores oxalic acid crystals in the vacuoles of leaf cells.

ODD FACT

The rhizomes of rhubarb, *Rheum rhaponticum*, contain druse crystals. These crystals are aggregates of calcium oxalate.

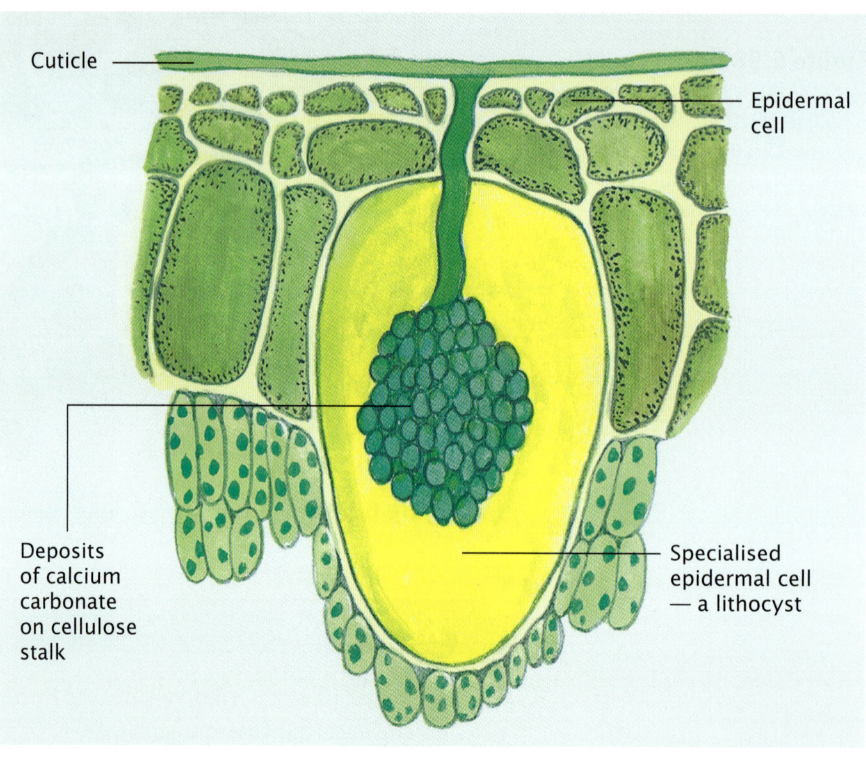

Figure 6.60 Deposits of calcium carbonate form on cellulose stalks in specialised and enlarged epidermal cells, called lithocysts, in the leaf of the india rubber plant, *Ficus elastica*.

The modular nature of plants enables them to cope very well with stored wastes. Some plant modules can be lost without detriment to the plant. Plants shed their leaves or other parts, and the accumulated wastes are, in effect, removed from the body of the plant. Before leaf fall, ions such as potassium and phosphorus that are required by the plant, may be selectively moved from the leaves to other parts of the plant. Unwanted ions such as sodium and chloride remain in the leaves that are shed by the plant.

Some plants have special features that enable them to eliminate substances that have been absorbed by the roots along with water. This is not true excretion but does demonstrate that plants are able to remove unwanted substances. Many plants, such as mangroves, have their roots immersed in salt water for much of the time. These plants must absorb water to replace that lost by transpiration. Although the roots of these plants filter out about 80 per cent of the salt in the water, a significant amount of unwanted salt is still absorbed by the roots and must be eliminated by the plant.

ODD FACT

The leaves of many species of mangrove and saltmarsh plants have salt-secreting glands, which control the salt content of the leaves. Secretion of salt by mangroves is not excretion because the salt is not a waste product of metabolism carried out within the mangrove plants.

Comparing plants with mammals

The transport systems of plants and mammals have some similarities. They are both made up of a network of fine tubes that penetrate all the tissues and come close to every cell of the multicellular organism. They each supply living tissues with the nutrients necessary to sustain life. However, there are important differences between the transport systems of the two groups of organisms.

- Mammals require the blood to provide a continuous supply of oxygen. Compare the relative need of oxygen by some mammal and plant cells (see table 6.6). The presence of the heart as a pumping station in the blood circulatory system ensures a constant steady flow of blood. Plants, on the other hand, lack an equivalent organ.
- Mammals have special pigmented cells in the blood. Haemoglobin in red blood cells carries oxygen in the blood close to tissues where the oxygen is required. The presence of haemoglobin results in the blood carrying much more oxygen than if the animal relied only on oxygen in solution. The increase in oxygen caters for the high energy needs of cells. Plants have no such cells or pigment in the transport systems, as they do not have the same needs for mass movement of oxygen. Most of their oxygen needs are met by diffusion or by the oxygen produced in cells during photosynthesis.
- The bulk intake of food and oxygen by mammals leads to large amounts of waste being produced that must be transported to specialised organs. In particular, carbon dioxide is transported to lungs and nitrogenous wastes are transported to kidneys for removal. Plants produce their own food and have no need for disposal. Carbon dioxide produced in respiration is used in photosynthesis, and oxygen produced in photosynthesis is used in aerobic respiration. Excess oxygen diffuses into the atmosphere.

ODD FACT

Plants rely on energy from the Sun to facilitate the movement of material from one site to another.

Table 6.6 Oxygen requirement in living organisms (one microlitre (μL) = one-millionth of a litre). Note that, in mammals, the rate of oxygen consumption decreases with increasing body size.

Organism	*Example*	*Oygen requirement (μL per gram body mass per hour)*
mammals	mouse (*Mus musculus*), at rest	2500
	cat (*Felix catus*), at rest	680
	dog (*Canis familiaris*), at rest	330
	human (*Homo sapiens*), at rest/active	200/4000
	elephant (*Loxodonta africana*), at rest	148
plants	typical plant cell	25
	barley grain, dry	0.06
	barley grain, germinating	108

KEY IDEAS

- Plants do not have excretory organs equivalent to those found in animals.
- Plants are able to store wastes in ways that are not detrimental to the plant.

QUICK-CHECK

31 Explain how the modular structure of plants overcomes their lack of excretory organs.

32 What is one similarity between the transport systems of plants and mammals, and what is one difference?

BIOCHALLENGE

1 Examine the simplified diagram of a portion of a mammalian blood circulatory system. Arrows indicate the direction of blood flow.

a Name a metabolic waste transported in vessel A.

b Name a high-energy molecule transported from the liver by vessel B.

c What important molecule does blood in vessel C deliver to the liver?

d What is the name of vessel D and what particular characteristic does it have that is *not* shared by all veins?

e Name one compound removed by the kidney before blood enters vessel E.

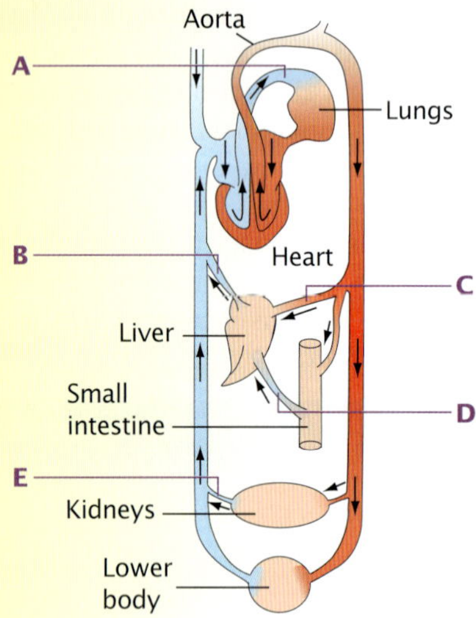

2 A scientist, investigating the loss of water from leaves, made measurements of stomatal openings in a range of different conditions, each over 24-hour periods. The different conditions were:

I a sunny humid day

II a sunny dry day

III a cloudy day.

The results obtained were graphed and are presented below.

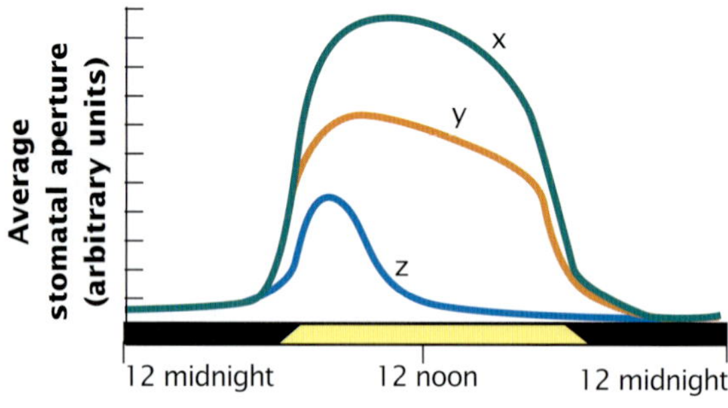

a The plants used were all of the same strain, raised under identical conditions. For what reason would the scientist choose plants satisfying these criteria?

b For several days before the experiments as well as during the experiments, all plants were given the same amount of watering. Why was the water treatment for all groups identical?

c Which graphs of X, Y and Z represent the results obtained for the three different conditions? Outline the reasons for your choices.

3 A plant was grown for one hour in a nutrient solution containing radioactive phosphorus. It was then removed, placed in a non-radioactive nutrient and monitored for six hours. Special photographic techniques were used to obtain images that tracked the radioactive material in the plant. The darker the image, the more radioactive material at that site.

Consider the following illustrations showing the plant at two different stages and answer the questions below.

a Where is most of the radioactivity located after one hour incubation?

b What other plant tissue shows radioactivity after one hour incubation? Explain your answer.

c Note the even distribution of radioactivity in young leaves that have grown during six hours. How would you account for this distribution? What tissues are involved?

(a) One hour after incubation in radioactive nutrient

(b) Six hours after being taken out of radioactive nutrient

CHAPTER REVIEW

Key words

alveolar duct
alveoli
ammonia
angioplasty
aorta
atrium
bladder
blood circulatory system
Bowman's capsule
bronchi (sing. bronchus)
carbon dioxide
chemical energy
closed circulatory system
cohesion
countercurrent
diffusion
ergastic substances
erythrocytes
excretion
flaccid
gaseous exchange
gills
glomerulus
glycogen
guard cells
heart
hypertonic
hypotonic
isotonic
kidneys
lamellae
larynx
lenticels
leucocytes
light energy
loop of Henle
lungs
lymphatic system
Malpighian tubules
nephrons
nitrogenous wastes
open circulatory system
operculum
phagocytic
pharynx
phloem
producers
pulmonary artery
pulmonary veins
respiratory system
sphincters
spiracles
stem cells
stoma
tissue fluid
trachea
tracheoles
translocation
tubule
turgid
urea
ureter
urethra
uric acid
urinary system
urine
vascular tissue
veins
ventricle
venules
xylem

Questions

1 *Making connections between concepts* ▸ Use at least eight of the chapter key words to construct a concept map.

2 *Analysing and commenting on an opinion* ▸ It has been suggested that, because human and fish circulatory systems are both 'closed' systems, they are equally efficient at circulating blood around the body. Comment on the validity of this idea.

3 *Applying your understanding to a new concept* ▸ Figure 6.61 shows the circulatory system of a fish. 'A' and 'V' indicate the auricle and ventricle of the heart. The arrows show the direction of flow of blood in the system. Consider the regions M, N, O, P and Q.

a Which of the regions would you classify as artery and which as vein?

b How would the concentration of oxygen in the blood in region P compare with that in regions M and Q? Explain.

c How would the blood pressure in regions P, Q and M compare with each other? Explain.

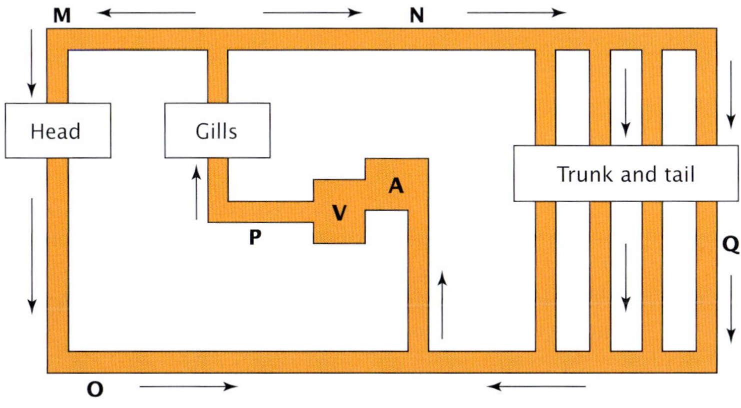

Figure 6.61

4 ***Interpreting biological information*** ▸ Two electrocardiograms (ECGs) are shown in figure 6.17 (see page 138). The person with the normal ECG had a heart rate of 56 beats per minute. The abnormal ECG was of a person with ischaemia, a condition in which a reduced oxygen supply has weakened heart muscles. This person had a heart rate of 137 beats per minute. Suggest why the ischaemic heart had a much higher rate than the normal heart.

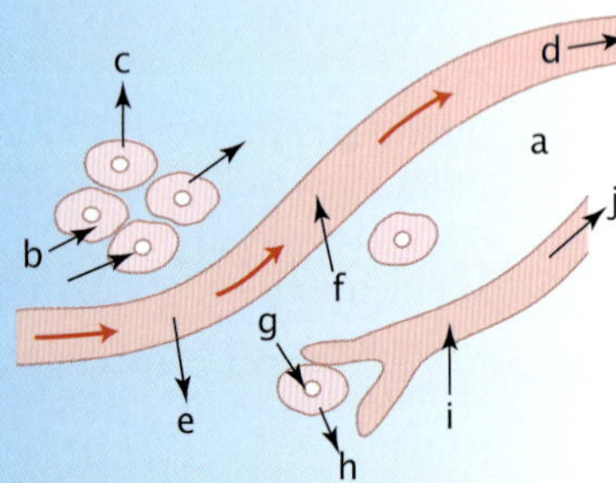

Figure 6.62

5 ***Applying biological understanding*** ▸ Figure 6.62 shows a few animal tissue cells close to a capillary network. The arrows show the direction of movement of various materials. Which of the following materials could each of the letters, a to j inclusive, represent?

carbon dioxide	tissue fluid	oxygen	lymph
nutrients	blood	nitrogenous waste	

6 ***Applying biological understanding*** ▸ Explain the following statements.

a Increasing your rate of exercise increases your rate of breathing.

b After sleeping with blocked nasal passages, you wake with a dry mouth.

c In humans, the surface area of lung alveoli is far greater than the surface area of the skin.

d Carbon dioxide is sometimes used in fumigation for an insect infestation in a confined space.

e Countercurrent flow is important in the oxygenation of fish blood.

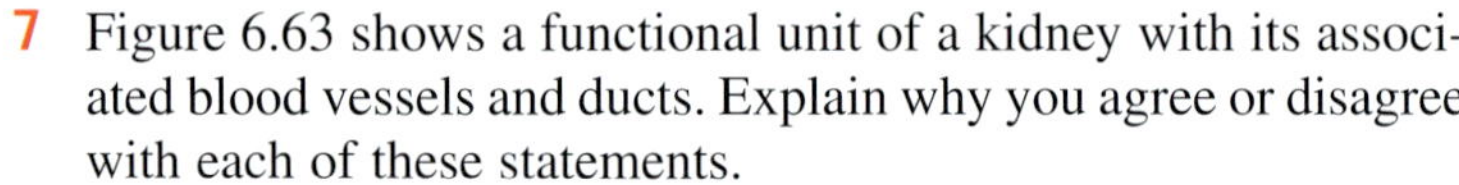

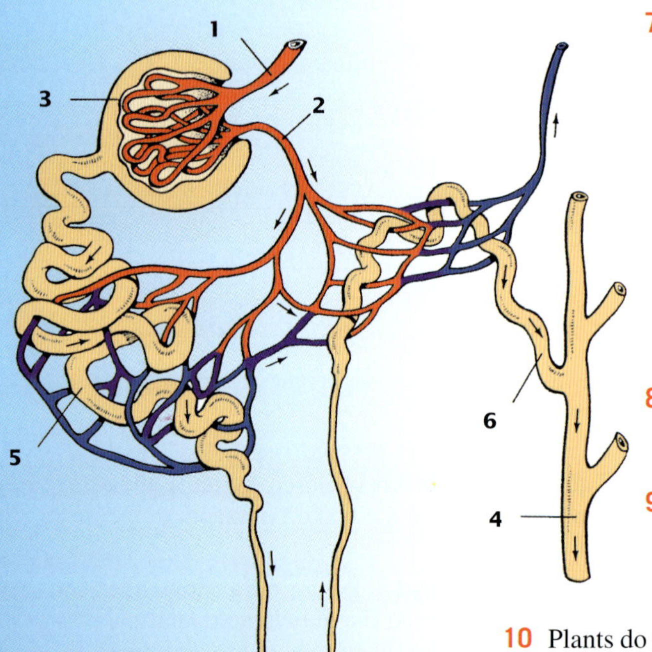

Figure 6.63

7 Figure 6.63 shows a functional unit of a kidney with its associated blood vessels and ducts. Explain why you agree or disagree with each of these statements.

a The blood pressure in vessel 1 is higher than the blood pressure in vessel 2.

b The concentration of urea is higher at 1 than at 4.

c There would be more protein per unit volume in 1 than in 5.

d The concentration of water in 5 is the same as in 6.

e Glucose is unable to pass the membranes at 3.

f Identify each of the numbered parts.

8 ***Applying understanding*** ▸ The more photosynthesis products are transported to a fruit, the larger the fruit grows. How could you encourage a fruit tree to produce larger fruits?

9 ***Applying understanding*** ▸ When a plant is transplanted, root hairs are often damaged. Before transplanting, a gardener will often remove many of the leaves. Explain how these two events are related.

10 Plants do not have the same requirement that most animals have for complex excretory systems. Write two or three sentences to substantiate this claim.

11 ***Applying biological understanding*** ▸ When a tree is ringbarked, phloem is removed from a ring around the trunk, but xylem is left intact (see figure 6.64). Explain the relationship between these actions and what causes the tree's death.

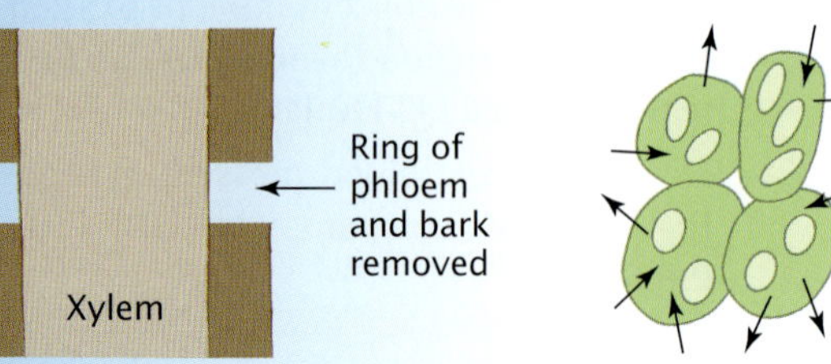

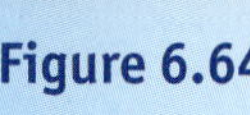

Figure 6.64

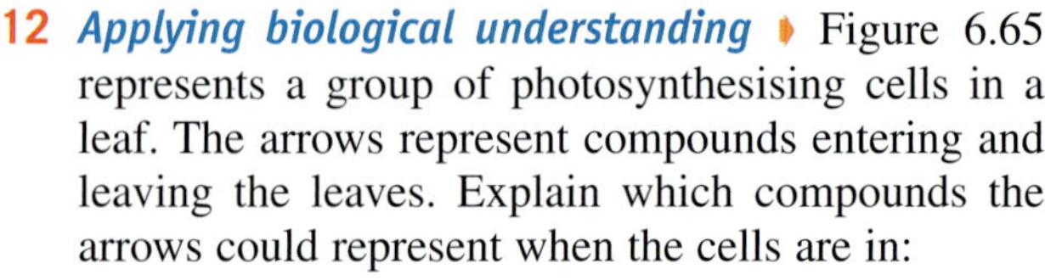

Figure 6.65

12 ***Applying biological understanding*** ▸ Figure 6.65 represents a group of photosynthesising cells in a leaf. The arrows represent compounds entering and leaving the leaves. Explain which compounds the arrows could represent when the cells are in:

a bright sunlight

b darkness.

13 *Analysing, evaluating and communicating* ▸ Comment on or explain the following statements.

a The leaves of many land plants have mechanisms to reduce the chance of wilting on a hot day.

b A leaf with a thin flat structure is likely to have a greater rate of photosynthesis than a round leaf with the same total volume.

c The epidermis of a root lacks the waterproof cuticle found on the epidermis of most leaves.

d Lenticels in the skin of fruit make an important contribution to the survival of living cells in the flesh of the fruit.

e Some plants growing in waterlogged soils may die through lack of oxygen.

14 *Interpreting and applying biological principles* ▸ A bird has a series of air sacs as well as two lungs. The air sacs are elastic and expand and contract rather like bellows. Gases do not diffuse through the internal surfaces of the sacs. Refer to figure 6.66; the arrows show the direction of air flow.

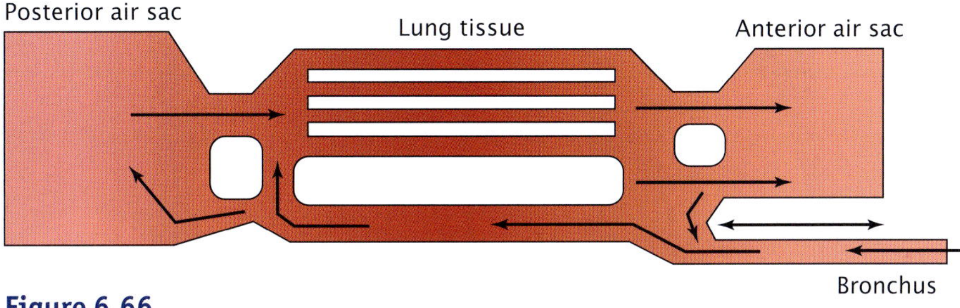

Figure 6.66

When a bird breathes in, air is drawn into the posterior air sacs which expand. Some air already in the sacs is pushed toward the lungs; this causes a push on air in the lungs. When the bird breathes out, air is forced out of the anterior air sac into the bronchus and then out of the body. Air is drawn from the lungs into the anterior air sac; this drawing effect results in air in the posterior sac being drawn into the lungs. Birds have a unique ventilating system in that air is continually pushed and drawn across the lung tissue.

a Explain how such a system increases the efficiency of the lung as an excretory organ.

b Suggest the likely advantage of such a system for birds.

c Explain how the movement of air in relation to lung tissue in a bird differs from that in humans.

15 *Using the web* ▸ Go to www.jaconline.com.au/natureofbiology/natbiol1-3e and click on the 'Robotic surgery' weblink for this chapter. From the left-hand side, select the option 'Clinical Specialties & Services'. Then enter 'robotic surgery' in the Search bar and click 'Go'. Select 'First-of-its-kind Pediatric Robotic Surgery a Success'. Read the story about Daniel Pham.

a About 300 children with the same condition as Daniel are born in the United States each year. What is the condition and why is it important to correct it surgically as soon as possible after it is diagnosed?

b In the discussion of robotic surgery on page 129, it stated that the robotics equipment virtually removes any effect from even a slight tremor in a surgeon's hands. Daniel's surgeon, Professor Albanese, explains how this is done. Write a sentence or two in your own words to explain how the robotic system prevents a slight tremor in a surgeon's hands from becoming a slight tremor in the movement of the surgical instruments.

7 Reproduction

KEY KNOWLEDGE

This chapter is designed to enable students to:

- understand the concepts of asexual and sexual reproduction
- describe how reproduction occurs in unicellular and multicellular organisms.

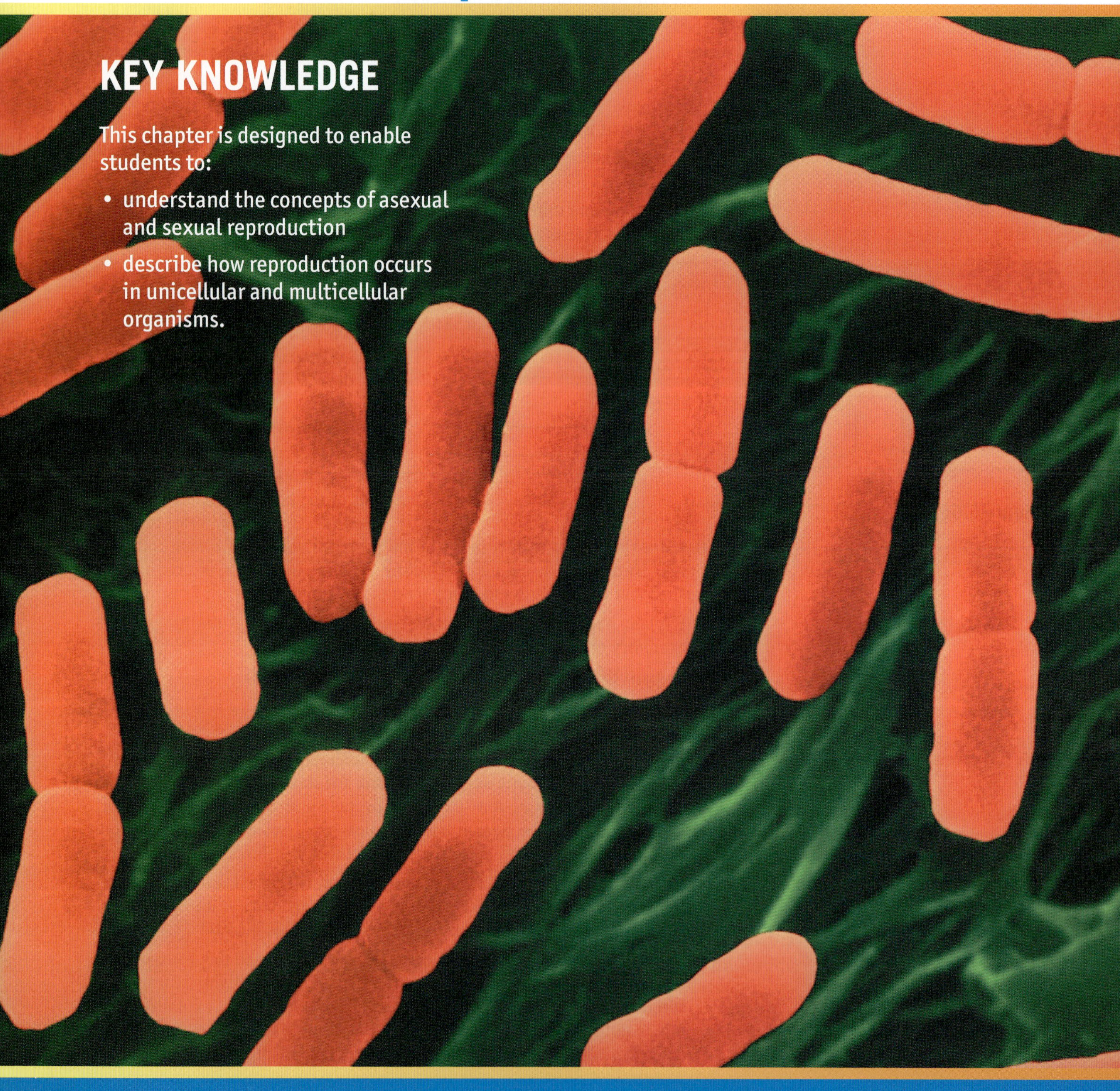

Figure 7.1 Reproduction is one of the defining characteristics of all living organisms, from unicellular prokaryotic organisms, such as bacteria, to multicellular eukaryotic plants and animals, such as giant mountain ash trees and blue whales. This is a scanning electron microscope image, taken at a magnification of 22 810 times, that shows reproduction in a bacterial species. Reproduction in all living organisms produces offspring that ensure continuity of existence of a species over time.

In this chapter, we will explore various types of reproduction and examine some of the different mechanisms, strategies and systems of reproduction that exist in the living world.

Births and birth rates

A ship called *Bee* arrived in the port of Geelong on 18 April 1857 after a four-month journey from England. Among the 459 passengers, mainly assisted migrants, was a young married couple, James and Sarah Minter.

Earlier in England, in May 1852, James, then 22, and Sarah, then 17, married. At first they lived in the south Yorkshire town of Barnsley where James worked as a handloom linen weaver. Their first child was born in June 1853 but died from pneumonia about a year later. Within a month of the death of their first baby, Sarah was again pregnant. Their second child was born in May 1855 but died nine months later in March 1856. Five months later, Sarah was again pregnant and during that pregnancy, she and her husband emigrated to Australia.

Soon after arriving on the *Bee* at the port of Geelong, Sarah gave birth to her third child in May 1857. The couple then moved to Ballarat where their fourth child was born in April 1859. Their fifth and sixth children were born near Maldon in April 1861 and in April 1863, but the sixth baby died aged 10 months. The family — the couple and three surviving children — then moved to Sandhurst (now Bendigo) where they lived at first in a tent on land off Sheepwash Road and later in a house at the same location. Their next five children (seventh to eleventh) were born in Sandhurst with birth dates as follows: January 1865, May 1867, April 1869, April 1871 and February 1873. Apart from the eleventh baby, which was stillborn, these babies survived beyond infancy. James and Sarah's twelfth baby, Emma, was born on 24 April 1874 at Sandhurst. Two days after the birth of this baby, Sarah, aged 40, died from childbirth (puerperal) fever. She was buried at White Hills Cemetery in Bendigo on 26 April 1874. Seven days later, baby Emma died and was buried in the same plot.

In terms of family size, intervals between births and infant mortality, this story of settlers living in country regions of Australia at that time is not unusual. An example of the change in average family size is shown in figure 7.2 with formal photographs for (a) a nineteenth-century family and (b) a twenty-first-century family.

Figure 7.2 Changing average family sizes. Formal family portraits taken **(a)** in the early to mid 1880s and **(b)** around 2005. The late nineteenth-century portrait shows two parents and their 11 surviving children. Contrast this with the twenty-first-century portrait that shows two parents with their two children.

(a)

(b)

One study has shown that if women breastfeed their babies five times per day for a total of at least 65 minutes, their ovarian activity is suppressed so that they are infertile.

The interval between successive births for Sarah Minter was usually about 24 months. The shortest interval was 14 months between the stillbirth (eleventh) and the twelfth baby, Emma. This occurred because breastfeeding did not take place after the stillbirth. Breastfeeding protects a woman from pregnancy by changing her hormone production to hormones concerned with the production and release of milk and suppressing those hormones concerned with ovulation. If no breast-feeding occurs or after a breastfed baby is weaned, hormone production changes to that for a normal ovarian cycle and fertility is restored.

Crude birth rates in Australia

Crude birth rates for Australia, expressed simply as the number of births in a given year per 1000 population, are shown in figure 7.3.

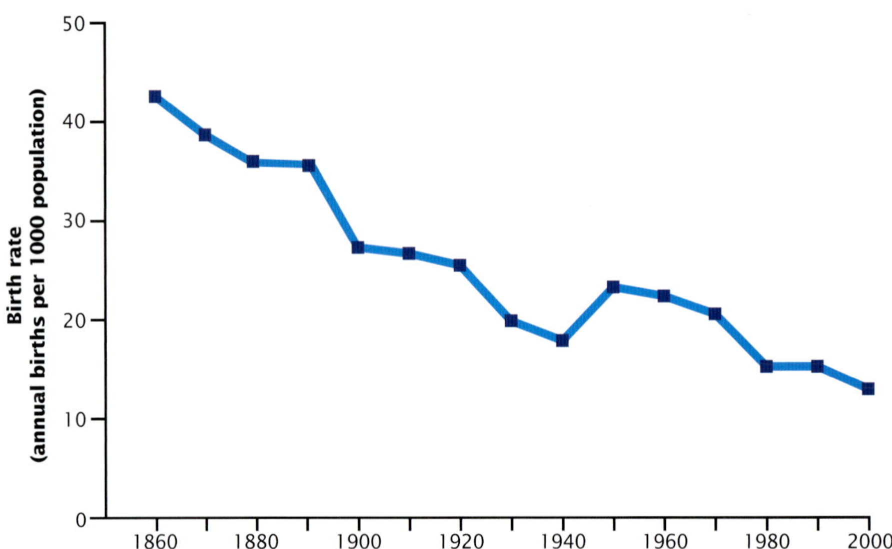

Figure 7.3 Trends in the crude birth rate (annual births per 1000 population) in Australia for given years in the period 1860–2000 (using data from ABS Cat. 3105.65.001, table 42). Suggest a possible reason for the increase in birth rate in 1950 compared with 1940. What factors have contributed to the overall decline in birth rates?

Over the period shown, a significant decrease in birth rates is apparent. The decrease is due to social factors, such as greater participation by women in the workforce and the later average age of marrying and starting families, and to technological factors, such as the availability from the late 1960s and early 1970s of reliable oral **contraception**.

Human reproduction is just one example of reproduction in the world of living organisms. We will return to human reproduction later in this chapter (page 204), but first we will explore reproduction across a range of living organisms and see that the living world has many surprises in store. Reproduction, as expressed in humans, is not the only way!

ODD FACT

Crude birth rates for Germany, Japan, Somalia and Uganda in 2000 were 9.0, 10.0, 46.8 and 47.2 respectively.

Reproduction: making offspring

One of the characteristics of living organisms is their ability to reproduce according to genetic instructions within the organisms themselves. Biological reproduction is different from 'reproduction' of inanimate (non-living) objects, such as formation of crystals.

Reproduction can be identified as either:

- asexual, or
- sexual.

While the human species is restricted to sexual reproduction, both types of reproduction can and do occur in many **invertebrate** animal species and in many plant species.

The prefix 'a' in asexual means 'not'. Other comparable words include: asymmetrical, atypical and acentric. What do each of these mean?

Asexual reproduction involves a single parent organism and produces off-spring or daughter cells that are **clones** — they are identical copies of each other

and of the single parent. With the exception of microbes, this uniformity arises because offspring are produced by mitosis. This means that should one offspring be susceptible to an infection, all of its 'siblings' will also be susceptible.

Sexual reproduction involves genetic contributions in the form of **gametes** from two sources (such as eggs and sperm in animals). The offspring produced differ among themselves and from their parents (see figure 7.4). We will see later (page 199) that this variation arises as a result of the process of **meiosis**.

Table 7.1 identifies some differences between asexual and sexual reproduction.

Figure 7.4 Sexual reproduction results in offspring that vary genetically from each other and from their two parents. In the case of this litter of piglets, some of the genetic variation can be seen in the body colouring.

Table 7.1 Some differences between asexual and sexual reproduction

Feature	*Asexual reproduction*	*Sexual reproduction*
number of parents or parental contributions	• one	• two
processes involved	• binary fission (prokaryotes) • cell replication involving mitosis (eukaryotes)	• gamete production involving meiosis (eukaryotes)
fertilisation	• absent	• fusion of gametes required
offspring	• no genetic variability; offspring are clones of single parent	• offspring vary from parents and from each other
time for completion	• faster	• slower

KEY IDEAS

- Reproduction according to embedded genetic instructions is a characteristic of living organisms.
- Two types of reproduction can occur: asexual and sexual.
- Offspring produced through asexual reproduction are clones of the single parent and of each other.
- Offspring produced through sexual reproduction show variation.

QUICK-CHECK

1 How many parents (or parental contributions) are involved in sexual reproduction?
2 The many offspring of a particular animal were genetically identical to each other. Which process of reproduction will have produced them?
3 Refer to the pig litter shown in figure 7.4.
  a Are the offspring identical with each other?
  b In light of your answer to (a), what reasonable conclusion may be made about the type of reproduction involved?

Reproduction without sex

Reproduction is asexual when offspring are produced from a single parent. In all cases of asexual reproduction, the offspring are identical with each other and with their single parent cell or parent organism.

Examples of asexual reproduction in which a single parent produces identical offspring include:

- binary fission in prokaryotic microbes
- splitting in single-celled eukaryotic organisms
- spore formation in fungi
- natural cloning in animals, for example,
  - budding in sponges and corals
  - 'virgin birth' in insects
- vegetative reproduction in plants, as in runners, cuttings, rhizomes and suckers.

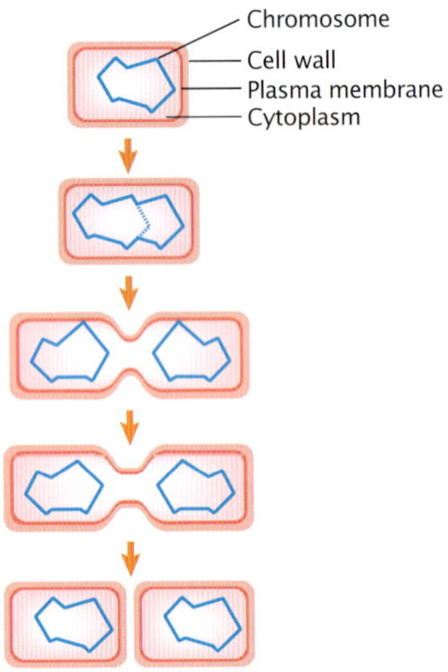

Prokaryotes: binary fission

Microbes such as bacteria consist of unicellular prokaryotic cells. Multiplication of bacterial cells occurs by an asexual process of reproduction known as **binary fission** (fission = 'splitting'; binary = 'two'). The binary fission of a bacterial cell involves:

- replication of the circular molecule of DNA of the bacterial cell
- attachment of the two DNA molecules to the plasma membrane
- lengthening of the cell
- physical division of the cell into two via a constriction across the middle of the cell, such that each new cell contains one circular molecule of DNA.

Figure 7.5 outlines this process.

Chromosome
Cell wall
Plasma membrane
Cytoplasm

Figure 7.5 Binary fission in a bacterial cell. Replication of the circular DNA molecule is followed by cell lengthening and then its division into two. Why is this process an example of asexual reproduction?

The process of asexual reproduction by binary fission in bacteria is simpler and faster than asexual reproduction in eukaryotic organisms. Asexual reproduction in eukaryotes involves the more complex process of **mitosis** (refer back to chapter 4) followed by division of the cytoplasm (cytokinesis). This process typically takes many hours to complete. Binary fission in bacterial cells can be completed in about 20 minutes at room temperature. This means that if resources are available, one bacterial cell, through successive binary fissions over an eight-hour period, could produce 16 million descendants! This is an example of **exponential growth** (discussed further in chapter 15) and it reminds us why a bacterial infection, if not treated, can have serious outcomes. This also reminds us why cooked meats left unrefrigerated can become a source of food poisoning from toxins produced by bacteria such as *Salmonella enteritidis*.

Figure 7.6 shows a cell of the bacterial species *Escherischia coli* dividing by binary fission.

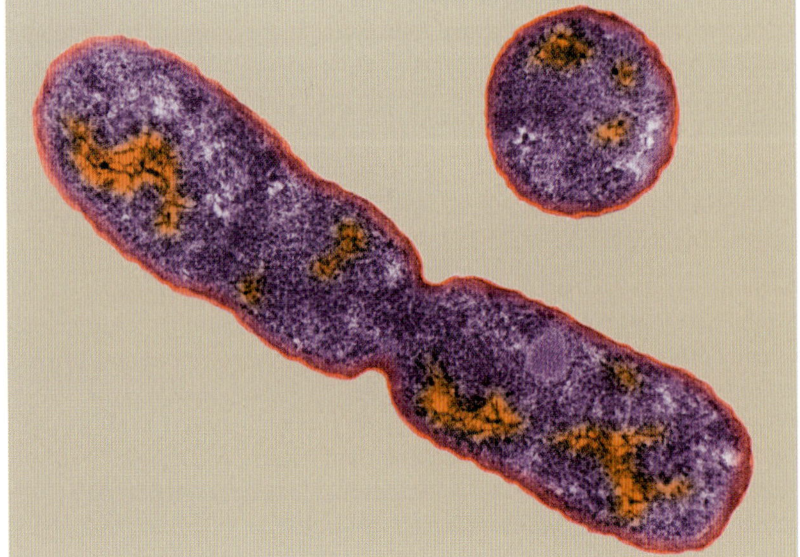

Figure 7.6 Cell of the bacterial species *E. coli* dividing by binary fission. This is a transmission electron micrograph at a magnification of 12 000 times. What has occurred in the cell prior to this stage? What is the circular object towards the upper right-hand corner of the image?

ODD FACT

In prokaryotes, the genetic material, DNA, is a naked circular molecule. In eukaryotes, the DNA is combined with proteins, greatly folded and organised into a number of chromosomes characteristic of each species.

Eukaryotes: asexual reproduction

Among the eukaryotic organisms — animals, plants, fungi and protists — many different forms of asexual reproduction occur. Let's look at some examples.

Let's split into two!

Some eukaryotic unicellular organisms, such as *Amoeba* (figure 7.7), *Euglena* and *Paramecium*, live in freshwater ponds and are less than pinhead-size. These unicellular organisms can reproduce asexually by splitting into two (see figure 7.8a and b). Just to confuse matters, this process is known as binary fission. However, the process of binary fission in these unicellular eukaryotes is different from that which occurs in bacteria. In eukaryotes, the formation of new cells by binary division involves the process of mitosis.

ODD FACT

Amoebae can also undergo **multiple fission**. Mitosis occurs repeatedly and many nuclei form within a single cell (see figure 7.9, page 180). Each nucleus becomes enclosed within a small amount of cytoplasm and forms a spore. Spores can later develop into new amoebae.

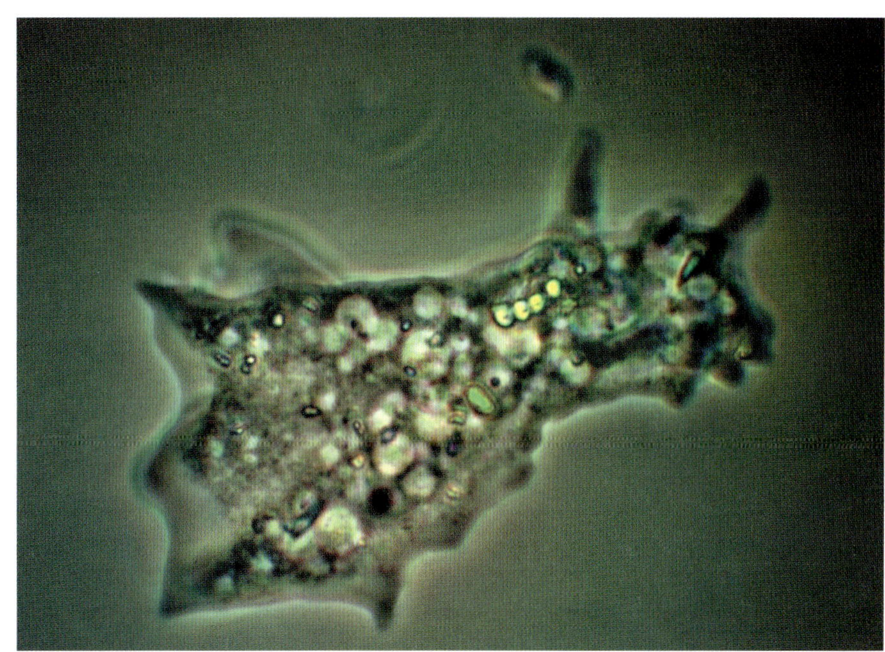

Figure 7.7 Phase contrast microscope image of *Amoeba* taken at 600 times magnification

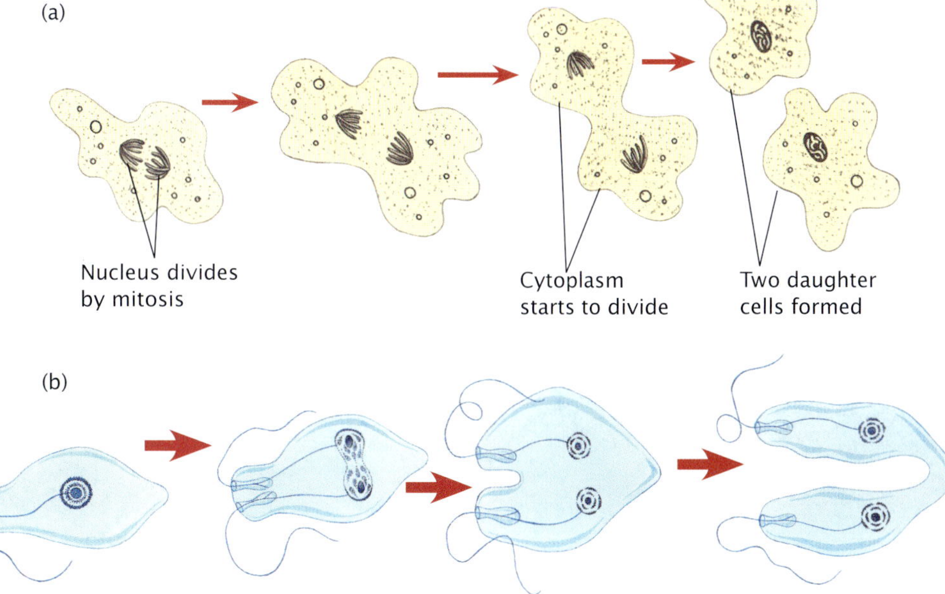

Figure 7.8 Binary fission or splitting in simple eukaryotes. **(a)** Splitting into two in *Amoeba* sp., a unicellular organism. The cytoplasm divides after the chromosomes have replicated and separated during mitosis. Will the daughter cells be identical to or different from each other? **(b)** Splitting into two along a longitudinal axis occurs in *Paramecium* sp., another unicellular eukaryote.

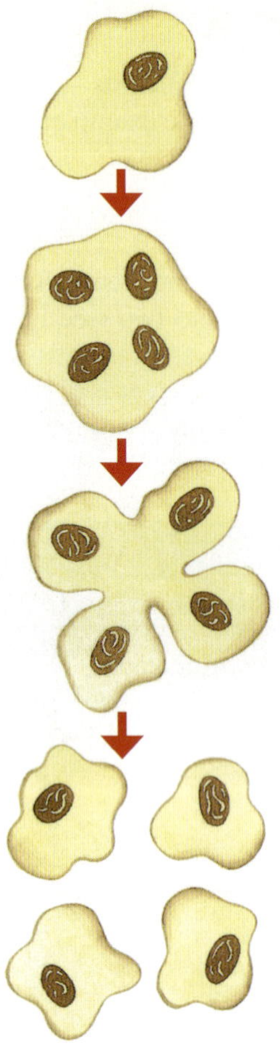

Figure 7.9 Multiple fission (Refer back to the Odd Fact on page 179.)

Simple multicellular animals, such as flatworms, anemones and coral polyps, can also reproduce asexually by splitting into two (see figure 7.10). Each of the parts then grows into a complete animal. This kind of splitting does not occur in other multicellular organisms because their structure is more complex and is built of many different tissues and organs.

Figure 7.10 A snakelocks anemone, *Anemonia sulcata*, reproducing by splitting in two. The name 'snakelocks' comes from the numerous non-retractable tentacles.

Budding to make more

Sponges are common in many marine habitats. Each sponge is made of thousands of cells but has no specialised organs or nervous system. Sponges are able to reproduce asexually from small groups of cells formed by mitosis that bud or break away from the main organism and are carried by currents to other locations where they settle and develop into new sponges. The small group of cells settles on some substrate; the cells reproduce by mitosis and develop into a new sponge. Other simple animals, such as *Hydra*, also undergo budding (see figure 7.11).

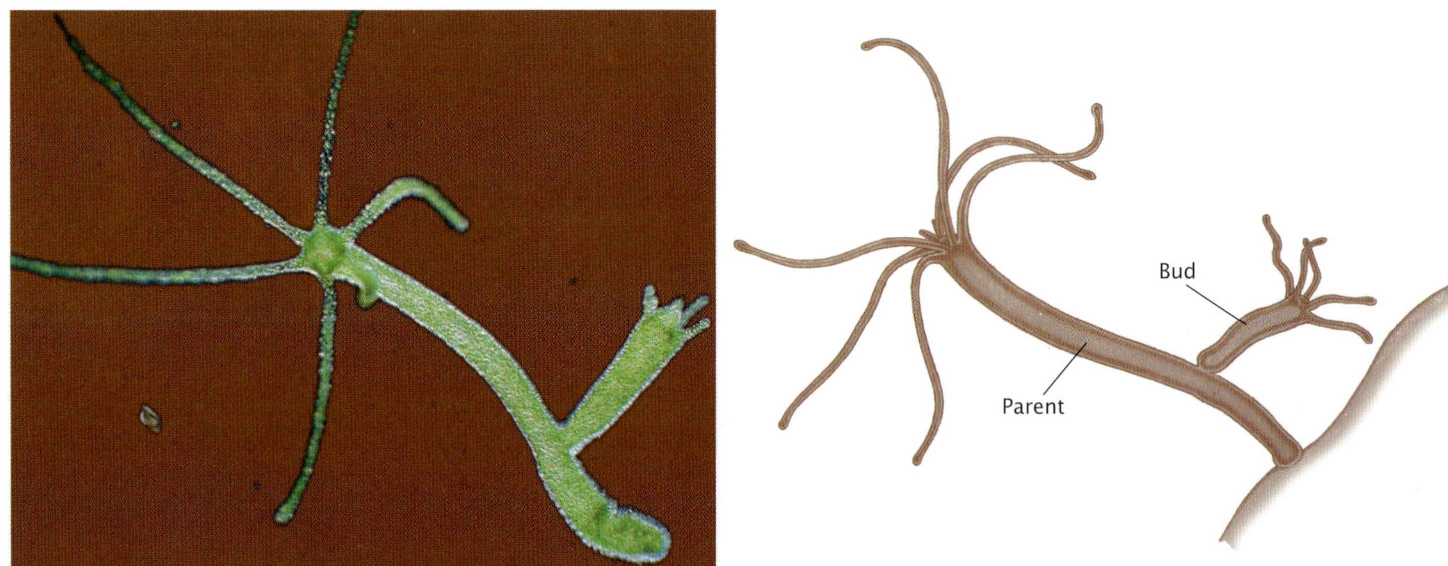

Figure 7.11 Asexual reproduction involving budding from one parent, occurs in *Hydra*.

ODD FACT

Coral polyps can also reproduce asexually. Some bud, others grow new polyps from fragments that break off from a parent polyp and some divide longitudinally to form two new polyps, genetically identical to each other.

Virgin birth in insects

One form of asexual reproduction in animals is **parthenogenesis**, which is also called virgin birth. Young are produced from unfertilised eggs — no sperm is necessary. For example, in certain seasons, aphids produce diploid eggs that develop into diploid individuals (see figure 7.12). The eggs are produced by mitosis and develop into offspring identical to the female parent.

Parthenogenesis is usually restricted to invertebrate animals. However, in 1958 a species of unisexual lizards was reported by a Russian scientist. A few genera of whiptail lizards are known to consist only of females. Each female produces eggs and each egg develops, without fertilisation, into a female identical to the mother.

Figure 7.12 Adult female aphid giving birth. The numerous offspring have developed asexually from unfertilised eggs. What is the name of this kind of reproduction?

Asexual reproduction in plants

Asexual reproduction is common in plants. In plants, asexual reproduction is also called **vegetative reproduction**.

Runners

Over 10 years, one strawberry plant (*Fragaria ananassa*) grew into the strawberry patch in figure 7.13. How did one small plant grow into such a large patch? Strawberry plants have runners, special stems that grow over the ground. The runner grows away from the parent plant and, at alternate nodes on the runner, new buds give rise to roots, leaves, flowers and fruit (see figure 7.14).

Another example of a plant that spreads by runners — this time in water — is the water hyacinth (*Eichhornia crassipes*) (see figure 7.15). This declared noxious weed infests wetlands, lakes and rivers in Australia.

Figure 7.13 In 10 years, one strawberry plant grew by asexual reproduction into this strawberry patch.

Figure 7.15 The water hyacinth

Figure 7.14 The strawberry (*Fragaria ananassa*) has runners, special stems that grow over the ground.

A variation of 'runners' occurs in blackberry plants (*Rubus* spp.) that propagate when their long stems (canes) bend over and shoots and roots grow from the point where the tips make contact with the ground.

As well as runners, other means of asexual reproduction include:

- cuttings
- rhizomes (underground stems)
- tubers (swollen underground stems)
- bulbs (underground structure with short stem and many closely packed, fleshy leaves).

Some of these structures are shown in figure 7.16.

(a)
Frond
Rhizome
Roots

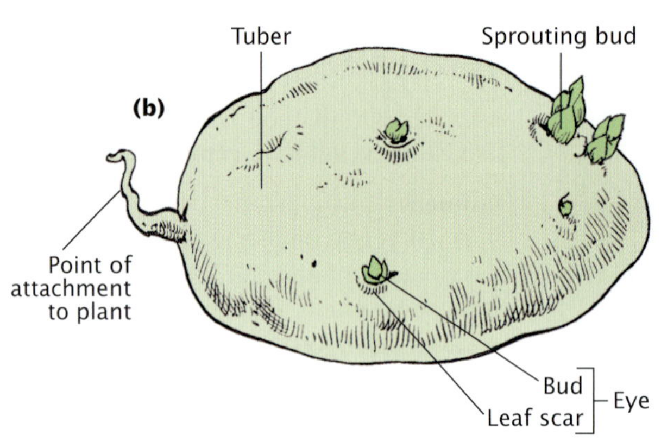

Figure 7.16 Asexual reproduction in plants includes **(a)** rhizomes as in bracken and some grasses, **(b)** tubers as in potatoes, and **(c)** bulbs as in onions. In each case, the new plants are genetically identical with the parent plant.

Most ferns are delicate and need moist, shady conditions to reproduce by sexual means. In contrast, austral bracken propagates asexually by rhizomes and thrives in exposed areas. Bracken is the only Australian native fern that is not protected by law.

Cuttings

With some plants, cuttings of shoots, roots or leaves can be used to produce clones. This is done when seed availability is limited or when the number of plants is low, as is the case for the rare Wollemi pine (*Wollemia nobilis*). Only a few specimens of this new species exist in the wild and new plants are now being propagated from cuttings.

Rhizomes

Some plants propagate through underground stems or rhizomes (see figure 7.16a). Buds sprout from the rhizome and produce new daughter plants. Plants that propagate by rhizomes include garden plants such as irises, grasses such as kikiyu grass (*Pennisetum clandestinum*) and couch grass (*Cynodon dactylon*), many reeds, and an Australian fern, the austral bracken (*Pteridium esculentum*) (see figure 7.17).

Figure 7.17 Austral bracken reproduces asexually from underground stems (rhizomes) when buds from the rhizome develop into new fronds. After a bushfire, austral bracken quickly becomes re-established in a burnt area. Can you suggest why?

ODD FACT

The suckering ability of blackberries, combined with their ability to spread by their canes taking root at their tips, contribute to the rapid spread of blackberry plants in some areas.

Suckers

Suckers are new shoots that arise from an underground root at some distance from a parent plant. Blackberry suckers can appear more than two metres from the parent plant.

Plantlets without sex

The fern *Asplenium bulbiferum*, which is native to Australia and New Zealand, can reproduce asexually. Figure 7.18 shows the small plantlets that arise from the fern frond. A similar process also occurs in the plant *Bryophyllum* sp. (refer back to page 86).

ODD FACT

In New Zealand, *Asplenium bulbiferum* is commonly known as hen and chickens fern. In Australia, the common name of the same fern is mother spleenwort.

Figure 7.18 Asexual development in the fern, *Asplenium bulbiferum*. Note the new plantlets on the fern fronds.

Technology: asexual reproduction

Reproductive technologies such as cloning involve methods of asexual reproduction in which the genetic information of new organisms comes from one 'parent' cell only. For many years, whole plants have been cloned by traditional methods, such as cuttings, but the horticultural industry is now using a newer technique of plant tissue culture to clone whole plants in large numbers. For mammals, however, artificial cloning of whole animals is a much newer area. The technique has been applied to dairy cattle with the aim of herd improvement and it has also been attempted with endangered mammals for conservation purposes. Mammalian cloning at present is limited because of high rates of embryo loss (fewer than five per cent of cloning attempts result in the birth of a cloned mammal) and, after birth, the survival rate of cloned mammals is low.

In the following sections, we will briefly explore technology use in artificial cloning of plants and mammals. This topic is discussed in more detail in *Nature of Biology Book 2, Third edition*.

Cloning plants from tissue culture

Using the technology of plant tissue culture in the laboratory, many identical copies or clones of a plant can be produced starting from a small amount of tissue from one plant. This technique is used with ornamental plants, such as orchids and carnations, and Australian native plants, such as bottlebrush (*Callistemon* spp.), the flannel flower (*Actinotus helianthi*) and various eucalypts (*Eucalyptus* spp.). It is also being used with endangered or very rare plants, such as the Wollemi pine (*Wollemia nobilis*). Figure 7.19a shows flannel flowers in tissue culture in a laboratory and figure 7.19b shows mass plantings of flannel flowers propagated by tissue culture.

(a)

(b)

Figure 7.19 **(a)** Flannel flowers in tissue culture, and **(b)** large-scale bed plantings of flowers propagated by tissue culture

Figure 7.20 Identical twins resulting from a natural process of embryo splitting. The frequency of twin births is about 1 in 80, but differs between ethnic groups.

Artificial cloning of mammals

Several artificial techniques for cloning mammals have been developed, including:

- cloning using embryo cells
- cloning using somatic cells.

Cloning using embryo cells

One form of artificial cloning in mammals involves artificially separating the cells of an early embryo produced through artificial insemination in a laboratory then transferring the separated cells into the uterus of surrogate mothers. This technique is an artificial form of the natural cloning that results in identical twins (see figure 7.20), triplets or larger multiple identical siblings in mammals.

ODD FACT

Dolly the cloned sheep was named after Dolly Parton, the country-and-western singer.

ODD FACT

cc (short for 'carbon copy'), the world's first cloned kitten, was born in December 2001 and was the only surviving clone from a total of 87 cloned embryos that were produced.

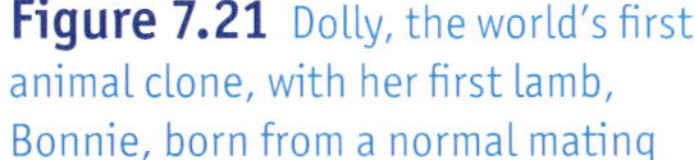

Figure 7.21 Dolly, the world's first animal clone, with her first lamb, Bonnie, born from a normal mating

Cloning using somatic nucleus

The somatic cells of a mammal are its body cells, such as skin, liver and kidney cells. A more experimental and dramatic technique of cloning of whole mammals has been developed. This first came to world attention with the birth, at 5 pm on 5 July 1996, of a lamb called 'Dolly' at the Roslin Institute in Scotland (see figure 7.21). Dolly was a cloned mammal and was produced by a radical new technique that involved the cloning of a somatic nucleus transferred to an egg cell.

Other mammalian species that have since been cloned include mice, rats, rabbits, pigs, mules, cattle, horses and cats. The technique has also been used to clone a banteng (*Bos javanicus*), an endangered species of wild cattle from south-east Asia, and a mouflon lamb, a rare breed of sheep.

Desirable or not, cloning of domestic pets is now available in the United States. A company called Genetic Savings and Clone has been set up for the artificial cloning of pets from cell samples provided by pet owners. In December 2004, media worldwide reported that this company had successfully cloned a kitten from a somatic cell of a 17-year-old pet male cat, Nicky, now dead. Dear Nicky's owner is reported to have paid US$50 000 for the cloned kitten (named Little Nicky) (see figure 7.22).

$50 000 KITTY — NICKY CLONED!

Figure 7.22 Little Nicky, the kitten who made headlines in December 2004

KEY IDEAS

- Asexual reproduction produces clones of a single parent or cell.
- In prokaryotes (bacteria), asexual reproduction occurs through binary fission.
- Asexual reproduction in eukaryotes involves the process of mitosis.
- Asexual reproduction in some eukaryotes, such as protists and simple animals, involves splitting and budding.
- Asexual reproduction in plants involves structures such as runners, rhizomes and bulbs.
- Plant clones can be produced through tissue culture.
- Artificial cloning of mammals has been achieved through separating embryo cells and through somatic cell cloning.

QUICK-CHECK

4 Identify the following statements as true or false.
   a Binary fission in an amoeba is identical to that in bacteria.
   b Two parents are involved in asexual reproduction.
   c Binary fission of a bacterial cell would be expected to produce daughter cells identical to the starting cell and to each other.
5 Starting with one bacterium, how many bacterial cells would be expected from six cycles of binary fission?
6 Name the key cellular process that is involved in all cases of asexual reproduction in eukaryotic organisms.
7 Identify two ways in which a plant can reproduce asexually.
8 What was the first mammal to be cloned by somatic cell cloning?

Sexual reproduction

Most commonly, sexual reproduction involves a male and a female parent, each of which makes a genetic contribution to each offspring. The genetic contribution from each parent is a single cell, either an egg or a sperm. Gamete is the general term for these cell types. Gametes are produced by meiosis and have only half the number of chromosomes present in each body cell of a multicellular organism (see figure 7.23). The number of chromosomes in a somatic cell is called the **diploid** number and for humans that is 46, denoted as $2n = 46$ since there are two sets of chromosomes. The number of chromosomes in a gamete is called the **haploid** number and for humans that is 23, denoted as $n = 23$ since there is only one set of chromosomes.

In animals, the gamete from a female is an **egg** that is produced in an **ovary**; the gamete from a male is a **sperm** and the sperm-producing organ is a **testis**. The gametes fuse to produce an offspring that will also mature to produce gametes.

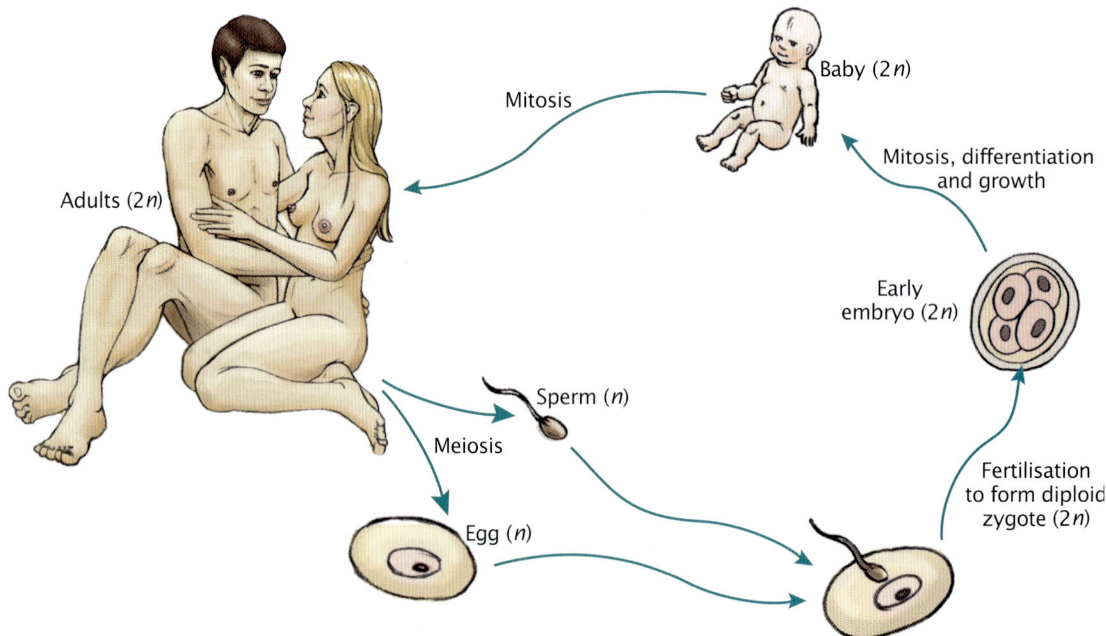

Figure 7.23 The life cycle of humans. In sexual reproduction, there are typically two parental contributions (egg and sperm) to each offspring. Note that apart from the gametes that are haploid, the life cycle is otherwise diploid.

In people and in familiar animals, such as domestic pets and farm animals, two different sexes produce the different gametes — males produce sperm and females produce eggs. However, this is not the case in many other kinds of animal (see pages 197–8) and it is rarely the case in flowering plants.

If you examine a flower, the most common situation is that it will have both pollen-producing organs (**stamens**) and egg-producing organs (**carpels**). Figure 7.24a shows the features of such a flower. About 80 per cent of flowering plants are of this type. In flowering plants, the female gamete is also called an egg and the egg-producing organ is the carpel (or **pistil** when individual carpels are fused into a single structure) and the male gamete is **pollen** produced by the stamen.

Cone-bearing plants, such as pines and firs, do not have flowers but they produce eggs and pollen on different types of cone (see figure 7.25). Ferns and mosses have neither flowers nor cones but they produce eggs and sperm in specialised sex organs at one stage of their life cycle.

(a)

Pistil (carpel)
Stigma
Style
Ovary
Pollen sacs
Anther
Filament
Stamen
Corolla
Petals
Calyx
Sepals
Ovule (egg)
Receptacle
Pedicel

(b)

Figure 7.24 (a) Parts of a typical flower. Note the presence of pollen-producing and egg-producing organs in the same flower. Such flowers are said to be perfect or hermaphrodite. **(b)** Magnolia flower (*Magnolia* sp.) showing many pollen-producing organs (stamens) surrounding numerous egg-producing carpels.

Figure 7.25 (a) Male cones of pine (*Pinus* sp.), and **(b)** mature female cones of pine (*Pinus* sp.), showing a young cone at left and a previous year mature cone at right

Life cycles of plants have two stages

The life cycle of a typical fern is shown in figure 7.26. The dominant part of the life cycle is a diploid fern but the life cycle also includes a tiny, independently living haploid plant that can survive only in moist environments. The diploid fern produces **spores** by meiosis. Spores usually develop on the underside of a fern frond in structures known as sori (singular = sorus) and these spores develop into tiny haploid plants. These tiny plants produce gametes by mitosis that, after fusing, develop into sporophytes and so the cycle is completed. Sperm can only move to fertilise an egg when liquid water is present.

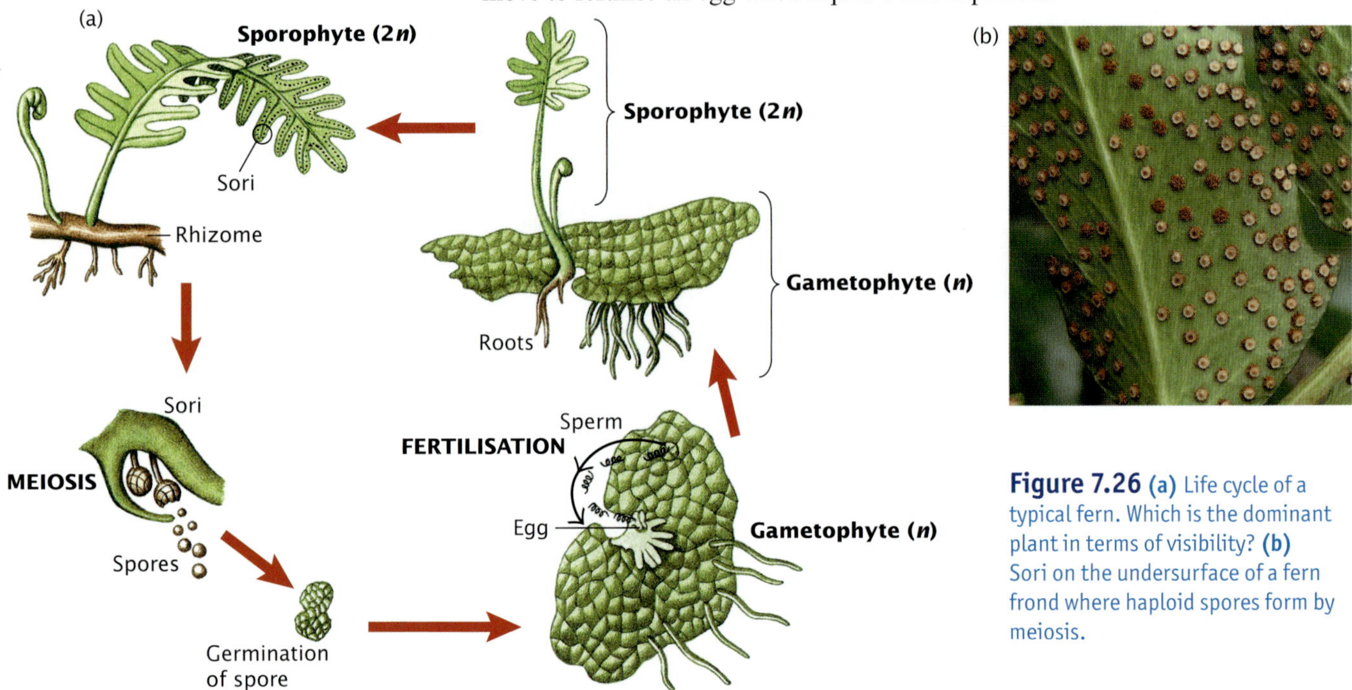

Figure 7.26 **(a)** Life cycle of a typical fern. Which is the dominant plant in terms of visibility? **(b)** Sori on the undersurface of a fern frond where haploid spores form by meiosis.

How do the life cycles of plants and animals compare? Generalised life cycles for both an animal and a plant are shown in figure 7.27. If you compare the generalised animal life cycle with that of a plant, you will see that the life cycle of plants is more complex because it includes both a diploid stage that produces

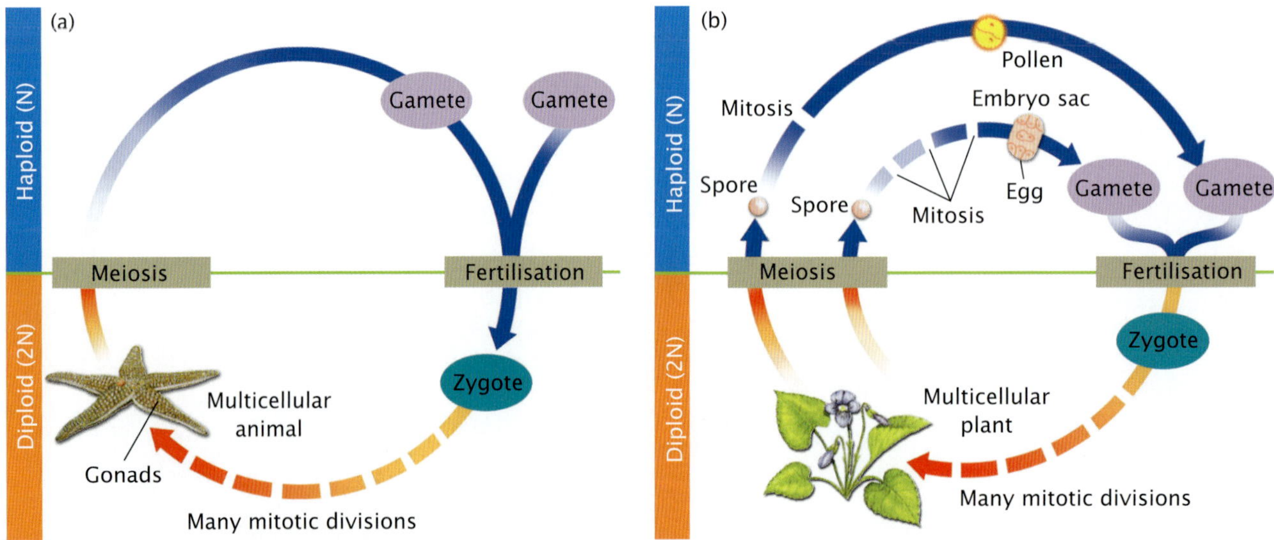

Figure 7.27 Generalised life cycle of **(a)** an animal and **(b)** a plant. Note that, in animals, the cells directly formed by meiosis are gametes. In contrast, in plants, the cells formed by meiosis are spores that develop into structures that form sperm (pollen) and eggs. What are these structures in flowering plants? in ferns?

spores and a haploid stage that produces gametes. The spore-producing stage is referred to as the **sporophyte** stage and the gamete-producing stage is called the **gametophyte**. In a plant life cycle, these diploid and haploid stages alternate, one giving rise to the other, and this is termed an **alternation of generations**.

In plant life cycles, however, the two stages in the life cycle are not equally prominent. One stage, either diploid or haploid, is the obvious recognisable plant. In mosses, the familiar green plant that we see and say 'There's a moss' is the gametophyte stage. In ferns, however, the familiar plant is the sporophyte stage and the gametophyte is tiny and inconspicuous. In conifers and flowering plants, the familiar plant is the sporophyte stage and the gametophyte stage is reduced to just a few cells that develop within the reproductive tissues of the sporophyte plant (see figure 7.28).

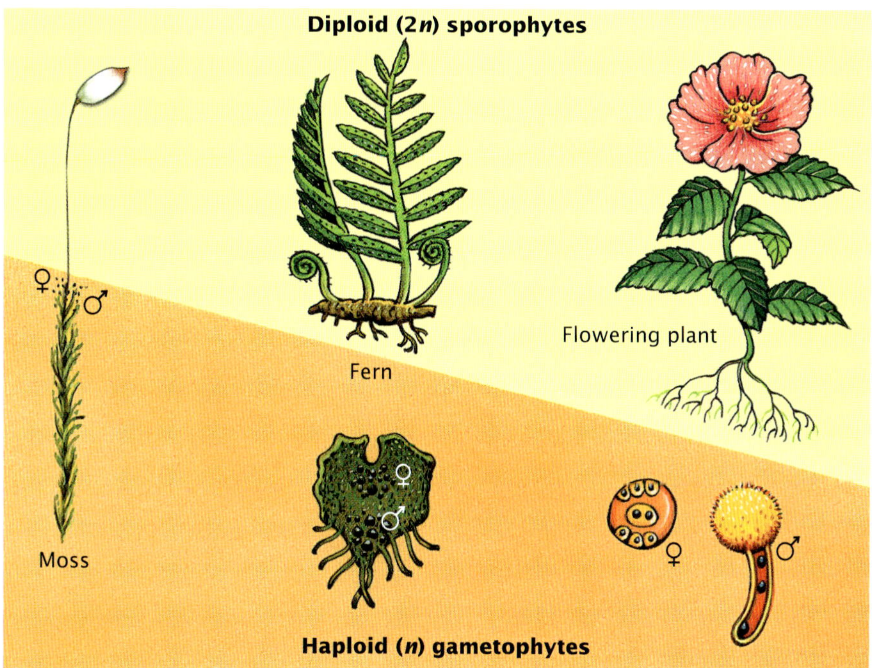

Figure 7.28 Alternation of diploid and haploid stages in a plant life cycle. Note that the dominant or 'obvious' stage in flowering plants, conifers and ferns is the diploid stage.

Animals: just what sex are you?

In some animal species, a single organism has both egg-producing and sperm-producing organs. These organisms are termed **hermaphrodites** and include the garden snail (*Helix aspersa*) (see figure 7.29) and the common earthworm (*Lumbricus terrestris*).

The term 'hermaphrodite' comes from Hermes, the male Greek messenger of the gods, and Aphrodite, the female Greek god of love and beauty.

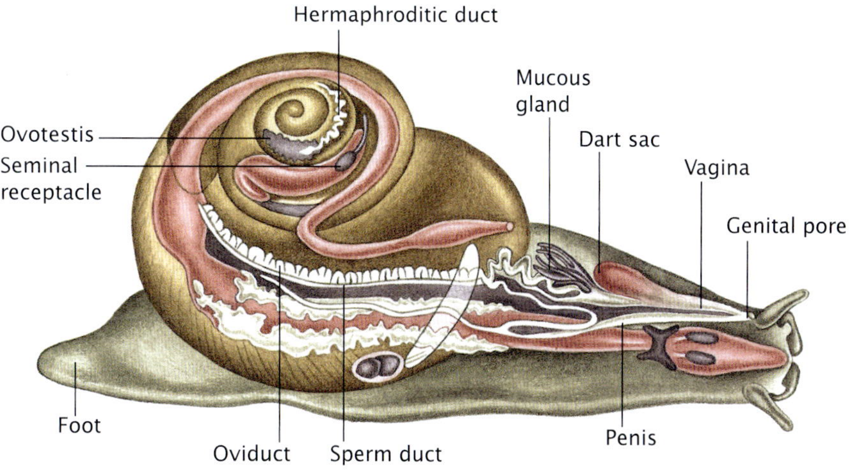

Figure 7.29 Internal anatomy of the garden snail. Note the presence of an organ labelled 'ovotestis' that produces both eggs and sperm.

Figure 7.30 Some animals can change their sex during their lifetime. They are called sequential hermaphrodites.

The common earthworm has about 100 body segments, with ovaries in segment 13 and testes in segments 10 and 11 (counting from the head). Sperm released through pores in segment 15 are exchanged between a mating pair of earthworms. They store the sperm in sacs in segments 9 and 10 prior to using them to fertilise their eggs that are then released through pores in segment 14.

When both sperm-producing and egg-producing organs are active at the same time in one organism, such as occurs in snails and earthworms, this is called **simultaneous (synchronous) hermaphrodism**. In contrast, some fish species found around coral reefs change sex and are known as **sequential hermaphrodites**. These fish start life as one sex and can transform to the other sex under certain conditions. Examples of sequential hermaphrodite organisms include various species of anemonefish (*Amphiprion* spp.) that live in groups of several males with one dominant female. If the dominant female fish dies or is removed, one large male fish then changes to become a functioning female (see figure 7.30). Other reef fish species, such as wrasses and some angelfish (*Centropyge* spp) live in groups comprising several females and one dominant male. If the male is removed, one of the large females then changes to a functional male.

Sex in flowering plants

We have already noted (see figure 7.24, page 187) that most flowering plants have both pollen-producing organs (stamens) and egg-producing organs (carpels). Less frequently, it may be observed that some flowers have either pollen-producing organs only or egg-producing organs only.

When both types of flower are present on the same plant, the plant is said to be **monoecious** (= 'one household') plant. Examples of monoecious flowering plants include pumpkin (*Cucurbita* spp.), corn (*Zea mays*), castor oil (*Ricinus communis*) and oaks (*Quercus* spp.).

(a)

(b)

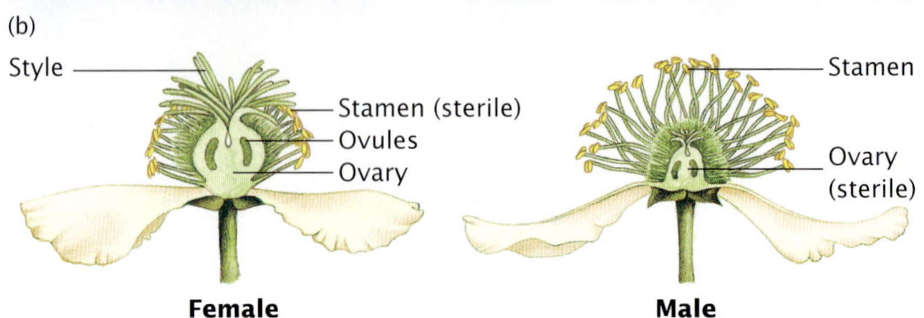

Figure 7.31 A single kiwifruit plant has either functional male or female flowers, but not both. **(a)** The photograph at top shows female flowers (at left) and male flowers (at right) of kiwifruit viewed from above (top row) and viewed in transverse section (bottom row). **(b)** Notice that the female kiwifruit flower (at left) has many prominent white styles, each forming part of a carpel. Below the styles can be seen stamens but they are sterile and do not produce functional pollen. The male kiwifruit flower (at right) has many pollen-producing stamens but the reduced carpels that are present do not produce eggs.

The flowering plant commonly known as Green Dragon (*Arisaema dracontium*) is dioecious. However, when young, it produces only 'male' flowers, but as it becomes larger, the plant also begins to produce 'female' flowers.

In other cases, the different flower types are present on different plants and the plant is said to be **dioecious** (= 'two households'). Examples of dioecious flowering plants that have either only functional 'female' or only 'male' flowers on the one plant include holly (*Ilex* spp.), date palm (*Phoenix dactylifera*), marijuana (*Cannabis* spp.), kiwifruit (*Actinidia deliciosa*) (see figure 7.31).

The dioecious condition is not restricted to flowering plants. Dioecious plants are also found among the conifers and other non-flowering plants, as for example, gingko (*Gingko biloba*).

KEY IDEAS

- Male gametes involved in sexual reproduction are sperm (and pollen) and female gametes are eggs.
- Plant life cycles have an alternation of generations, with a diploid spore-producing stage alternating with a gamete-producing stage.
- In some animals, egg and sperm production occur in separate female and male organisms.
- An hermaphrodite animal can produce both male and female gametes.
- Flowering plants show a variety of conditions regarding egg and pollen formation.

QUICK-CHECK

9 Identify the following statements as true or false.
   a A sequential hermaphrodite can change sex during its lifetime.
   b In animals, the male gamete is called sperm.
   c The term 'dioecious' refers to a flower that has both male and female reproductive organs.
   d In plants, the sporophyte stage produces spores by meiosis.

10 Give one example or the name of:
   a a monoecious plant
   b a simultaneous hermaphrodite animal
   c an organism that can be called a sequential hermaphrodite
   d one of the alternating generations in a plant life cycle.

11 Where might you find the following?
   a a carpel
   b an hermaphrodite animal

Getting gametes together

In sexual reproduction, two parental contributions (egg and sperm/pollen) fuse to produce a **zygote** that will then develop into an animal or a plant. Even in the case of hermaphrodite animals, the gametes typically come from two separate animals, rather than self-fertilisation occurring.

Fertilisation is the fusion of two gametes, an egg from a female parent and a sperm from a male parent. Some animals release their gametes (eggs and sperm) into the external environment so that fertilisation occurs outside the body of females that produce the eggs. This situation is termed **external fertilisation**.

On the other hand, in some animals, males deliver sperm directly into the reproductive tract of females so that fertilisation of eggs occurs inside the body of females that produce the eggs. This situation is termed **internal fertilisation**.

Tapeworms that live in the gut of an animal are hermaphrodites that self-fertilise.

External fertilisation in animals

ODD FACT

Mass release of gametes by coral polyps (or **spawning**) on the Great Barrier Reef occurs every year, three to six days after the full moon in October and November. For coral polyps of the Ningaloo Reef off the Western Australian coast, this mass spawning occurs in the period seven to ten days after the full moon in March.

When animals use external fertilisation, they produce very large numbers of gametes. These large numbers increase the chance of fertilisation but also mean much gamete wastage. Aquatic species that live in large bodies of water and depend on external fertilisation produce very large numbers of gametes. Oysters, for example, each produce about 500 million eggs in a single season.

Because sperm need a watery environment to swim to an egg, external fertilisation is limited to animals that either live in aquatic environments or reproduce in a watery environment. External fertilisation occurs in aquatic invertebrates such as coral polyps, bony fish and amphibians (frogs and toads). Developing embryos also require a watery environment for their development and this is provided by the aquatic conditions in which external fertilisation occurs.

External fertilisation can be a chancy process. Some species that use external fertilisation have developed strategies to increase the chance that eggs and sperm will meet and that fertilisation will occur.

Figure 7.32a shows the release of eggs by a colony of coral polyps, the animals that build coral reefs. All the polyps in one region release their eggs and sperm into the sea in a synchronised manner, creating a 'soup' of gametes (see figure 7.32b). The high concentration of gametes during spawning increases the likelihood of eggs and sperm colliding and fertilisation occurring. The typical life cycle of a coral polyp is shown in figure 7.33.

Figure 7.32 **(a)** Close-up of the release of gametes by a group of colonial animals known as coral polyps **(b)** Mass spawning of eggs and sperm bundles by coral polyps occurs at the same time (synchronously) each year. What possible benefit results from this synchrony?

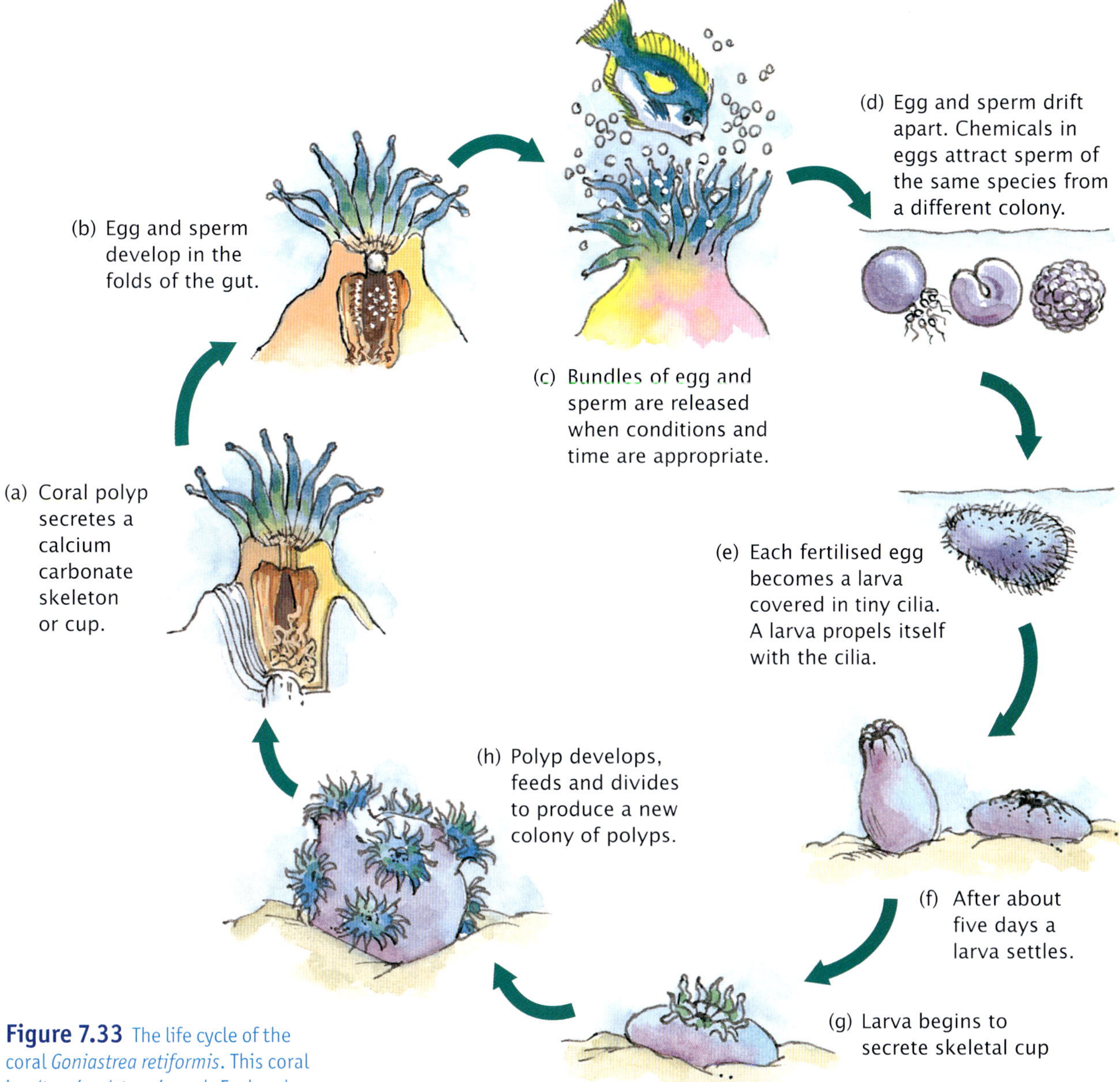

Figure 7.33 The life cycle of the coral *Goniastrea retiformis*. This coral is a 'true' or 'stony' coral. Each polyp lives in a protective cup of calcium carbonate which it has secreted.

ODD FACT

In one Australian frog species, fertilised eggs are swallowed by females and develop in their stomachs. They are regurgitated later as tiny froglets. In another Australian frog species, after fertilisation, the eggs are transferred into moist pouches on the hips of the male of the species where embryonic development occurs.

Just checking you out!

Fish depend on external fertilisation. If a female fish releases her eggs independently of the release of sperm by a male fish of the same species, then the chance of fertilisation of her eggs is extremely low or even nil.

Behaviours such as courtships before spawning can lead to an improved chance of fertilisation. Through a courtship display, a female can recognise that a male is a member of her species and, if she is ready to release her eggs, this occurs in close proximity to a male. In turn, he will release sperm nearby so that the likelihood of fertilisation is increased.

Hang on there!

Frogs and toads have external fertilisation. The chance of fertilisation is increased by a behavioural adaptation of males. When a female frog is ready to lay eggs she goes into a nearby pond or pool. A male tightly clasps her and remains on her back until she releases her eggs (see figure 7.34). As the female releases her

Figure 7.34 Ornate burrowing frogs (*Limnodynastes ornatus*). The male is clasping the female tightly and when the female releases her eggs into the water, the male releases sperm over the eggs. The eggs are suspended on a raft of bubbles.

eggs, the male is stimulated to release sperm over the eggs, fertilising them. The fertilised eggs then undergo embryonic development and give rise to larvae, known as tadpoles, that later metamorphose to frogs.

Internal fertilisation in animals

The watery environment in which aquatic animals live enables external fertilisation since it provides an environment in which eggs are protected from drying out and in which gametes can move to meet. However, with internal fertilisation, the chance of gametes meeting is greater and hence the chance of fertilisation is increased. So, it is not surprising that internal fertilisation occurs in some aquatic organisms.

Octopuses and sharks, for example, are marine organisms that have internal fertilisation. In the male octopus, one of his eight arms on one side is shorter than the other arms. This arm is specialised for the direct transfer of a parcel of sperm to a female. Figure 7.35 shows a male octopus that has mounted a female and inserted his specialised arm into her mantle cavity where he will deposit a package of sperm. In sharks, the male of the species has claspers that are appendages of his pectoral fins. The male shark inserts his claspers into the vagina of a female and forces his sperm inside her carried by water pressure that he generates.

Terrestrial animals do not live in an aquatic environment. Except for amphibians such as frogs and toads that return to water to get their gametes together, other terrestrial animals have internal fertilisation. We will briefly explore internal fertilisation in some terrestrial animals.

(a)

(b)

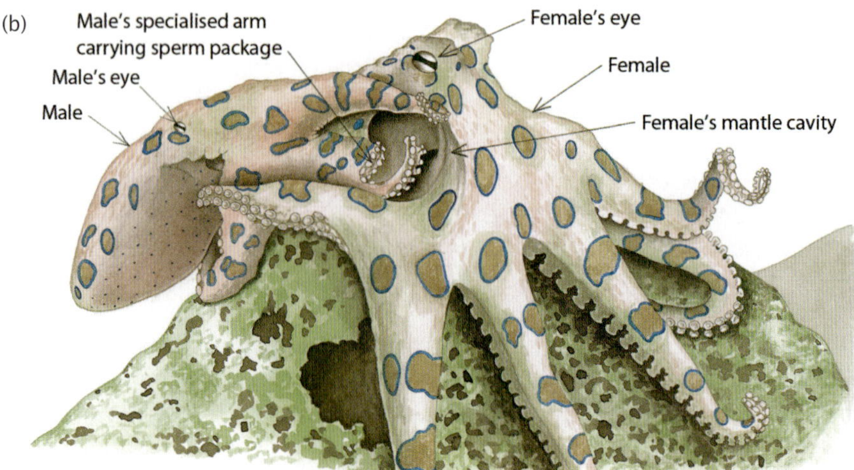

Figure 7.35 (a) Smaller male blue-ring octopus (*Hapalochlaena lunulata*) transferring a sperm package into the mantle cavity of a larger female (b) Diagram that shows the smaller male blue-ring octopus (at left) inserting his specialised arm carrying a sperm package into the mantle cavity of the larger female octopus (at right).

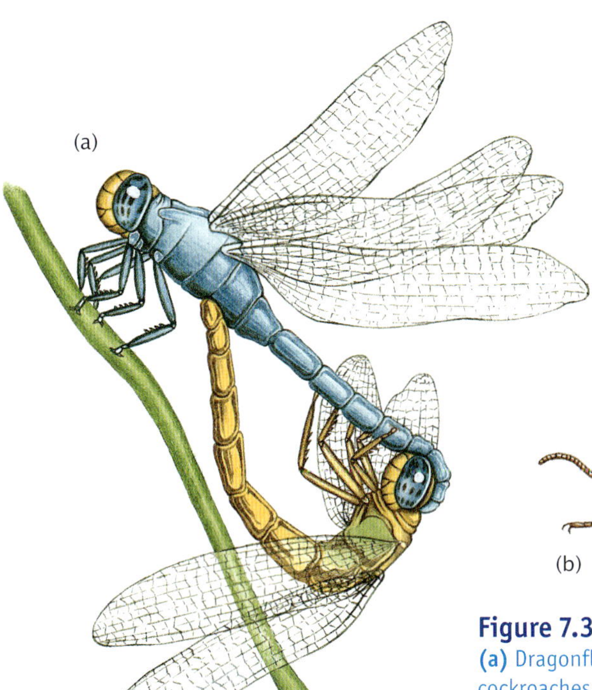

Insects: in the air and on land

Internal fertilisation occurs in insects. Complex genital structures at the end of the abdomen of a male insect enable him to couple with a female and transfer sperm packages into the reproductive tract of a female. Some insects, such as dragonflies, mate on the wing but most insects keep their feet (all twelve in a mating couple) on the ground (see figure 7.36).

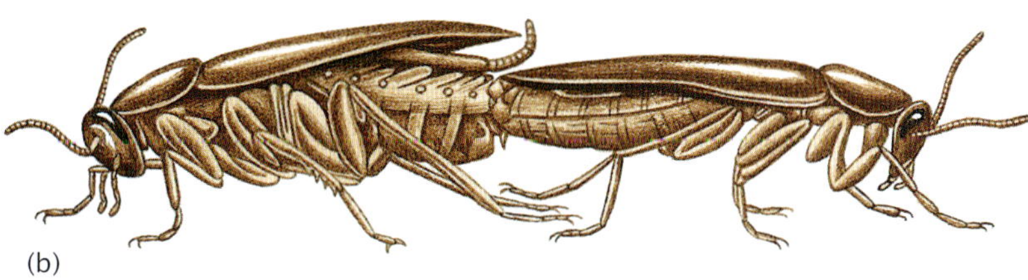

Figure 7.36 Insects use internal fertilisation to ensure meeting of their gametes. **(a)** Dragonflies can mate in flight or on narrow plant stems or leaves. **(b)** Mating of cockroaches on the ground

Reptiles: internal fertilisation

Males of one group of reptiles, the snakes, each have two penises, and each is called a hemipene.

Ancestral reptiles were the first vertebrates to evolve a male copulatory organ, the penis (see figure 7.37). The presence of a penis enabled reptiles to transfer sperm directly into the reproductive tract of a female. Apart from sharks, reptiles were the first vertebrates that did not need to release their gametes into water for fertilisation to occur. Internal fertilisation freed the ancestral terrestrial reptiles and their descendants from a need to return to water for reproduction.

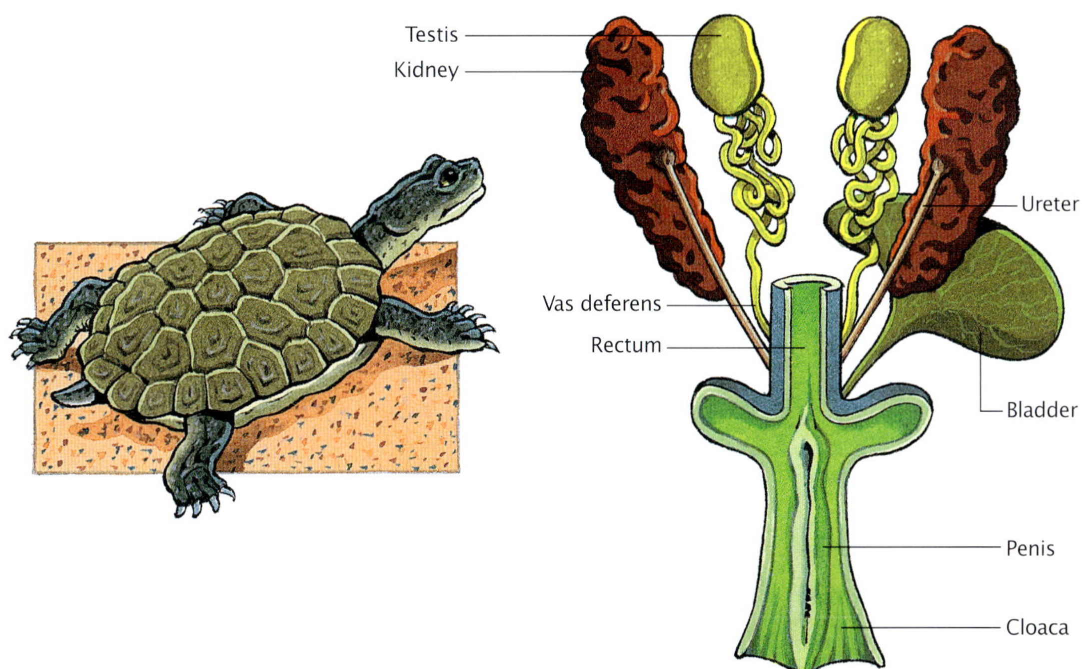

Figure 7.37 View of the urogenital organs of a male turtle. Note the presence of the penis. Ancestral reptiles were the first vertebrates in which this copulatory organ evolved.

For internal fertilisation, female animals produce a small number of eggs (or even a single egg) in each reproductive cycle. Less gamete wastage occurs and the chance of fertilisation for a given egg is quite high.

Another important evolutionary change that first appeared in ancestral reptiles was the egg with a protective outer shell, a series of internal membranes and a food supply for the developing embryo (see figure 12.18, page 385). This type of egg is a self-contained aquatic environment in which embryonic development occurs even when the egg is laid on land.

Birds do it!

Birds also have internal fertilisation but male birds do not have a penis. Instead, sperm transfer takes place through close contact of the urogenital openings (cloacas) of male and female birds. Like their reptilian ancestors, birds produce shelled eggs in which their embryos develop. The production of a shelled egg occurs as the fertilised egg passes through the genital tract of a female bird to the cloaca from where the egg is laid.

Mammals: internal fertilisation

Mammals have internal fertilisation.

- In monotreme mammals, such as the platypus and the echidna, the fertilised egg becomes enclosed within a shell and embryonic development occurs within the shelled egg (see figure 7.38).
- Female marsupials and placental mammals retain the fertilised eggs in their bodies and embryonic development proceeds within the uterus. Young marsupials are born at a very undeveloped stage. Young placental mammals are born at a much more developed stage.

Figure 7.38 Platypus eggs in the nest. The coin is an Australian 20-cent coin. What animal appears on this coin?

Pollination in plants

Reproduction in flowering plants and conifers involves two kinds of gametes: pollen and eggs.

The term 'pollination' commonly refers to the transfer of pollen from the anther of the flower of one plant to the stigma of the flower of another plant or to the transfer of pollen between one conifer and another. This process is known as **cross-pollination**. Fertilisation commences when pollen grains lodge on the **stigma** of a flower (or reach the female cone in the case of conifers).

More than 100 different varieties of the apple (*Malus domestica*) exist, such as Red Delicious, Gala and Granny Smith. Cross-pollination occurs in apples. If only one variety of apple tree is planted in an orchard, pollination will not occur. Orchardists must grow some trees of a second variety as well.

Figure 7.39 Cross-pollination involves pollen transfer between flowers of different plants of the same species.

Self-pollination refers to the transfer of pollen from the stamen of one flower to the stigma of the same flower or to another flower on the same plant. Self-pollination occurs, for example, in the opium poppy (*Papaver somniferum*) and the peach (*Prunus persica*).

For most flowering plants, cross-pollination is more common. Self-pollination is avoided in several ways, for example, by having eggs and pollen from flowers on the same plant maturing at different times. Self-pollination is also prevented by a chemical signal that stops the growth of the pollen tube when pollen from the same plant lands on the stigma of one of its flowers. This condition is termed **self-incompatibility**.

Pollen transfer between plants depends on various environmental agents: pollen may be wind-borne or carried by insects, birds or small mammals. We examine this in more detail in chapter 12.

Technology to assist in 'getting gametes together'

For animals that have external fertilisation, the gametes that meet and fuse are those that by chance are brought together by water currents. For animals that have internal fertilisation, the gametes that meet and fuse in fertilisation under natural conditions are restricted to those from a mating pair. For flowering plants, it is the 'chance' fertilisation that results from the arrival at a flower of either wind-blown pollen or a pollen-bearing insect. Technology has changed some aspects of this natural breeding through procedures such as:

- artificial insemination (AI) in livestock
- artificial pollination in plants.

These topics are explored in detail in *Nature of Biology Book 2, Third edition*, but we will look here at one example of the use of technology in artificial breeding — artificial spawning in scallops.

Artificial spawning of scallops

Culture of various animals in sea water, either in 'watery paddocks' offshore or in large tanks, is known as **mariculture**. Mariculture of various invertebrate animals has become a major industry with the artificial cultivation of commercial seafood species, such as scallops (see figure 7.40), oysters, shrimp and mussels.

Figure 7.40 The commercial or king scallop (*Pecten fumatus*) is widely distributed in temperate seas around Australia. Overharvesting and habitat destruction mean that this species is now not common in inshore waters.

Scallops are bivalve molluscs that have external fertilisation and successful mariculture depends on being able to initiate and control gamette release. Artificial spawning can be achieved by manipulating the environmental conditions in which scallops are cultured. Chinese scallops (*Chlamys farreri*) become sexually mature when the sea water temperature reaches 12–14°C.

For artificial spawning, scallops (mainly females with a few males) are collected from the sea when the temperature is 12–14°C and are placed in tanks. After a few hours, the scallops are removed from the water and left to dry out for several hours, after which they are returned to tanks containing warmer sea water (16–18°C).

The combination of drying out and being returned to warmer sea water stimulates the scallops to spawn and they release their eggs and sperm within a few minutes to half an hour. By separating the scallops into males and females, sperm and eggs can be collected separately and mixed to allow fertilisation to occur. The resulting larvae are free-swimming and are reared in tanks. When they reach the next stage of development, they are known as spats and are ready to settle.

Ropes or other inert supports are put in the tanks and the spats settle on them and develop further.

Eventually, they are transferred to the open sea in net cages where they mature and are harvested.

Some scallops, such as the bay scallop (*Argopectens irradians*), are hermaphrodites while others, such as the Chinese scallop (*Chlamys farreri*), have male and female individuals.

KEY IDEAS

- Sexual reproduction involves fusion of an egg and a sperm at fertilisation.
- In animals, fertilisation may be internal or external.
- Some species that use external fertilisation have developed strategies to increase the chance of fertilisation occurring.
- With internal fertilisation, there is a greater chance of gametes meeting and hence the chance of fertilisation is increased.
- Cross-pollination in plants is more common than self-pollination.
- Various natural agents assist the transfer of pollen between different plant species.
- Natural breeding can be altered by various technologies.

QUICK-CHECK

12 Identify the following statements as true or false.
  a All marine animals use external fertilisation.
  b Insects have internal fertilisation.
  c All vertebrates have internal fertilisation.
  d All hermaphrodite animals use self-fertilisation.
  e Cross-pollination in flowering plants is more common than self-pollination.
  f In flowering plants, fertilisation occurs when pollen grains fall onto the petal of a flower.
  g Mariculture refers to the artificial insemination of vertebrates.

13 Give one example of each of the following.
  a a mammal that lays eggs
  b a flowering plant that is self-pollinating

14 How is spawning initiated artificially in the Chinese scallop?

Meiosis: making gametes

Gamete production involves the process of meiosis and, in animals, this process occurs in the testes for sperm production and in the ovaries for egg production. Meiosis is the process by which one diploid ($2n$) cell with two sets of chromosomes produces four haploid (n) cells each containing one set of chromosomes. As well as halving the chromosome number, the process also reshuffles genes between pairs of chromosomes.

Let us first consider meiosis in terms of inputs and outputs only. It will be seen that:

- one cell entering meiosis with a chromosome number of $2n$ produces four cells each with n chromosomes (see figure 7.41a)
- one cell entering meiosis with two pairs of matching chromosomes produces four cells each with only one member of each chromosome pair (see figure 7.41b)
- one cell entering meiosis with two pairs of matching chromosomes produces four cells each with only one member of each chromosome pair, and with some reshuffling or exchange between matching chromosomes (see figure 7.41c).

(a) **Input**: 1 cell with $2n$ chromosomes

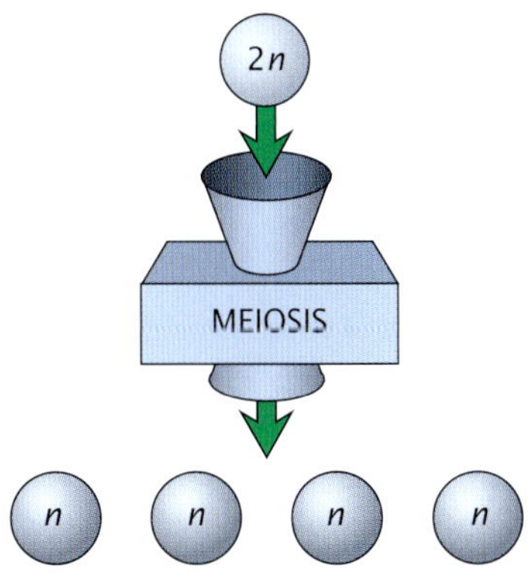

Output: 4 cells, each with n chromosomes

(b) **Input**: 1 cell with 2 pairs of homologous chromosomes

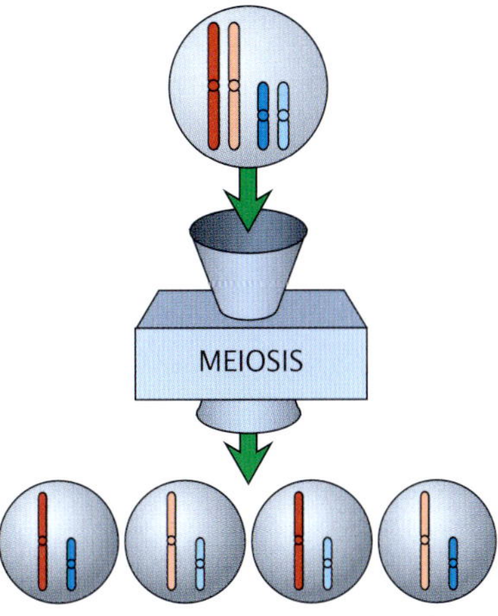

Output: 4 cells with one member only of each pair of homologues

(c) **Input**: 1 cell with 1 pair of homologous chromosomes with 2 linked genes

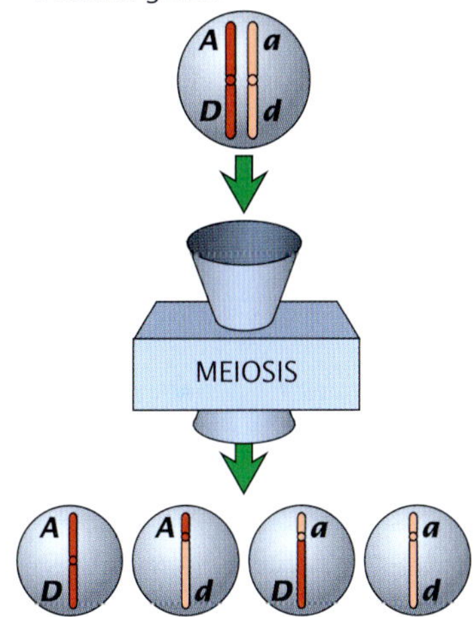

Output: 4 cells, some with 'splicing' of homologous chromosomes to create some new gene combinations

Figure 7.41 Meiosis: the halving and mixing machine **(a)** Meiosis halves the chromosome number. **(b)** The halving is precise — one member of each pair of chromosomes appears in the cell products. **(c)** During meiosis, exchange of segments can occur between homologous chromosomes. This is known as crossing over.

Details of what happens during meiosis between input and output are shown in figure 7.42, page 200. Look at this figure and notice that the chromosomes are duplicated immediately after stage 1. Can you identify where the reshuffling between matching chromosomes occurs? Can you identify where the chromosome number is reduced in each cell from diploid ($2n$) to haploid (n)? Compared with the starting cell, the halving of the chromosome number occurs at stage 6 (anaphase II). Each species has a constant chromosome number and the diploid number is restored at fertilisation when an egg fuses with a sperm:

$$\text{EGG } (n) \quad + \quad \text{SPERM } (n) \longrightarrow \text{ZYGOTE } (2n)$$

Table 4.2 (page 84) shows the diploid number in several species. Using this table, can you identify the number of chromosomes in the sperm of a cat, the egg of a dog, the embryo of a koala, the egg of a mulga tree?

Stages shown are:
1, **2**, and **3** prophase 1
4 metaphase 1 moving into anaphase 1
5 telophase 1 moving into prophase 2
6 anaphase 2
7 telophase 2

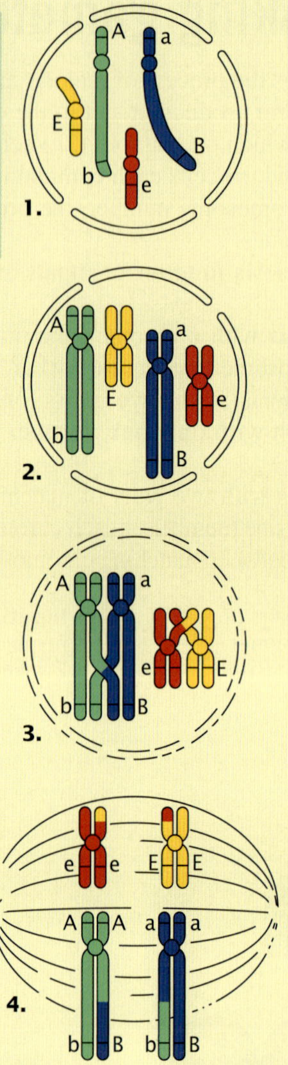

1. Starting point: cell with two pairs of single-stranded homologous chromosomes ($2n = 4$).

2. Early in meiosis, each chromosome replicates to become double stranded.

3. The homologous chromosomes align and pair closely or synapse. One or more exchanges or crossovers occur between a matching segment of one chromosome with a strand in its paired homologue. Crossing over produces new combinations of genetic instructions. The chromosomes then line up across the equator of the cell. Different arrangements are possible.

4. The homologous chromosomes separate (disjoin) from each other when their centromeres are pulled to opposite poles. The disjunction of each homologous pair is independent of any others. (For example, if the 'red' chromosome goes to the left-hand side, this in no way influences which of the longer chromosomes will go to the right-hand side.)

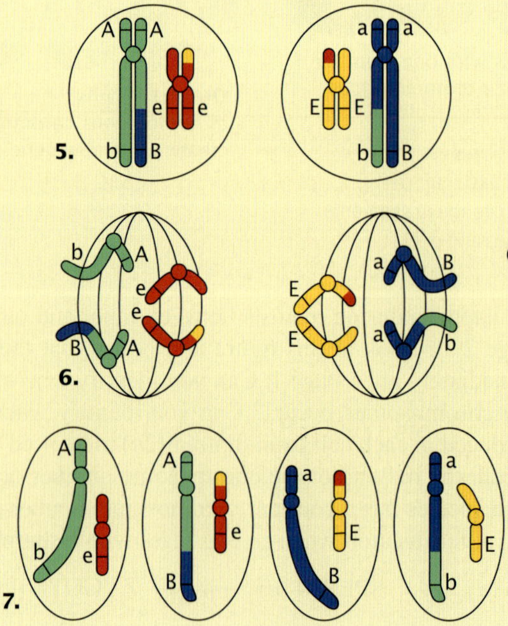

5. The resulting two products each have two double-stranded chromosomes, one long and one short.

6. The strands of each replicated chromosome then disjoin so that the single-stranded chromosome moves in opposite directions. The separation of the single-stranded copies of each chromosome is independent of that of other chromosomes.

7. End point: typically, four cells result from meiosis with each end product containing the haploid number of chromosomes. Having started from one cell ($2n = 4$), the process has produced four cells, each $n = 2$.

Figure 7.42 Stages in the process of meiosis, from the starting diploid cell to the final haploid gametes

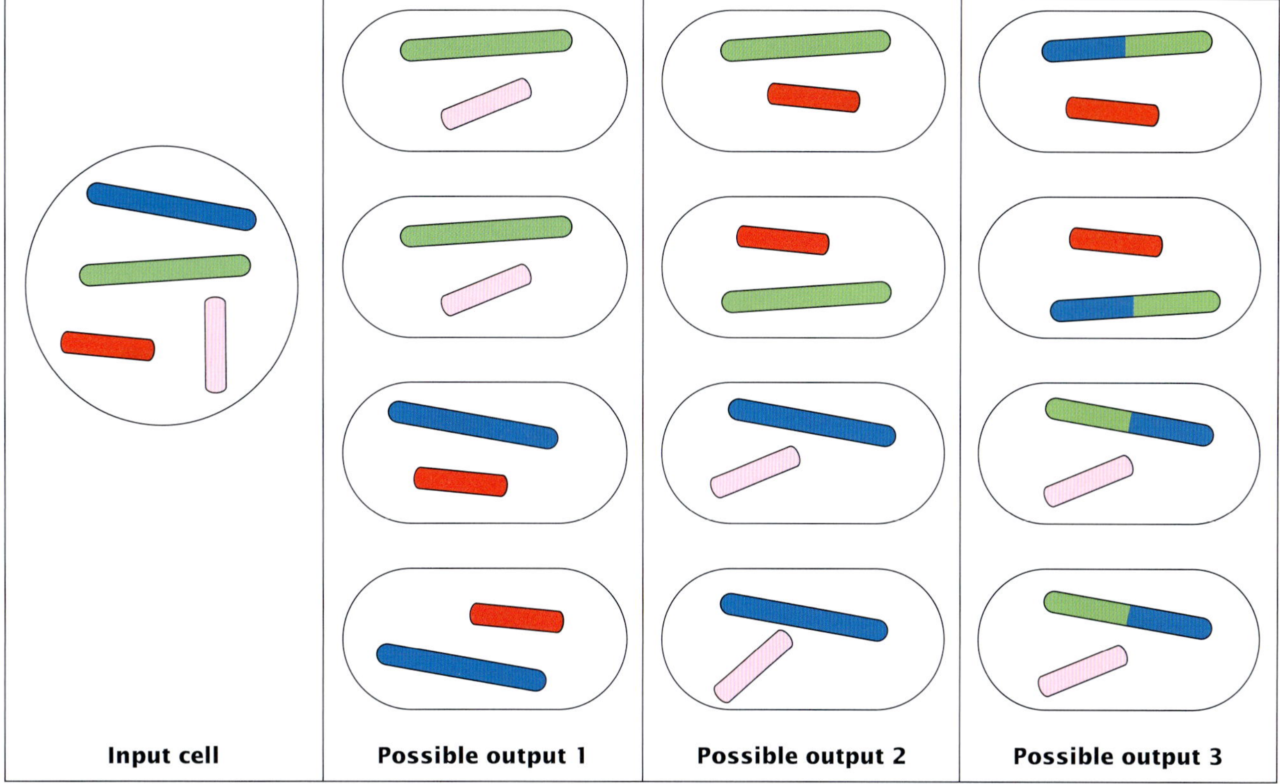

Figure 7.43 Variations in outputs of meiosis. Notice that, even with the same starting input, the outputs of meiosis can differ each time. In fact, just three of a very large number of outputs are shown here. As a result, meiosis generates variation in the gametes of one animal and in the spores of one plant.

The products of meiosis are not identical. Figure 7.43 shows some of the many possible outputs from meiosis — gametes in animals and spores in plants.

Let us look at the essential steps of reproduction, namely meiosis and fertilisation, in a flowering plant and then in humans. Let's start with a flowering plant.

Flowering plants: making eggs and sperm

Buried within the ovary of a flowering plant are **ovules**. Depending on the type of flowering plant, one ovary may contain one or many ovules. Within each ovule, one diploid cell (known as the megaspore mother cell) undergoes meiosis to produce four haploid cells called megaspores. This event normally takes place within the flower bud. Three of the megaspores disintegrate and the remaining one undergoes mitosis three times, but with no division of the cytoplasm, to produce a cell with eight nuclei. This structure then becomes organised into seven cells, one of which is the haploid egg cell and one of which has two nuclei, known as **polar nuclei** (see figure 7.44a). This seven-celled structure is enclosed within the ovule with the egg cell located close to an opening at the base of the ovule called the **micropyle**. By the time the flower is open and the stigma is receptive, this process is complete. We now have an egg!

Inside the anther of a flowering plant are many diploid cells (known as microspore mother cells). Each of these cells undergoes meiosis to produce four haploid cells called microspores. Each microspore divides to form a two-celled pollen grain with a distinctive external wall. The pollen is usually shed from the anthers of the flower at this stage, blown by the wind or carried by a vector. After it lands on the stigma of a flower, a pollen grain with its two cells germinates. One cell of the pollen forms a pollen tube that grows down the hollow style of the flower to the ovary. (If the pollen simply remained on the stigma, no fertilisation could occur because the ovule with its egg cell is totally enclosed within the ovary at the base of the carpel.) The second nucleus of the pollen grain divides to form two sperm nuclei (see figure 7.44b). We now have sperm!

ODD FACT

In some flowering plants, the pollen tube must grow a considerable distance to reach the micropyle. In maize, for example, the distance from the top of the stigma to the micropyle is about 15 centimetres.

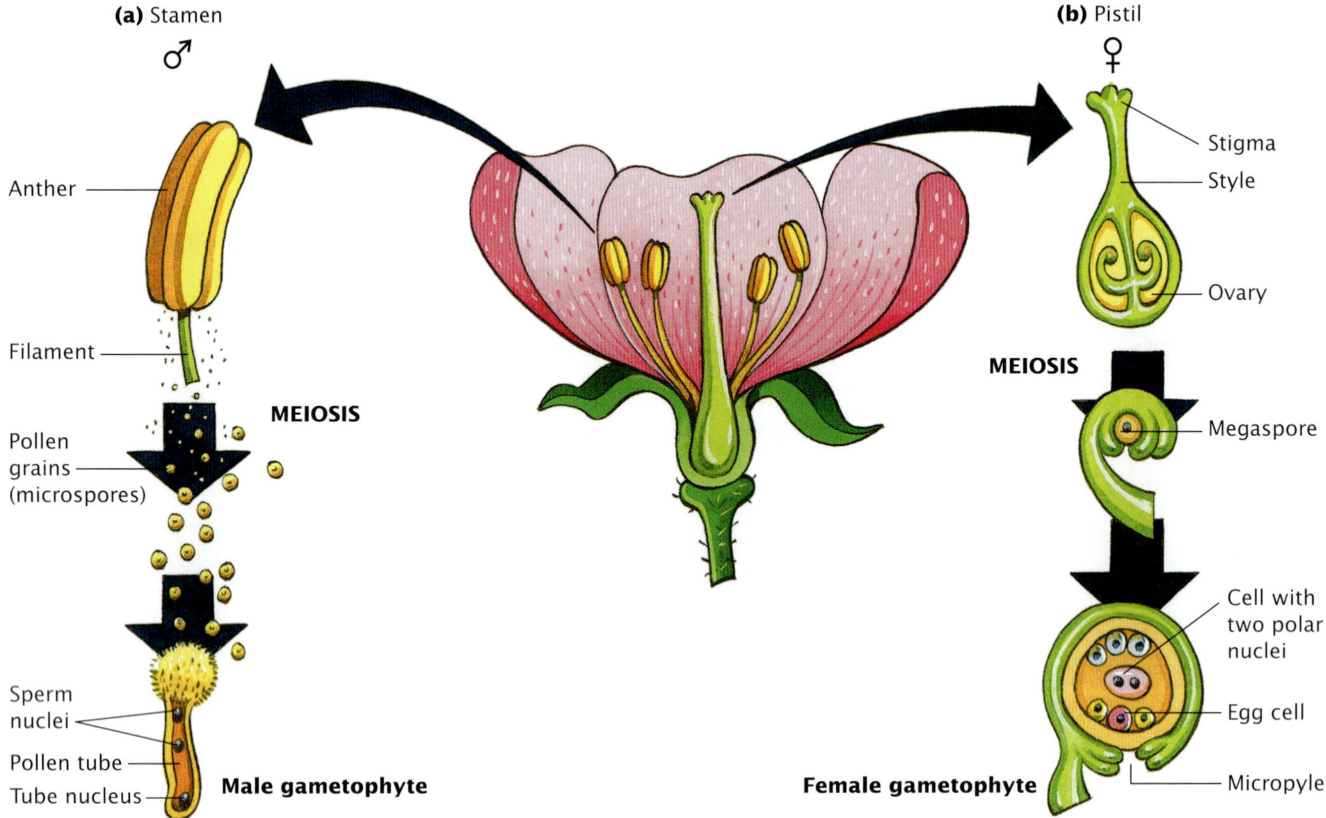

Figure 7.44 **(a)** Formation of the three-celled male gametophyte. Note the two sperm nuclei. Each will play a role in fertilisation. **(b)** Formation of the seven-celled female gametophyte. Note the egg cell near the micropyle and the central cell with its two polar nuclei. In flowering plants, this seven-celled structure is the female gametophyte stage. Unlike a fern gametophyte, which lives independently of the dominant fern sporophyte, the flowering plant gametophyte is enclosed within the ovule inside the ovary.

Flowering plants: fertilisation

Once the pollen tube reaches the micropyle of the ovule, one of the sperm nuclei moves inside where it fertilises the egg cell to produce a diploid zygote. The second sperm nucleus fuses with the two polar nuclei to make a triploid cell ($3n$) that will multiply to produce **endosperm** tissue that is a food store for the zygote during its development into an embryonic seedling. Flowering plants among all living organisms are unique in having the **double fertilisation**, with one sperm nucleus fertilising the egg and the other sperm nucleus fusing with the two polar nuclei.

Flowering plants: seed and fruit

After fertilisation, the flowers wither, petals and stamens fall and usually the style drops off. However, inside the ovary, each ovule (with the nutrient endosperm tissue and the developing embryo inside) enlarges to form a **seed**. A tough seed coat formed from the outer layers of the ovule protects the embryonic plant inside. A seed, then, is a fertilised ripened ovule containing a plant embryo and endosperm food store within a tough outer coat. As seeds are forming inside the ovary, the ovary wall itself thickens to become a **fruit**. Fruits of some flowering plants are the familiar soft and fleshy structures like tomatoes, olives, plums, peaches and cherries. In the cherry, for example, the outer skin of the fruit is the outer layer of the ovary, the soft flesh is derived from the middle layer of the ovary wall, and the hard stone is derived from the innermost layer of the ovary. Inside that stone is a seed! Figure 7.45 shows the formation of a pear fruit with its seeds.

ODD FACT

Sometimes a fruit includes other parts of a flower in addition to the ovary. In apples, for example, the ovary wall forms the scaly core of the apple that encloses the seeds. The edible flesh of an apple is formed from the receptacle of the flower.

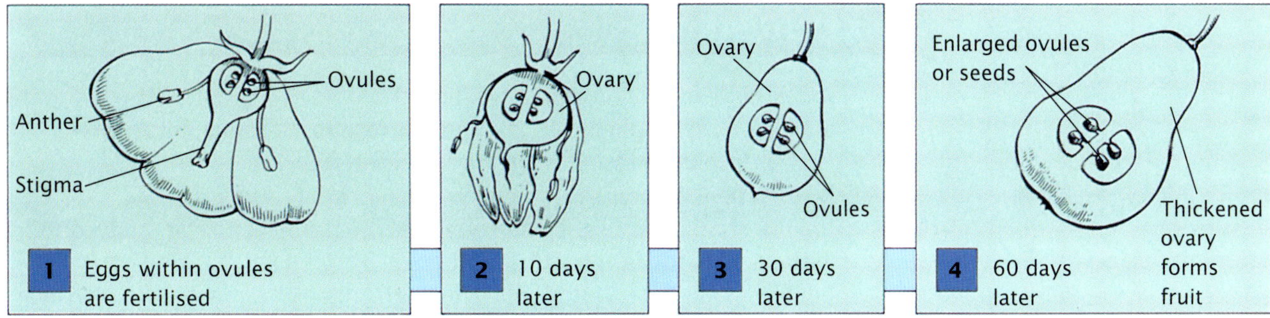

Figure 7.45 A pear is a fruit derived from one ovary containing several ovules. The general definition of a fruit is: the fleshy or dry, thickened and ripe ovary that encloses one or several seeds.

Some flowering plants, however, have fruits that are hard and dry. Table 7.2 gives examples of a few of the many kinds of fruits. Note that a grain is a fruit, not a seed. What are the outer layers of a grain made of — ovule wall or ovary wall?

Table 7.2 Some of the many kinds of fruit recognised by botanists. What survival advantage for a plant species might result from the production of fleshy fruit?

Fruit type		Examples
fleshy fruits	drupes	cherries, plums, apricots, peaches 
	berries	grapes, tomatoes
	aggregate fruit*	raspberries, strawberries, blackberries 
dry fruits	legumes (pods)	beans, peas, acacias 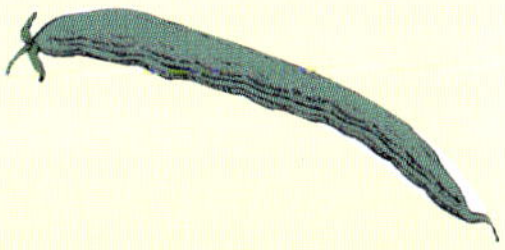
	capsules	poppies
	grains	wheat, oats, maize, rice 
	nuts	acorns (oaks), hazelnuts

*Aggregate fruit consist of a cluster of ripened ovaries.

Figure 7.46 If corn grains are heated, internal pressure from water vapour builds up and the kernel explodes. The white material that bursts out from the grain is the endosperm tissue that we call popcorn.

ODD FACT

Fruits of flowering plants are a major source of food for human populations, particularly cereal grains such as rice, wheat, oats and corn.

Normally, in flowering plants, fruits with seeds appear only after fertilisation. However, unexpectedly, some plants produce fruits without fertilisation of the eggs inside the ovules and, as a result, these fruits lack seeds. The formal term **parthenocarpy** refers to this condition.

KEY IDEAS

- Meiosis is involved in gamete production.
- Starting from a diploid cell ($2n$), meiosis results in four haploid (n) cells.
- Meiosis involves a reshuffling of genetic material.
- Fertilisation restores the diploid number of chromosomes.
- In flowering plants, fertilisation involves the fusion of an egg produced in the ovule with a sperm nucleus formed in the pollen grain.
- Double fertilisation occurs only in flowering plants.
- In flowering plants, the developing embryo is enclosed within a seed formed from the ovule.
- In flowering plants, one or more seeds are enclosed within fruit formed from the ovary.

QUICK-CHECK

15 Identify the following statements as true or false.
  a In meiosis, diploid gametes are produced from a haploid cell.
  b The endosperm in the seed of a flowering plant is triploid.
  c Pollination is the same as fertilisation.
16 An organism has a chromosome number $2n = 10$.
  a What is the origin of these 10 chromosomes?
  b How many chromosomes would be in one of its unfertilised eggs?
  c Would all eggs it produced be genetically identical?
17 An egg from an organism has four chromosomes.
  a Is this a haploid or a diploid cell?
  b How many chromosomes would be present in a somatic cell?
18 Domestic cats have 38 chromosomes. How many chromosomes would be present in a normal sperm cell of a male cat?
19 What is the difference between a seed and a fruit?
20 Give one example of each of the following:
  a a drupe
  b a berry.

Human reproduction

At the beginning of this chapter, you read the story of Sarah and James Minter. One striking feature of this story was the number of babies they had and the frequency of their births. Human reproductive patterns are very different today for several reasons, including advances in reproductive technology that help to prevent conception or overcome infertility. Let us look at the human reproductive system.

About 3 per cent of males are born with undescended testicles. During the first year after birth, about two-thirds of these undescended testes move into the scrotum. Surgery can be performed to correct the fault.

Human male reproductive system

The reproductive system of a human male is typical of male mammals (see figure 7.47). Two testes are located in a loose pouch of skin called the **scrotum**. The temperature within the abdominal cavity is usually about 37°C, which is too high for sperm to mature and survive. Because the testes hang outside this cavity they have a lower temperature. In addition, they have many sweat glands that assist in

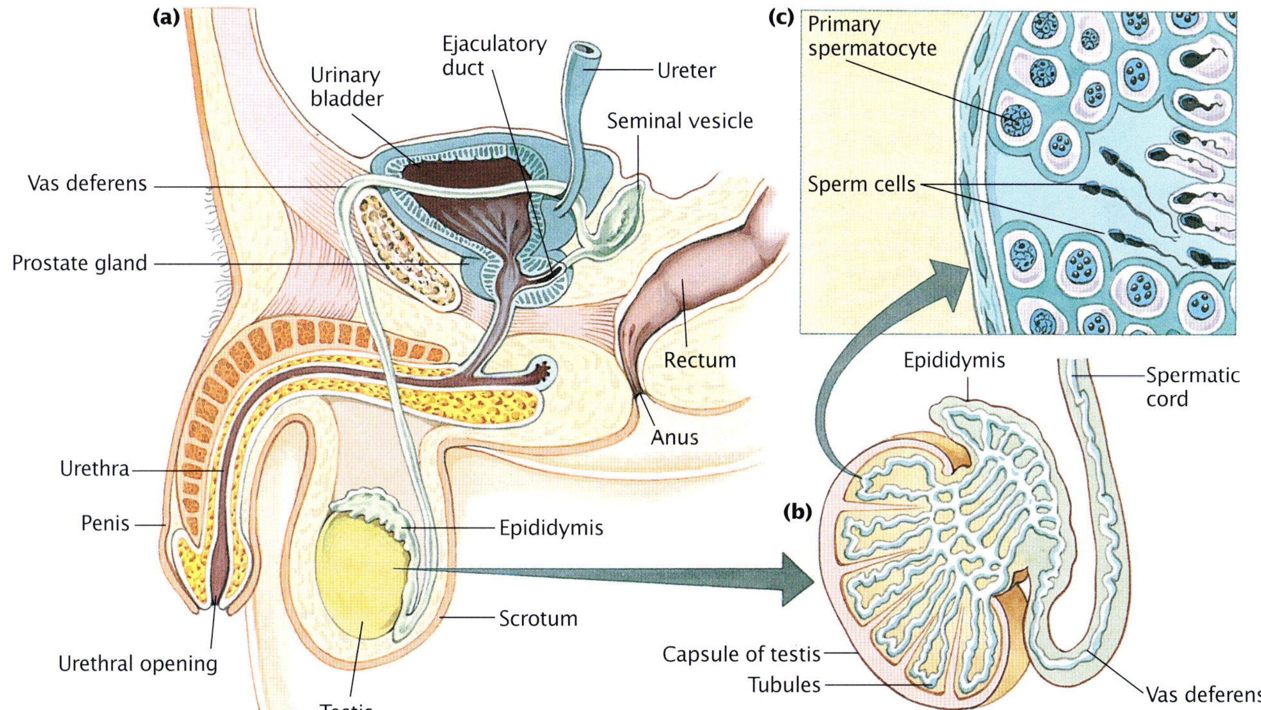

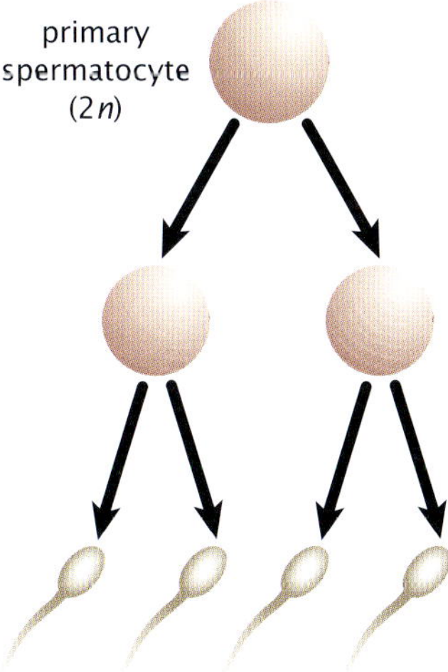

Figure 7.47 Reproductive system of the human male **(a)** Longitudinal section through the various parts **(b)** Longitudinal section through a testis to show the folded tubules in which meiosis occurs **(c)** Cross-section of a tubule showing details of sperm formation **(d)** Simplified diagram showing that each primary spermatocyte produces four haploid sperm by meiosis.

maintaining the testes' temperature at about 35.5°C, a temperature suitable for sperm development. Generally, the testes move into the scrotum several weeks before a male child's birth. If this does not occur, they will later fail to produce sperm because of the higher temperature in the body.

At **puberty**, **luteinising hormone** from the pituitary gland acts on the testes and the testes produce **testosterone**, the principal male sex hormone. When this occurs, the sex organs grow, muscles develop, the voice changes, a beard begins to grow and the testes begin to produce sperm and hormones.

In the human male, the time taken for the process involved in the conversion of an immature spermatogonium into a mature sperm is about 64 days.

Meiosis in the human male

Sperm are produced in tubules. The cells closest to the outer edge of the tubule wall are precursor sex cells. These cells remain dormant until puberty. At puberty, the precursor sex cells reproduce by mitosis to form primary spermatocytes. Primary spermatocytes form continuously in the testes after puberty, and each gives rise to four haploid sperm by meiosis (see figure 7.47d).

Each sperm can be thought of as a package of one set of human chromosomes organised ready for transmission to the next generation.

Mature sperm travel from a testis and are stored in the **epididymis** associated with the testis. From there, sperm move into a **vas deferens** which runs from each epididymis to the **urethra**. Sperm move into the urethra, along with a nutrient-rich fluid that has been secreted into the vas deferens from the seminal vesicle, Cowper's gland and the prostate gland. The fluid and the sperm together form **semen**.

During sexual arousal, blood enters the **penis** which becomes firm and erect. Semen moves through the urethra and leaves the body by ejaculation. Each ejaculation contains between 2.5 and 6 millilitres of semen and each millilitre contains 50 to 100 million sperm. Sperm are introduced into the vagina by the penis. During ejaculation, the outlet of the urinary bladder closes so that urine will not enter the urethra.

Human female reproductive system

Female gametes or eggs are produced in the ovaries. These are a pair of organs, each about four centimetres by two centimetres, located in the lower part of the abdomen (see figure 7.48). The mature ovaries produce hormones as well as eggs.

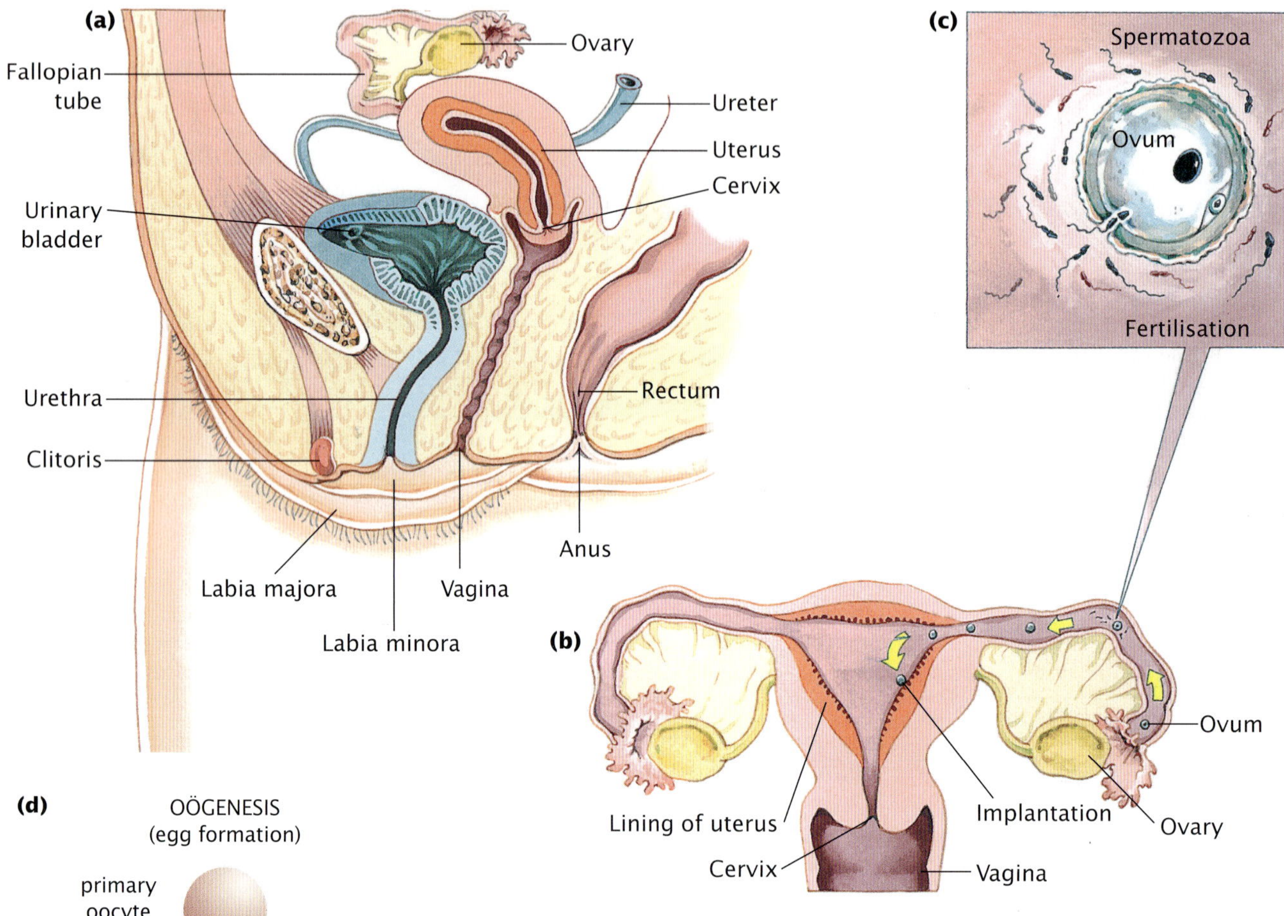

Figure 7.48 Reproductive system of the human female **(a)** Longitudinal section **(b)** Section through ovary, fallopian tube and uterus to show formation and passage of egg. Meiosis in an egg is completed only after fertilisation **(c)** which occurs in the fallopian tube. The diploid zygote that is formed by fertilisation develops into a small mass that moves along the fallopian tube into the uterus where it becomes implanted into the uterus wall. **(d)** Each primary oocyte (pronounced 'oh-oh-site') undergoes meiosis and gives rise to just a single cell or ovum.

Meiosis in the human female

Special cells present in the fetal ovary increase in number by mitosis during early fetal life and enlarge to form the potential egg cells or primary oocytes. It has been estimated that at birth each ovary contains about 200 000 such cells.

Although primary oocytes all start the process of meiosis in the fetal ovary, they stop very early in the meiotic cycle and remain dormant in the ovary until puberty. At puberty, in response to hormonal signals, an ovarian cycle begins. One of the potential egg cells increases in size and resumes the interrupted meiotic cycle begun many years before. Each primary ooctye is surrounded by a layer of follicle cells and, as it undergoes meiosis, the follicle in which it is located enlarges and a fluid-filled space appears. In the human ovary, a mature follicle can reach 10–13 millimetres in diameter.

The primary oocyte divides to give two cells of unequal size. The smaller nonfunctional cell, known as the first **polar body**, degenerates. The larger cell is released from the ovary at the time of ovulation when the mature follicle swells

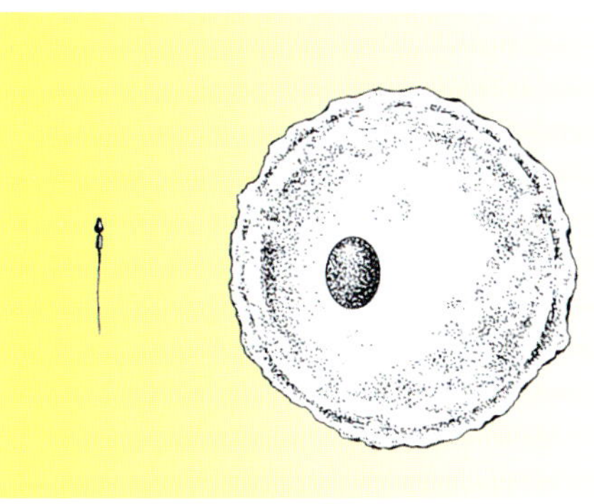

Figure 7.49 Relative sizes of sperm (left) and egg (right)

ODD FACT

In a mature female, the uterus is about 7.5 centimetres long, 5 centimetres wide, and 1.75 centimetres thick. The uterus enlarges greatly during pregnancy.

to form a bulge on the surface of the ovary and the bulge ruptures. Meiosis is still proceeding in this immature egg cell as it is released from the ovary into the fallopian tube. The empty follicle from which an egg cell is released becomes a **corpus luteum**.

About once a month after puberty, an immature egg is expelled from an ovary and moves into the funnel-like opening of a fallopian tube. Each fallopian tube is about 10 centimetres long and transports eggs to the **uterus**. The uterus is muscular and shaped like an inverted pear, and is situated between the urinary bladder and the rectum.

If an egg is fertilised, the second meiotic division is completed (see figure 7.48d), and the fertilised egg becomes embedded in the lining of the uterus. The single diploid cell that results from fertilisation is called a zygote and is just 0.1 mm long. The implantation of the zygote into the lining of the uterus is called **conception**. The zygote divides by mitosis to form an **embryo** which, within four weeks, is about 3 mm long with a body plan underway. When development is more advanced, the embryo becomes a fetus.

Blood vessels in the walls of the uterus supply the developing embryo with oxygen and nutrients. Later, the specialised organ known as the placenta serves that function. The narrow lower end of the uterus is called the **cervix**, which opens into the **vagina**. Sperm enter the female reproductive tract via the vagina. The vagina is the passageway for a baby during its birth. A female has separate openings for the urinary and reproductive systems.

The menstrual cycle

After puberty, a female produces an egg each month. Other changes also take place on a cyclic basis and include changes in hormone levels and the lining of the uterus. These changes are called the **menstrual cycle** (see figure 7.50).

Each immature egg is surrounded by a layer of cells — a **follicle**. Starting at puberty, a cycle begins with the secretion of **follicle stimulating hormone (FSH)** by the pituitary gland. Like all hormones, FSH is transported by the blood to all parts of the body. FSH acts in particular on the ovaries, and follicles begin to grow. Each month, about 20 eggs begin development but generally all but one degenerate.

ODD FACT

The first menstrual cycle, or menses, is called menarche, and the last cycle is called menopause. Menopause occurs because of a reduction in the amount of female sex hormones being produced.

The cells of the growing follicles secrete **oestrogen**, one of the female sex hormones. This hormone causes the thickening of the lining of the uterus in preparation to receive a fertilised egg. The increased secretion of oestrogen also stimulates the hypothalamus to produce more FSH and luteinising hormone (LH). The increase in LH causes the follicle to swell and erupt and the egg is released from the ovary, a process called **ovulation**. The egg is surrounded by a layer of follicle cells.

After ovulation, the follicle cells in the ovary form the corpus luteum that secretes the hormone **progesterone**. Progesterone acts on the uterus lining, preparing it for pregnancy. If fertilisation occurs, the corpus luteum persists and continues to produce progesterone. If fertilisation does not occur, the **ovum** and corpus luteum break down. This means the level of progesterone falls and the thick vascular lining of the uterus breaks down. Blood and tissue from the uterus wall are discharged through the vagina. This process lasts a few days and is called menstruation.

The production of FSH is suppressed while progesterone levels are high, but when the level of progesterone falls, FSH is again produced by the pituitary. The cycle begins again and more follicles begin to develop. The length of the cycle varies from female to female but on average lasts for about 28 days. The menstrual cycle continues, except during pregnancy, for between 30 and 40 years and, during this time, a female will produce and discharge a total of about 400 eggs from her ovaries.

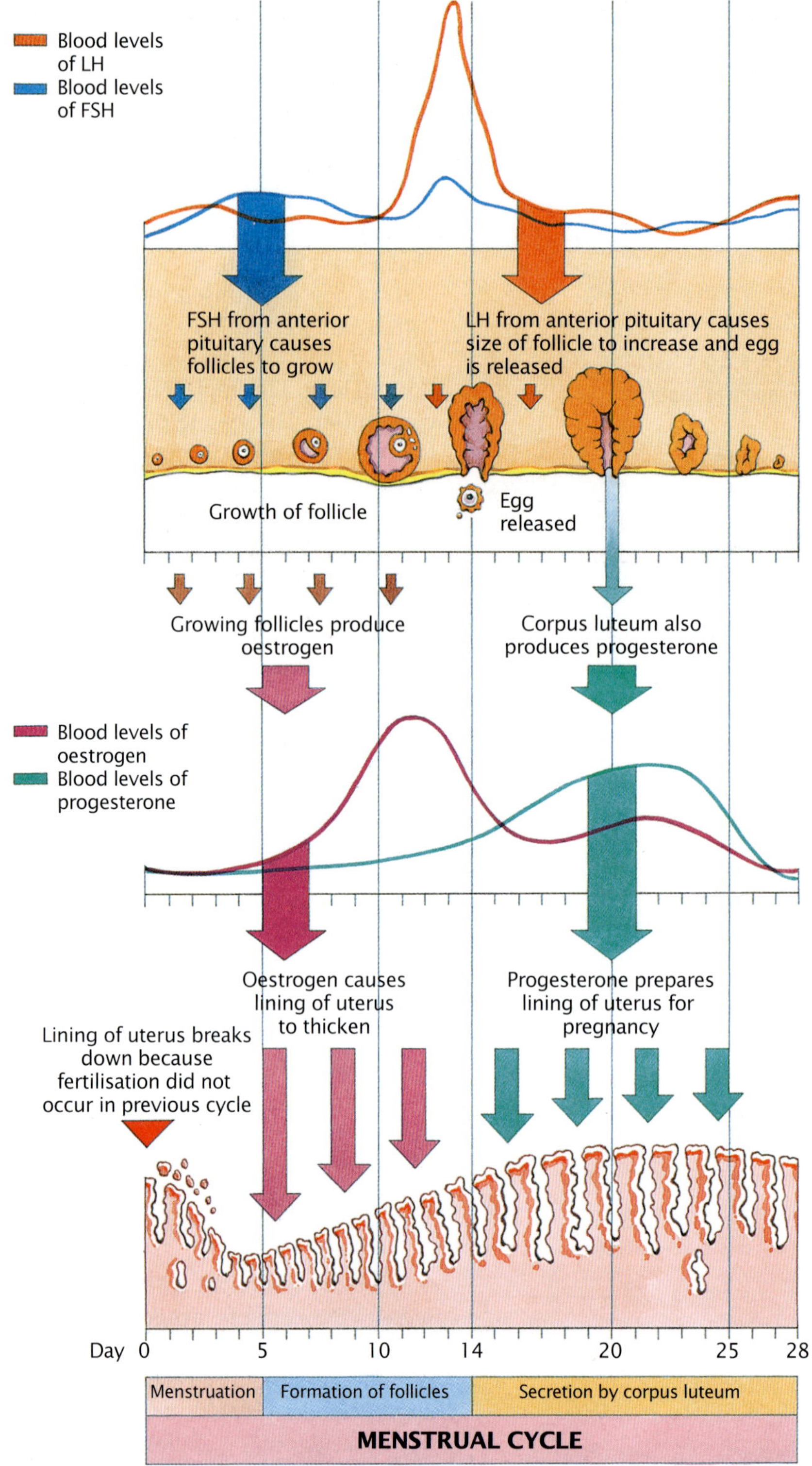

Figure 7.50 During the menstrual cycle, changes occur in the ovary and the lining of the uterus as the level of hormones in the blood changes. Follicle stimulating hormone (FSH) from the pituitary gland stimulates the growth of follicles in the ovary. Follicles produce oestrogen that stimulates the pituitary to produce luteinising hormone (LH). As LH increases, the size of the follicle increases until an egg is released. The corpus luteum formed in the follicle secretes progesterone that prepares the lining of the uterus for pregnancy. If fertilisation does not occur, the thick vascular lining of the uterus breaks down and is discharged from the body in the process called menstruation.

BIOLOGIST AT WORK

Dr Jan West — Biology researcher and teacher

In her biology lectures to students at Deakin University, Dr Jan West discusses reproduction in the context of her pet corgi who came into the world by means of artificial insemination. Here are some extracts from 'Lexie's story'.

I'm a Cardigan Welsh Corgi puppy called Lexie. My mother, Chelsea, lives in Australia and my father, Elmo, lives in California. My parents have never met. So, how did I come about?

In January 2000, semen was collected from Elmo who was tested to meet Australian Quarantine regulations. The semen was placed in vials, frozen in liquid nitrogen and shipped to Australia. The use of frozen semen has many advantages — it increases the breeding life of males and, as the animals do not move between countries, no quarantine time is needed. In addition, it increases the gene pool for rarer breeds.

Live sperm survives in the female reproductive tract for two to five days but frozen semen, once thawed, is viable for only 6 to 12 hours. To predict ovulation, Chelsea had daily blood tests to check her levels of the hormone, progesterone. Several days after ovulation had occurred, the semen was thawed and surgically implanted in Chelsea's uterus near the entrance to her oviducts. An egg was waiting there. One of the sperm that had been frozen for 1129 days arrived at this ovum and penetrated it. This egg was fertilised and would develop into me — Lexie. Two other eggs fertilised at the same time were to become my sister and brother.

Just over a week after fertilisation, meiotic cell divisions have increased my cell numbers and I am a hollow ball of cells about 2 millimetres in diameter.

Three weeks after fertilisation, I am 7 mm long and floating in amniotic fluid and the first signs of my eyes and ears have appeared. At four weeks after fertilisation, I am about 15 mm long and most of my internal organs have developed. After 5 weeks, the first signs of my sex are apparent. Both the sperm and the egg from which I originated carried an X chromosome, so I'm XX — a girl.

Figure 7.51 Dr Jan West in class with Lexie

After 6 to 7 weeks, I am about 90 mm in length and I have whiskers, toes and toe nails. I am starting to grow body hair and the shafts of my bones are forming. Between weeks 8 and 9, my footpads and coat are finished — a beautiful brindle and white coat.

Finally, after 63 days gestation, I'm born on 30 April 2003. My eyelids and ear canals stay closed for about 10 days after birth and my nose turns black. I have a well-developed sense of smell and a strong 'suck' reflex. When my mother's nipples are stimulated by my sucking, the pituitary gland in her brain releases the hormone oxytocin that triggers her mammary glands to release milk (figure 7.52). The milk is high in calories and nutrients which allows me and my siblings to grow into healthy adult Cardigan Welsh corgis.

Figure 7.52 Chelsea and her hungry new litter

KEY IDEAS

- Human eggs and sperm are produced in ovaries and testes respectively.
- From one starting cell undergoing meiosis, only one egg is produced by a human female but four sperm result.
- A human egg completes meiosis only after fertilisation.

QUICK-CHECK

21 Where does fertilisation of the human egg normally occur?
22 In which sex is there a single opening for both the reproductive and the urinary systems?
23 What is the difference between the terms conception and fertilisation?

BIOCHALLENGE

Carefully examine the images below. Each image relates to an aspect of reproduction. For each image, identify at least three key words (from the 'Key words' list on page 211) that can be associated with it. Try to use as many different relevant terms as possible.

1

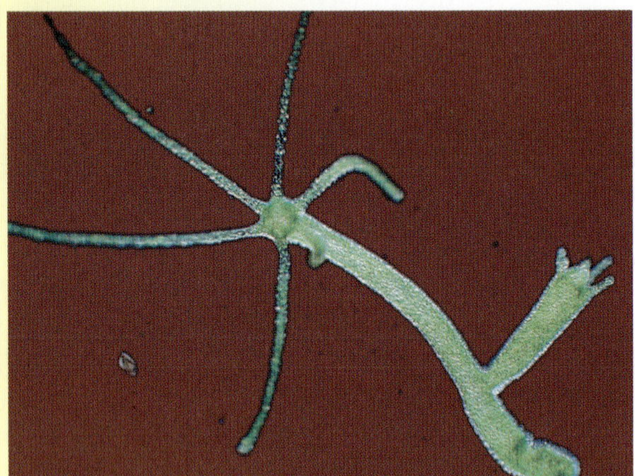

Hydra

2

Undersurface of fern frond

3

Clownfish

4

Single litter of kittens

5

Tomato

6

Lily flower

CHAPTER REVIEW

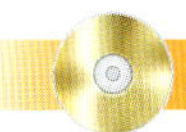

CROSSWORD

Key words

alternation of generations
asexual reproduction
binary fission
carpels
cervix
clones
conception
contraception
corpus luteum
cross-pollination
dioecious
diploid
double fertilisation
egg
embryo
endosperm
epididymis
exponential growth
external fertilisation
fertilisation
follicle
follicle stimulating hormone (FSH)
fruit
gametes
gametophyte
haploid
hermaphrodites
internal fertilisation
invertebrate
luteinising hormone
mariculture
meiosis
menstrual cycle
micropyle
mitosis
monoecious
multiple fission
oestrogen
ovary
ovulation
ovules
ovum
parthenocarpy
parthenogenesis
penis
pistil
polar body
polar nuclei
pollen
progesterone
puberty
scrotum
seed
self-incompatibility
self-pollination
semen
sequential hermaphrodites
sexual reproduction
simultaneous (synchronous) hermaphrodism
spawning
sperm
spores
sporophyte
stamens
stigma
testis
testosterone
urethra
uterus
vagina
vas deferens
vegetative reproduction
zygote

Questions

1 *Making connections* ▸ Use as many as possible of the chapter key words to construct a concept map.

2 *Applying your understanding* ▸ The horse (*Equus caballus*) has a diploid number of 64 while the diploid number in the closely related donkey (*Equus asinus*) is 62. Mules are the sterile offspring from the mating of a male donkey with a female horse. How many chromosomes would you expect in:

a an egg cell produced by a female horse?

b a sperm cell produced by a male donkey?

c a somatic cell from a mule?

3 For each of the following questions, identify two correct answers from the choices given.

a Which animals have internal fertilisation: crocodiles, frogs, koalas, cats?

b Which animals depend on the availability of water for breeding: cane toads, goannas, seagulls?

c Which animals produce shelled eggs: platypus, frogs, kangaroos, turtles, toads, penguins?

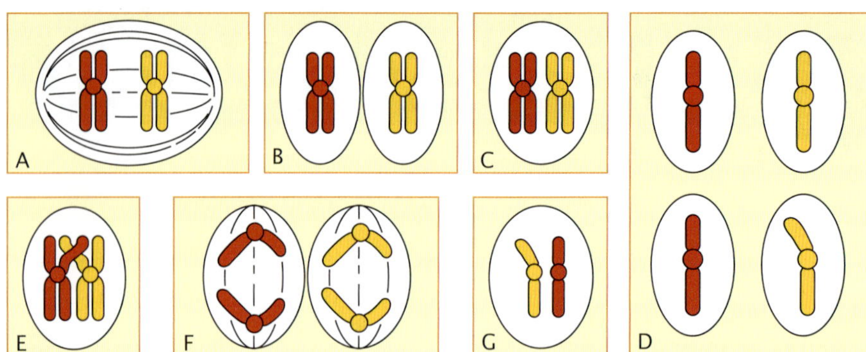

Figure 7.53

4 Figure 7.53 shows various stages during the process of meiosis. Examine the figure.

a List the stages in order from the beginning to the end of meiosis.

b What is the diploid number of this cell?

5 Plesiosaurs are extinct reptiles that lived in the sea. What prediction, if any, can be made about their manner of reproduction?

6 Consider the female reproductive system and associated hormones. Explain whether you agree or disagree with the following statements:

a If a woman's fallopian tubes are blocked, she can still produce eggs.

b If a woman's fallopian tubes are blocked, she can still become pregnant.

7 Consider the male reproductive system and associated hormones. Explain whether you agree or disagree with the following statements:

a Males born with undescended testicles are unlikely to produce sperm.

b A male produces more gametes in one meiotic cycle than does a female.

8 Indicate, giving a brief reason, whether or not each of the following might be easily seen as you walk around a botanic garden one weekend.

a a pine gametophyte

b a diploid human

c a megaspore mother cell of a flowering plant

d a fern gametophyte

9 ***Solving problems*** Refer back to figure 7.6 (page 178) which shows binary fission in the bacterial species, *E. coli*. Assume that the time for binary fission is an average of 20 minutes at 15°C.

a Starting with 10 bacteria, about how many bacteria would be present at the end of two hours of binary fission?

b The rate of binary fission doubles (or halves) for each rise (or fall) of 10 degrees in temperature. Assume that at 25°C, binary fission becomes twice as fast and is completed in 10 minutes. Assume that at 5°C, binary fission is slowed and requires 40 minutes for completion.

i At 25°C, starting with 10 bacterial cells, how many bacterial cells would be present after 2 hours?

ii At 5°C, starting with 10 bacterial cells, how many bacterial cells would be present after 2 hours?

c By what process do these new bacteria appear?

10 ***Apply your understanding*** Explain each of the following observations.

a Gametes from one fern gametophyte are identical in genetic content.

b For flowering plants, but for no other plants, the term 'double fertilisation' is used.

c The somatic cells of a blue whale have 40 chromosomes but their gametes have only 20.

d Cooked meat should be stored in refrigerators, not at room temperature.

e Ovulation stops during a woman's pregnancy.

f Ferns are generally not found in the same range of habitats as flowering plants.

11 *Discussion* ▸ Consider the following four observations.

I Orange groves often comprise trees that are clones because they have been produced by grafting shoots from the same plant (called a scion) onto healthy root stock.

II Fertilisation in oranges requires cross-pollination.

III Self-fertilisation is prevented by self-incompatibility reactions.

IV Under these circumstances, the orange trees produce fruit but the fruit differs from that produced when fertilisation occurs prior to fruit formation.

a What would be special about these oranges?

b Would such oranges be labelled as Valencia or Navel?

c Define the term 'parthenocarpy'.

d Why did cross-fertilisation not occur between the various orange trees in this grove?

e Is grafting an example of sexual or asexual reproduction?

12 Identify one significant difference between the members of the following pairs.

a Little Nicky (see page 185) and your cat

b sexual and asexual reproduction

c the gametophyte and the sporophyte stages of the life cycle of a plant

d cross-pollination and self-pollination

e a dioecious and a monoecious plant

13 Consider the following statements about internal fertilisation and identify them as true or false.

a Internal fertilisation occurs in all terrestrial animals.

b Internal fertilisation occurs in all marine animals.

c Internal fertilisation does not occur in hermaphrodite animals.

14 Identify the structure that assists sperm transfer to the female in internal fertilisation in the following species.

a tortoises

b sharks

c dogs

d octopuses

15 *Using the Web* ▸ In this question you will compare plant life cycles. Refer back to figure 7.26 (page 188) which shows the life cycle of a fern.

a Which is the dominant stage of the fern life cycle?

b Explain the term 'alternation of generations'.

c Go to www.jaconline.com.au/natureofbiology/natbiol1-3e and click on the 'Moss life cycle' link for this chapter. Examine the stylised life cycle of a moss. Compare the moss life cycle with that of the fern (figure 7.26) and:

i identify two differences

ii identify two similarities.

8 Ordering the living world

KEY KNOWLEDGE

This chapter is designed to enable students to:

- recognise the value of identification and scientific naming
- develop knowledge of the principles of classification
- recognise that biological classification contains a hierarchy of levels
- become aware that taxonomic systems are subject to change.

Figure 8.1 The living world contains millions of different kinds of organism. The science of taxonomy is concerned with organising this biodiversity into hierarchical groups based on the degrees of relationship between organisms.

In this chapter, we will explore the identification, scientific naming and classification of organisms, and examine how various technologies are extending our capabilities in these areas.

The seeds that poisoned

It is January 1697 and the Dutch sailors go ashore in a strange country. They notice palm-like plants with cones bearing large brightly coloured seeds (see figure 8.2a and b). The seeds are enticing and appear to be a delicious change to a monotonous seafaring diet. Some of the sailors eagerly sample the seeds. One of the sailors later writes about his experience:

> '. . . I ate five or six [seeds] and drank some of the water from one of the aforesaid holes, but after an interval of about three hours I and five more of the others who had also eaten the said fruit began to vomit so violently that there was hardly any distinction between death and us.' (from G. Schilder (ed.), *Voyage to the Great South Land: Willem de Vlamingh*, translated by C. de Heer, Royal Australian Historical Society, quoted in J. Kenny, *Before the First Fleet*, Kangaroo Press, Kenthurst, NSW, 1995, p. 76)

This event took place on the banks of the Swan River in Western Australia (see figure 8.2c) and the sailors were part of the crew of the *Nyptang*, one of the ships in an expedition led by the Dutch navigator, Willem de Vlamingh.

Figure 8.2 (a) Palm-like plants (in the foreground) of the kind seen by Dutch sailors in 1697. This particular kind is found only in Western Australia. **(b)** The attractive but poisonous seeds like those eaten by the Dutch sailors. **(c)** Location map of the approximate landing site of Vlamingh's crew in 1697

It is August 1770 and some British sailors pass the time on the banks of the Endeavour River in northern Queensland. They are part of the crew of the *Endeavour* voyage of James Cook and their ship has been badly damaged on a coral reef. It is now beached while repairs are made. The sailors are in a strange land where the plants and animals are unfamiliar to their eyes. They notice many palm-like plants growing nearby (see figure 8.3a, page 216), some of which have large brownish seeds described as '*nutts about the size of a large chestnut and rounder*' (from the *Endeavour Journal of Joseph Banks*, Vol. 2, p. 115) (see figure 8.3b). Thinking that the seeds are edible, some sailors try them. A few hours later, these men are suffering from painful stomach cramps and are vomiting explosively. The ship's captain later writes in his journal that the sailors had thought the seeds to be edible and:

> '... those who made the experiment paid dear for their knowledge of the contrary, for they [the seeds] operated as an emetic and cathartic with great violence ...' (from Hawksworth, *An account of the voyages for making discoveries in the Southern Hemisphere*, Vol. IV, Third edition, London, 1785, p. 195).

Figure 8.3 **(a)** Palm-like plant and **(b)** the seeds of the kind eaten by Cook's sailors. They are not palms but are members of a more ancient group of plants. This particular kind is common along the warm temperate and tropical regions of the east coast of Australia. **(c)** Map showing the location in Queensland where the *Endeavour* was beached in 1770

ODD FACT

Starch-rich cycad seeds were an important part of the diet of Aboriginal people in some regions of Australia. Prior to eating them, Aboriginal people used techniques such as roasting the seeds and then soaking the kernels for many days to wash out the poisons.

The plants involved in the poisoning of both the Dutch and the British sailors were unknown to both groups of sailors. While these plants are palm-like, they are not palms. They belong to a group known as **cycads** (pronounced 'sigh-cads') that occur naturally in tropical and warm temperate regions. Modern cycads are the living survivors of a very ancient group of plants that existed before the dinosaurs and the flowering plants evolved. Worldwide, nearly 300 different kinds of cycad have been identified and all produce toxic compounds. Eating fresh cycad seeds produces severe gastroenteritis and, if sufficient seeds are eaten, death can result. In fact, pigs were taken on board the *Endeavour* in Tahiti as a source of fresh meat and, when they were fed cycad seeds, several of them died.

Identification is important

The native vegetation of Australia consists of many different kinds of plants. More than 20 000 different kinds of **vascular plants** (for example, flowering plants, cone-bearing plants and ferns) have been identified in Australia. Many of these plants have edible parts, such as roots, seeds, fruits and leaves; only a relatively small number of native plants are poisonous (see figure 8.4).

Figure 8.4 Australian native plants include both edible and poisonous plants. Not many plants are poisonous but, if eaten, these can have serious or even fatal effects for people and for farm animals. The attractive plant with red pea flowers is commonly called crinkle-leaf poison. What does this suggest about this plant?

Knowing the difference between poisonous and harmless plants is critical. This was particularly important for the early European settlers in Australia who farmed sheep or cattle in pastures where the vegetation was unfamiliar. Their stock sickened and often died after eating poisonous native vegetation; for example, cattle sometimes developed an incurable condition known as 'staggers' that usually led to death. It was later found that this condition occurred when cattle ate the leaves and other parts of cycads.

As well as cycads, many other plants can be the source of poisoning of cattle and sheep. In Western Australia, for example, a group occurs of 43 closely related shrubs with pea flowers. Of these shrubs, 27 are poisonous and have toxic flowers, seeds and leaves. (These shrubs are commonly known by names such as crinkle-leaf poison, narrow-leaf poison and wallflower poison.) Over time, farmers learned to identify poisonous plants and distinguish them from closely related non-toxic plants that could provide good grazing for their herds and flocks. Identification matters!

ODD FACT

The poisonous agent in the toxic Western Australian pea-flowered shrubs is mono-fluoro-acetate, more commonly known as 1080 (ten-eighty). While 1080 is toxic for introduced cattle, sheep and the like, it is not toxic for native mammals of the region where these plants naturally occur.

The native animal life of Australia consists of many different kinds. Nearly 150 different kinds of marsupial mammal (for example, kangaroos and wallabies), nearly 800 different kinds of bird and more than 750 reptiles (for example, snakes and lizards) have been identified. Most of these occur nowhere else in the world. Among the reptiles are about 130 different kinds of snake that include non-venomous blind snakes, pythons that constrict their prey, and elapid snakes, such as taipans and tiger snakes, whose venom acts on the nervous systems of their prey. Knowing the difference between deadly and harmless snakes is critical. Identification matters!

It is important that living organisms can be accurately identified for many reasons, including:

- personal safety — to distinguish harmless from harmful organisms (see figure 8.5)
- quarantine — to recognise animals or plants and their products that are banned imports; for example, imported cargo unloaded from planes and ships can have unwanted 'living passengers'
- medicine — to identify particular kinds of infectious bacteria or fungi so that effective drug treatment can be prescribed
- conservation — to recognise endangered kinds of plants and animals so that their habitats can be preserved
- forensics — to identify plant or animal material that may be used in the identification and conviction of a law-breaker; for example, **palynology** is the study and identification of pollen. Pollen grains from different kinds of plants have distinctive shapes so that pollen found on a person's clothes can indicate where that person has been. Evidence of this type can link a person to the site of a crime.
- agriculture and horticulture — to identify pests of crops so that effective control measures can be introduced; for example, greenhouse whitefly, a serious pest of greenhouse crops, can be controlled by a particular wasp that locates young whitefly nymphs and kills them (see figure 8.6, page 218; also chapter 13).

Figure 8.5 Staff at the Australian Reptile Park prepare some tarantula spiders for transport to the Melbourne Museum's exotic animal quarantine facility for display. Museums play an important role in identification and preservation of species.

(a)

Figure 8.6 The tiny wasp, *Encarsia formosa* (less than 1 millimetre long), is used in the control of greenhouse pests, such as whitefly. **(a)** A wasp with black head and thorax and orange abdomen lays an egg into a whitefly nymph that is located under a clear scale. When the egg hatches, the wasp larva will eat the whitefly nymph. The black scale at left has already been attacked by a wasp. What will emerge from under this black scale — a wasp or a whitefly? **(b)** Wasp eggs are provided on cards to growers who locate them in their greenhouses where the wasps later hatch and begin their search for whitefly nymphs. What will be on the cards?

Material for identification

It is not always necessary to have an entire organism in order to identify it. Identification can be based on a study of:

- a whole specimen (actual or recorded as an image or as a verbal description)
- part of a specimen and macroscopic fragments
- microscopic fragments
- genetic material
- indirect evidence.

In each case, identification will depend on precise observations of the available material. The next step may be either:

- recognition of a particular set of features in the unknown organism that allow it to be identified as related to a known **species**
- careful comparison of the unknown with a database of relevant material from known specimens until a match is made: 'It looks like this one!'.

Whole specimens

Identification can be based on a whole specimen that is living, preserved, fossilised or recorded as an image. Artists were key members of the scientific parties who came to Australia in the 1700s and 1800s, and they drew images of the plants and animals of the southern continent, such as that shown in figure 8.7.

Figure 8.7 Drawing of the King Island emu by the French artist, Charles Le Sueur, who was on a French expedition to Australia in 1801–1804

Photographs and electronic images can record features of organisms that assist in their identification and verbal descriptions are useful. The following description was written in Sydney in 1808. Can you identify the organism?

> 'It is commonly about two feet [61 cm] long and one [30 cm] high; in girth about one foot and a half [45 cm]; it is covered with fine soft fur, lead coloured on the back, and white on the belly. The ears are short, erect and pointed … it bears no small resemblance to the bear in the fore part of its body; it has no tail; its posture for the most part is sitting.'

Bits and pieces: macroscopic, microscopic and molecular

Identification may depend on a view of part of a specimen as, for example, the tail flukes of a whale (see figure 8.8).

In other cases, **macroscopic** fragments of a specimen may be available, such as feathers, hairs, teeth in a lower jawbone, shells or fruits. This type of material can be used to identify an organism, often to the level of species.

Figure 8.8 Distinctive tails of three whale species (left to right): right, sperm and humpback whales. As well as recognising different species, particular nicks and marks on the outer edge of the tail flukes are sufficiently distinctive and permanent to allow individual whales to be identified.

Identification sometimes relies on specimens that must be examined at a **microscopic** level. Because plant cell walls are resistant to chemical breakdown, plant cells may be identified after passage through the digestive tract of an animal. This means that fragments of material taken from an animal's gut may be sufficient to identify the plants that it eats.

Microscopic examination of plant material, either living or fossilised, can reveal distinctive features, such as cell wall patterns, leaf vein patterns, pollen grain appearance and seed shape. These provide valuable clues to the identification of the plants. Databases of diagnostic features of many plant genera and species are available. Features of an unknown species can be compared with features of known specimens on the database until a match is made. Bingo! The unknown plant material is identified. Even burnt wood (charcoal) can be identified.

Animal cells are, in general, less hardy than plant cells but parts of some animals, such as hair, are resistant to chemical breakdown. Hair and fur from different mammals is distinctive. The shape of hairs in cross-section and the arrangement of scales on the outer surface of a hair fibre can be used to identify the species from which the sample came. A photographic database of hairs of Australian native mammals exists, and can be used to compare the unknown hair with hair from various specimens (see figure 8.9). Study of hairs found in the faeces (**scats**) produced by carnivorous mammals and in **pellets** regurgitated by **raptors**, such as owls, eagles and hawks, can identify their prey.

ODD FACT

If a bird is sucked into the air intake of a jet engine, little remains of it apart from feathers. It is possible, however, to identify a bird species from a single feather.

Increasingly, identification is based on molecular analysis of genetic material. For most organisms, this is DNA (deoxyribonucleic acid). For identification, samples of genetic material are examined for the presence of known markers that are distinctive in particular species.

Genetic material for identification can be isolated from hair follicles, blood stains, scats (see the box on page 221) and from any cellular or tissue fragments.

(a)

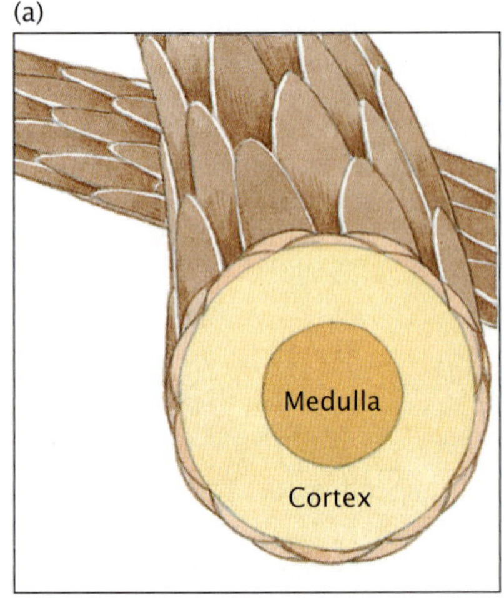

(b)

Cross-section through hairs

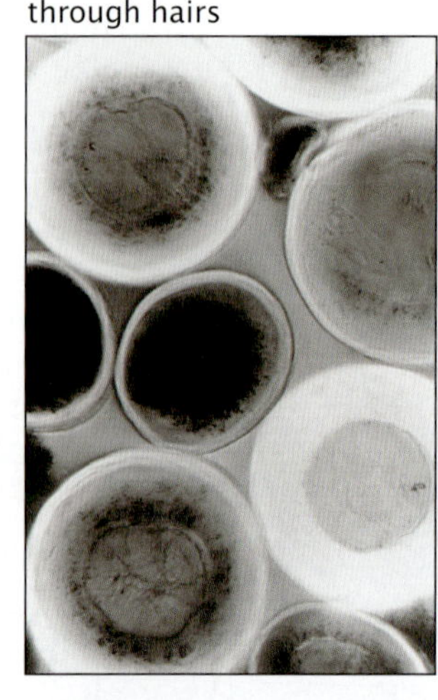

Pattern on outer layer

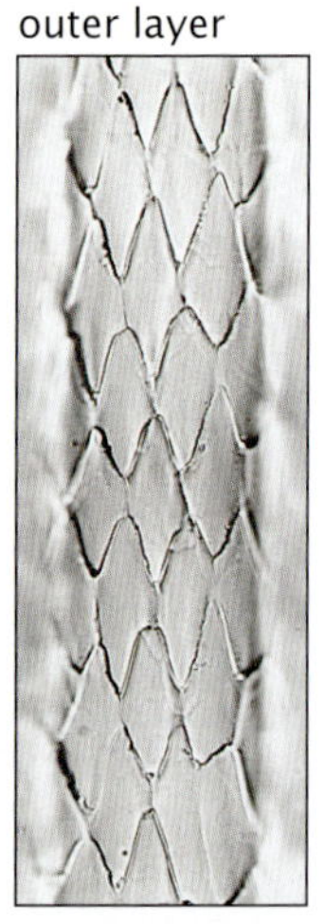

(c)

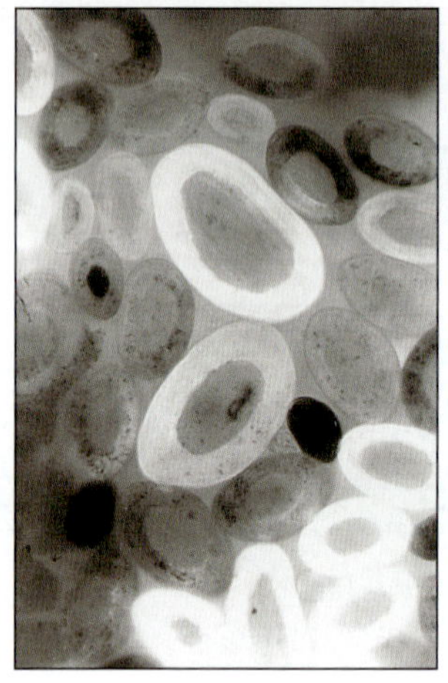

Figure 8.9 **(a)** Structure of typical hair including a cross-section. Different patterns seen on outer layer of mid-point of hairs from two species, **(b)** the spotted-tailed quoll (*Dasyurus maculatus*) and **(c)** the eastern grey kangaroo (*Macropus giganteus*). What differences are apparent?

(a)

Indirect evidence

Identification can be based on indirect evidence of an organism. In the case of mammals and birds, these might include tracks, burrows or nests, scratchings, scats and calls or songs. Figure 8.10 shows the distinctive tracks of two animals: the dingo (*Canis familiaris*) and the koala (*Phascolarctos cinereus*).

Field guides

Field guides (see figure 8.11) contain pictorial and verbal descriptions of organisms common to nominated areas. Identification is carried out by matching an unknown organism to a corresponding specimen in the guide. Field guides are commonly used for identification by bird watchers and by amateur naturalists.

Increasingly, computer technology is being used to assist in the process of identification. Images and text descriptions are now available on CD-ROMs which can hold very large amounts of data. For example, CD-ROMs are available to assist in the identification of families of flowering plants, tropical rainforest trees and beetle larvae.

(b)

Figure 8.10 Impressions of front and rear foot of **(a)** a koala **(b)** a dingo (smaller tracks)

Figure 8.11 Examples of field guides, both printed and on CD-ROM. Can you suggest a limitation of field guides for identification?

The World Wide Web is also a source of data that can be used to assist in identification; for example, the website of Sydney's Royal Botanic Gardens includes a PlantNet site which has special pages of information on the identification of cycads (go to www.jaconline.com.au/natureofbiology/natbiol1-3e and click on the 'Cycad Pages' weblink for this chapter).

SCATTY CLUES FOR IDENTIFICATION

It is claimed by some local residents that at night a cat-like beast roams Bodmin Moor in Cornwall, England, and is the cause of strange deaths among cattle and sheep. The most recent investigations have involved study of the scats (faeces) that are supposed to have been left by the 'beast' which many locals assert is some sort of 'big cat', such as a leopard or puma.

When faeces are produced, they carry, on their outer surface, cells dragged from the lining of the digestive tract. These cells can be washed from the surface of the faeces and the DNA present in the cells can be separated out and examined. DNA from different species can be distinguished and scats are thus a valuable source of cells from which DNA can be extracted and used for identification. It is important that care is taken to obtain cells only from the outside of the scat.

Figure 8.12

Scats claimed to have been produced by the Bodmin Moor beast have been analysed. The first sample was collected by a person who said he had seen the 'beast'. The DNA from this sample was found to match that of both dog and sheep. This result was interpreted as being due to a dog defecating on sheep droppings. The DNA from another sample collected at a different time showed that it was from a domestic cat.

To date, local rumours that the beast of Bodmin Moor is one of the 'big cats', remain unsubstantiated.

In another instance, a scat was claimed to be that of the Abominable Snowman. When analysed, the DNA isolated from this scat showed that it was from a fox!

Faeces can also be used to identify the number of individuals in a population. Not only can DNA from different species be distinguished, but also that from different individuals. In one American study, DNA obtained from faeces was used to estimate the number of coyotes in the population in a given area. DNA from such sources can also be used to identify the sex of individual members of the species.

Reference collections

Museums and herbaria (singular = **herbarium**) are institutions where **reference collections** of animal species and plant species respectively are held. If an unknown specimen of an animal or plant is to be identified, this process may involve comparison of the unknown specimen with reference specimens held in museums (for animal species) and in herbaria (for plant species).

Reference collections are held in Australian museums and other institutions. Some important reference collections are:

- the fish collection at the Australian Museum in Sydney, which contains almost one million specimens (see the box on page 223)
- the native plant collection, which includes the world's largest collection of eucalypts, at the Australian National Herbarium which is maintained in Canberra by the CSIRO Division of Plant Industry.

Keys

'Dichotomous' comes from the Greek dichotomia, *which means 'a cutting into two'.*

A means of identifying specimens is through the use of **keys**. Keys involve making decisions about the presence (or absence) of certain features in the specimen to be identified (see figure 8.13). When each decision involves choosing between just two alternatives, the key is called a **dichotomous key**.

The simple key in figure 8.13 shows how seven groups of animals can be identified using a dichotomous key. Each decision involves a choice between easily recognised features.

DECISION	1a	Hard outer covering or shell over part or all of body	Go to 2
	or		
	1b	No hard outer covering	Go to 5
DECISION	2a	Single shell	Go to 3
	or		
	2b	Shell comprises more than one valve	Go to 4
DECISION	3a	Rounded shell with circular opening	**snail**
	or		
	3b	Cylindrical shell with long opening	**cone shell**
DECISION	4a	Both valves convex	**mussel**
	or		
	4b	One valve flat, and one convex	**scallop**
DECISION	5a	Tentacles or 'arms' present	Go to 6
	or		
	5b	No tentacles	**slug**
DECISION	6a	Eight 'arms' present	**octopus**
	or		
	6b	Ten arms present	**squid**

Figure 8.13 A dichotomous key for identifying various groups within the phylum Mollusca. What feature is the basis for the first decision? What is one limitation of a key?

BIOLOGIST AT WORK

Mark McGrouther — collection manager

'Mention the words "marine biologist" and most people think of diving in crystal clear waters on the Great Barrier Reef. My role as collection manager of the Fish Research Collection which is kept by the Australian Museum does involve some exciting fieldwork, but there are also many duties that do not fit the popular image of a marine biologist.

'The collection manager's role is to manage and develop the fish collection of the Australian Museum. It is a fascinating job with roles as diverse as supplying fishes for research, participating in fieldwork, preparing pages for posting on the World Wide Web and answering enquiries from the public about strange fishes.

'The collection contains over a million specimens and serves many functions. One of the most important is scientific research. Researchers who wish to study the anatomy, taxonomy, diet or reproductive biology of a type of fish can use specimens from the collection. In this way, it is used like a library with specimens being loaned to ichthyologists world wide. Just about every fish you can imagine is stored in the collection, both marine and freshwater, from tiny fish eggs to great white sharks.

Figure 8.14 Mark McGrouther

'So where do all these fish come from? Some are brought to the museum by the public, who find them washed up on a beach, or caught on hook and line. Others are collected by professional fishers or other research institutes, and many are collected by staff during fieldwork.

'Fieldwork can vary from one day assessing the fauna of a local stream to a month or more away in remote locations assessing fish biodiversity. Staff have participated in expeditions to many countries including French Polynesia, Vanuatu, the Solomon Islands, Papua New Guinea and Madagascar.

'The collection needs to be large enough to provide adequate representation of each species, by covering variations in form, size, age, location, depth ranges and other variables. Of course, amassing a large collection and storing it is not an end in itself. The collection has to be utilised and well documented. The entire collection is recorded in computer databases. These data are constantly being updated and made available to researchers.

'Considerable effort has gone into posting information on the World Wide Web. The Australian Museum's fish pages contain a host of information which caters to the interests of scientists and the public. It seems everyone has a passion for fish.'

Access the website at: www.amonline.net.au/fishes

KEY IDEAS

- Accurate identification of a species is important for many purposes, including conservation, medicine and forensic investigation.
- Identification may be based on direct or indirect evidence of various kinds.
- Identification can be made from a variety of material ranging from whole organisms to microscopic fragments.

QUICK-CHECK

1 Identify the following as true or false.
  a Australia has more than 800 different kinds of bird.
  b It is necessary to have an entire organism in order to identify it.
  c Reference collections are held in museums and herbaria.
2 Give one example of why accurate identification to species level is important.
3 List two different materials through which a species might be identified.

Identification involves naming

Human learning involves putting labels on classes of objects in the world around us. As toddlers, we begin to learn to classify objects in the world around us and to give a distinctive verbal label (for example, 'dog', 'book', 'apple', 'chair') to each class. Young toddlers can recognise a large, fawn great Dane, a small, black and white Jack Russell terrier, a golden labrador and a black greyhound as having an essential 'dogness', in spite of differences in shape, size and colour. Toddlers know that cats are not dogs (see figure 8.15). Through labels or names, we can make sense of the world around us and can communicate our thoughts to others in a concise manner.

When we name something, we are identifying some of its essential and distinguishing characteristics. If something has the label of 'bird', you know that it has feathers, not fur.

ODD FACT

On seeing lions, tigers, leopards or other big cats at the zoo, young children readily recognise their essential 'catness' by calling 'Puss, puss, puss' to them.

Figure 8.15 Learning involves identifying classes of objects in the world around us with verbal labels.

Two-part names and rules

Putting labels or names on living organisms is an important area of biology. Each different kind or species of living organism presently known has a two-part (binomial) scientific name (see figure 8.16). You and every other person have the scientific name *Homo sapiens*, while your cat and every other domestic cat is *Felis catus*. The cycad species that poisoned the Dutch sailors in 1697

Figure 8.16 Each kind of organism has a unique scientific name. How many parts in each name?

(figure 8.2, page 215) has the scientific name of *Macrozamia riedlei*, while a different kind of cycad with the scientific name of *Cycas media* (figure 8.3, page 216) poisoned the British sailors in 1770.

The naming of various species occurs according to rules. Rules for the naming of new animal and new plant species appear in publications called the International Code of Zoological Nomenclature (ICZN) and the International Code for Botanical Nomenclature (ICBN) respectively. But, what is a species? The box headed 'What is a species?' on pages 230–1 explains how different species can be recognised.

Linnaeus' binomial system

Let us now look in more detail at the binomial system of naming species that was introduced by the Swedish biologist, Carl von Linné (1707–1778), also known by the Latinised form of his name, as Carolus Linnaeus.

In 1758, Linnaeus (see figure 8.17) introduced a new and uniform way of naming different kinds of organism by giving each species a two-part scientific name. This is the **binomial system of naming** (from the Latin *binomius*, meaning having two names). Linnaeus' important contribution to biology was to apply the use of binomial scientific names to all organisms in a systematic way.

The first part of a binomial scientific name is the **generic name** or the name of the genus to which the organism belongs; the generic name always begins with a capital letter. The second part always begins with a lowercase letter and identifies the particular member that belongs to the genus. This second part is known as the **specific name**. This binomial scheme for naming species continues to be used by all biologists throughout the world today.

Figure 8.17 Carolus Linnaeus is shown on this Linnean Society Silver Medal.

The Linnean binomial system replaced a number of clumsy systems of naming organisms that had developed during medieval times. An early **polynomial system** gave names to organisms that were descriptions. For example, the common buttercup was once known by the polynomial: *Ranunculus calycibus retroflexus, pedunculis fulcratus, caule erecto, foliis compositus*. This name described the plant as follows: buttercup with sepals bent back, with supporting flower stalk, with straight main stem, and with composite leaves. Linnaeus gave the common buttercup its binomial scientific name, *Ranunculus repens*.

Another significant feature of the Linnean binomial system is that binomial names are informative about the relationships between organisms. Where close relationships exist between organisms, the Linnean binomial system identifies this feature by giving them the same generic name. For example, Linnaeus gave the scientific name *Cervus elaphus* to the red deer and called the closely related fallow deer *Cervus dama*. Their close relationship is advertised by their having the common generic name *Cervus*.

The specific part of a scientific name does not denote any degree of relationship. So the organism *Balaenoptera musculus* is not at all closely related to *Mus musculus* — the former is the blue whale and the latter is the house mouse!

ODD FACT

By the time of his death in 1778, Linnaeus had described and given binomial scientific names to more than 9000 species of plants, nearly 2000 mollusc species, 2100 insect species and 477 fish species.

ODD FACT

Sometimes a person may know the name of the genus to which an organism belongs, for example, a tree belonging to genus *Eucalyptus*, but may not be sure of the specific name. In this case, the tree can be identified simply as *Eucalyptus* sp.

Figure 8.18 Sign in a university garden in China

Advantages of scientific names

Scientific names look more complex than common names. Wouldn't it be easier to stick with common names? Surely words like 'cat', 'koala' and 'numbat' are preferable to tongue twisters such as *Felis catus*, *Phascolarctus cinereus* and *Myrmecobius fasciatus*!

In spite of their tongue-twisting nature, scientific or species names have several advantages compared with common names.

1. **Common names vary from language to language, but scientific names are universal**. The domestic cat is the English *cat*, the Italian *gatto*, the German *Katze*, the French *chat* and the Hebrew *chatul*. In contrast, the scientific name, *Felis catus*, is recognised and used by all biologists, regardless of nationality. Figure 8.18 shows a sign from a university garden in China.
2. **The same common name is sometimes used to label different species**. People who speak the same language may mean different things when they use the same common name. Look at figure 8.19. Which is the robin? The answer depends on the nationality of the person answering the question. The common name 'robin' is used by Australians to refer to small flycatchers, such as the scarlet robin, *Petroica multicolor*. In contrast, the name 'robin' is used by Americans to refer to a different and larger bird, with a dull orange breast. The scientific name of this larger bird is *Turdus migratorius*.

Figure 8.19 The same common name 'robin' is used both by Australians and Americans for different kinds of bird. Australian robins **(a)** belong to the *genus Petroica*. Is this true of American robins **(b)**?

In Western Australia, the common name 'prickly Moses' is used for one kind of wattle, *Acacia pulchella*. In New South Wales, the same common name refers to a different species, *Acacia ulicifolia*. In Victoria and Tasmania, the common name 'prickly Moses' refers to yet a different species of wattle, *Acacia verticillata*. A scientific name refers to one kind of organism only.

3. **Scientific names give an indication of the degree of relatedness of different organisms**. Closely related organisms share the same generic name, for example, *Macropus*. So, there is *Macropus rufus*, the red wallaby, and its close relatives, including *Macropus giganteus*, the eastern grey kangaroo and *Macropus fulginosus*, the western grey kangaroo. These three kinds of kangaroo have similarities in physical structure (morphology), in functioning (physiology and biochemistry) and in their evolution (phylogeny). This shared similarity is expressed in the first part of their scientific names — the generic name, *Macropus*.

The fish *Argyrosomus japonicus* is found around the Australian coast. In Victoria, it is known as the mulloway, in South Australia as the butterfish, in Queensland and New South Wales as the jewfish, and in Western Australia as the kingfish or river kingfish.

Figure 8.20 Spotted-tailed quoll (*Dasyurus maculatus*). What common name did early settlers give to this animal? Why might this common name be used?

In contrast, common names cannot be used as a guide to the degree of relatedness of different kinds of organism. Consider the long-tailed mouse (*Pseudomys higginsi*), the spinifex hopping mouse (*Notomys alexis*) and the heath rat (*Pseudomys shortridgei*). Their common names suggest, perhaps, that the two mice are the most closely related. However, the heath rat and the long-tailed mouse are the most closely related pair.

4. **Common names may be misleading, suggesting relationships that are not valid.** Some common names given to organisms are misleading. In spite of their common names, cuttlefish are not fish, they are molluscs; likewise, starfish are *not* fish but echinoderms; reindeer moss is not a moss but a lichen and the sea tulip is not a plant but an animal.
   Early European settlers gave Australian animals common names that suggested relationships that did not exist. Figure 8.20 shows the animal that was given the common name of 'tiger cat' by early European settlers. This animal, the spotted-tailed quoll (*Dasyurus maculatus*) is not a close relative to the domestic cat (*Felis catus*), and it does not even belong to the same family or order as the domestic cat.

Sources of scientific names

A person seeing a scientific name, such as *Phascolarctos cinereus*, might say: 'It looks foreign to me!'

Many of the words used in scientific names come from Greek and Latin words. The scientific name of the koala is built from the Greek words *phaskolos* meaning 'pouch', *arktos* meaning 'bear', and the Latin word *ciner* meaning 'ash-coloured'. So *Phascolarctos cinereus*, the scientific name given to the koala in 1816, simply means 'ash-coloured, pouched bear'.

Some scientific names are shown below, along with their meanings:

Macropus rufus	=	big-foot (Gk), red (L)
Acrobates pygmaeus	=	acrobat (Gk), pygmy (L)
Thylacinus cynocephalus	=	pouched-dog (Gk), dog-head (Gk)
Ornithorhynchus anatinus	=	bird-snout (Gk), duck-like (L).

The big-footed red animal is the red kangaroo. Try to match the other scientific names to the Australian native animals shown in figure 8.21.

Other languages have contributed to scientific names; for example, the name of the plant genus *Pandanus* comes from the Malay language.

Figure 8.21 Which of these animals might have a scientific name that means 'bird-snout, duck-like', 'pouched-dog, dog head' or 'acrobat, pygmy?

ODD FACT

Scientific names sometimes have an extra bit, such as the messmate tree (*Eucalyptus obliqua* L'Hér), the koala (*Phascolarctos cinereus* Goldfuss) and the blue whale (*Balaenoptera musculus* Linnaeus — often shown as simply 'L'). This addition identifies the name of the person who first described the organism.

The generic names of some Australian native marsupial mammals are derived from Aboriginal words or phrases; for example the genus *Bettongia* is named from an Aboriginal word 'bettong' meaning small wallaby. Four marsupial species now have the generic name *Bettongia* including the brush-tailed bettong, *B. penicillata*, and the burrowing bettong, *B. lesueur*.

Scientific names of organisms may come in part from the names of people involved with their discovery or description. Names are usually 'Latinised' by adding an appropriate suffix. Some examples are given in table 8.1.

Table 8.1 Parts of scientific names based on people

Scientific name	*Common name*	*Source of name(s)*
Banksia serrata	saw-leafed banksia	Joseph Banks, English naturalist on Cook's *Endeavour* voyage
Wollemia nobilis	Wollemi pine	Dave Noble, who discovered this species in 1994 (see page 234)

Level of identification and naming

Look at the organisms in figure 8.22a and b. The most precise level of identification is to give each organism its binomial scientific name. This identification would be: 'Organism 1 is the golden wattle, *Acacia pycnantha*, and organism 2 is the clown anemonefish, *Amphiprion percula*'. However, identification can also be less precise and identify larger and more inclusive classification levels. Look at figure 8.22c. The statement 'That is a banksia' in fact identifies this plant to the level of genus since this is the saw banksia, *Banksia serrata*. Statements such as: 'This is a flowering plant' or 'That is a fish' identify organisms more generally, since, for example, about 250 000 different species of flowering plant have so far been identified and more than 21 000 bony fish.

(a) **1**

(b) **2**

(c) **3**

Figure 8.22 Can you identify these organisms?

KEY IDEAS

- Identifying classes of objects involves giving them names.
- Each different kind of organism or species has a unique scientific name.
- Scientific names are binomial: the first part is the generic name; the second part is the specific name.
- Identification can be at several levels and the most precise identification is at species level.

QUICK-CHECK

4 Identify the following as true or false.
   a The systematic use of the binomial system of naming organisms was introduced by Linnaeus.
   b Each different species has a unique scientific name.
   c Some, but not all, domestic cats have the scientific name *Felis catus*.
   d Two different species could have the same scientific name.
   e A species found in New Guinea must have the same scientific name as the same species found in northern Australia.

5 If a black labrador dog has the scientific name *Canis domestica*, what scientific name would be given to a white highland terrier?

6 Two organisms are similar in appearance. Will they have the same generic name?

Putting order into the living world

Rainforests cover less than 10 per cent of Earth's land surface but contain about half of our planet's species.

All the different species of living organisms (animals, plants, fungi, protists and microbes) comprise the biological diversity (**biodiversity**) of planet Earth. The species is the basic unit of the living world and the information on pages 230–1 explains how species are defined. To date, around 1.7 million different species have been identified and their description and scientific names have been published.

The biologists who specialise in identifying, naming, describing and classifying organisms are called **taxonomists**. Their area of study is called **taxonomy** and it is part of a broader area of study known as **systematics** that aims to describe relationships between different groups of organisms and to understand the evolutionary history of life on Earth. Taxonomists tend to specialise in one particular group of organisms. Read on page 232 about Bruce Maslin, a taxonomist and the world's expert on plants of the genus *Acacia*.

In the period 1863–78, the first published collection of Australian plants, entitled *Flora Australiensis*, listed 8000 plants. Since then, more than 12 000 new species have been described.

Describing each different kind of organism and giving each a unique scientific name makes an enormous amount of data. Imagine that the description and the name of each of the 1.7 million different species already identified were put into a book of 'Life on Earth', with one page for each different species. Such a book would be a multi-volume series that would form a stack more than 75 metres high! However, the total number of different kinds of living organisms on planet Earth is far more than the number so far identified and named. Various estimates of the total number of existing species range from 10 million up to perhaps 100 million different kinds. In addition, the species living today are just a tiny fraction of possibly hundreds of millions of different species that have existed over the eons since life first appeared on this planet. These species, now extinct and occurring only as **fossils**, could also be added to the book of 'Life on Earth' — in the 'past life' section.

WHAT IS A SPECIES?

Species can be defined in different ways, including:
1. classic definition — the use of structural similarities
2. biological definition — the ability to interbreed
3. modern definition — the use of DNA.

Classic definition of species

The classic definition of a species is based solely on similarities in appearance. If two organisms look sufficiently similar, they are defined as the same species; if they look sufficiently different, they are defined as different species.

One problem with this definition is that different species may look virtually identical. For example, the snow petrel (*Pagodroma nivea*) and the white tern (*Gygis alba*) are very similar in appearance but are different species. Butterflies and moths of different species can also be very similar in appearance. Another problem is that members of the same species may show enough variation to cause them to be mistakenly identified as different species, for example:

- different sexes of the same species may vary markedly, as seen in the *Electus* parrot (see figure 8.23a)

Figure 8.23 (a) The two sexes of the *Electus* parrot (*Electus roratus*) differ greatly. The adult male (left) is mainly bright green in colour while the female (right) is red and purple. For many years these two sexes were mistakenly thought to be different species. **(b)** The green python (*Chondropython viridis*) shows developmental or age variation. Shown here are an adult python that is lime green and a juvenile that is typically bright yellow. **(c)** The dark and the white variants of adult southern giant petrels (*Macronectes giganteus*)

- members of the same species at different ages may show striking variation, as seen in the green python (*Chondropython viridis*) (see figure 8.23b)
- members of the same species, regardless of sex and age, may look different because of inherited variation. For example, the southern giant petrel (*Macronectes giganteus*) has two variants, the common dark and a rarer white (see figure 8.23c).

The classic definition of a species is useful for fossil species. When two fossil organisms are very similar, they are identified as members of the same species. Figure 8.24 shows three fossil trilobites that are recognised as different species because of significant differences in aspects of their size and shape. Likewise, the classic definition can apply to organisms that reproduce only asexually. (Why?)

Figure 8.24 Three different trilobite species identified because of differences in their shape and structure: **(a)** *Elrathia kingii* (top) and *Agnostus pisiformis* (bottom); **(b)** *Phacops africanus*

Biological definition of species

The biological definition of species identifies a species as members of a group of similar organisms that are capable of interbreeding under natural conditions to produce viable and fertile offspring. This definition allows two different species to be recognised even when they appear superficially similar.

Lions (*Panthera leo*) and tigers (*Panthera tigris*) can mate but they fail the 'biological test' because (i) their offspring, known as ligers, are infertile and (ii) such matings occur only under captive conditions.

The biological definition of a species has limited application. It cannot be used for species that reproduce asexually.

Modern definition of species

Species are now recognised through criteria such as:

- the number and shape of chromosomes present in their cells
- molecular data.

This procedure can assist in separating, as distinct, two different species that look the same!

In the case of microbes, molecular analysis is used to identify various species. To be the same species, the DNA of the microbes concerned must match to a high degree.

This test is very useful for microbes that cannot be cultured in the laboratory. More recently, it has been proposed that all species be DNA barcoded (see pages 235–6).

BIOLOGIST AT WORK

Bruce Maslin — taxonomist and Acacia *specialist*

Bruce Maslin is a principal scientist with the Department of Conservation and Land Management in Perth, Western Australia. He is a botanist and expert on the plants of the genus *Acacia*, commonly known as wattles.

Worldwide, over 1300 different kinds (species) of wattle occur, with about 1000 in Australia, making this the largest group of woody plants in this country. Acacias have great symbolic significance to Australians because wattle (*Acacia pycnantha*) is our national floral emblem, the predominant colours of wattles (green and gold) are our national colours often worn by athletes, and the Order of Australia medal, our most important honour, is based on a single wattle blossom.

Bruce has spent many years studying the taxonomy of wattles and practical applications of his knowledge include finding ways to incorporate wattles into landcare, conservation and economic activities.

Bruce completed a Bachelor of Science (Honours) and a Master of Science degree at the University of Western Australia, but his specialist knowledge of Australian wattles has developed from 'on the job' experience. Western Australia, particularly the south-west, contains one of the world's richest floras.

As a taxonomist, Bruce classifies and formally names wattles as well as studying their evolutionary relationships. He observes plants in the field and collects specimens that are preserved in botanical museums (herbaria). He gathers data on the shape, size and form (morphology) of different kinds of wattle from herbarium collections and also uses data published by other taxonomists in scientific journals. Bruce's research is part of a worldwide study of *Acacias* carried out by scientists in many countries.

Bruce has developed and named around 200 previously unknown species of wattle. Scientific names are the principal 'hooks' by which information about organisms is stored and retrieved. Names enable people to communicate information and this, in turn, provides the foundation for applied areas of biology, including forestry, horticulture and conservation. Because accurate identification is important, Bruce has developed interactive electronic tools (keys) that guide users to the correct names for various wattles and provide drawings, colour photographs, distribution maps and written notes about each kind. Go to www.worldwidewattle.com (either click 'Info Gallery' then 'Identification' or click 'Identify a wattle').

Bruce's taxonomic work is used to develop conservation strategies. The geographic distribution and the conditions under which each kind of *Acacia* lives are documented so that rare and geographically restricted kinds have been identified. Bruce's department and other organisations responsible for land management can use this information to formally declare (gazette) these endangered kinds so that they and their habitats are protected.

Bruce's taxonomic work is also used in applied areas. Australian *Acacias* are used for a wide range of environmental, social and commercial purposes worldwide. *Acacia mearnsii* (black wattle) is an important source of tannin for the leather industry and is cultivated in South Africa, Brazil, China and Vietnam. Blackwood wattle (*Acacia melanoxylon*) is used for making fine furniture in Australia and is also cultivated in New Zealand, South Africa and Chile. Coojong (*Acacia saligna*) is used extensively for fodder and soil conservation purposes in North Africa (see figure 8.25b). Australian Aborigines used wattles for medicines, weapons and as a food, as for example, seeds of wirewood (*Acacia coriacea*) (see figure 8.25c). Around 50 species of *Acacia* have potential for cultivation in southern Australia as part of the fight against increasing salinity, a serious environmental threat in Australia.

Figure 8.25 (a) Bruce Maslin, botanist and taxonomist **(b)** Coojong (*Acacia saligna*), one of the most promising species for use in the fight against salinity in southern Australia **(c)** pods and seeds of wirewood (*Acacia coriacea*)

New species continue to be found

New species, both living and fossil, are constantly being discovered. In the three-year period from 2002 to 2004, many discoveries of new living species were made. New terrestrial species included a flightless bird species in the Philippines, an owl species in Sri Lanka, three monkey species, one in India and two in the Brazilian rainforest, and more than 100 new frog species in a Sri Lankan rainforest. New marine species included fish species in Antarctica and a new kind of bright red deep-sea jellyfish in the eastern Pacific. Examples of newly discovered extinct species were the 15 fossil species found in Antarctica in 2004 that included new kinds of shellfish, sea stars, sea lilies and a coral.

Many discoveries of new species have been made in remote and largely unexplored areas, such as Antarctica, in tropical rainforests and deep under the ocean surface. Some new species, however, have been found in areas not far from major cities — in 2002 a new species of centipede was discovered in New York's Central Park. There was also an exciting discovery not far from Sydney, as described in 'Biology in the workplace' on page 234.

ODD FACT

Among new species discovered are the world's smallest deer (found in mountains in south-east Asia in 1999) and, in 2004, the world's largest cockroach (found in an Indonesian cave) and the world's smallest fish (found in waters around the Great Barrier Reef).

Technology use to discover and identify

Technology has been an important tool in the discovery and identification of new species, for example:

- Both manned submersibles (discussed in chapter 13) and remotely operated submersible vehicles (see figure 8.26) have revealed new species in the ocean depths.
- Global positioning satellite technology allows scientists to identify positions with high precision so that the location of new discoveries can be accurately mapped (a handy tool when you are wandering through a large, unmapped rainforest!).
- Digital recordings of the calls of birds or monkeys have been the first clues to the existence of a new species.
- DNA technology is used to identify microbial species that cannot be cultured (grown) in laboratories.

Greg Pio for MBARI © 1997

Figure 8.26 ROV *Tiburon*, a remotely operated submersible vehicle of the Monterey Bay Aquarium Research Institute, is lowered into the ocean to undertake exploratory research. It can operate to a depth of up to four kilometres below the ocean surface and is equipped with various sensors and cameras which these scientists control with remote technology on the ship.

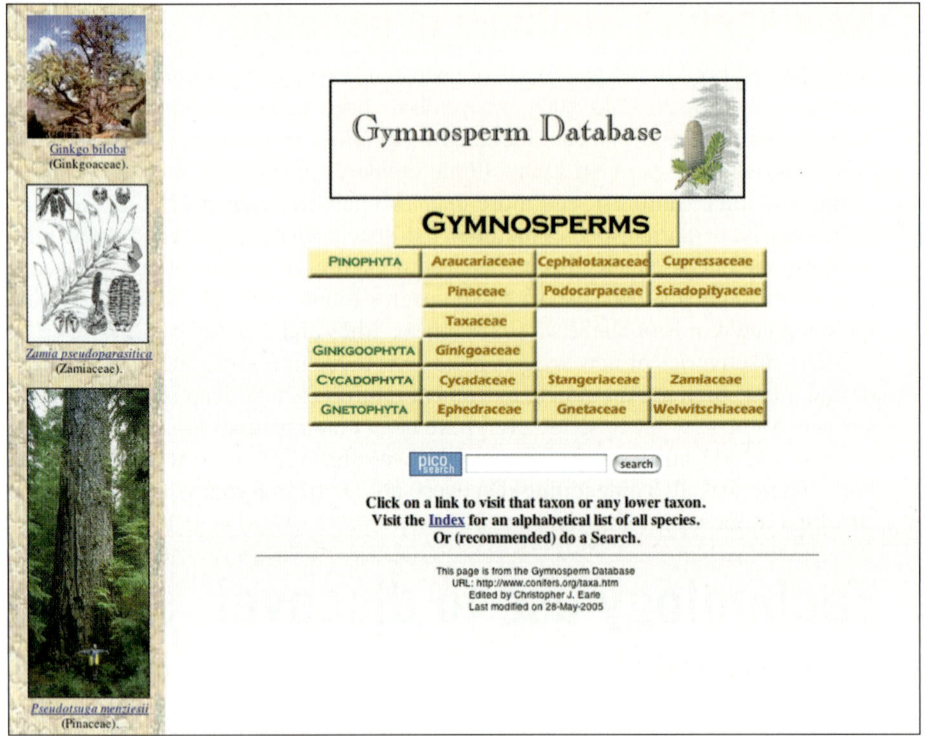

Figure 8.27 Computer technology enables the storage of large amounts of data relating to Earth's biodiversity. This example is the Gymnosperm Database at www.conifers.org/taxa.htm

Computer technology allows scientists to store, retrieve and organise the enormous amounts of data concerning Earth's biodiversity. The application of computer technology to the handling of biological data is the specialist field of biological informatics, more commonly called **bioinformatics**. Figure 8.27 shows a website that holds an electronic database of all the world's known gymnosperms (cone-bearing plants).

It is not always the use of high technology that results in the discovery of new species — sometimes it is keen observation and curiosity. Read the story below of Dave Noble, a ranger with the New South Wales National Parks and Wildlife Service, who made an exciting discovery of a new plant species.

BIOLOGY IN THE WORKPLACE

Dave Noble — ranger

'I am a ranger for the National Parks and Wildlife Service of New South Wales. I work in the Gardens of Stone and Wollemi National Parks about 100 kilometres to the west of Sydney. My job involves doing a variety of duties including fighting fires, picking up rubbish, search and rescue, wildlife surveys and writing reports. To get a job as a ranger, I completed a degree in park management at Charles Sturt University in Albury. After many interviews and lots of temporary work, I succeeded in getting a permanent position.

'I am very much an outdoors person and enjoy climbing, caving, bushwalking and canyoning. As I am a keen botanist, I am always looking at the plants around me. It was the combination of my outdoor activities and interest in plants that led me to discover the Wollemi Pine in 1994. I was walking down a canyon in a remote part of Wollemi National Park when I saw a plant that I didn't recognise. I was interested to find out what type of plant it was and took a small cutting home.

Figure 8.28 Dave Noble

'After looking through several books I gave up and decided to take it to a good friend and experienced botanist, Wyn Jones. Wyn didn't seem impressed and said it was some kind of fern. I remarked it was a large tree and left the specimen with him for a couple of days. We later visited the site to collect further specimens. The tree became known as the Wollemi pine (*Wollemia nobilis*); it was not just a new species but a whole new living genus dating from the time of the Gondwana supercontinent.

'I enjoy my work as a ranger and the surprise aspect of the job. You never know where you will end up when you turn up for work.'

In October 2005, over 100 'first generation' pines, grown from cuttings from the original trees discovered by Dave Noble, were auctioned to the public. The auction yielded more than $1 million.

Figure 8.29 Just a thought! In fact, each DNA barcode is a sequence of about 650 bases in the DNA of a particular gene.

Identifying species: DNA barcoding

We cannot know what species exist on Earth until they have been identified, but we do know that:

- a very large number of Earth's living species have yet to be discovered, named and identified
- species are becoming extinct at a very rapid rate through factors such as habitat destruction and many may be lost before they have been discovered
- the process of identifying, describing and naming species through traditional methods is a time-consuming process.

In light of these facts, one group of scientists is using DNA technology as a new and fast way of identifying new species of any kind — this new technology is called **DNA barcoding**.

Supermarkets keep track of their stock and of their sales through a barcode system called Universal Product Code. Barcodes on supermarket items are machine-readable digital labels, typically a series of stripes that uniquely identifies each different type of product. Barcoding is a convenient, cheap and fast way of knowing what stock is on hand, what has been sold and what needs to be re-ordered, particularly when very large numbers of different items are involved.

In a similar way, this group of scientists proposes that all species could be identified using a short stretch of their genetic material, DNA, that would serve as a unique identifier for the species. The barcode is a particular sequence of bases (A, T, C and G) present in DNA (see figure 8.29).

For animals, the barcode is a sequence of 650 bases from the **COI** gene that is found in the mitochondria of all eukaryote cells. For example, the unique barcode for the fish *Antigonia capros* is as follows:

GCACTCCTAGGAGATGACCAAATCTACAATGTTGTAGTTACAGCACATGCCTTTGTAATAATTTTCTTTATAG-
TAATACCAATTATAATTGGAGGATTTGGAAACTGACTAATTCCTTTAATGATTGGAGCCCCCGATATAGCATTC-
CCCCGAATGAACAATATGAGCTTCTGACTACTTCCACCCTCTTTTTTACTTCTCCTTGCCTCTTCTATAGTAGAAGCAG-
GGGCGGGCACTGGATGAACAGTTTACCCCCCTCTAGCTGGGAACCTGGCCCATGCCGGGGCATCAGTTGACTTA
ACAATTTTTTCTCTCCACTTAGCAGGGATTTCCTCAATCCTTGGGGCCATCAACTTTATCACAACTATTATTAATAT-
GAAACCTCCCGCTATTTCCCAGTACCAAACTCCCCTGTTTGTTTGAGCAGTACTAATTACTGCAGTTCTTCTTCTC-
CTCTCCCTTCCCGTCCTTGCTGCCGGAATTACAATACTTCTTACAGACCGAAACTTGAACACCACCTTCTTTGAC-
CCAGCCGGAGGAGGAGACCCGATTCTTTATCAACATCTATTCTGATTTTTTGGG

Use of a different gene is being considered for plant species.

Instead of the letters A, T, C and G, a DNA barcode can also be displayed as a pattern of coloured bands where A = green, T = red, C = blue and G = black. With colours, the distinctive patterns for various animal species are quite apparent (see figure 8.30, page 236). Applications of DNA barcoding include:

- identifying and separating animal species that are presently regarded as a single species because of similarity in appearance, and
- identifying whether a specimen is a new or an existing species.

The validity of DNA barcoding as a tool for identification is being tested. In one test reported in 2004, scientists took museum samples and showed that each of the 260 different bird species had a unique DNA barcode. This test also showed that, in one case, a bird species was in fact two distinct species.

Further testing to identify the limitations of DNA barcoding is continuing. In future, it is possible that each species will have:

- its binomial name, based on a study of structural and behavioural features using traditional taxonomy
- a DNA barcode, based on its unique genetic makeup.

Unlike traditional taxonomy, DNA barcoding can be done much more quickly, and will speed up species identification. Not all scientists agree that DNA barcoding can deliver reliable identification but the technique is being applied, particularly in animals, and a database of DNA barcodes has been established.

In April 2004, the Barcode of Life Initiative was started by a group of major natural history museums and herbariums. This project produces DNA barcodes for each of the specimens in their collections. Access the site at www.barcodinglife.com.

ODD FACT

DNA barcoding requires only small samples that can be taken from living specimens or from museum specimens and can be used on fragments of biological material, such as hair, skin cells and stomach contents.

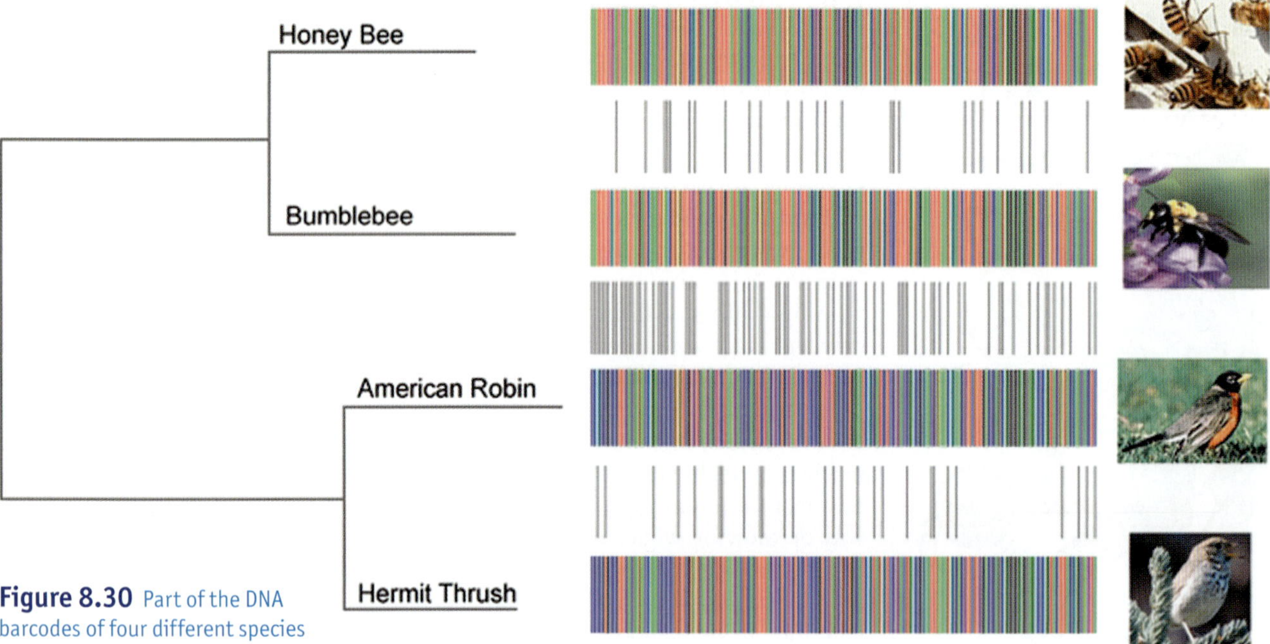

Figure 8.30 Part of the DNA barcodes of four different species (from the Public Library of Science, www.plosbiology.org). Each barcode is unique for the species. What do the colours represent?

KEY IDEAS

- Earth's biodiversity is variously estimated to comprise from 10 million to 100 million different species.
- Discovery of new species continues.
- Taxonomy is the science of systematically naming, describing and organising living things into related groups.
- Species can be defined in various ways, with each definition having different applications and shortcomings.
- DNA barcoding is a new tool for use in the identification of species.

QUICK-CHECK

7 Identify the following as true or false.
   a All living organisms on Earth have been identified and named.
   b The classic definition of a species is useful for fossil species.
8 Species can be defined in different ways. Which of the ways is most useful in the case of a fossil specimen?
9 List two ways that technology has assisted the discovery of new species.
10 Give one example of the successful use of DNA barcoding in species identification.

Classification: forming groups

The biodiversity of planet Earth comprises millions of different species. How can this enormous amount of biological data be organised in a meaningful way? This is done through the process of **classification**, which involves:

1. naming and describing each different kind of organism or species
2. organising closely related species into groups
3. combining these groups to form larger, more inclusive groups.

Taxonomy comes from the Greek taxis *= 'arrangement, order', and* nomos *= 'law'. Classification comes from the Latin* classis *= 'a class', and* facere *= 'to make'.*

We have already seen how species are named (pages 224–8) and we will now explore how species are organised into groups. As we saw earlier on page 229, the rules and principles of classification are part of the special area of biology known as taxonomy, the science of systematically identifying and describing species and organising them in related groups. Let us look first at the principles of classification and then at biological classification.

Principles of classification

Consider the following people: Shane is sorting stamps to put in a new album, Rosie is adding some new releases to the display of CDs in her music store and Tracey wants to locate a biology book from a library. A botanist, Tran, looks at a plant and says 'That's a fern'. What do these people have in common?

Each of these people is involved with an aspect of classification.

- Shane is *creating* a classification scheme by separating his stamps into groups by country of origin. The basis on which groups are formed is known as a **criterion** — in this case, 'country of origin'.
- Rosie, the music storekeeper, is *adding* new items to an existing classification scheme by putting new CDs into a display, previously arranged in separate areas by type of music (classical, rock, etc.) and, within each area, alphabetically by name of the composer or performer(s).
- Tracey is *retrieving* an existing book from a library organised according to the Dewey classification scheme. She goes to the 500s section for biology.

The classification of stamps, CDs and books involves their physical separation into groups. In contrast, the classification used by Tran, the botanist, is an intellectual exercise in which mental categories of objects are formed.

Tran organises plants into different groups according to the presence or absence of various features. Seeing a particular green plant having distinctive fronds with spores on their undersurface, Tran identifies it as belonging to the category labelled 'fern'. Tran uses a series of decisions, known as a dichotomous key (see figure 8.31), to organise the plants into groups.

Figure 8.31 These plants are organised into four major groups using a dichotomous key as follows:

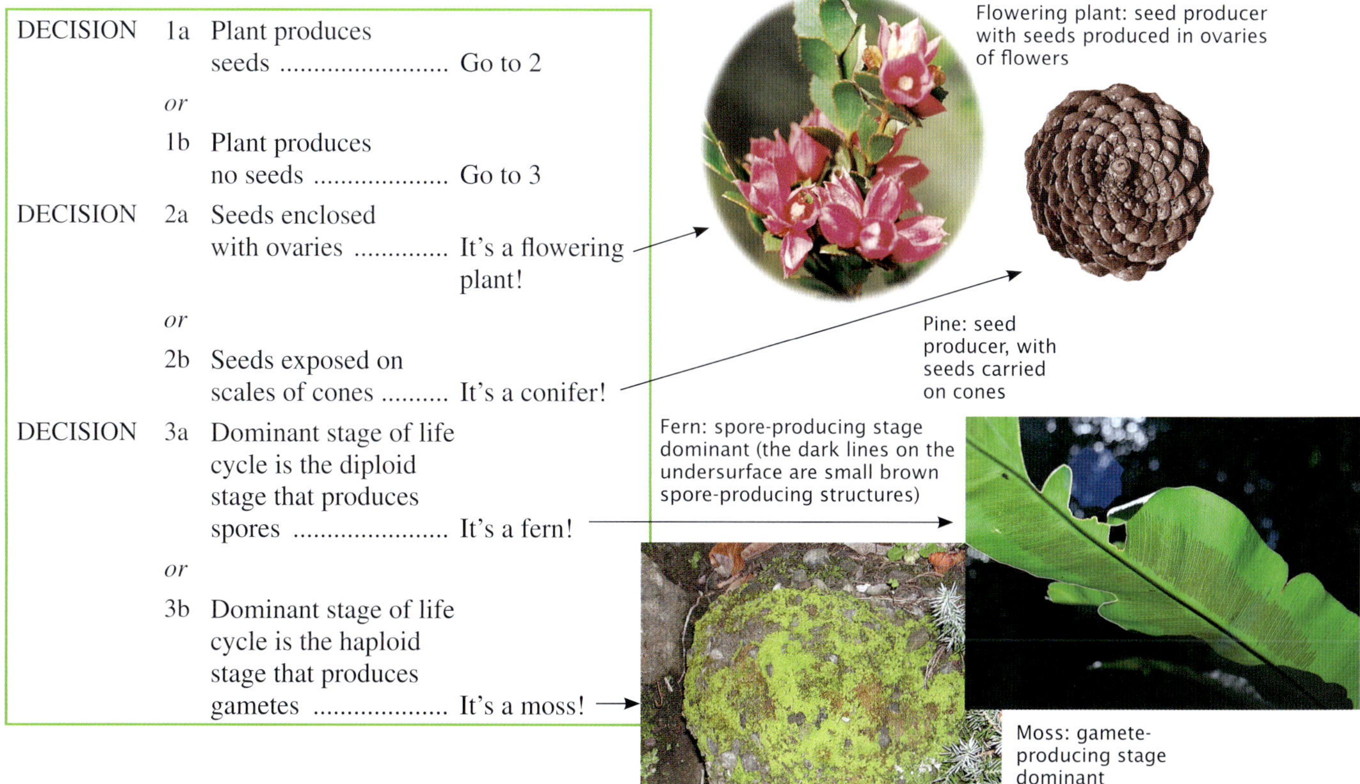

Features of classification schemes

- *Classification schemes can vary depending on their function.*
  There is no single correct way of classifying a group of items. Classification schemes can use different criteria to organise the same set of items. A plant nursery owner might classify plants on the criterion of size and group them into 'trees', 'shrubs', 'creepers' and 'herbaceous plants'. In contrast, indigenous tribespeople might classify them as 'edible', 'non-edible' or 'poisonous'.
- *Usefulness of classification schemes depends on the criteria selected.*
  Useful classification schemes involve criteria that are:
  — objective (rather than subjective)
  — meaningful (rather than arbitrary).
  An objective criterion is one that has the same meaning for different people; for example, with books, an objective criterion is 'subject matter' while a subjective criterion is 'interest level of contents'. Use of objective criteria in classification gives reproducible and predictable results, no matter who uses it. A meaningful criterion conveys useful information about what members of a particular group have in common, in contrast to an arbitrary criterion. For classical CDs, 'composer' is a meaningful criterion that provides useful information in contrast to an arbitrary criterion such as 'colour of CD label'.
- *Classification schemes are not fixed but can change as new information becomes available.*
- *Classification schemes can be single or multi-level.*
  Shane organised his stamps into a single-level scheme using one criterion. Rosie used two criteria and produced a multi-level scheme (see figure 8.32). A multi-level scheme is also called a hierarchical scheme. We will see later (pages 239–42) that biological classification involves a hierarchical scheme.

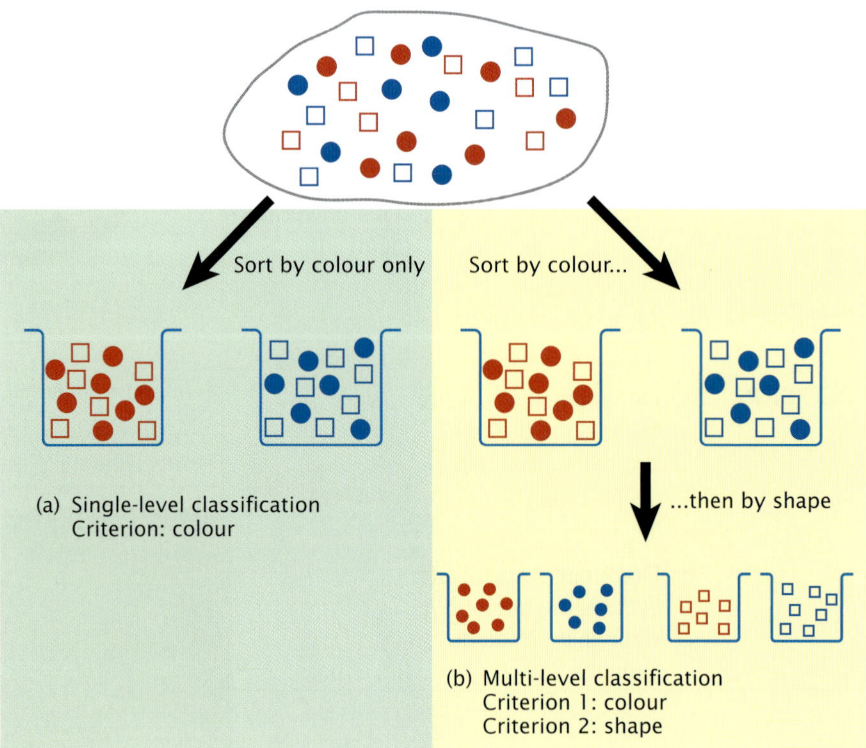

Figure 8.32 (a) Single-level classification schemes use one criterion. **(b)** Multi-level classification schemes use two or more criteria. How many criteria are used in this scheme?

ODD FACT

The Great Herball, a book published in 1526, has the following classification of fungi: 'Musheroons: There be two manners of them; one manner is deadly and slayeth them that eateth them and be called tode stools, and the other doeth not'. What criterion was used to group the fungi?

Figure 8.33 Knowledge of classification allows biologists to make predictions about a kind of organism — but *not* about its pet name!

Benefits of classification

Think about trying to find a phone number if the entries in a phone book were not organised alphabetically. Think about finding a book in a library if the books were just shelved in order of date of delivery. Think about communicating information about biodiversity if every organism, both living and fossil, was treated as an individual, unrelated item: aardvark, ammonite, armadillo, artichoke, avocado, and so on, to zebra, zebu, zinnia, zokor, zorro.

Classifying or organising different kinds of organism into a smaller number of groups using objective and meaningful criteria produces benefits.

- It is easier to deal with a smaller number of groups than a very large number of separate items. For example, the group 'family Felidae' refers to all members of the cat family including lions, leopards, cheetahs, jaguars, pumas, lynxes, ocelots and domestic cats.
- Classification can provide information. For example, what are zokors? Zokors are members of the order Rodentia and this label conveys the information that zokors are mammals closely related to mice and rats.
- New items can be added in a predictable way. For example, when a new kind of organism is discovered, it is compared with known groups and is classified as part of an existing group with which it shares key similarities.
- Information about items can be easily retrieved. For example, if a scientist wishes to access data about the group of invertebrate animals with soft bodies enclosed within paired shells of equal size, this can be done by accessing data about the animal group known as class Bivalvia.
- Predictions can be made about an item based on knowledge of its classification (see figure 8.33). If, for example, biologists read that a newly discovered fossil specimen has been classified as a member of the class Mammalia, they can confidently predict that its skull will show features characteristic of all members of that group, including a lower jawbone made of a single bone and teeth differentiated into various types. Examine figure 8.34. Can you identify the mammalian skull?

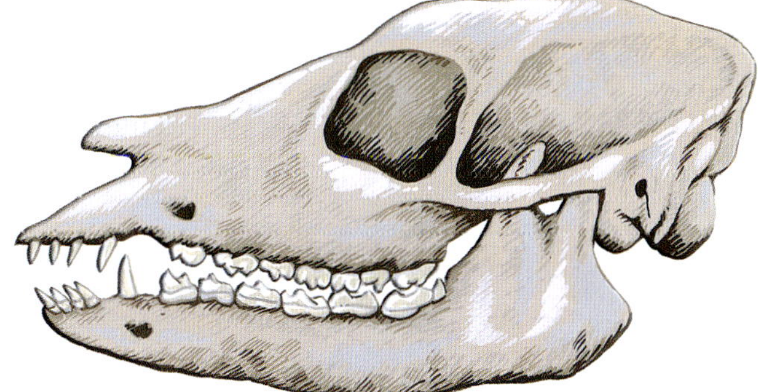

Figure 8.34 Two fossil skulls. Only one can be classified as belonging to the group known as Class Mammalia. Can you select which? (The other belongs to the group known as Class Reptilia.) You are a mammal. What prediction can you make about your lower jawbone?

Biological classification: forming a hierarchy

The system of biological classification forms a nested hierarchy of levels from species to phylum (see figure 8.35). Any level of classification — species, genus and so on — can be called a **taxon** (plural = taxa). The species level is the least inclusive and contains just one kind of organism. In contrast, the phylum level may contain a large number of organisms, as for example, the phylum Mollusca contains more than 50 000 different species.

The closer the evolutionary relationship between two organisms, the more similar their classification. Two organisms that belong to the same genus are more closely related than two organisms that share only the same family membership.

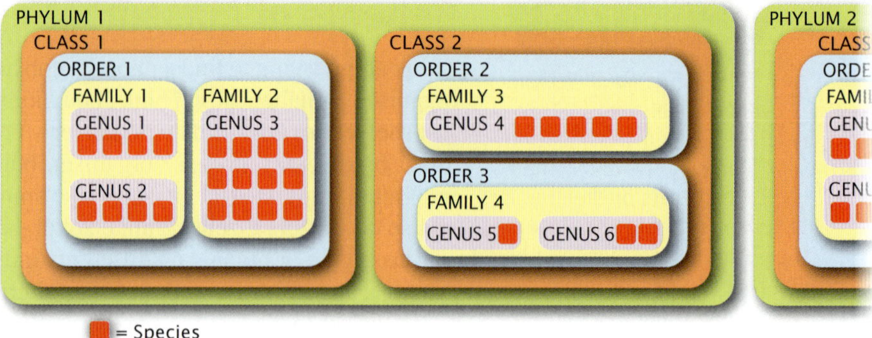

Figure 8.35 Classification schemes for organisms form a hierarchy.

Each different kind of organism, living or extinct, is known as a species, which is the basic level of classification. Examples of species include:

domestic cat (*Felis catus*)
lion (*Panthera leo*)
polar bear (*Ursus maritimus*)
emu (*Dromaius novaehollandiae*)
tiger (*Panthera tigris*)
great white shark (*Carcharodon carcharias*)
chimp (*Pan troglodytes*)
dog (*Canis familiaris*)

Figure 8.36 shows key features of their classification. Note that:

- Closely related species form a group known as a **genus** (plural: genera). Lions and tigers are both members of the genus *Panthera*; the other species belong to different genera.
- Groups of related genera form a group known as a **family**. Cats, lions and tigers are all members of the family Felidae. The others belong to different families: family Canidae for dogs, family Ursidae for bears and family Hominidae for chimps.
- Related families form a group known as an **order**. With the exception of chimps that are members of order Primates, the other five species are members of order Carnivora.
- Groups of related orders form a group known as a **class**. Six of the species are members of class Mammalia (with fur/hair and producing milk for their young).
- Related classes form a group known as a **phylum**. Mammals, birds and sharks are some of the classes that are grouped into phylum Chordata (see figure 8.37, page 241). What are the other classes?

Figure 8.36 The hierarchical (multi-level) system of classification of organisms. Observe how, as one moves to higher levels in the hierarchy, each group becomes more inclusive and contains more species.

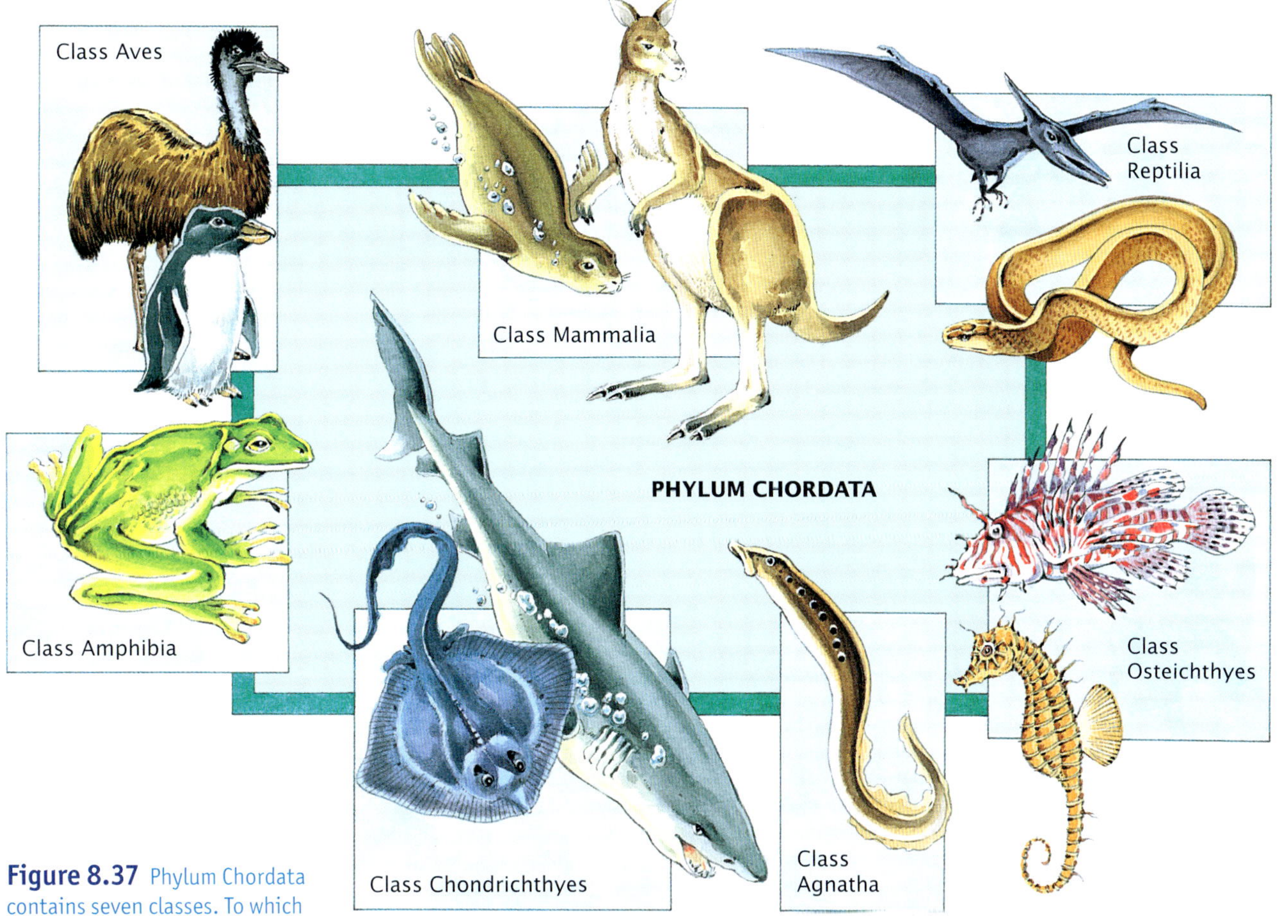

Figure 8.37 Phylum Chordata contains seven classes. To which class do you belong?

Beyond the level of phylum are the broadest and most inclusive groups — kingdoms and domains — that we will examine later (pages 245–50).

The classification of five animal species, including an extinct species of ammonite, is shown in table 8.2.

Classification also applies to plants and table 8.3 (page 242) shows the identification and classification of the peppermint gum.

Table 8.2 Classification of five animal species. Based on this information, which two are most closely related?

	Human	*Horse*	*Black rhinoceros*	*Dog*	*Ammonite*
phylum	Chordata	Chordata	Chordata	Chordata	Mollusca
class	Mammalia	Mammalia	Mammalia	Mammalia	Cephalopoda
order	Primates	Perrisodactyla	Perrisodactyla	Carnivora	Ammonitida
family	Hominidae	Equidae	Rhinoceritidae	Canidae	Dactylioceratiidae
genus	*Homo*	*Equus*	*Diceros*	*Canis*	*Dactylioceras*
species	*Homo sapiens*	*Equus caballus*	*Diceros bicornis*	*Canis familiaris*	*Dactylioceras commune*

Table 8.3 Classification of peppermint gum, one of the eucalypts whose leaves are eaten by koalas. What kind of fruit would *Eucalyptus marginata* have?

division	Spermophyta	It produces seeds.
class	Magnoliidae	It is a flowering plant having an embryo with two seed leaves or cotyledons.
order	Myrtales	It belongs to a group that includes some members with dry fruit and some with fleshy fruits; water-conducting tissue (xylem) in the stem is internal to the sugar-conducting tissue (phloem).
family	Myrtaceae	It produces aromatic oil from glands in the leaves, and has dry woody fruits, known as nuts.
genus	*Eucalyptus*	It is one of the eucalypts with each flower bud covered by a woody cap.
species	*Eucalyptus radiata*	It is the common peppermint gum tree, with grey fibrous bark and narrow leaves.

KEY IDEAS

- Classification is the process of organising objects or data into groups according to one or more criteria.
- Classification schemes may be single-level or multi-level (hierarchical) depending on whether one or more criteria are used.
- Taxonomists organise Earth's biodiversity into a series of taxa that form a hierarchical system.
- Biological classification begins with the species level and then moves through more inclusive groups such as families and phyla.

QUICK-CHECK

11 Identify the following as true or false.
  a Classification gives reproducible results when an objective criterion is used.
  b A phylum is a more inclusive group than a family.
  c A species consists of just one particular kind of organism.
  d Taxonomy is the study of the biology of particular species.
  e Organisms in the same order must be members of the same genus.
  f A taxonomist stated that she had published details of two new taxa so this must mean that she published details of two new species.

12 Identify two benefits that can result from classification.

13 You are told that an organism is a member of class Bivalvia. List one prediction that you can confidently make about this organism.

14 A plant fossil is found to have many features in common with modern eucalypts. Into which family should this fossil plant be placed?

Classification: how are groups formed?

Appendix C (pages 544–9) gives some detail of this complex picture and lists the nine phyla that include the majority of animal species.

Taxonomists have organised all the known animal species presently living on Earth into nearly 40 phyla. These animal phyla are subdivided into about 80 different classes and nearly 400 orders. All the plant species on Earth have also been organised into phyla (also known as divisions). Among the plants, there are more than 250 000 different species of flowering plants and taxonomists have classified them into two classes, more than 200 orders and more than 500 families.

How are decisions made about what should be included within the various groupings? Organisms that are directly descended from a common ancestor are placed in the same genus. But which genera should be grouped into the same family? What other mammals should be included with dogs in the family Canidae? in the order Carnivora? Biologists wish to group organisms according to inferred degrees of evolutionary relationship. One approach to identifying the relationships between various organisms and deciding which organisms should be included in particular groups, such as genus, family and order, is **cladistics**.

Figure 8.38 Which two animals appear more closely related? The most closely related pair is the echidna and the platypus.

Cladistics: recognising relationships

Cladistics asks the question 'How many derived features do they share?' to identify which organisms should be included in particular genus, family or order.

Characters seen in a particular group of organisms can be identified as primitive or derived. For a particular group, **primitive characters** are features that were present in their common ancestor and so appear in all members of the group. In a group of mammals, for example, a primitive or ancestral feature is the presence of fur. Why? Because this feature was present in the common ancestor of all mammals. **Derived characters** are advanced or modified features that evolved later and appear *in some members only* of the group in question. In a group of mammals, for example, a derived feature is the presence of nails on fingers and toes. Why? The hands and feet of the common ancestor of all mammals had claws so this is the primitive state; nails evolved later in one group of mammals, the primates, so this is the derived state.

In cladistics, only *shared* derived features can be used. (Derived features that are present in one member only of a group being studied cannot be used in cladistics.)

Cladistics can also be used to identify the relationships between genera within a family, as for example, Family Equidae, the horse family. Using various derived features, living and extinct horses can be related, as shown in figure 8.51 in the Chapter review, page 254.

The result of a cladistics analysis is shown in a diagram known as a **cladogram** (see figure 8.39). Branching points (forks) occur on the cladogram each time a derived character appears in some members of the group. A cladogram shows the evolutionary relationship between particular organisms based on the derived characters that they share. Various groups containing all organisms descended from a common ancestor are formed and these groups are used as the basis for various levels of classification.

Cladograms show evolutionary relationships but do *not* give the evolutionary history of the organisms (or groups) since they all appear at the end of the branches and no ancestral organisms are identified.

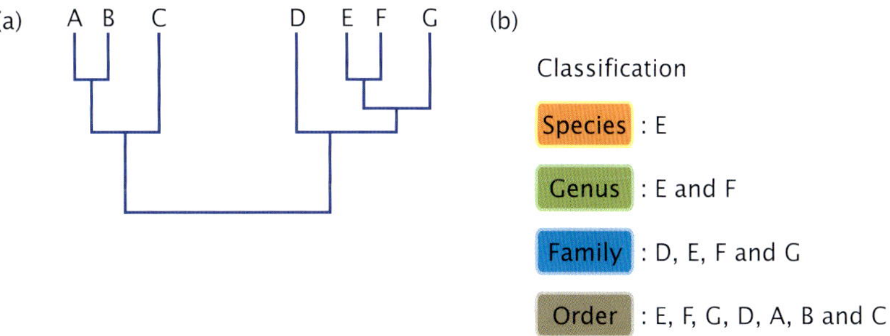

Figure 8.39 **(a)** A cladogram showing the presumed relationship **(b)** From this cladogram, classification groups can be formed. Organisms grouped in the same genus are the most closely related, for example, E and F. Families are formed from the next most closely related organisms, for example, E, F and G.

Classifications can change

Figure 8.40 These possums, once thought to be a single species, are now classified as two species — **(a)** the short-eared possum in the north (*Trichosurus caninus*) and **(b)** the mountain brush-tailed possum in the south (*Trichosurus cunninghamii*). This was first reported in the *Australian Journal of Zoology*, Volume 50, page 369, published in 2002.

Biological classifications can change. Organisms originally identified as belonging to a single species may later, as a result of further study, be split into two different species. Traditional classifications were based on structure and organisms were grouped because of similarities in their structures. Major changes in classification can occur as a result of **DNA sequencing** and **protein sequencing**. Technological advances in the last decade have made it possible to sequence DNA segments, specific genes and proteins very quickly and cheaply. Sequence data are held in a database that can be accessed by scientists worldwide and have become new tools in classification.

One example of a changed classification involves the Australian mountain brush-tailed possum, found in forests in eastern Australia from Queensland to Victoria (see figure 8.40). It was originally believed that this possum was a single species with the scientific name *Trichosurus caninus*. Based on extensive studies of the structural and molecular features of these possums, it was concluded in late 2002 that the northern and the southern populations were in fact two different species. The northern species has retained the scientific name *Trichosurus caninus* but has been given the common name of the short-eared possum. The Victorian species now has the scientific name *Trichosurus cunninghamii* and will now commonly be known as the mountain brush-tailed possum. One species has become two!

Tree shrews are a group of small mammals (see figure 8.41) that live in tropical rainforests in Asia; about 30 different species have been identified. Oriental tree shrews are not shrews and most do not live in trees! Tree shrews were originally classified as members of the order Insectivora that includes hedgehogs, moles and shrews (real ones!). In 1945, tree shrews were reclassified as members of the order Primates that includes lemurs, monkeys and apes. However, DNA technology showed later that tree shrews are not closely related to primates and they have now been reclassified as the only members of a new group known as order Scandentia.

Figure 8.41 One of the Asian tree shrew species, *Tupaia glis*. Two major changes have occurred in their classification. To what order do they now belong?

Classification: the big picture

As we move from species to the higher levels of genus, family, order, class and phylum, each classification category becomes more inclusive and there are fewer categories at each level. For example, currently more than 4600 species of mammal have been identified. These are included in 26 orders and then these are included in one class, namely class Mammalia. In fact, the living world is organised into even fewer, but more inclusive groups, above the level of phylum. Let us now look at the highest levels of classification, those above the level of phylum.

Figure 8.42 Sea pens belong to the class Anthozoa (= 'flower animals'), a group that also includes anemones and corals. This sea pen (*Sarcoptilus grandis*) is a sessile (non-motile) marine animal. It occurs in eastern Australian seas and grows to a maximum length of about 40 cm. Why does this not fit the Linnean definition of an animal?

The entire biodiversity of planet Earth has been organised by various biologists at different times into a small number of major groups. The most important are:

- the two-kingdom system of Linnaeus
- the three-kingdom system of Haeckel
- the five-kingdom system of Whittaker
- the three-domain system of Woese.

In addition to introducing a standardised way of naming species (refer back to page 225), the Swedish biologist Carolus Linnaeus (1707–1778) classified the living world into two kingdoms — vegetables (plants) and animals — so identifying every species as either a plant or an animal. Linnaeus stated that members of the Animal Kingdom 'have life, sensation and the power of locomotion', while members of the Plant Kingdom 'have life and not sensation'. While the Linnean two-kingdom system lasted for about one hundred years, it had shortcomings. Organisms such as bacteria and single-celled eukaryotes like *Amoeba* did not fit well into the two-kingdom system. In addition, some animals and plants are atypical, such as sea pens which are immobile (sessile) animals (see figure 8.42).

In 1866, a German biologist, Ernst Haeckel (1834–1919), proposed a three-kingdom system that comprised Kingdom Animalia (animals), Kingdom Plantae (plants and fungi) and Kingdom Protista (everything else). The Protista kingdom created by Haeckel included bacteria and single-celled eukaryotes.

In 1969, an American biologist, Robert H. Whittaker, proposed a five-kingdom system (see figure 8.43). Whittaker created two new kingdoms. One was Kingdom Fungi that separated fungi from plants, putting them into their own grouping, and the other was Kingdom Monera that encompassed all prokaryotic organisms, such as bacteria. Kingdom Protista, as was the case for Haeckel, was a 'left-over' group that accommodated those unicellular eukaryotic organisms, such as *Amoeba* and *Paramecium*, that were not plants, animals, fungi or bacteria.

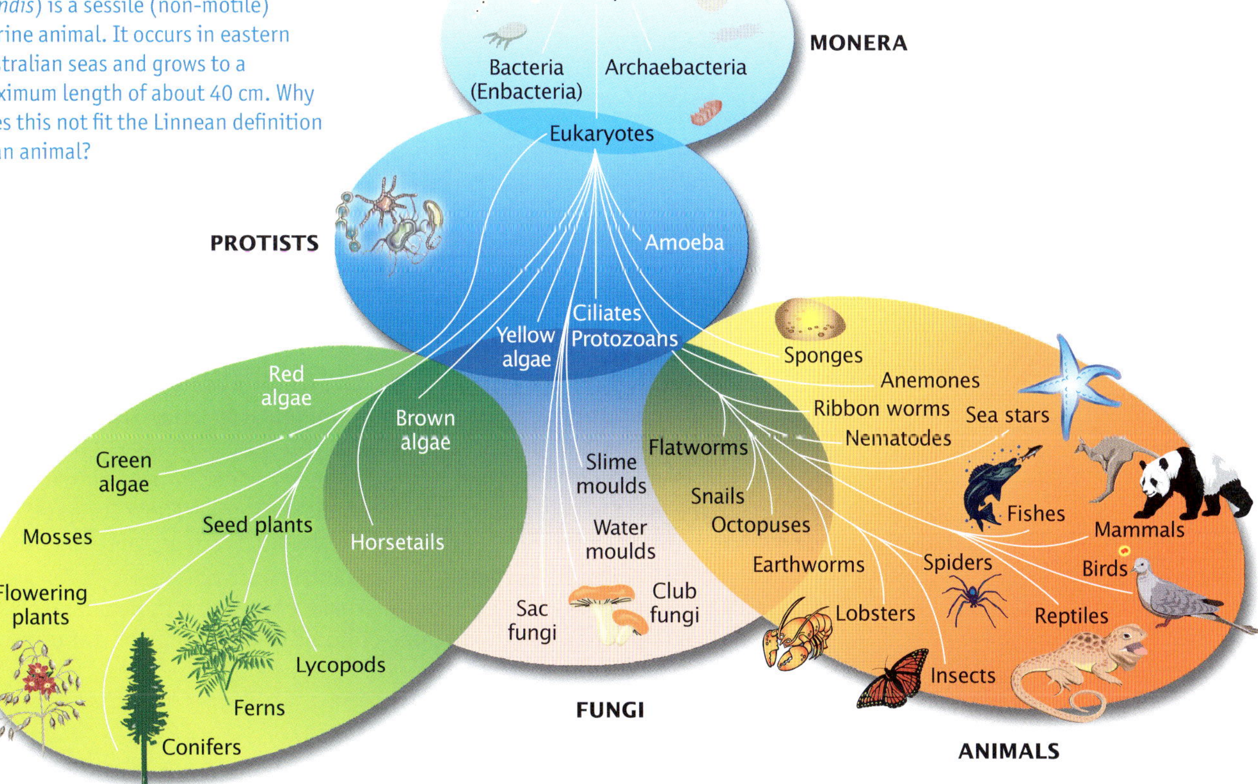

Figure 8.43 The biodiversity of Earth organised into a 'five kingdom' scheme

Table 8.4 gives more detail of the five-kingdom scheme and compares the different kingdoms in terms of some key features. Which kingdom has members that display the greatest variety of obtaining energy and matter for living? Which kingdom has members that have cellulose cell walls? In how many ways do fungi differ from plants?

Table 8.4 The five-kingdom system proposed by Whittaker in 1969

	FIVE KINGDOMS				
Name	***Monera***	***Protista***	***Fungi***	***Plantae***	***Animalia***
Examples	• cyanobacteria • gram-positive bacteria	• Amoeba • Paramecium • Volvox	• moulds • yeasts • toadstools	• mosses • ferns • cycads	• sponges • lobsters • birds
Cell type	prokaryotic	eukaryotic	eukaryotic	eukaryotic	eukaryotic
Other cell features	cell walls with complex carbohydrates		cell walls with chitin	cell walls with cellulose	no cell walls
Cell number	• unicellular	• unicellular • colonial • multicellular	• unicellular • multicellular	• multicellular	• multicellular
Reproduction	• binary fission	• asexual: mitosis • sexual: meiosis	• asexual: mitosis • sexual: meiosis	• asexual: mitosis • sexual: meiosis	• asexual: mitosis • sexual: meiosis
How energy and matter for life are gained	• autotrophic — photosynthesis — chemosynthesis • heterotrophic	• autotrophic — photosynthesis • heterotrophic — ingestion	• heterotrophic — absorption	• autotrophic — photosynthesis	• heterotrophic — ingestion

Domains: a new level of classification

In 1990, a more radical classification for all living organisms was proposed. Earlier schemes used structural features (morphological criteria), such as whether the organisms were built of prokaryotic or eukaryotic cells, to classify organisms. In 1990, however, Carl Woese and his colleagues created a new classification scheme for the living world that used molecular criteria. They compared the order of bases in particular genes, such as ribosomal RNA genes and other genes that code for specific proteins found in all organisms.

Based on these molecular comparisons, Woese proposed a radical new classification system that organised living things into different groups or **domains**:

- **Bacteria**
- **Archaea**
- **Eukarya**.

Domains are a higher level of classification than kingdoms. Each domain is subdivided into lineages or kingdoms shown as twigs branching from the main lines (see figure 8.44).

Figure 8.44 The three domains of life based on the molecular classification scheme of Woese. Some of the lineages (kingdoms) within each domain are identified. This tree was produced by comparing the base sequences of key genes, including particular ribosomal RNA genes, from various kinds of organism.

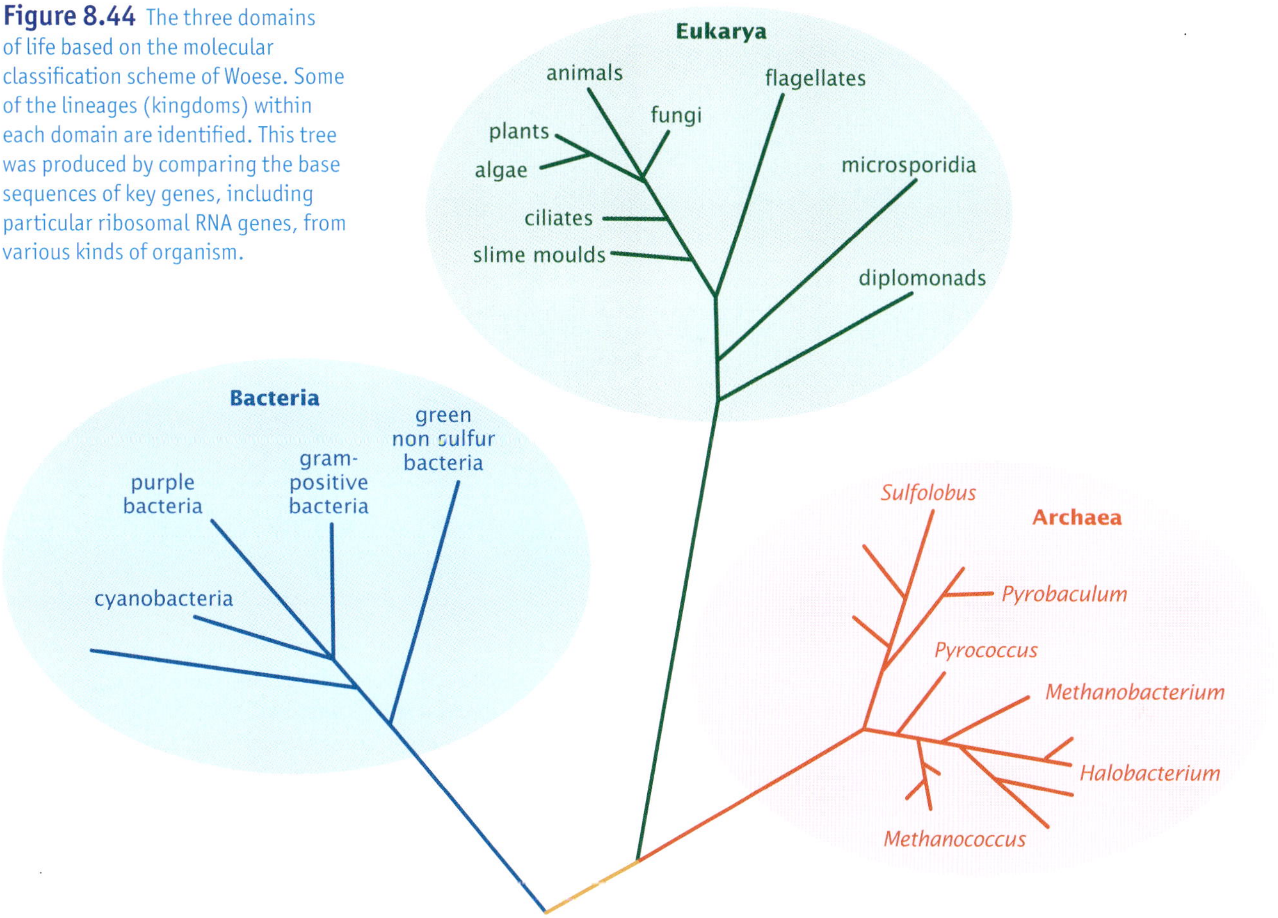

The Woese molecular classification upset the traditional separation of life into just two major groups, prokaryotes and eukaryotes.

The Woese classification system can assist in reconstructing major events in the evolution of life. Comparisons of the same gene from different organisms allow us to infer that, after the first forms of life evolved, a split occurred, perhaps at about 3700 Myr ago, to produce two distinct lines, the Bacteria line (shown in figure 8.44 in blue) and a second line (shown in orange). The second line later split to produce the Archaea line (shown in red) and the Eukarya line (shown in green).

Carl Woese's three-domain system of classification was revolutionary. This system recognised that, while all microbes were prokaryotic, not all microbes were bacteria. Molecular studies showed that, among the microbes, two major groups could be distinguished — typical bacteria that form the Bacteria domain and different microbes that form the Archaean domain. Archaeans differ from bacteria in some features of their structure and in their biochemistry; for example, the cell membranes of archaeans differ from those of bacteria and, unlike bacteria, archaeans are not sensitive to antibiotics. The third domain in the Woese classification system puts all other organisms — animals, plants, fungi, protists — as part of the Eukarya domain.

ODD FACT

Deep-sea vents have a variety of chemosynthetic bacteria that produce energy by different oxidation reactions. These include sulfur-oxidising bacteria, nitrifying bacteria that oxidise ammonia to nitrite (or nitrite to nitrate), bacteria that oxidise methane (natural gas, CH_4) and carbon monoxide (CO) and bacteria that oxidise ferrous iron.

Bacteria domain

The Bacteria domain comprises thousands of bacterial species and includes groups such as:

- **cyanobacteria** (see figure 8.45) that are members of a group that was the first form of life to carry out photosynthesis; for example, species in the genus *Nostoc* and *Anabaena*

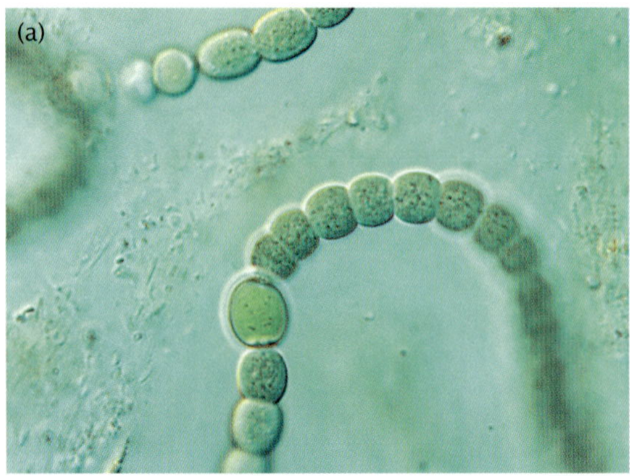

Figure 8.45 Cyanobacteria are photosynthetic members of the Bacteria domain. **(a)** One kind of cyanobacteria forms chains. Each cell is microscopic but large masses of cyanobacteria can form green scum on waterways. **(b)** Stromatolites at Shark Bay in Western Australia. These layered structures are built up over long periods by mats of cyanobacteria that trap sediment. Only the outer layer consists of living cyanobacteria.

- **nitrogen-fixing bacteria** that convert atmospheric nitrogen to ammonium ions (NH_4^+); for example, *Azotobacter* sp., *Frankia* sp.
- **sulfur-oxidising bacteria** whose energy comes from reactions with sulfur compounds; for example, *Thiobacillus* sp.
- **gram-positive bacteria** that include species that cause many diseases of humans and other mammals; for example, *Clostridium* sp., *Staphylococcus aureus*.

Methanogenic archaeans, such as *Methanobrevibacter ruminanitum*, live in the stomach (rumen) of certain herbivores. Methane from this source is a major contributor to greenhouse gases.

Figure 8.46

The Archaea domain

Most archaeans are **extremophiles** which means that they thrive in extreme environments, such as salt lakes, water more acidic than vinegar, water more alkaline than cloudy ammonia, oxygen-free mud of marshes and swamps, and hot springs and hot lakes (see figure 8.47). The scientific names of some archaeans give a clue to the extreme environments in which they live. Can you suggest where archaean species from the genus *Acidolobus* might be found? What about the genera *Thermococcus* and *Vulcanisaela*?

Figure 8.47 Champagne Pool, New Zealand, is a hot mineral pool and is the type of extreme environment in which archaean microbes can thrive. The orange and yellow colours at the edge of this lake are produced by living microbes.

Examples of archaean groups include:

- extreme **halophilic** ('salt-loving') archaeans that live in salt lakes, including the Dead Sea; some kinds even grow on the surface of salted fish!
- methane-producing archaeans, also known as **methanogens**, that produce methane from carbon dioxide and hydrogen and can survive only in oxygen-free environments, such as swamps and the stomachs of ruminant mammals like cattle and sheep (see the Odd fact)
- **thermophilic** ('heat loving') archaeans that live in hot springs.

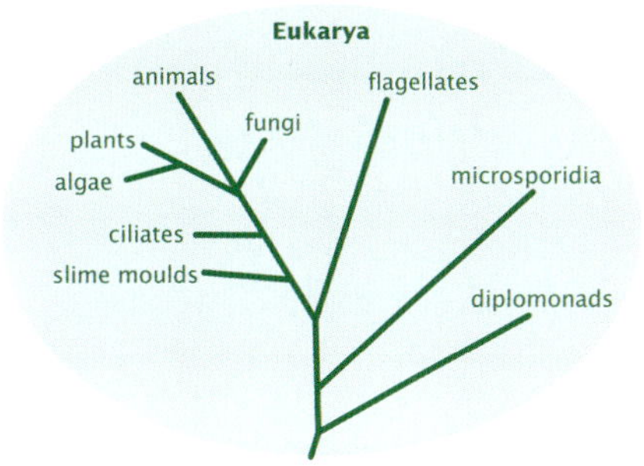

The Eukarya domain

Pat your cat or dog, look at the plants and insects in a garden, watch the fish in an aquarium, see the growth of green mould on bread. Look around you — almost all the living organisms that you see are members of the Eukarya domain. This domain comprises all animals, plants, fungi and the various lineages of protists (see figure 8.48). The other domains, Bacteria and Archaea, are almost all microscopic and usually 'out-of-sight'.

Eukarya
animals
fungi
flagellates
plants
algae
microsporidia
ciliates
slime moulds
diplomonads

Figure 8.48 The Eukarya domain comprises many lineages.

Figure 8.49 A tiny fraction of the biodiversity of planet Earth today

KEY IDEAS

- Decisions about the grouping of organisms into various classification categories can be made using different approaches.
- The most commonly used method to identify related organisms for classification purposes is cladistics.
- Classifications are not fixed but can change.
- Levels of classification above phylum are kingdoms and domains.
- The three-domain system of classification introduced in the 1990s organised all living organisms into three domains: Bacteria, Archaea and Eukarya.

QUICK-CHECK

15 Identify the following as true or false.
   a If two species share an immediate common ancestor, they are classified as members of the same genus.
   b Cladistics can be used to form classification groups.
   c Woese's three-domain system of classification uses molecular comparisons.
   d Fungi are members of the Archaea domain.

16 List two members of each of the following classification groups:
   a the Bacteria domain
   b the Archaea domain
   c the Eukarya domain.

17 Identify a classification scheme that:
   a groups all microbes in Kingdom Protista
   b separates prokaryotes into two different taxa.

BIOCHALLENGE

Taxonomists use computer technology in their challenging task of organising Earth's biodiversity into a hierarchical classification. Through the World Wide Web, taxonomists can access databases that hold information about various groups of organisms and check on the descriptions and names of various species.

Three propositions appear below which may or may not be correct. Your task is to investigate the validity of at least one proposition. As you proceed, you will become aware of the diversity of living things and how classification organises this biodiversity into inclusive groups.

PROPOSITION 1: 'Surely there is not much diversity in fish and certainly a herring is just a herring!' True or false?

a Go to www.jaconline.com.au/natureofbiology/natbiol1-3e and access the 'Phylum Chordata' weblink for this chapter. The website is located in the United Kingdom. A table of the various classes of phylum Chordata will appear on the screen.

b Click on the Class Osteichthyes (bony fish) and a list of the different Orders of bony fish will appear on screen. How many orders make up the Class Osteichthyes?

c Click on the order Clupeiformes (herrings).

d Back to proposition 1: true or false?

PROPOSITION 2: 'Cycads — there are just a few species in one genus, I believe!' True or false?

a Go to www.jaconline.com.au/natureofbiology/natbiol1-3e and access the 'Cycad Pages' weblink for this chapter.

b From the menu, select 'Identification and Classification'. A table will appear on the screen.

c Look at the classification table (just order, family and genus). So, what about proposition 2: true or false?

d Go to the genus column, click on genus *Cycas* and check the first section of text. How many different species in this genus?

e Scroll down to see the list of species in this genus. Click on *C. cairnsiana* and note some information about this species, such as the date of first description and the derivation of the specific part of the scientific name

f Scan the formal description of this species. Just get a 'feel' for how taxonomists describe a species.

PROPOSITION 3: 'The world of bacteria is more diverse (and so is organised into more genera) than the world of archaeans.' True or false?

a Go to www.jaconline.com.au/natureofbiology/natbiol1-3e and access the 'Bacterial names' weblink for this chapter. This website, located in France, goes to species level.

b Scroll down and select 'Genera of the domain (or empire) of Archaea'. Note the number of different genera in this domain.

c Go to the genus *Pyrolobus* and click on it; this will take you to a list of species in that genus and, for each species, details of the scientific publication in which the first description of the species appeared.

d Look at the entry for the archaean species, *Pyrolobus fumarii*. What is interesting about this species?

e Return to the starting page and select 'Genera of the domain (or empire) of Bacteria'. Note the large number of different genera (don't bother counting!).

f Click on the genus *Anoxybacillus* and go to the list of species in that bacterial genus. Where was the first specimen of *Anoxybacillus flavithermis* found?

g Back to proposition 3: which group is more diverse, Bacteria or Archaea?

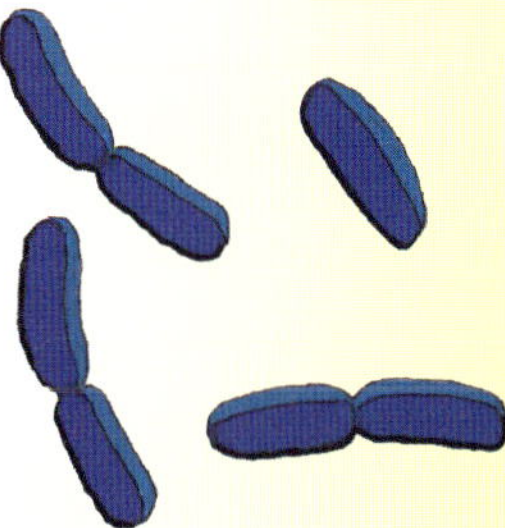

CHAPTER REVIEW

CROSSWORD

Key words

Archaea
Bacteria
binomial system of naming
biodiversity
bioinformatics
cladistics
cladogram
class
classification
criterion
cyanobacteria
cycads
derived characters
dichotomous key
DNA barcoding
DNA sequencing
domains
Eukarya
extremophiles
family
field guides
fossils
generic name
genus
gram-positive bacteria
halophilic
herbarium
keys
macroscopic
methanogens
microscopic
nitrogen-fixing bacteria
order
palynology
pellets
phylum
polynomial system
primitive characters
protein sequencing
raptors
reference collections
scats
species
specific name
sulfur-oxidising bacteria
systematics
taxon
taxonomists
taxonomy
thermophilic
vascular plants

Questions

1 ***Making connections*** ▸ Construct a concept map using at least eight of the chapter key words.

2 ***Applying understanding*** ▸ Consider the following items: cat, dog, cow, canary, budgerigar, kangaroo, emu, wallaby, platypus, elephant, tiger and polar bear.

a Identify an objective criterion to divide these items into groups.

b How would you test whether your criterion is objective?

3 ***Communicating your understanding*** ▸ Horses and donkeys can interbreed but their offspring, known as mules, are infertile or sterile. A student stated that, because they can interbreed, horses and donkeys must be members of the same species. Do you agree? Explain.

4 ***Analysing information*** ▸ Three different organisms (D, E and F) belong to Kingdom Animalia.

Animals D and E belong to the same class, but to different orders.

Animals E and F belong to the same genus but are different species.

Classify each of the following statements as either 'true', 'false' or 'can't say'.

a The degree of relationship between organisms D, E and F is equal.

b The first part of the scientific names of animals D and E must be identical.

c Animals D and F must belong to the same class.

d Animals E and F must belong to the same order.

5 ***Evaluating alternatives*** ▸ For this question, use the 'five kingdom' classification scheme.

Statement	Possible kingdom(s)
Organism X is built of cells.	
These cells each have a distinct nucleus bounded by a nuclear envelope.	
These cells have cell walls.	
These cell walls contain chitin.	

a Consider each statement in the table at left about organism X in turn and indicate what conclusions can be made about the kingdom(s) to which it might belong.

b At which point may a definite decision be made about the kingdom to which organism X belongs?

c What prediction can be made about how organism X obtains the energy and matter required for living?

6 ***Applying understanding*** ▸ Photographs of two different birds, A and B, are selected from a book *Birds of the World*. Photograph A is shown to a biologist who says: 'It's a magpie. Its scientific name is *Pica pica*.' Photograph B is shown to a second biologist who says: 'It's a magpie. Its scientific name is *Gynorhina tibicen*.' Could both biologists be correct? Explain.

7 ***Using scientific conventions*** ▸ Examine the following list showing some whale species.

Common name	Scientific name	Common name	Scientific name
fin whale	*Balaenoptera physalus*	southern right	*Eubalaena australis*
humpback	*Megaptera novaeangliae*	sei whale	*Balaenoptera borealis*

a How many different species are shown in this list?

b How many different genera do these whales represent?

c Identify the two most closely related whales.

8 ***Analysing information and making predictions*** ▸ Figure 8.50a shows a creeper-like plant (P). The undersurface of one of its leaves is shown in figure 8.50b. A different plant (Q) is shown in figure 8.50c.

(a) Plant P

(b) Undersurface of leaf of Plant P

(c) Plant Q

Figure 8.50

a Refer back to the dichotomous key (figure 8.31 on page 237). On the basis of the information in figure 8.50a only, can plant P be identified?

b Using the information in figure 8.50b, can plant P be identified?

c To what level of classification have you identified this plant? (Refer if necessary to appendix C on page 544.)

d List two resources you might use to make a more precise identification.

e Now consider the plant Q. You are told that it is a member of the same group as the plant P. In light of this, what prediction can you confidently make about its classification?

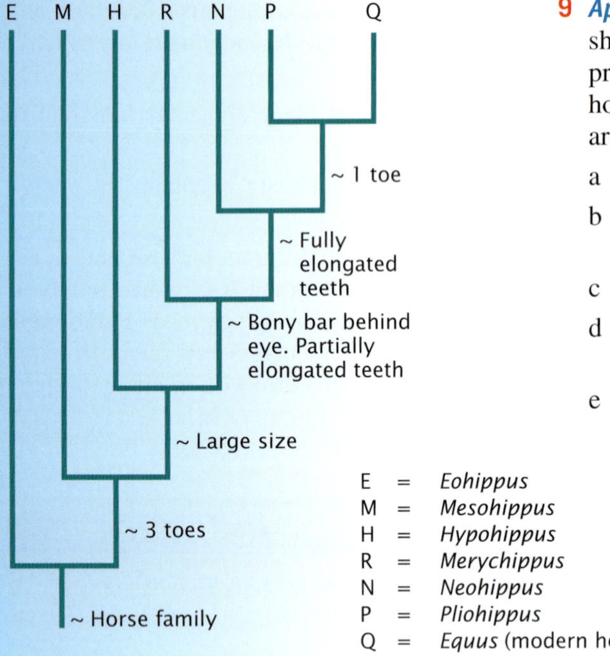

Figure 8.51

9 ***Applying your understanding*** The diagram in figure 8.51 shows the relationship between various horse species as produced by a cladistics study (refer back to page 243). Of the horse species concerned, one is living (*Equus*) and the other six are extinct.

a What name is given to this type of diagram?

b Identify one primitive (ancestral) character that would apply to this group.

c Are primitive characters used in cladistics study? Explain.

d Does this diagram show the evolutionary history of the horses?

e What shared derived feature was used to group species P and Q?

10 ***Discussion question*** Consider that a previously unknown kind of monkey is discovered in tropical forest. Identify at least three steps that might be taken in order to decide whether this monkey is a new species within an existing genus or whether it should be assigned to a new genus.

11 ***Using the Web*** Go to www.jaconline.com.au/natureofbiology/natbiol1-3e and access the 'Animalia' weblink for this chapter. The site lists the phyla of kingdom Animalia. By clicking on a phylum name, you can obtain information about it. In some cases, you will be given a list of the classes contained within a phylum and you can then drill down into orders, by clicking on a class.

From data in this website, match the numbered entries in the 'Feature' column of the following table with the appropriate name in the 'Phylum' column.

Phylum	Feature
Placozoa	(1) has 4 classes including class Cubozoa (box jellyfish)
Chordata	(2) the last phylum to be established
Cycliophora	(3) has more extinct fossil species than living species
Brachiopoda	(4) contains the class Pterygota (true insects) that has nearly 30 orders
Cnidaria	(5) has about 25 000 species organised into three classes
Platyhelminthes	(6) contains the most primitive multicellular animal discovered to date
Uniramia	(7) includes classes that contain birds, reptiles and amphibians

2

UNIT

ORGANISMS AND THEIR ENVIRONMENT

AREA OF STUDY 1

Adaptations of organisms

9 Habitats, environment and survival

KEY KNOWLEDGE

This chapter is designed to enable students to:

- develop a knowledge and understanding of habitats and the environments that exist in them
- identify major factors that produce the environments of various habitats and distinguish between biotic and abiotic factors
- give examples of structural adaptations that equip organisms for survival under particular environmental conditions
- explore technologies used to record environmental factors and track distribution of species within their habitats.

Figure 9.1 Here we see two contrasting environments. Above, a hot arid desert with its sparse community of plants, animals and microbes that can survive under the prevailing environmental conditions. Inset, a tropical rainforest with its rich community of many plant, animal, fungi and microbial species. In this chapter, we will examine a range of living spaces or habitats and identify the major factors that determine the environments in those habitats. We will also explore features of plants and animals that equip them for survival under various environmental conditions, with a particular emphasis on the Australian environment and the survival strategies seen in native plant and animal populations.

Giant monsters of the deep

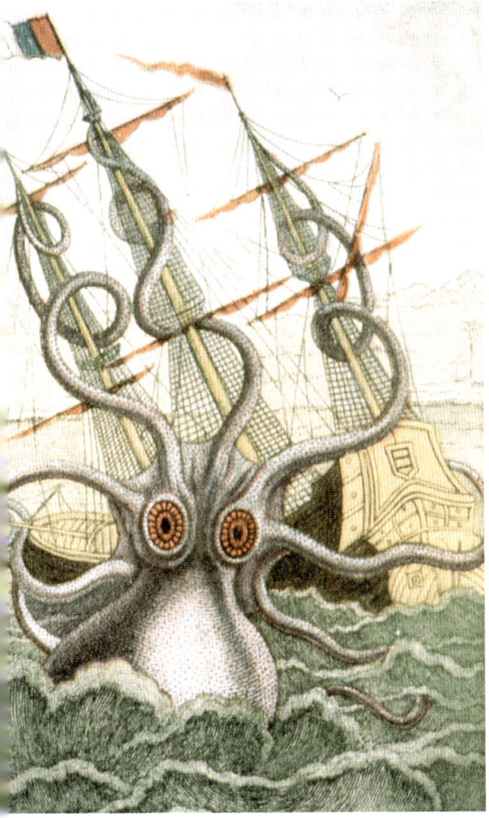

In 1830, the British poet Alfred Lord Tennyson wrote about a mythical sea creature, the Kraken:

> Below the thunders of the upper deep,
> Far far beneath in the abysmal sea,
> His ancient, dreamless, uninvaded sleep
> The Kraken sleepeth …

Stories of deep-sea creatures that could rise from the ocean depths to overwhelm a sailing ship are part of maritime legends. The Kraken was depicted as a creature with many arms that could reach to the top of the tallest mast of a sailing ship and engulf the ship, causing it to capsize with the loss of the crew (see figure 9.2).

While Krakens do not exist, giant squid do exist — in fact, several species of them. In 1888, a giant squid was washed up dead on a beach near Wellington, New Zealand. This squid was reported to have a length of about 18 metres (but some stretching of the tentacles of the dead creature may have occurred). The length of one species, the giant squid (*Architeuthis dux*), including its tentacles, is about 13 metres for males (nearly two bus lengths) and slightly shorter for females. Another species, the giant cranch squid (*Mesonychoteutis hamiltoni*), also known as the colossal squid, has an estimated mature length of more than 20 metres. (Pace that distance out.)

Figure 9.2 The legend of the Kraken is illustrated in this nineteenth-century print showing a giant octopus attacking a galleon.

Would you like to catch a living colossal squid? You will certainly not do this on land. You will not catch one as you fish from a pier or on a sandy beach. To catch a colossal squid, you need to go to the open ocean over the edge of the Antarctic continental shelf and descend into the squid's living space — a space that can extend from the surface to a depth of up to one kilometre.

In September 2005, a team of Japanese scientists announced some amazing research results. They had successfully, and for the first time ever, filmed a living giant squid in its deep, dark ocean environment. Search the web (enter the terms 'giant squid Japan') to access their story and view their fascinating pictures that unlock some of the mysteries surrounding this mysterious creature.

ODD FACT

One of the smallest squid is the southern pygmy squid (*Idiosepius notoides*). This squid is just two centimetres long when mature and lives in seas off Tasmania.

The giant squid lives in the cold ocean waters as a free-swimming predator, eating several hundred kilograms of food daily — perhaps a 100 kg Patagonian toothfish or two! By rapidly forcing water from a funnel below its head, the squid can suddenly jet-propel itself at speeds of more than 30 kilometres per hour. The giant squid has two long tentacles equipped with hooks at their clubbed ends that it uses to capture its prey (see figure 9.3c). After capturing its prey, the squid secures it using the suckers on its eight shorter arms and moves the prey to its mouth where its powerful parrot-like beak tears the prey into pieces to pass into its digestive tract (see figure 9.3b).

Figure 9.3 **(a)** The external anatomy of a typical giant squid, with its ten appendages — two longer tentacles and eight arms. **(b)** The powerful beak at the squid's mouth is capable of severing a steel cable. **(c)** The squid has hooks at the clubbed ends of the tentacles.

Figure 9.4 This 12-metre giant squid, washed up on Tasmania's west coast in February 2001, is being examined by a marine scientist on its arrival at the Melbourne Museum.

In their cold, deep-ocean living space, giant squid are efficient, streamlined and fast-moving predators. In warm shallow water, however, they become incapacitated and, out of the water, they become a motionless, collapsed blob since their body weight is no longer supported by water. Contrary to the maritime legend, no giant squid could raise its long tentacles out of the water to the top of a ship's mast — in the air, the squid could not support its tentacles. (Unlike a squid, you have an endoskeleton made of strong bones which, with their attached muscles, provide support for your limbs.)

The places where different species live and reproduce are many and varied. For the colossal squid, it is the open ocean around Antarctica. Other living spaces, such as coral reefs, open forests and hummock grasslands, are characterised by the presence of different species that live and reproduce there. In each case, members of the species concerned are equipped for survival but in their particular living space — the difference between a giant squid swimming in cold ocean waters and the lifeless squid beached on a shore is a graphic reminder of this.

In the following sections, we will look at factors that determine the environments in various habitats and explore how different species survive under those conditions.

Habitat: where an organism lives

Organisms are found in locations or natural settings where they can obtain the energy and matter required for living and where conditions are suitable for their survival and reproduction. The location or place where an organism lives at a given time is known as its **habitat** — this may be considered to be the organism's 'address'.

The habitats of different kinds of organisms vary greatly. For example, the typical habitat of porcupine grass (*Triodia irritans*) is a dry sandy area in inland Australia. The brown alga known as Neptune's necklace (*Hormosira banksii*) typically lives in the intertidal (or littoral) zone between low- and high-tide marks on rocky shorelines of south-eastern Australia. The habitat of the ragged-finned lionfish (*Pterois antennata*) is in the tropical waters around coral reefs in the Great Barrier Reef (see figure 9.5).

Figure 9.5 The habitat of the ragged-finned lionfish (*Pterois antennata*) is in the tropical waters of the Great Barrier Reef.

Bacteria of the species *Dienococcus radiourans* have been found living in the water surrounding nuclear reactors. These bacteria can survive levels of nuclear radiation that would be lethal to other organisms.

Figure 9.6 Some microbes live comfortably in so-called 'extreme environments'.

In our human perception, some living spaces are hostile and we call these '**extreme environments**'. You will not find familiar organisms living in extreme environments, but they are places that some hardy bacterial and archaean microbes call home. These unlikely places include salt lakes, mineral-laden water of hot springs, under boulders on glaciers, in oxygen-starved mud swamps and in acidic waters (see figure 9.6).

The size of habitats or living spaces varies for different organisms. White sharks (*Carcharodon carcharias*), a protected species, have as their living space an enormous volume of ocean where they feed and reproduce. In contrast, tiny mites live on the skin of a budgerigar. Some microbes have the human gut as their living space, such as *Escherischia coli* bacteria.

Many habitats can be described in general terms as being either:

- terrestrial (for example, deserts, grasslands, tropical rainforests), or
- aquatic:
  - — freshwater, for example, lakes, rivers, ponds
  - — marine, for example, coral reefs, sandy coastal seafloors, coastal seas and open ocean waters
  - — estuarine, for example, river mouths that are affected by tidal movements.

In reality, if we look at the various organisms living in the same habitat, we will see that each occupies or uses different parts of that habitat. In an open forest habitat, for example, one insect species might live under the bark of a particular kind of tree, while another insect species might be found in the leaf litter on the forest floor. One kind of bird might feed and nest in the upper canopy of the forest, a second bird species might feed and shelter in shrubs and low bushes, while a third might be a flightless ground dweller feeding and nesting in the litter layer on the forest floor. The more localised part of a general habitat where an organism lives is its **microhabitat**.

Look at figure 9.7 that shows a view of a small pond. Within this one pond, several microhabitats exist; for example, one microhabitat is the shallow water shaded by vegetation, another microhabitat is the muddy area at the bottom of the pond.

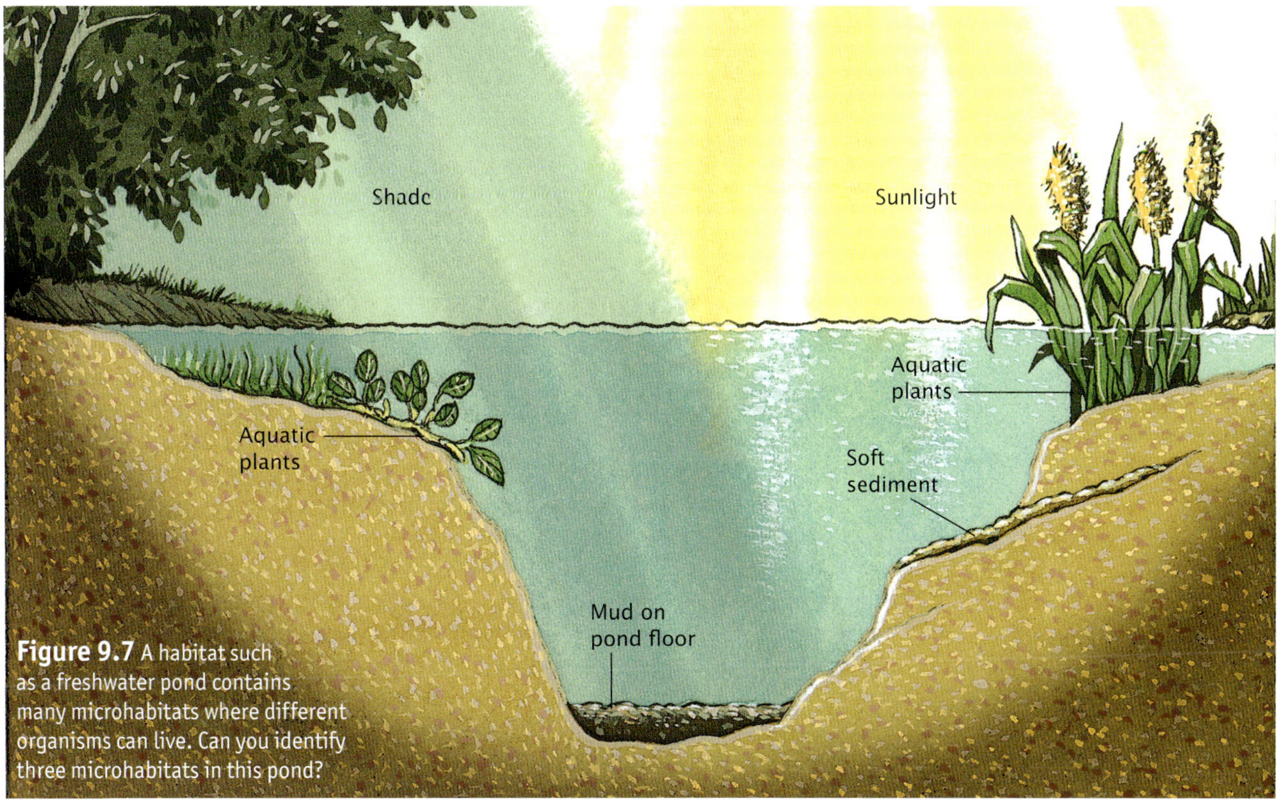

Figure 9.7 A habitat such as a freshwater pond contains many microhabitats where different organisms can live. Can you identify three microhabitats in this pond?

Habitats are not uniform

A habitat provides the resources that are needed for the life of a particular organism. For a typical animal, these resources include food, water, shelter, nesting and breeding sites. For a typical plant, the resources include sunlight, space, water and mineral nutrients. The distribution of resources in a habitat is not uniform and various resources may occur in different parts of habitat. Some areas of an animal's habitat may be where it feeds. Other parts of the same habitat may be where the animal shelters or, in the case of a female, where it lays its eggs or gives birth to its young.

The habitat of the tammar wallaby (*Macropus eugenii*) (see figure 9.8) provides an example of how the distribution of resources, such as food and shelter, in a habitat is not uniform. The general habitat of the tammar wallaby is shrubland scattered over large areas of south-west Western Australia. Within this habitat, the wallabies find shelter during the day in dense clusters or thickets of a native plant known as heartleaf poison (*Gastrolobium bilobum*) that provide protection from predators. At night, the wallabies leave the thickets and move to nearby open grassy areas where they feed, returning to their heartleaf thicket shelter before dawn. So the habitat of tammar wallabies has some areas that are rich in food resources (the open grassy regions) and other areas that provide shelter from predators (the heartleaf thickets).

ODD FACT

The heartleaf poison plants that provide shelter for tammar wallabies have bacteria that fix nitrogen from the air, converting it to nitrites. These nitrites enrich the soil and promote the growth of grasses that provide food for the wallabies. Find out more about this and other species of Western Australian flora at http://florabase.calm.wa.gov.au

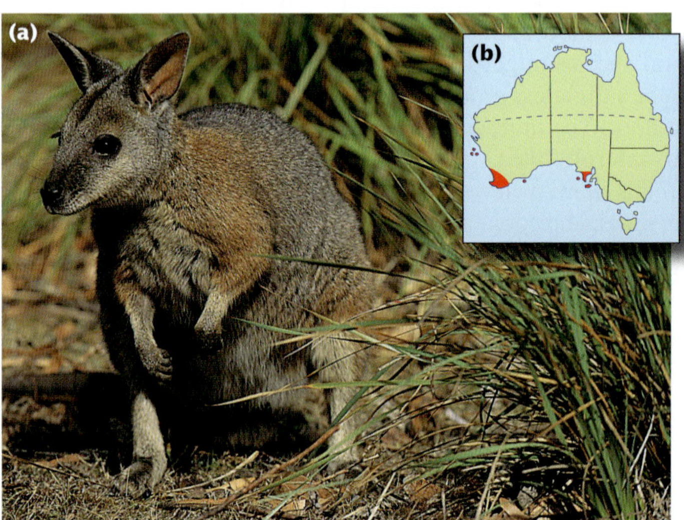

Figure 9.8 **(a)** The tammar wallaby is one of the smallest wallabies. **(b)** The tammar wallaby's general habitat is the south-west area of Western Australia. **(c)** Heartleaf poison (*Gastrolobium bilobum*) provides the tammar wallaby with shelter from predators.

Range: the distribution of habitats

The geographic area that encloses all the habitats where a species lives denotes the **range** or distribution map of that species. Ranges are commonly shown as maps similar to the inset map in figure 9.8b. It shows the distribution of the tammar wallaby and, as you would expect, the distribution of heartleaf poison — a species that is an essential part of the tammar wallaby's habitat — coincides with this range.

Within its range, a species may be plentiful or rare. A large range does not mean that a species is common. A small range does not mean that a species is very rare. The range of one small carnivorous marsupial, known as the kultarr (*Antechinomys laniger*) is large, but this species is rarely sighted within this range (see figure 9.9). In contrast, the current range of the banded hare-wallaby (*Lagostrophus fasciatus*) is restricted to two small islands, Bernier and Dorre Islands in Shark Bay, Western Australia. On those islands, hare-wallabies are common because predators are absent.

(a)

(b)

Figure 9.9 **(a)** The kultarr (*Antechinomys laniger*) is a small carnivorous marsupial mammal. **(b)** The kultarr's large range

Over time, the range of a species may increase or decrease. Many of our native Australian species have a shrinking range, as for example, the numbat (*Myrmecobius fasciatus*) (see figure 9.10). In contrast, many introduced species have increased their range from their point of introduction, as for example, cane toads (*Bufo marinus*). Since cane toads were introduced to northern Queensland in 1935, they have spread and their changing range is shown in figure 9.11.

(a)

(b)

200 years ago
Present range

Figure 9.10 **(a)** Young numbats (*Myrmecobius fasciatus*). **(b)** The past and present ranges of the numbat. Factors that have contributed to this shrinking range include predation by introduced feral predators, such as foxes and cats. Can you suggest other reasons?

(a)

(b)

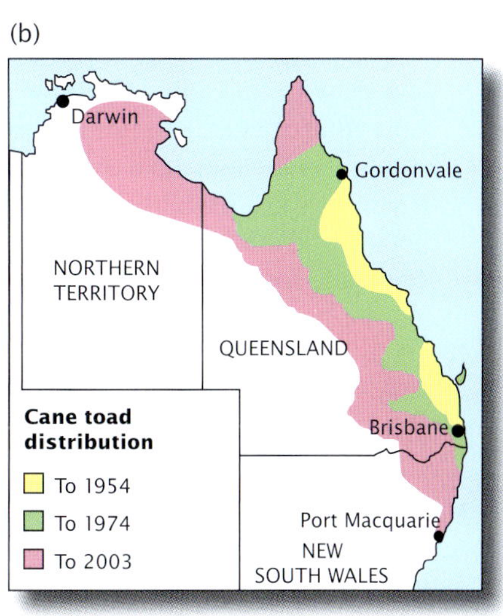

Figure 9.11 **(a)** The cane toad (*Bufo marinus*) was introduced to Australia in 1935 in an attempt to control the beetles that were damaging sugarcane crops. **(b)** Map showing the changing distribution of the cane toad. The rate of increase is about 35 kilometres per year. Can you suggest reasons for the expanding range of this introduced species?

ABSENCE FROM A REGION: WHAT DOES IT MEAN?

A biologist observes that a particular kind of organism is absent from a region. Does this mean that the conditions in that region are unsuitable for the species? This is sometimes true, but there may be other reasons. The following are some examples.

Unsuitable environment

The range of delicate common filmy fern (*Hymenophyllum cupressiforme*) does not include the sandy deserts of inland Australia because these ferns cannot survive the physical conditions there. The microhabitat of filmy ferns, which have small, flat, green fronds a few centimetres long but only one cell thick, is on wet surfaces or on fallen tree trunks in a sheltered gully in a wet sclerophyll forest.

Geographic barriers

The Australian brush-tailed possum (*Trichosurus vulpecula*) does not naturally occur in New Zealand. Many habitats in New Zealand have environmental conditions in which these possums can survive and reproduce. Some Australian brush-tailed possums were introduced into New Zealand in 1837 and large populations of these possums now inhabit forests, farmlands and suburban gardens — they are now considered a pest.

The absence of the Australian brush-tailed possum from New Zealand until 1837 was due to the geographic barrier of the Tasman Sea.

Competition

A species may be absent from a habitat because of **competition** from another species. One species (A) that is living in a habitat may successfully occupy the same living space and use the same resources that would be needed by a second species (B). Competition from species A would be expected to prevent species B from becoming established in this habitat.

Moving between habitats

So far, we have discussed the habitat of an organism in terms of one location. Some species, however, may move in a predictable way between very widely separated habitats on an annual (yearly) or 'once-in-a-lifetime' basis. Such animals are said to be **migratory**. Migrations are typically associated with access to food resources or to breeding sites or to movement from unfavourable seasonal conditions, such as extreme cold in winter.

Species with an annual migration in their life cycle include:

- short-tailed shearwaters or muttonbirds (*Puffinus tenurirostris*). These birds spend the Australian summer on islands in the Bass Strait where they breed, but then migrate to Japan, Siberia and Alaska for the northern hemisphere summer feeding in the seas (see figure 11.13, page 348).
- humpback whales (*Megaptera novaeangliae*). Some populations of whales spend the southern hemisphere summer feeding in Antarctic waters and courtship and mating also occurs during this time. The whales then migrate to temperate waters off the east and west coasts of Australia during the southern hemisphere winter (see figure 9.12). It is there that female humpback whales give birth to their single calves.

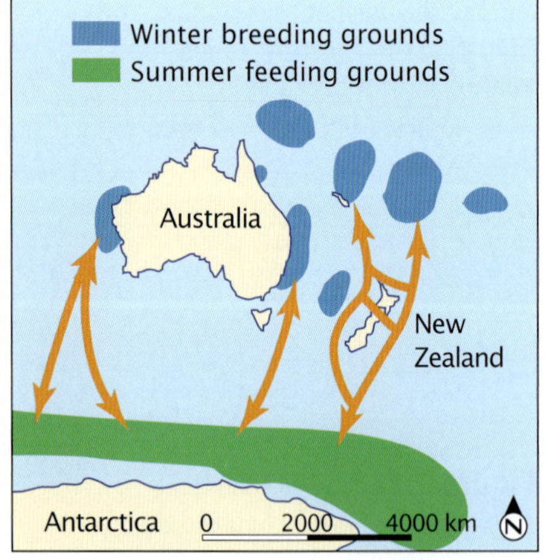

Figure 9.12 Some of the migration routes of humpback whales. Worldwide, other populations exist that migrate from Antarctica to South Africa and to South America and from the Arctic waters to the temperate regions of the Pacific and the Atlantic.

Species with 'once-in-a-lifetime' migration in their life cycle include:

- short-finned eels (*Anguilla australis*). These eels spend most of their lives in freshwater lakes and rivers in eastern and south-eastern Australia. When they are sexually mature, the eels move to the sea and travel several thousand kilometres to their spawning grounds in deep ocean waters near New Caledonia where the eels breed and die. Larvae hatch from the eggs and are carried by ocean currents back to the Australian coast, where the larvae change into elvers. The elvers swim up rivers and move into lakes where they feed and grow over several years. Once mature, this next generation of eels will make their 'once-in-a-lifetime' migration to the same ocean spawning grounds.
- bogong moths (*Agrotis infusa*). The larvae (caterpillars) feed in southern Queensland and northern New South Wales in winter months, pupate in early spring and emerge as moths. The moths migrate to the Snowy

Mountains and shelter in rock crevices at altitudes above 1500 metres, where they remain in a non-feeding and low energy use state (known as **diapause**) (see figure 9.13). In autumn, the moths return north and mate. The eggs laid by female moths before they die will hatch as the next generation of larvae.

Figure 9.13 **(a)** Bogong moths (*Agrotis infusa*) in the Snowy Mountains over winter. Since they do not feed, how might they survive? **(b)** The migration route of the moths

KEY IDEAS

- A habitat identifies the general area where an organism lives. A microhabitat identifies a particular area within a habitat.
- Within a habitat, the resources needed for living are not uniformly distributed.
- The range or distribution map of a species is the geographic area that encloses all the habitats where organisms of that species live.
- Migratory organisms move between widely separated habitats, either annually or once in their life cycle.

QUICK-CHECK

1. Identify the following statements as true or false.
   a. A species with a large range must be common.
   b. Humpback whales migrate annually during their lifetimes.
2. What is meant by the term 'habitat'?
3. In their habitat, where do tammar wallabies find (a) shelter and (b) food?
4. List three different general kinds of habitat.

Technology as a tool in biology

Keeping records of the kinds of vegetation in various terrestrial habitats that extend over large areas and recording any changes in the vegetation is not a task that can be done by one person in the field with a notebook. Nor can such a person readily monitor the movements of animals within a habitat over time.

The use of technology is an important tool in the study of habitats, particularly those that cover large areas. In the following sections, we will explore how analysis of satellite images enables scientists at a distance from a habitat to monitor the types of vegetation in that habitat and to detect changes over time. In addition, other applications of technology enable scientists to monitor movements of animals in their habitats.

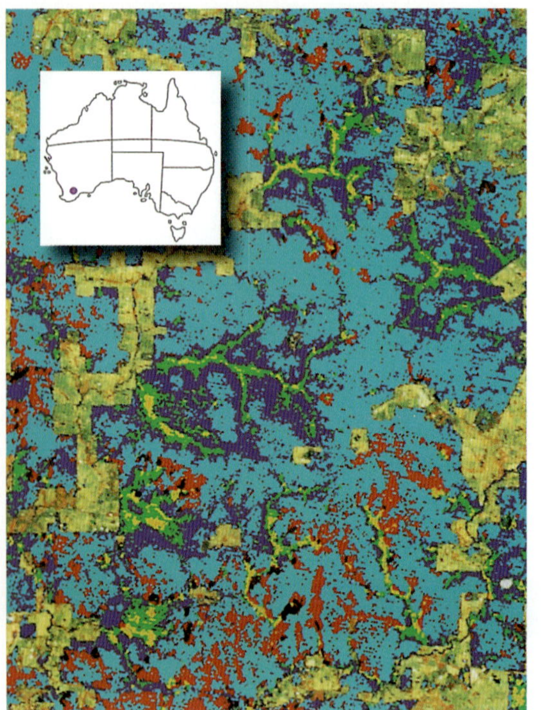

Monitoring vegetation in large-scale habitats

As we saw on page 260, the survival of tammar wallabies in their habitat depends on heartleaf poison plants that provide daytime shelter. Actions to conserve the tammar wallaby include recording the location of heartleaf thickets and monitoring the changes in the distribution of these plants as old thickets die and as new thickets regenerate after programs of controlled burning.

How can the distribution of heartleaf thickets over the large area of south-west Western Australia be measured and changes monitored? Direct field observations, which are adequate for studies of small areas, cannot be used over an area as large as this. Conventional techniques, such as aerial photography, are of no use because they cannot distinguish heartleaf plants from other kinds of plant. The answer? Use of **remote sensing** systems on an orbiting satellite, such as Landsat, which regularly passes over open forest areas in south-west Western Australia. The satellite carries sensors that can detect the distinctive 'signatures' of different kinds of vegetation, including heartleaf. Data collected by these satellite sensors are converted to colour-coded images that show the distribution of heartleaf thickets, as seen in figure 9.14. Further details about the sensors on Landsat are given in the Biotech box that follows.

Figure 9.14 Landsat image of a region of south-west Western Australia. In this colour-coded image, heartleaf thickets appear in red. Other colours represent different types of vegetation; for example, the dark blue regions show forest, blue regions show shrubland, green regions show woodland, and yellow regions show other plants.

BIOTECH

Eyes in the sky

We see objects in the world around us by detecting the radiation that they emit or reflect as visible light (wavelengths 0.4 to 0.7 μm). Just about every object also emits some infra-red (IR) radiation with wavelengths ranging from 0.8 μm to about 100 μm. We cannot see this radiation but we can feel it as heat. Satellites gather data about the Earth through their **thematic mapper (TM)** sensors, which detect radiation that is emitted or reflected from the land or sea surface. The intensity of radiation in particular spectral bands is recorded and each band provides specific information, as described in table 9.1.

One of the latest Earth-observing satellites is Landsat 7, which was launched in April 1999, and orbits at an altitude of 705 km. Every 99 minutes, Landsat circles the Earth, gathering data from 185 kilometre-wide strips of the Earth's surface. These data are relayed to a collection station on Earth. On each orbit, Landsat reorients itself slightly so that, after 16 days, it commences a repeat pattern of orbits.

Data from various bands are combined to produce false-colour computer images, such as that shown in figure 9.14 above.

Table 9.1 Data collection by Landsat thematic mapper sensors. Which band might differentiate heartleaf from other plants?

Band	*Range (μm)*	*Application*
1 visible blue-green	0.45–0.52	differentiates between soil and vegetation, as well as deciduous and evergreen trees
2 visible green-yellow	0.52–0.60	detects green reflectance of healthy vegetation
3 visible red	0.63–0.69	detects absorption by chlorophyll and can distinguish various plant species
4 near infra-red	0.75–0.90	surveys biomass
5 middle infra-red	1.55–1.75	measures vegetation moisture
6 thermal infra-red	10.40–12.60	measures surface temperature
7 middle infra-red	2.08–2.35	measures plant heat stress and soil moisture

ODD FACT

The most accurate means of monitoring movements of large animals uses **global positioning system (GPS) tracking** which depends on receivers in tags fitted on animals that pick up signals transmitted from special satellites.

Monitoring animal movements within habitats

Some animals move over very large distances within their habitat. For terrestrial habitats, radio tracking devices can be used to identify the positions of animals as they move within their habitats in search of food or a mate. Several **radio tracking** or **telemetry** techniques are currently in use. These techniques are restricted to animals of sufficient size to allow a particular tracking device to be fitted to them without it interfering with their normal activities.

The oldest technique for tracking animals in terrestrial habitats is **very high frequency (VHF) radio tracking** that was introduced in 1963. The procedure involves capturing the animal to be tracked and fitting it with a transmission unit built into a collar or the like. The scientist who tracks the animal carries a unit that comprises an antenna, a power source and a signal receiver with an audio or visual indicator of signal reception. The scientist moves around the habitat on foot or in a vehicle and picks up signals transmitted by the animal in question.

Later, by fitting tags with more powerful transmitters to animals, **satellite tracking** became possible. This technology enables animals to be tracked both across large areas of land and in the ocean. Each tag transmits a unique code to a special satellite network that can fix the position of the tag (and the tagged animal) on Earth. This information is then sent from the satellite to a ground station from where scientists can obtain the results. Unlike VHF tracking, satellite tracking does not require people to be in a habitat in order to track an animal and the technique is particularly useful for marine animals. The accuracy of results from satellite tracking is limited to several hundred metres. In the case of marine animals, the signal does not pass through sea water and is detected only when the animal comes to the ocean surface.

ODD FACT

A satellite tag on a male white shark, Bruce, transmitted data for an eight-month period. After being tagged in March 2004, the shark moved from South Australian waters to southern Queensland waters and then back to Bass Strait, covering more than 6000 kilometres over that time.

Scientists from CSIRO Marine Research use satellite tracking with marine animals including white sharks. A shark is captured, a tag is pinned to its dorsal fin and the shark is then released (see figure 9.15). Each time a tagged shark comes to the surface and its dorsal fin breaks from the water, the tag transmits to the satellite network a code unique to that tag and that shark. This means that each shark can be tracked within its habitat.

(a) (b) (c)

Figure 9.15 **(a)** White shark is manoeuvred onto a platform at the back of a boat **(b)** The shark is tagged. **(c)** Researchers from CSIRO check the position of the shark fitted with a satellite tracking tag.

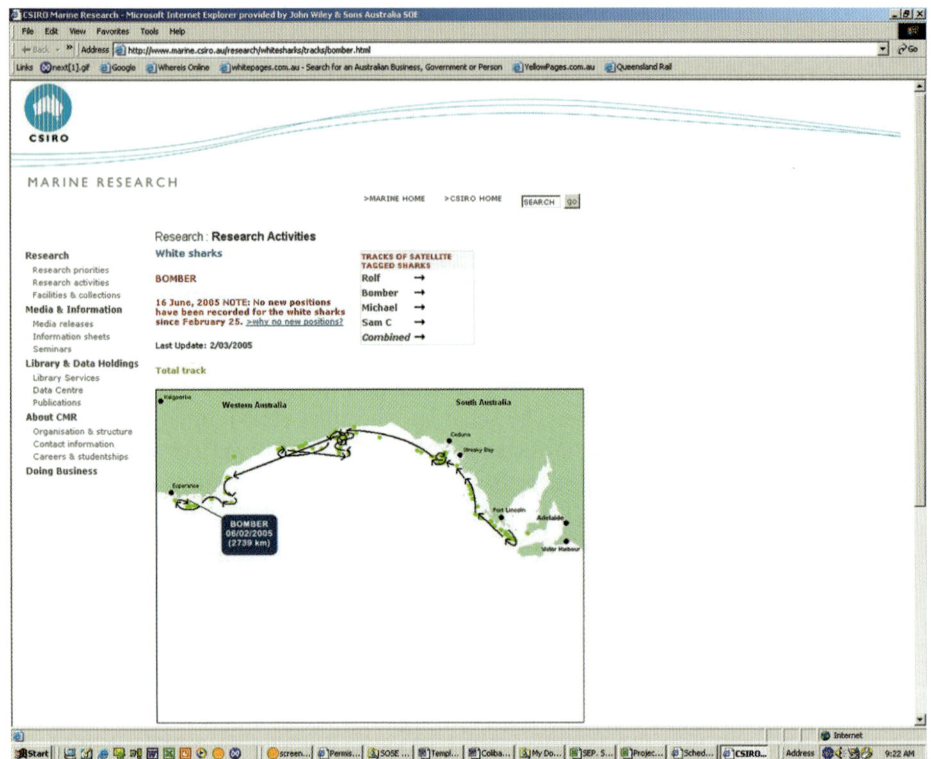

Figure 9.16 Bomber, a white shark, was tagged on 11 November 2004 off North Neptune Island and its movements were monitored for several months.

Does a shark stay in the same region where it was captured and tagged? Satellite tagging gives the answer. Figure 9.16 shows the web page on which the track of the white shark known as Bomber was recorded. This website at www.cmar.csiro.au records where, on which date and at what time of day tagged sharks surfaced most recently, and their movements through their habitat.

As well as tags that transmit data on the position of a white shark when the shark surfaces, another kind of tag is used to gather data on a shark's movements and its environment — this tag is called a **pop-off archival tag (PAT)**. These tags are programmed to collect data every few minutes over a given period. At the end of that time, the tag detaches from the shark and pops to the surface. Once at the surface, the tag transmits its data store to the satellite network.

Data collected by PATs include information about the shark's environment, such as water temperature and light levels, and data on the shark's movements, such as the depth at which the shark is swimming at particular times. Data from PATs help scientists to understand the daily behaviour of white sharks. It has been found, for example, that, during the day when they are near seal colonies, white sharks swim closer to the water surface than at night when they swim at deeper levels. Can you suggest a possible explanation?

KEY IDEAS

- Satellite technology is used to monitor vegetation in habitats that cover large areas.
- Various tracking devices have been developed to monitor the movement of animals in their habitats.
- Tags attached to animals can record data on animals' interactions with their environment.

QUICK-CHECK

5 Identify one use of satellite images in monitoring habitats.
6 Identify two different animals whose movements in their habitats have been monitored using satellite tracking.
7 What is a PAT?

BIOLOGIST AT WORK

Dr John Holland — tracking threatened wildlife with technology

John Holland is a senior lecturer in the Institute of Natural Resources at Massey University in New Zealand. He has established a wildlife tracking program that combines small transmitters with orbiting and geostationary satellites to track the daily movements of endangered terrestrial, aquatic and bird species in Australia, Africa and New Zealand. He says: 'It's a joy to work with people from many disciplines and cultures. We get together to combine our skills and passions and make a concrete contribution to saving endangered species'.

African elephant (*Loxodonta africana*)

Despite all we have learned over the years, wildlife managers have repeatedly stressed that they still know little about the precise movements of elephants and that this sort of information is essential to ensure sound management practices.

We are tracking an African elephant using a global positioning system (GPS) unit that is incorporated into a collar and attached to the elephant. The GPS fixes on a location point by responding to signals from a geostationary satellite that covers the whole of Africa, Australia and the Middle East and a constellation of orbiting satellites.

Figure 9.17 African elephant

Saltwater crocodile (*Crocodylus porosus*)

Up until about 1960, Australia's estuarine crocodiles were almost hunted to extinction. After being declared a protected species in 1971, their numbers in the Northern Territory increased from 3000 to 70 000. As the populations recover they are moving back into their ancient habitats that have become increasingly occupied by humans. This has led to ongoing management problems for park managers who still know little about the extent of their territories and home ranges. Due to the crocodiles' aggressiveness and the hostile environments in which they live, they have proved difficult to study using conventional methods and satellite tracking overcomes this problem.

Figure 9.18 The tracking device attached to the crocodile transmits data on its location and movements.

A transmitter was attached to the nuchal shield of a 4.2 metre crocodile (see figure 9.18) at a location near Darwin. The data collected are being sent on a daily basis to a computer at Massey University for analysis and dissemination to the Australian team members (you can read more about this project at www.croctrack.org.nz).

New Zealand falcon (*Falco novaeseelandiae*)

The New Zealand falcon is a threatened species under mounting pressure. To save the species, we need to know more about their home ranges and where they go if they leave their natal area.

A 17-gram solar powered platform transmitter terminal was harnessed to a 540-gram female bush falcon. The location signals have shown the falcon to be a very sedentary animal operating within a three-kilometre radius for most of the year. Over a three-year period that it has been tracked, its nest site has changed each year but has remained within its home range. The tiny transmitter has not affected the hunting or breeding behaviour of the falcon at all.

Figure 9.19 Dr John Holland checks the tiny transmitter before releasing and tracking the falcon's movements.

Who lives in a habitat?

Members of one species do not live in isolation in a habitat. Members of the different species that share one habitat form the living **community** of that habitat. Are these different species in competition for the limited resources of a habitat? Read the following case study.

CASE STUDY

Life in a lake community

If you visit a freshwater lake habitat, such as Cherry Lake in the Victorian suburb of Altona, you may notice many bird species busily feeding. These different species feed in different ways, often on different foods, and they occupy different feeding spaces, as shown in table 9.2.

Chestnut teal ducks and black swans both feed on submerged plants in Cherry Lake. Do these two species make the same use of this food resource? Chestnut teal ducks feed within a preferred range of water depths. They do not feed in deeper water, where they cannot reach the submerged vegetation, or in very shallow water. Black swans (see figure 9.20) have longer necks than the ducks and feed in deeper waters. The two species make different use of the food resource in their habitat by feeding in waters of different depths.

A **resource use** graph can show how the black swans and the chestnut teal ducks use different resources in a habitat. In such a graph, the amount of use of a resource is plotted against a relevant variable. Figure 9.21 plots the degree of use of food resources against water depth for both species.

Where various species differ in their use of food and space resources available in a habitat, they are said to occupy different **niches** in the community.

Figure 9.20 Black swans use their long necks to find food in deeper waters.

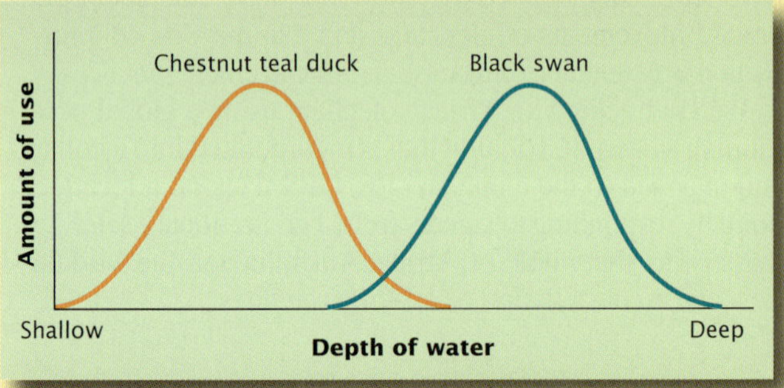

Figure 9.21 Use of food resources plotted against water depth for chestnut teal ducks and black swans

Table 9.2 Feeding habits of bird species living in a freshwater habitat

Species	Food	Method of feeding	Location of food
maned duck *Chernonetta jubata*	vegetation	grazing	grassy areas beside lake
sharp-tailed sandpiper *Calidris acuminata*	worms and small crustaceans	probing with beaks	mud at lake edge
chestnut teal duck *Anas castanea*	vegetation at bottom of lake	'bottom-up' grazing	shallow water
black swan *Cygnus atratus*	vegetation at bottom of lake	grazing	deeper water
pink-eared duck *Malacorhynchus membranaceus*	small plants and animals	filtering	water surface
musk duck *Biziura lobata*	small organisms near lake bottom	diving	deepest water
white-faced heron *Ardea novaehollandiae*	fish	thrusting with beak	shallow water
Australian pelican *Pelecanus conspicillatus*	fish	scooping with bill	deep water
welcome swallow *Hirundo neoxena*	insects	catching on wing	above lake surface
clamorous reed warbler *Acrocephalus stentoreus*	insects	catching with bill	in reeds at lake edge

Definition of niche

Table 9.3 Terms (and meanings) describing feeding niches of animal species in a community

Where it lives
terrestrial: on the ground
aquatic: in water
marine: in seas
arboreal: in trees
fossorial: digging underground
What it eats
herbivorous: plant-eating
frugivorous: fruit-eating
carnivorous: animal-eating
insectivorous: insect-eating
omnivorous: plant- and animal-eating
When it feeds
diurnal: by day
nocturnal: by night
crepuscular: at dawn and dusk

The term 'niche' refers to the 'way of life of a species'. Some other short definitions of niche include:

- the role or profession of a species in a community
- the way of life of an organism
- the status or role of an organism in its habitat.

An analogy of the concept of niche may be made using the human population of a city. Within the city, groups may be identified in terms of their different ways of life or roles, such as hardware retailer, educator, surgeon and security guard. Each group can be subdivided into more specific roles. For example, the educator role includes: pre-school teacher, secondary biology teacher and university Arts lecturer.

In the same way, the many species living in a biological community have different ways of life or niches. For each animal species, a description of its niche might include where it lives and feeds, what it eats, when it feeds and the environmental conditions it tolerates. Some terms that describe feeding niches of animal species in a community are shown in table 9.3.

A niche may be expressed in words, for example: 'its niche is that of a leaf-eating herbivore that feeds by day in the canopy of a tropical rainforest'. The same niche could be shown with several resource use graphs using variables such as 'time of day', and 'height above ground'.

Structural features equip various species to occupy different niches in a habitat. For example, in birds that feed on mudflats, their different beak lengths allow them to exploit different feeding niches, as shown in figure 9.22.

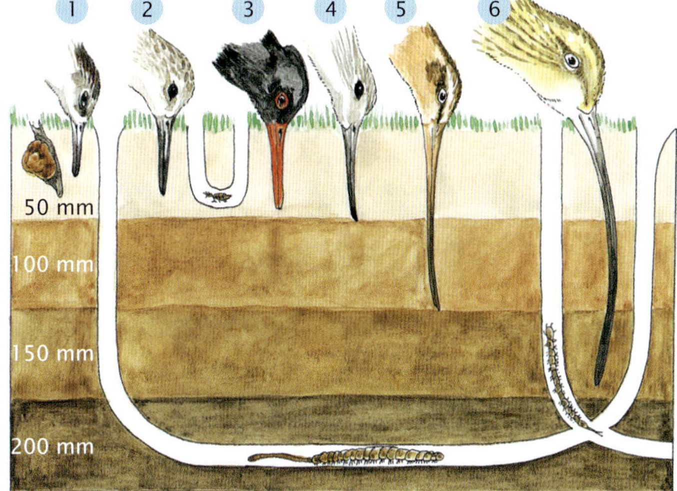

Figure 9.22 Different beak lengths — different foods!

Two into one will not go

Because they are not in direct competition for the same food and space resources at the same time, different species in a community in a natural habitat typically have either zero or a low degree of **niche overlap**.

In general, two or more different species cannot occupy the same niche in the same community for an extended period. These species will be in competition with each other and, over time, the species that can more efficiently exploit the food and space resources will displace the other from the habitat. Some native species have been eliminated from a region after exotic species were introduced which occupied the same niche. In semi-arid habitats in Australia, introduced rabbits competed more efficiently than small native herbivorous mammals for the same food and space resources and displaced these native plant-eating species.

Many species are able to occupy a broader niche than they actually do. In general, what stops them from extending their niche is competition from other species.

KEY IDEAS

- The niche of a species identifies its way of life or role in a community.
- A niche can be identified in terms of the degree of use of resources.
- Competition is avoided when various species in a community occupy niches that differ in how they use resources, such as food and space.
- A high degree of niche overlap is expected to lead to the elimination of one of two species.

QUICK-CHECK

8 Two species of mammals living in the same habitat both eat insects. Based on this information alone, can it be concluded that the mammals have identical feeding niches? Explain your answer.
9 List three different feeding niches in a freshwater lake habitat.
10 Would it be reasonable to predict that several species living naturally in a natural community would have a high degree of niche overlap?
11 In which of the following situations would competition between species be expected to be greatest?
 a species C and D feed on grass seeds by day
 b species E feeds on grass seeds and species F feeds on seed in a tree
 c species G feeds on grass seeds by day and species H feeds on grass seeds at night

Environment: what's it like there?

The physical, chemical and biological conditions that exist in a habitat make up the environment. The environment at the surface of a dry claypan during the day might be described as 'hot, brightly lit, wind-swept and arid'. However, the environment in a cavity 30 centimetres below the surface of the claypan that is the summer habitat of a water-holding frog might be described as 'cool, dark, still and moist'.

An environment may also be defined as:

- a collective term for the conditions in which an organism lives
- the sum total of external influences acting on an organism
- the external surroundings in which an organism lives and which are influenced by abiotic and biotic factors (see below).

Environmental factors

The various factors that produce the particular conditions in a habitat are called environmental factors. They can be divided into **biotic** and **abiotic** factors (see figure 9.23).

- Biotic or living factors relate to other living organisms in the environment and include factors such as presence of predators, presence of parasites, competition between members of one species.
- Abiotic factors are non-living factors relating to aspects of soil, water, light, temperature, topography, as, for example, light intensity, slope of land, aspect of slope (whether the slope faces north, south, east or west), rainfall.

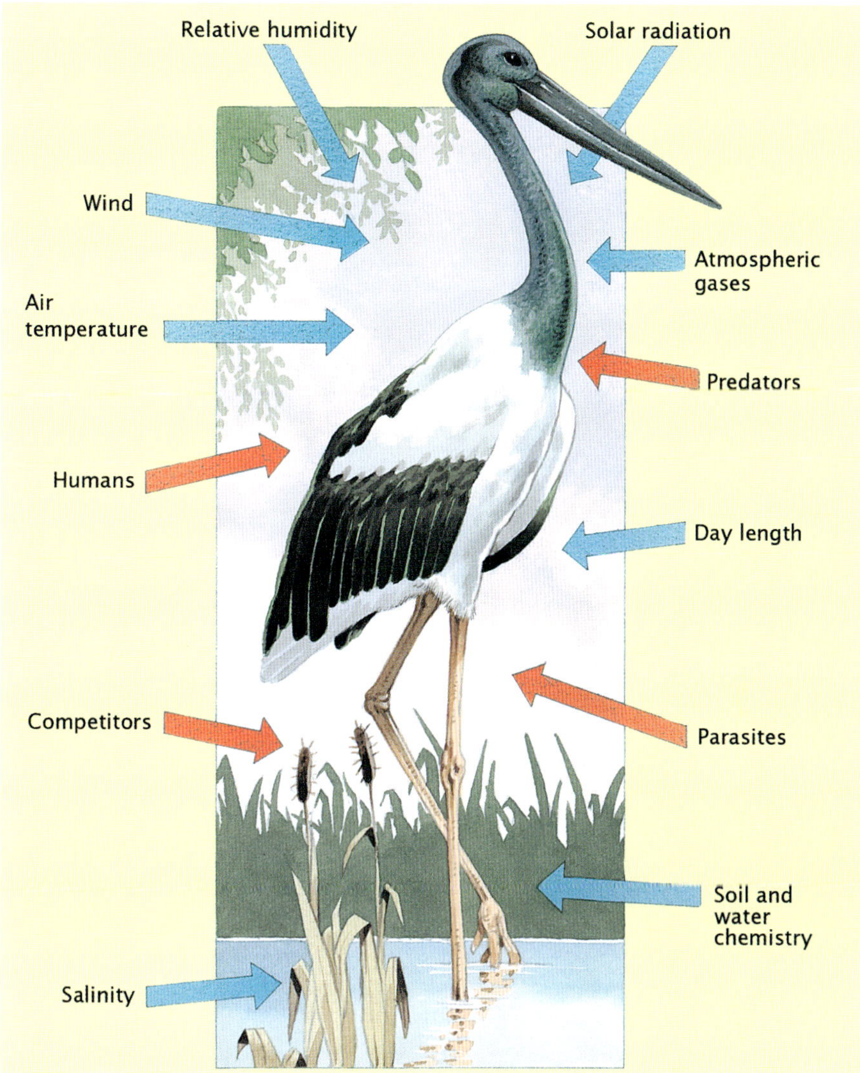

Figure 9.23 Biotic and abiotic environmental factors that contribute to the environment in the habitat of the black-necked stork (*Ephippiorhynchus asiaticus*), also known as the jabiru. Biotic factors are shown with red arrows and abiotic factors are shown by blue arrows. Would each factor remain constant over a year?

Terrestrial and aquatic environments differ in some key environmental factors. Rocky coastlines are exposed to strong wave action and currents (see figure 9.24) — these abiotic factors are absent from environments such as forests. Table 9.4 shows water-related environmental factors in terrestrial and aquatic habitats.

Figure 9.24 Part of the rocky ocean coast at Botany Bay National Park. The algae and shelled animals attached to the rocks are exposed to powerful breaking waves and backwash. Abiotic factors relating to wave action are important in shaping this environment but are absent from terrestrial environments.

Figure 9.25 Sea water absorbs light.

Table 9.4 Some water-related environmental factors in different habitats

Terrestrial habitat, such as open forest	*Aquatic (marine) habitat, such as coastal sea*	*Aquatic (freshwater) habitat, such as river*
• annual rainfall	• salinity of water	• rate of current flow
• seasonal pattern of rainfall	• water temperature	• pH of water
• relative humidity	• dissolved oxygen levels	• water temperature
• rate of run-off	• dissolved nutrients	• dissolved oxygen levels
• soil water	• tidal movements	• dissolved nutrients
• rate of drainage	• wave action	• frequency of flooding
	• clarity of water	• input of waste water

The environment in a habitat is produced by the action and interactions of several environmental factors. When biologists study a habitat, they cannot study all environmental factors. Instead, they select factors that appear to be most important. The mountain brush-tailed possum (*Trichosurus cunninghamii*) and the greater glider (*Petauroides volans*) live in eucalypt forest habitats in highland areas of south-eastern Australia. In one study, biologists showed that the numbers of possums and gliders were greatest in parts of the forest habitat that had the most trees with hollows.

Air and sea water compared

Air and sea water have very different properties and, in turn, terrestrial and marine habitats have very different environments. These different environments have a direct influence on the types of organisms that have evolved and can live and reproduce successfully in each type of environment. Table 9.5 outlines some of the significant differences between air and sea water and summarises how these differences can affect the communities of living organisms that live in the terrestrial or marine environments.

Figure 9.26 **(a)** A mixed phytoplankton bloom suspended in sea water. **(b)** Barnacles, such as this grey barnacle (*Tetraclitella purpurescens*), are filter feeders. Notice how the buoyancy of the water supports the delicate feeding arms of the barnacle. **(c)** When exposed to air, barnacles are tightly enclosed inside hard plates.

Table 9.5 Differences between air and sea water and some consequences for life in terrestrial and marine habitats

Difference	Effect on living community
Light absorption: sea water absorbs light much more strongly than air. Depending on water clarity, light entering water reaches to depths up to 200 metres before it is completely absorbed. Light absorption is selective, with most of the violet and red-orange light being absorbed within the first few metres.	At depth, there is no light (see figure 9.25). Photosynthetic organisms, such as phytoplankton, algae and plant life, are restricted to living in the zone of water at or near the ocean surface (see figure 9.26a and e); this is called the light or the **photic zone**.
Density: sea water is much more dense than air — in fact, more than 800 times more dense.	Unlike air, sea water contains large populations of tiny organisms that float or are suspended in the water, such as zooplankton and phytoplankton. Unlike air, sea water has many kinds of animals that feed by filtering suspended organisms or food particles from the water; these are called **filter feeders** (see figure 9.26b, c and d).
Gravity: the effect of gravity on organisms is far greater in air than in sea water and water provides significant buoyancy.	Compared with terrestrial organisms of the same mass, organisms living in marine habitats require less supporting tissue or supporting organs (such as exoskeletons).
Oxygen content: sea water holds much less oxygen than air and even less as temperature increases. air: 210 mL oxygen per litre; water (0°C): 8.0 mL per litre; water (20°C): 5.4 mL per litre	Marine organisms have developed extensive features to supply sufficient oxygen to them, such as respiratory surfaces with a high surface-area-to-volume ratio.
Sound conduction: sea water conducts sounds faster than air.	Sound is typically more useful than vision for marine animals for purposes of communication — think about whale songs!
Risk of desiccation: air has a variable and relatively low water content.	Unlike aquatic organisms, terrestrial organisms are at constant risk of water loss (**desiccation**).
Electrical resistivity: sea water has a much higher electrical resistivity than air.	Unlike any terrestrial animals, some marine animals have organs that can detect minute electrical activity from other animals in the water, such as sharks and rays (see figure 9.26f).

(d) Another filter feeder is the sea cucumber (*Pseudocolochirus axiologus*) with its feeding arms extended. **(e)** Brown algae (*Phyllospora comosa*) depend on photosynthesis for their survival. They are restricted to areas of the sea where light can reach. **(f)** This grey reef shark (*Carcharhinus amblyrhynchos*) has electro-receptive organs located in various areas of its snout. The pores are filled with a jelly that conducts electricity. What role do the electro-receptors serve?

Describing environmental conditions

An environment can be described in **qualitative** terms, such as 'warm and humid' where 'warm' refers to air temperature and 'humid' refers to the water vapour content of the air. An environment may also be described in **quantitative** terms when a numeric value is stated, such as, 'The air temperature is 23°C'.

When biologists measure environmental factors in a habitat, they typically express the values in quantitative terms. This is because quantitative values have precise meanings and are universally recognised. Measuring values of environmental factors can be achieved with a variety of instruments (see table 9.6).

Some environmental factors can be measured without sophisticated instruments, for example, water clarity can be measured by lowering a black-and-white disc into water on a weighted string and recording the depth (length of string) at which the disc can no longer be seen.

Table 9.6 Some instruments used in measuring some environmental factors. Several of these instruments can be combined into a single device.

Environmental factor	*Value units*	*Instrument*
air or water temperature	degrees Celsius	thermometer
wind speed	metres per second	anemometer
relative humidity of air	percentage humidity*	hygrometer
air pressure	hectopascals	barometer
light intensity	lumen per square metre	light meter
soil water content	gram water per gram dry soil	moisture meter
water salinity	gram per litre	conductivity meter

* The amount of water in air at a particular temperature compared with what it can hold.

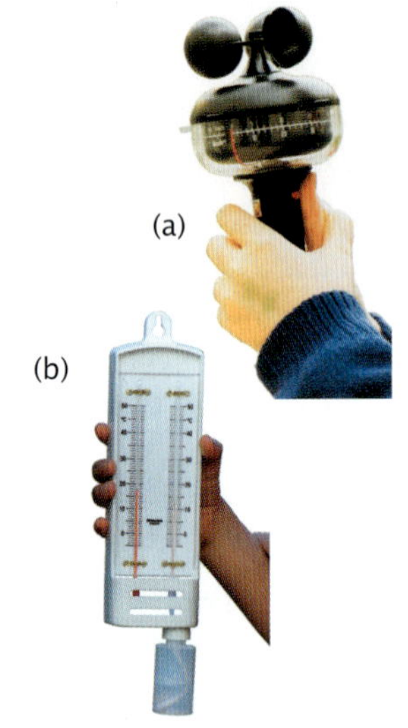

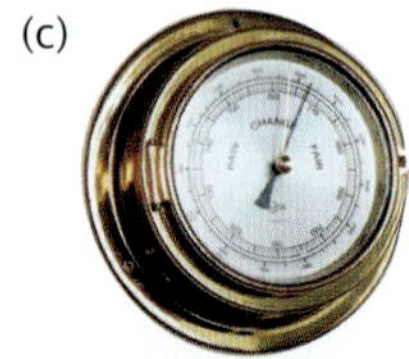

Figure 9.27 Some instruments that are used in measuring environmental factors in a habitat: **(a)** an anemometer measures wind speed **(b)** a hygrometer measures relative humidity **(c)** a barometer measures air pressure.

Comparing habitats

Let us look at the value of some environmental factors in two different marine habitats, one at high latitudes (closer to one of the poles) and the other tropical (close to the equator). Table 9.7 compares the two habitats. The same environmental factors can be identified in both habitats but the factors have different values in each. The range of values for the environmental factors listed is greater in the higher latitude marine habitat. This indicates that the tropical marine habitat is a more stable environment.

Table 9.7 Values of some environmental factors in two different marine habitats. Can you identify some quantitative values? some qualitative values?

Environmental factor	*High latitude marine habitat*	*Tropical marine habitat*
water temperature — winter — summer	 0°C 15°C	 24°C 28°C
day length	very much longer in summer than in winter	slightly longer in summer than in winter
solar radiation in winter	25 per cent of level in summer	75 per cent of level in summer
water clarity	1.5 to 6.0 metre	7.5 to 30 metre
wind speed	0 to 32 km/hr	24 to 32 km/hr
wave height	0 to 3.0 metre	0.6 to 1.8 metre

Environmental factors interact to create particular environmental conditions. For example, low rainfall, high temperatures and high wind speeds together produce a high evaporation rate and a dry environment — drier than one with identical rainfall and temperature but zero wind speed.

Micro-environments

A person standing in a forest habitat looks at a thermometer and says: 'It is 17°C'. This statement does not necessarily apply to all locations in that habitat. Just as many microhabitats can be identified in a habitat, many **micro-environments** can be identified within an environment.

Marked differences in environmental conditions may be seen, even in adjacent sites. For example, on a sunny day with a dry breeze, the temperature and humidity on the exposed upper surface of a leaf can differ from those on the lower surface. The two sides of the leaf are different micro-environments. In the Arctic in winter, the air temperature above snow-covered ground may be many degrees colder than the temperature within the snow layer (see figure 9.28).

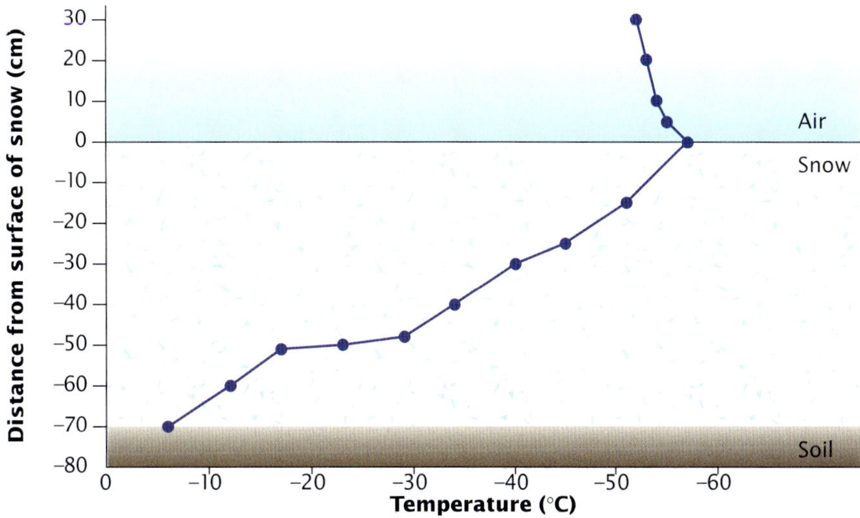

Figure 9.28 Temperature variations in air and snow. Is the temperature higher at the air–snow boundary or the snow–soil boundary?

ODD FACT

Some bacteria have a very high temperature tolerance. The bacterial species known as *Sulfolobus acidocaldarius* survives temperatures in boiling hot sulfur springs. This species of bacterium dies from cold at temperatures below 55°C.

Tolerance range

The distribution map of each species is affected by environmental factors. Every organism has a **tolerance range** for environmental factors, such as temperature, desiccation, oxygen concentration, light intensity and ultraviolet exposure. A tolerance range identifies the variation within which organisms can survive. Figure 9.29 shows the tolerance range for a fish species in terms of water temperature. The extremes of this range are the tolerance limits.

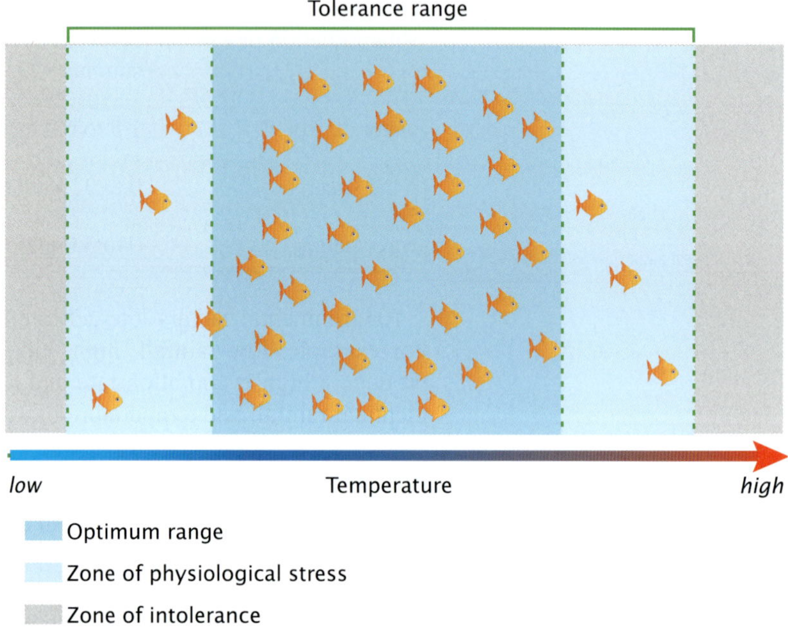

Figure 9.29 Tolerance range in terms of temperature for a fish species. Notice that, as water temperature moves closer to the tolerance limits, fewer fish are found. This is the so-called zone of physiological stress. What happens beyond the tolerance limits?

If an environmental factor has a value above or below the range of tolerance of an organism, that organism will not survive unless it can escape from or somehow compensate for the change. In some species, migration is one such escape behaviour and, in others, it is retreat underground.

Tolerance ranges differ for species and are influenced by structural, physiological and behavioural features of organisms. For example, the cold tolerance of various mammals is influenced by structural features such as fur density, shape of the body (see figure 9.30) and extent of insulating fat deposits, and by their behaviours, such as hibernating ('coping-with-it' strategy).

Figure 9.30 Apart from their differences in fur thickness, these two fox species differ in their ear sizes. The smaller ears have a lower surface-area-to-volume (SA:V) ratio than the larger ears. Which fox would be better able to conserve its body heat and tolerate lower temperatures? Why? (Surface-area-to-volume (SA:V) ratio is discussed in chapter 2, see page 26.)

Reef-building coral polyps live in warm, clear shallow seas, but if the temperature of the water falls below 18°C, the polyps die. What is the lower end of the temperature tolerance range for coral polyps?

Who survives where?

The kinds of organism found in various habitats are determined by their ability to survive in the prevailing environmental conditions. For example, organisms that can survive in the littoral (inter-tidal) zone are determined by the limits of their tolerance to exposure to the air and resistance to desiccation.

The upper limit of distribution of a barnacle species (see an example in figure 9.26b, page 272) is determined by its tolerance to desiccation. Barnacles tolerate periods of exposure to air by sealing themselves off from the air. They are found in regions of the littoral zone but cannot survive in the splash zone since this region is never submerged and barnacles feed under water.

Any condition which approaches or exceeds the limits of tolerance for an organism is said to be a **limiting factor** for that organism. Terrestrial and aquatic environments can differ in terms of their limiting factors.

Table 9.8 shows environmental factors that influence which kinds of organism can survive in various habitats. Those species that can survive under certain environmental conditions have tolerance ranges that accommodate those conditions.

Table 9.8 Examples of limiting factors in various habitats. Only one example of a limiting factor is given for each environment.

Habitat	*Key environmental factor*	*Comment*
floor of tropical rainforest	light intensity	Low light intensity limits the kinds of plants that can survive.
desert	water availability	Limited water supply means that only plants able to tolerate desiccation can survive.
littoral zone	desiccation	Exposure to air and sun limits types of organism that survive.
polar region	temperature	Low temperatures limit the types of organism that are found.
stagnant pond	dissolved oxygen levels	Low dissolved oxygen levels limit the types of organism that can live there.

In the next sections, we look at the key features of the Australian environment then examine survival of organisms in particular environments.

KEY IDEAS

- External agents whose actions produce a particular habitat are called environmental factors.
- Environmental factors include both biotic and abiotic factors.
- Environmental factors may be described in qualitative or quantitative terms.
- Various instruments enable environmental factors to be measured.
- Within a habitat, micro-environments can be identified.
- Tolerance range refers to the extent of variation in an environmental factor within which a particular kind of organism can survive.
- Organisms vary in their tolerance ranges.

QUICK-CHECK

12 Identify the following as true or false.
   a Algae (seaweeds) can be found in habitats at the ocean depths.
   b 'It is 35°C' gives a qualitative measure of an environmental factor.
   c Wind speed is an example of an abiotic environmental factor.
13 One plant species (P) grows equally well in soils with both high and low dissolved salt concentrations. A second plant species (Q) dies if the salt concentration rises above a low value. Which species has the broader tolerance range in terms of the salinity of the soil?
14 How would you measure the relative humidity of air?
15 What instrument would you use to measure the salinity of water?
16 What instrument would you use to measure wind speed?
17 List two differences between sea water and terrestrial habitats.

The Australian environment

Dorothea McKellar captured the essence of the Australian environment in her poem, *My Country*, published in 1904. The second verse is as follows:

> I love a sunburnt country,
> A land of sweeping plains,
> Of ragged mountain ranges,
> Of droughts and flooding rains.
> I love her far horizons,
> I love her jewel-sea,
> Her beauty and her terror —
> The wide brown land for me!

Sunburnt country, droughts, flooding rains ... Let us look at the Australian environment in terms of the following factors:

- average maximum temperatures — how hot is it?
- average annual rainfall — how dry is it?
- reliability of annual rainfall and incidence of drought
- soil nutrient levels.

How hot is it?

The average daily temperatures over Australia in January are shown in figure 9.31a and those for July in figure 9.31b. When would you plan a holiday to central Australia? Notice that the maximum temperature varies considerably across the country and that there is also a seasonal difference, with maxima in mid-summer (January) being much higher than in mid-winter (July).

Average annual rainfall in western Tasmania is more than 1200 mm but evaporation is less than 800 mm. Average annual rainfall in parts of central Australia is less than 200 mm while evaporation rate is more than 4000 mm.

How dry is it?

The Australian continent has a lower rainfall than any other inhabited continent. Almost two-thirds of the continent is arid (less than 250 mm annual average rainfall) or semi-arid (between 250 mm and 500 mm). Figure 9.32a shows the average annual rainfall over Australia.

In southern Australia (south of 35°S), most rainfall occurs in the winter months and the summer is typically a period of drought. In northern Australia (north of 25°S), rainfall occurs principally in the period from November to April, called 'the wet', and the period from May to October is known as 'the dry'.

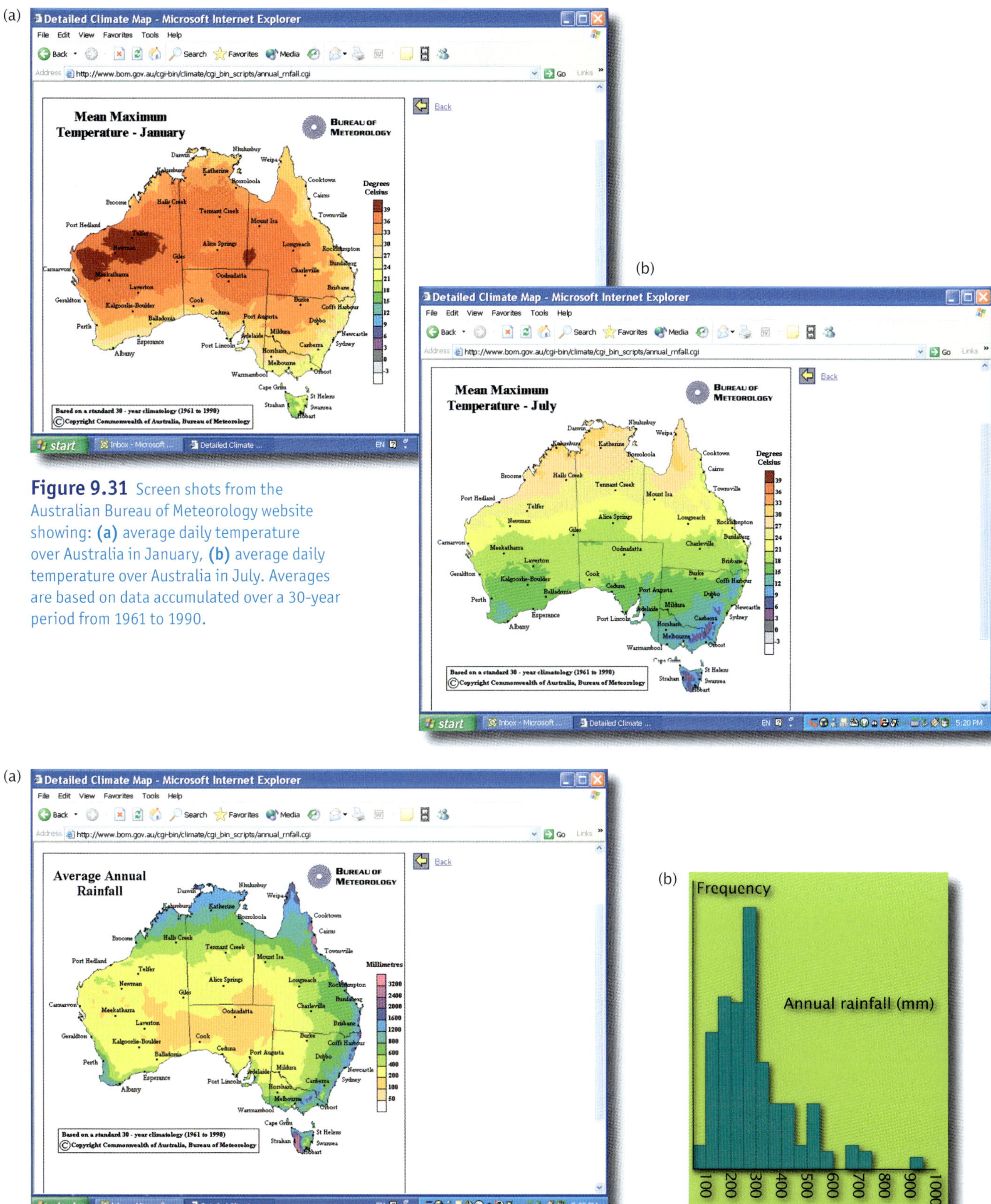

Figure 9.31 Screen shots from the Australian Bureau of Meteorology website showing: **(a)** average daily temperature over Australia in January, **(b)** average daily temperature over Australia in July. Averages are based on data accumulated over a 30-year period from 1961 to 1990.

Figure 9.32 **(a)** Screen shot from the Bureau of Meteorology website showing average annual rainfall. Values are based on the average over the 30-year period from 1961 to 1990. **(b)** Graph showing how often annual rainfalls of different sizes have occurred over a 113-year period at Alice Springs. What is the most common annual rainfall? (*Source: Ecos*, No. 73, Spring 1992, p. 16)

The driest area in Australia is in the region of Lake Eyre in South Australia where the annual average rainfall is about 100 millimetres.

Because inland Australia is a region of low rainfall, high temperatures and very high evaporation rates (more than 3200 mm per year), no surface water exists, apart from periods after rare heavy rain when temporary creeks and ponds are created. These bodies of water do not last long because of the high evaporation rates. They are termed **ephemeral** (= transient, temporary).

The rainfall of inland Australia is not reliable and commonly deviates from the average. The extremes of rainfall are well illustrated in parts of the arid 'red heart' of Australia. These areas are dominated by red soils with sparse vegetation. On occasions, however, the rainfall in these arid areas is well above average. Heavy rains bring about a remarkable change by triggering a sudden and rapid burst of germination of dormant seeds and the arid red heart of Australia is transformed into green (see figure 9.33).

When the rainfall in an area fails, regardless of whether this is an area of high or low rainfall, **drought** is said to exist. Drought conditions occur when the total monthly rainfall over a period of at least nine to twelve months falls below the lowest ten per cent of all values recorded. Figure 9.34 shows the incidence of drought for Australia.

Figure 9.33 An area of desert around Uluru in central Australia is quickly transformed by rain.

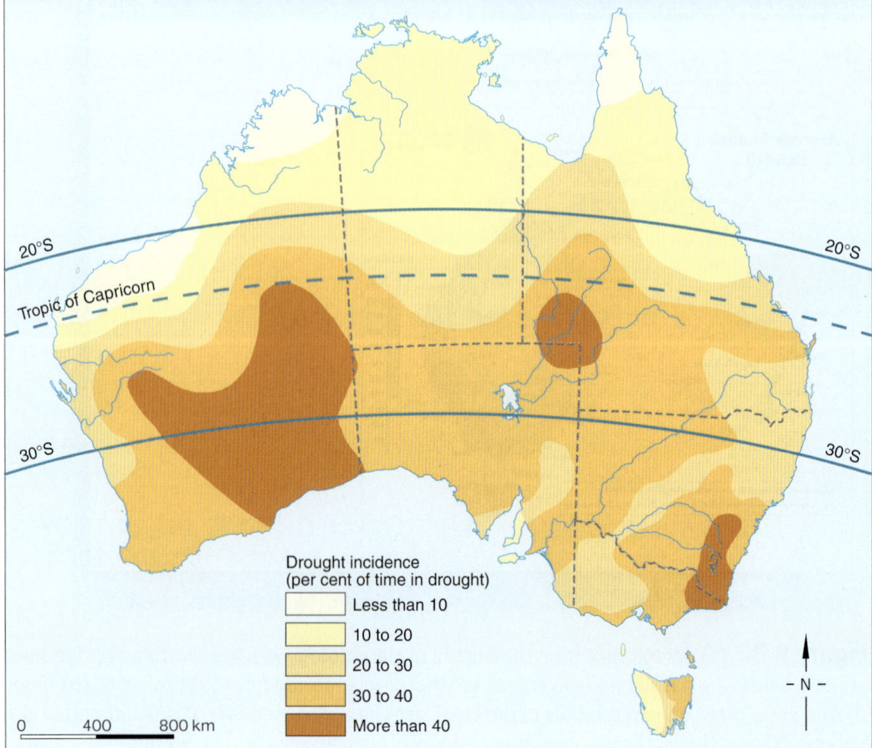

Figure 9.34 Incidence of drought over Australia. Where are drought conditions more likely to occur — in northern Australia or in central Australia?

ODD FACT

The annual average rainfall and number of rainy days per year for some Australian cities is as follows: Sydney 1214 mm over 148 days; Melbourne 655 mm over 143 days; Brisbane 1151 mm over 123 days; Cairns 2032 mm over 155 days; and Alice Springs 285 mm over 44 days.

El Niño and La Niña

The weather in inland Australia is highly variable from one year to the next. Every two to seven years, events over the Pacific Ocean affect the Australian weather. These events are known as **El Niño** and **La Niña**.

During an El Niño event, rainfall over Australia is lower than average and many areas, including inland Australia, are affected by drought. This effect results from weaker easterly trade winds that cause the surface temperature of the Pacific Ocean to rise so that the rainfall pattern is located over the central Pacific (see figure 9.35). In contrast, during a La Niña event, trade winds are stronger, the rainfall pattern is shifted to the east and higher than average rainfall, tropical cyclones and floods occur in eastern Australia.

Figure 9.35 Key features of an El Niño event. Note that the major rainfall zone occurs over the Pacific Ocean.

The onset of an El Niño event can be predicted by changes in the SOI (Southern Oscillation Index) that measures the difference in air pressure between Tahiti and Darwin ($SOI = P_T - P_D$) (see figure 9.36). When this index becomes negative, drought conditions are likely over eastern Australia; when this index becomes positive, rainfall is likely to be above average.

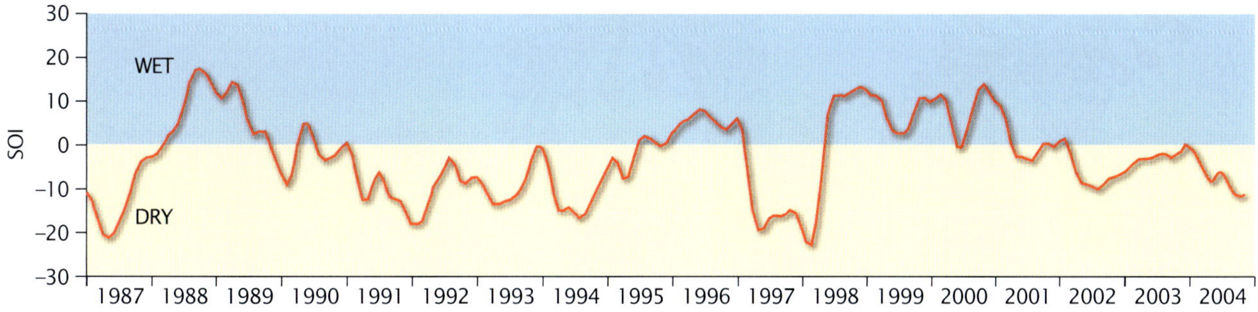

Figure 9.36 Fluctuations in the Southern Oscillation Index (SOI) in the period from 1987–2004 (*Source:* Adapted from graphs of Bureau of Meteorology, Australia)

How rich are the soils?

Much of the Australian continent consists of very ancient rocks. Soils derived from these ancient rocks have low levels of mineral nutrients. Why? Over long periods, mineral nutrients are lost from soil when they are dissolved and washed away, a process known as **leaching**. Soils are enriched again by volcanic eruptions that bring mineral-rich rocks to the surface that break down to produce high-nutrient soils. The ancient soils over much of Australia, however, have not been enriched by volcanic activity and so are depleted of mineral nutrients such as phosphates and nitrates.

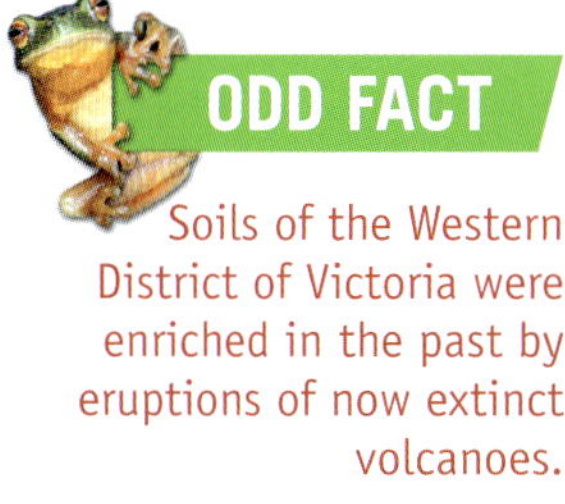

Soils of the Western District of Victoria were enriched in the past by eruptions of now extinct volcanoes.

Plants of several genera, such as genus *Gastrolobium* (for example, heartleaf poison) and genus *Acacia* (for example, mulga) have nitrogen-fixing bacteria living in nodules on their roots that enable these plants to survive on low-nutrient soils (see figure 9.37). Another example of survival on low-nutrient soils is seen in carnivorous plants, such as sundews. These plants have spade-shaped leaves with thin projections, each with a sticky tip (see figure 9.38) that trap small insects. Insects are a source of nitrogen compounds such as proteins, that can be used by these plants and enable them to thrive in nitrate-deficient environments.

Figure 9.37 Nodules on roots of clover plant where nitrogen-fixing bacteria live

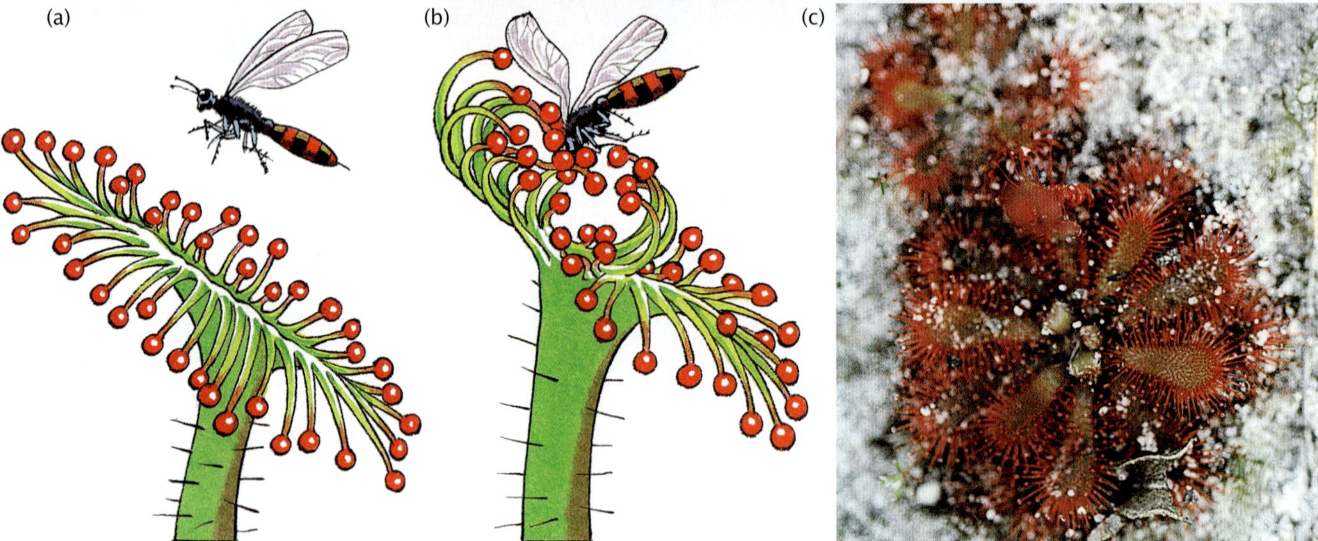

Figure 9.38 **(a)** Part of a sundew plant showing the leaves in the receptive state before stimulation **(b)** The plant after stimulation, showing the tentacles curved inwards. The plant traps small insects with its sticky-tipped projections that surround each leaf **(c)** Plant of one sundew species (*Drosera spathulata*)

KEY IDEA

- Much of the Australian environment is distinguished by high summer temperatures, low and erratic rainfalls and mineral-poor soils.

QUICK-CHECK

18 Identify each of the following as true or false.
   a Permanent surface water is present in arid Australia.
   b Arid regions are defined as having less than 250 mm annual average rainfall.
   c Rainfall over central Australia is low but reliable.
   d Some regions of Australia are commonly in drought.

Surviving in the Australian environment

Australia is the driest inhabited continent on Earth and it has been arid for millions of years. Many of the native plants and animals of this country have evolved in arid and semi-arid conditions of inland Australia and successfully survive and reproduce in this environment. These plants and animals possess features that enable them to cope under these conditions with the demands of life. Any genetically controlled features that may assist survival and reproduction of organisms in their specific environments are called **adaptations**. Adaptations for survival in the conditions prevailing in a particular environment may be structural, physiological or behavioural features. For example, adaptations for life in a hot, arid environment include those that minimise water loss:

- structural features, such as a thick waxy cuticle on the leaves of desert plants
- physiological features, such as the production of very concentrated urine
- behavioural features, such as sheltering by day and feeding by night.

Features that assist survival and reproduction must be identified in relation to a particular set of environmental conditions. A feature that is an advantage in one environment may be a disadvantage in a different environment. Gills are an excellent feature for survival in an underwater environment but are a death sentence on land.

Plants: what grows where?

Shrubs and trees are both woody plants, but shrubs are typically less than two metres high, while trees are taller. Trees are classified as tall (height greater than 30 metres), medium (10 to 30 metres) and low (less than 10 metres).

Let us look first at different types of vegetation that are found in Australia. To do this, we do not identify all the individual plant species. Instead, we identify vegetation type or structure, as for example, forests, grasslands or woodlands.

Both woodlands and forests are characterised by the presence of trees. Forests, however, differ from woodlands in terms of light penetration or how much of the sky is covered by the upper canopy of foliage (seen, for example, when looking up from below). For a forest, the coverage is from 30 per cent to 100 per cent of the sky but, for a woodland, the coverage is less than 30 per cent. Forests that have a coverage of 70 per cent to 100 per cent are termed **closed forests**, as for example, rainforests, while those with a lesser coverage are termed **open forests**, as for example, tall open eucalypt forests.

Different vegetation types can be further defined by climate, by use of labels such as tropical, temperate, semi-arid or arid preceding a vegetation type; for example, tropical rainforest. Different vegetation types are also defined in terms of the dominant family, genus or species in the plant community, for example:

- forest dominated by eucalypts
- shrubland dominated by acacia
- hummock grassland dominated by spinifex.

Rainforests cannot be defined in terms of a dominant plant genus. Why? Several hundreds of plant genera grow in rainforest and no one genus dominates.

Figure 9.39 shows the distribution of major native vegetation types in Australia.

In terms of area, the dominant vegetation type in Australia is **hummock grassland** that covers nearly 25 per cent of the Australian land surface. Hummock grasslands are found in arid inland areas of Australia and are dominated by spinifex grasses (*Triodia* spp.), as for example porcupine grass (*T. irritans*) (see figure 9.40a, page 285). Do not think of these grasslands in terms of your front lawn at home. Spinifex is a drought-resistant grass that looks nothing like the green grasses of a well watered and manicured suburban garden. Hummock grasslands are important habitats for native mammals and reptiles of the arid zone.

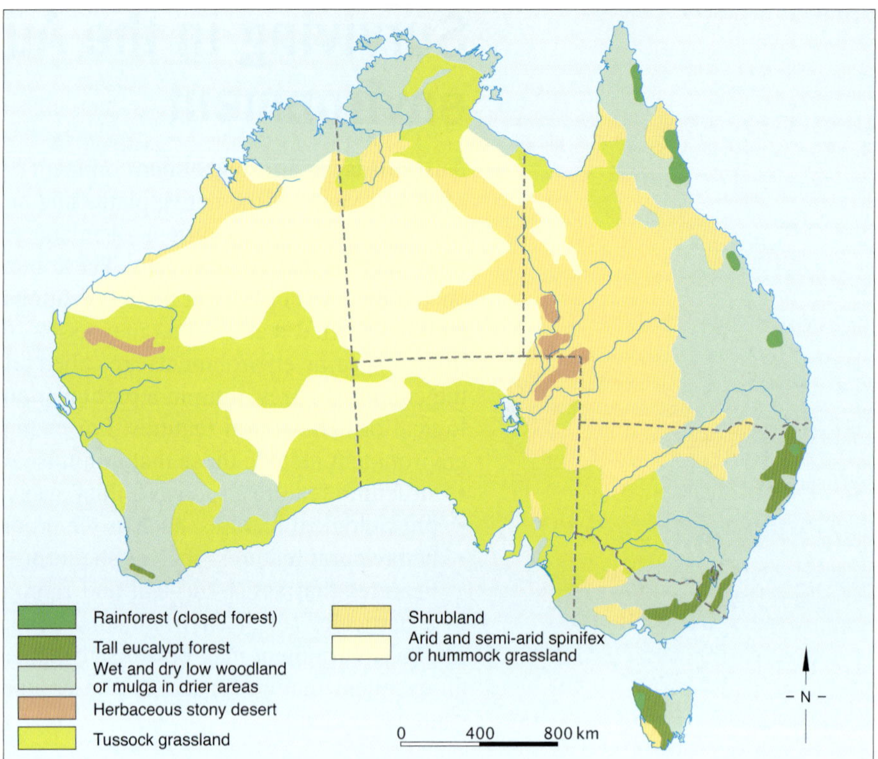

Figure 9.39 Major vegetation types in Australia

What does chenopod mean? The two genera Atriplex *and* Maireana *belong to the family Chenopodiaceae and so the plants of both genera are termed 'chenopods'.*

Acacia shrublands are also found in arid inland areas of Australia (see figure 9.40b). Acacia shrublands cover more than eight per cent of Australia's land surface and are typically dominated by mulga (*Acacia aneura*), one of the many hundreds of *Acacia* species that are native to Australia.

A third vegetation type found in Australia's arid and semi-arid areas are **chenopod shrublands** that cover about six per cent of Australia's land surface. Chenopod shrublands are dominated by saltbushes (belonging to the genus *Atriplex*) and bluebushes (belonging to the genus *Maireana*). Both groups of plants are salt-resistant and drought-resistant.

Another major vegetation type covering about nine per cent of Australia's land mass is **eucalypt woodlands**. These woodlands occur in areas of higher rainfall than the vegetation types mentioned above.

Vegetation types that cover very small areas in Australia include rainforests (see figure 9.40c) and heath. Rainforests, for example, now cover less than one per cent of Australia's land surface.

Compare figures 9.40a, b and c. You can see that the vegetation found in the hot and arid inland of Australia is very different from the vegetation of the high rainfall areas of Australia. In an arid environment the limiting factor is availability of water and, as a result, the vegetation is sparse, individual plants are well separated and the growth rates of individual plants are low. In contrast, in a rainforest, plant growth is luxuriant, plants are crowded and growth rates are fast. In a rainforest, the limiting factor is light (see figure 9.41).

Patterns in plant distribution

Why do rainforests occur in some regions of Australia while other regions of the country are covered by acacia shrubland? Why do hummock grasslands dominate such a large area of Australia? The distribution of various vegetation types is influenced by environmental factors mainly related to climate but also by other factors including soil type (for example, clay, sand) and soil salinity.

Figure 9.40 **(a)** Hummock grasslands dominate much of inland Australia. Spinifex grass exists as circular clumps, known as hummocks that, over time, grow outwards forming larger circles. **(b)** Acacia shrublands are dominated by mulga (*Acacia aneura*). These shrublands cover just over 8 per cent of the land surface. **(c)** Rainforests grow in high rainfall areas of tropical or temperate regions of Australia. What is the major limiting factor in this environment?

Figure 9.41 Light is a limiting factor in rainforests. What is a limiting factor in arid environments?

Plants adapted to a particular environment typically have a range or distribution map that includes areas that have particular environmental conditions. For example, hummock grasslands consist of plants that are adapted to living and reproducing in the arid and the semi-arid regions of Australia. Not surprisingly, the distribution map of hummock grasslands covers the arid and semi-arid regions of Australia that have the highest summer temperatures and the lowest rainfall. Table 9.9 summarises how the distribution of vegetation types is linked to aspects of climate and soil.

Table 9.9 Patterns in the distribution of various types of vegetation in Australia

Vegetation type	*Climate*
• hummock grasslands • mulga shrublands	arid: lowest and erratic rainfall, high evaporation rates, high temperature
• chenopod shrublands	arid and semi-arid: low rainfall, high temperature, salty or alkaline soils
• tussock grasslands • dominated by *Astrebla* spp.	semi-arid: annual rainfall between 200 and 500 mm, clay soils
• tropical grasslands dominated by *Sorghum* spp.	tropical: summer monsoons and winter drought
• mallee woodlands	temperate: intermediate rainfall, poor soil
• eucalypt forests	temperate: high rainfall, poor soil
• rainforests	tropical or temperate: high and reliable rainfall, rich soil

Plants: surviving in arid environments

Plant species that survive and reproduce in arid environments in Australia, such as porcupine grass (*Triodia irritans*) and mulga (*Acacia aneura*), show adaptations that equip them to

- maximise water uptake
- minimise water loss
- produce drought-resistant seeds.

Maximising water uptake

ODD FACT

Desert plants ofNorth, Central and South America are succulent and fleshy and are known as cacti (members of the family Cactaceae). In contrast, Australia has no native cacti.

The part of a plant that takes up water is the root system. In arid areas of Australia, some trees growing along dry creek beds produce long, unbranched roots that penetrate to moist soil at or near the watertable. Once moisture is reached, the major root branches and forms lateral roots. Plants that produce these deep roots are called **water tappers** and their major root can grow to depths of 30 metres. The part of the root that is located in the upper dry soil is covered by a corky waterproof layer that prevents water loss.

Other plants growing in arid regions develop extensive root systems that spread out horizontally, far beyond the tree canopy but just below the soil surface. In this case, the plant takes up water from an extensive area around it.

Reducing water loss

Transpiration is the loss of water vapour by evaporation from moist surfaces inside the plant. This loss of water vapour occurs through pores, known as **stomata** (singular: stoma), that are typically present on the lower surface of plant leaves. The higher the wind speed and the higher the temperature of the leaf, the greater the rate of water loss.

Some adaptations seen in plants that reduce water loss include:

- reduced density of stomata on leaf surfaces — fewer stomata per unit area of leaf surface means that water loss by transpiration is reduced
- restriction of period of opening of the stomata to night time when temperatures are lower — lower temperatures mean less transpiration

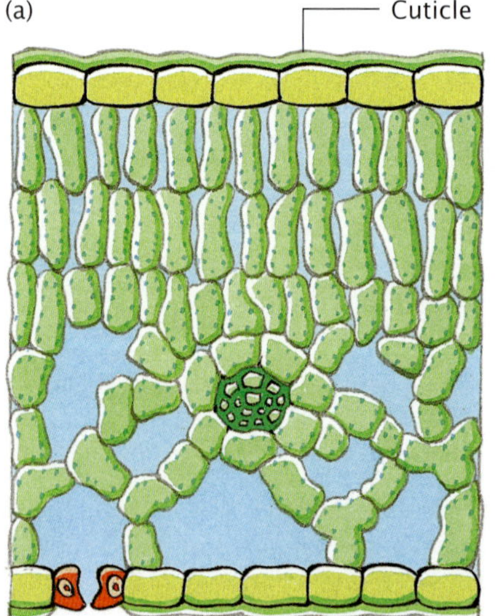

Figure 9.42 **(a)** Transverse section (TS) of a 'typical' leaf with stomatal openings flush with the lower leaf surface. Guard cells surrounding the stoma are shaded red. **(b)** Transverse section of a leaf from a plant with its stomata sunk below the leaf surface. Note also the hairs on the lower leaf surface and the thick cuticle on the upper surface.

- location of stomata in pits below the leaf surface rather than at the surface — sunken stomata create a region of higher humidity that slows water loss
- presence of a thick waxy cuticle on the leaf surface — this waterproofs the leaf surface so that loss of water vapour can occur only via the stomata
- presence of hair on the leaf surface — this slows the airflow over the leaf surface and reduces the loss of water vapour
- small leaves — this reduces the area from which water vapour can be lost
- glossy leaves — reflect the sun's heat, stay cool and reduce water loss
- infolding of leaves — this slows the airflow over the leaf surface and creates a region of higher humidity within the leaf.

Figure 9.42a shows a typical leaf with its stomata flush with the leaf surface. Contrast this with figure 9.42b that shows a leaf with adaptations, such as sunken stomata and the presence of hairs that minimise water loss.

Producing seeds for survival

Populations of some herbaceous flowering plants can survive in arid regions of Australia. These plants germinate from seeds, then flower and produce new seeds in a very short period. Plants that complete their life cycles in just two to three weeks are said to be *ephemeral*.

Because they produce drought-resistant seeds, populations of ephemeral plants can survive in arid regions. The outer coats of the seeds of these plants contain a water-soluble chemical that inhibits seed germination. So, dry conditions = no germination! When heavy rains fall, this chemical is dissolved away and the seeds then germinate to produce seedlings. Shortly after, the new plants produce flowers in a synchronised display (see figure 9.43). The plants soon die, but not before they have produced seeds that will lie dormant until the next heavy rains.

ODD FACT

Plants have different life spans. **Annuals** germinate, flower, set seed and die over one year. **Biennials** germinate and grow in the first year and flower, set seed and die in their second year. **Perennials** grow and reproduce over many years.

Figure 9.43 A magnificent display of flowers of ephemeral plants in the Australian desert. What event caused this blooming of the desert? Populations of these plant species exist most of the time as seeds, not plants!

Mulgas: trees of the arid inland

Acacia shrublands of the arid inland of Australia are dominated by mulga (*Acacia aneura*) that can exist either as trees or as small shrubs (refer back to figure 9.40b, page 285). Mulga trees have many features or adaptations that equip them for survival in arid conditions (see figure 9.44).

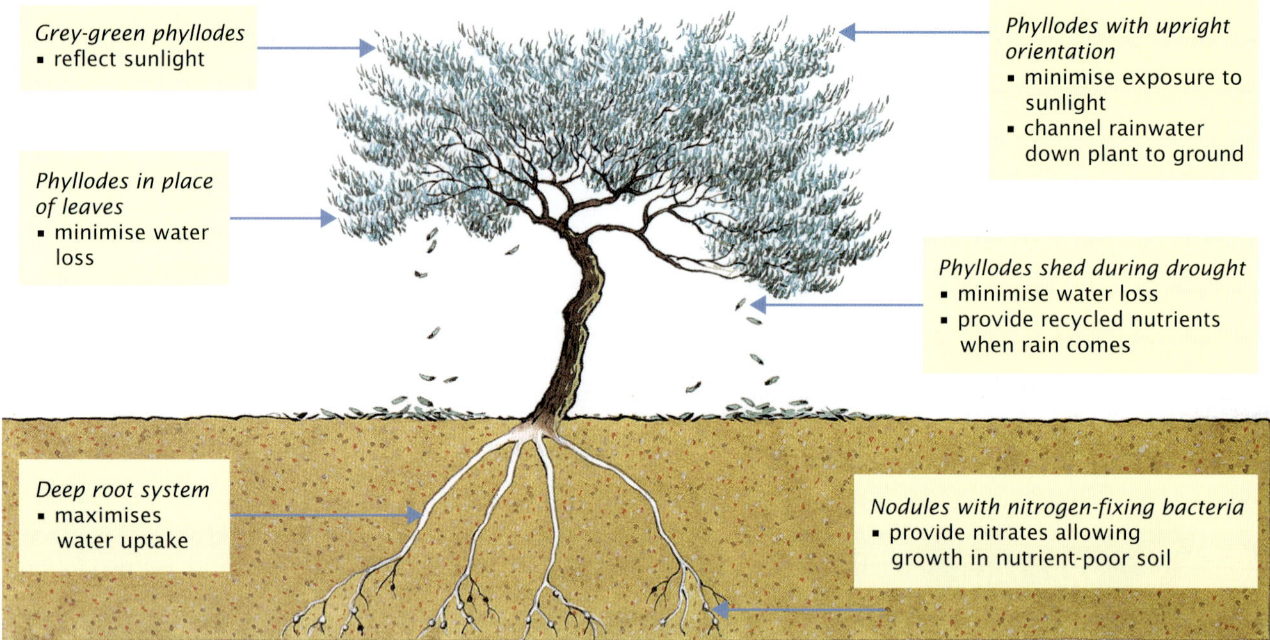

Figure 9.44 Some of the adaptations of the mulga tree. The vertical orientation of sparse foliage of mulga ensures that the little rain that falls is directed to the roots of the plant.

The root system of a mulga tree is concentrated around the base of the tree. When rain falls, it is caught by the upward-pointing leaves of this tree and funnelled down the branches to the centre of the tree. From there, the water falls to the ground around the trunk where the root system is most concentrated.

Mulga trees grow in regions where the rainfall is low and unreliable because they are drought-resistant and can survive a year or more without water. In dry years, a mulga tree does not produce any flowers. If, however, heavy rains fall in the summer, a mulga produces flowers and, if rains occur in the following winter, seeds are formed. The seeds germinate to produce seedlings the following summer and require rain to survive. Notice that a pattern of rainfall over three seasons (summer-winter-summer) is required for a new generation of mulgas to be produced. This pattern of rainfall occurs during a La Niña event.

Saltbushes

Soils of some areas of the hot, arid inland contain high concentrations of salt. Many species of salt-tolerant plants live in this environment, such as species of saltbush (*Atriplex* spp.) (see figure 9.45) and bluebush (*Maireana* spp.). These plants are also drought-resistant.

Figure 9.45 The saltbush (*Atriplex cinerea*) is salt-tolerant and drought-resistant.

Saltbushes grow in soils that are too salty for many other plant species. They can survive because they excrete the dissolved salt that is taken up by their roots from cells in their leaves. As a result, the leaves of saltbushes are covered in fine salt crystals. As well as being an excretory product, these salt crystals reflect the sun's heat and contribute to keeping the plants from overheating.

Saltbushes have structural adaptations to conserve water. Their leaves:

- have sunken stomata
- are covered in hairs
- are oriented so that they expose a minimal surface to the sun's rays.

Saltbushes produce seeds that have high concentrations of salt in their outer coats and this salt prevents germination. Saltbush seeds germinate only after the salt has been washed out after heavy rainfall. As soon as the salt inhibition is removed, the seeds germinate and new seedlings quickly become established. The salt inhibition of germination means the next generation of saltbush plants appears in times of good rainfall when their chance of survival is maximised.

Australian fauna: what survives where?

Different habitats provide different environmental conditions. Animal species that are regarded as 'successful' in a particular environment are those in which:
(1) individual animals survive to reproductive age and
(2) mature animals reproduce to give rise to sufficient numbers of offspring to ensure survival of the next generation.

Animals: surviving in arid environments

The presence of free water in arid and semi-arid areas is often temporary. Creeks exist only during the wet season and, in the dry season, water courses become dry creek beds. When rains fall, ephemeral (short-lived) lakes and pools are created. How do species that depend on free water survive the long dry periods between the shorter temporary wet periods?

Survive by flight

Some species cope with lack of surface water by emigrating from the drought-affected areas to areas where lakes and rivers exist. For example, banded stilts (*Cladorhynchus leucocephalus*) live near salt lakes in inland Australia and rely on these lakes for the brine shrimps that are their main food source (see figure 9.46). When one salt lake dries up, these birds simply fly to another salt lake.

Many animal species and all plant species, however, *cannot* use a 'get-up-and-go' strategy in periods of drought.

Figure 9.46 A group of banded stilts. What strategy do they use when their salt lake habitat dries up?

Figure 9.47 **(a)** The trilling frog (*Neobatrachus centralis*) is so called because of its high-pitched trill. **(b)** Distribution map of the trilling frog in Australia

Survive by dormancy

Some frog species live in arid inland Australia. Frogs typically live in moist surroundings and need a body of water in which to reproduce. How do they survive long periods of drought in the inland?

Some frog species that live near and breed in ephemeral waterholes respond in an amazing manner when the waterholes begin to dry out. The frogs burrow deeply into the soft mud at the bottom of their waterholes. Once underground at depths of up to 30 centimetres, the burrowing frogs, such as the trilling frog (*Neobatrachus centralis*) (see figure 9.47), make a chamber that they seal with a mucous secretion. The frogs then go into an inactive state known as **dormancy** in which breathing rates and heart rates are minimal and energy needs are very greatly reduced. Their low energy requirements are met from their fat reserves. Read the account written by two explorers about burrowing frogs:

> One day during the dry season we came to a small clay-pan bordered with withered shrubs … It looked about the most unlikely spot imaginable to search for frogs, as there was not a drop of surface water or anything moist within many miles …
>
> The ground was as hard as a rock and we had to cut it away with a hatchet, but, sure enough, about a foot [30 cm] below the surface, we came upon a little spherical chamber, about three inches [76 mm] in diameter, in which lay a dirty yellow frog. Its body was shaped like an orange … with its head and legs drawn up so as to occupy as little room as possible. The walls of its burrow were moist and slimy … Since then we have found plenty of these frogs, all safely buried in hard ground.
>
> (*Source:* W. B. Spencer and F. J. Gillen, *Across Australia,* 1912)

Frogs of the inland remain buried and are protected from desiccation until the next rains come — this may be a wait of one or two years. The frogs come out of their dormant state only when soaking rains fall and soil moisture rises. Once activated, the frogs return to the surface to feed and breed in temporary pools. The completion of the life cycle is very fast. Within days of being laid, eggs undergo embryonic development, hatch and the resulting tadpoles metamorphose to produce small frogs. These new populations of frogs feed on larvae of crustaceans and insects that have also hatched from dormant eggs.

Other animal species survive extended periods of drought by sealing themselves off from the drying conditions. For example, the univalve (one-shell) freshwater mollusc (*Coxiella striata*) seals itself inside its shell by closing the shell opening with a hard lid (**operculum**). These inland molluscs must stay sealed tightly in their shells for months or years.

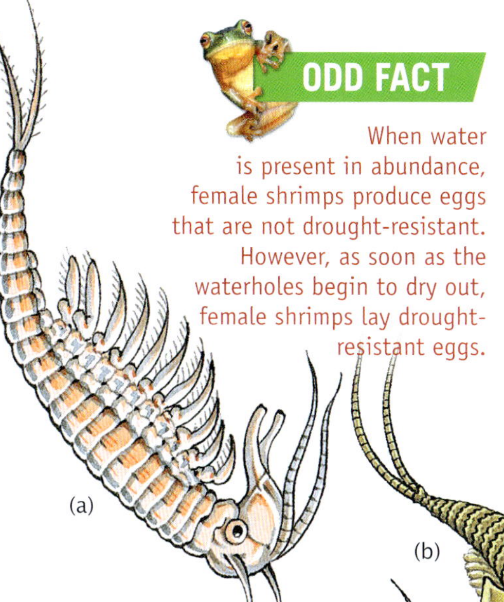

ODD FACT

When water is present in abundance, female shrimps produce eggs that are not drought-resistant. However, as soon as the waterholes begin to dry out, female shrimps lay drought-resistant eggs.

Figure 9.48 **(a)** A fairy shrimp (*Branchinella* sp.) about 3–4 cm in length. Fairy shrimps swim with their legs uppermost. **(b)** A shield shrimp (*Triops australiensis*) about 1.5 cm in length. How does this species survive drought?

Survive by the next generation

Some species are unable to survive long dry periods and, under these circumstances, adult members of the species die. In this case, the species survives through their offspring. This occurs in the case of crustacean species, such as fairy shrimps and shield shrimps. How?

As waterholes begin to dry out, fairy shrimps and shield shrimps (see figure 9.48) produce drought-resistant eggs. By the time the water has gone, all the adult shrimps are dead but the fertilised eggs that they have left can withstand desiccation for long periods. These drought-resistant eggs are in a state of dormancy and they can lie in the dust of dry waterholes for more than 20 years. When the drought breaks and the waterholes temporarily refill, embryonic development begins. Within just a few days, development is complete and the waterhole contains the next generation of shrimps that mature quickly and reproduce.

Survive without water

The fawn hopping mouse (*Notomys cervinus*) is a placental mammal found in most arid areas of Australia (see figure 9.49). The fawn hopping mouse can survive without drinking any water, even when its diet consists only of dry seeds. Water is needed for living cells. The hopping mouse, like other desert mice, can make its own water from the process of cellular respiration:

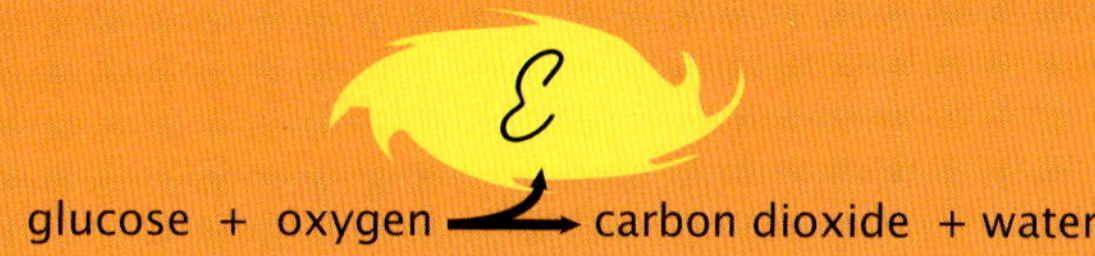

glucose + oxygen → carbon dioxide + water (with release of energy, $\mathcal{E}$)

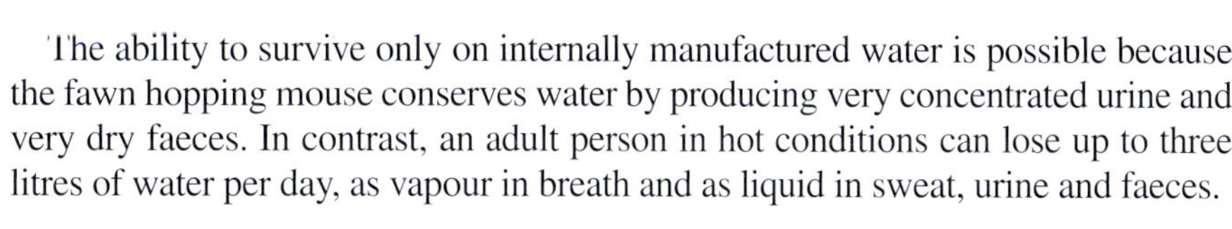

The ability to survive only on internally manufactured water is possible because the fawn hopping mouse conserves water by producing very concentrated urine and very dry faeces. In contrast, an adult person in hot conditions can lose up to three litres of water per day, as vapour in breath and as liquid in sweat, urine and faeces.

Figure 9.49 The fawn hopping mouse lives in the most arid area of Australia. It shelters by day in a moist, cool burrow and makes its own water through cellular respiration. At some times of the year, the fawn hopping-mouse can obtain some additional water from moist food items such as insects and green shoots.

KEY IDEAS

- The distribution map of various types of vegetation in Australia corresponds to areas with particular environmental conditions.
- Features of an organism can be identified as adaptations if they assist survival and reproduction.
- Adaptations may be structural, physiological or behavioural.

QUICK-CHECK

19 Identify two structural adaptations of mulga for an arid environment.
20 What is the dominant vegetation in terrestrial Australia?
21 Explain briefly how germination of the seeds of saltbush plants occurs only when conditions are favourable.
22 Identify two survival strategies for animals in arid inland Australia.

BIOCHALLENGE

A

Central Australian inland

B

Rainforest

C

Mallee region of western Victoria

D

Great Barrier Reef, Queensland

The images above show four contrasting habitats from locations around Australia:

1. a For each habitat, list three factors that would be important in determining its environment.
   b Identify each factor as either biotic or abiotic.
2. List three adjectives that could apply as part of a qualitative description of the environment in each habitat.
3. Identify one major limiting factor for the living organisms in each habitat.
4. For habitats A and D, identify one factor that would have no influence on the environment in one habitat but could be important in the environment of the other habitat.
5. In which habitat would you expect to find:
   a filter-feeding organisms?
   b large organisms without a supporting skeleton, internal or external?
   c organisms adapted to low light intensities?

CHAPTER REVIEW

CROSSWORD

Key words

abiotic
acacia shrublands
adaptation
annuals
biennials
biotic
chenopod shrublands
closed forests
community
competition
desiccation
diapause
dormancy
drought
El Niño
ephemeral
eucalypt woodlands
extreme environments
filter feeders
global positioning system (GPS) tracking
habitat
hummock grassland
La Niña
leaching
limiting factor
micro-environments
microhabitat
migratory
niche overlap
niches
open forests
operculum
perennials
photic zone
pop-off archival tag (PAT)
qualitative
quantitative
radio tracking
range
remote sensing
resource use
satellite tracking
stomata
telemetry
thematic mapper (TM)
tolerance range
transpiration
very high frequency (VHF) radio tracking
water tappers

Questions

1 *Making connections* ▸ Draw a concept map using a selection of the key words from this chapter. If necessary, add other concepts that you need to complete your map.

2 *Communicating your understanding* ▸ Dusky horseshoe bats (*Hipposideros ater*) are found by day roosting in dark caves or old mines. At dusk, the bats leave their roosts and fly to open woodland areas where they feed on insects.

a Identify a biotic and an abiotic factor in the bat's environment.

b Imagine you are the Minister for Conservation and have the task of introducing legislation to conserve dusky horseshoe bats. What measures might you suggest?

3 *Applying your understanding in a new context* ▸ Imagine that you must plan the accommodation in a zoo for rare terrestrial mammals.

a List five environmental factors about the natural habitats of these mammals that you might need to consider.

b How would your choice of factors be affected if the mammals concerned were rare aquatic mammals?

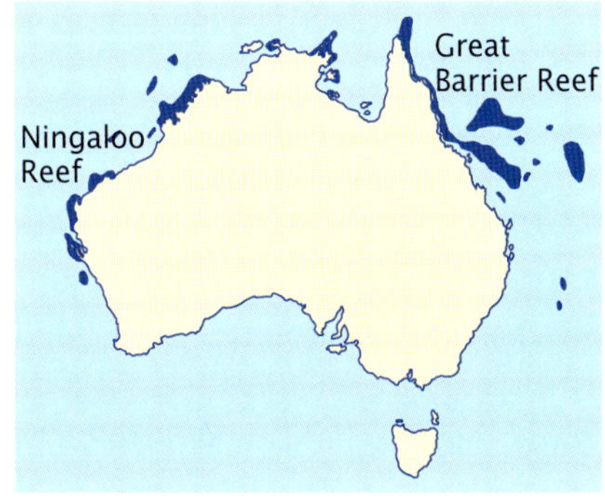

Figure 9.50

4 *Analysing and evaluating information* ▸ The map at left (figure 9.50) shows the distribution of coral reefs in Australian waters. From the information given in this map, what conclusions might be drawn about the possible physical surroundings in which coral-producing animals are able to survive and reproduce with regard to:

a water depth?

b water temperature?

5 ***Analysing information*** ▸ Figure 9.51 shows the present ranges of the western pygmy possum (*Cercartetus concinnus*) and the western grey kangaroo (*Macropus fulginosus*).

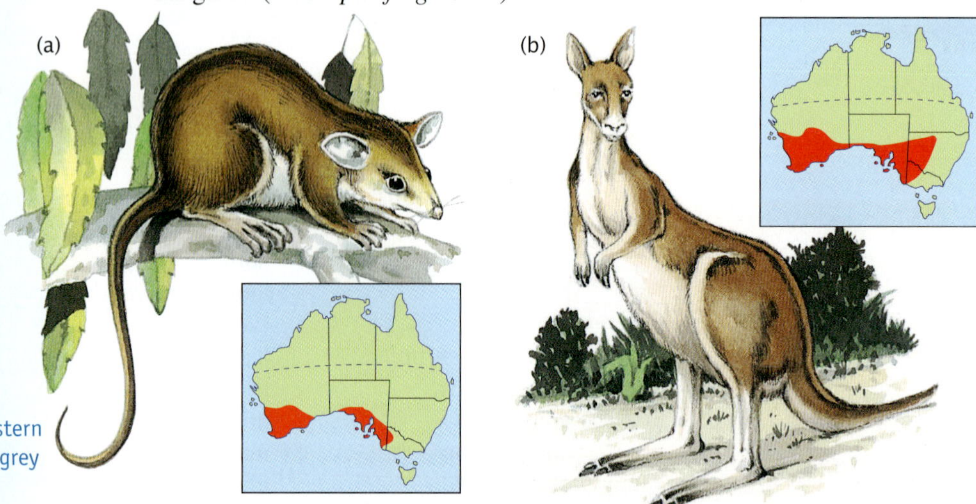

Figure 9.51 Range of **(a)** western pygmy possum and **(b)** western grey kangaroo

a Identify each range as continuous or broken.

b Suggest two reasons why the western pygmy possum is absent from the region around part of the Great Australian Bight.

c Could a person travel to any part of the range shown in figure 9.51b and expect to see western grey kangaroos? Explain.

6 ***Applying and communicating your understanding*** ▸ The community living in a rocky reef habitat along Australia's south-east coast includes populations of several species of sea star. Table 9.10 shows the typical depth range for each sea star population and its food source.

Table 9.10

Species	*Typical depth range*	*Food*
common sea star *Patirella calcar*	intertidal to 4 m	algae, molluscs, detritus
velvet sea star *Petricia vernicina*	low tide to 30 m	sponges
firebrick sea star *Asterodiscides truncatus*	20–400 m	sponges, detritus

a Velvet sea stars and common sea stars are often found in the same region of a rocky reef, just below the water surface. Do they occupy the same niche? Explain.

b The velvet sea star and the firebrick sea star both feed on sponges. Is it reasonable to conclude that one species will displace the other?

c Draw a resource use graph for the common sea star and the velvet sea star using the variable 'depth of water'.

7 ***Interpreting information and communicating ideas*** ▸ Refer to table 9.2 (on page 268) and answer the following questions:

a Do chestnut teal ducks and black swans compete for the same food resource? Explain.

b How do sharp-tailed sandpipers and musk ducks avoid competition?

c Both white-faced herons and Australian pelicans feed on fish. How can both survive successfully in the same habitat?

d How would the shapes of the beaks of sharp-tailed sandpipers and white-faced herons be expected to differ?

8 *Solving problems* ▸

a Imagine that you had to design a tag for use in tracking birds. Suggest three design features that you would consider important to incorporate into your design and give a brief explanation of your selection.

b Imagine you have to identify whether sharks in three different areas are part of the same population that moves during the year or are three separate populations. Identify how you might solve this problem.

9 *Applying understanding in a new context* ▸ Suggest an explanation in terms of tolerance range for the following observations.

a Brown trout (*Salmo trutta*) are found in cold, fast-flowing mountain streams but are absent from warm, sluggish waters.

b A plant in a large pot on a verandah was growing well in fresh potting mix and was watered regularly. The plant was moved indoors and continued to be watered regularly, but it died after a few weeks.

10 *Using scientific terminology and applying understanding* ▸ Identify one significant difference between the members of the following pairs.

a habitat and microhabitat

b oxygen content of air and of water

c terrestrial and aquatic environment

11 *Analysing and communicating ideas* ▸ Identify the type of environment, either aquatic or terrestrial, in which you would expect to find each of the following. Briefly explain your choice.

a gills as a structural adaptation

b presence of a thick cuticle on leaves

c absence of a supporting endoskeleton

12 *Using the web* ▸ Go to www.jaconline.com.au/natureofbiology/natbiol1-3e and click on the 'Species distributions' weblink for this chapter. Scroll down the page to the section headed 'Tolerance and Preference' and answer the following questions.

a What is the difference between tolerance range and preference range?

b Which is wider, tolerance range or preference range?

c Note the terms used to describe the breadth of tolerance: eury- = 'broad'; steno- = 'narrow'. So to describe, for example, temperature tolerance, the terms eurythermal and stenothermal can be used. Identify the meaning of each of the following terms:

i euryhygric

ii stenohalic.

d Examine the diagram that shows the tolerance range of two eurythermal species.

i What is the tolerance range of the species at left?

ii What is the tolerance range of the species shown at right?

e Examine the diagram that shows the tolerance range of the three steno-hygric species. What is the tolerance range in percentage water content for the middle species?

f Using the style of the diagrams above, show the tolerance ranges in terms of soil moisture content for the three plant species listed, namely coastal redwood, black oak and alder.

13 ***Discussion question*** (Note: Answer this only after completing question 12 as it requires some of the information from that question.)

Particular species are known as indicator species. This is because they can indicate the state of the local environment in the ecosystem they currently occupy and their decline in abundance can give an early warning of degradation in particular environmental conditions in an ecosystem. Examples of indicator species include corals that can signal water temperature fluctuations and degradation of water quality from runoff from land.

a Would these species be expected to be eury- or steno- in terms of their tolerance ranges of the environmental conditions for which they act as an indicator?

b Could one species be steno- for one environmental factor but eury- for another?

c Trout can only survive within a narrow range in terms of both water temperature and water quality. Could trout serve as indicator species? If so, what signal could they give?

d Convert the following true statement to plain English: Salmon are euryhalic and stenothermal.

10 Physiological adaptations for survival

KEY KNOWLEDGE

This chapter is designed to enable students to:

- relate physiological adaptations of organisms to their abilities to survive in particular environmental conditions
- identify the ways in which the nervous system contributes to the survival of animals
- understand the role that hormones play in the physiological adaptations of organisms
- understand how particular adaptations of organisms contribute to their regulation of water balance and control of body temperature.

Figure 10.1 The tarrkawarra, *Notomys alexis*, is a placental mammal that lives in the desert and semi-desert regions of central and western Australia. It is a small, nocturnal animal weighing from 20 to 40 grams, and lives in well-insulated burrows in sandy soils and established sandhills. Hence, it avoids the heat of the day. Its diet can include seeds, roots, shoots and insects. *Notomys alexis* can survive without drinking water and produces the most concentrated urine recorded for any known mammal. Note the elongated hind legs. It has a hopping gait and can travel at speeds of up to 4.5 metres per second. In this chapter, we will consider physiological adaptations that contribute to the survival of organisms, particularly relating to water balance and the control of body temperature.

Water balance in the tarrkawarra

Free water means water available for an animal to use, including to drink.

Many desert rodents (rats and mice) in Australia and other parts of the world live in extremely hot conditions with very little **free water**. They are at particular risk of becoming dehydrated and yet they survive. How do they manage their water balance? We will consider this question for the tarrkawarra (*Notomys alexis*) also called the spinifex hopping mouse, which lives in desert and semi-desert regions of central and western Australia.

Water sources for the tarrkawarra

Food

The main food for the tarrkawarra is dry seeds. The amount of water these contain depends on the **humidity** of the air in which the seeds are found. The relative humidity at night is greater than that during the day. The nocturnal habits of *Notomys* result in the animal collecting seeds at a time when the water content is likely to be at its highest. In addition, seed is stored in the burrows in which *Notomys* lives. The burrows are more than a metre deep, well insulated and have a relatively high humidity because animals huddle together there during the day. Seeds stored in burrows also have a greater water content than seeds collected from a plant. *Notomys* also eats green leafy shoots and insects when they are available but can gain weight on a diet containing dry seed only.

Metabolic or oxidation water

When carbohydrate and fatty foods are oxidised in an animal's body, the main end products are carbon dioxide and water. This oxidation water or metabolic water is used by *Notomys*.

Free water

Although free water may appear to be absent most of the time in desert environments, it can be present. There may be rainwater, and dew occurs after cold desert nights. *Notomys* does drink free water if it is available but does survive without it. A summary of the sources of water for *Notomys* is shown in figure 10.2.

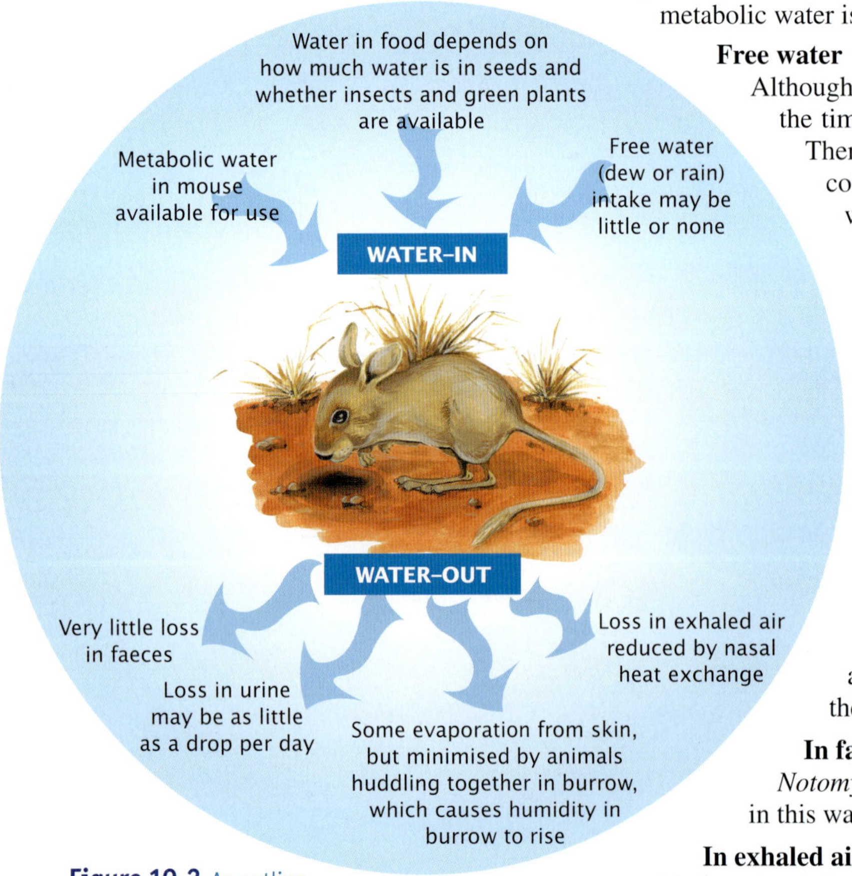

Water loss by the tarrkawarra

From the skin

Although *Notomys* has no sweat glands, some water is lost by diffusion through the skin. Evaporation from the skin occurs but this is minimised. During hot days, animals stay in their burrows huddled together. Air surrounding the group increases in humidity and has the effect of reducing water loss from the skin.

In faeces

Notomys faeces are very dry and little water is lost in this way.

In exhaled air

Air that moves from the lungs to the surrounding atmosphere is saturated with water vapour. This could result in significant water loss. In *Notomys*, a special heat exchange system in the nasal passages reduces that

Water in food depends on how much water is in seeds and whether insects and green plants are available
Metabolic water in mouse available for use
Free water (dew or rain) intake may be little or none
WATER-IN
WATER-OUT
Very little loss in faeces
Loss in urine may be as little as a drop per day
Some evaporation from skin, but minimised by animals huddling together in burrow, which causes humidity in burrow to rise
Loss in exhaled air reduced by nasal heat exchange

Figure 10.2 An outline of how *Notomys alexis*, or tarrkawarra, achieves a water balance. For survival, water-in must balance water-out.

loss. The temperature of air entering the body is lower than body temperature and so nasal passages are cooled as air enters. Warm air exhaled from the lungs passes over these cooled areas and is also cooled. Exhaled air is at a lower temperature than body temperature. As the air is cooled, some of the water vapour from the lungs recondenses on the walls of the nasal passages. Hence, not all the water vapour that leaves the lungs leaves the body.

In urine

Mammals must produce urine to be able to excrete their nitrogenous waste: urea (see chapter 6, page 150). Oxidation of proteins results in urea, as well as carbon dioxide and water. *Notomys* must remove this urea. *Notomys* produces the most concentrated urine recorded for any mammal, even several times more concentrated than that produced by North American desert rodents. Although some water loss occurs through the kidneys, it is clear that the kidneys are a significant site of water conservation in *Notomys*.

ODD FACT

The most concentrated urine that humans can produce is five times the concentration of dissolved material in their plasma. The white rat can produce ten times the concentration of dissolved material in their plasma, and *Notomys* 25 times the concentration of dissolved material in their plasma.

In milk for the young

Female *Notomys* feed their young with milk. The loss of water through having to feed young is balanced to some extent by a mother drinking the urine her young produce. The water in urine is recycled. It has been estimated that a female with suckling young requires only one millilitre of water per day. This water for lactation is obtained from fresh green food, rainwater or dew.

Although *Notomys* and other desert rodents live in very dry areas with little free water, their structural, behavioural and physiological characteristics enable them to survive in those harsh environments.

In this chapter, we discuss physiological adaptations of organisms that contribute to their survival. In particular, we consider those associated with maintaining water balance and control of body temperature.

External and internal environments

The **external environment** of a person can vary greatly — hot one day, cold the next, high humidity one day, dry the next — even over short periods of time. In spite of this great variation in the external environment, the living cells that make up a person exist in a relatively unchanging and stable internal environment — they have a limited **tolerance range**. So, in healthy people, whether they are eating or fasting, their blood **glucose** level is kept within a narrow range (3.6 to 6.8 mmol per L) and, regardless of the weather conditions, their core body temperature is kept within a narrow range of about 36.1 to 37.8°C (see figure 10.3).

Figure 10.3 The external environment may be different but the internal environment stays the same.

Homeostasis: keeping within the tolerance range

Homeostasis = staying the same

In good health, the **internal environment** in which body cells function is relatively constant. The condition of a relatively stable internal environment, maintained within narrow limits, is called **homeostasis**. When deviations (increases or decreases) occur in the internal environment of a healthy organism, mechanisms usually act to restore values to the 'normal' state. However, factors including infection, trauma, exposure to toxic substances or extreme conditions such as immersion in icy water, auto-immune diseases and inherited disorders may lead to a failure of homeostasis.

Homeostasis is critical to the survival of an organism. Uncontrolled and prolonged diarrhoea in young babies can result in a loss of the body's ability to compensate for changes in blood pH. This loss of homeostasis is a life-threatening situation. Table 10.1 shows some of the major variables that are subject to homeostasis.

The term 'homeostasis' was first used in 1932 by the physiologist W. B. Cannon to refer to the relatively constant internal environment.

Homeostasis is also important for plants. They maintain the right balance of water and have special features to absorb water and minimise water loss. Special transporting tissue carries salts and nutrients and distributes water throughout the plant. Plant systems and structures move gases, such as oxygen and carbon dioxide, and a variety of plant hormones to appropriate parts. (We investigate the responses of plants to temperature change more fully on pages 322–5.)

Table 10.1 Summary of major variables that are subject to homeostasis in humans

Variable	*Normal tolerance range*	*Comments*
temperature	36.1°C–37.8°C	Temperature of internal cells of the body is called the **core temperature**
blood glucose	3.6–6.8 mmol per L	Blood glucose is typically maintained within narrow limits regardless of diet
water	Daily intake must balance daily loss	Body tissues vary in their water content. Bone contains about 20% water and blood about 80% water. In prolonged dehydration, fluid moves from cells and tissue fluid into the body.
ions, e.g. plasma Ca^{2+}	2.3–2.4 mmol per L	Specific ions are required by some tissues.
pH of arterial blood	7.4	This pH is necessary for enzyme action and nerve cells.
blood pressure — arterial Diastolic (relaxed) Systolic (contracted)	 13.3 kPa (1000 mm Hg) 5.33 kPa (40 mm Hg)	Transport of blood depends on maintenance of an adequate blood volume and pressure.
urea (nitrogen containing wastes) in plasma	<7 mmol per L	Waste products of cellular processes must be removed by kidneys to prevent toxic effects on cells.

Body systems contribute to homeostasis

You will learn more about homeostasis in your studies for Unit 3, covered in Nature of Biology Book 2, third edition.

Various mechanisms monitor conditions inside the body and, when change is detected, body systems react to restore the balance. In the human organism, cells form tissues and systems that play an essential role in homeostasis. With the exception of the reproductive system, all body systems play a part in homeostasis.

The endocrine (hormonal) and nervous systems are the major systems responsible for the control and coordination of homeostasis. In addition, various types of behaviour of organisms contribute to the maintenance of homeostasis.

KEY IDEAS

- Tissue fluid and plasma form the internal environment of body cells.
- Homeostasis is the condition of a relatively stable internal environment.
- Homeostasis can be disrupted by agents such as disease and trauma.
- Most body systems play various roles in homeostasis.

QUICK-CHECK

1 Which is more variable: your external environment or the internal environment of your cells?
2 Define the term 'homeostasis' using words suitable for a non-biology student.
3 List three variables that are subject to homeostasis.
4 What are the two major body systems responsible for maintaining homeostasis?

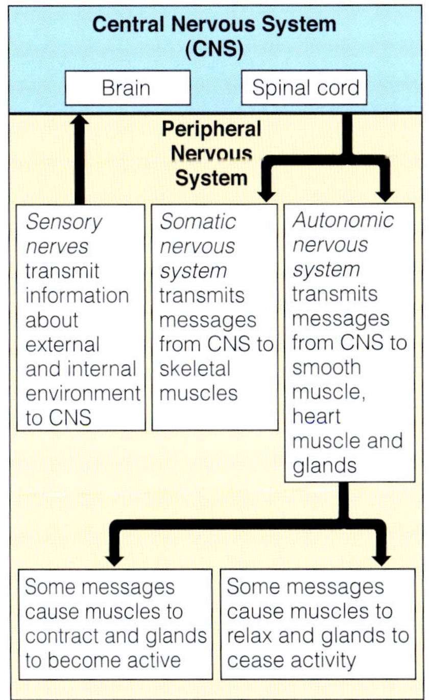

Figure 10.4 The nervous system coordinates the action of various kinds of muscles and glands in the human body. While different parts of the system are recognised, it should be noted that the system acts as a whole. Which part is responsible for the automatic control of the body's glands?

Nerves and hormones: detecting and responding to change

Continuous monitoring of variables and response to change occurs in the human body. The nervous system and hormonal system are the two main controlling systems in the body. They play major roles in the detection and response processes that are integral parts of homeostasis. In the majority of cases of maintaining homeostasis, both systems interact. In many situations, the nervous system stimulates the release of hormones. We will look first at the role of the nervous system.

Structure of nervous system

The **nervous control system** is composed of the brain, **spinal cord** and all the **nerve cells** connecting these to other parts of the body (see figure 10.4). The brain and spinal cord form the **central nervous system (CNS)**. All other nerve cells, in whole or part, that lie outside the central nervous system form the **peripheral nervous system (PNS)**.

Nerve cells

Nerve cells are the basic units of the nervous system. Nerve cells are also known as **neurons**. A typical neuron has a **cell body** which contains the nucleus.

Extensions arise from the cell body of a neuron. The extension that carries information away from the cell body to another neuron or tissue is known as an **axon**. In the human body, axons vary in length from a few millimetres to over a metre, and they may branch. Connecting and effector neurons also have extensions known as **dendrites**. Dendrites are highly branched extensions of the cell body that receive information from other neurons and carry information towards the cell body. A typical motor neuron is shown in figure 10.5a. A group of many axons bound together is called a nerve.

Three basic kinds of neurons are found in the nervous system.

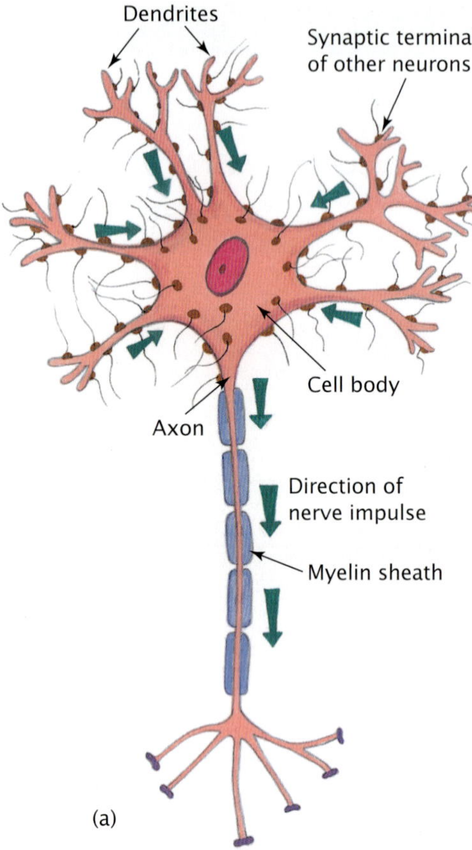

- **Affector neurons** may have one or more receptors that detect change in either the external or internal environment. Information detected is transmitted as an electrical impulse to the CNS by the affector neuron.
- **Effector neurons** carry impulses away from the CNS to muscle cells or glands and cause them to respond.
- **Connecting neurons** are typically located in the CNS and link sensory and effector neurons (see figure 10.5b).

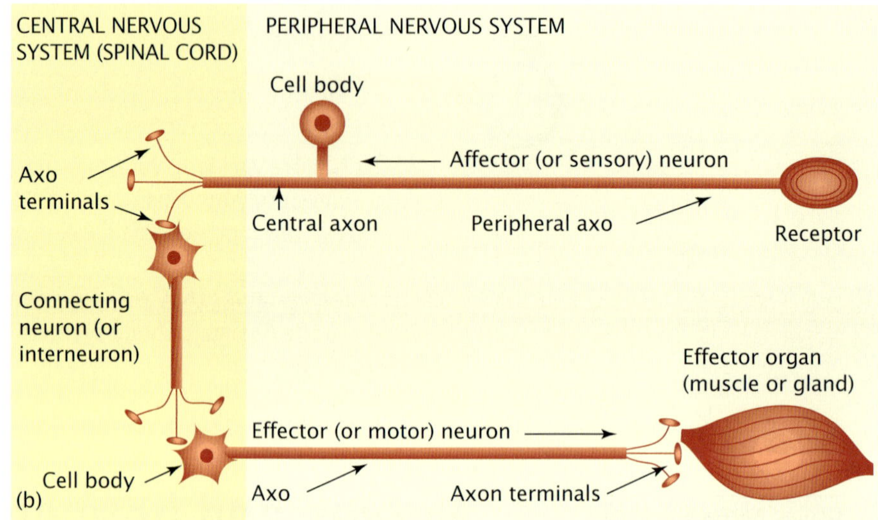

Figure 10.5 **(a)** A typical motor neuron. Note the cell body containing the nucleus, with many branching projections, called dendrites, and the single axon that ends with many synaptic terminals that allow it to communicate with other neurons. **(b)** Relationship between different kinds of neurons. Which type of neuron is located completely within the CNS? The junction between two neurons is called a synapse.

ODD FACT

In a person, the longest axons are extensions of neurons whose cell bodies are located in the lower spinal cord and whose axons reach to toes. In an adult, these extend the length of the leg.

Only a few neurons may be involved in an action such as a reflex arc, for example, when a painful stimulus to the foot causes the leg to lift away, as illustrated in figure 10.6. In more complex actions, many connecting neurons may be involved.

Major sense organs: sensing our environment

Our survival depends on our ability to monitor our external environment, particularly in situations where noxious (harmful) stimuli exist. We receive information all the time about the external environment from receptors. Some receptors in the human body are concentrated into a small area and organised into a structure called a sense organ, such as the eye. In contrast, others are distributed more diffusely over the body surface, such as heat, pain and touch receptors.

Receptors detect specific sensory information from the external environment, such as sound waves which are detected by receptors in your ears. Receptors then encode information about the stimulus into electrical signals that are carried as nerve impulses to your brain. When nerve impulses reach the brain, the brain stimulates effectors to produce a response.

The response to a stimulus may be might be a deliberate action, called a *voluntary* response, as in the case of hearing a car approaching and responding by waiting before crossing the road. In other cases, a response may be an automatic *reflex* response, as in the reaction to stepping on a tack (refer to figure 10.6). This ability to respond to stimuli in the external environment is an important part of our survival mechanisms.

ODD FACT

Putting your foot into it! At the ends of their feet, flies have hairs containing chemoreceptors that can detect the presence of sugar, salt and water. So information that we obtain through receptors on our tongues is obtained by flies through their feet.

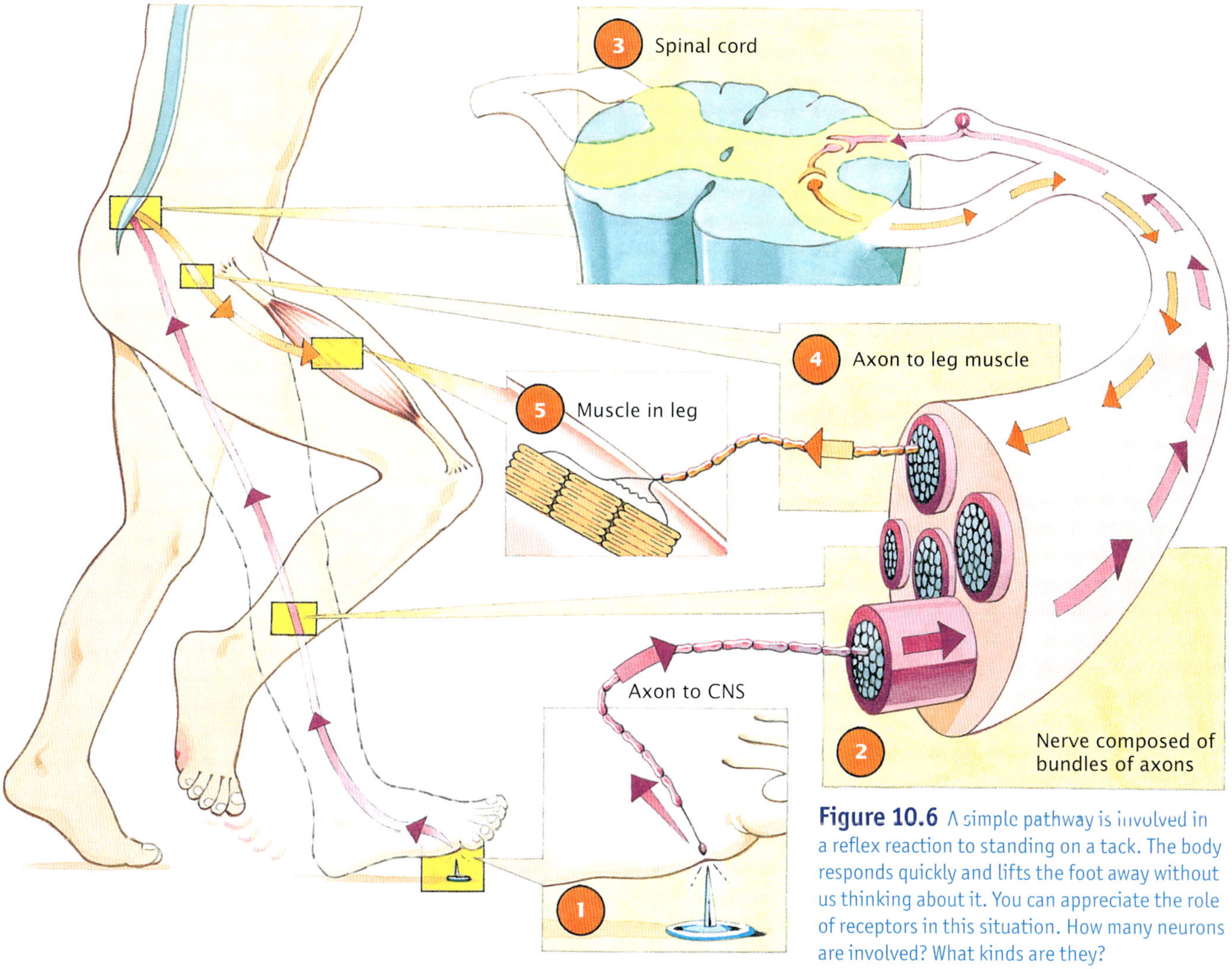

3 Spinal cord

4 Axon to leg muscle

5 Muscle in leg

Axon to CNS

2 Nerve composed of bundles of axons

1

Figure 10.6 A simple pathway is involved in a reflex reaction to standing on a tack. The body responds quickly and lifts the foot away without us thinking about it. You can appreciate the role of receptors in this situation. How many neurons are involved? What kinds are they?

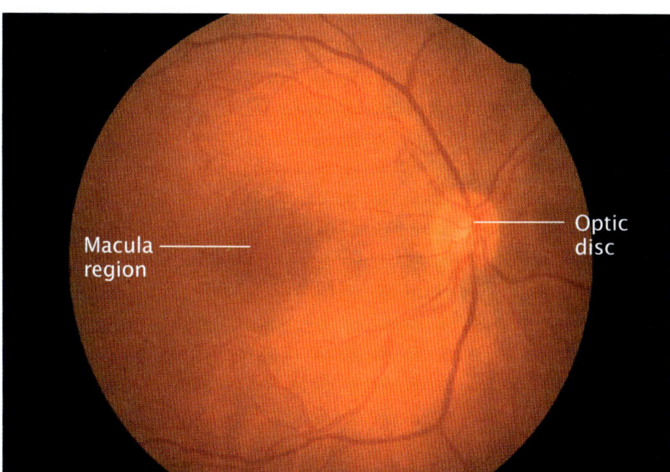

Figure 10.7 **(a)** A photographic image of part of the retina that lines the back of the eye. The majority of cone cells are packed in the macula area. The blood vessels branch and radiate out from the head of the optic nerve.

Human light receptors

Visual stimulus in the form of light enters a human eye through the cornea and passes through the lens where it is focused onto the retina (see figure 10.7b, page 304). The retina contains two kinds of photoreceptors, known as rods and cones, which contain light-sensitive pigments. Fibres from the rods and cones lead to the optic nerve that leaves the back of the eye and carries coded information in the form of a nerve impulse to the brain.

Cone cells function in high light intensities, and can detect colour and detail. Cones are most concentrated in the central region of the retina (see figure 10.7a), which provides us with our central vision which is used when a person looks straight at an object.

Rod cells detect light of low intensity and can detect movement of an object. Rods do not distinguish colour or detail, and occur at the highest concentration in the outer areas of the retina.

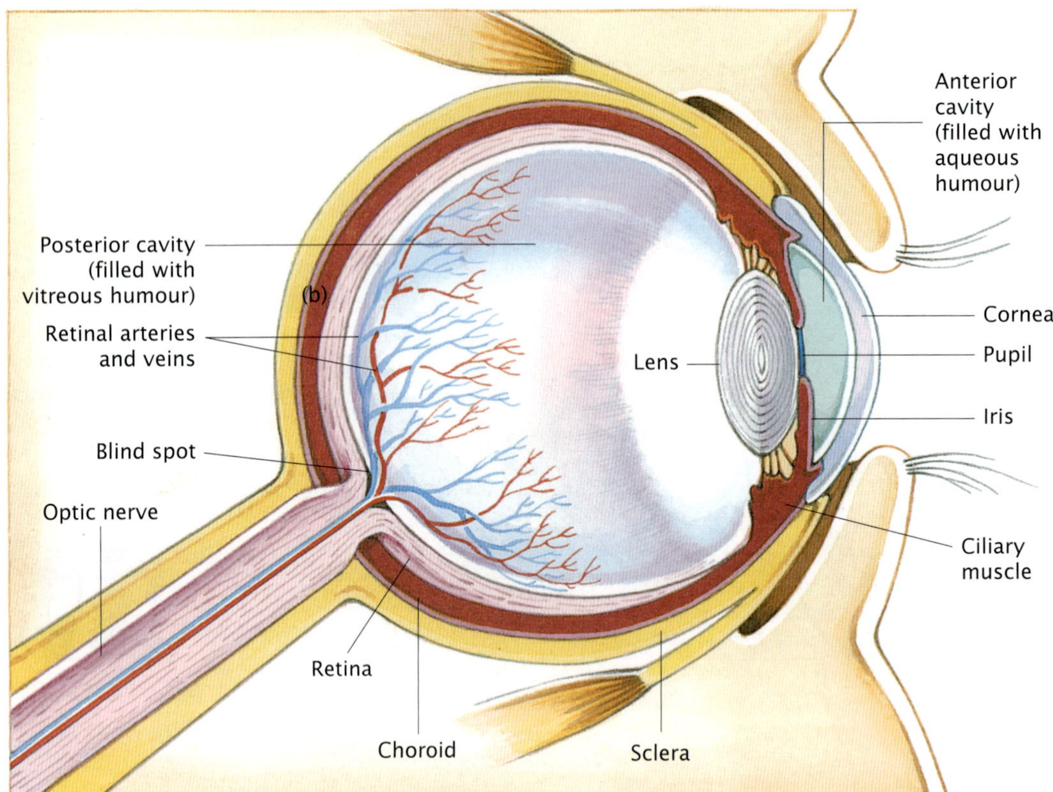

Figure 10.7 (b) The human eye

Human taste receptors

In humans, taste receptors are located in taste buds located on the tongue. Each taste bud is a collection of about 50 receptor cells (see figure 10.8a). Nerves from these receptors transmit impulses that carry encoded information about the taste of dissolved substance that enter the mouth. This information is decoded and interpreted in the brain.

Taste receptors can detect chemical substances that are in solution in the watery saliva of the mouth. Five basic tastes are identified, namely: sour, salt, bitter, sweet and umami. Umami is a taste sensation produced by monosodium glutamate (MSG) and other glutamates found in fermented foods.

Scientists have found that the traditional tongue map (see figure 10.8b) is now wrong. All taste buds detect all five basic tastes. In fact, the first taste maps produced (by D. P. Hänig in 1901) clearly show all four tastes identified at the same time over the same area of the tongue (see figure 10.8c).

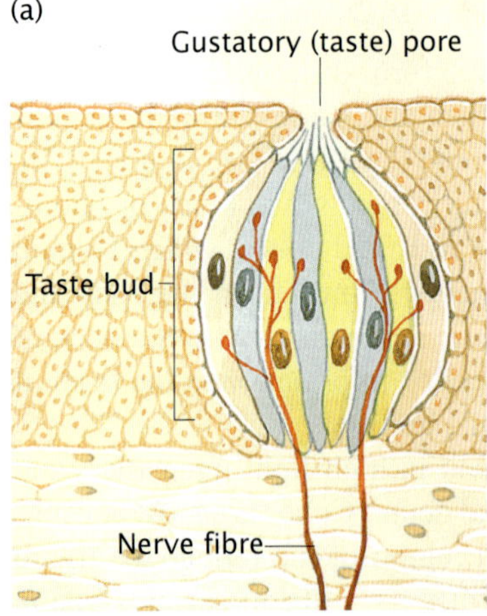

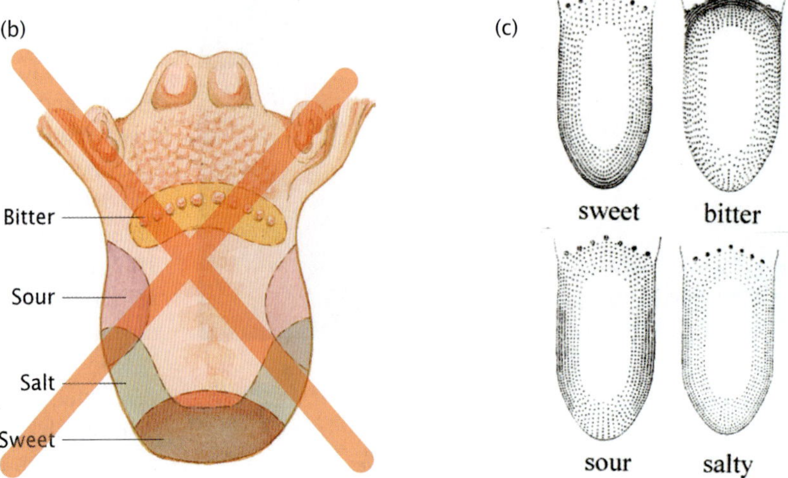

Figure 10.8 (a) Collection of sensory cells that form a taste bud **(b)** Traditional taste maps such as this are now known to be incorrect. **(c)** Taste maps produced by D. P. Hänig, 1901

Human olfactory receptors

If a person is asked 'What's that smell?', the person may sniff the air, drawing it sharply up through the nose into contact with olfactory receptors located on bony outgrowths inside the basal cavity (see figure 10.9). In humans, olfactory receptors are nerve cells, and the fibres from these cells form the olfactory nerve that leads to the brain. We smell something when vapours consisting of small lipid-soluble molecules bind to receptors, triggering an impulse that travels to the brain where it is perceived.

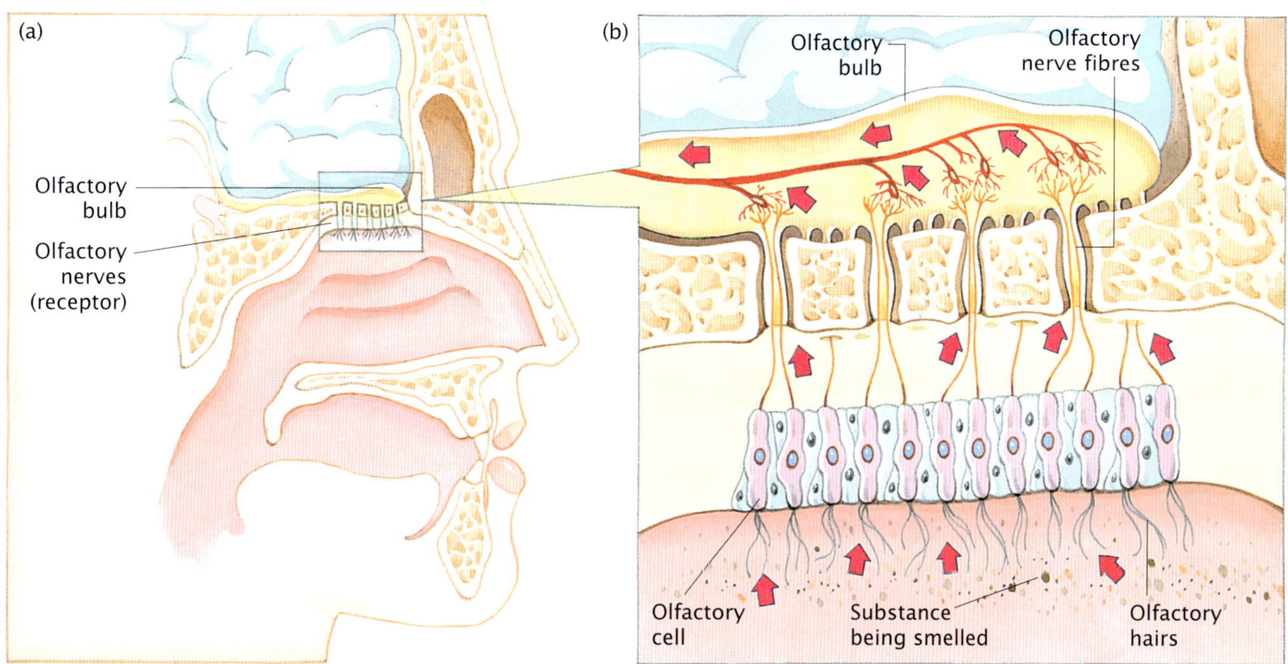

Figure 10.9 **(a)** Cross-section of human nose showing location of olfactory receptors **(b)** Information detected by olfactory cells is transmitted to the brain via the olfactory bulb. Why do pharmaceutical drugs taken through the nose act so rapidly?

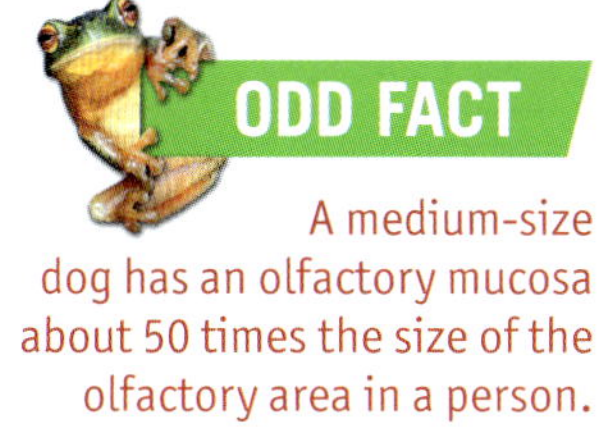

A medium-size dog has an olfactory mucosa about 50 times the size of the olfactory area in a person.

Olfactory receptors in the nose can detect substances at a concentration 10 000 times less than that required for detection by taste receptors. People vary in their smell sensitivity, and some people are unable to detect odours that are readily detected by others. Can you smell the scent of freesia flowers? Some people cannot. Differences exist between the sexes in terms of their sensitivity and women are, on average, a thousand times more sensitive to the odour of steroid-type substances than men.

The 'taste' of food

The 'taste' of many foods is a complex sensation, and comes from the combination of several sensory inputs. These sensory inputs include olfactory stimuli arising from the odour of food before and while it is in the mouth, tactile stimuli arising from the texture of the food, gustatory stimuli arising from the taste of the dissolved food, and temperature stimuli such as the heat or coldness of the food. We often use the senses of smell and taste to determine if food is edible. We may reject food if these senses inform us that the food has 'gone off', or spoiled.

Touch and other tactile senses

Receptors to detect stimuli that produce sensations of touch, pressure, temperature and pain are distributed over the entire skin surface (see figure 10.10). In order to stimulate tactile receptors, an object must make physical contact with the external body surface.

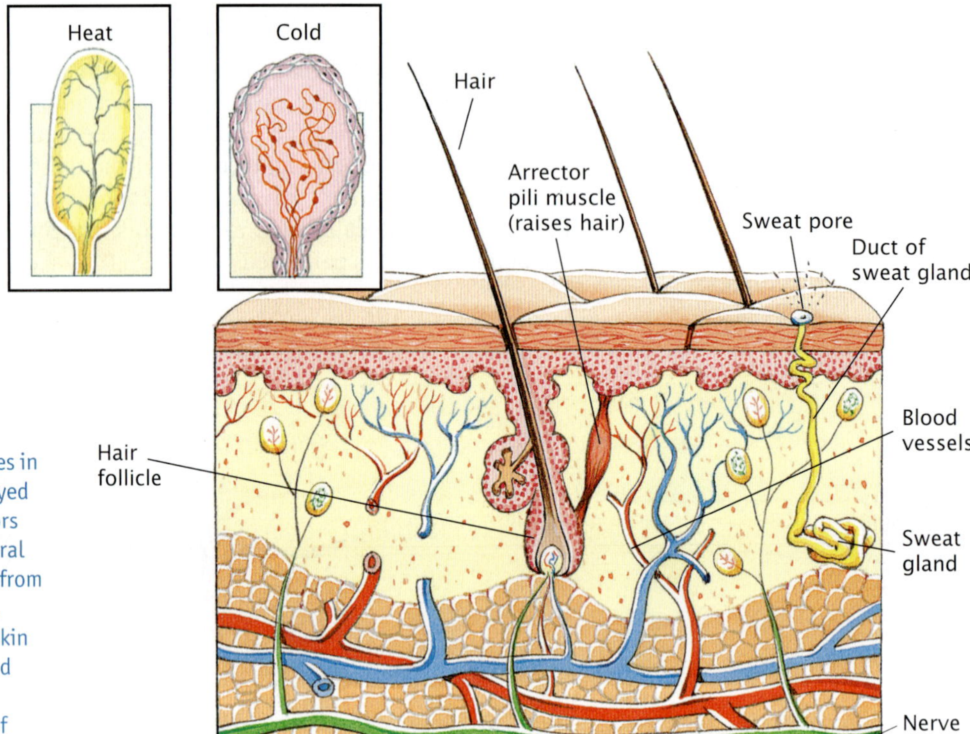

Figure 10.10 Heat and cold sensors in the skin detect changes in temperature, information is relayed to the hypothalamus and effectors act to dilate or constrict peripheral arterioles. Evaporation of water from sweat, secreted by sweat glands, reduces the temperature of the skin by evaporative cooling (discussed on page 313). Many mammals are insulated against extremes of temperature by fat and hair.

After a limb has been amputated or lost in an accident, a sensation of pain from the missing limb can still occur. This 'phantom pain' is probably due to stimulation of remaining nerve fibres in the stump of the limb. These fibres were once part of a nerve pathway leading from nociceptors in the limb to the brain.

ODD FACT

A canal, known as the Eustachian tube, leads from the middle ear to the back of the throat. The lower end of this canal is normally closed by a valve but swallowing or yawning opens it. When a lift descends rapidly, the pressure on either side of the eardrum becomes unequal, and the eardrum is pushed inwards. A person's ears 'pop' when air rushes into the middle ear through the Eustachian tube, equalising the air pressure.

Whiskers and bristles around the face of many mammals have touch receptors at their base, and these whiskers act as extensions of the body surface and increase the mammals' ability to collect information about their external surroundings through the sense of touch.

Would it be an advantage to have no pain receptors? We often think of pain as harmful and undesirable, but pain is a valuable sensation that alerts us to the fact that a stimulus causing tissue damage is occurring, and so enables us to avoid further damage.

Ears and hearing

In mammals, birds, reptiles and amphibians, sound receptors are concentrated in a sense organ known as the ear.

The ears of all mammals share a common structure that can be illustrated by the human ear (see figure 10.11). Three regions are commonly identified in the human ear:

- The outer ear consists of an external ear, made of cartilage, that leads into an ear canal, about 2.5 centimetres long. This canal ends in a delicate membrane (eardrum). The outer ear gathers sound waves.
- The middle ear is an air-filled cavity that contains three tiny bones that are joined by elastic ligaments. Sound waves cause the eardrum to vibrate, and this vibration is then conducted across the middle ear by these three bones to the inner ear. The force of the vibration is magnified because it is transmitted from a relatively large area of the eardrum to a much smaller area in the inner ear. The middle ear magnifies the sound vibrations.
- The inner ear consists of a small coiled structure, known as the **cochlea**, which is filled with fluid. Vibrations that reach the inner ear produce pressure waves in this fluid. The sound receptors are minute hair cells located on a membrane inside the cochlea. Information about the sound stimulus is encoded into nerve impulses and sent to the brain for interpretation. The inner ear receives the sound stimulus.

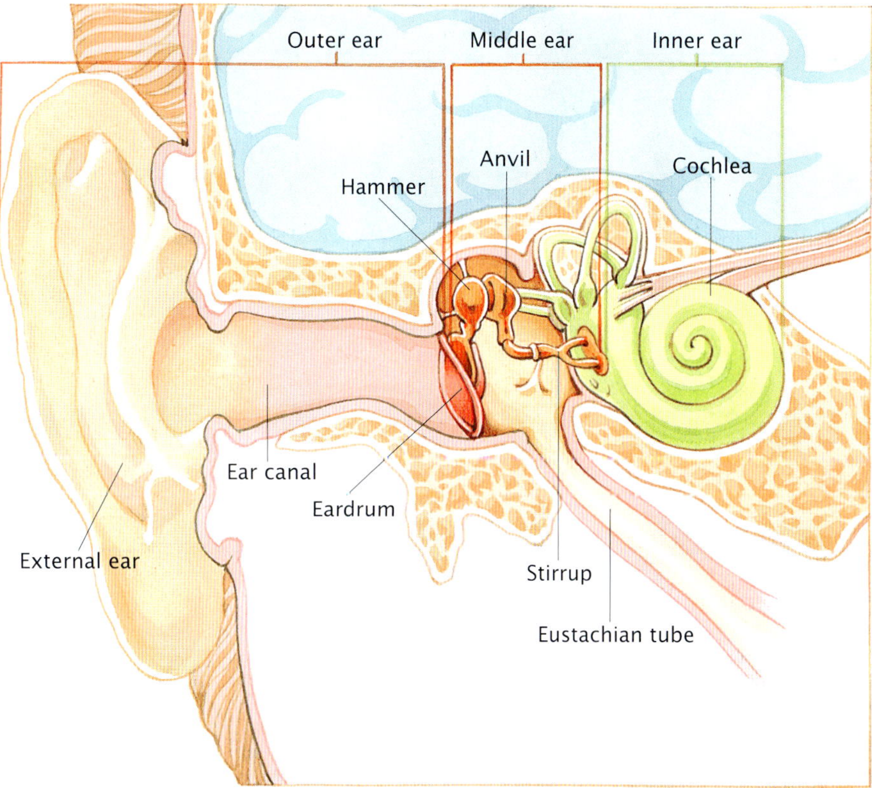

Figure 10.11 Section showing structure of human ear

The mammalian ear is also concerned with maintaining balance but the structures responsible are not discussed here.

The external ear varies in size and shape in different mammals, and is absent in some. Movement of the external ears helps gather sound waves and localise the direction from which a sound is coming. Because we cannot move our external ears in the way that some animals are able to, we sometimes move our heads to localise the direction from which a sound is coming.

Hormones — chemical regulators

The hormonal system is the other major controlling system in the body alongside the nervous system. The hormonal system, also called the **endocrine system**, produces hormones that, as does the nervous system, help maintain homeostasis.

A hormone only acts on cells that have particular receptors for that hormone.

Hormones are chemicals produced in special structures called **endocrine glands** (see figure 10.12, page 308). They are transported to other parts of the body through the bloodstream and act on other organs and tissues of the body.

Some organs can have other functions as well as an endocrine function. The pancreas produces digestive enzymes that are secreted via a duct into the small intestine. The pancreas also contains special cells that produce insulin which is secreted directly into the bloodstream. All hormones are secreted directly into the bloodstream. Because of this, endocrine glands are also called ductless glands.

When a vital factor of the internal environment of cells changes, such as temperature or water balance, the change is detected by either the hormonal or nervous system. Information about the direction of change is transported, either by chemical or nerve messages, to other parts of the body. These messages lead to another change in the vital factor, reverse to the initial one, so that the factor is now restored to within the normal range. For example, if your body temperature rises beyond the normal range, events take place to reduce the temperature back to within the normal range.

Such a system is called a **negative feedback** system. You will study such systems in greater detail in your future studies of Biology.

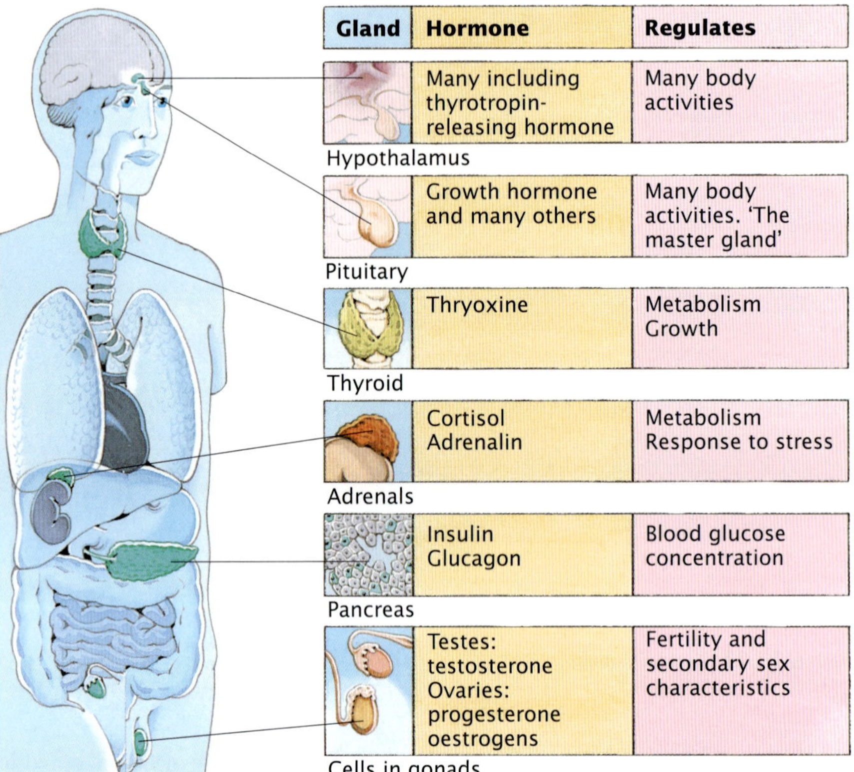

Gland	Hormone	Regulates
Hypothalamus	Many including thyrotropin-releasing hormone	Many body activities
Pituitary	Growth hormone and many others	Many body activities. 'The master gland'
Thyroid	Thryoxine	Metabolism Growth
Adrenals	Cortisol Adrenalin	Metabolism Response to stress
Pancreas	Insulin Glucagon	Blood glucose concentration
Cells in gonads	Testes: testosterone Ovaries: progesterone oestrogens	Fertility and secondary sex characteristics

Figure 10.12 The endocrine system: its main glands, the hormones they produce and their actions

KEY IDEAS

- The nervous control system is made up of the central nervous system and the peripheral nervous system.
- Nerve cells or neurons are the basic structure of nervous tissue.
- Neurons transmit nerve impulses.
- Major sense organs continually monitor our external environment.
- Special glands, called endocrine glands, produce chemicals called hormones which are secreted directly into the bloodstream.

QUICK-CHECK

5 Distinguish between the members of the following pairs:
   a CNS and PNS
   b sensory neuron and effector neuron
   c axon and dendrite
   d receptor and effector organ.

6 List:
   a the major sense organs
   b the five taste sensations.

7 What are the main glands involved in the endocrine system?

Detecting temperature change

External temperature change

Skin is the barrier between our body and the external environment and can be two or three degrees below core body temperature. Core body temperature is maintained at about 37°C but changes in the external temperature cause changes in the temperature of exposed skin. Such change is detected by two kinds of temperature receptors in the skin (see figure 10.10, page 306). One kind of receptor detects the cooling of the skin and the other detects warming.

- If there is a reduction in skin temperature, the cold receptors register this change by increasing the rate at which they discharge electrical information along affector neurons.
- If there is an increase in skin temperature, the heat receptors increase their rate of discharge of electrical information along the affector neurons.

The numbers of different kinds of temperature receptors vary in different parts of the skin. There can be up to ten times more cold than heat receptors present.

Affector, or sensory, neurons transmit impulses from skin temperature receptors to the **hypothalamus** in the brain (see figure 10.13).

ODD FACT

Although temperature receptors are generally represented as in figure 10.10, page 306, some scientists have suggested that they may be free nerve endings.

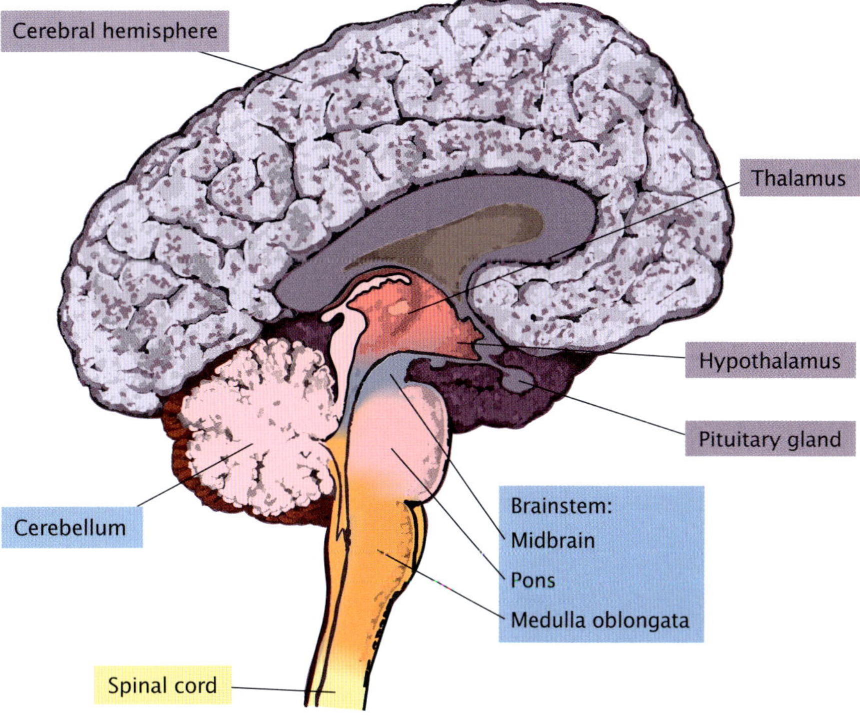

Figure10.13 Longitudinal section through the midline of the brain showing the relative positions of the hypothalamus and the anterior and posterior parts of the pituitary gland

Internal temperature change

Changes in core body temperature from 37°C are detected by a number of temperature receptors deep within the body. The most important of these are large numbers of temperature-sensitive receptors in the hypothalamus of the brain. The majority of these receptors, about three-quarters of the total, are sensitive to heat. The remaining one-quarter are sensitive to cold.

Other deep body temperature receptors are situated near the spinal cord, around large veins and in parts of the digestive system. All the receptors from these regions transmit impulses via affector neurons to the hypothalamus.

The hypothalamus, therefore, not only detects changes in core body temperature itself but also receives information about changes in body temperature from different areas of the skin and other body regions. The hypothalamus serves as the 'temperature control centre' of the body.

ODD FACT

No single value for the 'normal' human body temperature exists. A normal range is usually stated: oral temperature (36.1 to 37.2°C); rectal temperature (36.1 to 37.8°C). Cats have higher normal body temperatures (about 39°C) than people. On average, the body temperature of birds is about 41°C.

Maintaining core temperature

An example of homeostasis — in which nervous and hormonal systems interact — is the maintenance of a stable core body temperature.

Mammals keep their body temperatures within a narrow range. At any stage, a mammal is producing heat energy and losing heat energy. To maintain a stable core temperature, heat gain must balance heat loss (see table 10.2). In humans, this balance results in a core temperature of about 37°C.

Table 10.2 Some mechanisms of heat gain and heat loss

Ways of gaining heat	*Ways of losing heat*
basic metabolic processes	evaporation of sweat
shivering	panting
exercise or other muscular activity	convection
radiation and conduction to the body	radiation and conduction from the body

Losing heat

Heat can be lost by the body through **radiation**, **conduction**, **convection** and **evaporation** (see figure 10.14).

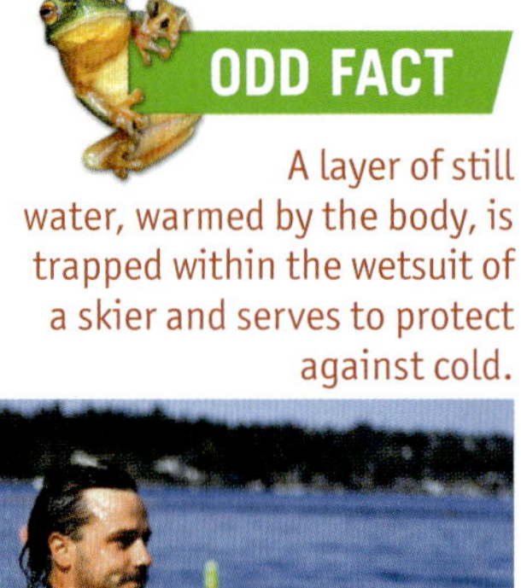

ODD FACT

A layer of still water, warmed by the body, is trapped within the wetsuit of a skier and serves to protect against cold.

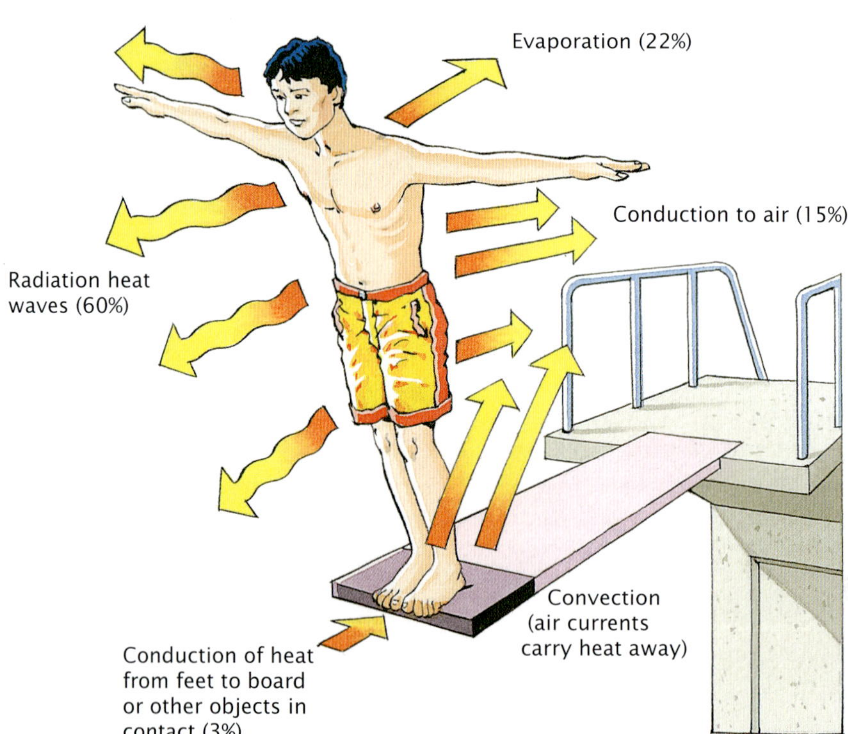

Figure 10.14 A person loses heat to the environment by radiation, conduction, convection and evaporation.

Radiation

Heat, in the form of infra-red heat rays, radiates from the body in all directions. Radiation accounts for about 60 per cent of heat lost from a person. Radiant heat can also be absorbed by a person from other objects if they are at a higher temperature than the skin of the person.

Conduction and convection

Direct contact with objects, and the transfer of heat into them, accounts for about three per cent loss of heat from the body. More heat loss (about 15 per cent) is due to conduction of heat to the air surrounding a person. Conduction of heat from the body into air occurs if the temperature of the air is less than that of the body but transfer ceases if air temperature becomes higher than that of the skin.

If the air surrounding a body is moving, air currents, or convection, carry the heat away from the body and more heat is lost. If a still layer of air surrounds the body, loss by conduction and convection is reduced. Water surrounding a body is generally moving continually, so the loss of heat by conduction and convection is greater than when a body is in air. Heat is conducted away from the body 24 times faster by water than air. Also, a particular volume of water is able to carry 2700 times the amount of heat carried by the same volume of air.

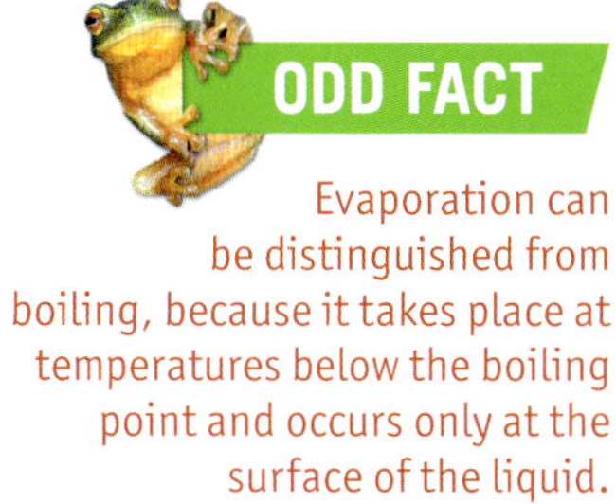

Evaporation can be distinguished from boiling, because it takes place at temperatures below the boiling point and occurs only at the surface of the liquid.

Evaporation

About 22 per cent of heat loss from a human body occurs through the skin and lungs when water evaporates from those surfaces. The evaporation of water requires heat which is provided by the body. Even if a person is not sweating, water still evaporates from the skin.

Temperature sensors, including the hypothalamus, continually monitor any loss of heat. If a reduction of core temperature is detected, the hypothalamus responds by directing events that reverse the reduction.

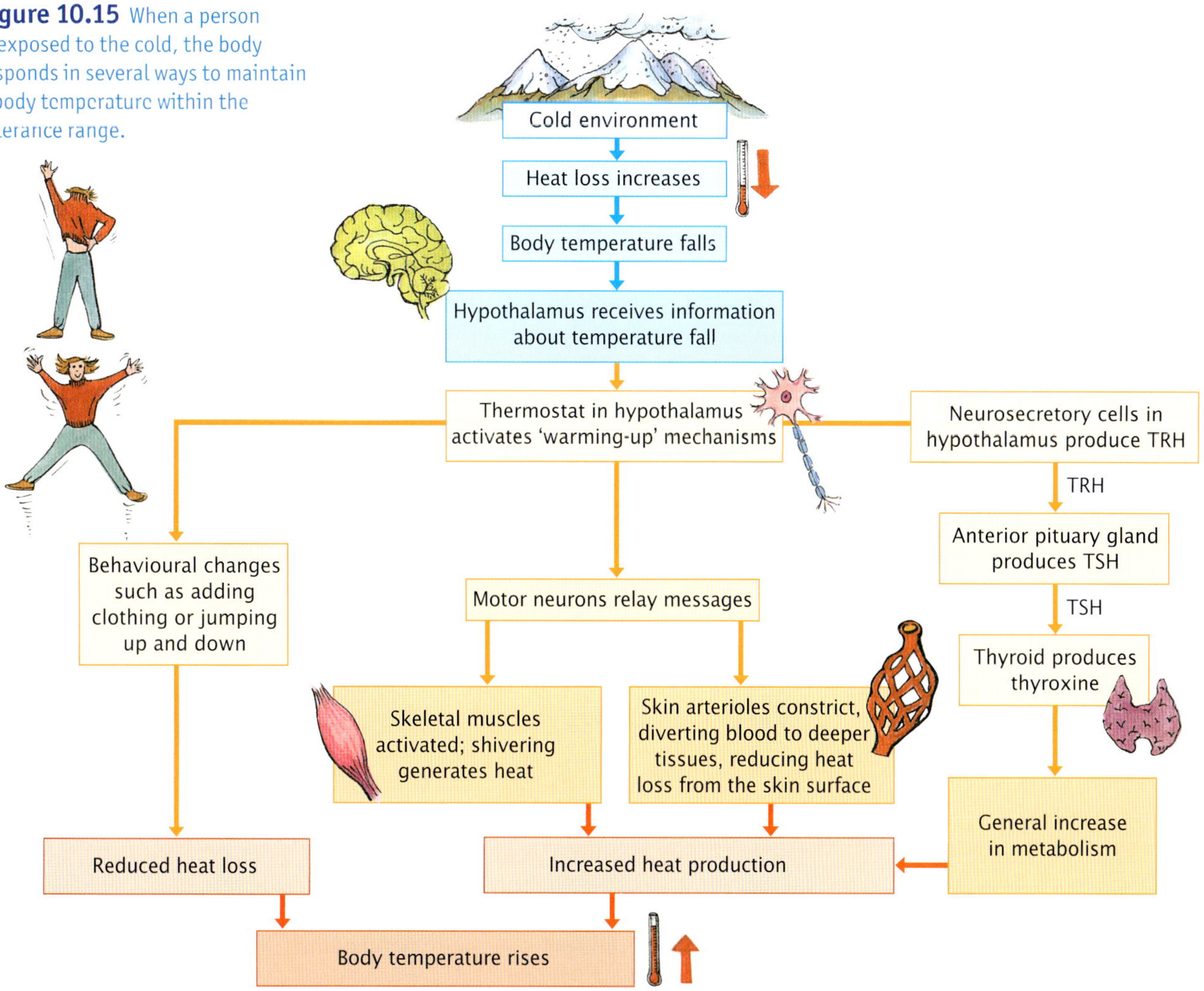

Figure 10.15 When a person is exposed to the cold, the body responds in several ways to maintain a body temperature within the tolerance range.

Cold environment
Heat loss increases
Body temperature falls
Hypothalamus receives information about temperature fall
Thermostat in hypothalamus activates 'warming-up' mechanisms
Neurosecretory cells in hypothalamus produce TRH
TRH
Anterior pituary gland produces TSH
TSH
Thyroid produces thyroxine
General increase in metabolism
Behavioural changes such as adding clothing or jumping up and down
Motor neurons relay messages
Skeletal muscles activated; shivering generates heat
Skin arterioles constrict, diverting blood to deeper tissues, reducing heat loss from the skin surface
Reduced heat loss
Increased heat production
Body temperature rises

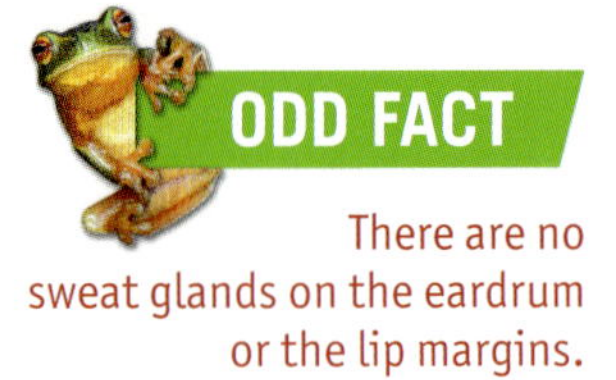

There are no sweat glands on the eardrum or the lip margins.

Gaining heat and reducing heat loss

The hypothalamus initiates two kinds of responses to balance any loss of heat. Some responses generate heat, others reduce the rate at which heat is lost from the body (see figure 10.15 on page 311).

Heat production by shivering

Shivering is the alternate contraction and relaxation of small muscle groups and is an involuntary action. The hypothalamus contains a centre for shivering that activates somatic motor neurons that control muscles in the upper limbs and body trunk. When muscles shiver, almost all of the energy of contraction is converted into heat energy.

Although maximum shivering can produce significant amounts of additional heat for a body — up to five times what is normally required — it cannot be sustained for long because it drains the energy reserves of the muscle tissue.

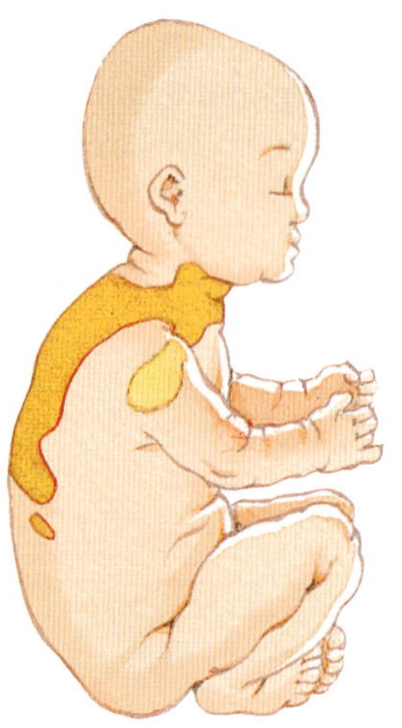

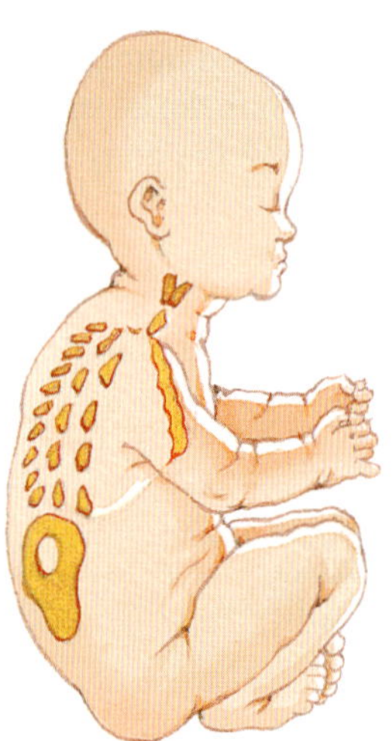

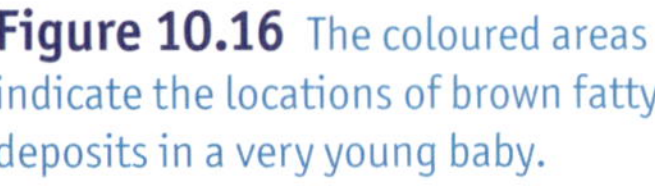

Figure 10.16 The coloured areas indicate the locations of brown fatty deposits in a very young baby.

Heat produced by metabolism

Metabolic processes in the body produce heat. Neurosecretory cells in the hypothalamus produce thyrotropin-releasing hormone (TRH). This hormone is transported to the anterior **pituitary** (see figure 10.12 on page 308) where it stimulates the secretion of thyroid-stimulating hormone (TSH). Thyroid-stimulating hormone is transported in the blood to the thyroid, which in turn increases its output of thyroxine. Thyroxine is a hormone that increases the metabolic rate of all cells of the body, resulting in an increase in heat production.

Motor nerves from the hypothalamus also cause the medulla of adrenal glands to secrete adrenaline and noradrenaline. These hormones increase the basal metabolic rate, particularly in skeletal muscles and brown fat, a special kind of fat of particular importance in young babies (see figure 10.16). Young babies have a relatively thin layer of insulating fat on their bodies and have poorly developed shivering mechanisms. Because of brown fat metabolism, a baby produces about five times as much heat (per unit body weight) from metabolic pathways as an adult.

Constriction of blood flow in skin

When cold is detected, neurons in the hypothalamus send impulses via the sympathetic nervous system to peripheral blood vessels in the skin. The impulses cause arterioles to constrict. This constriction reduces the surface area across which heat transfer can occur and reduces the amount of blood flow close to the skin. Hence heat lost via the skin surface is reduced. Heat is retained within the body.

Piloerection

Piloerection means 'hair standing on end'. Although it is not important for conservation of heat in humans, erection of hair is important for most mammals. A layer of air is trapped in the erect hair or fur and acts as an insulation layer between the skin of the animal and the external environment. Sympathetic motor neurons convey impulses from the hypothalamus to muscle at the base of each hair (see figure 10.10 on page 306) causing the muscle to contract and the hair to become erect. Although the hypothalamus reacts to decrease heat loss as outlined above, and restore normal core body temperature, it must also ensure that the body doesn't overheat beyond 37°C.

Cooling off

When the core body temperature of a person is likely to exceed 37°C, either from internal activity or from external factors, the hypothalamus acts to increase heat loss. Motor neurones from the hypothalamus send impulses through the parasympathetic nervous system and dilation of arterioles occurs. This results in a greater surface area across which heat exchange can take place, and in greater blood flow through the skin and so a greater loss of heat can occur. Similarly, activation of muscle by the shivering centre in the hypothalamus is reduced, metabolic activity is reduced and metabolic heat production is reduced. Metabolic heat production is also reduced because the thyroid produces less thyroxine. This is because neurosecretory cells of the hypothalamus reduce their secretion of TRH to the anterior pituitary which in turn reduces its secretion of TSH.

Nerve impulses also cause sweat glands to be activated (see figure 10.10 on page 306). Liquid sweat on the skin changes to a gas at body temperature by evaporation. The evaporation of sweat requires heat energy and so the body is cooled. When liquid water evaporates, energy is needed to change its state from liquid to gas. The evaporation of one millilitre of liquid sweat from a person requires about 2500 joules of energy. This is more than the amount of heat energy produced by burning a match. The heat energy needed to evaporate sweat is taken from the body of the person, thus the body is cooled. Cooling achieved in this way is called **evaporative cooling**.

An adult person has about two million sweat glands. In a very hot environment, a person can lose up to three litres of liquid water an hour just through sweating. Evaporation of this sweat results in cooling at the rate of about 2.5 kilojoules per millilitre (this does not include sweat that drops off).

So the hypothalamus plays a most significant role in the maintenance of core body temperature. It receives information about temperature change via sensory nerves from various parts of the body as well as detecting temperature change itself. The messages sent out by the hypothalamus in response depend on whether the information is that the temperature is higher or lower than 37°C.

Read the 'Biologist at work' box on page 315. It tells of the development of an ice-cooling jacket worn by athletes to reduce dangers from over-heating.

Behavioural activities

The events described in the previous paragraphs occur automatically. You do not have to think about them. In addition, animals may change their behaviour to reduce heat loss (or heat gain). A person may increase levels of activity, for example, by jumping up and down, thus generating heat from the muscle activity. The person may shelter or move indoors to be out of the cold. The person may turn on a radiator or put on some warm clothing. Clothing traps a layer of air — a good insulator — which reduces loss of heat by convection currents. The person may wrap their arms around the body to reduce the surface area from which heat can be lost.

Behaviours to cool down

Figure 10.17 Evaporation of saliva from the wallaby's paws uses heat energy. The heat energy is taken from the wallaby's body, which cools as a result.

Look at figure 10.17. In hot weather, kangaroos and wallabies often lick their front legs. What benefit might this behaviour produce? Energy is required to change water from a liquid to a gas state during evaporation. If this heat energy is taken from the tissues underlying the moist surface of the wallaby's front leg, the result is a cooling by removal of heat.

What other behavioural changes help to reduce body heat? Have you observed a cat or a dog sleeping in a hot location or in a cold location? The animal typically arranges itself differently depending on the temperature of its surroundings. These arrangements affect its exposed surface area.

For example, consider an adult cat. The volume of the cat is constant, but its exposed surface area can be minimised if the cat curls up in a ball. In contrast, its surface area increases if the cat stretches out. What arrangement, curled up or stretched out, would be expected to minimise heat loss?

In the cold, the cat usually assumes a tightly-coiled pose, with its limbs and tail drawn closely into its body. In this case, the animal is minimising its exposed body surface, and so reduces its heat loss. In contrast, in a hot situation, the animal usually assumes a stretched-out pose with its limbs and tail stretched away from its body. In this case, the animal is maximising the exposed body surface area and permitting maximum heat loss.

What types of behaviour do you carry out on a very hot day to help maintain a stable temperature?

Big or small: which stays warm more easily?

A tammar wallaby (*Macropus eugenii*) has a body mass about 800 times that of a long-tailed planigale (*Planigale ingrami*) (see figure 10.18). Which animal finds it easier to stay warm? To answer this question, we must consider not only heat production, but also heat loss.

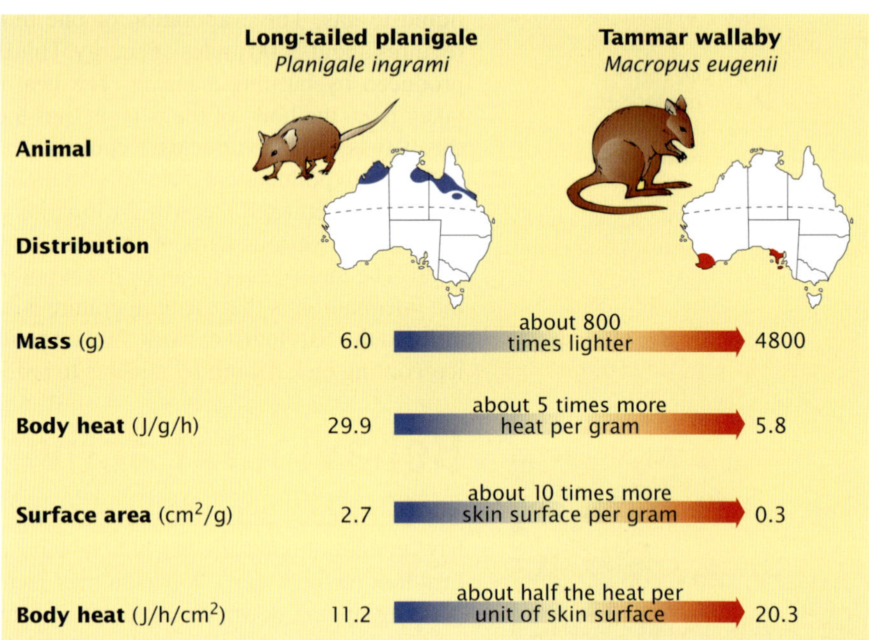

Figure 10.18 Although the wallaby produces less heat per unit mass than the planigale, it has a smaller surface area per unit mass across which heat can be lost and maintains its body temperature more easily than smaller animals.

The rate of heat production per unit body mass in a planigale is much (about five times) greater than in a wallaby. However, compared with the planigale, the larger wallaby has much less skin surface per unit body mass from which heat is lost. So when heat production is expressed in relation to units of skin surface, the wallaby produces more heat (20.3 joule/hour/cm^2) than the planigale (11.2 joule/hour/cm^2). As a result, the larger animal, the wallaby, maintains its body temperature more easily than the tiny planigale. Is it surprising then that the tiny planigale is found in northern regions of Australia?

As the body mass of a mammal gets larger, the rate of heat production for each unit of its body mass gets smaller. A similar relationship holds for other animals, including birds.

Tolerance of different environments

People cannot survive in freezing cold water for extended periods. They die. Yet other mammals, for example whales, swim in such waters. This contradiction of death or survival for organisms is true across the range of temperatures found in the various habitats on earth. The total temperature range across which living

BIOLOGIST AT WORK

Associate Professor Martin Thompson — exercise physiologist

Martin Thompson is an exercise physiologist who works in the University of Sydney's School of Exercise and Sport Science. He previously worked in the Environmental Physiology Unit of the Medical Research Council in London (United Kingdom). He writes:

'I have had a long-standing interest in the nature of fatigue that limits sports performance. In my quest to learn more, and desire to travel overseas, I completed a Master of Science degree in Human Biology at Loughborough University of Technology (UK) and a Doctor of Philosophy (PhD) degree at the London School of Hygiene and Tropical Medicine, University of London.

Figure 10.19 Associate Professor Martin Thompson

'I became interested in thermoregulation, that is, how creatures adjust their behaviour and physiological responses to maintain body temperature within narrow limits for survival. I was fascinated by comparative biology stories of how the Thompson gazelle (no relation!) could sometimes outrun the jaws of the cheetah because of its superior thermoregulatory mechanisms. I set about investigating the thermoregulatory responses of a group of marathon runners who undertook a series of treadmill-running experiments within a laboratory where the air temperature and humidity could be well controlled. This so-called 'climate chamber' allowed me to simulate a wide range of climatic conditions, and changing the treadmill velocity allowed me to alter the intensity of the exercise. Thus, the effect of different environmental heat loads and different metabolic heat loads could be studied in combination.

'This research extended the concept of a "prescriptive zone" where body (core) temperature during light to moderate exercise was shown to be dependent upon the exercise intensity and independent of the air (ambient) temperature. Under such conditions, the heat produced by the exercising muscles was dissipated by the evaporation of sweat and increase in blood flow towards the skin. A temperature gradient is established between the hotter parts of the body, the skin surface and the environment. Thus, thermal equilibrium was well maintained. However, in hot conditions, when simulating marathon-racing exercise intensity, the athletes invariably experienced extreme fatigue that was associated with an inability to maintain thermoregulatory control. Profuse sweating resulted in dehydration and the cardiovascular system was severely challenged with heart rate steadily drifting upward towards its maximum throughout the experiment. How then, could athletes improve their capacity to compete in hot environmental conditions?

'This question was of particular interest to the Australian Olympic team preparing for the 1996 Atlanta Olympic Games. In addressing this question, I led a team investigating heat acclimatisation strategies, immediate pre-competition cooling and textiles used in sports clothing that would enhance heat loss. For cooling of the skin, we developed a lightweight neoprene jacket that was close fitting and had pockets adjacent to the skin where plastic bags of crushed ice could be inserted. Our trials in the climatic chambers of the CSIRO demonstrated the dramatic skin cooling effect of the "ice cooling jacket": a reduced heart rate at sub-maximal exercise and a reduced rate of rise in core temperature. A cooling jacket was issued to all the athletes competing at outdoor venues at Atlanta and the feedback was most positive.

'Our current research is concerned with thermal tolerance and heatstroke. Heatstroke is a potentially fatal condition brought on by a failure in the body's thermoregulatory mechanisms. We have been investigating the collapse of runners in the annual City to Surf race. We are particularly interested in the increased synthesis of so-called heat shock proteins in response to overheating and their role in increasing thermal tolerance.

'I find this work is forever presenting interesting challenges. It has very applied outcomes and it is particularly gratifying to see ideas tested in the laboratory and subsequently translated into practice.'

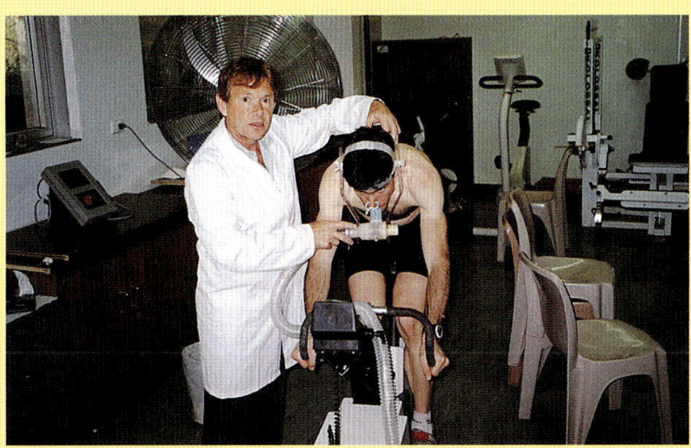

Figure 10.20 Monitoring a cyclist's performance

organisms may be found is quite large — it extends from sub-zero temperatures of the Antarctic to hot water springs. However, each species of organism is usually restricted in its tolerance range to a comparatively narrow range of temperatures. Humans are an exception to this restriction because of an ability to manipulate the environment in a way not possible by other organisms. In this section, we will examine a number of habitats and consider the strategies organisms adopt and the characteristics they have that enable survival in the temperatures encountered.

Heat source: external or internal?

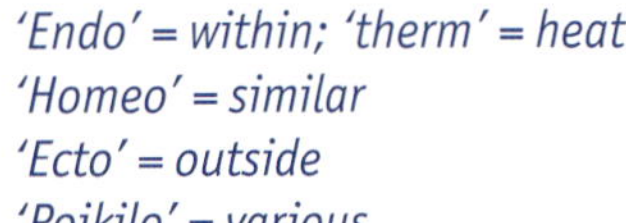

Visit an Arctic region on a wildlife safari and you may see various species of mammals, such as polar bears and seals. However, you will not see reptiles, such as snakes and lizards. Reptiles are typically found in temperate and tropical environments. Experiencing the warmth of the reptile house at a zoo is a reminder that reptiles cannot survive in Arctic or Antarctic habitats as they are **ectothermic**; that is, they depend on external sources of heat to generate their body warmth.

Reptiles cannot maintain constant body temperatures and their temperatures fluctuate with changes in the temperature of their external environment. Organisms with fluctuating temperatures are described as **poikilothermic**. In a cold environment, the body temperature of a reptile falls and the chemical reactions necessary for it to maintain its living state slow down. If this low temperature persists for too long, the reptile dies.

Figure 10.21 Ectotherms use the warmth of the sun to obtain body heat. Endotherms use the energy from reactions involving ingested food.

To warm up, a snake absorbs heat by exposing as much as possible of its body surface to the warmth of the sun. To prevent overheating, a snake seeks shade or shelters in a burrow. While this is an energy-cheap way of obtaining body heat, the price is that the body temperature of a snake can vary greatly. So, as the temperature in its external surroundings changes over a 24-hour period, a snake's body temperature varies and may range from 18°C to 40°C.

Mammals and birds occupy habitats in tropical, temperate and polar environments because they are **endothermic**: that is, they have an in-built source of body heat — their internal energy-releasing (exergonic) chemical reactions that produce heat energy. As a result, mammals and birds maintain fairly constant body temperatures, regardless of temperature fluctuations in their external environments. Such organisms with constant temperatures are described as **homeothermic**.

ODD FACT

Crocodiles, which are ectotherms, eat about two to three times their body mass in food each year. In contrast, lions, which are endotherms, eat about 20 times their body mass in a given year.

Figure 10.22 The yellow-bellied glider, an Australian mammal, uses an internal source of heat for controlling body temperature.

Using internally generated heat is an energy-intensive means of producing body heat. Typically a mammal uses up to 80 per cent of the energy that it obtains from feeding to mantain its core body temperature, but this means that its temperature stays within a very narrow range. For example, yellow-bellied gliders (*Petaurus australis*), with their high energy use, feed every day on insects and spiders (see figure 10.22). Red-bellied black snakes (*Pseudechis porphyriacus*), on the other hand, capture prey once every few weeks during warm weather and do not feed during winter.

KEY IDEAS

- The hormonal and nervous systems interact to maintain a stable internal temperature in mammals.
- The thermostat of the body is situated in the hypothalamus and in humans has a set point that ensures a core body temperature of about 37 °C.
- The ease with which mammals keep warm is related to their exposed surface area.
- The source of body heat in ectotherms, such as reptiles, is external and comes from the heat of the sun.
- The source of body heat in endotherms, such as mammals and birds, is internal heat generated from chemical reactions.

QUICK-CHECK

8 List the events that occur in the human body if the core temperature threatens to rise above the set point of the hypothalamus.
9 List the events that occur in the human body if the core temperature threatens to drop below the set point of the hypothalamus.
10 Explain why evaporation of sweat is an important factor in maintaining a stable core temperature on a very hot day.
11 Explain how behaviour of an individual can contribute to the maintenance of a stable core temperature.
12 What is the source of body heat for an ectotherm?

Animals surviving on land

Organisms that live successfully in a particular habitat would be expected to show particular **structural**, **behavioural** or **physiological features** or characteristics that assist them to survive in the range of environmental conditions that exist in that habitat. Features that appear to equip organisms for survival within a range of environmental conditions in their habitats are sometimes called **adaptations**. One of the challenges for a terrestrial animal (living on land) is to survive the fluctuating temperatures from day to night and from season to season.

Surviving in the heat

The spinifex hopping mouse, *Notomys alexis*, or the tarrkawarra (see figure 10.1, page 297), lives in a well-insulated burrow to avoid the heat of the day. However, like all mammals and birds, a tarrkawarra is an endotherm. It is able to generate heat to maintain its body temperature from an internal source — its body metabolism. Its body temperature stays fairly constant, day and night, regardless of how the temperature of the external environment changes.

Snakes bask in the sun

In contrast to mammals and birds, snakes and other reptiles are ectotherms. They obtain their body heat from external sources. The temperature of snakes and other reptiles typically varies by day and night as the temperature of their external environment changes. Their behaviour reflects their dependence on the

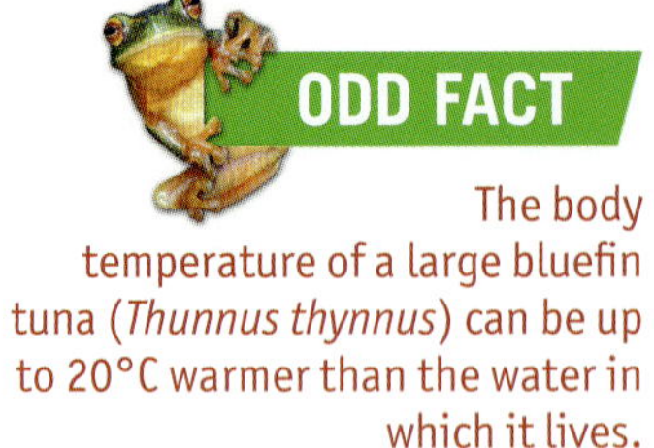

The body temperature of a large bluefin tuna (*Thunnus thynnus*) can be up to 20°C warmer than the water in which it lives.

sun for heat and, not surprisingly, they are often seen basking in the sun (see figure 10.23). It is not correct to call them 'cold-blooded'. The body temperature of an ectotherm can get just as high as that of an endotherm. However, the body temperature of an ectotherm is dependent on that of the environment and fluctuates with the temperature of the environment.

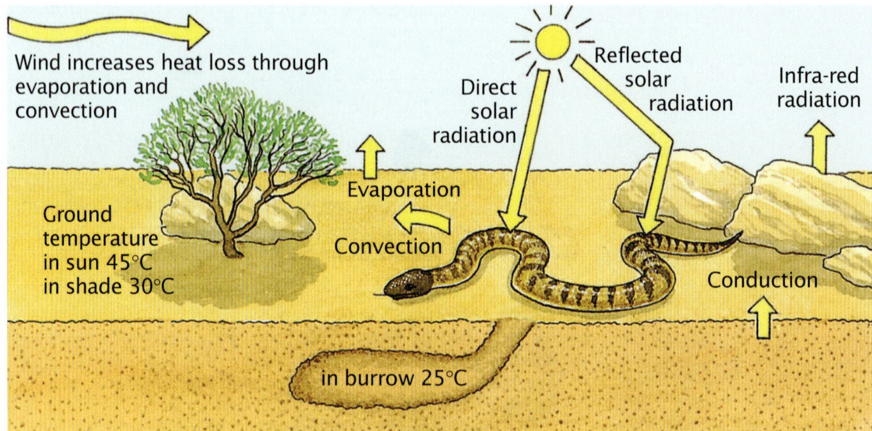

Figure 10.23 A snake can gain and lose heat from its environment in several different ways. Changes in its body temperature occur through changes in behavioural and physiological characteristics. The snake moves into the shade if it is in danger of overheating but exposes itself to the sun to increase its body temperature.

Because they often cool down, ectotherms such as snakes tend to be sluggish. Their hearts are unable to pump sufficient blood to supply the oxygen needed for vigorous activity. Snakes and other ectotherms use anaerobic respiration during muscular activity. Because lactate builds up during anaerobic activity, the time spent in strenuous activity is limited. In general, snakes escape to cover when danger is near.

The rate at which a snake heats up as it basks in the sun is influenced by a number of factors. Snakes move in and out of the shade and vary their exposure to sun (see figure 10.24). Some change the shape of the body. A flattened body exposes a greater surface area to the sun. The amount of wind and the air and ground temperatures of an area also have an impact. Physiological features include an increase in blood flow in vessels close to the skin as a snake basks in the sun. As a result, more heat is absorbed and transported to the inner body tissues and organs.

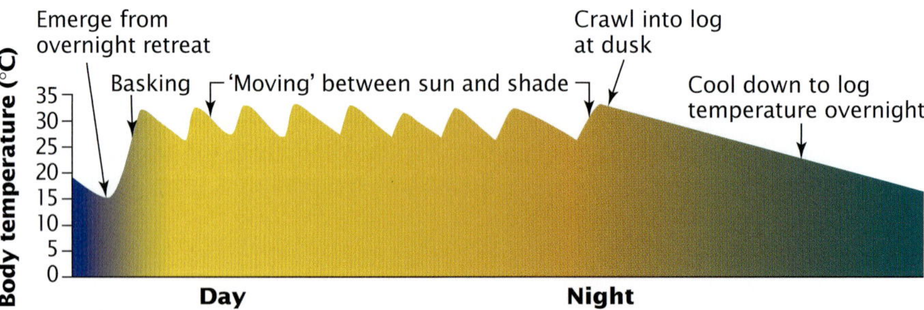

Figure 10.24 During the day, the black snake, *Pseudechis porphyriacus*, regulates its body temperature by moving in and out of the shade. The body temperature drops overnight.

Surviving in the cold

Ice can damage or kill

Processes that are essential for life include chemical reactions that take place between substances that are dissolved in liquid water — that is, in solution. These processes cannot take place in solid water (ice). If all the liquid water in a living organism were replaced by solid water, life would be destroyed. When ice forms, the solid water expands. If cells freeze, the expanding ice crystals rupture the cell membranes and kill the cells.

DEATH IN A BUSHFIRE

The photograph in figure 10.25 appeared on the front page of *The Age* on Friday 4 September 1998. The caption read: 'Twelve metres between life and death: two identical fire tankers — one burnt out, another unscathed — in the blackened bush near Linton'. Five men survived in the tanker on the right-hand side, while five men died in the truck on the left-hand side.

Figure 10.25 One fire truck was burnt out and the other was unscathed in a serious bushfire in 1998.

Distance was not the important factor with the firefighters and their trucks shown in the photograph.

The truck containing the firefighters that survived the fire had a reserve of water. Two of the crew huddled under a fire blanket in the cabin. The remaining three firefighters in the back of the unburnt truck turned small water hoses on themselves and then pointed them skywards so that water rained over the whole of their truck. Note the unburnt vegetation near the unburnt truck. This vegetation was also protected by the veil of water sprayed over the truck.

The men who perished in the fire were in a truck with no water. The change in fire direction happened so quickly there was no time for the men to get to the other truck. The fierceness of the fire, which is indicated by the complete absence of living vegetation near the burnt truck, meant that radiant heat would have been extreme and death inevitable for the unprotected. Water in the front truck was insufficient to protect the second truck.

A key issue identified was that low-water-level warning devices should be installed in all Country Fire Authority tankers.

Figure 10.26 Radiant heat can kill. Remember, as soon as you become aware of a fire, cover up to survive.

Surviving a bushfire

Bushfires are an integral part of the Australian bush. The Ash Wednesday fires of 16 February 1983 caused the death of 47 Victorians and 26 South Australians. How can your chance of survival be increased if you are in such danger? Remember that, apart from the flames themselves, it is the level of radiant heat that kills.

Make sure you are well clothed and take cover. Wear protective clothing to reduce your exposure to radiant heat. Wear long pants and a long sleeved shirt or light pullover. Natural fibres such as light wool or close-weave cotton are best. Wear solid footwear, preferably leather, and cover your head with an appropriate hat. Remember — **cover up to survive**.

Take cover inside your house. You will be protected from the radiant heat. Shut windows and doors. This ensures your supply of oxygen and prevents embers from blowing into the house.

You run the risk of becoming dehydrated in a bushfire. Drink water often even if you don't feel thirsty. Avoid alcohol and fizzy drinks.

If caught on the road in a car DO NOT get out and run. Stay in the car until the fire passes. Park the car with lights on and the engine running in a clear area away from vegetation, especially any that is dry. Close the windows and vents and get as low as you can within the car and cover yourself with a woollen blanket (see figure 10.27).

Figure 10.27 In the country and other fire-prone areas always carry woollen blankets in your car. They will help protect you from radiant heat.

Radiant heat can be the killer. It can lead to heat exhaustion, heart failure and dehydration. Some people have died from asphyxiation (lack of oxygen) during a bushfire. Why do you think this occurs?

You may wish to visit the Country Fire Authority website at www.cfa.vic.gov.au.

ODD FACT

What is frostbite? At temperatures below freezing, body parts such as hands, feet, nose, chin and ears are at risk of damage from the cold. Sometimes just the skin freezes. In more severe cases, the skin and underlying tissues become frozen. If ice crystals form, the affected part of the body can be permanently damaged. Gangrene may result from damage to the blood supply. In this case, amputation of the frostbitten part may be necessary.

Many living things can exist on land in Antarctica or the Arctic. During winter, the air temperatures fall well below the freezing point of pure water. How do living things survive in these low temperatures? Organisms have special features or behaviours that enable them to survive extremely low temperatures.

Pure water freezes at 0°C, but water with dissolved material in it has a lower freezing point than this. For example, a very concentrated salt solution (280 grams per litre or 4.8M) starts to freeze only when the temperature falls to about –18°C. One strategy used by some living things to assist their survival in very low temperatures is to produce **antifreeze** substances. For example, some insects, fishes, frogs and turtles can survive in regions that have low temperatures during winter. These animals make antifreeze substances such as glycerol, amino acids and sugars, or mixtures of substances, at the start of the freezing season. These antifreeze substances are released into their body fluids. The presence of these dissolved substances lowers the freezing point of their body fluids to well below that of the surrounding water temperatures. This means that the body fluids of these organisms stay liquid.

Some frogs and toads burrow underground to avoid freezing temperatures.

Birds and mammals living in Antarctica or the Arctic use another strategy to protect themselves from the damaging effects of low temperatures. Birds and mammals convert chemical energy present in their food into heat energy. This internal supply of heat keeps the body temperatures of these birds and mammals well above the freezing point of pure water. This heat is retained by excellent insulation: mammals have **insulating layers** of fat under the skin and thick fur and birds have layers of feathers. Would you expect that these Antarctic animals would need to eat more or less than animals of comparable size living in temperate conditions?

Burramys *has a long sleep*

The mountain pygmy possum, *Burramys parvus*, is the only Australian mammal that lives permanently in alpine regions. Its distribution is limited to two small areas (see figure 10.28), one in Kosciuszko National Park of New South Wales and the other near Mount Hotham in Victoria.

Burramys has both behavioural and physiological features that enable it to survive the low winter temperatures of its alpine environment. It collects and hides seeds and fruits for use during winter. Unlike other pygmy possums, *Burramys* has no storage of fat in its tail. At low temperatures during winter, *Burramys* goes into a torpor that is equivalent to hibernation. When mammals hibernate, their heartbeat slows down considerably and their breathing rate drops. Body metabolism is significantly reduced and their body temperature drops. In captivity, *Burramys* can hibernate at about 6°C and remains in that state for three to seven days at a time. Normal body temperature is around 36.1°C and during hibernation drops to that of the environment. The body metabolism of *Burramys* in hibernation ranges between 0.6 per cent and 3.9 per cent of normal metabolic rate of an active *Burramys* at 6°C. Hibernation and the reduced metabolic rate for periods means that the amount of food required by an animal, overall, to survive the winter period is reduced.

(a)

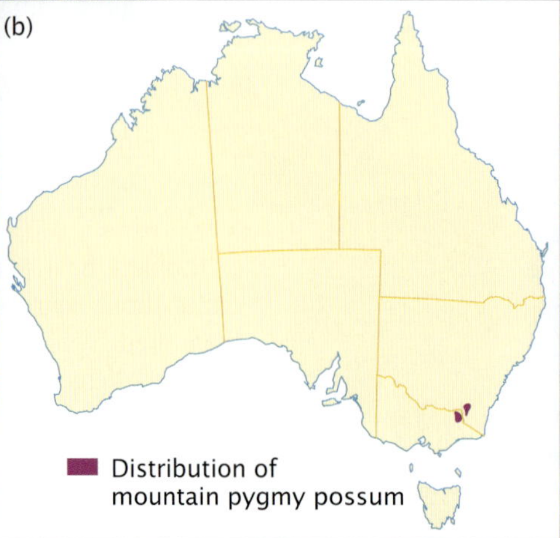

Figure 10.28 **(a)** Mountain pygmy possum, *Burramys parvus* and **(b)** its distribution, limited to two small areas

Animals surviving in water

ODD FACT

Of approximately 30 000 species of fish, most are exothermic but a few, including marlin (*Maikaira* spp.) and tuna (*Thunnus* spp.), are endothermic.

The range of temperature variation in an aquatic environment is far less than in a terrestial environment. However, organisms that live in water must be able to meet many of the same needs as organisms that live on land. Temperatures must be appropriate, and prey must have a reasonable chance of escaping their predators, just as predators must have a reasonable chance of catching their prey. In addition, there may be situations that are unique to a watery environment.

Whales and other aquatic mammals

Some marine animals breathe air. The time they can stay under water is determined by the amount of oxygen they are able to carry in their lungs or store in other body tissues. Mammals such as elephant seals, *Mirounga leonina*, and sperm whales, *Physeter macrocephalus*, that dive to great depths, are able to do so because they have special characteristics that increase their oxygen-carrying capacity. For example, they have a much higher concentration of red blood cells in their blood than many other mammals.

Whales and dolphins (order Cetacea) are mammals that spend their entire lives in water. Like all mammals, they are endothermic and they breathe air, and so must come to the surface every so often. The females give birth to young that they suckle on milk secreted by mammary glands.

Most land mammals have an insulating fur coat that assists in the regulation of the body temperature. Whales and dolphins rely on an insulating layer of fat or blubber below the skin. This layer may be up to 50 centimetres thick and can vary with the different seasons. Cetaceans maintain a stable body temperature of 36–37°C in an environment that is usually less than 25°C and may be as low as 10°C. In addition to blubber under the skin, fat may also be deposited around organs and tissues such as the liver and muscles, and in bone in the form of oil. These deposits can make up to half of the body weight of an animal.

Countercurrent systems to warm blood

Whales and dolphins also maintain their body temperature by using a **countercurrent exchange** system (see figure 10.29). There is a fine network of vascular tissue within the fins, tail flukes and other appendages. An outgoing artery is paired with an incoming vein. Blood coming from the body core to the skin is warm. Blood flowing from the skin back to the body core has been cooled. In this countercurrent exchange system, heat in the blood coming from the core flows to the blood that is returning from the skin to the body core. This warms the blood flowing in from the skin and so prevents the venous blood from cooling the internal organs and muscles. At the same time, the blood moving out to the skin is cooled and so the loss of heat across the skin is reduced.

Heat is readily lost from appendages such as hands and feet. Whales and dolphins have few protruding parts (fins and tail flukes). This means that they have a relatively small surface area-to-volume ratio and heat loss across the skin is further minimised. These features enable large whales to live in the cold waters of the Antarctic Ocean.

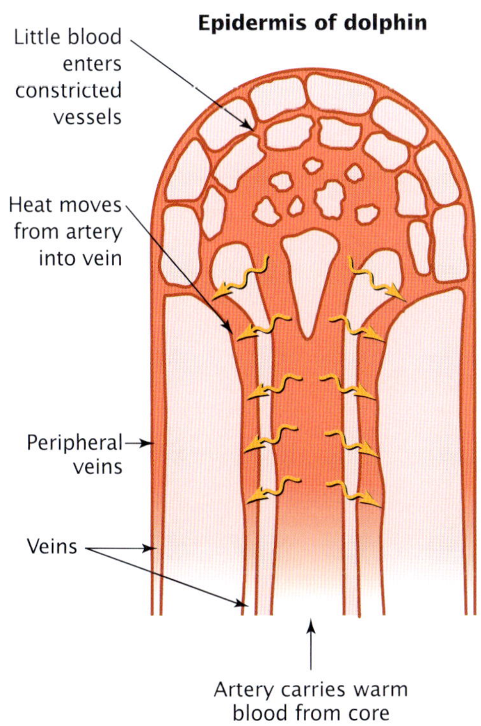

Figure 10.29 A countercurrent exchange system in the skin of dolphins. When the animal needs to conserve heat, the outermost blood vessels contract, little blood flows and heat loss from these vessels is reduced. In addition, heat flows from the warm blood coming from the core of the body into the cooler blood that is returning to the body from the skin.

KEY IDEAS

- Organisms have structural, behavioural and physiological adaptations that equip them to survive in their habitat.
- Extremes of environmental temperature can overwhelm the homeostatic mechanisms for body temperature control in an organism.
- Some animals go into torpor or hibernate to survive sub-zero temperatures.
- There is less temperature variation in aquatic environments than in terrestrial environments.
- If the water in cells freezes, the cells are killed.
- Animals that remain at cold latitudes during winter have adaptations that prevent cells from freezing.

QUICK-CHECK

13 Give one structural and one behavioural adaptation of a particular mammal that enables it to survive in hot, dry regions.
14 Give one structural and one behavioural adaptation of a particular animal that enables it to survive in a cold environment.
15 List two actions that increase your chance of survival in a bushfire.
16 Many desert mammals have countercurrent systems leading to the brain. What is the purpose of such a system and its structure?
17 Why is ice lethal to cells when water is a necessary and significant component in all living tissue?
18 How do hibernating animals survive their drop in body temperature?

Plant responses to temperature change

An enzyme in a plant cell has the same general characteristics as an enzyme in an animal cell, including an optimal temperature at which the enzyme has maximum efficiency. Plants tend to maintain their temperature within an optimal range to ensure optimal metabolic action and to minimise damage that can occur at extremes of heat and cold.

Plants in a hot environment

Green plants depend on radiant energy from the sun to carry out photosynthesis. However, only a small fraction of energy absorbed is used. To prevent overheating, a plant must lose much of the radiant energy it absorbs. A plant does this in the following ways (see figure 10.30):

- *Radiation* — a plant radiates heat to objects in its environment.
- *Transpiration* — plants are cooled when heat within them is used to evaporate water from cell surfaces. The water vapour formed exits a plant mainly through the stomata of leaves, with some loss from the cuticle, in a process called **transpiration**. If water loss continues, guard cells become less turgid and stomata close. Excessive water loss can cause plant death.

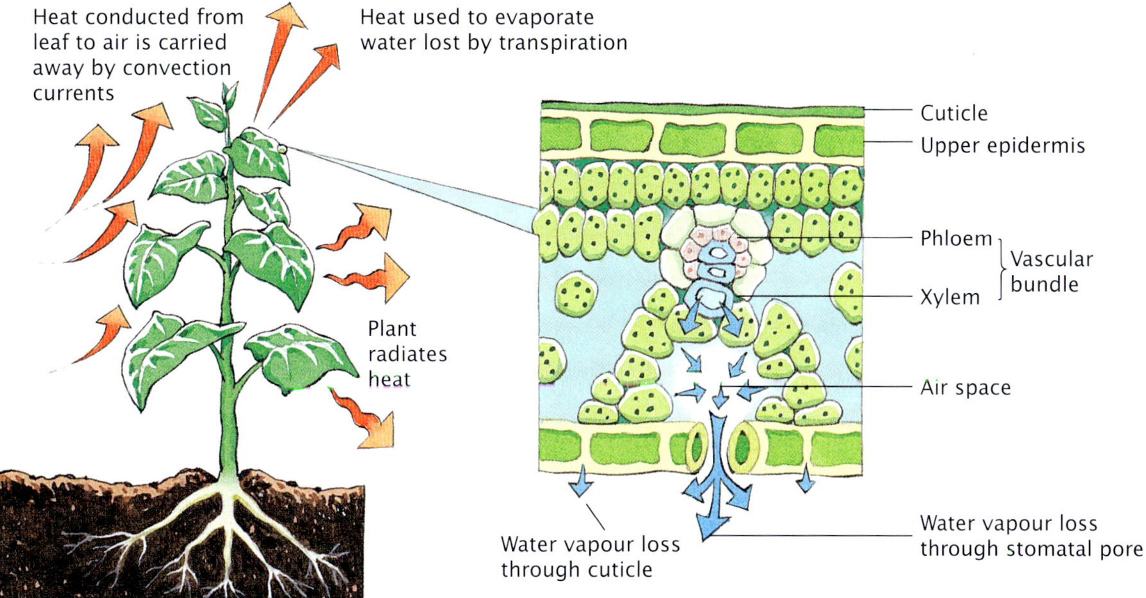

Figure 10.30 A plant must lose much of the heat it absorbs from the sun. It does this by radiation, conduction and convection and by evaporating water.

- *Convection* — air surrounding a plant becomes heated and hence is less dense than air further away from the plant. The heated air rises, carrying heat away from the plant.

Other factors that affect heat loss from, or heat gain by, a plant are as follows:

- *Leaf shape* — leaves are thinnest where the two surfaces of a leaf come together and lose most heat from that region. The larger the ratio of edge length to surface area of a leaf, the faster the leaf will be cooled. In figure 10.31, leaf A has a larger ratio of edge length to surface area than leaf B and will cool more quickly than leaf B.
- *Heat-shock proteins* — plants in temperate climates produce special proteins, called heat-shock proteins, at about 40°C. It is thought that these proteins may protect enzymes and other proteins in some way so that they are not denatured as the temperature rises.
- *Leaf orientation* — in hot weather, the leaves of some plants orient themselves so that a minimum surface area is exposed to the direct rays of the sun. Leaves hang so that their flat blade surfaces are parallel with the rays of the sun, and less radiant energy from the sun falls on the leaf (see figure 10.32). The leaves of many eucalypts orient themselves in this way.

Figure 10.31 Which leaf has the greater ratio of edge length to surface area? Which leaf will cool more quickly?

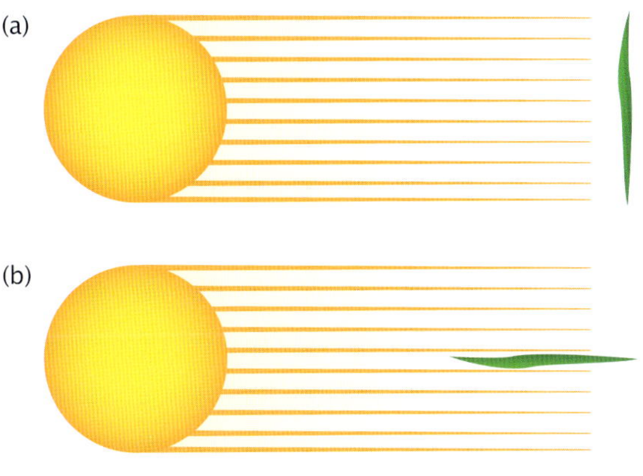

Figure 10.32 An aerial view looking down on two leaves. In **(a)**, a leaf has maximum exposure to the sun's rays. In **(b)**, the leaves minimise the radiant heat they receive by hanging vertically with the leaf blade surface parallel with the sun's rays. Compare the radiant heat falling on each of the leaves.

Figure 10.33 Boab trees, *Adansonia gregorii*, survive in very hot climates by storing water in a swollen bottle-shaped trunk and dropping their leaves in summer. At what time of year was this photo taken?

- *Structure* — one Australian native species that survives well in a hot environment is the boab tree (*Adansonia gregorii*) which grows to about 15 metres high. The boab or bottle tree has an unusually thick, bottle-shaped trunk (see figure 10.33) that can be up to five metres in diameter. The thick trunk is a structural adaptation for water storage. Young trees have a dense crown of leaves but older trees are more sparse and spreading. Boab trees are deciduous and they shed all of their leaves during the very hot summer months. This means that the tree has no stomata during the summer and a far smaller surface area through which water can be lost. The tree also has a far smaller surface area through which heat can be absorbed.
- *Leaf fall* — although eucalypts are 'evergreens', they do drop leaves continually. In very hot areas, some eucalypts increase their leaf fall during the dry season, thus decreasing the surface area through which heat may be gained and water vapour lost through transpiration.

Plants in a cold environment

Many plants survive in sub-zero temperatures without being damaged by these extremely low temperatures. Unlike animals, plants do not produce an 'anti-freeze'. They gradually become resistant to the potential danger of ice forming in their tissues as the temperature falls below 0°C. How does this occur?

Remember that water is transported through plants in very fine xylem vessels and is subjected to a number of forces. These forces affect the way in which water behaves in plants in freezing temperatures. As the temperature surrounding the plant drops below freezing, ice forms suddenly in the spaces outside the living cells of the plant. The inside of the cells doesn't freeze because the concentration of ions in the cytosol is greater than the concentration outside the cell. The cytosol has a lower freezing point.

Because ice has formed, the concentration of water inside the living cells is higher than the concentration outside and so water moves out of the cells. The ice crystals outside the cells grow (see figure 10.34). The movement of water out of the cells increases the ion concentration inside the cells and so lowers their freezing point even further. The living cells are then able to withstand further drops in the external temperature because the more concentrated cytosol acts as an anti-freeze. The ice crystals grow between the cells and do not damage the cell membranes which are pliable and bend under pressure of the ice. Many species of trees are able to withstand extremely low temperatures (see table 10.3).

The coldest recorded temperature in Australia was –23°C at Charlotte Pass near Mount Kosciuzko in June 1944.

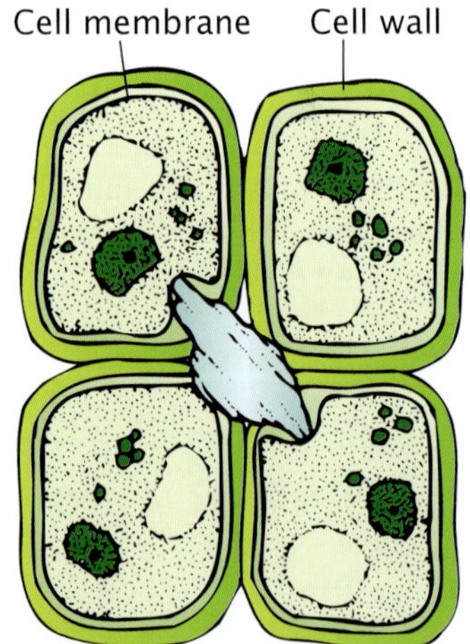

Figure 10.34 Ice formation in living plant tissue. Water leaves cells and adds to ice crystals growing in the spaces between the cells. Although the ice punctures cell walls, the cell membranes are merely pushed inward and the cells remain intact.

Table 10.3 Lethal temperatures for some trees. Many species of trees can tolerate extremely low temperatures before they are killed. The temperatures at which the living tissue in a tree is killed influences the latitudes at which it can grow. Which tree is most likely to be found in the northerly latitudes of Canada?

Species	*Temperature (°C) at which killed*
redwood (*Sequoia sempervirens*)	–15
southern magnolia (*Magnolia grandiflora*)	–15 to –20
swamp chestnut oak (*Quercus michauxi*)	–20
American beech (*Fagus grandifolia*)	–41
sugar maple (*Acer saccharum*)	–42 to –43
black cottonwood (*Populus trichocarpa*)	–60
balsam fir (*Abies balsamea*)	–80

Ultimately, if there is an excessive drop in the surrounding temperature, ice crystals will form inside the cells which will die and so the tree may die. It has been suggested that an excessive drop in temperature damages the protein molecules that form part of the cell membranes so that ions can leak out of the cell.

Australia does not experience the sustained extremes of cold temperatures found in many other countries and low temperature is rarely a limiting factor for plant growth. Growth of native plants in Australia is determined by whether a plant has the adaptations to survive the various altitude zones and their associated temperatures. Some plants, particularly exotic garden plants, may be killed or damaged by an unusually severe frost.

KEY IDEAS

- Plants maintain their temperature within an optimal range.
- Many plant characteristics facilitate heat loss, others minimise heat uptake.
- Ice formation between plant cells facilitates plant survival in sub-zero temperatures.

QUICK-CHECK

19 What is meant by antifreeze in animals and plants and why are these substances important?
20 Explain two features of plants that enable them to reduce heat uptake.
21 Explain the dangers of evaporative cooling for a plant.

Water balance in living organisms

At the start of this chapter, we considered water balance in the tarrkawarra. In the following section, we will consider water balance in a range of other organisms.

Water balance in mammals

Kidneys — organs for water balance control in vertebrates

As we considered in chapter 6 (pages 151–4), the **kidneys** are the organs that control water balance of the body for all vertebrates. Kidneys eliminate nitrogenous wastes from the body while simultaneously ensuring water balance. The kidneys play a major role in stabilising the internal environment of the body. They are vital organs; you would survive for only a few days without them.

Water balance and blood pressure

Water conservation in the body is ultimately associated with the maintenance of blood pressure because, as water varies, blood pressure also varies. Increased water raises blood pressure; decreased water lowers blood pressure. The processes of **osmoregulation** and blood pressure control interact. The two significant compounds involved are an **antidiuretic hormone (ADH)**, also called **vasopressin**, and **renin**.

Vasopressin in action

Vasopressin is an antidiuretic hormone (ADH) produced by neurosecretory cells (see figure 10.35) in the hypothalamus of the brain. Blood contains very little vasopressin if the body contains sufficient water. Neurosecretory cells are activated when **osmoreceptors** in the hypothalamus detect a rise in blood solutes. This means there is a drop in water concentration in the blood that could be due to either insufficient intake of water or excessive sweating or diarrhoea. Vasopressin flows through the axons of the neurosecretory cells to the posterior pituitary gland, where it is released into the bloodstream. It may also be stored.

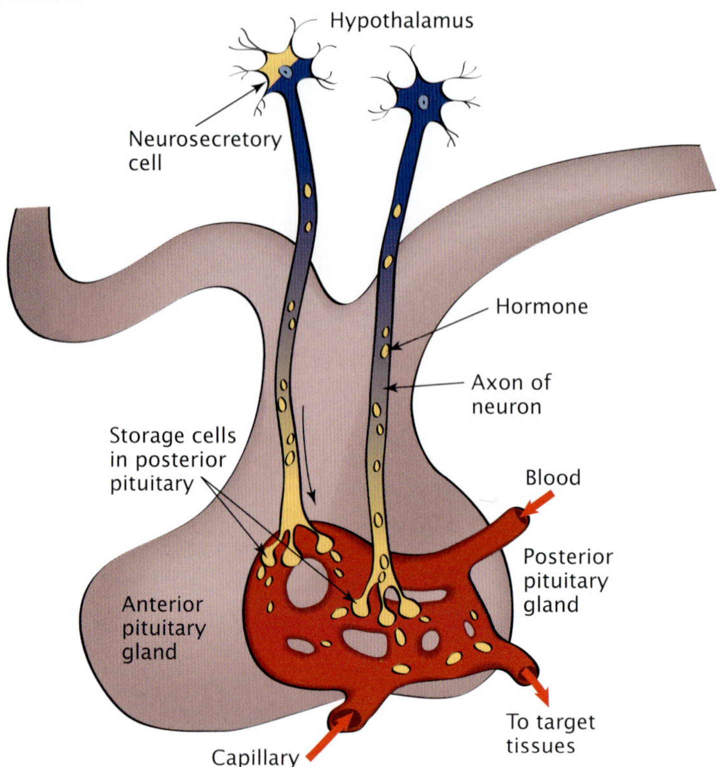

Figure 10.35 Neurosecretory cells in the hypothalamus in the brain produce antidiuretic hormone (ADH) or vasopressin. Vasopressin flows through axons to the posterior pituitary gland, where it is stored or secreted into the bloodstream.

Vasopressin is transported to the kidneys where it increases the permeability of distal tubules and collecting ducts to water. The amount of water reabsorbed by these areas increases and the concentration of solutes in the blood declines. Negative feedback then leads to a decreased secretion of vasopressin.

The osmoreceptors also generate a sensation of thirst in the body when they detect a rise in blood solutes. Increased drinking also acts as a feedback mechanism leading to a reduced secretion of vasopressin.

Aldosterone is essential for life. If it is absent, excessive amounts of sodium are excreted and death occurs within a few days. Treatment with injections of the hormone is possible. Drinking a salt solution also replaces excreted salt in cases of reduced production of the hormone.

Renin in action

When dehydration begins, the blood volume decreases and blood pressure falls.

Renin is secreted within the kidneys when blood pressure falls. A fall in blood pressure reduces glomerular filtration. Pressure-sensitive receptors in the kidney register this and afferent arteriole cells secrete renin. Renin initiates chemical reactions that cause the adrenal cortex of adrenal glands to release **aldosterone**. Aldosterone acts on nephron distal tubules, sodium ions are actively reabsorbed from the tubules, water follows and hence blood pressure rises.

The way in which vasopressin and renin act in water conservation and blood pressure control is outlined in figure 10.36.

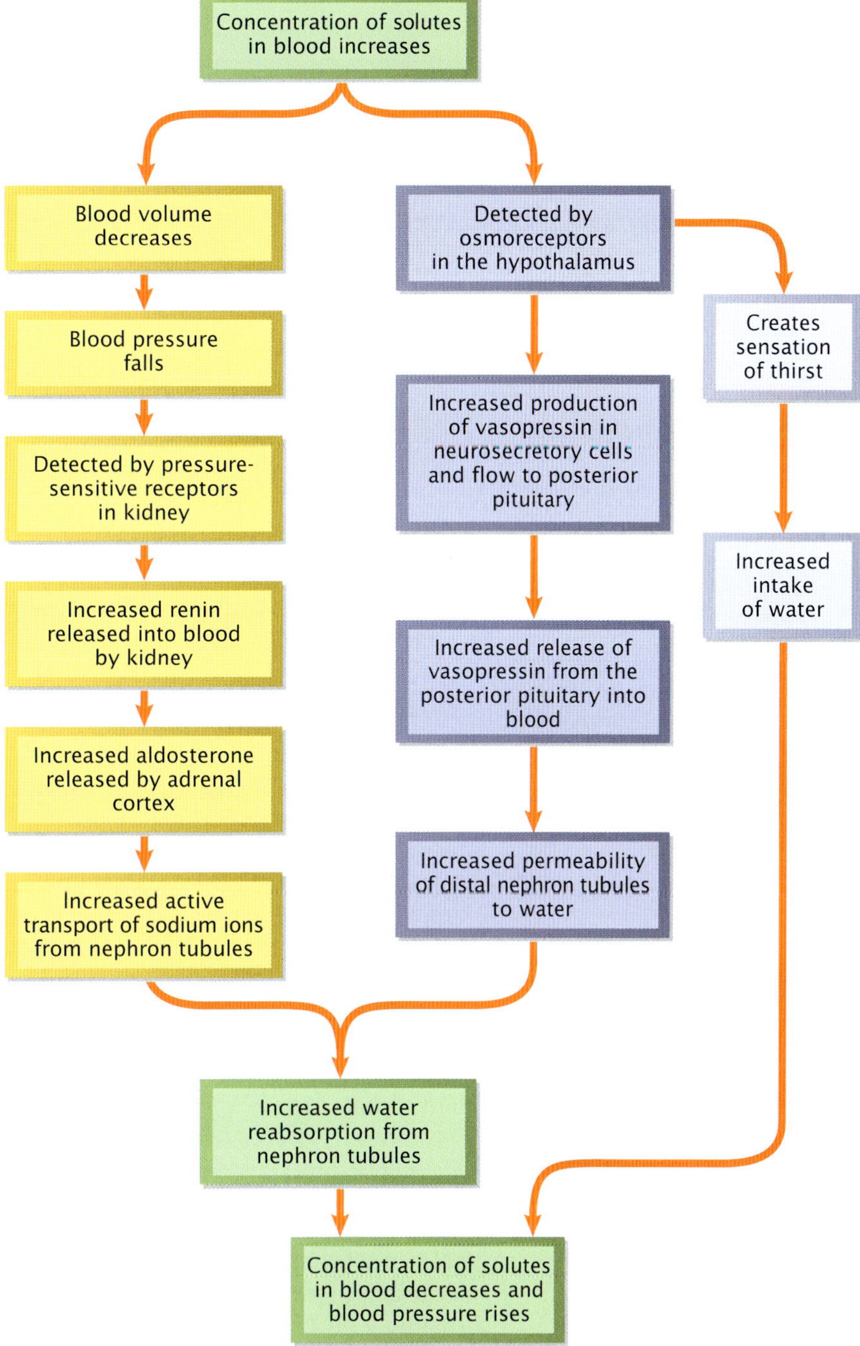

Figure 10.36 A simplified flow chart showing how vasopressin and renin both play a role in water conservation

Water balance in *Amoeba*

The contents of single-celled organisms such as *Amoeba* are at a higher concentration than the surrounding fluid in which they live. Water enters across the cell membrane by **osmosis**. If this continues unchecked, the cell will burst. Contractile vacuoles in the cytosol accumulate excess water that is then expelled from the cell. Thus, water balance within the cell is maintained. Waste products of metabolism can diffuse across the membrane.

Water balance in fish

We have already considered excretion and the role played by water in fish (see chapter 6, pages 154–5). A brief summary of osmoregulation in fishes is given in table 10.4.

Table 10.4 Summary of osmoregulation in freshwater and marine (saltwater) fish

Freshwater fish	*Marine (saltwater) fish*
Tissues hypertonic to surroundings	Tissues hypotonic to surroundings
Concentration gradient results in a loss of salts and an uptake of water	Concentration gradient results in a loss of water and an uptake of salts
Fish must counter these changes to maintain homeostasis	Fish must counter these changes to maintain homeostasis
1. Does not drink.	1. Drinks sea water
2. Kidney contains glomeruli and secretes copious amounts of very dilute urine that contains ammonia. Tubules actively reabsorb NaCl.	2. Minimal urine produced. Kidneys lack glomeruli. Tubules actively secrete $MgSO_4$.
3. Gill membranes permeable to water.	3. Gill membranes are relatively impermeable to water.
4. Gills actively absorb ions. Some ammonia leaves gills at the same time.	4. Gills actively secrete sodium from chloride cells; chloride ions follow.

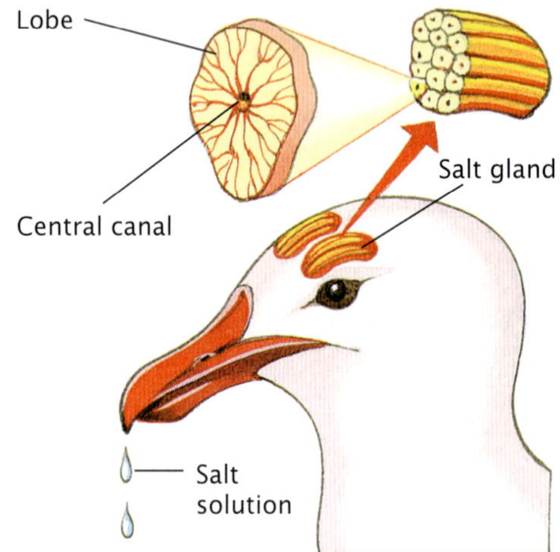

Figure 10.37 Salt drains from glands above the eyes into a canal that runs into the nose.

Water balance in seabirds

All birds excrete their nitrogenous waste as uric acid, the most efficient mode of excretion with regard to water requirements. However, because of a need to conserve water, seabirds have a particular problem — with salt. They take in relatively large amounts of salt in the food they eat and the water they drink. Water is reabsorbed in the cloaca so that uric acid is 3000 times more concentrated in urine than it is in blood. The salt level in urine, however, is relatively low. The high levels of salt in the blood must be excreted by an alternative route if water balance in the body is to be maintained.

Seabirds have salt glands above the eyes (see figure 10.37). The size of the salt glands can vary depending on the amount of salt in a bird's diet. Salt solution, twice the concentration of sea water, runs from these glands into the nostrils. A bird appears to have a 'runny nose' as the salt solution drips away.

Water balance in reptiles

In aquatic reptiles such as turtles and crocodiles, that have plenty of water available, nitrogenous waste is usually ammonia and urea. Terrestrial reptiles such as goannas (see figure 10.38) generally need to conserve water and their main excretory products are uric acid and urate salts. These are highly insoluble and little water is needed to eliminate them from the body.

Goanna kidneys have large numbers of kidney tubules (see figure 10.39a). If a goanna is dehydrated or has an excess of ions, the number of active tubules is significantly reduced. This reduces the amount of filtrate produced and so conserves water. As the water taken in increases, the number of tubules activated also increases. The kidney plays a significant role in water conservation in the goanna.

Figure 10.38 A goanna is a lizard in the family *Varanidae*. It is also referred to as a 'monitor' and 'varanid'.

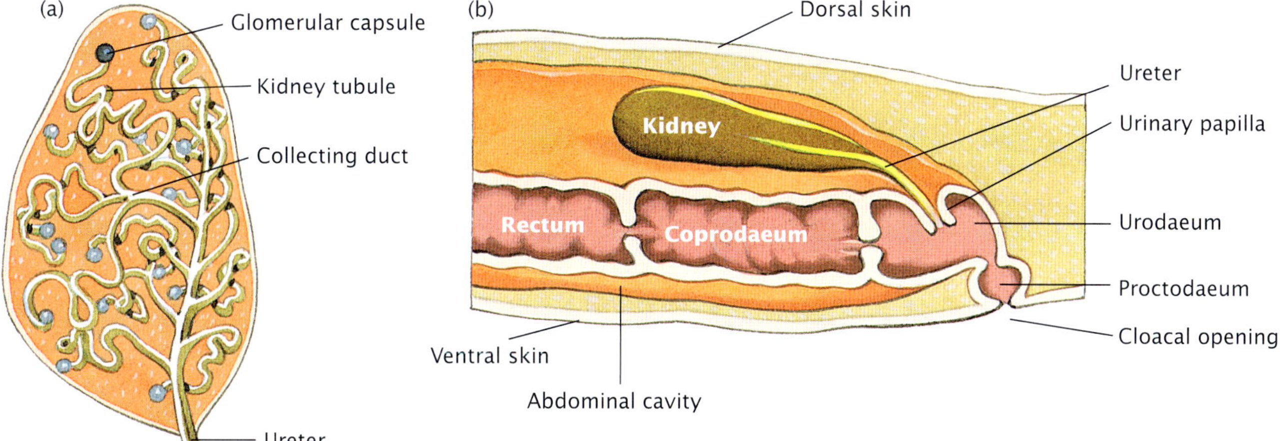

Figure 10.39 **(a)** Section through a goanna kidney showing kidney tubules **(b)** Longitudinal section through hind end of goanna showing entry of kidney ducts into the cloaca

A goanna has a cloaca, the terminal part of the gut into which kidney ducts open (see figure 10.39b). Urine moves from the kidney into the urodaeum then into the coprodaeum which has highly folded walls. Water is reabsorbed by the folded wall of the coprodaeum. In addition, urine is acidified in this region. Urate salts become insoluble and precipitate and can be excreted in solid form. The outcome is that more water can be reabsorbed from the walls of the coprodaeum. Goanna nitrogenous wastes leave the body by the cloacal opening.

An excretory pellet from a goanna with adequate water is only 48 per cent water. This represents a water loss of only 2.7 mL for every gram of nitrogen excreted. Compare this with the urine of the tarrkawarra which represents a loss of 7 mL water for each gram of nitrogen excreted.

Some reptiles have salt-secreting glands. In goannas these glands are in the nasal capsules, in turtles behind the eyes and in some crocodiles, the glands are on the tongue. All these glands are similar in structure and have densely packed secretory tubules radiating from a central duct (see figure 10.40). Note the structure of cells lining the tubules.

The interlocking folds of adjacent cells and large numbers of mitochondria are typical of cells with high metabolic activity. Both of these characteristics are essential for the production of secretions that have a higher concentration than that of blood. A concentrated solution of salt moves from cells into the central duct and drains from the nose, eyes or mouth to the outside.

Figure 10.40 **(a)** Goanna salt gland with its many secretory tubules radiating from a central duct. **(b)** Cell from secretory tubules. What do folds in adjacent cell walls and large numbers of mitochondria indicate?

Water balance in amphibians

Frogs live in fresh water. Because the concentration of ions in their tissues and fluids is higher than that of the surrounding environment, water continually diffuses by osmosis through the permeable skin. Also, salts continually diffuse across the frog's skin from the internal to the external environment (figure 10.41).

To counter these problems, frogs produce large quantities of dilute urine which contains urea and ammonia. Also they actively transport sodium and chloride ions across the skin from the surrounding water into the body to balance diffusion loss from the body.

In dry air, a frog produces concentrated urine and hence reduces water loss but still excretes large amounts of urea. Frogs also have cloaca.

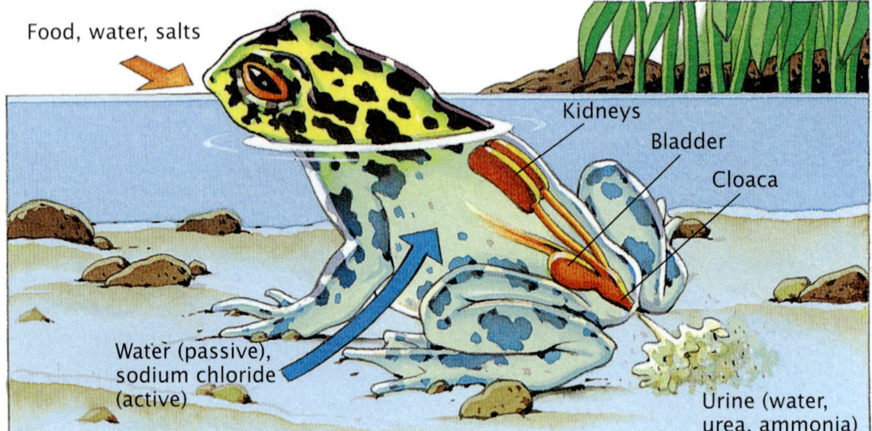

Figure 10.41 Note the input and output of water and salts in a frog. Why is active transport required for the movement of ions from water into the frog?

KEY IDEAS

- Water in living organisms must be maintained at a relatively stable level.
- Water loss from and water gain by an organism occur in many ways.
- Kidneys are essential organs for water balance.
- Kidneys eliminate nitrogenous wastes from the body at the same time as maintaining water balance.
- Vasopressin, renin and aldosterone are important compounds in the control of water balance and blood pressure in humans.

QUICK-CHECK

22 Where are vasopressin and renin formed and what is the role of each in water balance?
23 Explain how a variation in water balance causes a variation in blood pressure.
24 What are the three main kinds of nitrogenous wastes excreted by animals? What is the relative need for water in each case?

Water balance in plants

Also refer to pages 286–9 for information relevant for Australia.

Water makes up about 90–95 per cent of the living tissues of plants. Plants often grow in situations where they are continually losing water. Plants cannot move around and search for water. They have features that help them obtain and retain sufficient water for their cells to operate effectively. Under conditions of water shortage, plants maximise their opportunity to obtain and conserve water and at the same time minimise loss. They do these things in a number of special ways.

ODD FACT

The roots of creosote bushes (*Larrea divaricata*) produce a poison that prevents other plants, including other creosote bushes, from growing too close. The result of this is that creosote bushes are well spaced out, with roots that have no competition for the water in their area.

Stoma = singular
Stomata = plural

Doorways in the leaf: stomata

Most leaves are covered by a waterproof layer, the **cuticle**, through which relatively little water is lost. Most water is lost from a leaf through its stomatal pores (see figure 10.42). The walls of the mesophyll cells are moist and the air spaces around them contain water vapour. When **stomata** are open, gases, including water vapour, are able to diffuse in or out of the leaf. If water vapour moves through the stomata into the air surrounding the leaf, water evaporates from the moist surfaces of the mesophyll cells to replace that lost from the air spaces.

As water moves out of the mesophyll cells to replace that lost from their surfaces, water moves from small xylem vessels in the leaf into the mesophyll cells. In effect, water vapour moving out through stomata sets up a chain reaction which results in water moving through xylem vessels as a result of a pulling or sucking movement of the water ahead.

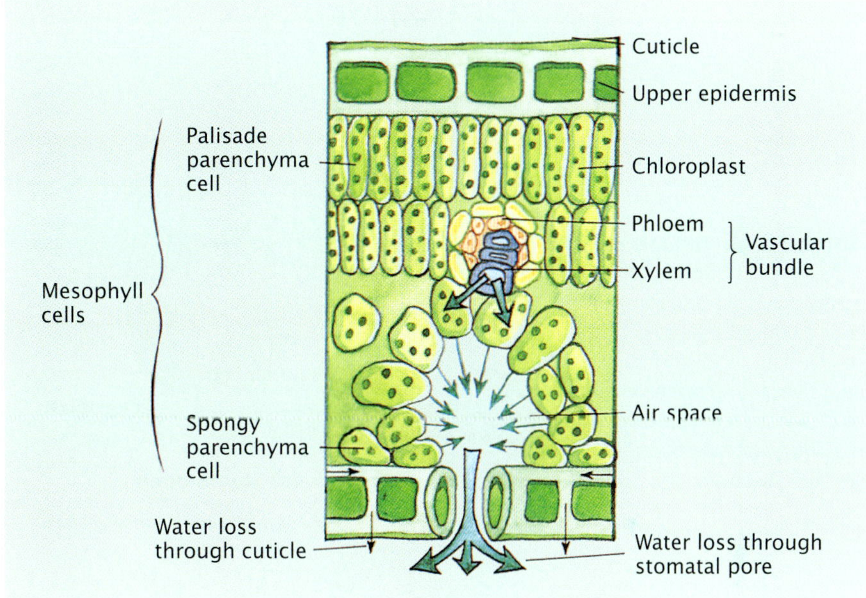

Figure 10.42 In a leaf, water moves from xylem into surrounding mesophyll cells. As water vapour moves out of the leaf through stomata, water evaporates from the moist surfaces of the mesophyll cells. Most water is lost from a leaf through stomata, but a little evaporates from the cuticle.

Transpiration: loss of water vapour

The movement of water from the roots, through the stem, and to the leaves where it may pass through the stomata as water vapour is called the **transpiration stream**. The loss of water vapour from a plant is called transpiration and occurs mainly through the stomata with some loss through the cuticle. Up to 98 per cent of water absorbed by a plant can be lost through transpiration. Only 2–5 per cent is retained within the plant.

You may have seen wilting plants, particularly on a hot day. Wilting can occur if the rate of loss of water vapour from a plant is greater than the rate of uptake of water by the plant. The rate at which transpiration occurs is influenced by a number of factors. Some of these relate to structural features of the plant itself while others are features of the surrounding environment.

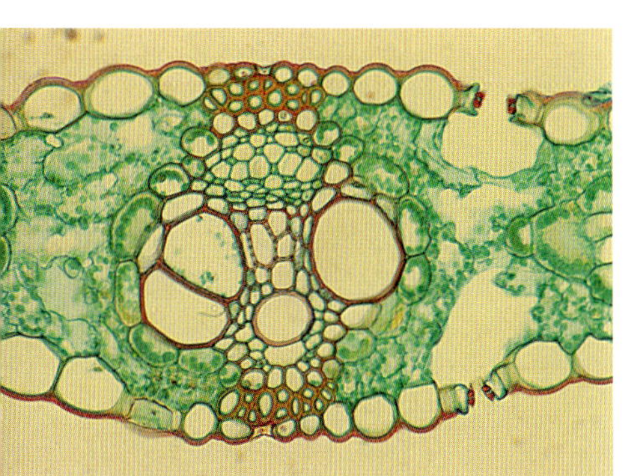

Figure 10.43 Transverse section (× 200) through a monocotyledon leaf. Note a stoma (arrows) on each leaf surface, and the large vascular bundle comprising large thick-walled cells of the xylem tissue and thin-walled cells of phloem.

Most water is lost through the stomata

Green plants require carbon dioxide to carry out photosynthesis. Although some carbon dioxide comes from respiration of cells inside a plant, this is insufficient to serve all the needs of a plant. Stomatal pores allow carbon dioxide to enter. Inevitably, stomatal pores that allow the entry of carbon dioxide also allow the exit of water vapour. Leaves contain many stomata (see figures 10.43 and pages 164–6).

ODD FACT

The 'dew drops' at the tip of some leaves in early morning in fact come from within the plant. The drops are excess water absorbed by roots when stomata are closed. It is pushed through the plant and out through hydathodes, permanently open pores.

In chapter 6, we mentioned that ions play a role in the control of stomatal aperture. How does this occur? Ions, particularly potassium ions, migrate into guard cells from adjacent cells as carbon dioxide concentration declines (see figure 10.44) when photosynthesis commences as light becomes available. As the concentration of ions rises, greater water absorption by the guard cells occurs and stomata open. When potassium ions move out of guard cells, water also moves out. This may occur at the end of a day. Reduced light means less photosynthesis. Carbon dioxide accumulates making the cell more acid. Potassium ions move out of the guard cells and this leads to water also leaving the guard cells. The guard cells become flaccid and stomatal pores close.

When a plant is in a situation during the day where water availability is reduced, and if water leaves the guard cells, this also leads to stomatal closure.

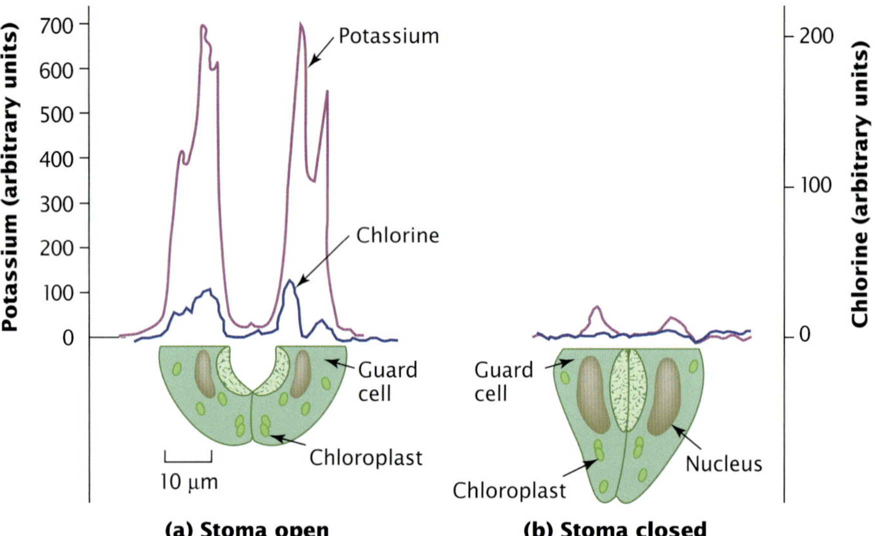

Figure 10.44 The distribution of potassium and chloride ions in guard cells of open and closed stomata. When guard cells are turgid **(a)**, they contain relatively high levels of potassium and chlorine ions. When ions leave the leaf **(b)** water is also lost, the guard cells become flaccid and stomata close.

Leaf structures help reduce water loss

The waterproof outer layer of a leaf, the cuticle, reduces water loss. The thickness of the cuticle is just one of the ways in which the structure of a leaf can vary depending on its particular environmental conditions (see figure 10.45). The thinner the cuticle, the more transpiration occurs. Plants that live in dry areas are known as **xerophytes** and show a variety of specialised features.

The cuticle is a noncellular layer of waxy material called cutin.

(a)

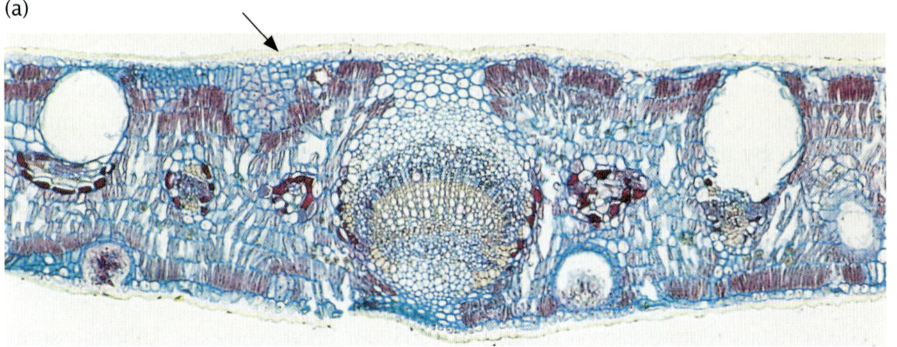

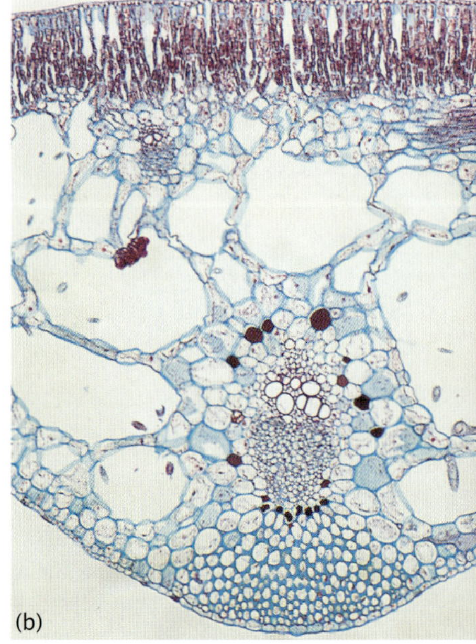

(b)

Figure 10.45 Transverse sections through leaves from two different species of plants: **(a)** *Eucalyptus globulus* and **(b)** waterlily, *Nymphaea* sp. The two leaves are at the same magnification. Note the thick cuticle (arrow) of the *Eucalyptus* leaf and apparent lack of cuticle on the waterlilly leaf. The thicker the cuticle, the less transpiration occurs.

Sunken stomata

In many plants, stomata are at the same level in the leaf surface as other cells. However, some plants, such as the fig (*Ficus carica*), have stomata that are sunk in pits on the leaf (see figure 10.46). A small pocket of air is trapped beside each stomal pore. This forms a barrier between the air spaces inside the leaf and the air that circulates around the leaf, and assists in the reduction of transpiration. Fig leaves also have two or more layers of epidermal cells that store water.

The shape of a leaf is important in relation to the amount of water it loses through transpiration. The edges of some flat leaves curl over and form a protective layer above the stomata in order to reduce transpiration rate. How do you think such behaviour helps to reduce water loss from a leaf? Many leaves have hairs or other projections on their surfaces. It has been suggested that these also serve to trap a layer of air around the plant and so contribute to a reduction in transpiration rate. The results of recent experiments indicate that this may not always be the case.

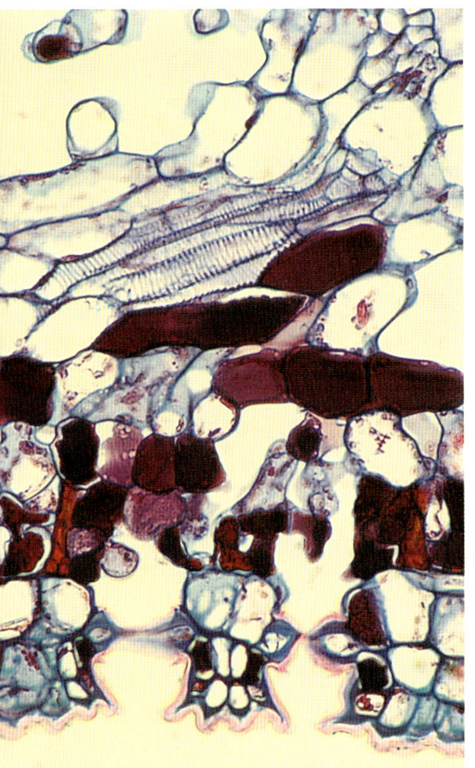

Figure 10.46 Some leaves have sunken stomata. Section (× 200) through a fig leaf (*Ficus carica*) showing sunken stomata. Note the xylem tissue above centre of the photograph. Suggest why sunken stomata may be of benefit to a plant.

Rolled up leaves

Figure 10.47 shows a rolled-up leaf of marram grass, *Ammophila arenaria*. Although this is not an Australian species, some related species in Australia have the same characteristics. These leaves have a number of features to restrict water loss, including:

- hinge cells that lose turgor if water is lost and cause the leaf to curl, creating a humid chamber for the stomata
- stomata on only one side of the leaf so that when the leaf curls, no stomata are directly exposed to the environment
- stomata located in 'folds' of the leaf so that they are shielded from air currents even when the leaf is unrolled
- a thickened cuticle on the surface that is exposed when the leaf curls.

Note the hairs on the upper epidermis of the leaf in figure 10.47b.

(a)

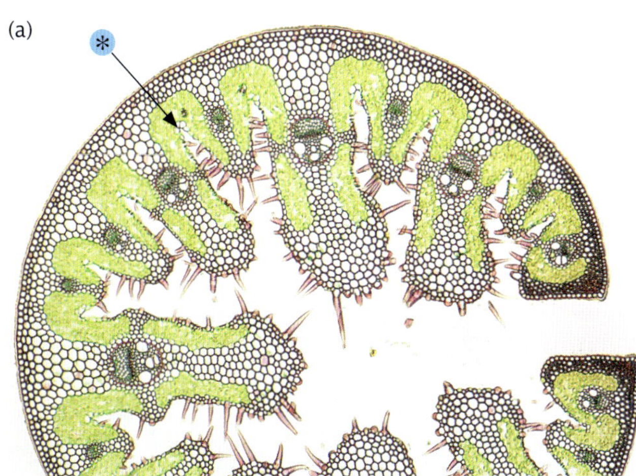

(b)

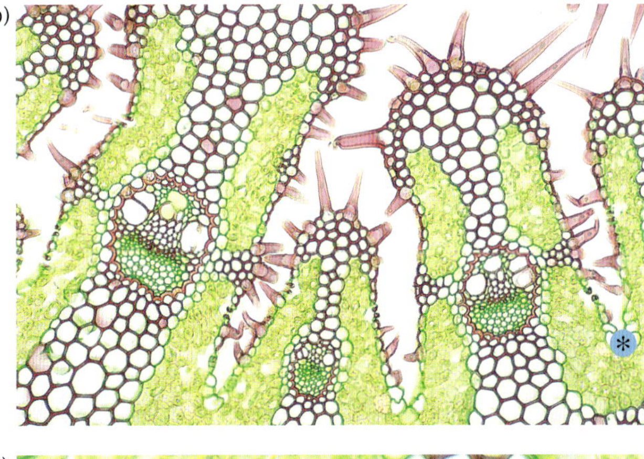

(c)

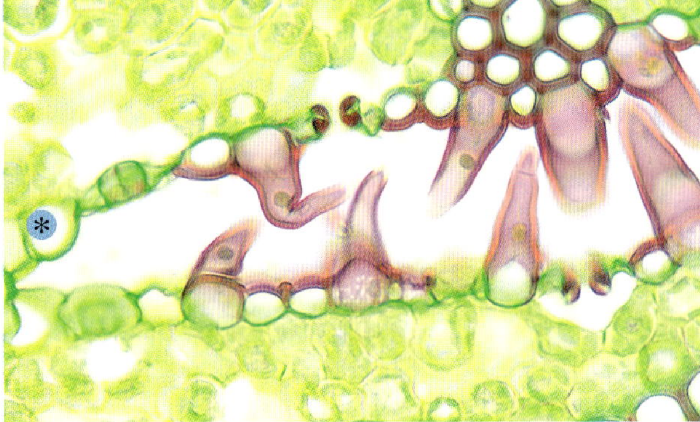

Figure 10.47 Transverse section of marram grass, *Ammophila arenaria*, at three different, increasing, magnifications. **(a)** Water loss in hinge cells (marked by ✻) causes the leaf to curl. Note the thickness of the cuticles on the upper surface (thin) and lower surface (thick). **(b)** Note the vascular tissue, photosynthetic tissue, thickened sclerenchyma cells and hair cells. **(c)** Can you identify stomata, photosynthetic tissue and hair cells in the upper epidermal layer?

ODD FACT

Some cacti avoid dehydration by having 90 per cent of their mass below ground, and extending their tops above the surface only when conditions are favourable.

Succulents

Some leaves have cells with very large vacuoles that act as water stores for the plant. Some succulent plants have very reduced or even no leaves and very succulent stems that store water, often for many years (see figure 10.48). Cacti have this characteristic. They can also absorb surface water efficiently because of their extensive shallow roots. They have very thick cuticles and few stomata, so the amount of water lost because of transpiration is low. There are no true cacti — that is, no plants belonging to the family *Cactaceae* — that are native to Australia. However, Australia does have 'drought succulents' — plants that are adapted to living in drought areas.

Figure 10.48 Cacti have succulent stems, few or no leaves and extensive shallow roots. Note the succulent stems of prickly pear cactus, *Opunta stricta*, which can store large quantities of nutrients and water. Also note the berries. These contain seeds that can be distributed over large areas by birds and other animals.

Cylindrical leaves

Hakea is an Australian plant whose leaves show many characteristics that enable a plant to conserve water (see figure 6.56, page 166). These leaf characteristics include:

- narrow and cylindrical
- epidermis with thick cuticle
- concentrically arranged palisade tissue, interspersed with dumbbell shaped, thick-walled sclereid cells
- sunken stomata (arrows).

How does each of these characteristics contribute to the reduction of water loss by a *Hakea* plant?

No leaves?

Acacias are wattles, a common plant in Australia. Although some adult *Acacia* species have leaves, most have **phyllodes** (see figure 10.49). Phyllodes may look like leaves but they are actually flattened petioles. A petiole is a leaf stem. Phyllodes carry out photosynthesis for the plant but lack the stomata of true leaves. This reduces the water such a plant will lose through transpiration.

Figure 10.49 Portion of an *Acacia* plant showing the true leaves of the plant and a developing phyllode. The leaves will eventually fall from the plant. (Refer also to figure 9.44, page 288.)

Factors affecting transpiration

Humid air is air that contains relatively high levels of water vapour. Water vapour moves out of stomatal pores in a leaf much more quickly when the leaf is surrounded by dry air than when the leaf is surrounded by humid air. This is because there is a much greater difference between the concentration of water vapour inside and outside a leaf when a leaf is surrounded by dry air (see figure 10.50). The more humid a day then the less water vapour is lost from a leaf.

The transpiration rate is much greater on a hot windy day than on a hot still day. On a still day, a boundary of still air surrounds the leaf. Water vapour leaving stomatal pores tends to stay close to the leaf, keeping the humidity of the boundary layer similar to the humidity inside the leaf. On a windy day, water vapour that diffuses out of stomatal pores is immediately blown away from the leaf; there will still be a significant difference in humidity inside and outside the leaf, and more water vapour will diffuse from the leaf. The windier a day is, then the higher is the transpiration rate.

ODD FACT

Saturated air at 10 °C contains nine milligrams water per litre of air and at 25 °C contains 23 milligrams of water per litre of air.

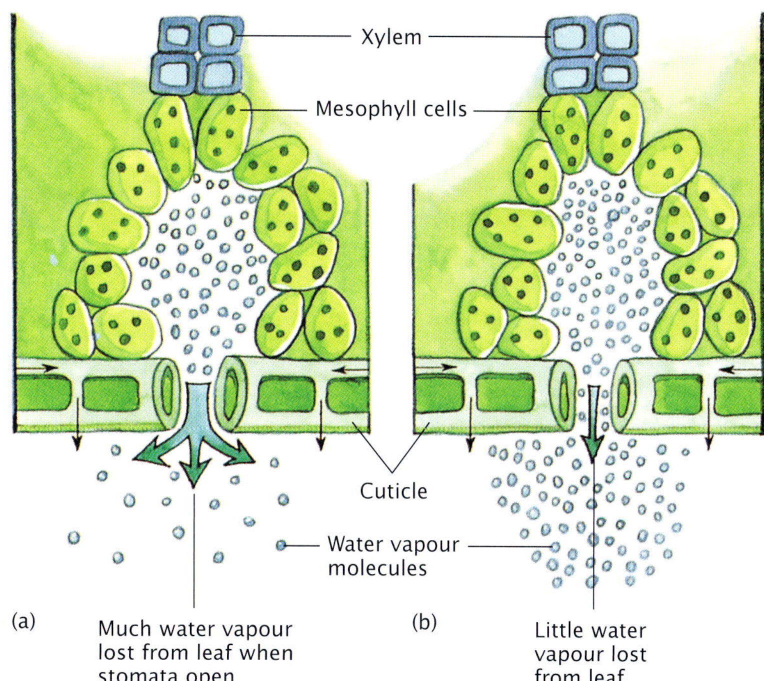

Figure 10.50 The amount of water vapour lost through stomata of a leaf depends on the relative amounts of water vapour inside and outside the leaf when stomata are open: **(a)** on a dry day, **(b)** on a humid day.

ODD FACT

In summer, flowers with large petals may lose so much water through their petals that stomata close, thus significantly reducing the rate of photosynthesis.

Air temperature is another factor that influences the transpiration rate of a plant. The higher the temperature, the greater will be the amount of water lost from the plant. As water evaporates from the surfaces of a plant, particularly the leaves, heat is required so the evaporating process cools the surface of the leaf.

Stomatal pores close if excessive water loss occurs. As long as there is sufficient water in the soil to replace the water that is being lost by a plant, stomata stay open.

KEY IDEAS

- Up to 98 per cent of water absorbed by plants is lost through transpiration.
- The transpiration stream carries water and ions to all parts of a plant.
- Plant factors and environmental factors influence the rate of water loss from a plant.

QUICK-CHECK

25 Why is water loss from a plant inevitable?
26 List the regions and cells through which water passes as it moves from the soil into a plant, to the leaves and to the atmosphere.
27 What causes stomatal pores to open?
28 Give two special structures that reduce water loss from a plant.
29 Explain how atmospheric humidity influences water loss from a plant.
30 List three characteristics of xerophytic plants and explain the role each plays in water conservation.

BIOCHALLENGE

The flow chart below represents the events that occur when the internal temperature of mammals becomes raised. Some of the boxes in the flow chart have been completed but eight are empty. Eight terms or phrases are given below the flow chart.

1. Complete the flow chart by placing each of the eight terms/phrases in an appropriate space in the chart.
2. Describe how dilation of skin arterioles contributes to a fall in internal body temperature.
3. Describe how sweating contributes to a fall in internal body temperature.
4. Describe how a reduced rate of cell metabolism contributes to a fall in internal body temperature.
5. Name a typical human behaviour that contributes to a fall in internal body temperature.
6. Name a behaviour of a particular Australian mammal that contributes to a fall in internal body temperature.

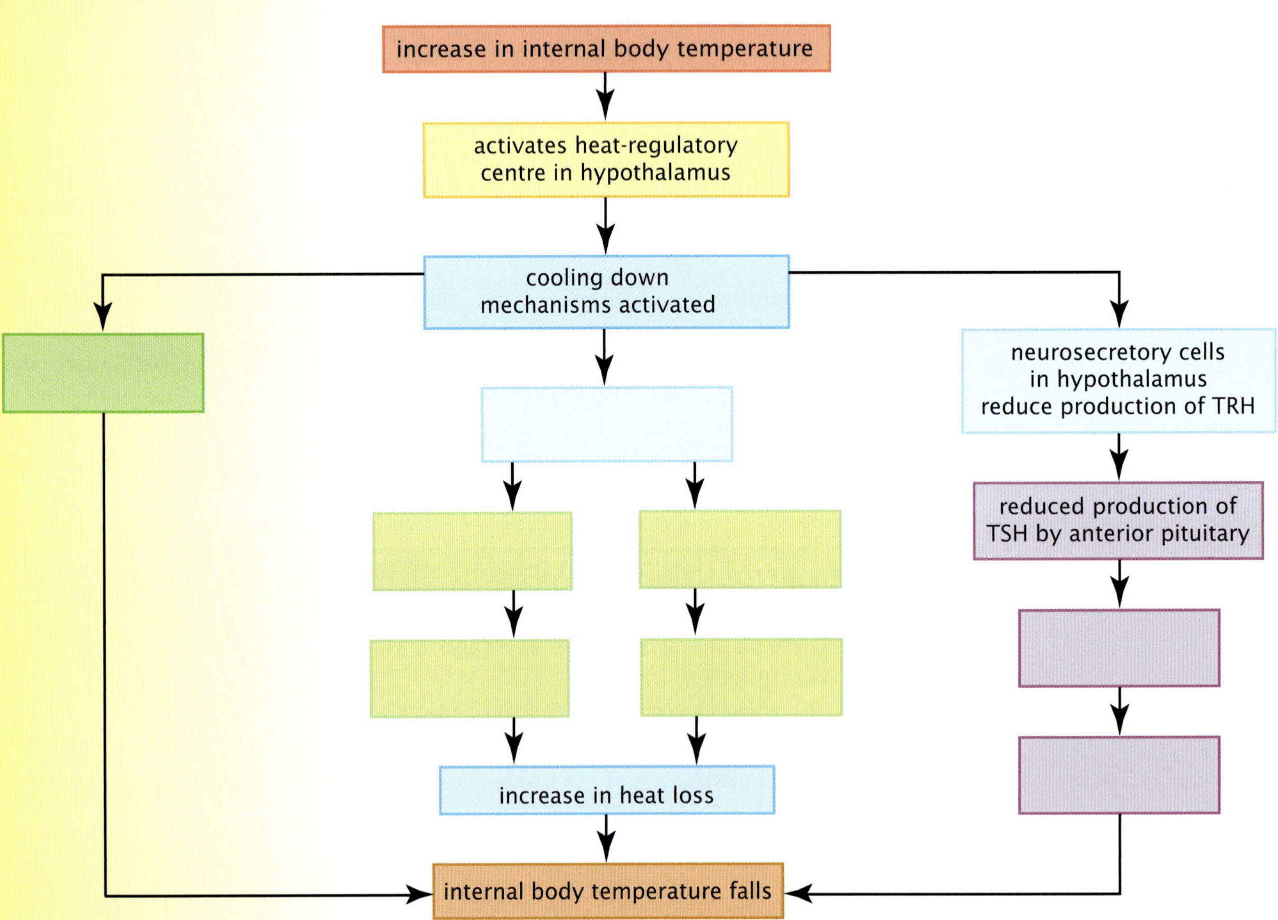

Figure 10.51

reduced rate of cell metabolism
skin blood vessels
sweat glands
neurones relay messages to
behavioural activities
sweating
reduced thyroxin from thyroid
dilation of skin arterioles

CHAPTER REVIEW

Key words

CROSSWORD

adaptations
affector neurons
aldosterone
antidiuretic hormone (ADH)
antifreeze
axon
behavioural features
cell body
central nervous system (CNS)
cochlea
conduction
connecting neurons
convection
core temperature
countercurrent exchange
cuticle
dendrites
ectothermic
effector neurons
endocrine glands
endocrine system
endothermic
evaporation
evaporative cooling
external environment
free water
glucose
homeostasis
homeothermic
hormones
humidity
hypothalamus
insulating layers
internal environment
kidneys
negative feedback
nerve cells
nervous control system
neurons
osmoreceptors
osmoregulation
osmosis
peripheral nervous system (PNS)
phyllodes
physiological features
pituitary
poikilothermic
radiation
renin
spinal cord
stomata
structural features
tolerance range
transpiration
transpiration stream
vasopressin
xerophytes

Questions

1 *Making connections* ▸ Use at least eight of the key words in this chapter and draw a concept map. You may use other words in drawing up your map.

2 *Applying your understanding in a new context* ▸ Some homes have central heating. When the air reaches a certain temperature, a thermostat turns the heat off. When the temperature drops below a certain level, a thermostat turns the heater on. Explain whether you think this is similar to, or different from, the control of internal body temperature.

3 *Applying your understanding* ▸

a Which photoreceptors are involved in a colour vision defect?

b Which photoreceptors would have a higher density in a mammal that is active at night?

c What major differences exist between touch and auditory receptors in terms of their distribution?

d Why do blind people 'read' braille with their fingertips rather than the palms of their hands?

e Which part of the mammalian ear:

i gathers sound waves from the external environment?

ii converts sound vibrations to nerve impulses?

iii connects the eardrum to the inner ear?

4 *Analysing information and communicating ideas* ▸ Refer to figure 10.8 on page 304. Scientists now believe that all taste buds detect all five basic tastes. Initial taste maps of the tongue drawn in 1901 (figure 10.8c) show identical

distribution for the four tastes known at the time and yet, for almost a century since then, incorrect taste maps have been presented in textbooks.

Suggest how such a mistake could be made by large numbers of people over such a long period.

5 ***Applying your understanding*** ▸ Suggest an explanation for the following observations:

a The young of tiny bats such as little bent-wing bats (*Miniopterus australis*) are always observed to be huddled very tightly together in large groups on the walls of caves, rather than being widely separated from each other.

b The smallest penguins, the fairy penguin (*Eudyptula minor*), are found in temperate climates in southern Australian states. The largest penguins, the Emperor penguin (*Aptenodytes forsteri*), live in Antarctica.

c Cougars or mountain lions (*Puma concolor*) in the northern regions of North America are, on average, larger than those in the southern regions. Of what advantage might this size difference be?

Emperor penguins can live for up to 70–80 years.

6 ***Applying and communicating your understanding*** ▸ Explain how each of the following features assists a plant to survive in a very hot environment:

a Desert plants generally have deeply penetrating root systems.

b Succulent plants (that store water) have stomata that open only at night.

c Some plants have special cells, called hinge cells, on one surface of their leaves that also has stomata. When hinge cells lose water, the leaf rolls up with the hinge cells on the inside of the rolled leaf.

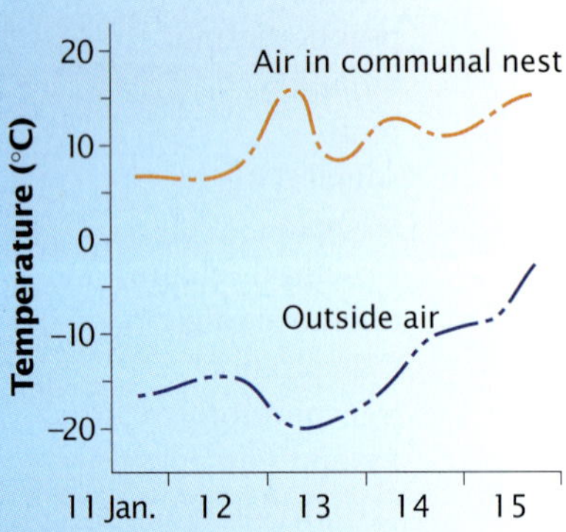

Figure 10.52

7 ***Interpreting data and communicating ideas*** ▸ Many small animals that are solitary over summer tend to become social during winter and often construct nests under the snow. The temperature inside a communal nest of beavers was compared with the temperature of the outside air. The results are shown in figure 10.52.

a What is the maximum difference between the temperature of the beaver nest and the temperature of the outside air?

b Suggest what causes this difference in temperature.

8 ***Applying your understanding*** ▸ Look at figure 10.29 on page 321. Assume a dolphin needed to lose heat. What changes would occur in the countercurrent exchange system to facilitate that loss?

9 ***Applying your understanding*** ▸ Identify a physiological characteristic that assists each of the following to maintain water balance.

a humans (on a hot day)

b sea birds

c the tarrkawarra (*Notomys alexis*)

10 ***Applying your understanding in new contexts*** ▸ Air in the Antarctic is relatively dry. Antarctic explorers can become dehydrated relatively quickly. Explain the relationship between these statements. How can Antarctic explorers reduce the chance of dehydration?

11 ***Inquiring scientifically and communicating ideas*** ▸ A stream of air was blown over a leafy plant growing in a well-illuminated glass chamber and well supplied with water. The temperature of the chamber was kept constant. The amount of carbon dioxide in the air was measured as it entered the chamber and as it left. The rate of water loss from the plant and stomatal aperture were also measured throughout the experiment.

a Explain where most of the water loss from the plant would have occurred. Would the kind of plant used make any difference to your answer?

b In what directions, in or out of the plant, would carbon dioxide and oxygen move? Across what area/s would the movement occur?

c In what way, if any, would photosynthesis in the leaf impact on carbon dioxide in the air blown over the leaf?

d Comment on the suggestion that if photosynthesis and respiration in the plant were equal, there would be no carbon dioxide in the air leaving the chamber.

12 ***Applying your understanding***

a Explain why most stems and leaves on plants have a waterproof cuticle and yet roots do not.

b When cuttings of plants are first potted, they have no roots and may wilt. Wilting is prevented if the pot is enclosed in a plastic bag and shaded from sunlight. Explain why this treatment prevents wilting.

13 ***Analysing information*** In an experiment to investigate the impact of different water content in plant cells, two similar pieces of potato were set up as shown in figure 10.53.

a Explain what has happened with regard to water movement and content in potato chip A.

b Explain what has happened with regard to water movement and content in potato chip B.

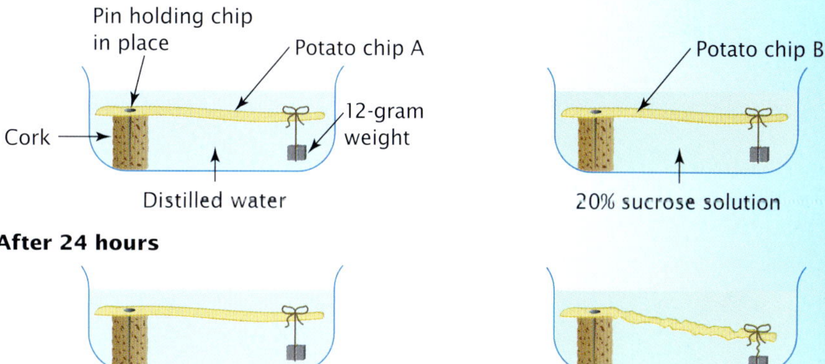

Figure 10.53

14 ***Analysing and synthesising data*** In an investigation of the rate of water transpiration and absorption in a plant over four-hourly periods on a summer's day, the following measurements were made.

Time	*Water (g) absorbed in 4 hours*	*Water (g) transpired in 4 hours*	*Amount of water in leaf (as ratio to dry mass)*
4 am	6	1	7.2
8 am	7	9	6.0
12 noon	15	21	5.5
4 pm	23	30	3.0
8 pm	13	10	3.8
midnight	8	3	7.5

a Use suitable scales to draw the following graphs:

i amount of water absorbed against time

ii amount of water transpired against time.

b At what times did maximum absorption and transpiration occur?

c Suggest why both transpiration and absorption drop during the night.

d What is the relationship between transpiration and absorption and the amount of water in a leaf?

15 ***Using the web*** Desert plants have physical and behavioural features that enable them to survive in those regions. Go to www.jaconline.com.au/natureofbiology/natbiol1-3e and click on the 'Desert plants' weblink for this chapter. Read the section on desert plant adaptations.

a What are the two broad groups of desert plants that survive because of special physical features?

b What is the special feature of each group you identified in part a?

c What is the significance of the shallow but extensive root systems that some desert plants have?

d What is a significant feature contributing to the survival of phreatophytes such as the mesquite tree and how does the feature contribute to survival?

e What are the features of an ephemeral plant?

f One Australian plant that often grows as an ephemeral is Sturt's desert pea, *Swainsona formosa*, the floral emblem of South Australia (see figure 16.50, page 535). The hard seed coat inhibits germination and for garden planting needs to be given special treatment, such as gentle rubbing between sandpaper. Suggest what causes the seed to germinate naturally.

11 Behaviours for survival

KEY KNOWLEDGE

This chapter is designed to enable students to:

- gain an overview of behavioural adaptations in plants and animals
- distinguish between innate and learned behaviours in animals
- recognise the importance of behaviours to the survival and reproduction of individuals and groups
- understand that the complexity of animal behaviours contrasts with the reduced array of behaviours shown by plants.

Figure 11.1 A dominant male splendid fairy wren, *Malurus splendens*, showing the bright blue plumage that he has during the breeding season. Both females and non-breeding males are dull brown. These birds live in social groups of up to 12 birds, comprising one dominant male, one breeding female and subordinate non-breeding birds of both sexes and young from the previous year's breeding. The group carries out cooperative breeding. All members of the group carry out parent-like care of the young in the nest or a brood arising from several clutches of three to four eggs over a breeding season. Many Australian birds exhibit cooperative behaviour.

In this chapter, we examine some of the behaviours of animals and plants and how behaviour relates to survival of individuals and groups.

Welcome to my bower

Figure 11.2 The male satin bowerbird, *Ptilonorhynchus violaceus*, puts the finishing touches to his bower in preparation for courtship.

Male birds take a variety of actions to attract a female to mate. Some build elaborate structures to attract females for courtship and a fascinating case in Australia is the male satin bowerbird, *Ptilonorhynchus violaceus* (see figure 11.2).

The male bowerbird builds a bower, often near a fallen log or moss-covered rock. To do this, he covers the ground with thin twigs to form a platform about 30 cm long. He then interweaves twigs to form two parallel walls about 30 cm high. The walls often bend inwards and meet to make an arch. The bird paints the walls with a mixture of vegetable juices or charcoal mixed with saliva and then decorates them with shells, feathers, flowers, pebbles, bleached bones, berries and leaves. Anything that glints or has a bright colour may be chosen, but blue and yellowish-green coloured decorations are favoured. Human-made items, such as bottle tops, buttons, straws and other pieces of plastic, are often included as bower decorations (see figure 11.2).

A female is attracted to the bower during courtship, which is followed by mating. The female leaves the bower to build her nest and care for the young, but the male remains near the bower and displays to other females.

Building a bower is part of the courtship **behaviour** of a bowerbird and is an important part of reproduction of the species.

Behaviour in an organism is some action that occurs in response to a particular stimulus or a range of stimuli. In the case of the male bowerbird, the stimulus is the urge to mate, and his actions in response to this stimulus are to build a bower to attract a female.

Actions by plants most often take the form of some predictable change in growth — for example, a plant may grow towards a light. The action, or movement, generally involves only part of the plant and relies on chemical coordination.

Behaviour in animals is far more complex than that of plants and often results in action that involves movement of the whole organism. For example, a small bird may perceive danger and fly away from a hawk in the same area. Such an event involves some part of the nervous system — such as eyes, ears, the sense of touch, taste or smell — to detect the impending danger and then involves both the nervous and hormonal systems to cause action to avoid the danger. In the example of the male bowerbird, nerves and hormones play a role in his response to his urge to mate.

In this chapter, we examine some of the behaviours of plants and animals and consider the significance of these behaviours for the fitness of the individuals involved, and hence the survival chances of individuals and groups.

Although behaviour will be discussed under a number of separate headings, you should become aware that most situations involve an interaction of more than one kind of behaviour.

Behaviour in animals

The study of animal behaviour is called **ethology**. Three men, Nikolaas Tinbergen (1907–1988) from the Netherlands, Konrad Lorenz (1903–1989) from Austria, and Karl von Frisch (1886–1982) from Germany, played a significant role in observing animal behaviour in nature. In 1973, they received the Nobel Prize in Physiology and Medicine in recognition of their work.

This chapter is mainly concerned with the question 'What is the function of a particular behaviour?' and does not deal with the mechanisms underlying a particular behaviour — although this is a fascinating topic that may interest you in future biology studies.

Innate versus learned behaviour

All male bowerbirds build bowers. When a behaviour is essentially the same in all members of a species (in this case the male members), the behaviour is called **innate** or **inborn behaviour**. It is genetically controlled. A baby suckles its mother's breast shortly after birth. Sucking is an innate behaviour. What are some other innate behaviours that people show?

Figure 11.3 A very young baby has the innate ability to make sounds but must learn to make sounds that have meaning to older people.

There are a number of different innate behaviours (see figure 11.3). You probably see many examples of individual animals showing innate behaviour every day: a spider spins its web and a bird calls from a nearby tree before it flies off. Birds do not have to learn to fly; spiders do not learn to spin webs. In other cases, groups of animals participate in innate behaviour. Flocks of birds migrate for the winter, lions stalk their prey in groups and bees work together to care for their hive.

Now consider these questions: What behaviours have you had to learn? Can you swim? Can you ride a bicycle? When did you learn to read? When did you learn to write? **Learned behaviours** are those that develop or change as a result of experience. Learning may take place because we are shown how to do something. We may copy someone else, or we may learn how to do something on a trial-and-error basis; that is, we keep trying to do something in different ways until we find one that works (see figure 11.3). When a behaviour is changed as a result of experience, we say that learning has occurred. Learning enables an animal to adapt to change.

Figure 11.4 outlines a number of innate and learned behaviours. First, we will examine a number of innate behaviours and their importance, then consider the role that learned behaviours contribute to the survival of individuals and groups.

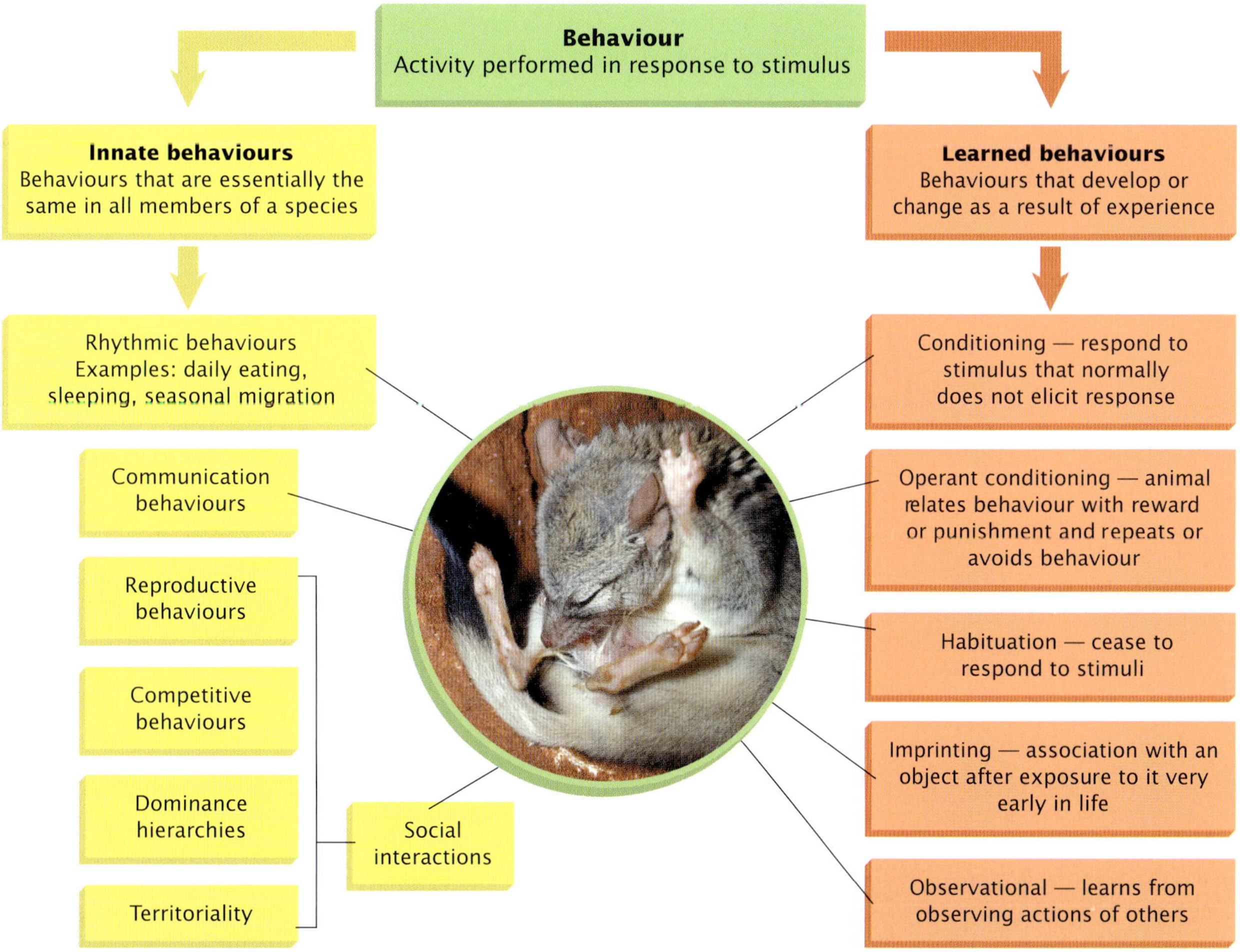

Behaviour
Activity performed in response to stimulus

Innate behaviours
Behaviours that are essentially the same in all members of a species

Rhythmic behaviours
Examples: daily eating, sleeping, seasonal migration

Communication behaviours

Reproductive behaviours

Competitive behaviours

Dominance hierarchies

Territoriality

Social interactions

Learned behaviours
Behaviours that develop or change as a result of experience

Conditioning — respond to stimulus that normally does not elicit response

Operant conditioning — animal relates behaviour with reward or punishment and repeats or avoids behaviour

Habituation — cease to respond to stimuli

Imprinting — association with an object after exposure to it very early in life

Observational — learns from observing actions of others

Figure 11.4 Many facets of behaviour can be studied about a single animal. Some of these will be inborn or innate and be similar to the behaviour shown by other members of the species. Innate behaviour may be modified by experience. Other behaviours are learned and may vary from individual to individual. (The animal grooming itself here is a kowari, *Dasyuroides byrnei*.)

The study of animal behaviour in the field has concentrated on innate behaviour. Laboratory studies of animal behaviour have tended to concentrate on learned behaviours.

Some innate behaviours are rhythmic

Animals repeat behaviours at regular intervals. For example, they eat and sleep. These are called **rhythmic behaviours**. Different species of animals may follow different patterns of rhythmic behaviour. Some sleep during the day while others sleep at night. Differences in these daily rhythmic behaviours may be a significant factor in the ability of a group of animals to exploit the resources within the area in which they live and avoid competition from other species.

Rhythmic behaviours are regulated by both internal and external factors. The internal factor is often referred to as the biological clock. The external factor is generally light.

Feeding behaviour

Feeding behaviour is an example of a rhythmic behaviour. Feeding may be:

- on an individual basis, as in the examples of the spiders' behaviour (page 345) and the baleen whale's feeding techniques (pages 346–7)
- on a group basis, in which members of a species cooperate in some way to improve their chances of obtaining their nutrients.

An example of group feeding behaviour is seen in the cooperative behaviour of common dolphins (*Delphinus delphis*). A pod of dolphins swimming in a bay detects a shoal of fish. This encounter represents a chance for the dolphins to feed. They surround the fish, with some dolphins forming a U-shape and others positioning themselves opposite the open side of the U. The dolphins gradually close in, herding the fish into a dense shoal so that the fish are more easily caught and eaten, as shown in figure 11.5.

Figure 11.5 Cooperative hunting in dolphins

Migration

Rhythmic behaviours may also be seasonal. A common example of this is **migration**. Most animals move around from time to time. However, some species move thousands of kilometres at regular intervals in their lives. The movement of large numbers of animals over long distances from one area to another area, and their subsequent return to their original home, is called migration.

See chapter 9, pages 262–3; for more information on migration, and pages 265–7 for more on satellite tracking of migratory animals.

Animals that migrate usually use the same route each time. The migration routes of animals have been mapped by tagging the animals or fitting them with small radio transmitters. A very diverse range of animals migrate including many species of bird, fish, eel, insect and mammal. The migration cycle of many birds and land animals often takes place annually. Other animals, for example salmon and eels, may take several years to complete their migration cycle.

Animals usually migrate because of the approach of winter and the disappearance of food in an area. Also, winter temperatures may be fatal to a particular species. Migration is an important survival strategy (see page 348).

STRANDS OF SILKY DEATH

Silk plays a major role in the life of a spider. Many innate behaviours in spiders involve the use of silk — in feeding and in reproduction.

Spiders have silk-producing glands that open to the exterior via spinnerets at the rear ventral surface of the abdomen. Spiders pull fluid silk from the spinnerets using their hind legs. As the silk is released, it hardens into a solid thread.

Some spiders build elaborate vertically oriented webs of silk that they locate in the flight paths of their insect prey (see figure 11.6).

Some species of spider do not build elaborate webs but still use silk to catch their prey, for example, the bolas or angler spiders. At night, a bolas spider hangs from a silk harness and dangles a strand of silk about five to seven centimetres long from its front legs. Along the length of this silk strand are sticky globules of gum. The spider releases a chemical that mimics the chemical signals produced by female *Noctuid* moths. This chemical diffuses into the surroundings and male moths are attracted to the area. When a moth approaches, the waiting spider begins to swing its gum-laden strand of silk around in circles, as shown in figure 11.7. When the silk strand contacts a moth, the prey is trapped by the gum. The spider pulls in the strand with prey attached, bites the moth to immobilise it and wraps it in a different kind of silk for storage.

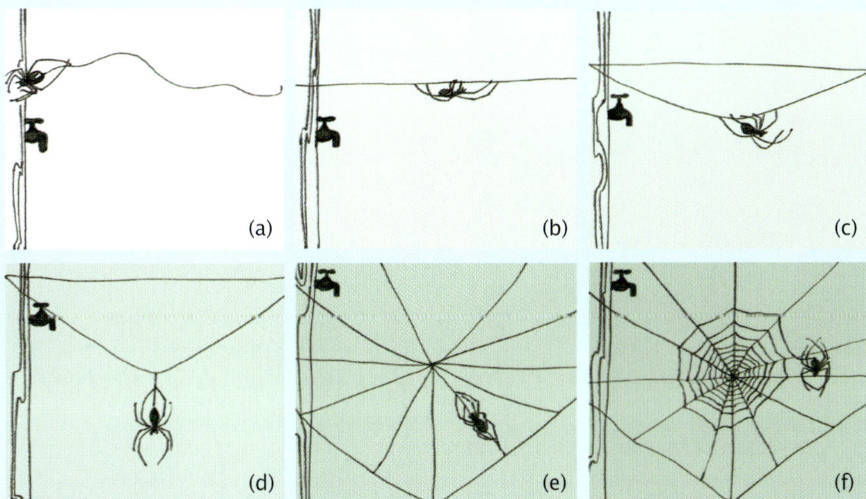

Figure 11.6 Stages in the building of an orb web: **(a)**–**(b)** the 'bridge' is established and strengthened; **(c)**–**(d)** a Y-frame is put in place; **(e)** radials are added; then **(f)** a spiral is completed. Not all spiders build elaborate vertical webs. Huntsman spiders capture their prey by pursuit; funnel-web spiders ambush prey that approaches their burrow entrances.

Figure 11.7 The swirling strand of gum-laden silk can catch a moth.

WHALES AND THEIR FEEDING BEHAVIOURS

The many species of whale are classified into two major groups, as shown in table 11.1.

The diet of toothed whales may include squid and cuttlefish and schooling fish, such as pilchards and anchovies. In addition, killer whales attack and eat seals, penguins, dolphins and, when hunting as a coordinated group, will attack and feed on whales much larger than themselves.

Baleen whales feed on fish schools, squid and marine invertebrates. The largest baleen whale, the blue whale, feeds on krill and must ingest enormous quantities to meet its energy needs.

Different species of baleen whale feed using various techniques: 'gulping' and 'skimming'. With the 'gulping' technique, a whale lunges open-mouthed into a shoal of fish or krill and engulfs an enormous volume of food-laden water. The volume of water engulfed is greatly increased in those baleen whales which have throat grooves. Throat grooves are longitudinal pleated areas of skin extending from the lower jaw to the navel which can be expanded during 'gulping-style' feeding (see figure 11.8).

Once the baleen whale has engulfed a mixture of food and water, what next? If we looked in the mouth of a baleen whale we would see hanging down from the upper jaw many long thin strands (plates) of tough flexible material, known as baleen, that form an effective sieving system. The baleen plates form parallel rows and, as they wear, the baleen on the inner edge of each plate becomes frayed into many thin hairs.

Figure 11.8 Grooves on the throat of a humpback whale. What role do they play in feeding?

Table 11.1 Body mass and length of some whales

Whale species	*Average mass (kg)*	*Length (m)*
suborder Odontoceti (toothed whales):		
common dolphin (*Delphinus delphis*)	80	1.5–2.2
bottle-nosed dolphin (*Tursiops truncatus*)	200	2.3–3.8
sperm whale (*Physeter catadon*)	60 000 (male) 45 000 (female)	11–20 8–17
killer whale (*Orcinus orca*)	5000 (male) 3000 (female)	5–9 4–7
suborder Mysteceti (baleen whales):		
humpback (*Megaptera novaeangliae*)	65 000	11–19
fin whale (*Balaenoptera physalus*)	80 000	17–25
southern right whale (*Eubalaena australis*)	80 000	15–16
blue whale (*Balaenoptera musculus*)	150 000	21–33

Just as a coffee plunger separates the coffee grounds from the coffee fluid, the baleen plates are used to separate food (fish and krill) from sea water. Filtering occurs when the whale closes its mouth, pushes its tongue up and forces water out the sides of its mouth through the baleen plates, as shown in figures 11.9 and 11.10. The separated food trapped by the baleen plates is then swallowed. In the 'skimming' technique, a baleen whale feeds continuously as it moves on or close to the surface of the water with an open mouth. This movement directs a continuous stream of water into its mouth and food is retained by the baleen plates.

Another apparent feeding technique observed in humpback whales is 'bubble netting' — an unusual technique that they apparently use to concentrate or herd scattered fish. In bubble netting, one (or more) whales swims below a school of fish. The whale then ascends in a spiral track around the fish and moves more closely inward. As the whale moves upward and inward, it exhales air that forms columns of tiny bubbles around the fish, crowding them together (see figure 11.11a). The fish are trapped within the cage of bubbles and move closer together. Once the fish are concentrated, the humpbacks can swim through the mass of prey and feed using the 'gulping' technique (see figure 11.11b).

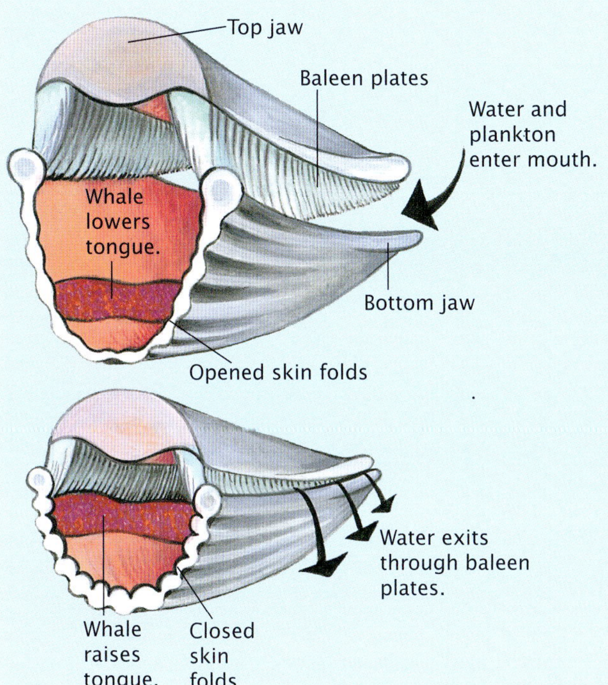

Figure 11.9 Filtering occurs when the tongue is pushed upward, forcing water out through the baleen plates.

Figure 11.10 Baleen plates form a dense filter through which water can be pushed out and food retained.

(a)

(b)

Figure 11.11 Bubble netting in humpback whales is illustrated above. How does this technique assist the humpback whales to get the energy they need for living?

Figure 11.12 The mutton-bird *Puffinus tenuirostris*. A young mutton-bird banded in South Australia travelled 13 000 kilometres along its migration route when it was sighted six weeks later.

Birds on the wing

More than one-third of the world's species of birds migrate.

The white-throated needletail swift, *Hirundapus caudacutus*, migrates to Australia each year from northern Asia. These birds breed in eastern Siberia, Mongolia and Japan and migrate to Australia during late spring. Towards the end of summer they fly north again.

Mutton-birds (short-tailed shearwaters, *Puffinus tenuirostris*) (figure 11.12) breed in south-eastern Australia in the summer months each year. As winter approaches, millions of them migrate to the north Pacific (see the map, figure 11.13) and return to Australia the following spring.

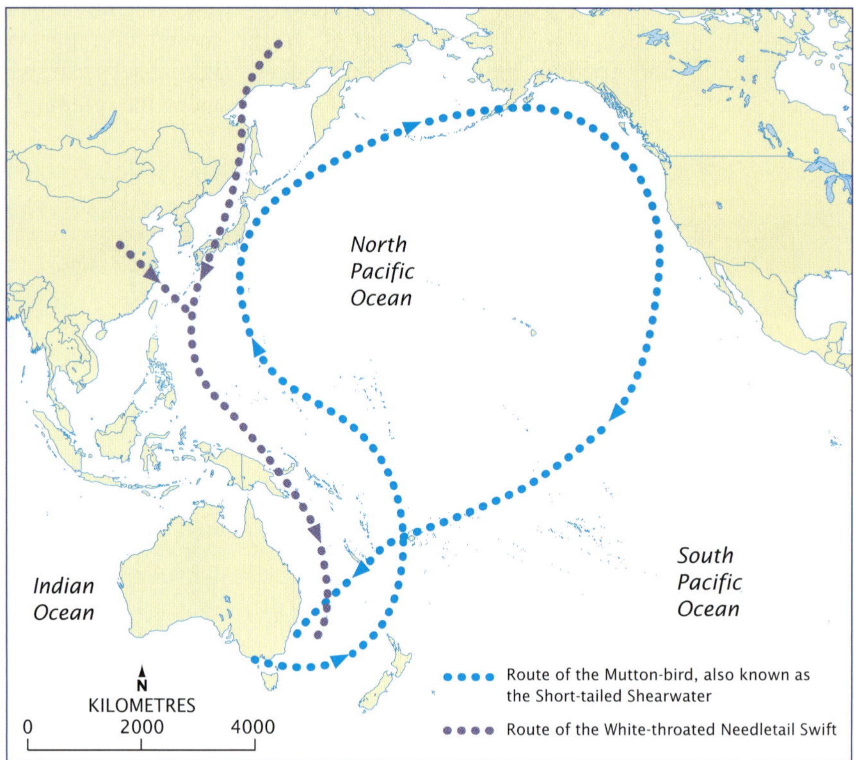

Figure 11.13 The migration paths of the two birds. How does this innate and rhythmic behaviour contribute to survival?

Why do birds migrate?

Birds are active animals that need a constant supply of food. In many regions, as the summer ends and winter approaches, temperatures fall and food supplies decline. Birds migrate from such regions to warmer climates to find a more abundant food supply. The white-throated needletail swift feeds on insects that it catches while on the wing. The swifts would starve to death if they remained in the winter snows of their breeding grounds. Migratory behaviour enhances survival.

How do birds know where to fly?

Young birds migrating for the first time don't need to be guided by their parents. They are born 'knowing' to migrate. For example, young mutton-birds don't leave on their long migration flight until about two weeks after the parents have left, and yet they 'know where to go'. How do birds find their way over these immense distances around the world, often flying high in the sky without any landmarks to guide them? They must be born with the ability to navigate.

It is thought that birds use the sun and stars as compasses. In addition, they use the Earth's magnetic field to determine the direction they need to fly. Magnetite, a magnetic iron-oxide compound, has been found in the tissues of some birds and it is thought that the presence of this compound is related to the bird's ability and helps the bird to use the Earth's magnetic field.

Communication behaviours

With few exceptions, including aspects of human **communication** via technology, communication in the Animal Kingdom is an innate behaviour. Communication can be through touch, posture, sound, visual display and chemical signals.

As with all behaviours, communication behaviour occurs in response to a stimulus. The important components of communication include:

- stimulus for the communication
- sender of a signal
- receiver to whom the signal is directed
- the kind of signal sent
- how the signal is sent
- the behaviour of the receiver
- the setting in which the communication occurs.

In building his bower, the male bowerbird whom we meet on page 342 was communicating via a visual display to female birds that he was a good potential mate (table 11.2).

Table 11.2 Communication between animals involves a number of components. The making of a bower by a male bowerbird is a behavioural response to a particular stimulus. His behaviour sends a signal to female bowerbirds.

Aspect of communication	*Particular case with bowerbird*
Stimulus	Desire to mate
Sender	Male bowerbird
Receiver to whom the signal is directed	Female ready to mate
Kind of signal sent	Appearance of carefully made and decorated bower
How the signal is sent	Visual image
Behaviour of the receiver	Attracted to the bower
Setting in which the communication occurs	Courtship behaviour

We will now consider aspects of communication for some different animals.

Messages to trick a competitor

In South America, white-winged-tanager-shrikes often perch high in trees and scan their environment. If a perched shrike sees a hawk, the bird makes a hawk alarm call which warns other birds to dive for safety or freeze. Other species of birds are at the same time competing for insects in the dense foliage. A white-winged-tanager-shrike also makes the hawk alarm call if it sees a competitor species chasing an insect (figure 11.14). The competitor reacts as if a hawk is close-by and the insect is left for the shrike.

Why does the competitor bird respond to the hawk alarm call when no hawk is nearby? If the bird didn't respond and a hawk really was nearby, the bird may be killed. The benefits of reacting seriously to each alarm call outweigh the cost of not reacting. By reacting to each alarm call, the shrike's competitors are more likely to survive and reproduce.

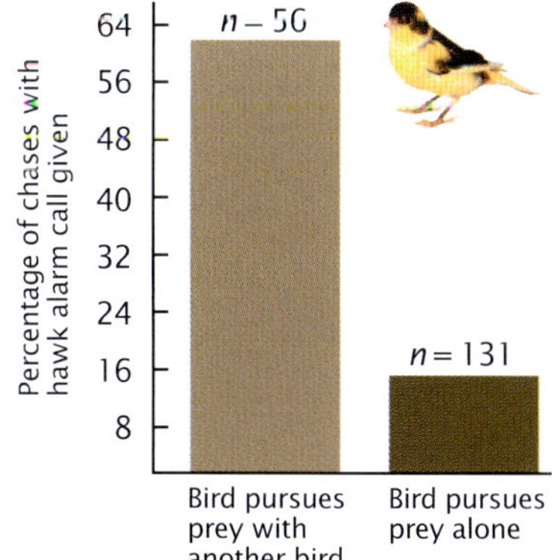

Figure 11.14 White-winged-tanager-shrikes are birds that cry 'wolf'. The shrike makes a hawk alarm call when a competitor chases an insect. This tricks the competitor into thinking that a hawk is nearby. The competitor flies away and leaves the insect for the calling bird.

The language of bees

Honeybees, *Apis mellifera*, live in groups. Each group lives in a hive which has a queen bee, male drones and workers. Worker bees bring food such as pollen and nectar back to the hive. In the 1940s Karl Von Frisch suggested that worker bees communicated the source of food to other workers by dancing in special ways depending on the position of the food source and its distance from the hive.

The honeybee dance

Soon after a worker bee finds a new food source and returns to the hive, a large number of bees appear at the food source. Von Frisch claimed that information related to the new food source was transmitted from the first worker bee to others by a 'waggle dance' performed on a vertical surface in the hive.

During the waggle dance, a worker bee moves in a figure-of-eight and waggles its abdomen from side to side as it moves across the centre of the eight (see figure 11.15a). The time a bee takes for this waggle segment is related to the distance of the food from the hive — the longer the dance, the further the food is from the hive. The angle that the straight centre across the eight makes to the vertical is related to the position of the food in relation to the position of the sun. A waggle dance performed at an angle of 20° to the left of the vertical means that the flight-path from the hive to the source of food is 20° to the left of the line to the sun (figure 11.15b). As the worker bee performs the dance, other workers gather around, often touching the dancer. The dancer also makes a sound different from the buzz that is made at other times.

Do bees also smell their food?

It is now accepted that smell is a factor in a bee's ability to find a new source of food. Worker bees that observe the waggle dance probably obtain information about the direction they should fly and the approximate distance they need to fly to reach the food. The exact position of the food is located by smell.

Why is it an advantage to the species for a single bee to be able to communicate the location of a new food source? The alternative would be for all workers to go out each day and look for food on a random basis. This could result in a lot of wasted energy. By having a rough idea of the direction and distance of the new food, bees can use their sense of smell to find the food in a much shorter time and so use less energy. This leaves more energy available for increasing the numbers in a hive, and so increases the chance of survival of the species.

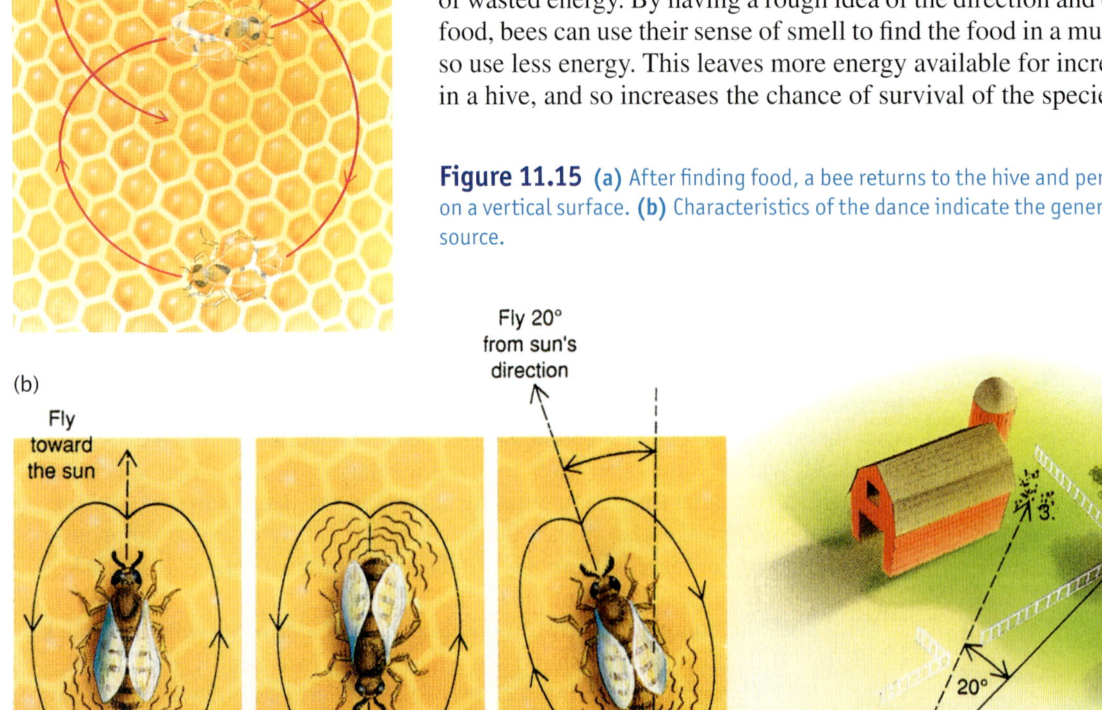

Figure 11.15 **(a)** After finding food, a bee returns to the hive and performs a waggle dance on a vertical surface. **(b)** Characteristics of the dance indicate the general position of the food source.

Communication under water

Sound is an important means of communication underwater. Sound is transmitted through water for long distances, whereas light can only penetrate a short distance below the surface. Underwater noises transmitted by whales and dolphins have been studied extensively and the patterns of the sounds recorded. The range of noises includes what we would call blips, clicks, groans, moans and snores which can last from a few seconds up to several hours.

ODD FACT

One whale in the Caribbean was heard to sing continuously for 22 hours.

Singing whales

Although the noises made by some groups of whales differ, in other cases, noises by different groups of the same species can have very similar patterns even though they live thousands of kilometres apart. Whales living off the coast of California sang the same song as whales living off Hawaii. Both groups changed their songs in the same way over two years. Although the sounds of a group are very similar overall, individuals can be identified on the basis of minor variations.

The noises of whales are continually being changed. They may be lengthened or shortened. New sounds may be introduced. When such changes occur, they generally take place in all whales in the group. It is not known why whales continually change their songs in this way.

Why do whales sing?

Some whales make sounds to locate and identify each other. Whales can respond to a call from another whale over 5 kilometres away by changing direction and swimming towards the calling whale. Other whales seem to communicate by sound to coordinate their feeding (as we discussed on pages 346–7) or simply to keep the group together. At other times sounds are used in courtship.

Singing during courtship

During courtship, a singing male humpback whale is generally alone. A lone male who joins a mother and her calf stops singing and becomes the female's principal escort. Mating may or may not occur. If the three come close to other singing males (figure 11.17), one of these may attempt to reach the female. If he does, he stops singing and jostles to get close to the female. The principal escort and the newly arrived male then fight over the female. They try to ram one another or thrust at each other with their tail flukes and blow bubbles underwater. During this aggression, the males make low grunts, high trumpet calls and sing snatches of songs that attract other singing males from as far away as 9 kilometres.

Figure 11.16 Some whales use sounds to communicate to each other when coordinating their feeding.

Figure 11.17 Courtship behaviour in whales. **(a)** An adult male stops singing when he joins a cow and calf and becomes a primary escort. **(b)** Other escort males join the group. **(c)** A secondary escort may physically challenge the primary escort after which the successful male remains with the female and her calf: other escorts leave the group.

The males that are attracted are called secondary escorts. Any one of them may challenge for the position of principal escort. The principal escort retains that position for up to eight hours but is eventually replaced by another. The other males lose interest, move away and start singing again. The mother and calf and new principal escort swim on again until the next encounter.

It appears that a male humpback makes a song during the mating season to advertise his sexual availability to females. Whales may be separated by several kilometres of water so it is important for reproduction that they are able to communicate over long distances.

Table 11.3 Sounds made by bottle-nosed dolphins during various behaviour situations

Behaviour	Sounds
Navigation, hunting	Clicks
Play-chase, pain	Squawk
Courtship, mating	Yelp
Threat	Buzz
Alarm, fright, distress	Squeaks, cracks, pops
Resting, predator nearby	Silence

Dolphins

Dolphins appear to communicate information to one another using various sounds (table 11.3). Dolphins also use some sounds to find their location and for navigation. The ability of some animals to use sound bounced off objects to determine their position is called **echolocation**. In addition, humans have trained dolphins to carry out many tasks in response to various sounds. You may have seen some of this taught behaviour if you have visited a sea park that has performing dolphins.

Figure 11.18 Volunteers try to help a pod of whales that became stranded on a beach at Busselton, Western Australia, in June 2005. Although the reason for mass stranding of whales and dolphins is unknown, it may be because members of a pod respond to and follow the distress calls of one of its members.

Overall, it is an advantage to the survival of a species of animals if they cooperate and respond to communication from other members of the same species. However, some of the mass strandings of dolphins and whales on beaches (figure 11.18) may occur because a pod responds to distress calls from a member of the pod.

Chemicals for communication

Chemicals called **pheromones** are used by some animals for communication. Some bee species communicate the location of a new food source by leaving a pheromone at the site. As the worker returns to the hive, it chews on grasses along the way and leaves a signal every few yards. After returning to the hive the worker is able to lead others back along the same track to the food source (see figure 11.19).

Figure 11.19

Pheromones are also used by many animals to attract members of the opposite sex. There are examples in virtually all animal groups. Dogs use chemical communication: a female dog about to ovulate comes 'on heat' and releases a pheromone into the urine to notify male dogs that she is ready for mating.

Some animal pheromones act over very short distances. For example, a stalked ciliate, *Vorticella*, secretes a pheromone which attracts another member of the species over about 1 mm. Other animal pheromones act over very long distances. The pheromone of a female gypsy moth can be detected by the large feather-like antennae of a male up to eight kilometres away. Pheromone communication makes it very easy for insects and other animals to locate a mate, even in sparsely populated areas.

Social and territorial interactions

Courtship, reproductive strategies and parenting behaviours are discussed in chapter 12.

Social interactions involve two or more individuals and may involve cooperation as in mating or ensuring a food source. Social interactions may also involve aggression and conflict as individuals or groups fight to defend territories or to select a mate. As mentioned earlier, the different kinds of behaviours interact with each other.

Courtship and other **reproductive behaviours** (discussed further in chapter 12) sometimes enable us to distinguish species that are difficult or impossible to distinguish on their appearance alone. Two distinct species of Australian ravens were identified only after quite different behaviours were observed (table 11.4).

Table 11.4 Reproductive behavioural differences between two species of ravens

Behaviour observed	*Corvus coronoides*	*Corvus mellori*
Courtship	Aerial chase	Ground promenade
Nest-site	In tree more than 13 m above ground, commanding good all-round view	In tree less than 13 m above ground or on ground, often hidden in thick cover
Commencement of breeding	Mid-July	Early August
Time for young in nest	45 days	36 days
Young leaving nest	Remain in territory for at least 4 months	Move, with parents to flock, after 2 or 3 days

Ravens of one species would be able to identify members of their own species by these subtle behaviour differences. This is important for successful mating and reproduction.

There are minor differences in the beak structure and length of neck feathers between the two species of raven so it is possible to distinguish them by careful observation of the physical characteristics.

Group organisation

We have already seen how some animals form groups of various sizes. The size of a group and the reason for which the animals group varies. Dolphins form groups to herd fish; jays form groups to raise their young. A couple may come together just to mate. Other animals, such as the honeybee, always live in groups. A group may be permanent or it may stay together until its numbers are such that it splits into two. For many animals, being a member of a group is safer than being alone.

Different kinds of groups are found in different species of animals. Groups may be formed to make collecting food easier, for protection, for shelter or for mating. Whatever the reason for grouping, it increases the chances of survival of members of the group and hence the species.

Castes

Ants, termites, some wasps and bees have a **caste system**. Each caste has a different structure and performs a different function in the group (figure 11.20). Ants have a reproductive caste, sterile workers, who collect food and maintain the living quarters, and soldiers who guard the group.

(c)

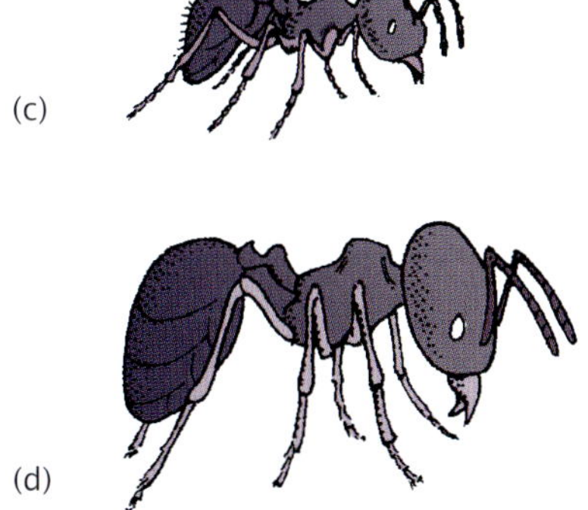

Figure 11.20 Ants, *Pheidole* sp, have a caste system. The **(a)** queen and **(b)** male are the reproductive castes; **(c)** sterile workers and **(d)** soldiers which guard the group.

Social hierarchies

Animals that live in a group will often fight to establish a **pecking order**. At first this may seem to be disruptive, but the animals fight to learn which others can beat them and which others they can beat. The term pecking order is used because the first experiments on social hierarchies were carried out using chickens. Chickens in a group peck other chickens that are below them in rank.

Once an animal learns that it will be beaten by particular members of the group, it no longer attempts to fight those members. The animal 'knows its place' in the group and so in the end the fighting in a group is actually reduced. Valuable resources will not be wasted on continued fighting.

The higher an animal is in the pecking order then the greater will be its access to food and other aspects of life in the group. Animals low in a pecking order are less likely to mate and have offspring than those high in the pecking order. The stronger animals in the group are therefore more likely to have offspring. These offspring are more likely to be the stronger members of the next generation.

Leadership

Baboons, *Papio ursinus*, have a complex **social hierarchy**. A baboon group usually contains several infants, juveniles, adult females and males (figure 11.21). The strongest male is usually the dominant male and leader of the group with other males in a rank order behind the dominant male. The dominant male defends the group against predators or strange males and protects a lower ranking member of the group who runs to him after a fight with another male. Sometimes two or three males form a ruling group within the troop. Members of a group are generally safer than solitary animals.

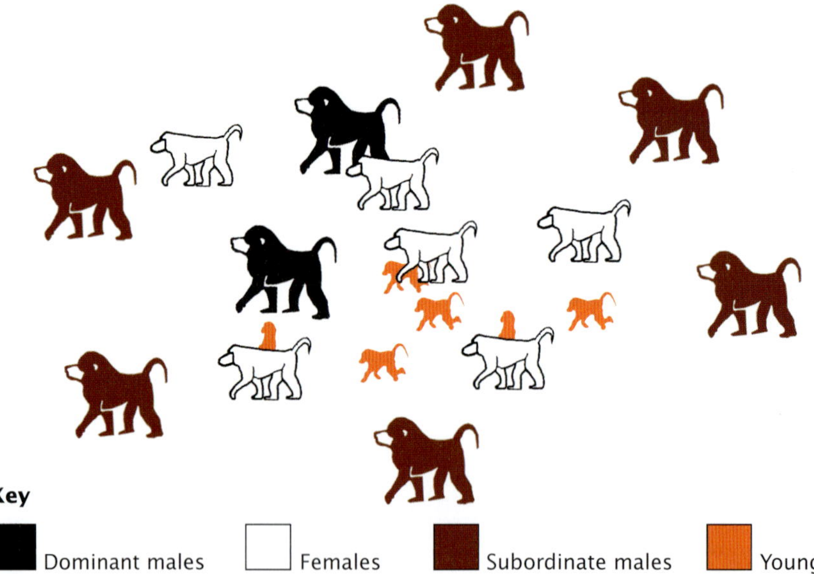

Figure 11.21 Baboons are social animals that live and hunt in large groups. The group is dominated by one or two males. Other males take up strategic positions at the front and rear of the group. If the group is attacked, all males move towards the aggressor.

ODD FACT

Black-headed gulls fight to establish individual territories for nest sites. Once the territories are established, the gulls do not intrude into a neighbour's territory. When the young hatch, the parents are free to look after them as they no longer have to defend their territory against other gulls.

When the group is on the move, the troop is led by strong adult males, childless females and young males. These are followed by the dominant male and females with very small infants. Young males bring up the rear of the moving troop.

The dominant male has priority with the females for mating. He is the male most likely to father the next generation of young. The stronger the parents, the more likely they are to have strong offspring. These are more likely to survive and produce the next generation, an important aspect for survival of the species as a whole.

Territorial behaviour

Have you ever been attacked by a magpie? If so, you probably walked through its territory during the nesting season. Animals select a territory and defend it against others for a number of reasons. They show **territorial behaviour**.

A DOG OWNER MUST BE THE LEADER OF THE PACK

By nature, dogs are pack animals that obey and respect the leader of the pack. The leader gains and holds its position by strength, courage, fighting skills and the ability to dominate all challengers. Periodically all dogs challenge their leader's position. If the challenge fails then a dog and leader continue in their respective positions.

The pack for a domesticated dog includes people. The owner of a dog must be the leader of the pack and make sure that the leadership role is maintained. If this occurs, the dog assumes that it will be defended by its leader and will relax. If an owner does not play a firm role as leader, a dog may see itself as the leader of the pack which in this case is the family and others who visit the home. If it sees its role as the leader, a domesticated dog will act to defend its pack.

Part of a dog's defence is to bite. It will bite to protect itself, its food and territory and other members of its pack. It will bite to protect its position as pack leader. Such a dog may readily bite a family (pack) member if it thinks its position as leader is being threatened by some behaviour of the family member. A dog may think its position is being threatened by a boisterous child and may bite the child even though the child is a member of the dog's own pack.

To make sure that their dogs do not bite people, both in the house and in public places, owners must train their dogs to understand that they are subordinate to all humans.

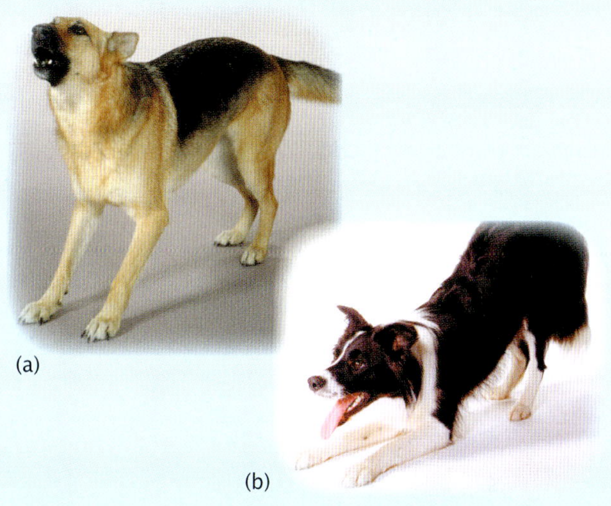

Figure 11.22 Different emotions produce different behaviours: **(a)** the hackles rise on an aggressive dog; **(b)** quite a different posture is assumed when a submissive dog greets its master.

ODD FACT

We tend to associate aggression with the large and obvious animals we see around us, and yet aggression also occurs with small, apparently placid animals.

Male bowerbirds defend their bowers from other males which try to steal ornaments or damage the bower. The better he can build and defend a bower, then the greater chance a male bowerbird has of finding a mate. Other animals will defend a territory because it provides sufficient food for the occupant (and possibly family) but insufficient to sustain additional members of the species.

Competition

Animals often compete with each other. Birds may compete for space on a cliff edge to build a nest, humpback whales may compete for a mate or two seagulls may squabble over a piece of food they have found. Such **competition** may lead to **aggression**.

As we have seen, while aggression between two individuals is common, such aggression is often related to establishing a position in the hierarchy within a group. Group dynamics is important for survival and reproduction in many species.

ODD FACT

Florida scrub jays often have other non-reproducing jays living in their territory and helping raise newly hatched young. Pairs with helpers have, on average, 2.06 fledglings. Those without helpers have 1.03 fledglings. Cooperation in these jays increases the chance of survival of the offspring.

Cooperation

Animals often **cooperate**. As we saw on page 344 (see figure 11.5), dolphins display cooperative feeding when they enclose a school of fish. Animals may also cooperate to kill an animal much larger than themselves. African hunting dogs and hyenas hunt in packs to kill a wildebeest or zebra. The combined weights of the predators enable them to overcome the prey. A single hyena is unlikely to be able to kill a wildebeest but, by cooperating, a small group of hyenas can ensure food for all. In this case cooperation is essential for food and survival.

Do innate behaviours remain constant?

Innate behaviours are genetically determined and are similar in all members of the species. However, an innate behaviour is not necessarily fully developed at birth and may be modified by learning. As individuals grow they have many experiences from which they learn. The new information can be used to modify or improve a previous behaviour.

When laughing gull chicks (*Larus atricilla*) feed, they follow a particular feeding pattern (figure 11.23). As the parent lowers its head, the chick begs for food first by pecking at the parent's red beak. The chick then grasps the beak and strokes it downward. The parent regurgitates partially digested food which is eaten by the chick. Uneaten food is reingested by the parent.

(a)

(b)

(c)

(d)

Figure 11.23 Feeding behaviour of a laughing gull chick (*Larus atricilla*). As the parent lowers its head **(a)** the chick begs for food by pecking at the parent's beak. **(b)** The chick strokes the beak downward. **(c)** The parent regurgitates partly digested food which is **(d)** eaten by the chick.

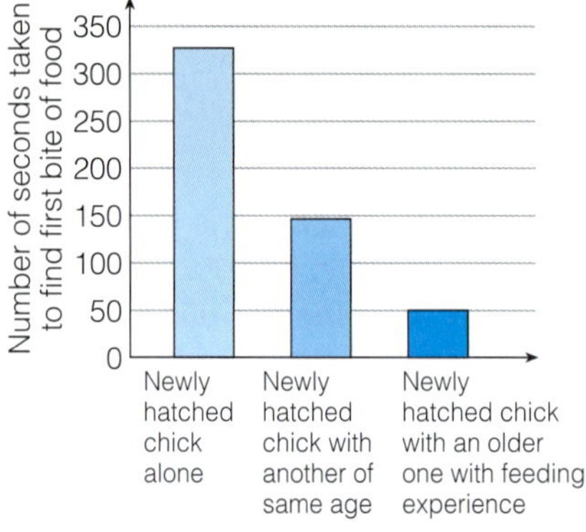

Figure 11.24 Time taken for newly hatched laughing gull chicks to take their first bite of food. The most time was taken by chicks on their own. Shortest time was taken by chicks accompanied by older chicks experienced in feeding.

Immediately after hatching, many of the chick's pecks are inaccurate. Only about a third of the pecks strike the parent's beak. On the second day about 50 per cent of the pecks are accurate, and by the third day a steady level of more than 75 per cent accuracy is achieved. A chick becomes more accurate in aiming its pecks as a result of practice. The more accurately a chick pecks, then the quicker a parent will respond to the request for food. This is important if a chick is to receive sufficient food for it to survive and grow.

Although feeding behaviour is innate, gull chicks must learn to recognise food. If a chick does not immediately recognise food that is dropped by the parent it must find food by trial-and-error. The chick pecks at any material in the nest until it eventually finds the food. The speed with which gull chicks learn to recognise food depends on whether or not they have another chick from which to learn (figure 11.24). Chicks take less time to recognise food if they have an older chick from which to learn.

Consider the bowerbirds once again. Building a bower is an innate behaviour and yet a male bird improves on the quality of the bowers it makes as it gets older. The better built and more decorated bowers are more likely to attract females. Improving the behaviour through experience increases the chance of mating success.

KEY IDEAS

- A behaviour is an activity carried out by an animal in response to a stimulus.
- Behaviours are important for the survival and reproduction of a species.
- Innate behaviours are those that are essentially the same in all members of a species.
- There are different kinds of innate behaviours.
- Innate behaviours may be refined through experience.

QUICK-CHECK

1 List three kinds of innate behaviour.
2 Copy and complete the gaps in the following table. The magpie is given as an example.

Animal	*Kind of innate behaviour*	*Stimulus for the behaviour*	*Response (the behaviour)*
magpie	territorial	intruders in territory	attacks to drive potential competitor away
newt		urge to mate	postures in front of female and raises crest
sea slug		chemical substance from starfish	flaps body up and down to move away
lion			male kills cub of another male
dolphins			herd fish
mud dauber wasp			builds a mud nest

3 If attacked, an opossum often feigns death — it 'plays possum'. How would you classify this behaviour? What is the stimulus for the behaviour? In what way does this behaviour contribute to the survival of the organism?
4 Mother cats sometimes give their kittens live mice to play with. Why might such play be an important behaviour for the kitten?

Learned behaviours

Some behaviours change as a result of maturation of the body of an organism due to hormonal and structural changes. These are not to be confused with learned behaviours which are those that develop or change as a result of experience. Learned behaviour has been summarised in figure 11.4 (page 343). An expansion of some of these follows.

Conditioning

A dog will salivate if meat powder is placed in its mouth. The food is a stimulus that triggers the salivation response. In 1904, a Russian physiologist, Ivan Pavlov (1849–1936), discovered that he could make a dog salivate by ringing a bell provided he had conditioned the dog to believe that it would receive meat powder if the bell was rung. He conditioned the dog by carrying out a series of experiments (figure 11.25).

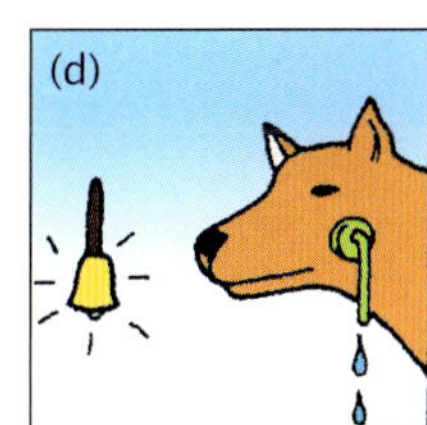

Figure 11.25 Pavlov conditioned a dog to believe that it would be fed if a bell was rung. **(a)** A bell on its own produced no response from the animal. **(b)** However, on exposure to food, the dog produced saliva. **(c)** When shown food at the same time as the bell was rung, the dog salivated. **(d)** After that experience, the dog salivated when the bell was rung. The dog was conditioned to believe that food would accompany the bell.

ODD FACT

Birds avoid taking the unpleasant tasting cinnabar caterpillars by learning to recognise their colour. After learning this, birds refuse to eat both edible and inedible caterpillars of the same colour as the cinnabar.

Initially, ringing a bell had no effect on the dog. Then, every time Pavlov rang a bell he gave meat powder to the dog. The dog salivated automatically in response to the food. After a time the dog salivated when the bell was rung even if no meat powder was provided. The dog had learned that the bell was associated with receiving food. This kind of associative learning is called **classical conditioning**. The reward follows the stimulus.

Another kind of conditioning is called **operant conditioning** or trial-and-error learning. The term operant learning comes from early behaviour experiments with rats. Caged rats learned by trial-and-error that if they operated a bar in their cage, they would receive food. The behaviour is a conditioned response that results in a reward. The reward follows the response. Animals generally have to be motivated to perform an activity that results in a reward. Usually, the motivation is hunger. In the case of the experimental rats the stimulus for the behaviour is hunger, the action is pressing the bar and the reward of food follows.

ODD FACT

Otters spend most of their time in the water lying on their backs with one or two feet sticking out of the water. This behaviour helps the otters to regulate their body temperature. Heat is lost from the skin surface of the large splayed feet if an animal is too hot, or heat is absorbed from the sun if the animal is too cold.

Figure 11.26 Young otters play with stones and shells. Eventually they are able to use the stone to break open various shellfish and sea urchins they collect for food.

Children and young animals at play often engage in trial-and-error learning. Greater activity in **play behaviour** as a child leads to a more successful adult because of the wider range of experiences that can be drawn upon.

An adult sea otter often lies on its back in the water with a stone on its chest (figure 11.26). The otter bangs abalone, mussels and other shelled molluscs against the stone to break their shells so that the flesh of the animal can be eaten. The stones can weigh up to a kilogram and are carried by the otters as they dive to collect their food. Young otters learn the technique by trial-and-error. They begin to hit their chests with their paws at about 5 weeks. Next, they hit their chests with a shell or some other object or food such as a crab. By about 9 to 10 weeks, the young otters try to hit two objects together. By about 20 weeks they can use a stone to open soft prey like mussels.

Habituation

A new teacher at a school was startled to hear a deafening noise. Her students showed no reaction and appeared not to have heard the noise. The noise was a large aeroplane coming in to land at a nearby airport. The students assured the teacher that within a short time she too would become accustomed to the noise, realise the plane was not going to come through the roof and would react as they had — as if the noise didn't exist. After a few days, the teacher had learned to behave in exactly the way the students predicted. The ability to 'get used to' a repeated stimulus, such as a noise, is called **habituation**.

Figure 11.27 Animals habituate in all kinds of situations.

Why is habituation important for animals?

In nature, animals are subjected to numerous stimuli to which they must respond. Animals will respond to a noise if they think the noise has been made by a predator. Other animals will respond to a shape overhead if the shape resembles that of a predatory bird, such as a hawk or an eagle. Such responses are critical for survival.

If animals react to every noise they hear or every shape they encounter then they would waste a lot of time in reacting to non-threatening and non-important situations. Habituation enables animals to distinguish the unimportant noises and shapes from those that are important. Habituation allows animals to ignore meaningless stimuli and save precious energy for other activities which are critical for survival.

Insight

English sparrows raised with canaries will learn the canaries' song.

Insight learning is related to the ability of an animal to apply past experience to solving a new problem without a trial-and-error period. Insight involves the use of reason. This type of behaviour is most highly developed in humans. Human behaviour often involves insight — not only do we use our own experience to solve problems but we also use the ideas of others that have been communicated to us in writing, speech or some other way.

It is often difficult to judge whether the new problem really is new. Early experiments into insight learning involved putting chimpanzees into a new situation in which a bunch of bananas was suspended from the ceiling. When the bunch could not be obtained by jumping, chimpanzees used boxes to build steps that enabled them to reach the bananas. Some scientists argued that the problem may not necessarily have been completely new. Previous play may have given the chimpanzees a similar experience.

Play forms an important part in developing problem-solving skills for later use.

Imprinting

ODD FACT

A male duckling reared by a female of a different species is unable to mate at maturity because he courts females of the different species who do not react to his advances.

If a duck egg of one species is hatched by a female duck of another species, the baby duck will follow its hatching mother away from the nest. Later, at maturity, it will try to mate only with a duck of the same species as the hatching mother. The behaviour of the young duck in following the mother shortly after hatching is important in fixing behaviour later in life.

Konrad Lorenz, an Austrian scientist, found that newly hatched ducks and geese follow the first moving object they see. In his experiments, Lorenz acted

Orphan lambs that are bottle-reared by a person become imprinted with their human keeper and will leave the flock to follow the keeper.

as the first moving object to young ducks and later found that they would follow him everywhere. They followed Lorenz by preference even when an adult duck was introduced. When the ducks matured, they courted Lorenz and other humans rather than other ducks. It is important for a young duck to follow its mother in nature (figure 11.28). The young duck forms a special attachment to its mother and at maturity it can recognise a suitable mating partner.

The formation of an attachment to something in the environment shortly after hatching or birth is called **imprinting**. The learning that takes place during imprinting is rapid and cannot be reversed. In experiments with mallard ducklings, the critical period for imprinting was the fifteenth hour after hatching. Imprinting fixes behaviour in relation to a species. Lorenz's ducks would follow other humans as well as himself. They would not follow other ducks.

Imprinting is important for the survival of many species in the wild. Mothers protect their young and may be aggressive towards other species. Without imprinting, the young may lack this protection and be far more vulnerable to attack.

Figure 11.28 Maned duck adults with young. Young ducklings imprint to their mother shortly after hatching. Imprinting is important because it fixes behaviour in relation to the species. If a mother duck is not available at the critical time for imprinting, young ducklings will imprint to other animals or moving objects.

Does imprinting occur in humans?

The first bond formation in humans has some similarities with imprinting. The first bond formation with an individual, often the mother, is important in setting the scene for the ability to form lasting friendly and loving relationships later in life.

Babies who are forced by circumstances to transfer their relationship from one person to another can suffer damage to their emotional structure which results in their reduced ability to form attachments with other people. Awareness of this has resulted in parents being allowed to spend much more time with their sick children in hospitals. Previously sick children who spent long periods in hospital ran the risk of emotional damage as they were forced to transfer their attachments from nurse to nurse as the nursing shifts changed.

Learning from animal behaviour

Although the study of animal behaviour has become increasingly popular in recent years, humans have relied on certain aspects of animal behaviour for centuries. Farmers used domesticated wolves, or dogs, to help them look after flocks of other domesticated animals. In doing this, farmers were exploiting the natural hunting behaviour of the dog.

Dogs are pack animals which obey the leader of the pack. The farmer is seen as the leader of the pack by his dogs and they are happy to drive the 'prey', in this case sheep, towards their leader. An understanding of animal behaviour has practical applications in agriculture and animal welfare. A better understanding of the natural behaviour of animals means that we can look after them better, while still using them as a resource.

If we understand the natural behaviour of animals, we can also analyse and interpret any change in behaviour. A change in animal behaviour can sometimes indicate a change in the environment. For example, a change in the distribution of different species of microorganisms in water may reflect the presence of a pollutant. We should also be able to predict more accurately the impact that an environmental change might have on the survival of a group of animals in an area. We need to be able to give answers, based on sound scientific study, to questions such as the extent to which feeding and nesting behaviour of fairy penguins in Southern Australia might be affected by a spill of oil in Western Port Bay, or the impact that logging would have on the native animals in a particular area.

ODD FACT

Animals do not recognise international boundaries so, in some cases, cooperation is needed between nations. For example, monarch butterflies will become extinct if the forests to which they migrate in winter are reduced below critical levels. Maintenance of their summer habitat alone would not enable them to survive.

The increase in the number of people studying animal behaviour also reflects a change in our outlook on life. Many people no longer see animals as a resource to be plundered as humans see fit. Animals form part of the aesthetics of the world in which we live and it is important that we preserve them for future generations. This is only possible if we have an understanding of animals' needs in terms of space for feeding, living and reproduction.

BIOLOGIST AT WORK

Dr Christopher Gleeson, Veterinarian

I am writing this from England where I have been working and travelling for the past 17 months. This is one of the many positive aspects of being a veterinarian — along with other Australian graduates, I have the opportunity to work in the United Kingdom and see many other parts of Europe. I have worked in small rural practices where I have been the only vet and in busy London clinics where I was one of 15 vets.

When I left school I did not have a particular career in mind and followed science because it was 'in the family'. My father is a veterinarian so I knew something about the work and had helped out at his practice many times. By the end of my first year of BSc/BA studies, I had decided on Veterinary Science as a career and entered the BVSc after completing the four-year BSc course. Eight years after finishing the VCE, I graduated in Veterinary Science from the University of Melbourne. None of this was time wasted. As one of several 'older' students, and as the vet science class was relatively small, I formed lifelong friendships while studying. The final two years are spent in residence or close to the clinic and I loved the clinical work. I soon decided that small animals and the lifestyle of a city vet were more for me than large animals.

Becoming a vet requires a commitment to intensive study, and the work can involve stress and long hours. Despite this, I have found my first few years of veterinary work very rewarding. The aspects of veterinary science that I enjoy most are the contacts with people while treating their pets and the variety of tasks, from surgery to consultations, to treatment and drug administration. Vets use a wide spectrum of disciplines, including microbiology, biochemistry and pharmacology and, unlike a doctor who treats one species, a veterinarian can see a dog, a cat, a bird, a snake in one block of consultations. Increasingly there are specialist practitioners in veterinary science: skin specialists, eye specialists, orthopaedic specialists, pathologists and even specialists in dog behaviour. Specialist roles facilitate further research and expand the pool of knowledge in animal medicine. They also allow vets to change direction and follow special interests in their careers.

Figure 11.29 Australian vet Chris Gleeson is gaining a broad range of veterinary experience in the UK.

KEY IDEAS

- Learned behaviours are those that develop or change as a result of experience.
- There are different kinds of learning behaviours.
- Learning enables an animal to adapt to change.
- Animals learn by observing other animals or by experimenting.

QUICK-CHECK

5 List three kinds of learned behaviour.
6 What type of learned behaviour is responsible for each of the following actions?
  a People living next to a fire station don't 'hear' the bells of the fire engine.
  b A dog stands on its hind legs and presses on a latch or handle to open a door.
  c A rat is fed against a false wall placed halfway along a passage. The rat continues to go to the halfway mark even when the false wall is removed and food is placed at the far end of the passage.
  d Birds on a seed platform ignore a cat that stands on its hind legs but is 20 cm short of the platform.
7 Outline a situation in which you have applied insight learning.
8 Chaffinches (birds) raised in isolation are able to sing simple 'songs'. They only sing the full song characteristic of the species if they hear adult chaffinches singing. What kind of behaviours are involved in chaffinch singing?

Plant behaviour

In the previous section, we considered animal behaviours in response to a range of stimuli. What stimuli impact on plants and how do plants respond as a result? In this section, we consider the impact of environmental stimuli on plants and their effect on plant growth. We also examine some rhythmic activities of plants in response to external stimuli.

Tropisms

External factors, such as light, gravity and touch, exert an influence on plant growth and development. The growth of a plant in response to a stimulus such as light or water is called a **tropism**.

When a plant grows towards a stimulus, the term positive tropism is used. Growing away from the stimulus is called negative tropism.

Light as a stimulus

If you have ever had a pot plant on a windowsill you may have noticed that the plant turns towards the light. You may have even kept turning the pot around so that the plant takes on a more even appearance. When a plant moves in response to light, it shows **phototropism**. When the movement is towards the light it is known as positive phototropism.

Although a coleoptile is sometimes called a stem, it is really a leaf.

Over the years many experiments have been carried out with young plants to study the effect of light on their growth. These experiments were carried out using the first leafy shoot of a growing seedling of the grass family. This first shoot is termed a coleoptile.

In an investigation of growing seedlings, Charles Darwin (1809–1882) found that he could prevent bending towards the light by covering their growing tips (see figure 11.30). Darwin concluded that the tip of the seedling produced an 'influence' that passed from the tip to the region below the tip where bending occurred.

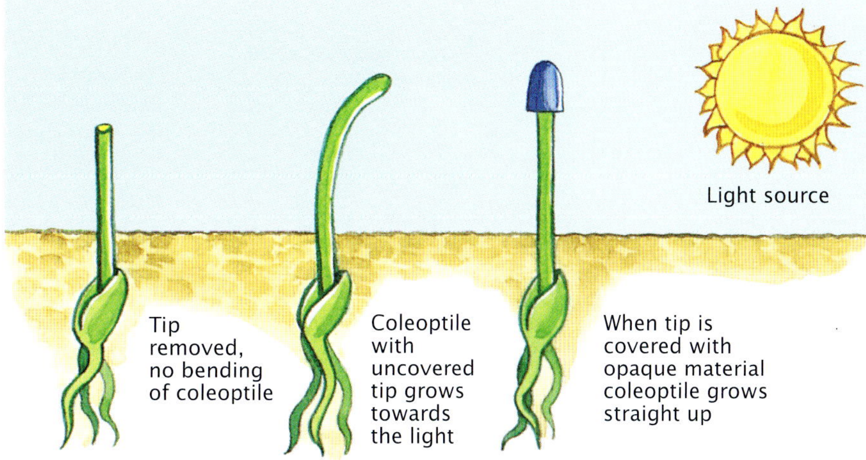

Figure 11.30 Growing coleoptiles bend towards a light source. Bending does not occur if the tip is shielded from the light. What conclusion can you draw from the observation that a coleoptile without its tip fails to respond to the light?

We now know that auxin, a plant hormone, is produced in the tip of a coleoptile and causes growth of cells in the coleoptile. The tip is the site of reception of the light stimulus. The growing region below the tip is the effector.

Auxin moves away from light

What causes a plant to bend towards a light? If a coleoptile is evenly illuminated, auxin is evenly distributed throughout the tip and the coleoptile grows straight up. If light is concentrated to one side of a coleoptile then auxin moves away from the light source to the darker side of the tip and becomes more concentrated in the cells in that region. The increased concentration of auxin in these cells means they grow more quickly than cells nearer the light (see figure 11.31). The uneven growth of cells results in bending of the coleoptile.

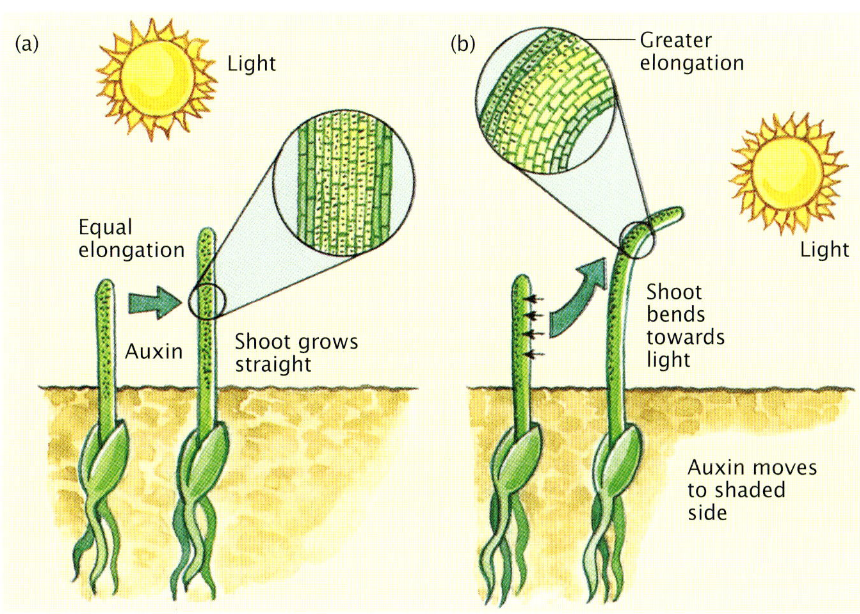

Figure 11.31 Distribution of auxin influences bending in a plant. **(a)** An even distribution of light results in an even distribution of auxin in the coleoptile which grows straight. **(b)** If light shines from one side, auxin moves to the shaded side and causes the elongation of cells in that area. The different rates of growth of cells on each side of the shoot result in bending.

The label 'auxins' refers to many hormones within plants. You will consider plant hormones in greater detail in your studies of biology next year.

Further evidence for the involvement of auxin in the bending of coleoptiles was obtained experimentally. When a block of agar containing auxin was placed to one side of the cut tip of a coleoptile placed in total darkness, the coleoptile bent away from the side with the auxin-agar block. Auxin in the agar had diffused into one side of the coleoptile and caused the cells to grow more quickly on that side.

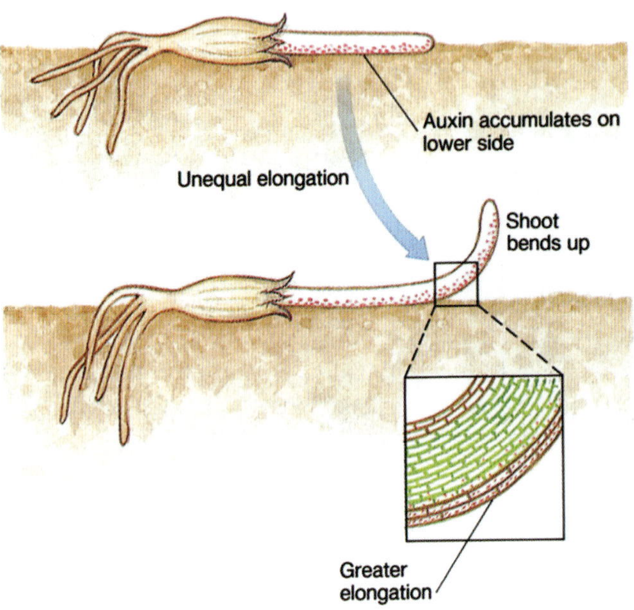

Figure 11.32 In a horizontal seedling, auxin accumulates along the lower horizontal part of the shoot causing the cells in that part to grow faster. Hence, the shoot turns upwards, away from gravity. In the roots, it is the upper part that grows faster so the root turns downwards towards gravity.

Gravity as a stimulus

If a germinating seed is oriented so that its root and shoot emerge horizontally from the seed, then after a short time the shoot turns upwards and the root turns downwards even if the seed is placed in total darkness. This directional growth occurs because the plant shows **geotropism**; it responds to gravity.

Auxin is produced in growing roots and shoots. Gravity causes most of the auxin produced to accumulate in the lower sides of the horizontal shoot and root (see figure 11.32). Auxin stimulates the growth of cells in shoots but inhibits the rate of growth of cells in roots. The uneven distribution of auxin in the cells of the horizontal seedling causes uneven growth of cells: the cells along the *lower* horizontal part of the shoot grow faster and hence the shoot turns upwards away from gravity; the cells along the *upper* horizontal part of the root grow faster and hence the root turns downwards towards gravity. Shoots show negative geotropism and roots show positive geotropism.

The response of plants to particular distributions of auxins enables plants to orient themselves to a more favourable position within their environment. Seedlings grow through the soil and shaded plants can grow in a way that their photosynthetic tissues are more likely to be exposed to increased sunlight.

Figure 11.33 Tendrils form when there is uneven growth on the inside and outside of the tendrils. Uneven growth occurs because of uneven distribution of plant hormone.

Up and away: climbing plants

Have you ever looked closely at a grapevine or some other climbing plant? You may have noticed the tendrils or twining parts of the plant that are wrapped around support structures or other plants (see figure 11.33). The twisting is caused by uneven growth of cells along the tendril as it comes into contact with other plants or structures.

A change in growth because of contact with another object is called **thigmotropism**. What do you think happens to the distribution of auxin when a growing part comes into contact with another object?

Rhythmic activities

Plant, like animals, exhibit a range of rhythmic behaviours.

Time as a stimulus

We tend to sleep during the night and stay awake during the day. Have you ever travelled overseas to a country in a very different time zone from Australia? Your body may have taken a while to get used to the different arrangement of night and day. Your internal clock, called a **biological clock**, must be reset to the new surroundings and this may take a few days to occur.

Plants also have biological clocks. For example, some plants have leaves that follow a pattern with a daily rhythm. The leaves are horizontal during the day but fall into a 'sleeping' position at night. The 'sleep' movements occur whether the plant is kept in normal daylight and dark or in 24 hours of light. The internal biological clock of the plant determines the 'sleep' pattern which recurs approximately every 24 hours.

An activity that follows a 24-hour cycle is called a **circadian rhythm** (or circadian cycle). Other plant activities that involve a circadian rhythm include the opening and closing of flowers, and nectar and perfume production. This kind of movement, in contrast to movement in response to light or gravity, is independent of the direction of the stimulus and is called **nastic movement**.

Solar tracking

The leaves and flowers of many plants are able to move during the day so that they are oriented either perpendicular (at right angle to) or parallel to the sun's direct rays. This ability is called **solar tracking**, or **heliotropism**. Sunflowers, *Helianthus annuus* (see figure 11.34) have that ability and so, in a field of sunflowers, all the flower heads and leaves face the same direction. Leaves that remain perpendicular to the sun's rays throughout the day, such as sunflower leaves, have a higher rate of photosynthesis than parallel-tracking or non-tracking leaves.

Timing of flowering

A hormone probably plays some role in determining when a plant flowers. Also of importance is the length of night and day to which a plant is exposed. The relative length of day and night is called the **photoperiod**. The response of plants to particular periods of light and dark is called **photoperiodism**.

Figure 11.34 Sunflowers exhibit solar tracking behaviour.

Photoperiodism differs in different plants. Some plants, such as chrysanthemums, flower only when the day length is shorter than some critical period. These are called **short-day plants** (see figure 11.35). Other plants, such as carnations, are **long-day plants**: they flower only after being exposed to day lengths that are longer than a certain critical minimum. Some plants, such as the dandelion, flower regardless of the length of day. These are called **day-neutral plants**.

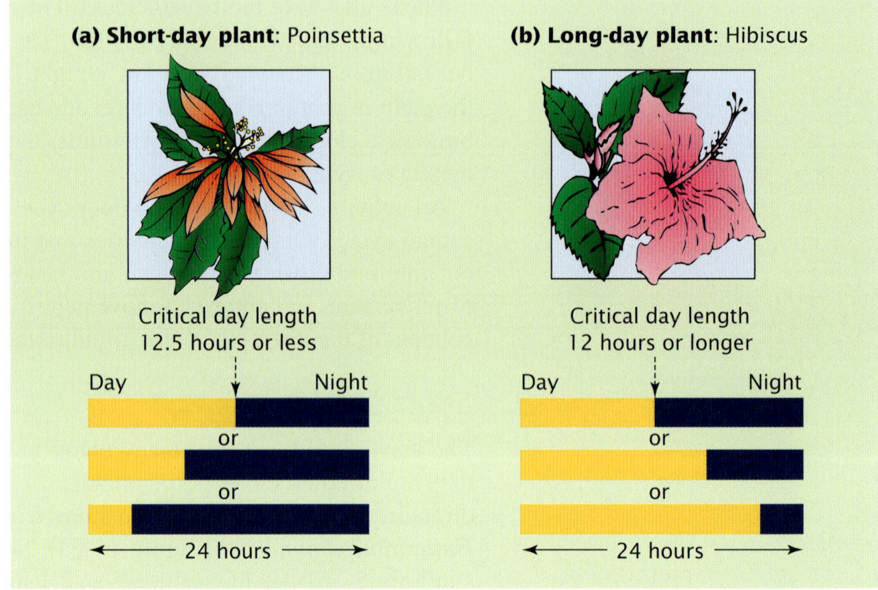

Figure 11.35 Different plants flower in response to different day and night lengths. **(a)** The poinsettia (*Euphorbia pulcherrima*) is a short-day flower and flowers only when the day length becomes less than 12.5 hours. **(b)** The hibiscus (*Hibiscus* spp.) is a long-day plant and flowers only when the day length becomes greater than 12 hours.

Examples of short-day, long-day and day-neutral plants are shown in table 11.5. How would you tend to classify plants that flower in spring?

Table 11.5 Some examples of short-day, long-day and day-neutral plants

Short-day	*Long-day*	*Day-neutral*
monocotyledons		
rice (*Oryza sativa*)	wheat (*Triticum aestivum*)	corn (*Zea mays*)
dicotyledons		
chrysanthemum (*Chrysanthemum* spp.)	cabbage (*Brassica oleracea*)	potato (*Solanum tuberosum*)
poinsettia (*Euphorbia pulcherrima*)	hibiscus (*Hibiscus syriacus*)	rhododendron (*Rhododendron* spp.)
violet (*Viola papilionacea*)	spinach (*Spinacea oleracea*)	tomato (*Lycopersicon esculentum*)

A grower who knows the day–night characteristics of a particular plant is able to encourage or delay flowering by controlling the periods of light and dark to which the plants are exposed. Why would this be an advantage to the grower?

Although the commonly used terms of short-day and long-day plants suggest that the exposure to light is important, we now know that the length of darkness is critical. In experiments with the cocklebur (*Xanthium strumarium*), it was shown that the length of the night is the critical factor that determines whether a plant flowers or not (see figure 11.36).

The cocklebur, a short-day plant, would flower only when the dark period of its cycle exceeded nine hours; the day length did not matter. When the plant was exposed to a brief flash of light during the dark period, however, the plant no longer flowered. Uninterrupted dark is important for flowering. Short-day plants are really long-night plants. Long-day plants are really short-night plants.

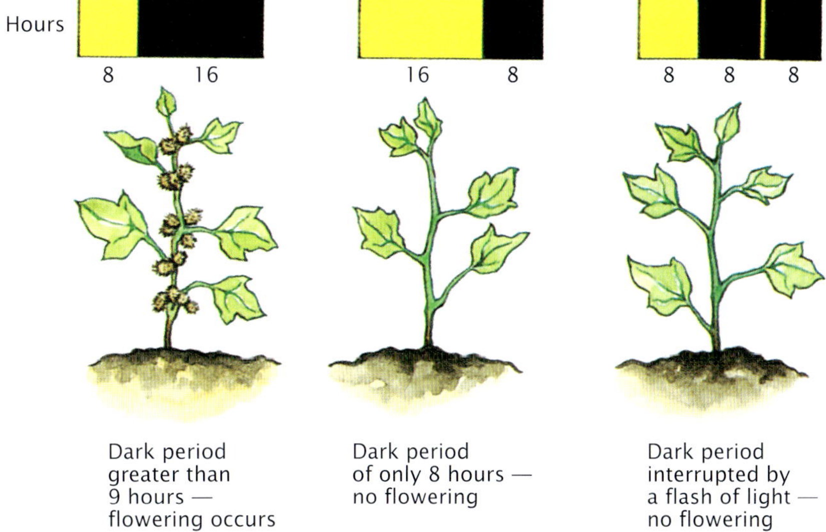

Figure 11.36 Examination of the effects of different day and night lengths on flowering of cocklebur (*Xanthium strumarium*). Flowering occurs only if darkness exceeds about nine hours. If the dark is interrupted, even with a brief flash of light, flowering does not occur.

How does a photoperiod exert its influence on flowering? It is thought that phytochromes, light-sensitive pigments in leaves, react to a photoperiod that is appropriate for the particular plant and as a result the leaves produce a hormone. The hormone travels to buds which then flower.

Photoperiod can also be important in breaking the dormancy of seeds and other parts of plants.

KEY IDEAS

- Factors external to a plant influence growth and development of the plant.
- The direction of response of a plant is sometimes influenced by the direction of the external stimulus.
- Day–night lengths are critical for some plant activities.

QUICK-CHECK

9 List two external factors that influence plant growth and describe what happens as a result of each influence.
10 How does a plant that is positively phototropic respond to light?
11 How does a plant that is negatively geotropic respond to gravity?
12 Name two activities of plants that show a circadian rhythm.
13 How would you classify a plant that flowered after being exposed to 14 hours light followed by 10 hours dark?

BIOCHALLENGE

A number of animal behaviours are outlined below. How would you classify the behaviour(s) in each case and of what advantage would each be for the individual or group involved?

A

Puffins, *Fratercula arctica*, that nest in burrows on the cliffs of Fair Isle, north of Scotland, gather in groups in the sea and spend some time together before they fly to the cliff faces.

B

When a colony of meerkats is active, moving in, out and around their burrows, one of the colony usually stands on a high point acting as a sentry.

C

Male peacocks (*Pavo cristatus*) raise their highly decorative tail feathers when females are close by.

D

Many brown algae, 'kelps', have holdfasts that secure them to the seafloor. Sea otters, *Lutra lutra*, living in the vicinity of kelps often wrap themselves in kelp before they sleep.

E

Laughing kookaburras, *Dacelo novaeguineae*, remain with their parents for two or three breeding seasons before they leave the family group. Females tend to leave before their male siblings.

F

Vultures are usually seen in numbers when there is a carcase nearby but they have not caused the death of the animal.

CHAPTER REVIEW

Key words

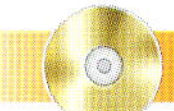

CROSSWORD

aggression
behaviour
biological clock
caste system
circadian rhythm
classical conditioning
communication
competition
cooperate
day-neutral plants
echolocation
ethology
geotropism
habituation
heliotropism
imprinting
innate or inborn behaviour
insight learning
learned behaviours
long-day plants
migration
nastic movement
operant conditioning
pecking order
pheromones
photoperiod
photoperiodism
phototropism
play behaviour
reproductive behaviours
rhythmic behaviours
short-day plants
social hierarchy
social interactions
solar tracking
territorial behaviour
thigmotropism
tropism

Questions

1 *Making connections* ▸ Use as many as possible of the key words from this chapter to make a concept map.

2 *Applying your understanding* ▸ State whether each of the following situations describes innate or learned behaviour:

a The ability to perform a task often requires a teacher.

b An action present in an animal with a very short life span

c An ability to perform the act develops only after practice.

d The action may be influenced by rewards.

e An ability which may not be apparent at birth but requires no practice

f An action that is genetically determined

g Webs of spiders of the same species are virtually identical.

3 *Applying and communicating understanding in a new context* ▸ How would you classify the general type of learning represented in each of the following situations? Explain your choice.

a You visit your grandmother for a week's holiday. On the first night you are kept awake by a wall clock that chimes at quarter-hour intervals. By the end of the week you have no difficulty in getting to sleep.

b A male puppy will only raise its leg on objects to urinate if it has lived with a male dog who urinates in that way.

c i A mother rings a bell each night to let her children know that the evening meal is ready and they should come in from play. After a few days, the family cat also comes home in response to the bell and is fed at the same time. Occasionally, when the family is going out and they wish to lock the cat up, they ring the bell in the middle of the day. The cat comes home, but is not fed.

ii Predict what might happen if the cat is brought home regularly by ringing the bell without feeding.

d A model bird made of red and white feathers was fiercely attacked when it was first placed near a robin nest in spring. After several days the robins ignored the model.

4 ***Analysing and interpreting information*** ▸ What is the general type of behaviour in each of the situations described below? What is the stimulus in each case that led to the behaviour?

a A cat arches its back, with its tail and hair erect.

b As a female bird approaches, a male lyrebird dances and sings on his mound with his tail well displayed over his head.

c Thousands of termites build a large nest.

d An old mirror stands against an outside wall of a building. A pair of birds, one male, one female, peck among the nearby grass and come close to the mirror. The male sees himself in the mirror, jumps up fluttering his wings and pecks at his own image in the mirror. He repeats the action several times.

5 ***Interpreting data and communicating ideas*** ▸ Sticklebacks are small freshwater fish. During the mating season male sticklebacks develop red bellies and each male sets up a territory and other males are chased away. Tinbergen experimented to establish what provoked the chasing response. He introduced different models into the range of a male stickleback and recorded what happened. The results are summarised in figure 11.37.

Which characteristic is important in provoking an aggressive response in male sticklebacks during the mating season? Explain your reason.

MODEL OF INTRUDER	BEHAVIOUR OF TERRITORY OWNER
(a)  Identical shape – no red belly	No effect
(b)  Identical shape – red belly	Strong aggression
(c) 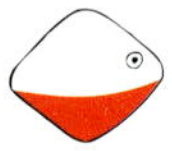 Different shape – red belly	Strong aggression
(d)  Roughly identical shape – red belly	Strong aggression
(e)  Different shape – red belly	Strong aggression

Figure 11.37 Summary of Tinbergen's experiments with sticklebacks

6 ***Interpreting and communicating biological information*** ▸ Two subspecies of deer mouse, *Peromyscus gracilis*, live in different habitats:

- Subspecies A lives in forests and brushy areas
- Subspecies B lives in open fields and meadows.

Experiments were carried out to investigate mouse choice of habitat. Each of the two subspecies was allowed to breed in a laboratory for several generations. Laboratory-raised deer mice, which had no direct experience with the natural environment, were allowed to choose between two habitats. The results are presented in the following table.

Habitat provided	*Preferred by*
Tree trunks with flat plates attached to the top to simulate shade	Subspecies A
Artificial grass made of stiff manila paper	Subspecies B

a Why were the deer mice allowed to breed in a laboratory for several generations before being given a choice of habitat?

b What kind of behaviour is habitat choice in deer mice? Explain your answer.

7 *Interpreting and communicating biological information* ▸ Ground squirrels give a special call when a falcon or hawk is seen. The caller squirrel and others hide for safety when this occurs. Ground squirrels give a different kind of call when a ground predator approaches. Observations were made to see if there was a difference in the calling behaviour of males and females for ground predators. The data are presented in the following table.

Category of squirrel	*Exposure to predator*	*Number of squirrels observed to give alarm call*
Males, more than 1 year old	67	12 (18%)
Females, more than 1 year old	358	106 (39%)

a Are male or female ground squirrels more likely to call the alarm when a ground predator approaches?

b Suggest reasons for the difference in calling pattern shown by male and female ground squirrels when a ground predator approaches.

Observations were made to determine whether special female caller squirrels operated to respond to aerial and ground predators. Data were collected for different groups of females and are shown in figure 11.38.

c Explain whether one group of females is more likely to sound a warning call at the approach of an aerial predator than another group of females.

d Explain whether one group of females is more likely to sound a warning call at the approach of a ground predator than another group of females.

e Suggest reasons for the difference in behaviour with the different kinds of predators.

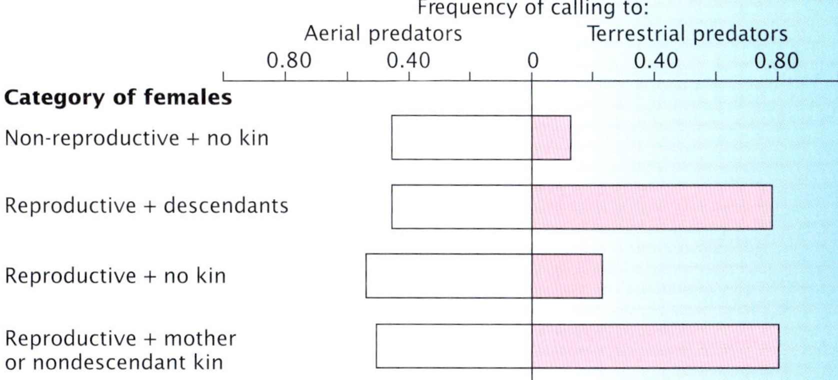

Figure 11.38 Frequency of alarm calling of different groups of female ground squirrel when threatened by aerial and ground predators. Reproductive females include the pregnant, lactating and those living with weaned offspring.

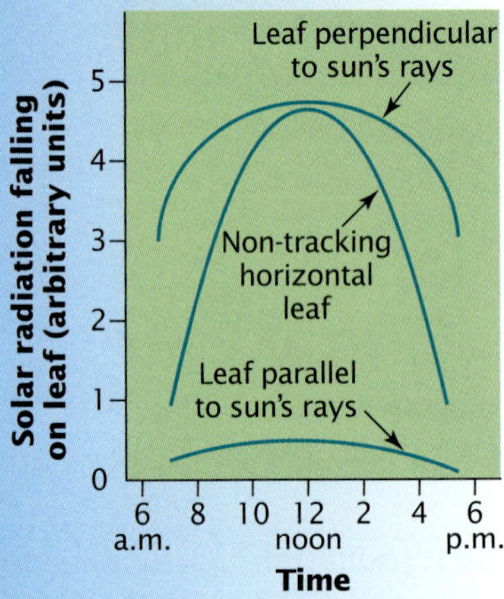

Figure 11.39 Solar radiation falling on three different kinds of leaf

8 ***Applying your understanding and communicating ideas*** Leaves may orientate differently to the direction of the sun's rays. Examine figure 11.39.

a What is the amount of solar radiation falling on each of the three leaves at 10 a.m.?

b Explain why more solar radiation falls on the leaf perpendicular to the sun's rays than on the other two leaves.

c Explain the difference between solar radiation falling on a non-tracking leaf at 8 a.m. and that falling at 12 noon.

d Of what advantage might it be for a leaf to remain parallel to the sun's rays?

9 ***Investigating scientifically*** You note that a particular kind of flower closes its petals at night. Design an experiment to investigate whether the flowers close their petals in response to light intensity or to temperature.

10 ***Analysing information and communicating ideas*** Blocks of agar containing various amounts of auxin were prepared. They were placed at each end of sections cut from coleoptiles or roots and left to stand for four hours. The results are shown in figure 11.40.

a Explain whether the experimental results support the claim that auxin can move both up and down a coleoptile.

b Do the data support the claim that auxin can move both up and down a root? Explain.

Coleoptile	Start of experiment	After 4 hours	Root	Start of experiment	After 4 hours
	100 0	33 60		100 0	45 45
	100 100	35 159		100 100	90 90
	100 250	34 305		100 250	167 167

Figure 11.40 The numbers in the agar blocks represent the relative amounts of auxin at the start and the end of the experiment.

11 ***Using the web*** Vampire bats, *Desmodus rotundus*, are extremely small animals, weighing between 15 and 50 gram. Their food consists of blood. They eat about half their body weight each night and, because of their low weight, run the risk of dying if they fail to feed on two consecutive nights.

Go to www.jaconline.com.au/natureofbiology/natbiol1-3e and click on the 'Vampire bats' weblink for this chapter. Click on the photograph for information about bats. Open and read the 'Fun facts' and the information below them.

a In which countries do vampire bats live?

b From which animal(s) do bats obtain their source of blood?

c Outline the cooperative feeding behaviour that occurs in vampire bat colonies that reduces the likelihood of a bat dying from starvation.

You may wish to investigate the behaviour of other animals by clicking on the appropriate image.

12 Reproductive strategies for survival

KEY KNOWLEDGE

This chapter is designed to enable you to:

- extend your knowledge of the variety of reproductive strategies in animals and plants
- develop understanding of how various strategies assist successful reproduction in particular environments.

Figure 12.1 A colourful marine mollusc, the orange spotted clown nudibranch (*Ceratosoma amoena*), has laid a cluster of many eggs on a green alga. This nudibranch is found in temperate waters of south-eastern Australia and of northern New Zealand. If the eggs are not eaten by predators, they will hatch and produce free-swimming larvae. Most larvae will not survive but a few may survive and mature to become sexually reproducing nudibranchs of the next generation. In this chapter, we will explore various aspects of reproduction, including strategies and modes that enable the survival of species in different habitats.

Sex at sea

The Great Barrier Reef (GBR) stretches about 2300 kilometres along the Queensland coast, from Gladstone in the south to the tip of Cape York in the north. The GBR is not a single reef but consists of many types of reef (platform, fringing, barrier, ribbon) that are located several tens of kilometres to about one hundred kilometres offshore. Reefs consist of hard corals that are produced by small **colonial** animals known as coral polyps. Each polyp in a colony builds a hard calcareous cup in which it sits, and together the colony produces the calcareous mass that we recognise as coral (see figure 12.2). Polyps are carnivorous animals and, when feeding, they extend their tentacles from the cups (see figure 12.3) and capture small passing prey. When not feeding, polyps retract into their protective cups.

If we take a dive beneath the turquoise waters surrounding a reef, we can explore the reef front. Here, we will meet some of many hundreds of different species of reef fish of different sizes, shapes and colours. If we swim to the shallow waters overlying the reef flats, we are likely to see sharks and rays, such as blacktip reef sharks (*Carcharhinus melanopterus*) and blue-spotted stingrays (*Taeniura lymna*). These marine **vertebrate** animals are just a few of the myriad of species that inhabit the tropical waters around and near the reefs, along with many **invertebrate** animals, including feather stars, giant clams, crabs, shrimp, squid, octopuses, sponges and, of course, the polyps that create the reefs.

Figure 12.2 Some of the many forms of coral, including boulders and plates. Other corals are convoluted and yet others are columnar or highly branched.

Figure 12.3 Coral is the product of tiny animals known as polyps. This photograph of coral (*Tubastrea* spp.) shows the polyp arms extended. (See chapter 13, page 423, for a photograph of a coral polyp that has captured its prey.)

Broadcast spawning: eggs floating freely

ODD FACT

Coral polyps, along with jellyfish, sea anemones and hydra, are members of phylum Cnidaria. Their distinctive body plan includes (i) a single opening to the body (ii) a three-layered body wall and (iii) tentacles covered by stinging cells.

Depending on the time of day and year, we may be lucky enough to see the reproductive activities of some of the animals of the GBR. For example, we may observe some kinds of reef fish gathering and simultaneously releasing their eggs and sperm into the water where fertilisation will occur. This process is known as **broadcast spawning** and usually occurs in regions of the reef affected by strong currents and when an outgoing tide is flowing. As a result, the fertilised eggs are quickly carried away from the reef and from the many predators that live in this habitat. Eggs of this type that float within the water column are termed **pelagic** eggs.

Examples of broadcast spawners include butterfly fish (*Chaetodon* spp.), coral trout (*Plectropomus leopardus*), boxfish (*Lactoria* spp.) and many species of wrasse. Some of these reef fish mate in pairs, such as boxfish. Some reef fish that engage in broadcast spawning, such as coral trouts and cardinalfish, gather in very large groups at one location (see figure 12.4) and release their eggs and sperm simultaneously in a so-called **mass spawning** event.

The term 'fish' refers to two different classes of animal. Reef fish have skeletons made of bone and belong to the Class Ostichthyes. Sharks and their close relatives, rays and skates, have skeletons made of cartilage and belong to the Class Chondrichthyes. The label 'elasmobranch' is often used for the latter group.

Mass spawning also occurs in coral polyps (refer back to page 192).

Figure 12.4 A large group of cardinalfish (family Apogonidae) gather for mass spawning among the coral of the Great Barrier Reef.

As well as reef fish, fish of the open ocean, such as mackerel, tuna and cod, are also broadcast spawners. Another group of broadcast spawners are freshwater fish that live in large flood-plain river systems, such as the deltas of the Amazon River and the Mekong River. Freshwater fish living in these habitats spawn either immediately before or during river floods. This is an optimal time for spawning because the flooding opens up large areas of new habitat and food sources for the larvae that will develop from the eggs.

Figure 12.5 These are newly hatched fish larva (termed 'alevins'). Their yolk sacs — the orange bubble-like shapes — are still present and when this food source shrinks and disappears, the juvenile fish will feed independently.

Broadcast spawners, such as reef fish, put all their energy into egg production and produce very large numbers of eggs. Any fertilised eggs, however, are immediately on their own as they drift away in the ocean currents. Embryonic development takes place within the membranes that enclose the egg. The original single egg cell divides many times, forming layers of cells and then organs. The egg yolk provides energy and nutrients to the developing embryo. After several days, a larva (immature fish) breaks free from the membranes that surround the egg. At this stage, the newly hatched larva is small, has little power of locomotion, still has some yolk reserves (see figure 12.5) and floats in the ocean as a temporary member of the so-called **zooplankton**. When the yolk supply is used, however, the larva must find its own food. After some time, the larva is transformed into a juvenile fish and later the juvenile becomes a sexually mature adult fish, a transition that may take years.

Demersal spawning

While most reef fish use broadcast spawning and produce pelagic eggs, some produce **demersal** eggs. Demersal eggs do not float and a female fish that produces such eggs may lay them in a 'nest' in the sand or in a crevice in the reef or she may attach them to some part of the reef surface. After being laid, the eggs are fertilised by sperm released by the male. Demersal spawning occurs, for example, in the anemone clownfish (*Amphiprion* spp.). The female clownfish lays her eggs, usually on coral, close to the anemone with which she shares living space. (What advantage might this location offer?) The male clownfish then releases sperm to fertilise the eggs. Unlike fish that use broadcast spawning, fish species that produce demersal eggs typically give some care to their eggs by guarding them until they hatch or keeping them clean of debris (see figure 12.6).

ODD FACT

Fish, after hatching, receive no parental care, in spite of impressions given by the film *Finding Nemo*.

Figure 12.6 Clownfish (*Amphiprion polymnus*) guarding fertilised demersal eggs. Both parents remain close to the eggs until they hatch, after which the larvae are 'on their own'.

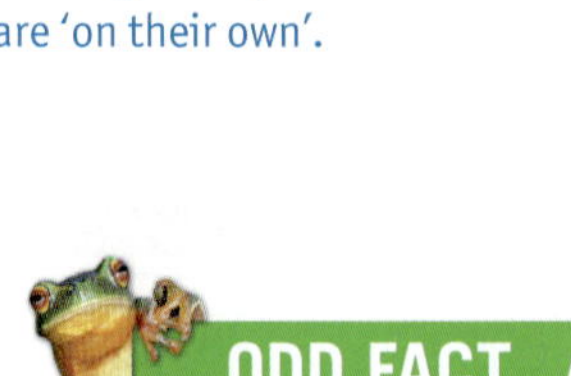

ODD FACT

Most demersal eggs hatch in the early evening after dusk, a period when predators are least active. How might this behavioural feature assist survival?

Because demersal spawners invest energy into guarding and cleaning their eggs, these fish produce fewer and larger eggs than broadcast spawners. The developing embryos within demersal eggs have a greater chance of survival than eggs that float in the sea unguarded. After demersal eggs have hatched, the larvae usually swim away and become part of the zooplankton community of the open seas.

Living organisms have evolved many different **reproductive strategies**. The strategies must be suitable for the environmental conditions in the habitats where the species live. In addition to egg and sperm production, there are other aspects to reproduction, such as displays by males to attract mates, size and number of eggs produced, care of eggs and offspring. Various species make different energy investments into these aspects; boxfish, for example, invest all their energy into producing many small eggs but none into the care of the eggs. In contrast, clownfish put energy into producing fewer larger eggs but also invest energy into care of the eggs. In the following sections, we explore overall strategies for successful reproduction and survival of species — the goal of the game of life.

KEY IDEAS

- Different reproductive strategies exist in the living world.
- Reproductive strategies seen in reef fish include broadcast spawning and demersal spawning.
- Energy investments by different species into various aspects of reproduction differ.

QUICK-CHECK

1 Identify the following as true or false.
   a All marine vertebrates have the same reproductive strategies.
   b Pelagic eggs are free-floating within the water.
   c An example of a fish that broadcasts its gametes is the boxfish.
   d Broadcast spawning always involves large groups of fish.
2 Identify a species of reef fish that:
   a puts all its reproductive energy into egg production
   b puts energy into egg production and egg care.
3 Which fish would be expected to produce more eggs: a broadcast spawner or an egg guarder?

A range of reproductive strategies

In the animal kingdom, reproductive strategies can differ markedly in terms of several major features including:

- type of reproduction
  - asexual
  - sexual
- gender system
  - sexes exist as separate male and female
  - hermaphrodite
  - parthenogenesis
- mode of fertilisation
  - external
  - internal
- mating system
  - single pair matings: monogamy
  - multiple matings of individuals (polygamy)
  - promiscuity
- numbers of offspring
  - r-selection
  - K-selection
- place of development and source of nutrition for the embryo
  - oviparity — egg laying
  - viviparity — giving birth to live young (several types)
- investment of parental care into offspring
  - nil
  - care by one or both parents
  - care by extended family.

In chapter 7 (pages 174–213), we discussed aspects of reproduction including asexual/sexual reproduction and gender systems. In the following sections, we explore other aspects of reproduction and mode of fertilisation.

Mating systems

First, let us explore mating systems using examples in particular from species of bird (Class Aves).

Monogamy: just two of us!

Monogamy is less common in mammals than in birds but occurs in several primate species, including gibbons (*Hylobates* spp.) and owl monkeys (*Aotus trivirgatus*).

Some bird species form 'pair bonds' in which one male mates with one female for one or more breeding seasons or for life. This situation is termed **monogamy**. It is estimated that most bird species (more than 90 per cent) are monogamous. Birds that form 'pair bonds' for one breeding season include emperor penguins (*Aptenodytes forsteri*) while birds that form lifelong pair bonds include various species of eagle, parrot and albatross.

The monogamous mating system is seen in species where the survival of the young requires the care of both parents for tasks (see figure 12.7a, page 378), such as:

- defending a nesting site
- incubating eggs
- feeding offspring.

The young of many, but not all, monogamous bird species are born featherless, blind and helpless (see figure 12.7b). Birds that produce this type of young are said to have an **altricial** development mode.

(a)

(b)

Figure 12.7 **(a)** A monogamous pair of princess parrots (*Polytelis alexandrae*) that form pair bonds and care for their young. **(b)** A newly hatched laughing kookaburra (*Dacelo novaeguineae*), an example of a bird with altricial development. Survival of such birds after hatching typically depends on both parents for protection, warmth and nutrition.

ODD FACT

Polygamy is often, but not always, seen in birds that have **precocial** young. On hatching, these birds are covered in down or in feathers, have strong legs, are alert and can feed themselves.

Polygamy

Some animal species are polygamous. In this case, either one male or one female has multiple partners during a breeding season. **Polygamy** is seen in bird species that live in habitats with plentiful and reliable food resources and where one parent can ensure survival of the young.

- When one male has multiple female partners during a breeding season, the situation is called **polygyny** (poly = 'many'; gunos = 'woman').
- When one female has multiple male partners during a breeding season, the situation is termed **polyandry** (poly = 'many'; andro = 'male').

Figure 12.8 Magpies are one example of a polygynous bird species. The particular form of polygyny in magpies is termed harem polygyny. Can you suggest why this term is used?

Polygyny: one male, many female mates

Polygyny occurs in about two per cent of bird species and two types are recognised: **harem polygyny** and **serial polygyny**.

Harem polygyny occurs when one dominant male lives with a group of females and mates with each of them during the breeding season. Harem polygyny occurs in magpies (*Gymnorhina tibicen*) (see figure 12.8). Magpies live in small groups and one male magpie mates with one or more females in his group. (Among mammals, harem polygyny is seen, for example, in lions and in gorillas.)

In serial polygyny, one male attracts passing females in turn for mating. To attract females, males must advertise themselves through visual or vocal signals and, as a result, the males of polygynous bird species often differ markedly in colour or feather length from the females. This is seen, for example, in

ODD FACT

Research shows that when female birds choose a partner from various males on display, they use signs such as size, symmetry and intensity of colouring. Asymmetry, for example, can indicate handicaps such as poor health, presence of parasites or abnormal internal organs.

ODD FACT

Scientists showed that the success of a male bowerbird in attracting females is reduced if some of the decorations near his bower are removed (reported by Gerald Borgia, in *Animal Behaviour*, 1985, Vol. 33, pages 266–71).

Figure 12.9 Note the prominent tail feathers of the male lyrebird. What does the dramatic display signify? What purpose does this serve?

Figure 12.10

the lyrebird (*Menura novaehollandiae*). Male lyrebirds use their tail feathers in dramatic visual displays to try to attract passing females to mate with them (see figure 12.9). A female lyrebird will select a male based on factors that advertise his genetic fitness, including the length and symmetry of his tail feathers.

Male lyrebirds make no contribution to the care of the eggs or the offspring. They put their energy into growing long tail feathers and putting on displays and contribute only sperm to the partnership. Female lyrebirds, in contrast, put their energy into egg production and into the care of offspring (see figure 12.10).

Another group of Australian native birds that exhibit serial polygyny are the bowerbirds. As we saw in chapter 11 (page 342), male satin bowerbirds (*Ptilonorhynchus violaceus*) interweave sticks and twigs to construct a structure known as a **bower** and decorate the areas at the ends of the bower with colourful objects. When a female approaches, the male puts on a vigorous display of singing and dancing outside his bower. If she is sufficiently impressed, she enters his bower and the pair mate. If not, the female moves on and inspects other displays until she chooses her mate. Each female bowerbird mates with only one male during a breeding season. She selects a male based on factors that advertise his genetic fitness — in this case, the quality of his display. One male, however, may mate with 20 or more females in turn; immediately after mating with one female, a male bowerbird looks for another mate.

Male bowerbirds in one region will not be equally successful in the mating game; some remain bachelors. A few male birds only will father most offspring. The female parent alone incubates the eggs and cares for the hatchlings. The male's energy is invested into building a bower, collecting decorations and in displaying.

Serial polygyny sometimes involves a process known as **lekking**. Lekking occurs when males gather in one communal area, known as a **lek**, and perform conspicuous visual or vocal displays in order to attract passing females for mating. Lekking occurs in bowerbirds and is also seen in hummingbirds, birds-of-paradise, Australian riflebirds and the New Zealand kakapo (*Strigops habroptilus*). Mammals, such as walruses and fallow deer, also engage in lekking to attract females.

Lekking using mobile phones? In 2000, researchers observed the behaviour of young people around tables in an English pub and noted that, as the number of males relative to the number of females at a table increased, the time spent by the males in fiddling with, displaying and checking their mobile phones increased (reported by J. E. Lycett and R. I. M. Dunbar, 'Mobile phones as lekking devices among human males', in *Human Nature*, Vol. 11, No. 1, 2000, pages 93–104).

Figure 12.11 Are mobile phones like the display of a lyrebird?

Polyandry: one female plays the field

Polyandry occurs when one female mates with two or more male partners during a breeding season. This mating system is rare and occurs in less than one per cent of bird species. In polyandrous bird species, the female is often larger and more brightly coloured than the male.

Australian bird species with a polyandrous mating system include the comb-crested jacana (*Irediparra gallinacea*) (see figure 12.12a) and the cassowary (*Casuarius casuarius*) (see figure 12.12b). A female jacana mates with a male and lays three to four eggs in a nest that he has built. She then leaves the male to incubate the eggs and protect the chicks and looks for another mate and will produce another clutch of eggs for him to mind. Likewise, after mating and laying a clutch of eggs, a female cassowary takes no part in the incubation of the eggs but moves on and mates with another male and produces the next clutch of eggs. The male is left to incubate the eggs over a period of 47 to 54 days and the chicks then stay with the father for many months.

Figure 12.12 **(a)** Young chicks of the jacana (*Irediparra gallinacea*), also known as the lily-trotter jacana, which has a precocial mode of development. Jacana chicks have very long toes and claws that equip them to walk on waterlily pads. Which parent incubates the eggs and protects young jacana? **(b)** Male cassowary (*Casuarius casuarius*) taking care of a chick and an unhatched egg. The female cassowary departs to mate with different males and produces several clutches of eggs each breeding season. What name is given to this type of mating system?

Promiscuous mating systems

In a promiscuous mating system, males and females within a social group engage in multiple and indiscriminate matings. In a promiscuous mating system, all males are about equally successful in producing offspring. How does this compare with individual male success in a polygynous mating system?

In general, promiscuous behaviour is more common in male birds than in females. This is probably because the investment by the male into each mating is just sperm and, if a mating is unsuccessful, there is little disadvantage to him. On the other hand, because of the higher energy cost of producing eggs, a female puts a greater investment of energy into each attempt at reproduction. It is more likely, then, that a female in a group will not mate promiscuously but will be more selective about her choice of partner.

KEY IDEAS

- Several mating systems occur in the animal kingdom.
- Monogamy and polygamy are examples of different mating systems.
- Several polygamous mating systems can be identified, including polyandry and polygyny.
- Polygamy is seen in bird species that live in habitats with plentiful and reliable food resources.

QUICK-CHECK

4 Identify the following as true or false.
   a A polygynous mating system exists when one male mates with several females during a breeding season.
   b Monogamy is more common among birds than polygamy.
   c Bird species that have precocial young are more likely to be monogamous than polygamous.
   d Lekking occurs in monogamous bird species.

5 Give one example of the following:
   a a monogamous bird species
   b a polygynous bird species
   c a bird species that has an altricial mode of development.

Offspring: how many? how often?

Two reproductive strategies are commonly identified in animals and these are:

- the 'quick-and-many' strategy, known as **r-selection**
- the 'slower-and-fewer' strategy, known as **K-selection**.

However, many species cannot be neatly 'pigeonholed' as using one or other of these strategies with species showing aspects of both r-selection and K-selection in their reproduction. The two reproductive strategies (r- and K-selection) are at contrasting ends of a continuum. (Further discussion of these strategies appears in chapter 15.)

Among plant species, similar strategies can also be identified, that is:

- shorter-lived annual plants often produce large numbers of minute seeds
- longer-lived plants often produce fewer but very much larger seeds with food reserves for the developing plant embryo.

Because of their high natural rates of population growth, r-selected species are suited to rapid spread in habitats where the population density is low. In contrast, because K-selected species have very slow rates of population increase, these species are suited to habitats where the population size is near the maximum that can be supported in a particular habitat.

ODD FACT

Organisms within the same broad category, for example 'lizards', can differ in their reproductive strategies. Among Australian lizards, for example, some species reproduce twice yearly while others reproduce only every second or third year. Likewise, the number of eggs laid at one time by different lizard species can vary from one to many dozens.

Quick-and-many: r-selection

The 'quick-and-many' strategy of reproduction is typical of species that reach sexual maturity quickly, produce large numbers of offspring or breed more frequently (or both) and may put little or no parental care into their offspring. Such species are said to have high **fecundity** where fecundity refers to the number of eggs produced by each female on an annual basis.

Offspring produced through the 'quick-and-many' strategy are produced in larger numbers but they have higher mortality rates so that few survive to adulthood. Species that use this 'quick-and-many' strategy include reef fish, squid, oysters, scallops, many insects, cane toads and, among mammals, mice and rabbits.

As well as reef fish, many other inhabitants of the Great Barrier Reef waters show aspects of r-selection.

- A common female octopus (*Octopus vulgaris*) lays from 100 000 to 400 000 eggs in clusters and deposits them in crevices in a coral reef. She keeps her eggs supplied with oxygen by squirting them with water and keeps them clean by gently wiping the eggs with her arms.
- Individual giant clams when mature have both male and female reproductive organs. (What is the scientific term for this? Check on page 189.) Giant clams

ODD FACT

One female scallop can produce more than 3 million eggs. A female squid may produce 30 egg cases, each containing several hundred eggs, in a breeding season.

Figure 12.13 A giant clam (*Tridacna gigas*) releasing its sperm. Each clam has both male and female reproductive organs. How is self-fertilisation prevented? How does this species disperse?

are **sedentary** or fixed on one location and, during the breeding season, they release large numbers of eggs and sperm into the water. To avoid self-fertilisation, a giant clam releases its sperm at a different time from when it releases its eggs (see figure 12.13). The fertilised eggs of the giant clam and the larvae that hatch from them are free-floating. Having free-floating larvae enables dispersal of giant clams to new habitats.

Slower-and-fewer: K-selection

The 'slower-and-fewer' strategy of reproduction is known as K-selection. Animal species that are K-selected mature slowly, breed later, produce fewer and larger offspring and may put extensive parental care into their offspring. Examples of species that use this strategy include some birds, gorillas, elephants, whales and humans. Because of their relatively longer lifespan, K-selected species typically have more than one breeding season during their lifetimes.

ODD FACT

The endangered bird, the Californian condor (*Gymnogyps californianus*), produces just one egg every two years. What challenges does this pose for conservation of this K-selected species?

K-selection at sea

In the late autumn and winter each year, humpback whales (*Megaptera novaeangliae*) migrate up the east coast of Australia to breeding grounds in tropical or semi-tropical waters. It is here that the whales mate and it is also here that pregnant females give birth during the southern hemisphere winter.

Humpback whales show many of the features of a K-selected species as follows:

- sexual maturity does not occur until whales are about five years old
- gestation (pregnancy) in humpback whales lasts about 11.5 months
- each female gives birth to just one calf every one or two years
- for an average of 10 months after its birth, a mother suckles her calf on milk
- the life span of humpback whales is up to 50 years.

Figure 12.14 A humpback whale and her calf. The long gestation (pregnancy), the single offspring per birth and the extended period of maternal care of the humpback whale are features of a K-selected species. The generic name (*Megaptera*) of humpback whales means 'big-winged', reflecting the fact that their flippers are the longest of any whale species.

KEY IDEAS

- Two reproduction strategies are r-selection and K-selection.
- r-selection is characterised by large numbers of offspring produced quickly.
- K-selection is characterised by fewer offspring produced more slowly.
- r- and K-selection are the extremes of a continuum and species may show features of both strategies.

QUICK-CHECK

6 Identify the following as true or false.
   a An r-selected species would be expected to have a higher rate of population growth than a K-selected species.
   b A typical K-selected species would be expected to reach sexual maturity very early.
   c Rabbits would be expected to produce more offspring over a given period than gorillas.

7 What term applies to the reproductive strategy of a species that breeds often and has many offspring?

8 Give an example of an r-selected species.

Eggs or liveborn young?

In the world of animals, different modes of producing offspring can be identified:

- **oviparity** — eggs are released by the mother so that embryos develop outside the mother's body with nutrients for the embryo coming from the egg yolk; animals that use this mode are said to be **oviparous** (ovum = 'egg'; parus = 'bearing')
- **viviparity** — embryos develop within the mother's body and are born as miniature copies of the adult; animals that use this mode are said to be **viviparous** (vivus = 'living'; parous = 'bearing'). In viviparous animals, nutrition of the developing embryo within the mother occurs in different ways.

Egg layers: the oviparous animals

Egg laying or oviparous vertebrate animals include some sharks, all rays, bony fish, amphibians (such as frogs and toads), most reptiles (turtles, crocodiles, goannas but only some snakes) and all birds (see figure 12.15). Unlike all other mammals, the monotreme mammals (platypus and echidna) are egg layers (refer back to figure 7.38 page 196). In oviparous animals, the size of the newly hatched young is determined by the yolk supply in the egg — the larger the yolk, the larger the young.

Figure 12.15 Crocodiles, like most other reptiles and all birds, are oviparous.

Figure 12.16 Spiral black egg cases of a Port Jackson shark attached to sea tulips. These egg cases are about 12 to 15 centimetres long and can occasionally be seen washed up on a beach. The egg case is soft when first laid but it later hardens. To see egg cases of some other sharks and rays, go to www.jaconline.com.au/natureofbiology/natbiol1-3e and click on the 'Shark egg cases' weblink for this chapter.

The eggs of bony fish and of amphibians (for example, frogs and toads) do not have shells and quickly dry out if they are not released into water or other situations with high moisture content. No problem for bony fish whose eggs are shed into their aquatic habitats! Amphibians live on land but must return to water for breeding. Embryonic development in both bony fish and amphibians occurs outside the female's body. The embryo obtains its nutrients entirely from the egg yolk and releases its wastes into the water.

About 25 per cent of sharks and all rays lay eggs and these eggs are covered with a leathery membrane (see figure 12.16). Oviparous sharks include the Port Jackson shark (*Heterodontus portusjacksoni*) that wedges her eggs into rock crevices where, after a period of 10 to 12 months, the young sharks (pups) hatch.

All birds and most reptiles (for example, all turtles, all crocodiles but only some snakes) are oviparous. Eggs of reptiles and of birds are relatively large because they contain large amounts of yolk. In reptiles and birds, eggs are released from the ovary and are fertilised internally. As each fertilised egg passes along the oviduct, it becomes coated by egg white (albumen), is then enclosed within thin membranes and is finally surrounded by a hard shell (see figure 12.17). When the egg is laid, the developing embryo consists of a tiny group of cells on the yolk surface. Embryonic development then occurs outside the mother's body and the source of nutrients for the embryo comes from the egg yolk.

The eggs of reptiles and birds are known as **amniote eggs**. Features of an amniote egg include:

- an outer shell that protects the egg contents and permits gas exchange; immediately under the shell is a thin membrane that also allows gas exchange
- a series of membranes inside the egg, including
  - a membrane, known as the **amnion**, that envelops the embryo and secretes a fluid that bathes the embryo, protecting it from mechanical damage and from drying out, a bit like a personal water-filled bath (see figure 12.18a)
  - a membrane, known as the allantois, that forms a sac in which wastes produced by the embryo are stored in solid form, a bit like a personal toilet (see figure 12.18a)
  - a membrane that forms the yolk sac, a bit like a personal pantry (see figure 12.18a).

At hatching, the baby reptile or bird breaks free from the egg (see figure 12.18b).

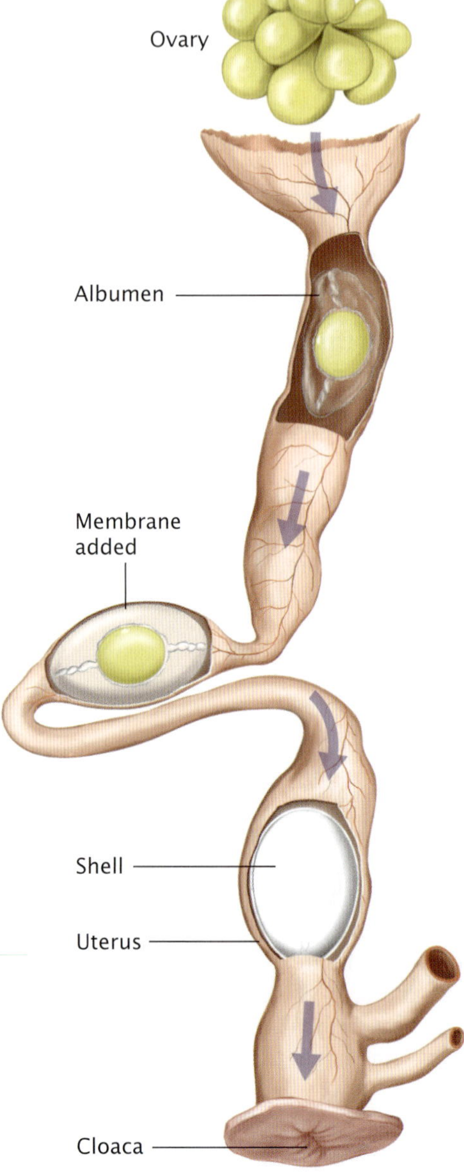

Figure 12.17 A typical bird's egg from the time of ovulation (release of the egg from the ovary) to when it is laid. After ovulation, the egg is fertilised and when it reaches the upper part of the oviduct, it gets a coating of egg white (albumen); at the middle of the oviduct, membranes are added to enclose the egg with its egg white and, finally, the shell gland adds a hard calcareous shell. (The process is similar in reptiles.) In birds, the process from ovulation to laying typically takes about 24 hours.

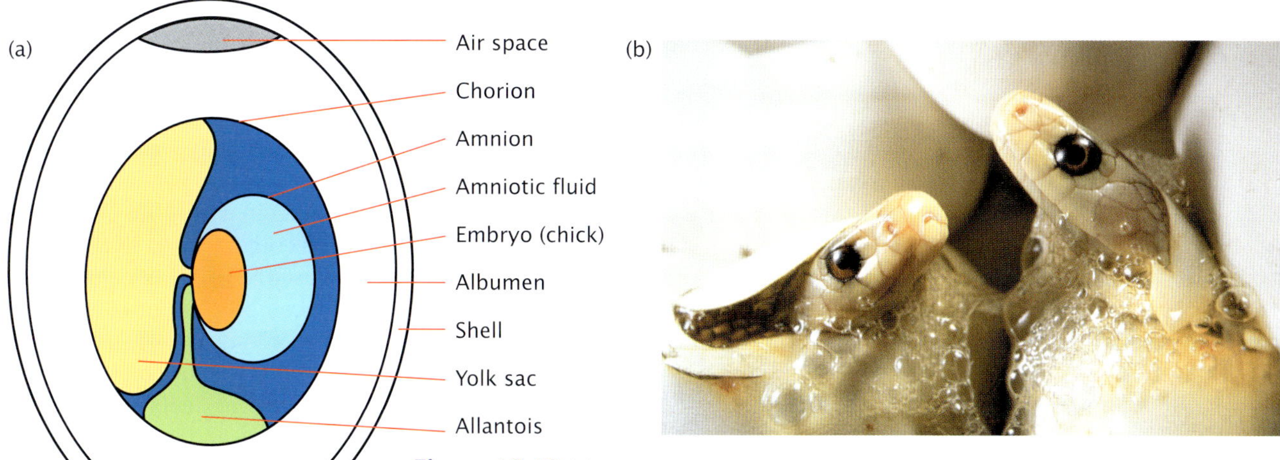

Figure 12.18 **(a)** Stylised diagram showing a bird embryo and the membranes within the egg. The eggs of reptiles and of monotreme mammals (platypus and echidna) are similar. This type of egg, known as an amniote egg, enables reproduction to be independent of water. **(b)** Baby taipans hatching from eggs. What was their source of nutrients while they were in the eggs? The taipan (*Oxyuranus scutellatus*) lays from 3 to 20 eggs in a clutch.

ODD FACT

Baby snakes of oviparous species break out of their eggs using a special egg tooth. This tooth is lost soon after a baby snake has hatched from its egg.

The amniote eggs of birds and of reptiles can be laid on land since they carry their own supply of water and are protected from **desiccation** (drying out). This adaptation has enabled reproduction in reptiles and birds to be independent of free water and has allowed these animals to occupy many terrestrial habitats. For example, the animals commonly found in habitats in the arid regions of Australia include many reptiles, such as snakes, lizards and goannas. The success of reptiles in these arid habitats is due in part to the fact that they produce amniote eggs.

In general, oviparous animals can either put energy into producing a smaller number of large eggs or into producing a large number of small eggs. It should be noted that:

- larger eggs have larger food stores in the form of yolk
- young that hatch from larger eggs have longer periods of embryonic development and are bigger at the time of hatching than young that hatch from small eggs
- the larger the egg size, the fewer the number of eggs laid.

Viviparity: eggs made but not laid!

ODD FACT

Egg sizes vary greatly between bird species with the largest egg (1600 grams) being that of the ostrich (*Struthio camelus*) and the smallest (0.3 gram) being that of hummingbirds.

Eggs with their yolk supply are a valuable source of nutrients for a developing embryo. However, this also makes eggs a target for many predators seeking a source of food and many are lost to predators. (Two hens' eggs, for example, are a popular breakfast selection even for humans!)

Instead of laying eggs, some vertebrate animals produce miniature live-born young and these include most sharks, some snakes and, except for monotremes, all mammals. Viviparity contributes to an increased chance of survival for a developing embryo by protecting the eggs within a female's body rather than leaving them exposed to possible loss from predators.

Three different types of viviparity can be identified, depending on the source of nutrients for the developing embryo, as follows:

(i) nourished by the yolk of the egg only (= egg yolk viviparity)
(ii) nourished via the mother's placenta (= placental viviparity)
(iii) nourished via other sources.

Egg yolk viviparity

A headache of terms! In older textbooks, egg yolk viviparity is known by the term 'ovo-viviparity'.

Some sharks and some snakes produce amniote eggs but do not lay these eggs. Instead of being laid, the eggs are retained in the mother's uterus. During this time, the developing embryos are nourished by the egg yolk only. After hatching inside the mother, the young are born as miniature adults. The size of these young is determined by the amount of yolk available to support development.

ODD FACT

Male snakes have two penises and each is termed a hemipene. During mating, a male snake inserts one of his hemipenes into the female's reproductive opening (cloaca). In some snakes, the hemipene has small spines that hold it in place during mating.

Egg yolk viviparity is seen in some shark species, as for example, the nurse shark (*Ginglymostoma cirratum*) that gives birth to 30 to 40 miniature sharks (known as pups) each about 30 cm long. About 20 per cent of snakes worldwide show egg yolk viviparity, including many Australian native snakes such as copperheads (*Austrelaps superbus*), tiger snakes (*Notechis* spp.) and the bardick (*Echiopsis curta*). Copperheads produce up to 20 young in a litter.

Many viviparous snake species live in habitats with cold climates. Researchers suggest that female snakes can better control the temperature of the developing embryos by retaining the eggs in their bodies (see figure 12.19). The majority of sea snakes, however, that live in tropical and semi-tropical waters are also viviparous and give birth to live young in the sea. As a result, these snakes do not have to come ashore to lay eggs — quite an advantage since sea snakes are very clumsy on land.

Figure 12.19 Do viviparous snakes control the temperature of the fertilised eggs they are carrying? Warming the eggs so that they hatch faster might be an advantage in a cold climate, especially if winter is coming.

Placental viviparity

Apart from monotremes, all mammals, including humans, cats, monkeys, dolphins and sheep, are viviparous and produce tiny pinhead-sized eggs with very little yolk. (How does this compare with eggs of birds and reptiles?) Each mammalian embryo is enclosed within a fluid-filled sac and develops within its mother's uterus. Nutrients are delivered from the maternal bloodstream via the **placenta** to the embryo. The placenta is a region where the bloodstreams of the embryo and the mother come into close contact allowing nutrients to be taken up by the embryo and its wastes removed. When the gestation period (pregnancy) is complete, the young are born.

ODD FACT

Embryos of grey nurse sharks (*Carcharias taurus*) develop teeth very early after hatching from eggs inside their mother's uterus. The largest unborn shark pup inside the mother's body kills its smaller siblings. By feeding on unfertilised eggs from the mother and also by feeding on its dead siblings, the single unborn shark in each of the mother's paired uteri continues to develop.

Viviparous with nutrients from other sources

Some species of sharks and rays give birth to a few large young (called pups). These baby sharks and rays are much larger than would be expected if their nutrients came only from the egg yolk. Such pups grow *after the eggs hatch but while they are still inside their mother*. So, there must be an additional source of nutrients available for these pups before birth. Several sources of nutrients have been identified as follows:

- Nutrition via a shark placenta. This is seen, for example, in the hammerhead shark (*Sphyrna zygaena*). After three to four months, hammerhead shark pups have used their yolk supplies. When this occurs, the membranes of the empty yolk sac become greatly folded, their blood supply increases and they become

Figure 12.20 Viviparous rays give birth to a few large young.

closely attached to the lining of the mother's uterus, forming a so-called shark placenta. The placenta enables nutrients to pass from the bloodstream of the mother to the bloodstream of the developing shark pup and for wastes to be removed. This placental structure maintains the developing pup for a further 7 to 12 months of gestation. Not surprisingly, shark pups whose nutrition is supplied first by the egg yolk and then via a placenta can be quite large at birth, up to one-third of the size of the mother.

- Feed them eggs! Some female sharks, such as the porbeagle shark (*Lamna nasus*), produce a supply of unfertilised yolky eggs as a source of food for the shark pups after their own egg yolk is used. This egg supply allows these shark pups to continue growing within their mothers' bodies so that at birth they can be up to one metre long.
- Feed them 'milk'. In some rays, such as the bat rays (*Myliobatus californica*), cells of the inner lining of the mother's uterus secrete a protein-rich fluid or 'uterine milk' to nourish the embryo after the egg yolk is used. This 'milk' allows the unborn ray to continue growing while still inside the mother's body.

KEY IDEAS

- Two major reproductive strategies are oviparity and viviparity.
- Oviparous species lay eggs and their young hatch from eggs outside the mother's body.
- Viviparous species give birth to live young that develop within the mother's body.
- Various kinds of viviparity occur that differ in how the embryo is nourished.
- Reproduction in fish and amphibians is dependent on the presence of water.
- Amniote eggs of reptiles and birds enable reproduction to occur away from water.

QUICK-CHECK

9. Identify the following statements as true or false.
   a Monotreme mammals are oviparous.
   b Some bird species are viviparous.
   c Amphibian eggs are examples of amniote eggs.
10. List two different strategies relating to reproduction in sharks.
11. Identify two significant differences between the egg of a toad and the egg of an eagle.
12. Name the source of nutrients for embryos of the following:
    a an oviparous snake
    b a domestic cat.
13. Of two shark species, which would be expected to give birth to the larger pups — a species with egg yolk viviparity or a species with viviparity involving a shark placenta? Explain.

Parental care or not?

Different animal species vary in the energy that they put into the care of their eggs or their young. Some species give no care to their eggs after laying or to their young, such as most reef fish, all turtles and most frogs and toads. In contrast, other animal species invest significant energy into guarding their eggs and into feeding and protecting their young (after hatching or after birth), including most bird species and all mammals.

Figure 12.21 A male seahorse, *Hippocampus brevicips*, gives 'birth' to young seahorses after incubating them in a pouch for about four to five weeks. A male incubates up to 100 young at a time. The young are nourished by the egg yolk.

Caring for eggs after laying

Some fish, such as the clownfish, guard their eggs after they are laid (refer back to page 376). Another example of a fish species that cares for its eggs is the seahorse (*Hippocampus* spp.). Unlike other fish, the female seahorse lays her eggs in a pouch in the male where they develop for about four to five weeks (see figure 12.21). When the young leave the male's pouch they are almost one centimetre long. By carrying the young in this way, the male seahorse has provided protection and increased the chance of survival of the young as they developed.

In a few oviparous snake species, the female guards her eggs until they hatch, as for example, the Australian diamond python (*Morelia spilota*) and the children's python (*Bothrochilus childreni*). After laying her clutch of eggs, the female coils her body around them until they hatch (see figure 12.22). If the temperature falls, she 'shivers' and these rapid muscular contractions raise both her temperature and that of the eggs she is incubating.

Figure 12.22 Eggs within the coils of a female children's python. How does this behaviour assist survival of the eggs to hatching?

Cephalopod (pronounced kef-al-o-pod) refers to a member of the class Cephalopoda of the phylum Mollusca.

Several invertebrate animals also look after their eggs after they are laid. Among the caregivers are different kinds of **cephalopods** (octopuses, squid, cuttles and argonauts). For example, reef-dwelling species of octopus attach their parcels of eggs to 'safe' locations, such as in a crevice in the reef, and the female octopus keeps them clean and keeps predators away. Cephalopods of the open ocean, such as argonauts (*Argonauta* spp.) use different strategies to care for their eggs, and female argonauts, for example, secrete paper-like 'shells' in which they place their eggs.

Spiders also care for their eggs. After laying their eggs, many spiders wrap them in a silken cocoon known as an egg sac. A female of the bird-dropping spider (*Celaenia excavata*) found in Victoria lays up to 200 eggs on a silken sheet that she shapes into a ball and coats with a waterproof fluid (see figure 12.23). Over time, the female produces a dozen or more such egg sacs and she remains close to them until the eggs inside hatch.

Wolf spiders (*Lycosa* spp.) provide remarkable care of their eggs and their young. The female carries her egg sacs with her and, after hatching, the miniature spiders are carried on their mother's back for some time.

Figure 12.23 A bird-dropping spider (at bottom centre) in a Melbourne garden with her four egg sacs. Each round egg sac contains many eggs. This spider gets its common name from the fact that it looks like bird dung. An interesting camouflage!

Parental care of young

In most bird species, both parents care for the hatchlings. In a few species, however, only one parent cares for the eggs and the young — this is the female parent in polygynous bird species, but it is the male in polyandrous species and the young show precocial development.

All mammals care for their young for some period after birth, principally by the female parent suckling her young on milk from her mammary glands (see figure 12.24). Mammals of the group known as the primates (monkeys and apes including humans) continue to care for their young for considerable periods after they are weaned.

Figure 12.24 A female mammal — a human — feeding her newborn young

Case studies in parental care

Emperor penguins (*Aptenodytes forsteri*)

These birds form monogamous pairs for one breeding season. In May, about 60 days after the pair has mated, the female Emperor penguin lays a single large egg. She quickly rolls the egg to her male partner for incubation and returns to the sea to feed. The male keeps the egg on his feet where it is kept warm by a thick flap of feather-covered skin known as a brood pouch (see figure 12.25) and he continues to do this for about nine weeks without feeding during the Antarctic winter. By clustering in tight groups, the males ensure added protection against the sub-freezing temperatures of the Antarctic winter. The females return in August and both male and female parents take turns to share the feeding of their young chick. By the Antarctic summer, the chick has fledged and begins an independent life in favourable conditions.

Figure 12.25 A male Emperor penguin with egg in brood pouch. The temperature in the male Emperor penguin's brood pouch is about 30°C while the temperature in the external environment can be −60°C.

Australian mallee fowl (*Leipoa ocellata*)

The male mallee fowl invests considerable energy into 'nest' building by constructing a large incubation mound for the eggs that will be laid by his mate. The mound is a large pile of sand that covers pieces of vegetation. The female mallee fowl invests her energy into producing from 12 to 24 eggs that she lays one at a time at intervals of several days. Each time the female is ready to lay an egg, she goes to the mound and the male opens the mound so that she can lay an egg on the vegetation and he then buries the egg, a task that involves moving several cubic metres of soil and vegetation (see also chapter 13, page 415).

The male alone looks after the incubation of the eggs for about seven weeks, keeping the eggs at a suitable temperature through a series of earth works (see figure 12.26). At hatching, mallee fowl chicks are partly feathered, able to move and feed themselves and receive no further parental care.

May–June: Plant matter is collected and placed in depression in mound and is moistened by rain. Damp vegetation is then covered to form mound.

July–October: Male mallee fowl detects temperature inside mound. When warm enough, he opens mound to let female lay eggs inside.

October–December: If mound becomes too warm from sun, male adds sand to protect eggs from sun's heat.

January–April: When heat production from decomposing plant matter stops, male removes sand to let in sun's heat to warm eggs.

Figure 12.26 The male mallee fowl maintains the eggs at a temperature of 33°C. Decomposition of the vegetation releases heat. In spring, if the decomposing vegetation generates too much heat, the male removes sand from the top of the mound. In summer, the male adds sand to the mound to insulate the eggs from the heat of the sun. In autumn, the male scoops sand from the mound to expose the interior to the sun's heat. In addition, the male makes adjustments during each day as part of maintaining the correct temperature.

Emus (*Dromaius novaehollandiae*)

In this polyandrous species, the male is the caregiver. He builds the nest and cares for the offspring. The female emu lays a clutch of 5 to 15 eggs over a four-week period and, after laying, she departs and may mate with another male and produce another clutch of eggs. Meanwhile, the male emu incubates the eggs for about 8 weeks until the chicks hatch. For many months after hatching, the chicks will continue to be guarded by the male parent (see figure 12.27).

Figure 12.27 A male emu with his chicks. Sometimes, smaller chicks from another brood may join his original family so that he becomes a super dad.

ODD FACT

The birth weight of the honey possum (*Tarsipes rostratus*), an Australian marsupial, is just 5 milligrams — this is the lightest birth weight of any mammal.

Reproductive strategies: Australian marsupials

The predominant Australian native mammals are marsupial mammals (or just **marsupials**). These mammals have evolved strategies that equip them for successful reproduction in the drought-prone and unpredictable environmental conditions of this continent.

Features of marsupial reproduction

The embryos of non-marsupial (placental) mammals have an extensive period of development in the mother's uterus. In contrast, marsupial embryos develop in the mother's uterus for a very short period only and are born at a very immature stage of development. Tammar wallabies (*Macropus eugenii*), for example, are born just 26 days after fertilisation. The tiny newborn is only 16 mm long and weighs just 0.4 gram — the mass of a bean. The newborn climbs into its mother's pouch where it attaches to a nipple and undergoes further development over an extended period (see figure 12.28).

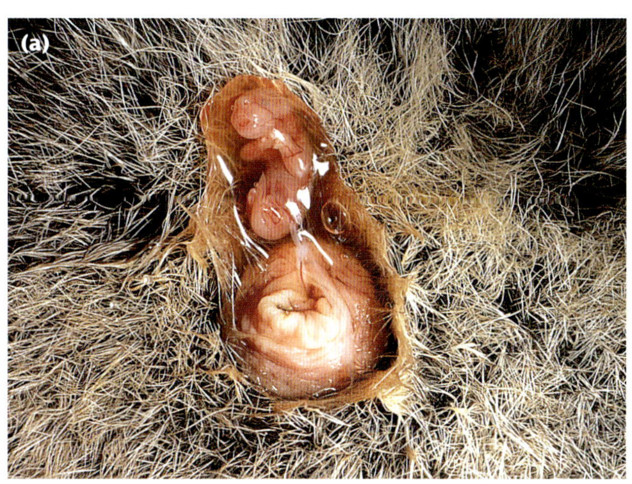

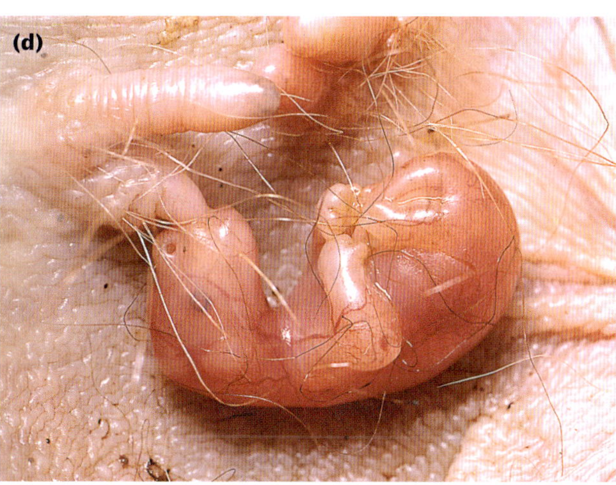

Figure 12.28 Birth of a tammar wallaby **(a)** The tammar wallaby is born enclosed in fetal membranes. **(b)** Freed of membranes, the young begins to climb with a swimming motion. **(c)** The newborn struggles through the fur to the pouch. **(d)** Inside the pouch, the young takes only a few minutes to locate a teat and attach.

Marsupials use a 'stay-on-the-nipple' approach to early development whereas placental mammals use a 'stay-in-the-uterus-for-longer' approach. All mammals — monotreme, marsupial and placental — produce milk for the nutrition of their young.

Reproductive strategies of kangaroos and wallabies include the ability to suspend embryonic development. Immediately after giving birth and having young in her pouch, a female kangaroo conceives again. However, her newly conceived embryo in her uterus does not undergo development but is held in a state of suspended animation. This arrested development continues as long as young are feeding on milk from the nipples in the mother's pouch.

Reproduction in kangaroos and wallabies is influenced by environmental conditions, such as drought. Let's look, for example, at reproduction of the red kangaroo (*Macropus rufus*).

CASE STUDY: THE RED KANGAROO

A male red kangaroo, *Macropus rufus*, becomes sexually mature at about two years, the female between 15 and 20 months. The female reproduces immediately she matures, provided local conditions are favourable. The egg she produces is fertilised and, after 33 days development, the small hairless young kangaroo weighing less than one gram is born. The young travels to the mother's pouch and becomes attached to one of the four teats there. Within a couple of days the female has mated again and a new embryo begins to develop. If the young in the pouch continues to survive, development of the embryo is arrested and remains dormant at the blastocyst stage, a small sphere of cells. If the young in the pouch dies or is lost, then the blastocyst resumes development and the new young kangaroo produced makes its way to the pouch.

The young kangaroo stays in the mother's pouch for about 200 days, always feeding from the same teat. As the time nears for the young to leave the pouch, the dormant blastocyst resumes development so that a new young is born as the older one leaves the pouch. The young that has left the pouch weighs four to five kilograms and is called the young-at-foot because it remains with the mother for about four months and continues to suckle her. The new arrival in the pouch must attach itself to a teat other than the one that the young-at-foot is using. The two teats provide different kinds of milk. The milk provided for the young-at-foot contains more fat than the milk for the pouch young.

A couple of days after the birth, the female mates again, the egg develops to the blastocyst stage, when once again development is arrested until the young in the pouch is about to leave.

So, at the one time, a female kangaroo can have a young-at-foot, a young in the pouch, and an embryo arrested at the blastocyst stage of development (see figure 12.29). This cycle continues throughout the reproductive life of the kangaroo as long as there is plenty of food available. In less favourable conditions, the young-at-foot is the first to die. If the pouch young also dies, then the embryo completes development and is born. Another egg is fertilised and arrested in development. If drought continues, then the pouch young generally dies in about two months. The embryo continues development and is born, and another egg is fertilised. This cycle continues every six to eight weeks until the drought ends. If the drought is extreme and extended, the female may stop reproducing. Once the drought is over, she mates and the reproductive cycle is resumed.

This pattern of reproduction in kangaroos has evolved in response to the selective pressure of changing environmental conditions in Australia. These conditions may include erratic rainfall and long periods of drought during El Niño weather events. Although death of individual joeys occurs in unfavourable conditions, the continuation of the population is favoured as new young can be produced by a female as soon as conditions become favourable.

Figure 12.29 At any one time, a female red kangaroo can have three offspring at different stages of development: a young-at-foot, a young in the pouch still on a teat, and an embryo arrested at the blastocyst stage of development.

KEY IDEAS

- The amount of parental care invested by parents into the care of their eggs or of their offspring varies between species.
- Among bird species, different patterns of parental care can be identified in terms of which parent(s) provide the care.
- The reproductive strategy of Australian marsupial mammals differs from that of placental mammals.
- Reproduction of Australian marsupials includes features such as suspended embryonic development that equips these species for life in drought-affected habitats.

QUICK-CHECK

14 Identify the following as true or false.
   a Only vertebrate animals provide parental care of eggs after laying.
   b All female mammals suckle their young with milk.
   c For mammals of the same adult size, a newborn placental mammal would be expected to be larger than a newborn marsupial.
   d The most common situation in birds is that care of the new hatchlings is provided by both parents.
15 Give an example of a bird species in which care of the new hatchlings is provided:
   a by the male parent alone
   b by neither parent — no parental care is provided.
16 Which kind of mammal (placental or marsupial) can suspend embryonic development?
17 Which mode of development (altricial or precocial) would be expected in birds that provide no parental care of their young?

How often do matings occur?

In many animal species, mating occurs just before each fertilisation event because sperm cells typically do not live long after they are released by a male, either into the environment (external fertilisation) or into the female's reproductive tract (internal fertilisation).

In some animals, however, sperm from one mating remains capable of fertilising eggs for a long period and can be stored by the female. In honeybees (*Apis mellifera*), for example, the queen bee is the only breeding female in a hive and she mates just once. The sperm that she receives from a male drone are stored in a special structure in her reproductive tract. Later, depending on the needs of the hive, the queen bee produces eggs that either remain unfertilised and hatch into male drones or are fertilised from her store of sperm and hatch into female workers.

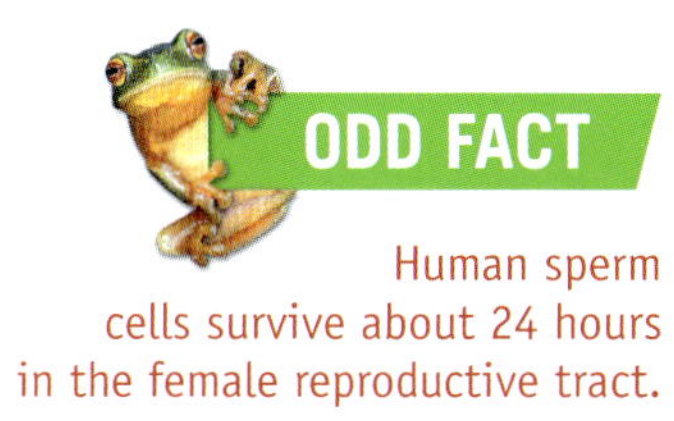

Human sperm cells survive about 24 hours in the female reproductive tract.

Patterns and times of breeding

For sexual reproduction, eggs and sperm must be available at the same time in members of a population. The **breeding season** is the period when mature members

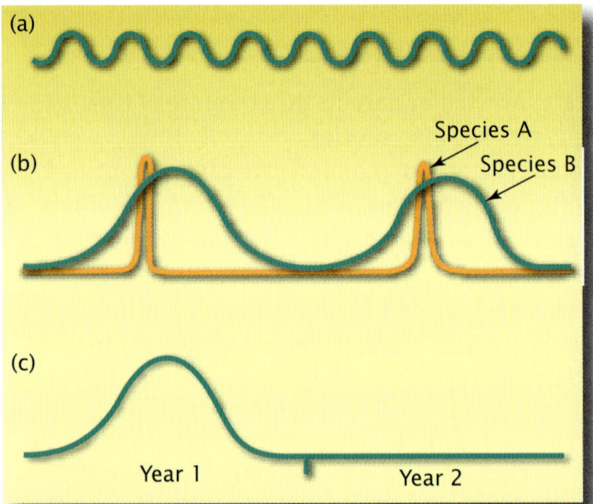

Figure 12.30 Different breeding patterns over a two-year period. **(a)** Pattern in population that breeds through the year **(b)** Annual cycle of breeding **(c)** Biennial cycle of breeding. What is the difference in breeding patterns between species A and species B?

of an animal population have sperm and eggs ready for release and fertilisation. This period is called the **spawning season** (for animals that have external fertilisation in an aquatic environment) or the **mating season** (for animals that have internal fertilisation). Producing sperm and eggs (or gametogenesis) is influenced by internal factors, such as hormone levels, and is also influenced by external factors, such as temperature and day length.

Different patterns of breeding are seen in various animal populations. Populations of species that live in tropical habitats, or at the ocean depths where environmental conditions do not vary much over the year, may show continuous reproduction throughout the year but at an overall low intensity (see figure 12.30a). However, most animal populations, including those that live in habitats where environmental conditions change over the year, have distinct and regular breeding seasons. The most common pattern is a single annual breeding season (see figure 12.30b). Yet another is the biennial pattern of breeding in which two years are required for individuals to mature and to spawn (see figure 12.30c).

Female mammals produce their eggs during a so-called **oestrous cycle** that varies in length in different species: 28 days in human females; 21 days in cows and horses; 4 to 6 days in mice and rats. (The oestrous cycle in human females is more commonly termed the **menstrual cycle**.) However, while some female mammals, such as humans and cattle, have regular oestrous cycles throughout the year, other mammalian species in the wild show different patterns, as follows:

- wolves, foxes and bears — one oestrous cycle per year
- deer, sheep and goats — several oestrous cycles within a short period, typically in autumn when hormonal changes are triggered by shorter day length
- hamsters and horses — several oestrous cycles but all within a short period, typically in spring when hormonal changes are triggered by longer day lengths.

Note that breeding and fertilisation in wild species is influenced by the environmental factor of day length. What advantage might this have for survival of offspring?

Unlike human females, the females of most mammalian species will accept males for mating only at limited times of the year. At these times, the females of non-human mammals are said to be in **oestrus** (commonly called 'heat'). These receptive periods are associated with the cycle of egg production and egg release from the ovary into the reproductive tract.

A female mammal in oestrus shows behavioural changes and, in some cases, also produces visual signals, for example, the coloured calluses on the rear of female baboons (*Papio* spp) that become more prominent during oestrus.

ODD FACT

A male deer produces sperm only during the rutting (breeding) season when his testes move into his scrotum. At other times, his testes are positioned in his abdominal cavity.

Once in a lifetime: mate then die

For some animal species, breeding occurs annually and one animal may breed year after year. Other animal species, however, reproduce only once in their lifetime.

Species that breed just once include the Australian short-finned eel (*Anguilla australis*) and Pacific salmon (*Oncorhynchus* spp). Short-finned eels live for many years in freshwater lakes and rivers in eastern Australia. However, near the ends of their lives, mature eels go on a one-way journey from their freshwater habitats to the sea and migrate to spawning (breeding) grounds in ocean waters near New Caledonia. Here, each female sheds a very large number of eggs that are fertilised by a male. The adult eels die, but the eel larvae that hatch from the eggs are carried back to the Australian coast by ocean currents. The larvae mature into elvers that return to the parental inland freshwater lakes and rivers.

ODD FACT

In some mammals, including members of the cat family, ferrets, camels and llamas, a female releases her egg(s) only *after* mating has occurred. No mating, no ovulation! This is called **induced ovulation** and is in contrast to the spontaneous ovulation that occurs in other mammals, such as cows, sheep, goats and humans.

Figure 12.31 A short-finned eel. How many times does it breed in its lifetime?

ODD FACT

The longest known spawning journey for a Chinook salmon is from the Bering Sea to Lake Teslin in Canada, a total of 3800 kilometres. These salmon swim about 65 kilometres per day.

Chinook salmon (*Oncorhynchus tshawytscha*) live for many years in the sea but migrate near the end of their lives to spawning grounds in inland freshwater lakes and rivers. Once at the spawning site, a female salmon lays a batch of eggs in a nest in the gravel at the bottom of a lake. Her eggs are fertilised by nearby males. She covers the eggs with gravel and then moves to another location where she lays another batch of eggs that are also fertilised. In all, one female salmon lays about 5000 eggs. The adult salmon die after breeding. After two months, the eggs hatch and the young salmon eventually return to the sea, but will finally return to the same freshwater spawning grounds to breed and die.

KEY IDEAS

- In most animal species, fertilisation must be shortly preceded by mating.
- In some animal species, including insects, sperm is long lived and the female can store sperm in her reproductive tract.
- Females of various mammalian species have regular oestrous cycles throughout the year.
- Reproduction may occur annually or may be a 'once-in-a-lifetime' event.
- Reproduction in wild species can be influenced by external environmental factors such as day length.

QUICK-CHECK

18 Identify the following statements as true or false.
   a A queen bee mates just once in her lifetime.
   b The pattern of breeding in animal species is influenced by environmental conditions.
   c Females of all mammalian species accept males for mating throughout the year.

19 Give an example of an animal species:
   a in which females release eggs only after mating
   b that has a single oestrous cycle each year
   c that breeds only once in its lifetime.

20 Members of a population of species A breed only during spring while members of a population of species B breed throughout the year. What reasonable prediction can be made about the environmental conditions in the habitats of species A and species B?

Plant reproduction

So far, we have looked at reproductive strategies in the animal kingdom. Reproduction, however, is a feature of all living organisms, including plants. In this section, we will examine one important aspect in the reproduction of flowering plants and in conifers, namely the process of pollen transfer or **pollination**. The reproductive cells in these plants are the eggs and pollen that are produced at particular times of the year (the flowering season for flowering plants). The eggs remain in place on the plants, either within the flowers of flowering plants or in the female cone of conifers. The pollen is the reproductive cell that must travel to the eggs. How does this occur?

Strategies to ensure pollen transfer

Because plants are permanently fixed in one location, they have evolved strategies to ensure that their pollen grains can be transferred. Pollen may be transferred between plants by the following means:

- blown by the wind
- carried by animals such as
  - insects (e.g. beetles, flies, gnats, bees and butterflies)
  - birds of many species
  - mammals (e.g. bats, small rodents and honey possums).

The pollen of all conifers and some flowering plants (e.g. grasses and several trees) is carried by the wind. These plants are said to be **wind pollinated**. Pollen that is carried by the wind is small, smooth and very light. Flowering plants that are wind pollinated:

- do *not* have obvious brightly coloured flowers (why not?)
- have their pollen-producing organs (stamens) exposed to the wind
- have feathery stigmas that can catch passing pollen grains.

Wind pollination is a chancy 'hit-and-miss' event and is affected by climate.

Most flowering plants are pollinated by insects, birds or mammals. Animals that are carriers of pollen are called **vectors**. Flowering plants attract and 'reward' their animal vectors. The attraction may be brightly coloured flowers and the reward may be sugar-rich nectar or protein-rich pollen (see figure 12.32).

Figure 12.32 Nectar attracts insects. To produce nectar, the plant must use energy. Is this a waste of energy?

Some marsupial mammals play an important role in transferring pollen of various Australian native plants. For example, the tiny honey possum (*Tarsipes rostratus*) transfers pollen among *Banksia* flowers (see figure 5.29 on page 120). These possums feed exclusively on the pollen and nectar of native plants in their heathland habitat in the south-west corner of Western Australia. The possums have long, brush-tipped tongues that they use to probe deeply into flowers and mop up nectar.

Other mammals that act as pollinators include flying foxes and other bats that are important in pollinating some rainforest plant species.

Clues to likely animal vectors in pollen transfers

Insect-pollinated flowers:

- are typically blue or purple or yellow as these colours are visible to insects (insects cannot see colours at the red-orange end of the spectrum)
- may have a shape that includes a 'platform' on which an insect can land
- may have a scent that mimics the odour of something that attracts the insect species concerned in pollination; for example, the odour of some plants is like rotting meat that attracts blowflies as vectors
- usually have nectaries inside the flower at its base that produce sugar-rich nectar that attracts insects
- often have white or yellow dots or lines that reflect ultraviolet light, producing bright dots or lines that signal the presence of the flowers to insects that can see UV radiation.

Bees do not visit flowers that make nectar with a low sugar content. Bees need a sugar concentration of at least 30 per cent for making honey and they visit only flowers with a sugar concentration of this value or greater.

Look at figure 12.33 that shows some pansy flowers. What agent of pollination is likely for these flowers?

The stamens of insect-pollinated flowers are often arranged so that the insects must push through them to reach the nectaries. As the insect pushes past the stamens, it becomes coated in pollen grains. The pollen grains of plants that are insect-pollinated are normally larger than those of wind-pollinated plants and are often sticky. Can you suggest why?

Flowers that are pollinated by nocturnal insects, such as moths, often have a strong scent to attract their pollinators and this scent is strongest at night.

Some plants that are pollinated by one insect species may have a shape that mimics the shape of the female of that species. This shape, combined with a

Figure 12.33 What is the likely agent of pollination of these pansy flowers?

Figure 12.34 Insects such as this honey bee (*Apis mellifera*) transfer pollen between the flowers of plants.

matching odour that is released when the pollen is ripe, attracts male insects to the flowers. The male insects attempt to mate with the flower and in doing so become covered in pollen — then, off they go to the next flower!

Greenhood orchids (*Pterostylis longifolia*), for example, are pollinated by male gnats. This orchid has a structural adaptation that ensures that the gnat picks up some pollen when it visits an orchid. The gnat appears to be lured to the plant by an attractive scent that he mistakes for a female gnat sending out messages that she is available to mate. The flower has a sensitive lip that snaps shut if touched. When a gnat lands on the lip and it snaps shut, the gnat has only one way of getting out of the flower. The exit path the gnat must take ensures that the gnat brushes against the anther. Pollen is transferred to the gnat which transports it to the next greenhood orchid flower it visits (see figure 12.35).

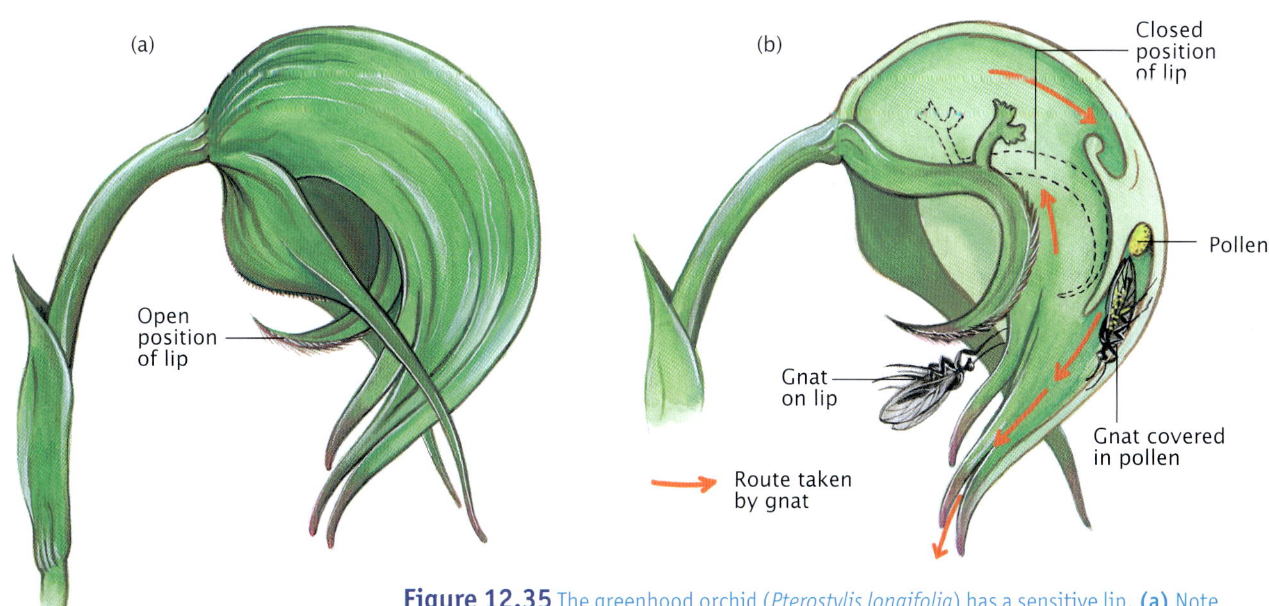

Figure 12.35 The greenhood orchid (*Pterostylis longifolia*) has a sensitive lip. **(a)** Note the open position of the lip and the hooded nature of the flower. **(b)** The lip flicks up into the hood when a gnat lands on it and the gnat is forced to leave the flower via a route that takes it past the pollen-laden anther. Pollen is transferred to the gnat. Note the closed position of the lip and the exit path the gnat must take.

Bird-pollinated flowers:

- are often, but not always, red or orange or yellow as these colours are recognised by birds as suitable sources of food
- generally are not scented since birds rely on vision, not smell, to detect possible food sources; birds do not have a strong sense of smell
- often have petals forming a tubular shape
- usually have nectaries inside the flower at its base.

As birds try to obtain the sugar-rich nectar from the nectaries, they dislodge pollen from the stamens inside the floral tube and this pollen coats the back of their heads. When the birds then fly to other plants of the same species and poke their heads into other flowers for nectar, they transfer pollen to the stigma of those flowers.

The flowers of many Australian native plants that are pollinated by insects or birds do not have large petals so that petal colour is not important in attracting bird or insect pollinators. In these plants (such as members of the following genera: *Eucalyptus*, *Grevillea*, *Hakea*, *Callistemon* and *Banksia*), the flower colour comes from parts such as stamens (see figure 12.36a and b).

Figure 12.36 Many Australian species have 'flowers' that differ from the typical flower. **(a)** *Grevillea* 'flowers' are actually composed of a number of flowers like those shown in **(b)**. The style is long and often protrudes from the joined ring of petals as shown in the unopened flower on the left-hand side. The anthers are attached to the petals and become exposed as the flower unfolds. Flowers of *Hakea* are similar to these. **(c)** The 'flower' of *Banksia* has flowers as shown in **(d)**. **(e)** The colour of *Eucalyptus* flowers is due to the coloured stamens. Does this flower (*Eucalyptus ficifolia*) have any petals?

ODD FACT

A dry fruit from a plant known in some parts of the world as devil's thorn has a 'drawing pin' shape that is an adaptation for dispersal on the hooves of mammals. Would you expect plants with this feature to be native to Australia?

Dispersing plant offspring

In flowering plants and in conifers, the embryo that results from fertilisation of the egg by pollen is enclosed in a seed. At some stage of the life cycle of these plants, however, there must be a mobile stage that enables the plant population to spread or disperse to new habitats. In flowering plants and in conifers, dispersal or spread of the next generation of plants occurs through the seed.

In flowering plants, seeds are enclosed in fruit. Either the ripe seeds are set free when the fruit opens (see figure 12.37a) or the seeds are retained within fruits and are dispersed (see figure 12.37b). In contrast, the seeds of conifers are not enclosed in fruit and are said to be naked.

(a)

(b)

Figure 12.37 Fruits may be dry or fleshy. **(a)** Dry fruits that open and shed the seeds that they contain. In this case, the seeds are the mobile stage that enable dispersal. **(b)** Fleshy fruits that contains seeds but do not open. The unit of dispersal in this case is the fruit plus its seeds.

Seeds or fruits may move or be moved by:

- sailing in the wind
- drifting on water currents
- hitchhiking *on* animals
- hitchhiking *inside* animals.

Seeds or dry fruits that disperse by sailing in the wind are lightweight so that they can be dislodged from their mother plant by even a slight puff of wind. They may be flat or have papery wings that cause them to glide on air currents (see figure 12.38a) or they may have parachutes of fine silky hairs that set them sailing even in a slight breeze (see figure 12.38b).

Seeds or fruit that drift on water currents have a waterproof coat and are buoyant. Seeds or fruit that hitchhike on animals have hooks or spines that assist the seeds in becoming attached to the fur of mammals or the feathers of birds (see figure 12.38c). Seeds that hitchhike inside animals, such as birds, are typically enclosed in brightly coloured berries or other soft fruits. This type of hitchhiking depends on the seed getting into the digestive tract of the animal concerned — in other words, by the fruit being eaten. An adaptation of these seeds is that they are resistant to the action of digestive enzymes (see figure 12.38d).

(a)

The agent that disperses plant seeds may be non-specific; for example, dry grass seeds can be dispersed on the coats of many kinds of native mammals. In other cases, the agent of dispersal is very specific. For example, the mistletoe bird (*Dicaeum hirundinaceum*), an Australian native bird, feeds exclusively on the soft berries of mistletoe plants that are parasites on trees or shrubs and is the only agent of dispersal of mistletoe seeds.

Figure 12.38 Adaptations for dispersal by various agents. **(a)** The honesty plant (*Lunaria annua*) has flat dry fruit, containing flat seeds. **(b)** The dandelion (*Taraxacum officinale*) with its head of many dry fruit each with a parachute of fine hairs. **(c)** Dry fruit, commonly called a burr, with barbs on the surface. **(d)** Seeds in emu dung.

KEY IDEAS

- Sexual reproduction in flowering plants and conifers involves pollination.
- Transfer of pollen involves vectors, such as animals or wind.
- Pollination in conifers is exclusively by the wind.
- Depending on the pollination vectors, flowers will show particular features.
- Seeds are the stage of the life cycle of conifers and of flowering plants that enable dispersal.

QUICK-CHECK

21 Identify the following statements as true or false.
   a Pollination of most flowering plants occurs through animals.
   b Pollen grains of insect-pollinated flowers would be expected to be smaller than those of wind-pollinated flowers.
   c All conifers are wind pollinated.
   d A flowering plant with a strong odour would be expected to be wind pollinated.

22 The flowers of a particular plant have prominent red petals but have no scent. What is the probable pollination vector?

23 List two adaptations that you would expect to observe in a wind-pollinated flowering plant.

24 Why is seed disperal important for the survival of plant species?

25 Identify two examples of adaptations of seeds that assist their dispersal.

BIOCHALLENGE

In 2002, scientists examined aspects of mating in red-shouldered widowbirds (*Euplectes axillaris*), a bird species that is a native of Africa. In this species, each male displays in his territory and mates with any females that he can succeed in attracting. After the female leaves, she later lays her eggs in one of many rough nests that he has constructed in his territory. Meanwhile, the male seeks to attract other females.

The scientists hypothesised that female widowbirds would prefer males with the longest tails. In order to test this hypothesis, they identified four groups of males: birds with tail lengths of 6 cm, 7 cm, 8 cm, and birds whose tails had been artificially extended to 22 cm. The scientists recorded the number of nests with eggs on the territory of each male and used this as a measure of his mating success. The results are shown in the following bar graph:

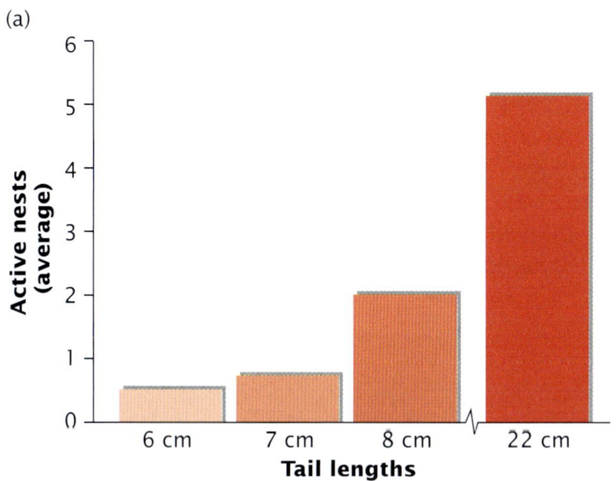

Figure 12.39 (a) Bar graph showing the average reproductive success of four groups of male red-shouldered widowbirds with tails of different lengths. The tail lengths of the four groups of birds are shown on the horizontal axis. (*Source:* Adapted from Pryke and Anderrson, 'A generalised female bias for long tails in a short-tailed widowbird', Proc.R.Soc.Lond.B (2002) 269, 2141–2146)

(b) A red-shouldered widowbird with an artificially lengthened tail

Consider the bar graph and answer the following questions:

a How was the mating success of each male widowbird measured?

b Which mating system appears to exist in this bird species?

c Which variable is shown on the vertical axis?

d Which variable is shown on the horizontal axis?

e In an experiment, the variable that is controlled by the researcher is known as the *independent variable*. The variable that is directly affected by the independent variable is known as the *dependent variable*. For this particular experiment, identify:
  i the independent variable
  ii the dependent variable.

f Complete the following sentence: The average number of nests with eggs in a male's territory is dependent on ____________.

g What hypothesis was tested by the scientists?

h Did the experimental results support or disprove the scientists' hypothesis? Explain.

i Suggest why female widowbirds might prefer males with longer tails.

j Over many generations, what might be predicted to happen to the average tail length of male widowbirds?

k What prediction could reasonably be made about the tail length and colour of female widowbirds? Explain.

CHAPTER REVIEW

CROSSWORD

Key words

altricial
amnion
amniote eggs
bird-pollinated
bower
breeding season
broadcast spawning
cephalopods
colonial
demersal
desiccation
fecundity
harem polygyny
induced ovulation
insect-pollinated
invertebrate
K-selection
lek
lekking
marsupials
mass spawning
mating season
menstrual cycle
monogamy
oestrous cycle
oestrus
oviparity
oviparous
pelagic
placenta
pollination
polyandry
polygamy
polygyny
precocial
r-selection
reproductive strategies
sedentary
serial polygyny
spawning season
vectors
vertebrate
viviparity
viviparous
wind pollinated
zooplankton

Questions

1 ***Making connections*** ➧ Use as many as possible of the words or terms above to create a concept map.

2 ***Applying your understanding and drawing conclusions*** ➧ Identify the probable mating system (as discussed on pages 377–80) in the following bird species. Briefly justify each of your answers.

Species A: Males of this species engage in elaborate displays.

Species B: Females produce a clutch of three to four eggs that hatch, producing blind, featherless and helpless young.

Species C: Females lay several clutches of eggs during the breeding season.

3 ***Applying understanding of scientific terminology*** ➧ Identify each of the following combinations of reproductive strategies as either logical or illogical and give a reason for your decision.

a a broadcast spawning animal that produces amniote eggs

b an animal that cares for its eggs after laying and which shows egg yolk viviparity

c a female bird that engages in multiple matings, produces several clutches of eggs during the breeding season and is an example of a monogamous bird species

d an oviparous snake that lays eggs

e a demersal spawning animal that lives in a terrestrial habitat

f a mammal with a single oestrus each breeding season that produces young throughout the year

g a viviparous shark species with placental nourishment that gives birth to fewer young per breeding season than a related shark species with egg yolk viviparity

4 *Making connections between concepts* ▸ For each of the following reproductive strategies, identify the environmental conditions in which this strategy could be used to advantage:

a broadcast spawning

b production of amniote eggs

c r-selected numbers of offspring.

5 *Using scientific terminology and drawing conclusions* ▸ One very unusual group of bony fish is known as the surfperches (family Embiotocidae). More than 20 different species are found mainly in seas off the west coast of North America. In contrast to almost all other species of marine bony fish, surfperches give birth to miniature baby fish and guard their young.

a What scientific term can be applied to this mode of offspring production in surfperches?

b Identify the scientific term that applies to the usual mode of offspring production in bony fish.

c Would you expect female surfperches to release more or less eggs in each breeding season compared with fish that have demersal spawning? Explain.

6 *Applying your understanding in new contexts* ▸ Suggest an explanation in biological terms for each of the following observations.

a The egg of a great dane bitch is very much smaller than the egg of a female budgerigar.

b Eggs of reptiles have a larger yolk supply than mammalian eggs.

c The period between ovulation and egg laying in oviparous sharks is much shorter than the period between ovulation and giving birth in some viviparous sharks.

d Many more species of reptile live in habitats in inland Australia than amphibian species.

e Flowers of grass species lack petals.

f Reef fish that engage in broadcast spawning produce very large numbers of eggs and sperm.

7 *Making comparisons* ▸ Two different habitats are listed below:

a aquatic: a tropical sea

b terrestrial: an arid scrubland that is subject to drought.

For each habitat, identify a combination of two reproductive strategies that could contribute to the successful reproduction of an animal species that lives in that habitat.

8 *Making comparisons* ▸ Identify one major difference between the members of the following pairs and suggest a reason in biological terms for this difference.

a a bird-pollinated plant and an insect-pollinated plant

b a wind-pollinated plant and an insect-pollinated plant

c a wind-dispersed fruit and an animal-dispersed fruit

d a dry fruit and a fleshy fruit

Hydrilla question: Hydrilla is dioecious. Female flowers come to the surface attached to the plant by an elongated thin 'stem' and open when they reach the surface. In contrast, the male flowers break free from the plant as buds and rise to the surface within an air bubble. When they reach the surface, the male flowers open explosively, scattering pollen.

9 ***Analysing information and communicating ideas*** ▸ A population of one particular animal (species M) has survived in a particular habitat for many generations. A student concluded that other animal species living in the same habitat would be expected to show the same reproductive strategies as species M. Do you agree with this student or not? Explain your choice.

10 ***Analysing information and communicating ideas*** ▸ Consider reproduction in marsupial and placental mammals.

a Identify a major difference between these two types of mammal.

b Should one strategy be regarded as more successful than the other?

c A third mammalian group are the monotremes. How does the reproductive strategy of monotremes differ from the other mammals?

d Construct a table to show similarities or differences between monotreme, marsupial and placental mammals in terms of the following items:

i fertilisation: internal or external?

ii oviparous or viviparous?

iii source of nutrition for developing embryo?

iv source of nutrition for young?

11 ***Making valid predictions*** ▸ Make a prediction, giving brief justification, about each of the following:

a the colour of flowers of an insect-pollinated plant

b the size of flowers of a wind-pollinated plant

c the presence of nectaries in flowers of a wind-pollinated plant

d the odour of flowers of a bird-pollinated plant.

12 ***Discussion question*** ▸ Some flowering plants that live in aquatic habitats are permanently submerged below the water, such as the water thyme (*Hydrilla verticillata*).

a Identify one major challenge regarding pollination in these species as compared with pollination in plants living in a terrestrial habitat.

b Identify a possible strategy for pollination in these fully submerged plants. Use your imagination, since you are asked to produce a possible answer, not the correct answer. (For your interest, the means of pollination in *Hydrilla* appears at the foot of page 403.)

13 ***Using the web*** ▸ Typically, fertilised frog eggs are suspended in a jelly-like mass in a freshwater pond and receive no parental care. However, some frog species put their eggs in less usual locations and the female frog cares for her developing young.

a Go to the websites listed below and find out about maternal care in:

i the Surinam toad (*Pipa pipa*) — go to www.jaconline.com.au/natureofbiology/natbiol1-3e and click on the 'Surinam toad' weblink for this chapter

ii the gastric-brooding frog (*Rheobatrachus* spp.) that is native to Australia — go to www.jaconline.com.au/natureofbiology/natbiol1-3e and click on the 'Gastric-brooding frog' weblink for this chapter.

b How would the numbers of eggs produced by a Surinam toad compare with the number produced by a cane toad (*Bufo marinus*) that gives no care to its eggs? Explain.

Figure 12.40 A female Surinam toad, *Pipa pipa*. Sadly, these frogs appear to be extinct since they were last seen in the mid-1980s.

UNIT 2

ORGANISMS AND THEIR ENVIRONMENT

AREA OF STUDY 2

Dynamic ecosystems

Chapter 13 Ecosystems and their living communities

Chapter 14 Flow of energy and cycling of matter

Chapter 15 Population dynamics

Chapter 16 Changes in ecosystems

13 Ecosystems and their living communities

KEY KNOWLEDGE

This chapter is designed to enable students to:

- identify the components of an ecosystem
- understand the nature of a living community
- become aware of various ecological roles in a community
- understand the variety of interactions that occur within an ecosystem.

Figure 13.1 Part of a living community in a marine habitat. Communities consist of populations of different organisms living in one habitat at the same time. Many interactions occur between different members of a community and between members of the community and their physical surroundings and the whole is known as an ecosystem. In this chapter we will explore the kinds of organisms found in ecosystems and some of the interactions and relationships that occur in ecosystems.

ODD FACT

The life-support module of *Alvin* is a sphere built of titanium. The small viewports (windows) are made of plexiglass more than 10 centimetres thick. The life-support module is maintained at a constant pressure of one atmosphere.

To the ocean depths

A pilot and two scientists crowd into the spherical life-support module of a deep submergence vehicle or submersible known as *Alvin* (see figure 13.2). Launched from a mother ship on the surface, *Alvin* descends to the ocean floor at 30 metres per minute. Down and down — the light becomes dim, then darkness at about 200 metres. At first, the passenger module is warm but, below 500 metres, the temperature drops to that in a coldroom. How long to reach the ocean floor? The ocean floor here lies nearly 2.5 kilometres below the surface and the journey down will last more than two hours.

Figure 13.2 The deep submersible vessel known as *Alvin*. The submersible carries lights and an external extendible wand that has a thermometer, an analyser (to identify chemicals present in water) and a 'sipper' (to collect small samples of water).

Cindy Lee Van Dover, an American scientist who has piloted *Alvin*, wrote:

> '*Alvin* is a safe boat, with an unsurpassed safety record. Even so, no pilot can work the seafloor without wondering what it would be like to find oneself trapped 4000 meters down with a short seventy-two hours to sort things out . . . In quiet talks among pilots I have since found we all share two great fears: getting trapped in a cave and being buried.'
>
> Van Dover, C. L. 1996, *The Octopus's Garden*, Helix Books (Addison-Wesley Publishing Company), p. 39.

At last, *Alvin* reaches the ocean floor, a region of total darkness, crushing pressure and frigid temperatures. The floor of the Atlantic and Pacific oceans is crossed by **mid-ocean ridges**. In these regions, the ocean floor is shaped into valleys, cliffs and seamounts (see figure 13.3). In the valley areas, new ocean crust is being formed as the plates of the Earth's crust spread apart. *Alvin* has reached such an area.

Here, cold sea water seeps into fissures in the Earth's crust and is heated by the underlying molten rock to temperatures up to 400°C. The water reacts with the hot rock and dissolves out minerals, such as iron sulfide. Because the superheated mineral-rich water expands, it is forced upwards and emerges from outlets in the ocean floor known as **hydrothermal vents** (see figure 13.4b). As the hot water comes into contact with the surrounding frigid ocean waters, minerals precipitate out of solution and form 'chimneys' around the vent openings. One kind of hydrothermal vent, known as a 'black smoker', emits streams of dark acidic water from a chimney made mainly of iron sulfide (see figure 13.4a). Could life exist in such a dark, high pressure, acidic and sulfurous environment?

Most of the vast expanses of the ocean floor and the deep ocean waters themselves are regions where the density of animals is very low because there is a great scarcity of food. (Why?) However, in 1977, scientists discovered it was a different story at the hydrothermal vents.

ODD FACT

For each 10 metres of descent, the pressure increases by one atmosphere so that, at a depth of 2.5 kilometres the pressure is a crushing 250 atmospheres.

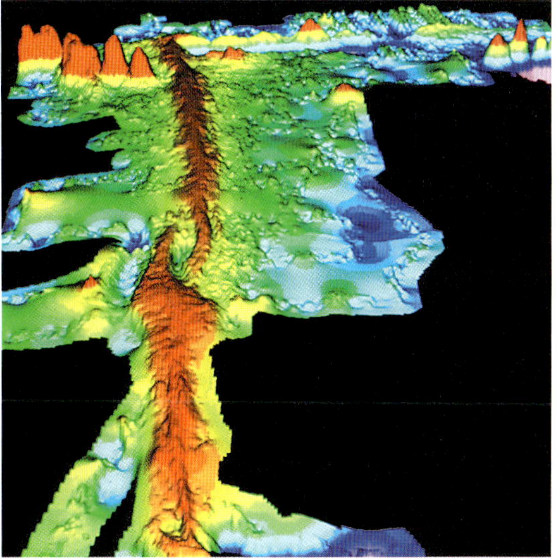

Figure 13.3 This figure shows a false-colour image of the ocean floor obtained by sonar. In the regions of the mid-ocean ridges, the ocean floor is not a smooth region such as we experience at the seaside. Instead, mounts, cliffs and ridges occur on the ocean floor.

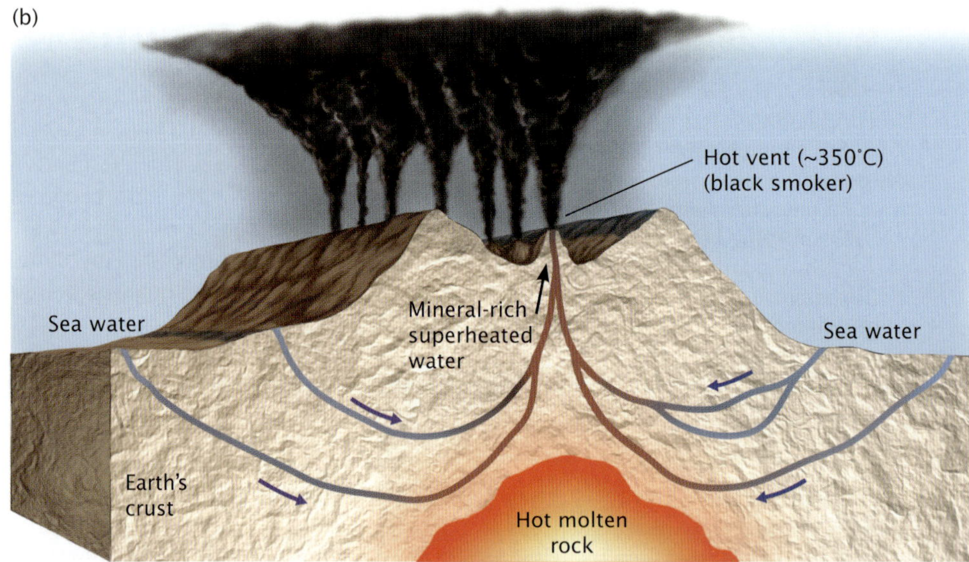

Figure 13.4 **(a)** 'Black smokers' at a hydrothermal vent. Note the 'chimneys' surrounding the escaping water. From what material are these chimneys made? **(b)** Sea water seeps down through rock and becomes heated when it comes near molten rock. This superheated water, rich in minerals, is forced up and escapes through 'black smokers'.

ODD FACT

At sea level, water boils (turns to steam) at a temperature of 100°C. As pressure increases, the boiling point of water increases and at the ocean depths, water is still liquid at temperatures of 400°C. What happens to the boiling point of water as pressure reduces at the top of a mountain?

In 1977, scientists aboard the submersible *Alvin* discovered a fantastic abundance of organisms around a hydrothermal vent about 2.5 kilometres below the surface of the Pacific Ocean near the Galapagos Islands (some examples are shown in figure 13.5). Unlike the very low density of organisms found in other parts of the ocean floor, the organisms around the vent were living at high densities. The vent organisms included white clams, mussels, blind white crabs, shrimp and various kinds of worm.

Among the most spectacular organisms found were giant tubeworms (*Riftia pachyptila*), two centimetres in diameter but up to 150 cm long; these worms are so different from other types of worm that they have been classified as the only members of a new phylum: Pogonophora. Less obvious but critically important were the enormous numbers of **sulfur bacteria** living around the vent. Since the discovery of the first hydrothermal vent in 1977, many other vents have been found and each has its own dense collection of living organisms.

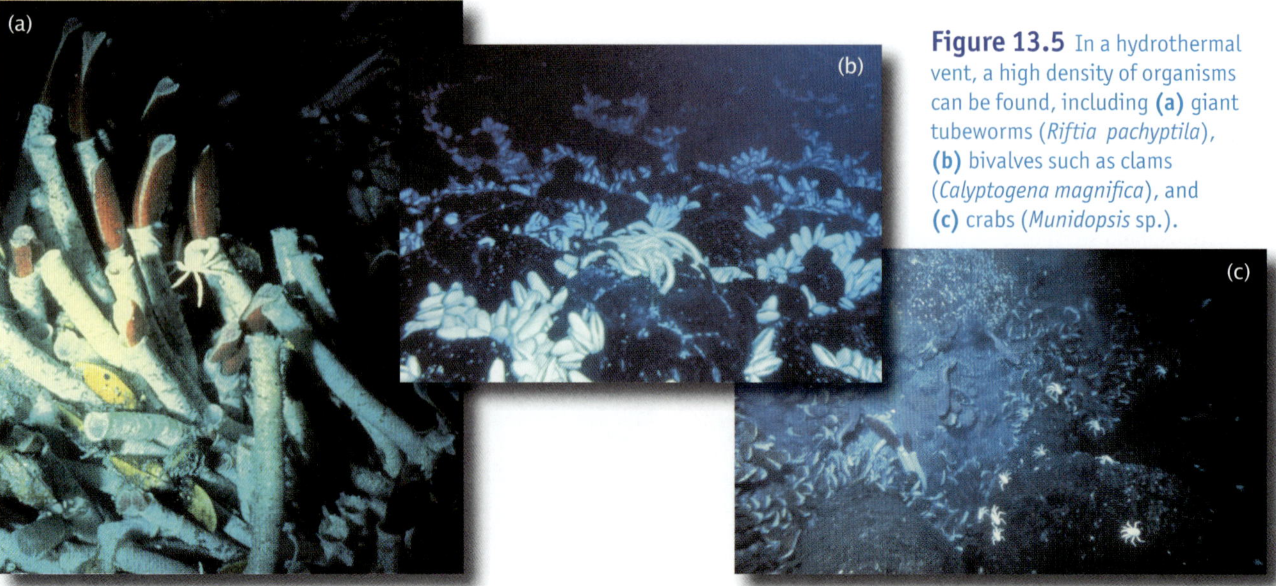

Figure 13.5 In a hydrothermal vent, a high density of organisms can be found, including **(a)** giant tubeworms (*Riftia pachyptila*), **(b)** bivalves such as clams (*Calyptogena magnifica*), and **(c)** crabs (*Munidopsis* sp.).

Giant tubeworms are part of a community

Giant tubeworms do not live in isolation but live around a hydrothermal vent along with many other kinds of living organism. All the different species living around a particular hydrothermal vent form a **community**.

A community consists of many populations

What is a community? Each community is made up of populations of various organisms living in the same location at the same time. All the different populations living in one location at a particular time form a community:

community 1 = population 1 + population 2 + population 3, and so on.

Each **population** consists of one species that may be a different species of animal, plant, fungus, protist or microbe.

Communities in different ecosystems can differ in their **diversity**. Diversity is not simply a measure of the number of different populations (or different species) present in a community.

A population is defined as all the individuals of one particular species living in the same area at the same time. So, we can talk about the population of giant tubeworms (*Riftia pachyptila*) in the hydrothermal vent community and the population of clams (*Calyptogena magnifica*) in the same community.

Just as the hydrothermal vent has its own living community, other habitats, such as coral reefs, sandy deserts and tussock grasslands, also have their own living communities. Different communities can be compared in terms of their diversity. When ecologists measure the diversity of a community, they consider two factors:

1. the richness or the number of different species present in the sample of the community, and
2. the evenness or the relative abundance of the different species in the sample.

As richness and evenness increase, the diversity of a community also increases.

The various populations found in the different communities differ according to the environmental conditions in the particular location, with members of the populations in each community showing features that equip them for survival and reproduction in the particular environmental conditions.

A community is one part of an ecosystem

The hydrothermal vent and its community is just one of a diversity of **ecosystems** that exist on planet Earth.

An ecosystem consists of a community, its physical surroundings and the interactions within and between them. An ecosystem is a complex level of organisation.

In studying biology, it is possible to focus on different levels of organisation. Some biologists focus on the structure and function of cells. Other biologists focus on whole organisms, others on populations. Different levels of biological organisation are shown in figure 13.6.

Figure 13.6 Different levels of biological organisation, from the most basic unit of life, the cell, to the complex unit of an ecosystem

The term ecology comes from the Greek oikos *meaning 'home' or 'place to live' and* logos *meaning 'study'.*

The study of ecosystems is the science known as **ecology**. In this chapter, we look at the living communities of some ecosystems, the ecological role of members of different populations and the interactions within communities and between communities and their physical surroundings. In chapter 14, we will look at the flow of energy through an ecosystem and the cycling of matter within an ecosystem. In chapter 15, we examine population dynamics and, in chapter 16, we will explore changes in ecosystems, both local and global, over time and space.

Introducing ecosystems

Each ecosystem includes a living part and a non-living part. The living part is a community that consists of the populations of various species that live in a given region. The non-living part consists of the physical surroundings. However, an ecosystem consists of more than living organisms and their non-living physical surroundings.

Look at some definitions of an ecosystem:

- a biological community living in a discrete region, the physical surroundings, *and the interactions that maintain the community*
- an assemblage of populations grouped into a community *and interacting with each other and with their local environment.*

Each definition conveys the idea that an ecosystem includes both a living community and the non-living physical surroundings, as well as the interactions both within the community and between the community and its non-living surroundings.

We can develop an understanding of the concept of an ecosystem using an analogy with a hockey game. A hockey game has a 'living part' made up of players, coaches, umpires and time keepers — these are like the living community. A hockey game also has its 'non-living part' that includes a pitch, line markings, hockey sticks, goal nets and scoreboard — these are like the non-living surroundings of an ecosystem. A hockey game also includes interactions that occur within the 'living part' and between the 'living part' and the 'non-living' surroundings. These are like the interactions occurring in an ecosystem.

Figure 13.7 Professor Eugene Odum (1913–2002) was a world-famous ecologist who received many honours for his research and wrote many scientific publications and books on ecology.

The continuation of an ecosystem depends on the intactness of the parts and on the interactions between them. A hockey game could not proceed normally if there were no goal keepers, or if the line markings were altered. Likewise, an ecosystem depends on its parts and may be destroyed if one part is removed or altered. This idea was expressed by Professor Eugene Odum, the world-famous ecologist (see figure 13.7), who wrote: 'An ecosystem is greater than the sum of its parts'. This is another way of saying that an ecosystem is a functioning system, not just living things and their non-living surroundings.

When you think about ecosystems, such as the Antarctic marine ecosystem or a tropical rainforest ecosystem, remember its three essential parts:

- a living community consisting of various species, some of which are microscopic
- the non-living surroundings and their environmental conditions
- interactions within the living community and between the community and the non-living surroundings.

Ecosystems can vary in size but must be large enough to allow the interactions that are necessary to maintain them. An ecosystem may be as small as a freshwater pond or a terrarium or as large as an extensive area of mulga scrubland in inland Australia.

Naming ecosystems

Ecosystems can be named in a variety of ways. One method is to name a terrestrial ecosystem in terms of its plant community and the growth form and structure of its dominant vegetation. Using this scheme, an ecosystem in inland Australia where widely-dispersed grasses are the dominant vegetation could be called 'an open grassland ecosystem'. An ecosystem in a dense forest in northern Queensland could be called 'a tall closed forest ecosystem'. One advantage of this scheme is that it does not require knowledge of the scientific names of the plants.

Ecosystems may also be named in terms of the growth form and structure of the dominant plants *plus* the name of the genus of those plants. This scheme can be used when the tallest plants belong predominantly to one genus, such as 'a tall open *Eucalyptus* forest ecosystem' in south-east Victoria.

KEY IDEAS

- A community is composed of many populations.
- A population consists of all the organisms of one species living in the same area at the same time.
- An ecosystem consists of a living community, its physical surroundings and the interactions both within the community and between the community and the physical surroundings.
- Ecosystems can be named in various ways.

QUICK-CHECK

1 Identify the following statements as true or false.
  a A population of animals is an example of an ecosystem.
  b The water in a lake is an example of an ecosystem.
  c Various populations living in the one location at the same time make up a community.
  d More species would be found in one community than in one population.
2 List the three essential components of an ecosystem.

Ecological communities

If we look at a variety of ecosystems, we find that the living community in each ecosystem differs. The various populations that make up each community have physical, biochemical and behavioural features that equip them for survival and reproduction in the particular environmental conditions of their ecosystem.

The community of a littoral zone

The **littoral (inter-tidal) zone** of a rocky seashore (see figure 13.8) has a living community that includes various populations of green and brown algae and sponges that are situated at the low tide mark. Higher up in the inter-tidal zone, animals including arthropods (such as barnacles and crabs), molluscs (such as periwinkles and mussels) and echinoderms (such as starfish) are found.

The littoral (or inter-tidal) zone is the narrow strip of coast that lies between the low-water mark (LWM) of low tide and the high-water mark (HWM) of high tide. The littoral zone is affected by the tides, being exposed at each low tide and submerged at each high tide.

The littoral zone can be a near-vertical cliff face; in other locations, it can be a near-horizontal rock platform.

Twice per month, at new moon and at full moon, spring tides occur when high tide is at its highest and low tide is at its lowest. At the first and third quarter phases of the moon, neap tides occur when the tidal movement is at its minimum.

Figure 13.8 (a) The inter-tidal or littoral zone of a rocky shoreline. The living community of this ecosystem is most apparent at low tide and includes many organisms, such as (b) barnacles (*Tetraclitella purpurascens*) and (c) starfish or seastar (*Nectria ocellata*). At the low tide mark, brown algae such as (d) the strapweed, *Phyllospora comosa*, may be seen.

The littoral zone is often subdivided into lower, middle and upper regions. Note that these are artificial subdivisions and there is no sharp boundary delimiting each region. The 'splash zone' lies above the upper region (see figure 13.9a).

The various species living in the littoral zone are exposed to a wide range of environmental conditions varying from total submergence to total exposure to the air. At low tide when the littoral zone is exposed, organisms must cope with drying air and heat from the sun. They may be exposed to fresh water during periods of heavy rain. At high tide, when the littoral zone is submerged, organisms are returned to conditions where they are affected by water currents.

The various species found in the littoral community are not distributed uniformly throughout the zone. This suggests that the various species differ in their tolerance to exposure to the air and risk of **desiccation** (drying out). One species may tend to be found higher up in the littoral zone than another species; for example, periwinkles are typically found much higher in the littoral zone than algae.

Few organisms are found in the 'splash zone'. For most of the time, this region is dry. Small amounts of moisture come from spray thrown up by crashing waves or blown by strong winds, especially at high tide. Periwinkles with squat spiral shells are often found clustered in rock crevices in the 'splash zone' (see figure 13.9b).

(a)

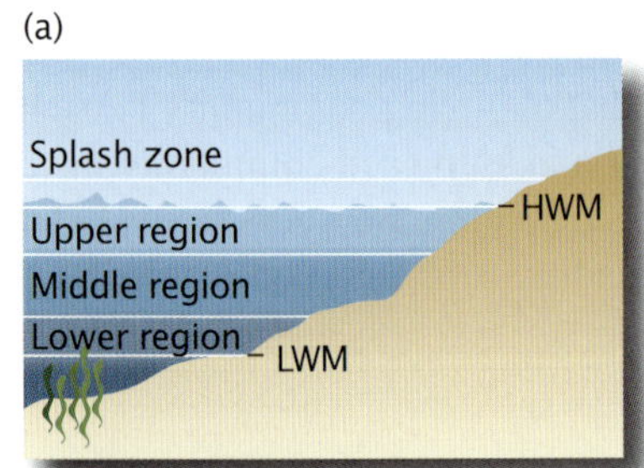

Figure 13.9 (a) The littoral zone can be subdivided into a number of regions: lower, middle and upper. The splash or spray zone lies above the upper region. Tide heights vary during the month. LWM = low-water mark; HWM = high-water mark. (b) Periwinkles are common in the splash zone of rocky shores on Australia's east coast. These periwinkles of family Neritidae are often seen clustered in depressions or along cracks in rocks.

ODD FACT

Barnacles begin life as free swimming larvae. The adults are sessile (fixed) and live adhering to rock surfaces. Barnacles are hermaphrodites, each animal having both male and female reproductive organs. Cross-fertilisation occurs in barnacles.

Several species of rock barnacles, including the six-plated barnacle (*Chthamalus antennatus*) and the surf barnacle (*Catomerus polymerus*), can be found in the littoral zone. During low-tide periods when they are exposed to the air, barnacles are protected against desiccation by the presence of hard valves that seal each barnacle into its moist chamber (see figure 13.8b).

The lowest region of the littoral zone has the least period of exposure during low tide periods. Various species of algae live near and below the low-tide mark, including leather kelp (*Eklonia radiata*) and strapweed (*Phyllospora comosa*) as shown in figure 13.8d.

The community of an open forest

The plant community living in one open forest (see figure 13.10) consists of:

- an upper storey made of the leafy tops (canopies) of various trees, such as the narrow-leaved peppermint (*Eucalyptus radiata*), the messmate (*E. obliqua*) and the manna gum (*E. viminalis*)
- a middle storey consisting of shrubs, including the black wattle (*Acacia mearnsii*)
- a ground cover of various herbs and hardy ferns.

Other members of this open forest community include various fungi and, within the soil, many microbe populations.

The open forest community includes many animal populations. Hidden in tree hollows are possums that are active at night. Various bird species feed in the forest, some kinds in the upper canopy, others in the middle storey and some are ground feeders. Tiny skinks sun themselves on rocks and scurry into the litter layer when disturbed. The litter layer and the underlying pockets of soil also contain populations of invertebrates, such as centipedes and beetles.

Figure 13.10 Part of an open forest ecosystem. Which members of its living community are most prominent? The living community of this ecosystem includes various species of *Eucalyptus* and wattles of the genus *Acacia* as well as many small plants and various animals. What component of an ecosystem cannot be easily shown in a photograph?

ODD FACT

The foliage of she-oaks (*Allocasuarina* spp.) does not consist of typical leaves. Fine green branches, known as **cladodes**, do the photosynthetic work of leaves. The true leaves are reduced to very small pointed scales at the nodes of these branches.

Figure 13.11 Jointed stems (cladodes) of a species of *Casuarina* sp.

The community of a Mallee ecosystem

If we visit the Little Desert National Park in north-west Victoria, we would find ourselves in the Mallee, a region of low rainfall, with sandy soils that in some areas are salty. The Little Desert National Park covers an area of 132 000 hectares. The living community in this Mallee ecosystem includes more than 670 species of native plants and more than 220 species of birds, including the endangered mallee fowl (*Leipoa ocellata*) (figure 13.12c).

The living community of this Mallee ecosystem includes many reptiles, including geckos, skinks, snakes and lizards, such as the shingleback lizard (*Trachydosaurus rugosus*) (figure 13.12b), many birds including the mallee fowl and the musk lorikeet (*Glossopsitta concinna*) (figure 13.12d), and many mammals including Mitchell's hopping mouse (*Notomys mitchelli*). Near waterholes or after rain, some of the frogs of this ecosystem can be seen and heard, such as the eastern banjo frog (*Limnodynastes dumerillii*). A diversity of plant species lives in a Mallee ecosystem, including many multi-stemmed eucalypts (figure 13.12a) such as green mallee (*Eucalyptus viridis*) and red mallee (*E. calycogona*). Other plants include the drooping she-oak (*Allocasuarina verticillata*) and the buloke (*Allocasuarina luehmannii*), and native conifers, for example, slender cypress pine (*Callitris preissii*) and Oyster Bay pine (*Callitris rhomboidea*). Invisible bacteria that are present in the soil form an important part of this ecosystem.

The box on page 415 tells the story of Whimpey who has worked for many years in the Mallee ecosystem of the Little Desert National Park.

(a)

Figure 13.12 Part of a Mallee ecosystem of the Victorian Little Desert. The living community includes various plants and animals. Note the multi-stemmed nature of the Mallee eucalypts that differs from the single trunk seen in most other eucalypts.

(b)

(c)

(d)

BIOLOGIST AT WORK

Whimpey's World in the Mallee

Ray 'Whimpey' Reichelt, a Nhill district local, and his wife Maureen started Little Desert Tours in 1970 as a way to protect and share with others the natural wonders of the Little Desert National Park.

Whimpey had a fascination with mallee fowls from childhood. In 1973, the Reichelts purchased 117 hectares of mallee scrub as a site for a mallee fowl sanctuary and an educational camp. In 1980, they started the Little Desert Lodge located on the northern edge of the park.

Whimpey was granted a wildlife shelter permit and, in 1984, commenced a captive breeding program for mallee fowl. He records detailed information on the breeding of mallee fowl and has an intimate knowledge of their behaviour. The mound of one pair of birds has data logging equipment installed at its base (shown in figure 13.13b) to monitor temperature and humidity in the egg chamber during the breeding cycle as well as external temperature, rainfall and humidity. The data that Whimpey collects on the birds' behaviour are analysed by scientists at Latrobe University Bendigo who are researching the life cycle and breeding patterns of mallee fowls.

(a) (b)

Figure 13.13
(a) Ray 'Whimpey' Reichelt **(b)** Whimpey can work with the birds at close range on their mounds.

The program also operates as an educational facility and allows visitors to watch the birds. People of all ages come to see Whimpey's famous rapport with the timid birds.

Whimpey has received both international and national recognition. He has been identified as one of Victoria's Living Treasures (1986). In 1997, the Little Desert Lodge hosted the Third International Megapode Symposium 'signifying the importance and value of Whimpey's ongoing study on the mallee fowl'. In 2003, Whimpey received an Order of Australia Medal (OAM) for services to nature conservation and the development of regional and eco-tourism in Victoria.

His 35 years of studying mallee fowl behaviour has been 'exciting and rewarding', says Whimpey. Now in his seventies, Whimpey is reluctant to hang up his trademark towelling hat but hopes he has succeeded in leaving 'something for our grandkids'.

Ecological groupings within an ecosystem

The living communities in different ecosystems, such as a littoral zone, an open forest and a Mallee ecosystem, are apparently very different in terms of the species that are present. However, there is a common pattern to each community. Members of every community can be identified as belonging to one of the following ecological groups: **producers** or **autotrophs**; **consumers** or **heterotrophs**; **decomposers**.

Producers or autotrophs

The community in every functioning ecosystem must contain some species that can be identified as producers or autotrophs. Producers or autotrophs are those members of an ecological community that can manufacture organic compounds, such as glucose, from simple inorganic compounds, such as carbon dioxide, using an abiotic energy source, such as sunlight. Through their energy-transforming actions, the producers in every ecosystem make chemical energy available in the form of organic compounds for their own use and also, directly or indirectly, for use by all other members of the living community of that ecosystem.

Details of the process of photosynthesis are given in chapter 5, pages 96–99.

The process by which most producers transform the radiant energy of sunlight into the chemical energy of sugars is known as **photosynthesis**:

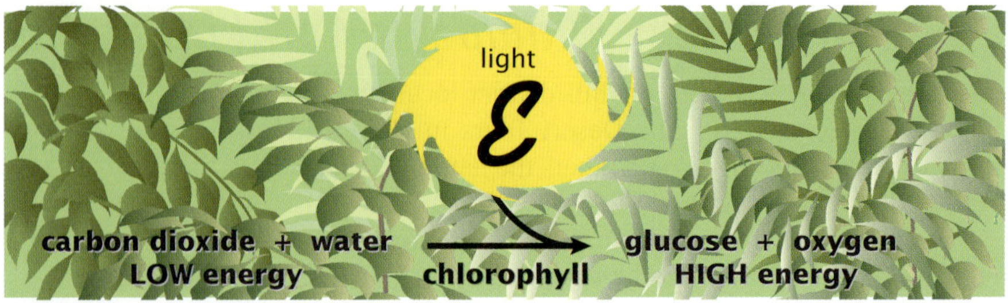

In aquatic ecosystems, such as seas, lakes and rivers, the producers are microscopic phytoplankton, macroscopic algae and seagrasses.

In terrestrial ecosystems, producer organisms include familiar green plants. These, in turn, include trees and grasses, other flowering plants, cone-bearing plants such as pines, and other kinds of plant such as ferns and mosses (see figure 13.14). All these producers convert the energy of sunlight to the chemical energy of glucose by the process of photosynthesis.

Figure 13.14 Producers from a range of ecosystems

Ecosystem: temperate marine kelp forest
Producers: algae, including the string kelp, *Macrosystic angustifolia*

Ecosystem: Antarctic marine ecosystem
Producers: many species of phytoplankton

Ecosystem: temperate closed forest
Producers: various woody flowering plants, ferns and mosses

When a cyanobacteria bloom occurs, waterways and rivers become covered in a mat of cyanobacteria that causes water quality to fall and, in some cases, to become toxic.

In some ecosystems, various species of bacteria, such as cyanobacteria, are also among the producers (see figure 13.15). Cyanobacteria have a form of chlorophyll and can carry out photosynthesis. While individual cells of cyanobacteria are microscopic, under certain favourable conditions the cell number can increase exponentially — this is a so-called 'bloom'.

Figure 13.15 Some bacteria, such as cyanobacteria, can transform the energy of sunlight to the chemical energy of sugars, such as glucose. Here we see a so-called 'bloom' of cyanobacteria (either *Anabaena* sp. or *Microcystis* sp.) in a river. Would you predict that cyanobacteria would possess a type of chlorophyll?

All of that leaves us with the interesting question: *If every functioning ecosystem must have producer organisms, what are the producers in the hydrothermal vent community?* Obviously, they cannot be any of the producers that are identified in figure 13.14 since the vents are areas of total darkness several kilometres below the ocean surface. No sunlight reaches more than about 200 metres down into the clearest waters. (We will explore this question in chapter 14.)

Consumers or heterotrophs

Consumers or heterotrophs are those members of a community that must obtain their energy by eating other organisms or parts of them. All animals are consumers and, in aquatic ecosystems, examples of consumer organisms include fish that graze on algae, sharks that eat fish and crabs that eat dead fish. In terrestrial ecosystems, consumer animals include wallabies that eat grass, koalas that eat leaves, snakes that eat small frogs, eagles that eat snakes, echidnas that eat ants, numbats that eat termites and dunnarts that eat insects (see figure 13.16).

(a)

(b)

Figure 13.16 Examples of consumer animals in a terrestrial ecosystem **(a)** The koala (*Phascolarctos cinereus*), an Australian marsupial, is a herbivore. It consumes the leaves of certain species of Eucalyptus. **(b)** The common dunnart (*Sminthopsis murina*) is a small Australian marsupial mammal. It is a carnivore and feeds mainly on insects.

ODD FACT

The dietary habits of the little crow, *Corvus bennetti*, which lives in inland Australia, show that its diet is typically composed of insects (48 per cent), plant material (26 per cent) and carrion (26 per cent).

Consumer organisms can be subdivided into the following groups:

- **herbivores** that eat plants, for example, wallabies and butterfly caterpillars
- **carnivores** that eat animals, for example, numbats and snakes
- **omnivores** that eat both plants and animals, for example, humans and crows
- **detritivores** that eat decomposing organic matter, such as rotting leaves, dung or decaying animal remains, for example, earthworms, dung beetles and crabs.

Fragments of dead leaves and wallaby faeces on a forest floor, pieces of rotting algae and dead starfish in a rock pool are all organic matter that contains chemical energy. Particles of organic matter like this are called **detritus**. The organisms known as detritivores use detritus as their source of chemical energy. Detritivores take in (ingest) this material and then absorb the products of digestion. Detritivores differ from decomposers (see below) in that decomposers first break down the organic matter outside their bodies by releasing enzymes and then they absorb some of the products.

Decomposers

Typical decomposer organisms in ecosystems are various species of fungi and bacteria (see figure 13.17). Decomposers are heterotrophs that also obtain their energy and nutrients from organic matter and, in their case, the 'food' is dead organic material. Decomposers differ from other consumers because, as they feed, decomposers chemically break down organic matter into simple inorganic forms or mineral nutrients, such as nitrate and phosphate. These mineral nutrients are returned to the environment and are recycled when they are taken up by producer organisms. So, decomposers convert the organic matter of dead organisms into a simple form that can be taken up by producers.

Figure 13.17 **(a)** Clusters of fruiting bodies of the fungi *Coprinus disseminatus*, found on rotting wood that is its food **(b)** Fungal growth on apples. What is happening?

Decomposers are important in breaking down wastes from consumers, such as faeces, shed skin, and the like. All these wastes contain organic matter and it is the action of decomposers that converts this matter to simple mineral nutrients.

Of the three groups described — producers, consumers and decomposers — only two are essential for the functioning of an ecosystem. Can you identify which two? One essential group comprises the organisms that capture an abiotic source of energy and transform it into organic matter that is available for the living community. Who are they? The second essential group is the one that returns organic matter to the environment in the form of mineral nutrients. Who are they?

ODD FACT

Guilds originated in medieval times as associations of craftsmen or tradesmen who made their living by practising similar trades, such as the guilds of goldsmiths, of silversmiths and of cloth merchants. Use of the term 'guild' in ecology was introduced in 1967.

Guilds: grouping organisms by function

As well as grouping the organisms in an ecosystem into producers, consumers or decomposers, organisms can also be grouped into **guilds**. The term 'guild' refers to groups of different species that exploit the same food resource in similar ways in an ecosystem. Examples of guilds include:

- various species of bird that eat insects from the leaves of trees
- various species of bird that eat seeds
- various species of reef fish that graze on algae
- various species of nectar-eating insects.

KEY IDEAS

- Each ecosystem has a living community that comprises various populations.
- The organisms in an ecosystem can be grouped into the major categories of producers, consumers and decomposers.
- Producers typically use the energy of sunlight to build organic compounds from simple inorganic materials.
- Consumers obtain their energy and nutrients from the organic matter of living or dead organisms.
- Decomposers break down organic matter to simple mineral nutrients.
- Consumers can be subdivided into various groups.
- Organisms can also be grouped into guilds.

QUICK-CHECK

3 Identify the following statements as true or false.
   a A producer organism in an ecosystem must be a flowering plant.
   b Different ecosystems have different living communities, each consisting of organisms that can survive in the particular environmental conditions.
   c Every functioning ecosystem must have producer organisms.
   d Decomposer organisms are important in breaking down organic matter to its more simple constituents.

4 Three ecosystems were identified in figure 13.14 (page 416). For each ecosystem, give an example of an organism that is likely to be identified as: (a) a producer organism; (b) a consumer organism; (c) a decomposer organism.

Interactions within ecosystems

In ecosystems, interactions are continually occurring:

- between the living community and its abiotic surroundings
- within the abiotic surroundings
- within the living community.

Interactions: between organisms and their surroundings

Interactions between organisms and their physical surroundings are occurring all the time in an ecosystem. In a terrestrial ecosystem, for example, animals take up water and oxygen from their surroundings and may use rocks for shade and protection. Plants capture the energy of sunlight and take up carbon dioxide from the atmosphere and water and mineral nutrients, such as nitrate and phosphate, from the soil. At the same time, members of the living community also release material to their physical surroundings (see figure 13.18).

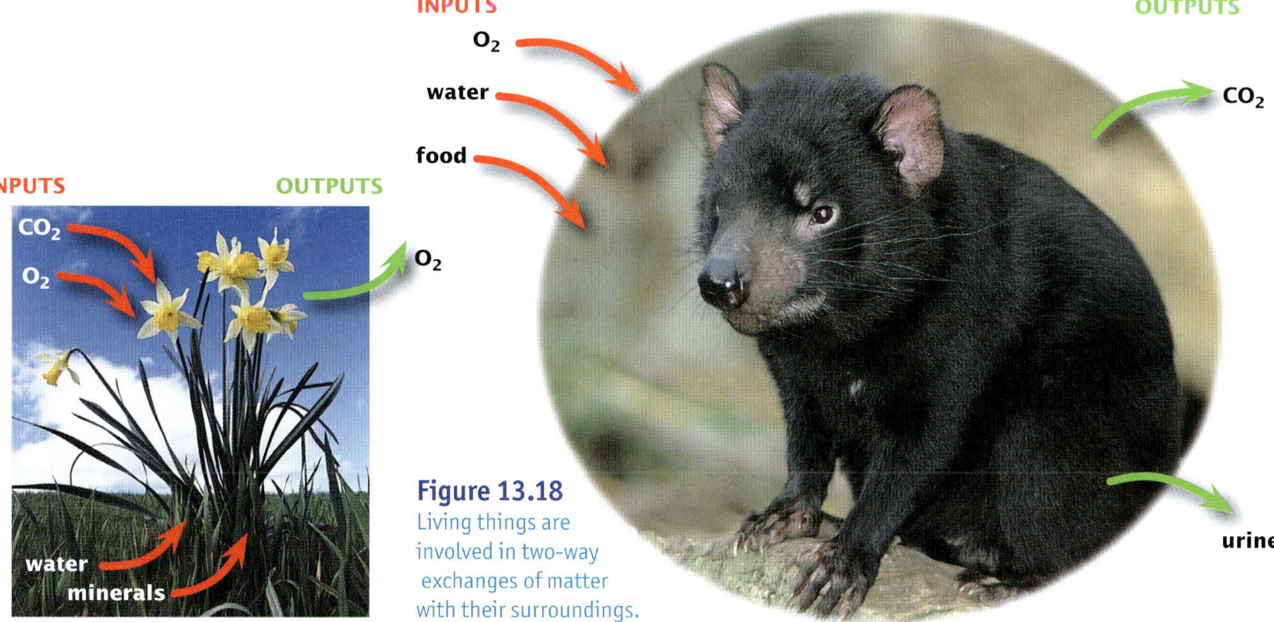

Figure 13.18 Living things are involved in two-way exchanges of matter with their surroundings.

Interactions between organisms and their surroundings can affect both the surroundings and the organisms. Shrubs growing in exposed coastal clifftop habitats, where prevailing strong and salty winds blow almost constantly from the sea, show the effect of their environment. In a habitat with different environmental conditions, shrubs such as the one shown in figure 13.19 would be expected to be symmetrical in shape and larger in size.

Figure 13.19 A wind-shorn tree

Plants growing in mineral-deficient soils show the effects of this environment by stunted growth and loss of green colour from their leaves. Factors such as atmospheric pollution can reduce the diversity of lichen populations growing in an area. Animals, too, are affected by their environments. The numbers and diversity of animals living in rivers are affected by water quality measures, including dissolved oxygen, clarity, pH as well as levels of industrial and agricultural contaminants.

Interactions within the abiotic surroundings

A sudden storm brings a downpour to a Mallee ecosystem causing localised flooding as creek beds fill and overflow and soil is washed away, causing erosion. A dry windstorm blasts across the land surface, picking up fine particles of topsoil and carrying them away from the region. Lack of rainfall over an extended period causes the water flow in rivers to fall and creeks to dry out and become a disconnected chain of shallow pools. High temperatures and high wind speed over an extended period produce an increase in the evaporation rate that causes the disappearance of surface water. Each of these is an example of an interaction between the components of the abiotic environment in a particular ecosystem. In the littoral zone of a rocky seashore, what interactions might occur between components of the abiotic surroundings?

Interactions within a living community

Interactions within the community of an ecosystem may involve members of the *same* species, or may involve members of *different* species. These interactions can be classified in various ways and include:

- competition
- predator–prey relationships
- parasitism
- mutualism
- commensalism.

Competition within and between species

In an ecosystem, certain resources can be in limited supply, such as food, shelter, moisture, territory for hunting and sites for breeding or nesting. In situations where resources are limited, organisms compete with each other for them.

Competition may be between members of the same species. For example, pairs of parrots of the same species compete for suitable hollows in old trees for nesting sites. Competition between members of the same species for resources is termed **intra-specific competition**.

Members of a population of one species also compete with members of populations of other species and this is termed **inter-specific competition**. For example, members of different plant populations in the same ecosystem compete for access to sunlight and different animal species may compete for the same food source. We will see later (page 432) that competition between members of different populations living in the same community is minimised through niche separation.

Competition occurs when one organism or one species is more efficient than another in gaining access to a limited resource, such as light, water or territory, for example:

- faster-growing seedlings will compete more efficiently in gaining access to limited light in a tropical rainforest than more slowly-growing members of the same species; the faster growers will quickly shade the slower growers from the light and deprive them of that resource
- when soil water is limited, a plant species with a more extensive root system will compete more efficiently for the available water than a different plant species with shallow roots, and deprive it of that resource.

Figure 13.20 shows an example of competition for space between two species of anemones. The competitive interaction between the two anemones is very visible to an observer. This is also the case for other competitive interactions, as for example, when a number of smaller birds of one species 'mob' a larger bird of another species that enters their nesting territory.

Figure 13.20 Anemones compete for space and food. **(a)** If an anemone encroaches too closely to another **(b)** the original occupant will inflate its tentacles and **(c)** release poisoned darts from stinging cells. The intruder may retaliate and return fire. **(d)** Eventually one of the anemones retires from the fight.

Competition between different species for food or for space does *not* always involve visible actions such as mobbing in birds, or displays of threat or actual fighting. Instead, competitive interactions may involve the use of chemicals. An example of this type of competition is the release of a chemical by one plant species that inhibits the germination or the growth of another species. Chemical inhibition of this type is termed **allelopathy**. The inhibitory chemicals are known as **allelochemicals** and they are made in various parts of a plant, such as roots, leaves or shoots. Plants known to produce allelochemicals include:

- crop plants such as barley (*Hordeum vulgare*), wheat (*Triticum* spp.), sorghum (*Sorghum bicolor*) and sweet potatoes (*Ipomoea batatas*)
- many species of pine.

The area under a pine tree that is covered with fallen needles (see figure 13.21) is bare of plant growth. Why? When the pine needles fall, they release allelochemicals.

ODD FACT

Some fungi produce chemicals that prevent the growth of bacteria. One such fungus is the green mould, *Penicillium notatum*, that is often seen growing on stale bread. The chemical that it produces is the antibiotic, penicillin.

Figure 13.21 Pine needles contain allelochemicals that inhibit the growth of weeds around the base of pine trees.

Chemicals are also important in other interactions, such as communication between members of the same species. For example, many female moths signal their readiness to mate by releasing chemicals called **pheromones**. Pheromones from a female moth spread through the air (that is, they are **volatile** substances) and attract males of the same species (see figure 13.22).

Figure 13.22

Predator–prey relationships

In an open forest ecosystem after rain, a frog leaps towards a shallow pond. It pauses momentarily beside a clump of dense vegetation. A sudden movement occurs as an eastern tiger snake (*Notechis scutatus*) strikes. The frog collapses, its muscles paralysed by neurotoxins in the snake's venom. The snake eats the frog that has become a source of nutrients and energy for the snake. This is just one example of the **predator–prey relationships** in ecosystems. A predator–prey relationship is one in which one species (the **predator**) kills and eats another living animal (the **prey**).

Carnivorous heterotrophs have structural, physiological and behavioural features that assist them to obtain food. These features are as varied as their food sources and include the web-building ability of spiders, claws and canine teeth of big cats, heat-sensitive pits of pythons, poison glands of snakes, visual acuity of eagles, speed of extension of a chameleon's tongue, tail-luring behaviour of death adders and cooperative hunting by dolphins.

Vacuuming, grasping, netting, ambushing, pursuing, piercing, filtering, tearing, engulfing, spearing, constricting, luring and biting — these are some of the different ways that carnivores capture and eat their living prey. Can you identify heterotrophs that obtain their food in some of these ways?

Think about the labels 'carnivore' or 'predator'. On land, these tend to call up an image of an animal, such as a lioness (*Panthera leo*) equipped with strong teeth, sharp claws and powerful muscles, which stalks its prey, pursues it over a short distance and then overpowers and kills it. However, a net-casting spider (*Deinopis subrufa*) is equally a carnivore that waits for its living prey to come to it to be snared on its web (see figure 13.23).

(a)

(b)

Figure 13.23 **(a)** A net-casting spider pulling out silk threads and crafting them into a net. **(b)** Spider with a completed net. In Australia, nine species of spider within the genus *Deinopis* are distributed across most states.

ODD FACT

The chameleon, a lizard native to Africa, can extend its tongue more than the length of its body (about 35 cm) in just 20 thousandths of a second to catch its insect prey.

If you were asked to name a predator of the seas, the powerful white shark (*Carcharodon carcharias*) that actively hunts its prey would probably spring to mind. However, the coral polyp (see figure 13.24b) which uses a 'sit-and-wait' strategy is also a carnivore, equipped not with teeth but with stinging cells (see figure 13.24a).

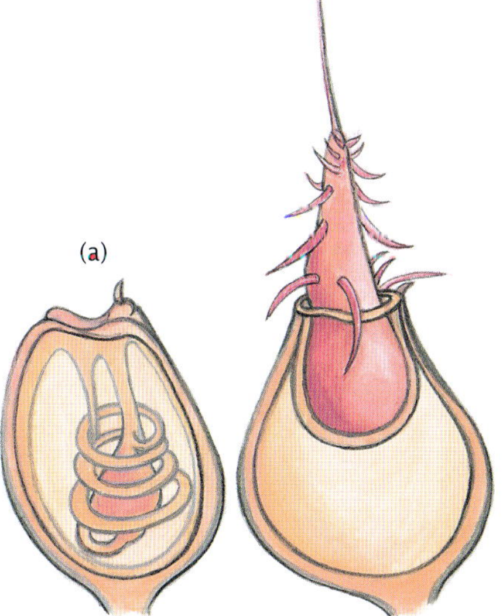

Figure 13.24 **(a)** Stinging cells or nematocysts (left: charged; right: discharged). The hollow coiled thread when discharged penetrates the prey and injects a toxin. **(b)** Coral polyps capture their prey, including fish, using the stinging cells on their 'arms'.

Carnivores come in all shapes and sizes and different species obtain their prey using different strategies. Let's look at three snake species. Snakes are a remarkable group of predators — legless but very efficient!

- The copperhead (*Austrelaps superbus*) lies in wait for its prey, such as a frog or a small mammal. When the prey comes within striking distance, the copperhead strikes, injecting its toxic venom.
- The desert death adder (*Acanthophis pyrrhus*) (see figure 13.25a) attracts its prey by using its tail as a lure. The death adder partly buries itself in sand or vegetation and wriggles its thin tail tip. When its prey is attracted by the 'grub' and comes close, the death adder strikes and injects its venom.
- The green python (*Chondropython viridis*) actively hunts its prey by night in trees (see figure 13.25b). Its prey includes bats, birds and tree-dwelling mammals. The python locates its prey in the darkness using its heat-sensitive pits, located mainly on the lower lip under the eye, and it kills its prey, not by toxic venom, but by constriction.

Figure 13.25 **(a)** The death adder has a short thick body but its tail tip is thin and is used as a lure to attract prey. Note how the snake positions its tail tip close to its head. What advantage is this behaviour? **(b)** A green python on the hunt. Like most members of the family Boidae, the green python has heat-sensitive pits that can detect temperature differences as small as 0.2°C between objects and their surroundings. By moving its head and responding to information from these sense organs on either side of its head, a python is able to locate a source of relative warmth such as a bird or a mammal. Can you see the pits along its lower lip?

Want to go swimming? You first!

Figure 13.26 tells an interesting story. Notice the Adelie penguins (*Pygoscelis adeliae*) that are lined up eager to enter the water but holding back. They must enter the water to feed on krill and small fish for themselves and for their chicks that are in a rookery away from the water. The reluctance of each penguin to enter the water is because of the leopard seal (*Hydrurga leptonyx*) on the ice. Leopard seals are major predators of Adelie penguins.

Figure 13.26 Adelie penguins waiting to enter the sea to feed. At left is one of their major predators — a leopard seal. The leopard seal on land is slow and cumbersome and the penguins are safe. In fact, they can come quite close to the seal. In the water, however, the leopard seal becomes a swift, agile and efficient predator of these penguins. Also present is a bird, a south polar skua (*Stercorarius maccormicki*), that will feed on fragments left by the leopard seal.

Prey are not always caught

Among reef fish that are hunted by predators, defence mechanisms include spines, thick and tough skins, mucous coatings, camouflage, eyespots and toxins.

In the living community of an ecosystem, predators are *not* always successful in obtaining their prey. Various prey species show structural, biochemical and behavioural features that reduce their chance of becoming a meal for a potential predator. Following are examples of some features that protect prey.

Structural features

- **Camouflage:** look like something else! Some insects in their natural surroundings look like green leaves, dead leaves or twigs, for example, the stick insect. Refer back to the bird-dung spider on page 389.
- **Mimicry:** look like something distasteful! Viceroy butterflies (*Limentitis archippus*) mimic or copy the colour and pattern of monarch butterflies (*Danaus plexippus*) that are distasteful to birds that prey on butterflies.

Figure 13.27 Meerkats on sentry duty while a pup is feeding

Behavioural features

- Stay still! Prey animals such as some rodents and birds reduce their chance of being eaten by staying still in the presence of predators.
- Keep a lookout! Meerkats gain protection from predators by having one member of their group act as a sentry or lookout when the rest of the group is feeding (see figure 13.27). The lookout signals the approach of a predator, such as an eagle, and the group immediately flees to shelter.
- Schooling — safety in numbers! Individual organisms in a large group, such as a school of fish (see figure 13.1, page 406), have a higher chance of not being eaten than one organism that is separated from the group.

Biochemical features

- Produce repellent or distasteful chemicals! Larvae of the monarch butterfly (*Danaus plexippus*) (see figure 13.28) feed on milkweeds that contain certain chemicals that cause certain predator birds to become sick. The larvae store these chemicals in their outer tissues and the chemicals are also present in adult butterflies. Predator birds that are affected by this chemical rapidly learn that the monarch butterflies are not good to eat. In fact, monarch butterflies advertise that they are distasteful with bright **warning colouration**.

Figure 13.28 A monarch butterfly in its caterpillar stage

Herbivore–plant relationships

One of the most common relationships seen in living communities is a **herbivore–plant relationship**. Herbivores are organisms that obtain their nutrients by eating plants. Herbivores include many mammals, such as kangaroos, koalas and cattle, but the most numerous herbivores are insects, such as butterfly larvae (caterpillars) (see figure 13.28 above), bugs, locusts, aphids and many species of beetle. Plants under attack from herbivores cannot run, hide or physically push them away. What can plants do?

Plants can protect themselves from damage by herbivores by physical means, such as thorns and spines, as seen in cacti, and also by means of stinging hairs, as in nettles. Various plant species also produce allelochemicals that either protect the plant from attack by herbivores or limit the damage done by them.

- Some plants produce chemicals that deter or poison insect herbivores, for example, some clovers (*Trifolium* spp.) produce cyanide.
- Some plants produce chemicals that interfere with the growth or development of insects, for example, an African plant known as bugleweed (*Ajuga remora*) produces a chemical that causes serious growth abnormalities in herbivorous insects.

Parasite–host relationships in animals

In a temperate forest, a wallaby hops through the undergrowth. A close examination shows this wallaby is carrying some 'passengers' in the form of ticks that are attached to the animal's face, near its eyes. The passengers in this case are adult female paralysis ticks (*Ixodes holocyclus*), which are native to Australia. The female ticks are noticeable because they are fully engorged after feeding on their host's blood (see figure 13.29).

Unfed Half-fed Fully fed (engorged)

Figure 13.29 Life-size representations of a tick. A female paralysis tick is smaller than a match head before it feeds on its host. As a tick feeds, it increases from about 3 mm in length to about 12 mm in length when it is fully engorged. Notice that a tick has eight legs and this makes it a member of the Class Arachnida that includes other eight-legged arthropods, such as mites and spiders.

This is just one example of the many **parasite–host relationships** that occur in ecosystems. In a parasite–host relationship, one kind of organism (the **parasite**) lives on or in another kind (the **host**) and feeds on it, typically without killing it, but the host suffers various negative effects in this relationship and only the parasite benefits. A variety of animals are parasitic, including insects, worms and crustaceans. (In addition, other kinds of organism — plants, fungi and microbes — can also be parasites.) It is estimated that parasites outnumber free-living species by about four to one.

Parasites that live *on* their host are called **exoparasites**, while those that live *inside* their host are termed **endoparasites**. Fleas, ticks and leeches are examples of exoparasites that feed on the blood of their host. Other external parasites are fungi, such as *Trichophyton rubrum*, a fungus that feeds on moist human skin causing tinea and athlete's foot.

Some exoparasites, such as fleas, live all of their lives on their host. Others, such as ticks, attach to and feed from their hosts at specific times only; at other times, they are off the host. The life cycle of the paralysis tick, for example, has three stages: larva, nymph and adult (see figure 13.30). Each stage must attach to a host and feed on the host's blood. Tick larvae hatch from eggs and must attach to a host and feed on its blood. After feeding, the larvae drop off the host to the ground where they moult and become tick nymphs. A nymph then attaches to a host and feeds, after which it drops to the ground and moults to become an adult. The adult must also attach to a host and feed. After an adult female tick has fed,

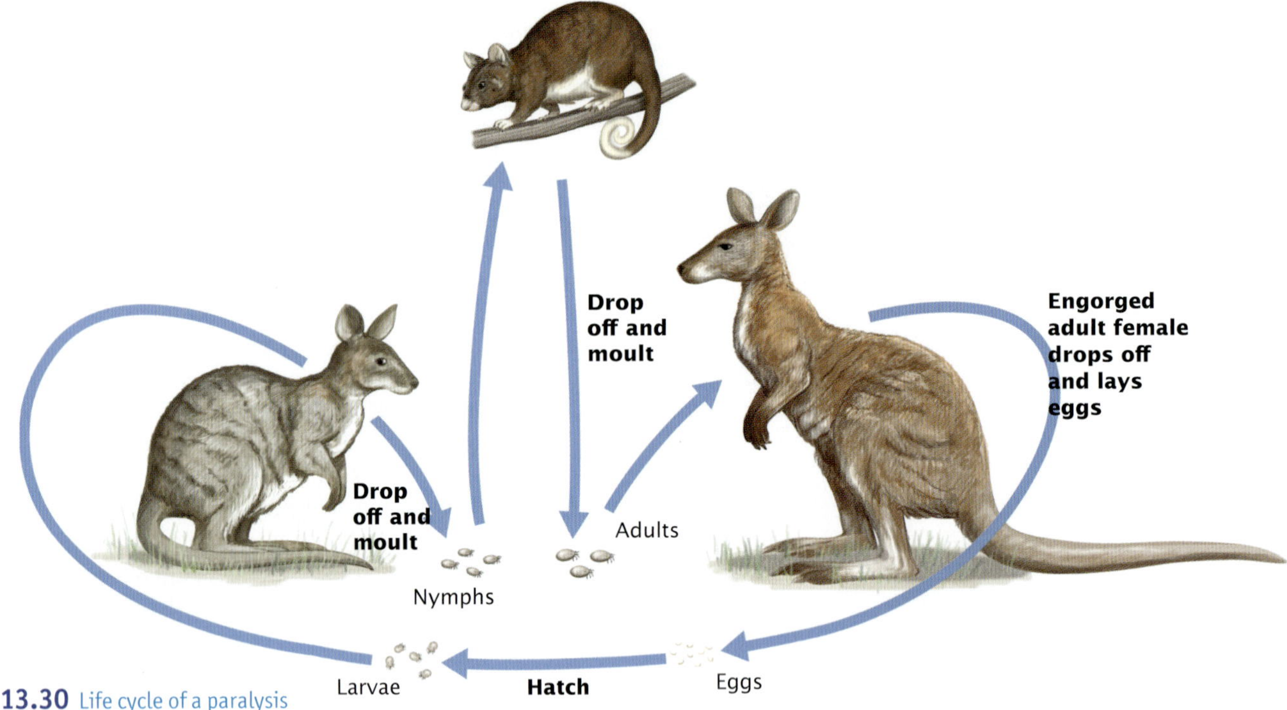

Figure 13.30 Life cycle of a paralysis tick. This is sometimes referred to as a three-host tick. Can you suggest why?

ODD FACT

Paralysis ticks produce venom that has adverse, even fatal, effects on children, cats and dogs, and also on the young of introduced species, such as cattle and horses. In contrast, native marsupial mammals, such as wallabies, that are the natural hosts of this tick are unaffected by the toxin.

Figure 13.31 Mouth of a wide-mouthed lamprey (*Geotria australia*) showing large teeth above the mouth and radiating plates with smaller teeth around the mouth

she drops off her host, lays several thousand eggs and dies. The eggs hatch after about two months.

Parasites are also found in freshwater and marine ecosystems. Parasitic lampreys have round sucking mouths with teeth arranged in circular rows (see figure 13.31). Lampreys attach themselves to their host's body using their sucker mouth and rasp the skin of the host fish and feed on its blood and tissues (figure 13.32). Other examples of parasite–host relationships include roundworms and tapeworms that are endoparasites in the guts of mammals, such as the beef tapeworm, *Taenia saginata* and the pork tapeworm, *T. solium*. A strange example of a marine parasite is described in the box on page 428.

Figure 13.32 Lamprey attached to its host

ODD FACT

A tiny parasitoid wasp (*Encarsia formosa*) is used in commercial greenhouses to control the greenhouse whitefly (*Trialeurodes vaporariorum*), a pest of tomato and other crops (see figure 8.6, page 218). An adult female *Encarsia* parasitoid lays 100 to 200 eggs each day with just one egg being laid in each greenfly larva.

Parasitoid: halfway between predator and parasite?

Parasitoids are a varied group of organisms, mainly small wasps and flies, that are like parasites. (The suffix '-oid' means 'like'.) Parasitoids kill their hosts that are usually another kind of insect. A predator–prey relationship is of short duration with the death of the prey occurring quickly. In contrast, a parasitoid–host relationship has a longer duration before the host dies. In addition, unlike parasites, it is only the adult female wasp or fly that is a parasitoid.

A parasitoid, such as an adult female wasp, lays one or more eggs on or in the body of her specific host. The host is usually the immature stage (larva or caterpillar) of another kind of insect such as a fly or butterfly (see figure 13.33). The parasitoid wasp larva that hatches from the egg then slowly eats the host from inside and, when the vital organs are eaten, the host finally dies.

Figure 13.33 Eggs of a parasitoid wasp on the caterpillar or larval stage of its host insect. What happens when the eggs hatch?

ODD FACT

The parasitic plant *Rafflesia arnoldii* was first reported to the scientific world in 1816, when it was discovered by Sir Stamford Raffles and Dr Joseph Arnold in Sumatra — hence the scientific name of this species. Raffles Hotel in Singapore is named after the same Raffles.

Parasite–host relationships in plants

Parasite–host relationships also exist in the Plant Kingdom. Two kinds of parasite–host relationships can be recognised, namely:

1. **holo-parasitism**, in which the parasite is totally dependent on the host plant for all its nutrients
2. **hemi-parasitism**, in which the parasite obtains some nutrients, such as water and minerals, from its host but makes some of its own food through photosynthesis.

Holo-parasitism

Holo-parasitism in plants is rare. One striking example of holo-parasitism is seen in 15 plant species belonging to the genus *Rafflesia*. These species occur only in the tropical rainforests of Borneo and Sumatra. *Rafflesia* plants are remarkable — they have no leaves, stems or roots and, most of the time, they grow as parasites hidden inside the tissues of one specific vine (*Tetrastigma* sp.). They obtain all their nutrients from their host vines.

At one stage of its life cycle, a *Rafflesia* parasite forms a bud on the roots of its host vine and, over a 12-month period, the bud progressively swells until finally it opens out into a single giant flower (see figure 13.34) that has the smell of rotting meat. (Why?) *Rafflesia* plants are dioecious and so the flowers are either male or female. The pollinators for *Rafflesia* flowers are flies and beetles that usually feed on dead animals (carrion).

Figure 13.34 Giant flower of *Rafflesia arnoldii*. This holo-parasitic species produces the largest single flower in the world, measuring about one metre in diameter and weighing about 10 kg. A flower lasts for about one week. What are the pollinating agents for this plant? What is its host?

PARASITISM CAN AFFECT HOSTS IN STRANGE WAYS

One species of barnacle (*Sacculina carcini*) has a remarkable life cycle. This barnacle is a parasite and its specific host is the green crab (*Carcinus maenus*). Most species of barnacle are free-living sedentary animals in the inter-tidal zones of rocky coasts. However, this barnacle species lives most of its life as an endoparasite within the body of its host crab. Many endoparasites are hermaphrodites and do not need a partner for reproduction. This is not the case for the *Sacculina* barnacle that has two separate sexes. So what is its story?

After hatching from their eggs, the larvae of the *Sacculina* barnacle are free-swimming and pass through a number of stages. Once they reach a particular stage of development, a female *Sacculina* larva will respond to the presence of a green crab. When she senses a green crab, the larva crawls onto it and finds a soft spot at a joint in one of the crab's legs. Here, the female larva injects some of her cells into the crab. The remainder of the larval material is discarded, while the injected material will develop into an adult female.

Figure 13.35 The *Sacculina* parasite grows to be a large lump (shown in yellow) on the underside of the crab.

The injected material moves into the soft tissue in the middle of the crab's body and the female parasite grows by sprouting many finger-like strands of tissue that spread throughout the crab's body and absorb nutrients from the crab's blood. The female *Sacculina* parasite continues to grow and becomes a distinct lump on the underside of the host crab located exactly where a non-parasitised crab would normally carry her own egg mass. This is the end of the story — unless a male *Sacculina* larva finds this parasitised host crab.

If a male *Sacculina* larva finds a parasitised green crab, he too injects himself into the crab. The injected male enters the body of the female *Sacculina* and remains there fused to her tissue. For the rest of their lives, the male *Sacculina* and the female *Sacculina* who houses him reproduce inside the green crab host and produce many broods of eggs during one season.

The presence of the parasite changes the shape, the physiology and the behaviour of the host green crab. When unaffected crabs reproduce during the mating season, parasitised crabs do not attempt to reproduce but simply continue feeding, ensuring that their internal parasites are nourished and can continue to reproduce. Host crabs, regardless of whether they were once unaffected males or females, care for the parasitic lumps on the under-surfaces of their bodies as if these lumps were their own egg masses. When the parasite eggs hatch and the *Sacculina* larvae are ready to emerge, a host crab goes into deep water as an unaffected crab would, forces the larvae out of their body and assists them to disperse as if they were green crab larvae.

Dodder has various common names throughout the world and these include strangleweed, devil's hair, hellbind, love vine, pull-down and devil's guts. Can you suggest why these names might have been given?

Species of another group of holo-parasites, known as dodder, have a worldwide distribution, including some species that occur in Australia. Dodders belong to the genus *Cuscuta* and our native species include the Australian dodder, *C. australis* and the Tasmanian dodder, *C. tasmanica*. Dodder plants have no leaves and are parasites on the stems of many different host plants, including crops such as clover, alfalfa and potato, and ornamental plants such as dahlias and petunias.

Dodders obtain the water and nutrients they need for survival and for reproduction from their host plants. Initially, a young dodder seedling has roots but as soon as it attaches to a host, the roots die and the dodder simply consists of a mass of thin intertwining yellowish stems (see figure 13.36a). At various points, there are thickenings in the dodder stems and these are the sites where the dodder parasite penetrates the tissues of the host plant (see figure 13.36b).

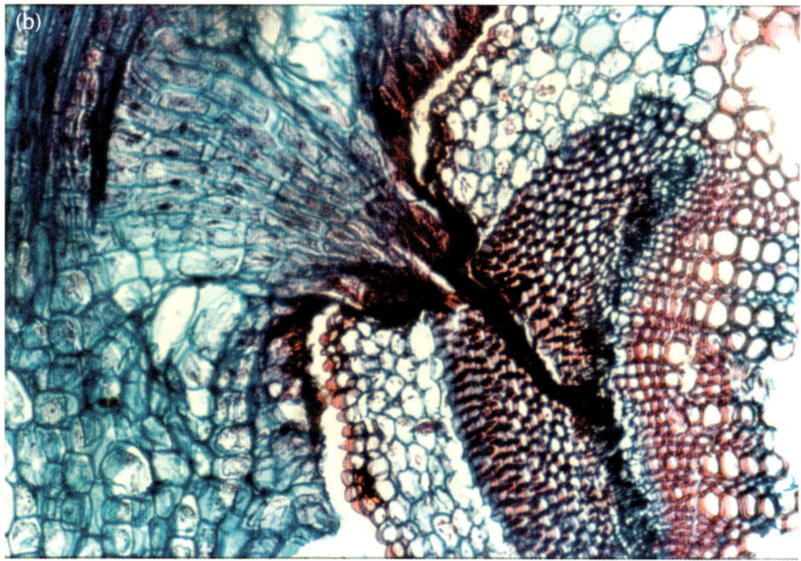

Figure 13.36 **(a)** Dodder parasite on its host. Note the intertwining yellow stems of the dodder and the occasional thickened regions of the dodder stem. What is happening here? **(b)** Photomicrograph showing tissue of one species of dodder (*Cuscuta campestris*) penetrating the tissue of its host plant (at right). The thin strand of parasite tissue that forms the connection with the host is known as a haustorium (plural: haustoria).

Hemi-parasitism

Hemi-parasitism ('hemi' = half) is best known to most people through plants known as mistletoes. Australia has many species of mistletoe that are parasitic on different host plants (see table 13.1). Mistletoes form connections (known as **haustoria**) with their host plants and the parasites obtain water and mineral nutrients from their hosts through the haustoria (see figure 13.37).

Table 13.1 Mistletoe species and their common hosts

Mistletoe species	*Hosts*
sheoak mistletoe (*Amyema cambagei*)	various sheoaks, especially *Casuarina cunninghamiana*
paperbark mistletoe (*Amyema gaudichaudii*)	various paperbarks especially *Melaleuca decora*
grey mistletoe (*Amyema quandong*)	*Acacia* species in dry woodlands
drooping mistletoe (*Amyema pendulum*)	*Eucalyptus* species in forests and woodland
Amylotheca dictyophleba	various rainforest tree species

Figure 13.37 Here we see a mistletoe stem (at right) and its host plant (at left). Note the connections between the parasite and its host. These connections (haustoria) are modified roots.

ODD FACT

The largest plant parasite in the world is the Christmas bush (*Nuytsia floribunda*) that is native to Western Australia. Could this plant be a parasite? In fact, this species is parasitic on the roots of other plants. Would it be a holo- or a hemi-parasite?

Figure 13.38 The Western Australian Christmas tree, *Nuytsia floribunda*

How do mistletoes reach their host plant since their seeds are large and fleshy? We get a clue from the word 'mistletoe' that comes from two Anglo-Saxon words: *mistel* = 'dung' and *tan* = 'twig'. This name arose from the mistaken belief that mistletoes spontaneously arose from bird dung on trees. We now know that seeds of many mistletoe species are dispersed by birds. For example, in Australia, mistletoe seeds are spread by the mistletoe bird (*Dicaeum hirudinaceum*) (see below).

Mutualism: both organisms benefit

Mutualism is a prolonged association of two different species in which both partners gain some benefit. Examples of mutualism include:

- mistletoe birds (*Dicaeum hirudinaceum*) and mistletoe plants. The birds depend on mistletoe fruits for food and, in turn, act as the dispersal agents for this plant. The birds eat the fruit but the sticky seed is not digested. It passes out in their excreta onto tree branches where it germinates. An interesting behaviour is that, before voiding excreta, the birds turn their bodies parallel to the branch on which they are perching so that their droppings plus seeds lodge on the branch rather than falling to the ground.
- fungi and algae that form lichens (see figure 13.39). The fungus species of the lichen takes up nutrients made by the alga and the alga appears to be protected from drying out within the dense fungal hyphae.
- fungi and certain plants. A dense network of fungal threads (hyphae) becomes associated with the fine roots of certain plants to form a structure known as a **mycorrhiza** (see figure 13.40). Plants with mycorrhiza are more efficient in the uptake of minerals, such as phosphate, from the soil when compared with plants that lack mycorrhiza. This is because the mycorrhiza increase the surface area of root systems. The fungal partner gains nutrients from the plant.

Figure 13.39 Lichen on a tree trunk. Which two organisms form the lichen partnership?

(a)

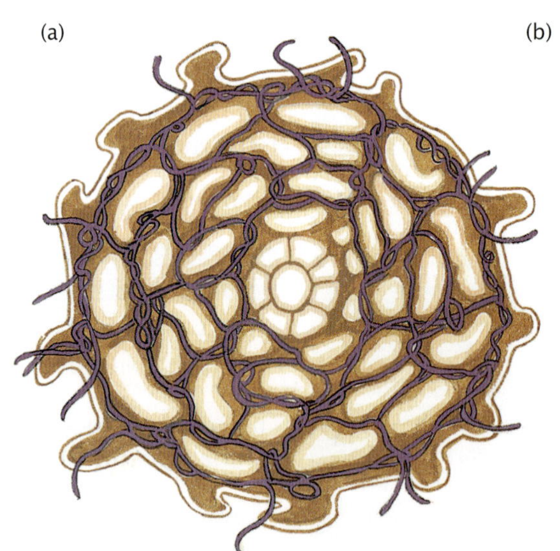

(b)

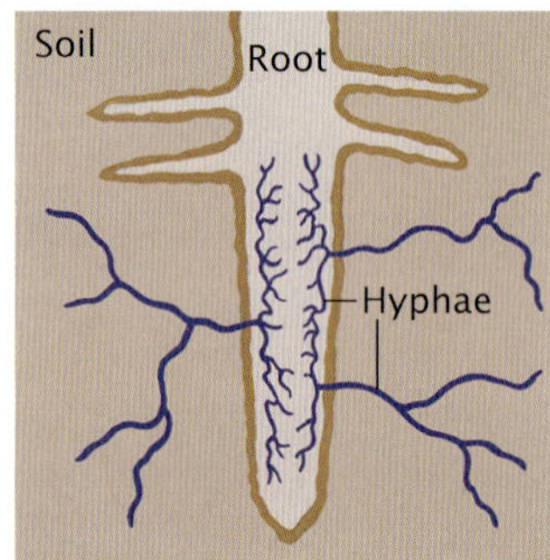

Figure 13.40 (a) Transverse section through a plant root showing thin threads (hyphae) of the associated fungus (b) Longitudinal section through root showing fungal hyphae

- **nitrogen-fixing bacteria** and certain plants. Plants require a source of nitrogen to build into compounds such as proteins and nucleic acids. Plants can use compounds such as ammonium ions (NH^{4+}) and nitrates (NO^{3-}) but cannot use nitrogen from the air. However, bacteria known as nitrogen-fixing bacteria can convert nitrogen from the air into useable nitrogen compounds. Several kinds of plants, including legumes (peas and beans) and trees and shrubs such as wattles, develop permanent association with nitrogen-fixing bacteria. These bacteria enter the roots and cause local swellings called nodules (see figure 13.41). Inside the nodules, the bacteria multiply. Because of the presence of the bacteria in their root nodules, these plants can grow in nitrogen-deficient soils.

Since the sixteenth century, farmers noted that growing crops of peas, beans and clover improved the quality of the soil. It was not until the twentieth century that this phenomenon was recognised as due to the action of nitrogen-fixing bacteria.

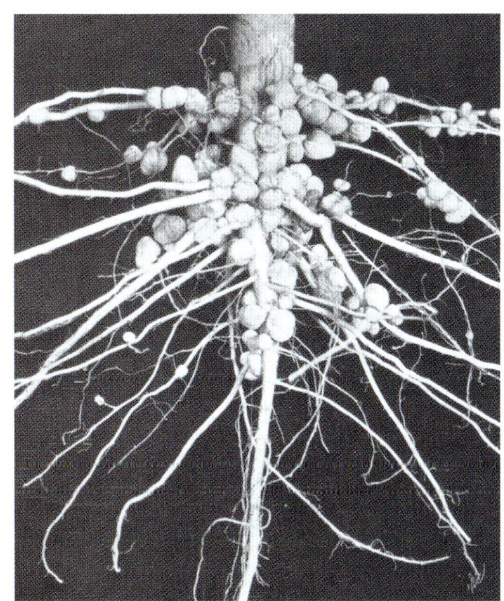

Figure 13.41 Root nodules contain large numbers of nitrogen-fixing bacteria. These nodules occur on the roots of several kinds of plants. How does each partner in this association benefit?

One stunning example of mutualism on the Great Barrier Reef involves small crabs of the genus *Trapezia* and a specific kind of coral (*Pocillophora damicornis*). The crab gains protection and small food particles from the coral polyps. The coral receives a benefit from the crab. Look at figure 13.42 and you will see a tiny *Trapezia* crab defending the coral polyps from being eaten by a crown-of-thorns starfish (*Acanthaster planci*). The crab repels the starfish by breaking its thorns.

Figure 13.42 *Trapezia* crab repelling a crown-of-thorns starfish from eating coral polyps

Commensalism

Commensalism (= at the same table) refers to the situation in which one member gains benefit and the other member neither suffers harm nor gains apparent benefit. An example of commensalism is seen with clown fish and sea anemones (see figure 13.43). The clownfish (*Amphiprion melanopus*) lives among the tentacles of the sea anemone and is unaffected by their stinging cells. The clownfish benefits by obtaining shelter and food scraps left by the anemone. The anemone appears to gain no benefit from the presence of the fish.

Figure 13.43 The clownfish (*Amphiprion melanopus*) lives among the tentacles of a sea anemone in tropical seas and is unaffected by the anemone's stinging cells.

Interactions such as parasitism, mutualism and commensalism are all examples of close associations between two species that have evolved over geological time.

Figure 13.44 shows an example of commensalism from about 400 million years ago. In the seas of that time lived many animals known as crinoids or sea-lilies. They were sessile animals fixed to the sea floor by long stalks. At the top of each crinoid stalk were branched arms surrounding its mouth. Many fossil crinoids have been found with small, shelled molluscs within their branched arms. These molluscs fed on wastes produced by the crinoid. Figure 13.44a shows the shell of one of these waste-eating animals (*Platyceras* sp.) within the arms of the crinoid (*Arthroacantha carpenteri*).

(a)

(b)

Figure 13.44 **(a)** Commensalism from nearly 400 million years ago! Inside the branched arms of this fossil crinoid can be seen part of a small shell. The animal that lived in this shell ate wastes produced by the crinoid. Part of the segmented stalk of the crinoid is also visible. Compare this crinoid fossil with the living specimen illustrated in (b). **(b)** Diagram of a living crinoid. Note the fine detail of the branched arms that wave about creating currents that carry food particles to the crinoid.

Interactions such as parasitism, mutualism and commensalism are sometimes grouped under the general term **symbiosis** (= living together) which is defined as a prolonged association in which there is benefit to at least one partner. Table 13.2 summarises these symbiotic relationships in terms of benefit or harm or neither to each of the species concerned.

Table 13.2 Summary of symbiotic relationships

Interaction	*Species 1*	*Species 2*
parasitism	parasite: benefits	host: harmed
mutualism	species 1: benefits	species 2: benefits
commensalism	species 1: benefits	species 2: neither harm done nor benefit gained

Minimising competition between species

In chapter 9, pages 268–9, we saw that, in some communities, competition between members of different species is minimised because they exploit different resources in different ways in the same habitat.

When various species in the same community differ in the use that they make of a resource, such as food or space, the various species are said to show **niche separation** — the greater the niche separation of two species, the smaller the level of competition between them. If, however, two species use the same resource in similar ways, they can be said to show **niche overlap** — the greater the niche overlap, the greater the intensity of competition between two species.

KEY IDEAS

- Interactions occur continuously between and within the various components of an ecosystem.
- Interactions between the living community and their physical surroundings may affect both the surroundings and members of the community.
- Interactions occurring within the living community of an ecosystem can be grouped into various kinds.

QUICK-CHECK

5 Identify the following statements as true or false.
  a In a parasite–host relationship, the host is always killed by the parasite.
  b A predator–prey relationship is an example of mutualism.
  c In lichens, the interacting partners are a fungus and an alga.
  d Endoparasites live on the outside of their hosts.
  e In mutualism, both partner organisms gain some benefits.

6 Give an example of each of the following.
  a a plant that is hemi-parasite
  b a fungus that is a parasite
  c a predator from a forest ecosystem
  d two partners with a relationship of mutualism.

BIOCHALLENGE

We have seen that, in an ecosystem, the members of the living community interact in various ways with each other.

1. Examine images A, B and C and for each image:
   a. describe the interaction that has occurred or is occurring using relevant biological terms
   b. for each member of the pair of organisms concerned, identify whether the interaction is beneficial, neutral or harmful
   c. give two other examples of the interaction shown in each image, one for a community in a terrestrial ecosystem and one for a community in an aquatic ecosystem.
2. Give two other examples of the interaction shown in C.
3. Identify two other kinds of interaction that can occur in the community of an ecosystem.

A

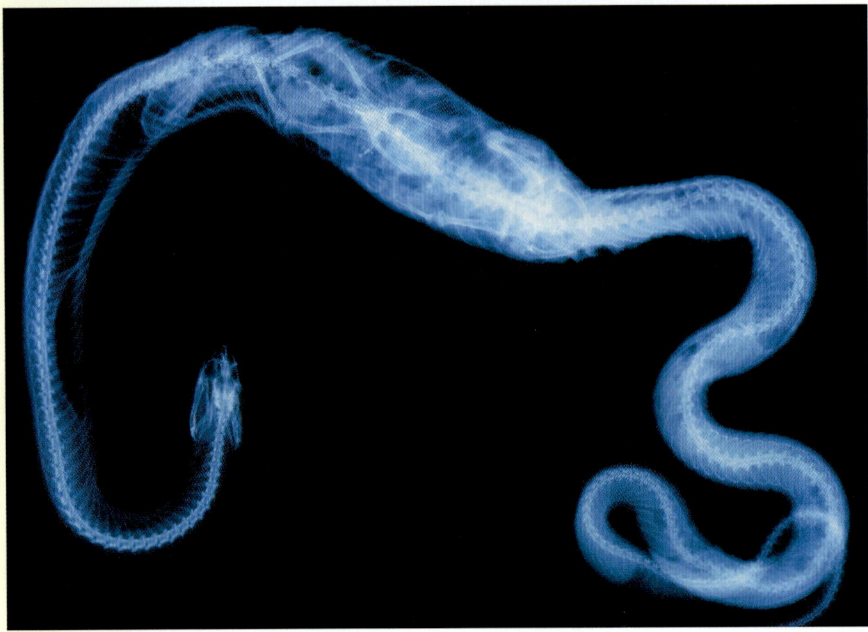

Two skeletons, a snake and a frog

B

Flower and honeybee

C

Tick on human skin

CHAPTER REVIEW

CROSSWORD

Key words

allelochemicals
allelopathy
autotrophs
camouflage
carnivores
cladodes
commensalism
community
competition
consumers
decomposers
desiccation
detritivores
detritus
diversity
ecology
ecosystems
endoparasites
exoparasites
guilds
haustoria (singular: haustorium)
hemi-parasitism
herbivore–plant relationship
herbivores
heterotrophs
holo-parasitism
host
hydrothermal vents
inter-specific competition
intra-specific competition
littoral (inter-tidal) zone
mid-ocean ridges
mimicry
mutualism
mycorrhiza
niche overlap
niche separation
nitrogen-fixing bacteria
omnivores
parasite
parasite–host relationships
parasitoids
pheromones
photosynthesis
population
predator
predator–prey relationship
prey
producers
sulfur bacteria
symbiosis
volatile
warning colouration

Questions

1 *Making connections* ▸ Draw a concept map using a selection of key words from this chapter. If necessary, add any other concepts that you may need to complete this map.

2 *Identifying differences* ▸ Identify one essential difference between the members of the following pairs:

a parasite and parasitoid

b host and prey

c predator and parasite

d symbiosis and commensalism.

3 *Analysing a situation* ▸ Could a fish tank with clean fresh water containing three different fish species be regarded as a miniature ecosystem? Explain.

4 *Analysing information and applying your understanding* ▸ In a forest habitat, the following four bird species feed by day on insects:

- White-throated tree creepers (*Cormobates leucophaea*) move *up* the tree trunks, locating insects in crevices.
- Varied sitellas (*Daphoenositta chrysoptera*) move *down* the tree trunks, locating insects in crevices.
- Crested shrikes (*Falcunculus frontatus*) locate insects by tearing bark off tree trunks.
- Bell miners (*Manorina melanophrys*) locate insects on leaves.

a Do these birds share (i) the same habitat? (ii) the same guild?

b Are these birds in direct competition for food?

c How is niche separation achieved between bell miners and shrikes?

d How might the direction of movement up or down a tree trunk be expected to result in different food sources?

5 ***Applying your understanding*** ▸ Suggest an explanation in biological terms for the following observations.

a A common fern produces a chemical that interferes with the development of insect larvae.

b A grower of tomatoes in a commercial glasshouse invests in the purchase of eggs of a particular parasitoid.

6 ***Developing a hypothesis*** ▸ Look at figure 13.45. Suggest a possible biological explanation for the pattern on this moth species. How would you test your hypothesis?

Figure 13.45 Can you explain the pattern on this moth?

7 ***Applying and communicating your understanding*** ▸ Competition can be achieved either by one species:

I preventing access to a resource by another, or

II making more efficient use of a resource than another.

a Identify each of the following situations in terms of alternatives I or II.

i A bird defends its territory from other members of the same species.

ii Larvae of two barnacle species (P and Q) both settle in the lower regions of an inter-tidal zone. The larvae of the P barnacles grow more quickly and crowd out the Q barnacles.

iii Some fungi produce a chemical that restricts bacterial growth.

iv Some members of a plant population grow more quickly and develop a more extensive root system than other members of the same population.

b For each example i–iv identify the competition as interspecific or intra-specific.

c Suggest the resource that is under competition.

8 ***Using the Web*** ▸ Go to www.jaconline.com.au/natureofbiology/natbiol1-3e and click on the 'Parasitic plants' weblink for this chapter.

a What is the major group of parasitic plants?

Click on 'General properties of parasitic plants'.

b What is the connection between a plant parasite and its host called?

c At what stage of its life cycle does a plant parasite establish its relationship with its host? Is the same true of animal parasites?

Return to the main page and click on 'Types of parasitic plants'.

d What is a root parasite?

e What is the host of *Nuytsia floribunda*? (Refer back to figure 13.38, page 430, to see Nuytsia.)

14 Flow of energy and cycling of matter

KEY KNOWLEDGE

This chapter is designed to enable you to:

- describe the ways in which energy flows through ecosystems
- recognise that matter cycles within ecosystems
- understand the key role of producers and decomposers in ecosystems
- become aware of trophic levels
- give examples of biogeochemical cycles.

Figure 14.1 Adelie penguins are part of a community whose members find the energy for living and the matter of their structure within an Antarctic marine ecosystem. Here we see a group of Adelie penguins leaping into the sea at the start of their quest for food that is their source of energy and matter. The penguins, in turn, are a source of energy for their predators. In this chapter, we will explore the flow of energy through ecosystems, starting from its capture as radiant energy by producer organisms and continuing with its transfer as chemical energy through organisms at various trophic levels. We will also explore the cycling of matter within ecosystems.

A day in the life of krill

It is late summer in Antarctica and the sun shines from a cloudless sky onto the clear blue waters. While these seas appear clear, they are in fact a concentrated soup of phytoplankton, which is a mixture of hundreds of different species of single-celled microscopic algae, such as diatoms, dinoflagellates and silicoflagellates. These tiny organisms possess coloured pigments that capture the energy of sunlight during the long daylight hours of an Antarctic summer. The **radiant energy** captured by the phytoplankton is transformed from non-material sunlight to **chemical energy** in carbohydrates, such as glucose. These floating phytoplankton themselves represent succulent morsels of chemical energy for hunters — such as krill.

Hunters of all sizes lurk in the Antarctic waters. The smallest are an army of zooplankton which comprise a biodiverse mixture — tiny protists, jellyfish, fish larvae, tunicates known as salp and various crustaceans, including copepods and krill (shrimp-like organisms). Antarctic krill (*Euphausia superba*) are the most abundant organisms in the zooplankton found in these waters. Individual krill reach an adult length of about 6 cm (see figure 14.2a) and gather in swarms so dense that at times the sea water becomes red (see figure 14.2b). These enormous aggregates or super-swarms can contain more than two million tonnes of krill spread over an area of 450 square kilometres.

Figure 14.2 (a) One *Euphausia superba* from a swarm of Antarctic krill. Notice the bristle strainer formed by the 'feeding' limbs around the anterior end. What function might it serve? **(b)** Part of a swarm of krill colouring the Antarctic waters

By day, the krill swarm typically stays in deep water. As the light dims, the swarm approaches the surface to feed on phytoplankton. Krill have five pairs of jointed limbs for swimming and other paired limbs, including six pairs of specialised 'feeding' limbs, located around their mouths, that are covered with long fine bristles — just like a built-in strainer! To capture their food, the krill enclose a quantity of water within their feeding limbs and strain it by squeezing the water through the bristle network. Small organisms, particularly phytoplankton, are trapped in the net formed by the bristles and this material is eaten. The radiant energy originally trapped by the phytoplankton has now been captured by the krill.

Suddenly, the krill swarm becomes the target of larger hunters. Adelie penguins (*Pygoscelis adeliae*) (see figure 14.3a) returning from a foraging trip have detected the swarm. They move in, swallowing the krill whole. Adelie penguins must obtain food, not only for themselves, but also for the young hatched in December that wait onshore in the creche area of the penguin rookery. The young penguins are close to fledging, the period when they will replace their fluffy down with adult feathers. Having taken as many krill as they can store in their crops, the Adelie

ODD FACT

In one breeding season, it was estimated that penguins delivered 18 485 kilograms of food to the colony and that this amount of food produced 412 fledged penguin chicks.

parents begin the return journey to their rookery. Here they will regurgitate from their crops a fishy stew, known as barf, to feed their young. One Adelie penguin, however, will not complete this journey. As the penguins approach the shore, a leopard seal (*Hydrurga leptonyx*) that has been waiting for the return of the penguins dives deep into the water. Suddenly changing direction, the seal accelerates upward, seizing one of the penguins. The leopard seal vigorously shakes the dying penguin, stripping the skin from its body. The seal now eats the exposed flesh and, in feeding, obtains the chemical energy it needs for living.

Figure 14.3 **(a)** Adelie penguins are consumers within the Antarctic marine ecosystem. Krill is a major food item for these penguins and so is a major source of chemical energy for them. From where do krill obtain their energy? **(b)** Adelie penguins themselves are food for higher-level consumers in the Antarctic marine ecosystem. Here we see a leopard seal that has caught a penguin. For this seal, the penguin simply represents part of the intake of chemical energy that is needed by the seal to maintain its own life.

ODD FACT

The term 'krill' is Norwegian for 'whale food' and refers to a group of more than 80 species of shrimp-like organisms whose habitats are the tropical, temperate and polar oceans of this planet. Antarctic krill are the most numerous of all krill species. Antarctic female krill spawn twice a year, on each occasion producing several thousand eggs that fall to the sea bed where they hatch.

Other animals also feed on the krill swarm. Crab-eater seals (*Lobodon carcinophagus*) sieve krill from the water through their multi-lobed teeth (see figure 14.4). (In spite of their name, they do *not* eat crabs, but feed mainly on krill.) The adult crab-eater seals that feed on the krill are survivors of a much larger group of crab-eater seal pups whose numbers were reduced in part by the feeding activities of hunters, such as leopard seals and killer whales (*Orcinus orca*).

The krill swarm re-aggregates after the attack by the Adelie penguins. It now becomes the subject of massive feeding activity by a pod of humpback whales (*Megaptera novaeangliae*) that circle the dense krill swarm at depth and then rise vertically through it, with open mouths, engulfing quantities of krill-rich water from which they filter the krill.

Figure 14.4 A crab-eater seal. Its unusual multi-lobed teeth (see inset) enable it to sieve krill from water.

Figure 14.5 Feeding relationships (food web) in an Antarctic ecosystem. Arrows denote the flow of chemical energy as one organism feeds on another kind. How does energy enter this ecosystem?

The krill swarm continues to feed on phytoplankton but the krill also become food for other animals, such as squid and fish. The squid, in turn, become food for flying birds (albatrosses, petrels), penguins and toothed whales. The fish are hunted by seals — the Weddell seal (*Leptonchotes weddellii*), the southern elephant seal (*Mirounga leonina*), the Ross seal (*Ommatophoca rossii*) and the Antarctic fur seal (*Arctocephalus gazella*). In spite of the feeding activities of so many hunters, the krill population over time is largely unaffected because of its high reproductive rate.

The living community in Antarctica in summer, which includes toothed and baleen whales, various sea birds, penguins and seals, depends on the seas for their food. A simplified version of the feeding relationships between these animals is shown in figure 14.5. Note the importance of krill as a direct and indirect food resource for this community.

Ecosystems require an energy source

Several features characterise all self-sustaining **ecosystems**. Among these characteristics, we can identify that every ecosystem must have a continual input of **energy**.

Imagine a city with no energy supplies — no electricity, no gas, no petroleum or diesel. Such a city would have no lighting, no artificial heating, no refrigeration, no industrial activity, and no mass transportation of people or goods. The city would be unable to operate and would cease to be recognised as a functioning city.

Just as the operation of a complex unit like a city requires an input of energy, an ecosystem requires an input of energy for its operation. Energy is not recycled in an ecosystem. Because of this, energy must continually be supplied to an ecosystem. What is the energy source for ecosystems?

The forms of energy that are particularly important in ecosystems are the radiant energy of sunlight and the chemical energy stored in organic matter — the latter is 'food'. To read more about energy, refer to the box 'Energy: what is it?' on page 442.

Sunlight: energy source for ecosystems

When we view a grassland or coastal seas bathed in sunlight, we are, in fact, observing energy coming into an ecosystem. The visible light that travels from a star known as the sun, across a distance of about 150 million kilometres, is the source of energy for almost all ecosystems and is the foundation for life on planet Earth.

Sunlight includes a range of wavelengths (see figure 14.6), such as visible light, with wavelengths in the range from about 0.4 to 0.7 micrometres (μm), and some longer wavelength infra-red (IR) and shorter wavelength ultraviolet (UV) rays.

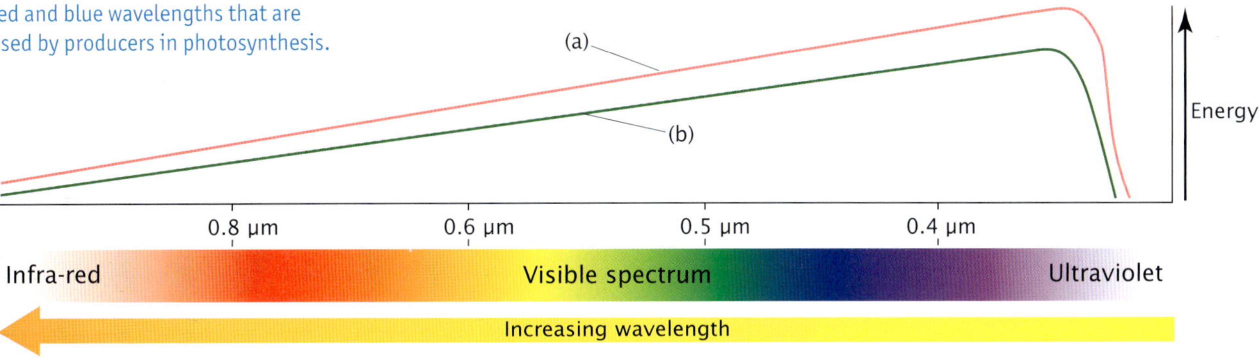

Figure 14.6 Solar radiation **(a)** outside the Earth's atmosphere and **(b)** at the Earth's surface. Solar radiation is in the ultraviolet (UV), visible and infra-red (IR) parts of the electromagnetic spectrum. As solar radiation passes through the atmosphere, much of the short wavelength UV radiation is absorbed by the ozone layer and water vapour in the atmosphere absorbs some of the IR radiation. Most of the solar energy is in the visible spectrum and it is the red and blue wavelengths that are used by producers in photosynthesis.

ENERGY: WHAT IS IT?

Energy is that which can change the motion, physical composition or temperature of an object. Energy exists in various forms which can be grouped into:

- forms of **kinetic energy** (or the energy of motion)
  - Thermal energy or heat is energy produced by the moving atoms or molecules in a substance; for example, steam.
  - Electrical energy is energy produced by the movement of electrons; for example, electricity, lightning.
  - Motion energy is energy of objects moving forward or rotating; for example, wind, speeding car, spinning turbine.
  - Radiant energy is energy of radiation that moves in waves; for example, sunlight, X-rays, UV, gamma rays.
- forms of **potential energy** (or stored energy).
  - Nuclear energy is energy stored in the nucleus of an atom that is released when nuclei are either split (fission) or combined (fusion).
  - Gravitational energy is energy stored in objects because of their position; for example: water stored in a dam (hydropower).
  - Stored mechanical energy is energy stored in objects under an applied force; for example, a compressed spring.
  - Chemical energy is energy stored in the bonds of molecules; for example: fuels, foods.

Of the various forms of energy in our surroundings, the radiant energy of sunlight is the principal direct and indirect source of energy for ecosystems.

Energy obeys several rules

Energy can be transformed — this means that energy can be converted from one form into another, but energy can neither be created nor destroyed.

Transformations of energy include:

- Solar collectors convert the radiant energy of sunlight into electrical energy (see figure 14.7a).
- Wind turbines convert the motion energy of wind to electrical energy (see figure 14.7b).
- Hydroelectric plants convert the gravitational energy of stored water to the motion energy of a turbine that is then converted to electrical energy.
- Heaters convert chemical energy of fuels, such as wood, gas, or electrical energy, to heat energy.
- Green plants convert the radiant energy of sunlight to the chemical energy of organic matter, such as sugars, through the process of photosynthesis (see figure 14.7c).
- Animals convert the chemical energy of food into the motion energy of running.

When energy is changed from one form to another, some of the energy is dissipated at each step as less useful energy, usually in the form of heat; this means that energy conversions cannot be 100 per cent efficient.

For example:

- About 10 per cent of the chemical energy in petrol is converted to the motion energy of a car.
- About 5 per cent of the electrical energy passing through the filament of a standard incandescent light bulb is converted to light; the remaining 90 per cent appears as heat. This is why a light bulb becomes very hot when it is 'on'. For a fluorescent light, the efficiency of conversion is about 20 per cent.
- Animals are typically about 25 per cent efficient in converting the chemical energy of food to other forms, such as motion energy.

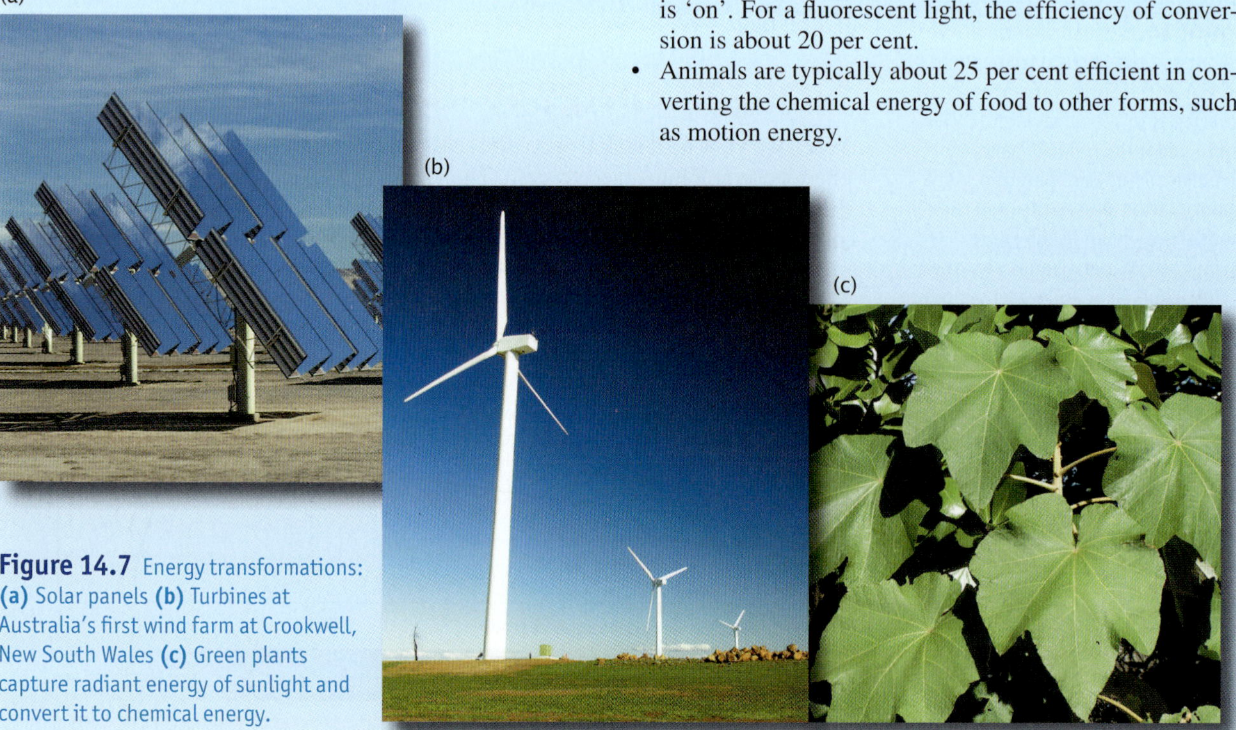

Figure 14.7 Energy transformations: **(a)** Solar panels **(b)** Turbines at Australia's first wind farm at Crookwell, New South Wales **(c)** Green plants capture radiant energy of sunlight and convert it to chemical energy.

Producers capture and transform energy

To check on producers and consumers, refer back to chapter 13, page 415.

No ecosystem can exist in a sustainable way without the presence of producer organisms. All producers are autotrophs. The **producers** in a marine Antarctic ecosystem are microscopic phytoplankton of many species, including diatoms (see figure 2.2, page 23). The producers in a tropical coral reef ecosystem are various algae and phytoplankton. In terrestrial ecosystems, the producers include flowering plants, conifers, ferns and mosses. Producer organisms can vary in size from a microscopic diatom to a very large organism like a mature mountain ash tree (*Eucalyptus regnans*) that can reach heights of up to 110 metres (see figure 16.46, page 531). All, however, are united in their ability to capture the radiant energy of sunlight and convert it to chemical energy stored in organic matter.

One group of producers visible on a global scale are the **phytoplankton** of the Antarctic. Studying phytoplankton across the oceans on a global scale has become possible through the use of satellite technology. Satellite-mounted instruments, such as SeaWiFS (see the Odd Fact), require just a few minutes to gather data on phytoplankton concentrations in the seas, when ship-based instruments would take about 10 years.

SeaWiFS (Sea-viewing Wide Field-of-View Sensor), an instrument carried on the Sea Star spacecraft orbiting 705 km above Earth, collects data on the concentration of phytoplankton. Go to www.jaconline.com.au/natureofbiology/natbiol1-3e and click on the 'SeaWiFS' weblink for this chapter to view an image of the concentration of phytoplankton around Tasmania.

Figure 14.8 is a satellite image of Antarctic waters in summer that shows the surface chlorophyll concentration. This chlorophyll is present in the phytoplankton. Note the high concentrations of chlorophyll (and hence of phytoplankton) in the waters over the continental shelf surrounding Antarctica. The productivity of these waters is high during summer but low during winter. The limiting factor in winter is sunlight.

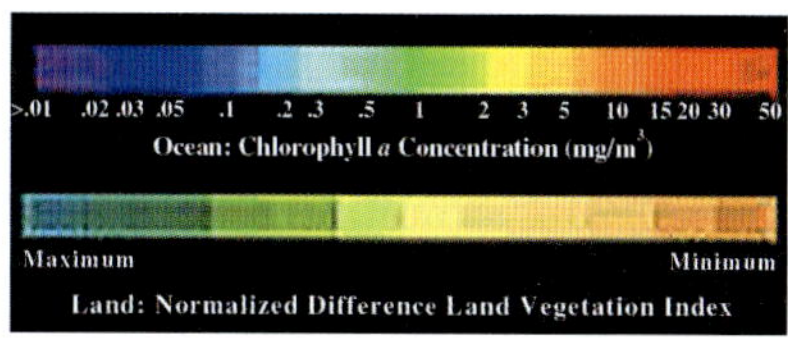

Figure 14.8 Colour-coded image of phytoplankton in the Southern Ocean around Antarctica. Purple areas have little phytoplankton; orange areas have the highest concentration of phytoplankton.

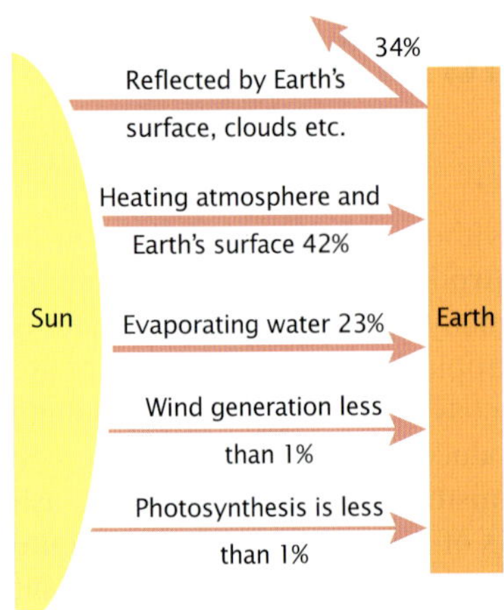

Figure 14.9 Solar energy coming into Earth has various fates. What percentage of this energy is immediately reflected out? In what form is the incoming energy? In what form is the major outgoing energy?

The incoming source of energy for life on planet Earth is the radiant energy of sunlight. Producers transform the radiant energy of sunlight into the chemical energy stored in organic matter through the process of photosynthesis (see chapter 13, page 415). Sunlight also provides the energy that drives winds, produces cyclones, evaporates water and drives the water cycle (see page 465) and warms the land and the atmosphere. On average, about one per cent of the incoming radiant energy of sunlight to Earth is captured by producers and converted to chemical energy (see figure 14.9).

So, the energy input to an ecosystem is in the form of the radiant energy of sunlight. It is the producer organisms, such as plants, that capture the radiant energy and transform it to chemical energy stored in organic matter, such as sugars.

Chemical energy is transferred from producers to consumers

The chemical energy made by producers is available both for the producer organisms themselves and also directly or indirectly for all consumer organisms in the living community of an ecosystem (see figure 14.10a). This chemical energy can be transferred from one organism to another.

- In a terrestrial ecosystem, a transfer of chemical energy occurs when herbivores eat plant material.
- In an aquatic ecosystem, chemical energy is transferred when herbivores eat phytoplankton or algae.

Chemical energy present in organic molecules is a concentrated (high-quality) form of energy that can be readily stored and transferred as required. In contrast, the radiant energy of sunlight is a dispersed form of energy that cannot be stored in that form and cannot be transferred (see figure 14.10b).

Figure 14.10 (a) Producer organisms, such as plants, capture radiant energy of sunlight and transform it to chemical energy of organic matter. Chemical energy in this form can be transferred from one organism to another. **(b)** What is easier to transfer — radiant energy or chemical energy? Consider this — it's more difficult to put light into a box than it is to put sugar in a bag or petrol in a tank!

Consumers at different feeding levels

- *Refer back to figures 13.16a (page 417) and 13.28 (page 425) to see some primary consumers.*
- *Refer back to figures 13.23 to 13.25 (pages 422–3) to check on some secondary consumers.*

All animals (mammals, birds, reptiles and invertebrates) are consumers.

- The consumers in a marine Antarctic ecosystem are many and varied and include fish, squid, seals and whales (refer back to figure 14.5, page 440).
- The consumers in a tropical coral reef ecosystem include fish, octopuses, sharks and starfish.
- In a terrestrial ecosystem, such as an open forest, the consumers include wallabies, snakes and insects.

Consumer organisms can vary in size from a tiny flea to a very large organism like a blue whale (*Balaenoptera musculus*) that can reach a length of about 25 metres and a mass of 110 tonnes.

As we saw in chapter 13 (page 415), all the consumers in an ecosystem can be classified in terms of the major source of their nutrition — plants, herbivorous animals or carnivorous animals. Consumers that mainly feed directly on the organic matter of producers are termed **primary consumers**, as, for example, leaf-eating caterpillars and other sap-sucking insects and herb-eating wallabies. Consumers that feed on primary consumers are termed **secondary consumers**, such as birds that eat caterpillars. Eagles that eat these carnivorous birds are termed **tertiary consumers** or top carnivores. Decomposers cannot be identified easily in terms of their major source of nutrition since they feed on the dead remains of plants and animals.

Trophic comes from the Greek trophikos *meaning 'relating to food'.*

Producers and the various consumers in an ecosystem can also be identified in terms of their 'feeding' level or **trophic level**. Producer organisms that make their own 'food' are said to be at the first feeding level or the first trophic level. Other trophic levels that can exist in ecosystems are shown in the following table 14.1.

Table 14.1 Different trophic levels may exist in an ecosystem. What level is occupied by a primary consumer? Why is it difficult to include decomposers in this table?

Trophic level	*Organisms at that level*	*Source of chemical energy or 'food'*
first	producers	make organic matter (food) from inorganic substances using energy of sunlight
second	primary consumers (herbivores)	eat plants or other producers
third	secondary consumers (carnivores)	eat plant-eaters
fourth	tertiary consumers (top carnivores)	eat predators

Figure 14.11 Sometimes herbivores or primary consumers themselves are not visible, but the results of their feeding activities can be seen.

Figure 14.12 (page 446) compares the general types of organisms that occupy the various trophic levels in an open forest ecosystem (a terrestrial ecosystem) as compared with a temperate coastal sea (an aquatic ecosystem).

When organisms feed, a transfer of chemical energy occurs from one organism (the eaten) to another (the eater). The eaters may be **detritivores**, such as a dung beetle that eats dung, or they may be **herbivores** or primary consumers, such as a koala that eats eucalypt leaves, or they may be **carnivores** or secondary consumers, such as a coral polyp that uses its stinging arms to snare fish. One group of consumers are the decomposers that feed on organic matter in dead organisms, both plants and animals.

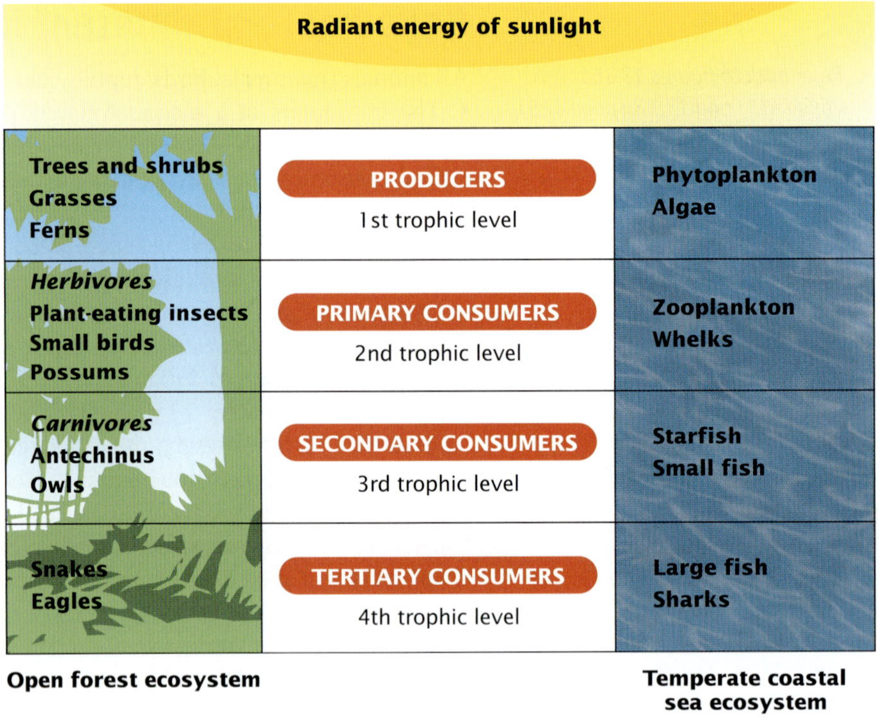

Radiant energy of sunlight

Open forest ecosystem		Temperate coastal sea ecosystem
Trees and shrubs Grasses Ferns	PRODUCERS 1st trophic level	Phytoplankton Algae
Herbivores Plant-eating insects Small birds Possums	PRIMARY CONSUMERS 2nd trophic level	Zooplankton Whelks
Carnivores Antechinus Owls	SECONDARY CONSUMERS 3rd trophic level	Starfish Small fish
Snakes Eagles	TERTIARY CONSUMERS 4th trophic level	Large fish Sharks

Figure 14.12 Comparison of producers and consumers in a terrestrial and an aquatic ecosystem. Which organisms are the producers in each? What is the energy source for organisms at the third trophic level?

KEY IDEAS

- Energy can exist in many forms.
- All ecosystems require an energy source.
- The energy input to ecosystems is typically the radiant energy of sunlight.
- Energy flows through an ecosystem and must be continually supplied.
- Energy is captured and brought into an ecosystem by producers.
- Producers transform radiant energy to chemical energy in organic matter.
- Chemical energy made by producers is available for themselves and for other members of the living community of an ecosystem.
- In an ecosystem, energy is transferred from producers to consumer members of a community in the form of chemical energy.
- Organisms in an ecosystem can be classified into different trophic levels.

QUICK-CHECK

1 Identify the following as true or false.
  a A primary consumer is a producer.
  b A predator in an ecosystem is at the third trophic level.
  c Energy moves through an ecosystem as the chemical energy of organic matter.
  d Producers can capture and transform radiant energy.
  e Energy is recycled within an ecosystem.

2 Give an example of an organism that is:
  a a producer in a terrestrial ecosystem
  b a producer in an aquatic ecosystem
  c a consumer of chemical energy
  d a tertiary consumer
  e an organism at the first trophic level.

Energy flows through an ecosystem

The energy input to an ecosystem is the radiant energy of sunlight that is transformed by producers into chemical energy stored in organic matter. Once transformed to chemical energy, the energy is transferred within an ecosystem through feeding activities. At each transfer, however, some energy is 'lost' from the ecosystem as heat energy.

Energy is lost at each transfer

All energy transfers in an ecosystem begin with producers that trap external energy and convert it to chemical energy in organic matter (food). Energy in this form can flow to primary consumers (herbivores), then to secondary consumers (carnivores) and then to higher order consumers (top predators). This energy transfer is not 100 per cent efficient and, at each step, only part of the energy entering can be transferred out, as shown in figure 14.13 (refer also to the box on page 442).

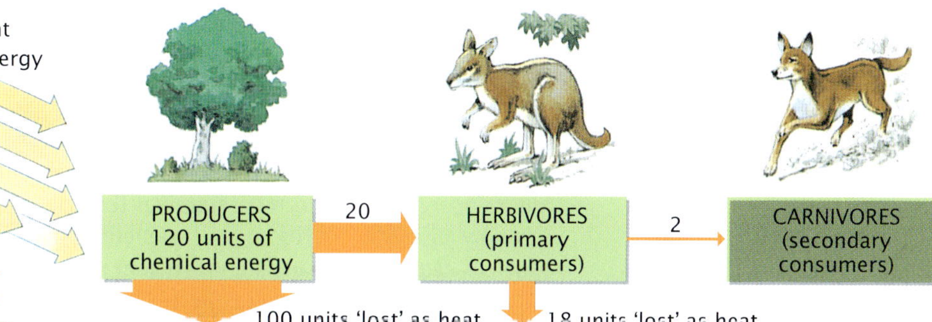

Figure 14.13 Energy flow in an ecosystem. The values are averages. Is the amount of energy that enters a trophic (feeding) level equal to the amount that flows to the next level?

Why is energy lost at each transfer?

What happens to the energy captured by producers and transformed to chemical energy? Most of the energy captured by producers is used for their own energy needs through the process of **cellular respiration** and is lost as heat. The remainder of this chemical energy is stored by producers in their own organic matter, as, for example, in new leaves, roots, and other tissues of flowering plants. This organic matter, such as leaves or fruits, in the tissues of producers is 'food' for consumers in an ecosystem (see figure 14.14) and it is only this chemical energy that can be transferred to consumers.

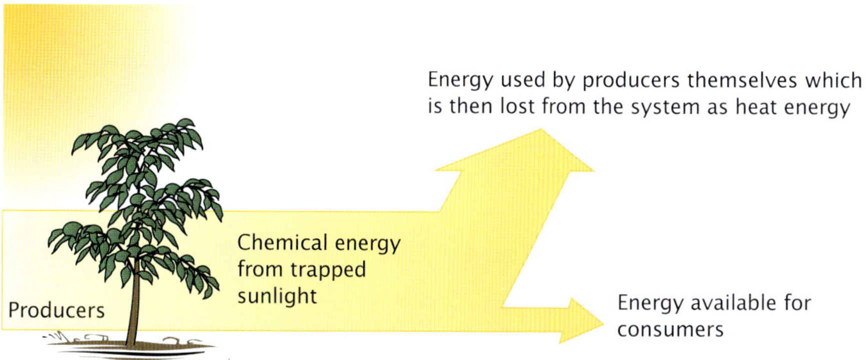

Figure 14.14 The chemical energy in sugars from sunlight energy trapped by producers is used mainly by the producers themselves for staying alive. A small amount of this energy is available to consumers in the ecosystem.

Unicellular producers such as phytoplankton use a lower proportion of the chemical energy that they make for their own needs as compared with complex multicellular producers such as a eucalypt tree. The latter have a large amount of non-photosynthetic tissues (roots, stems, flowers, fruit) that require chemical energy. It is estimated that, of radiant energy that they capture and transform to chemical energy, producer organisms use between 15 and 70 per cent for their own maintenance. (The higher percentages occur in those producers that are complex multicellular producers with large amounts of non-photosynthetic tissue.) The remainder is available to consumers.

Likewise, most of the chemical energy taken in by a herbivore when it eats leaves, fruits, seeds, nectar or plant sap is used by the herbivore itself for its own maintenance. Only a small fraction of the energy taken in by a herbivore is available to flow on to carnivores, and this is the energy that is stored in the muscle and other tissues of the herbivore. Where does this 'lost' energy go?

- Some is used by the herbivore for the energy-requiring processes that are part of being alive, such as active transport, muscle contraction, digestion, nerve transmission and hormone production. This can be described as the energy for 'staying alive and kicking' (see figure 14.15). In this use, chemical energy is converted to heat energy.
- Some is lost as organic matter in the faeces, urine and tissue that the herbivore egests, excretes or loses, for example, leaves that fall from mangrove trees, or skin cells shed by animals.

In beef cattle, for every 100 kilograms of pasture that they eat, they produce about four kilograms of meat.

Energy needs: mammals compared with others

Mammals, as well as birds, are warm-blooded (homeothermic) and keep their body temperatures within a narrow range. This is not the case for cold-blooded (poikilothermic) animals such as reptiles, fish and invertebrates like insects. As a result, more energy is needed by a mammal for heat production via cellular respiration than by an insect. Therefore, the production of new tissues from food by mammals, such as cattle and sheep, is less efficient than by fish, reptiles and insects.

A study of one mammalian herbivore showed that more than 30 per cent of the chemical energy in the ingested food is used via cellular respiration for 'staying alive' and so is finally lost as heat energy. More than 50 per cent of the energy in the food was lost in organic matter in faeces and urine. Only a small fraction of the food consumed appears as new tissues of the herbivore (see figure 14.16a).

In contrast, in insects, about 10 per cent of the chemical energy in the food they ingest will appear in insect tissues and be available for transfer to the next trophic level (see figure 14.16b).

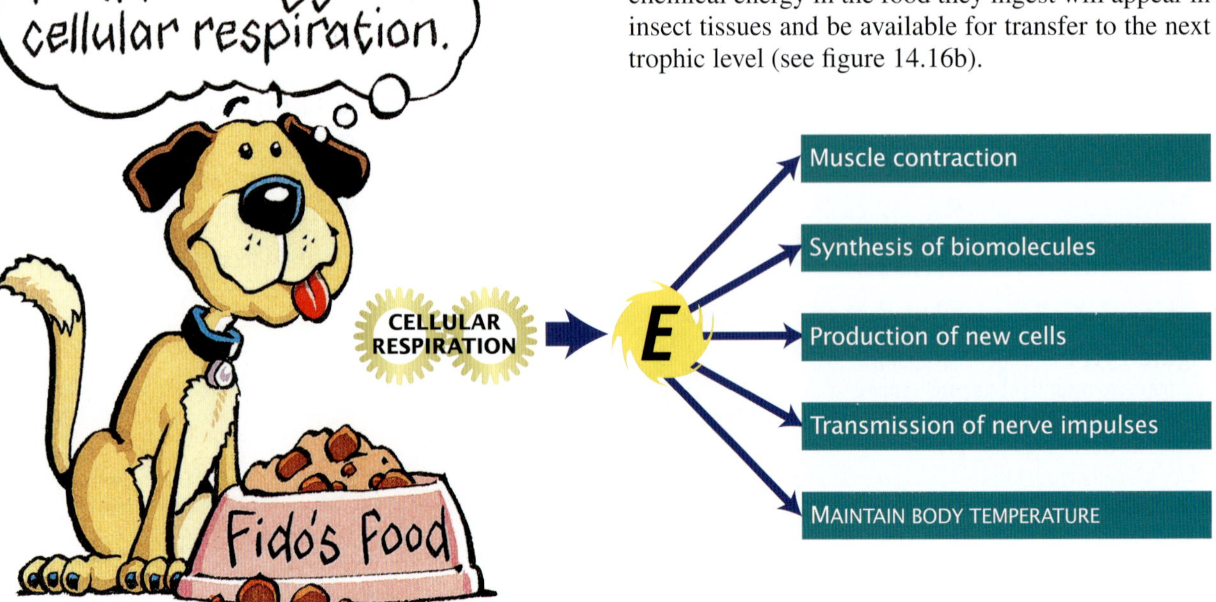

Figure 14.15 Mammals, including Fido, use energy from the food they ingest for many purposes. One key purpose is maintenance of body temperature within a narrow range.

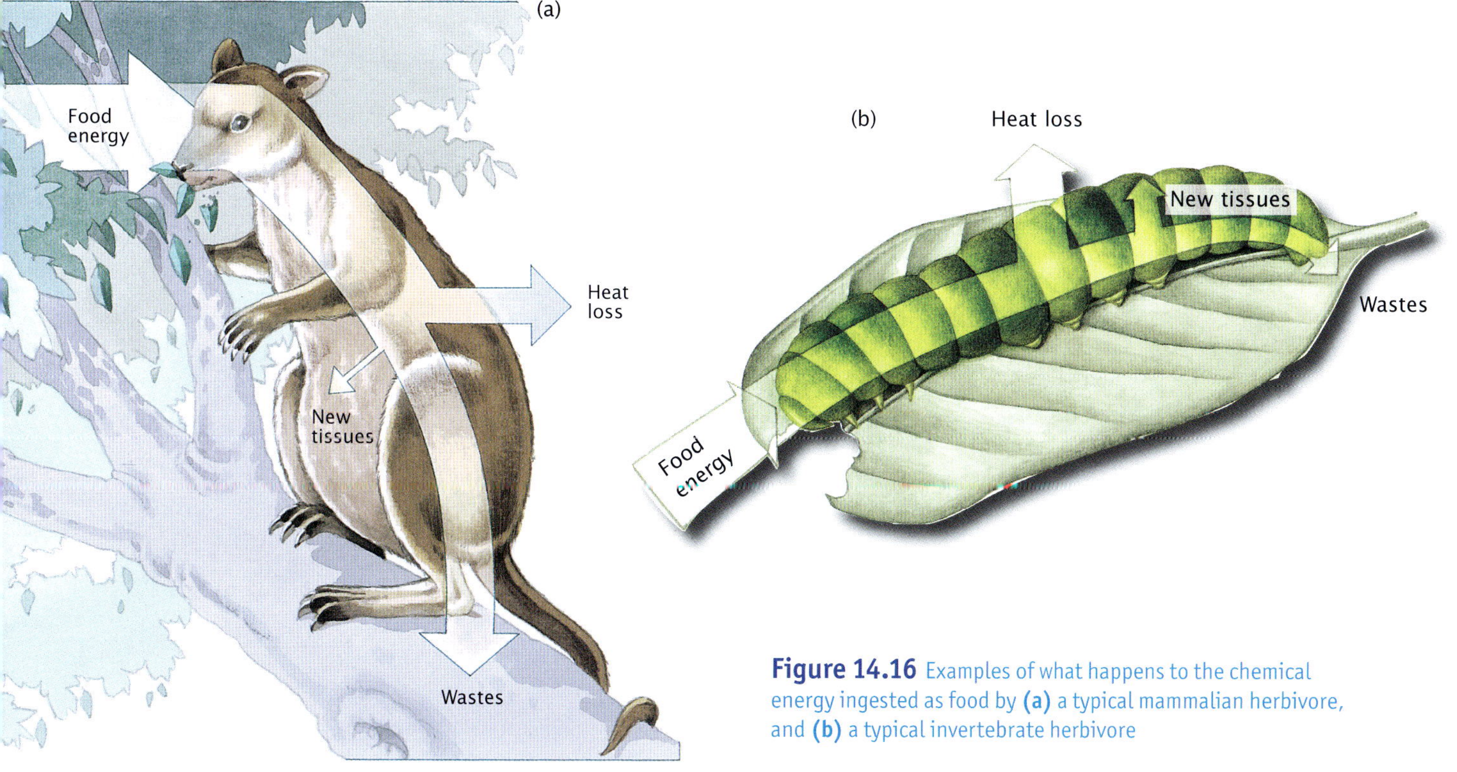

Figure 14.16 Examples of what happens to the chemical energy ingested as food by **(a)** a typical mammalian herbivore, and **(b)** a typical invertebrate herbivore

A rough rule-of-thumb

A rough 'rule-of-thumb' figure used by ecologists for the transfer of chemical energy between trophic levels is the 10-per-cent rule; that is, about 10 per cent of the energy going into one trophic level is available for transfer to the next trophic level in the form of organic matter in tissues for tissue production. We can show this by the following diagram (see figure 14.17).

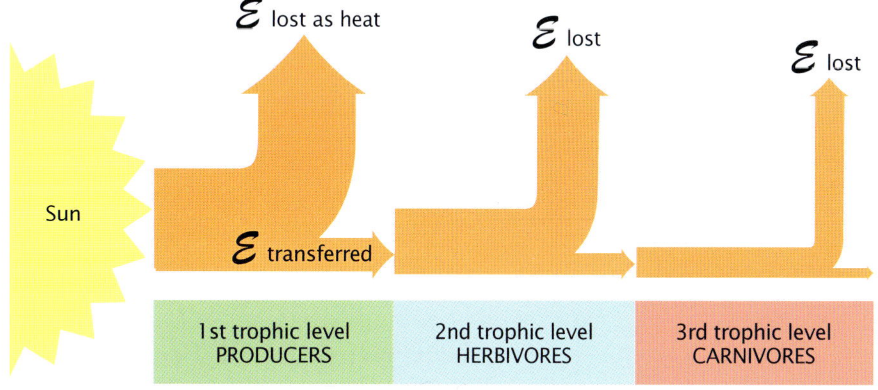

Figure 14.17 As chemical energy moves through organisms at different trophic levels in an ecosystem, some energy is 'lost' as heat energy. Note that the useful chemical energy leaving one trophic level is typically only 10 per cent of the energy entering that level.

Because loss of energy occurs at each trophic level, there is a limit to the number of transfers through trophic levels in an ecosystem.

We can use this rough rule-of-thumb to calculate the cost in energy terms of producing the organic matter in the tissues of organisms at different trophic levels. For example, we can estimate that 100 grams of plant material (leaves) can produce 10 grams of caterpillar tissue and that this amount of caterpillar tissue can produce just one gram of spider tissue. So, one unit of carnivorous spider tissue is 10 times more expensive to produce in energy terms than one unit of herbivorous caterpillar tissue.

WHAT IS THE ENERGY COST OF PRODUCING A CHICK?

There are significant challenges in answering the question, such as: How much chemical energy in food is required to rear an Adelie penguin chick from hatching to fledging? However, Australian scientists developed technology to help answer this question.

Adelie penguins (*Pygoscelis adeliae*) are the most common penguin species living in Antarctica. The major part of their diet is krill. It is important that the feeding behaviour of Adelie penguins is understood. To do this in a way that minimises the handling of penguins, a team led by Dr Knowles Kerry developed an automated penguin monitoring scheme (APMS) that is being used with a colony of Adelie penguins consisting of 1800 breeding pairs. The colony is located on Bechervaise Island near Mawson Station, Antarctica.

The APMS is an automated scheme that involves a weighbridge that penguins must cross when they enter or leave the colony (see figure 14.18). By measuring the mass of the penguin on exit from the colony and its mass on return from a feeding trip, an estimate can be made of the success of its feeding activity.

As well as recording their mass, the APMS also logs individual birds into and out of the colony. This outcome involves implanting an electronic identification tag under the skin along the lower backs of a selected sample of penguins. As the penguin crosses the weighbridge, a nearby antenna detects and records the specific tag. The direction that the penguin is travelling can be identified because the bird must cross two infra-red beams. The order in which the beams are cut identifies the direction.

Figure 14.18 An Adelie penguin returning to the colony via a weighbridge

By weighing particular penguins as they leave their breeding colony and as they return, it is possible to get an estimate of the mass of krill that these penguins are bringing back to the colony to feed their young. From these measurements, it was estimated that raising a penguin chick to fledging required 45 kg of food. The weight of an Adelie penguin at the time of fledging is about 3.1 kg.

Showing energy transfers

Chemical energy transfers in an ecosystem can be shown in various ways, such as:

- **food chains**
- **food webs**.

Food webs and food chains show the energy transfers or **energy flow** in an ecosystem by simply indicating who specifically eats whom. In both food chains and food webs, arrows show the direction of energy transfer, but neither shows the relative amounts of chemical energy at each transfer.

Showing energy flow: food chains and webs

The flow of chemical energy can involve more than two kinds of organism. In a mulga scrub ecosystem in the Flinders Ranges, part of the chemical energy present in the organic matter of grass is transferred to yellow-footed rock wallabies (*Petrogale xanthopus*) when they feed. In turn, some of this chemical energy is transferred to the wedge-tailed eagles (*Aquila audax*) that prey on young rock wallabies.

This one-way energy transfer can be shown as a simple diagram known as a food chain (see figure 14.19). Arrows show the direction of flow of chemical energy from the eaten to the eater.

Figure 14.19 A simple food chain in an ecosystem. What do the arrows denote?

The energy flow in an ecosystem is more complex than can be shown with a food chain. Food chains have certain limitations.

- Food chains may suggest that a particular kind of organism obtains its chemical energy from a single source, but this is not always the case. The flow of chemical energy to wedge-tailed eagles is not only from rock wallabies, but comes also from rabbits and small birds.
- Food chains may suggest that a particular kind of organism always occupies the same position in terms of energy flow. This is not always the case. Many organisms such as yellow-bellied gliders (*Petaurus australis*) obtain chemical energy from both producers (blossoms, sap, nectar) and consumers (insects and spiders).
- Food chains may suggest that chemical energy flows from a particular kind of organism to only one other kind of organism. This is not always the case. One kind of organism may be eaten by several other kinds. The chemical energy in grasses flows not only to rock wallabies, but also to feral goats and rabbits and many insect species in the same ecosystem.
- Food chains often do not show the energy flow from dead organisms, from parts of organisms or their waste products.

The flow of chemical energy in an ecosystem can also be shown using a representation known as a food web, such as the one provided in figure 14.20.

Figure 14.20 A food web showing the flow of chemical energy through the different kinds of organism in an ecosystem

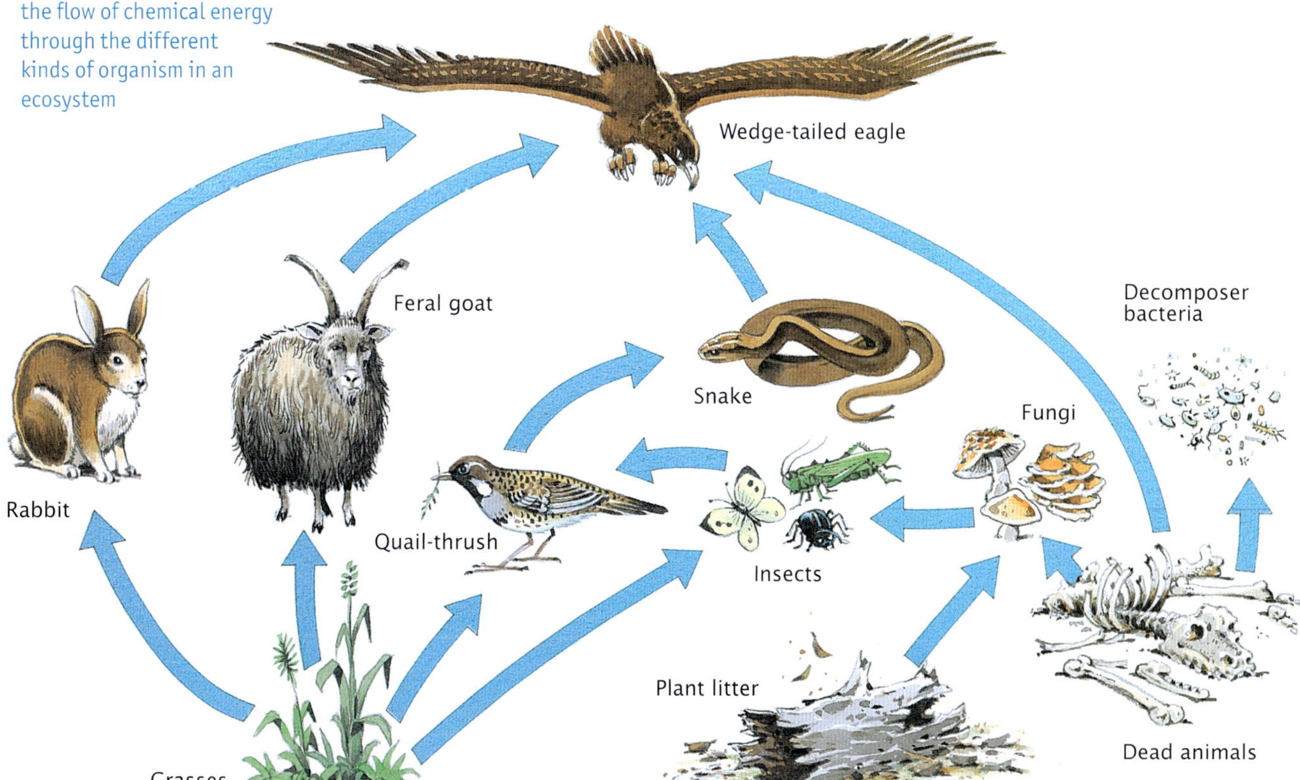

The producer organisms in an ecosystem are usually shown at the base of a food web diagram. Arrows show the direction of flow of chemical energy in an ecosystem from the eaten to the eater. In a food web, the flow of chemical energy from the organic matter of dead organisms can be included.

You will notice that a food web includes many food chains. Can you identify a food chain from grasses to wedge-tailed eagles through a primary and a secondary consumer?

Putting it together

Several important conclusions arise from the fact that energy is lost as heat energy at each trophic level in an ecosystem.

1. The number of trophic levels in ecosystems is limited, with many ecosystems having only three levels.
2. The higher the trophic level of organisms, the greater the energy cost of production of their organic matter. So, the production of carnivore organic matter requires more energy than the production of an equal amount of herbivore organic matter.
3. Energy must be continually supplied to an ecosystem, because it flows in a one-way direction and is not recycled.

People are primary consumers when they obtain chemical energy from eating cereal crops and are secondary consumers when they obtain chemical energy from eating beef. A consequence of conclusion 2 above is that larger human populations can be supported on cereal crops grown on a given area of land than can be supported on beef from cattle reared on this same area of crop.

KEY IDEAS

- Chemical energy from organisms at one trophic level passes to organisms at a higher trophic level, with loss of heat energy at each level.
- Only a small fraction of the energy taken in by consumers in their food appears as organic matter in their tissues.
- A rough 'rule-of-thumb' is that the fraction of chemical energy transferred between trophic levels is about one-tenth, or 10 per cent.
- Energy transfers within an ecosystem can be shown as food chains and food webs.

QUICK-CHECK

3 Identify the following as true or false.
  a Producers occupy the first trophic level in an ecosystem.
  b Energy is gained at each higher trophic level in an ecosystem.
  c Herbivorous insects produce a greater proportion of new tissue from their food compared with herbivorous mammals.

4 In an ecosystem, which energy flow would be greater?
  a energy flow from primary to secondary consumers
  b energy flow into producers

5 Identify two sources of 'lost' energy in a consumer organism.

6 Which has the higher energy cost of production: one gram of herbivore tissue or one gram of carnivore tissue?

Ecological pyramids

In the previous section, we saw that the flow of chemical energy through an ecosystem can be shown through the use of food chains and food webs. These representations do not give us any information about the numbers of different types of organisms involved. For example, the food web shown in figure 14.20 (page 451) gives no indication as to the number of grass plants or the number of wedge-tailed eagles in this ecosystem. We know that many producers (grasses) are needed to provide energy for just one herbivore and that many herbivores are needed to provide energy for just one carnivore.

Information about numbers of organisms at each trophic level and other aspects can be shown for an ecosystem through **ecological pyramids**. As you will see, some pyramids (numbers and biomass) can be far from a pyramid shape! In all the following cases, the pyramids are simplified by omitting detritivores and decomposers. (This has been done because detritivores and decomposers take in matter and energy from all trophic levels.)

The use of pyramids is a reminder that organisms at higher levels depend for their existence on energy flowing to them from organisms at lower levels. Remove the base of the pyramid and the structure is destroyed. Several kinds of pyramid can be drawn.

Pyramid of numbers

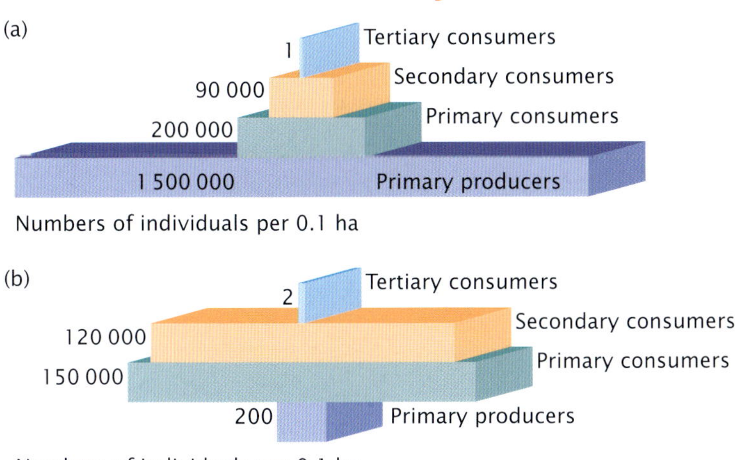

Figure 14.21 (a) Pyramid of numbers for an area of a grassland ecosystem (b) Pyramid of numbers for an area of woodland. Why is this pyramid 'inverted'?

A **pyramid of numbers** shows the number of organisms at each trophic level per unit area of an ecosystem. Numbers are denoted by the lengths of the horizontal bars. The lowest bar shows the number of producer organisms; the next bar shows the primary consumers, and so on (see figure 14.21a which shows an example for an area of a grassland ecosystem).

A pyramid of numbers does not take into account the relative sizes of the organisms. A forest ecosystem has a relatively small number of large trees that are the producers and a much larger number of primary consumers such as plant eating insects. When producers are much larger in relative size than consumers, an 'inverted' pyramid of numbers is produced (see figure 14.21b).

Pyramid of biomass

In an ecosystem, the total amount of matter present in organisms at each trophic level at a given time is referred to as **biomass**. (You can get a feel for the producer biomass in a terrestrial ecosystem by asking: 'How much green stuff can I see?'.) Producer biomass can be as a renewable fuel provided replanting and regrowth occurs to match the rate of harvesting.

Because biomass is an amount, it is measured in units of mass, such as kilograms or tonnes per unit area, and is typically measured after removal of water. (Why? Because water has no energy value.) Biomass is preferred to the use of numbers of organisms because individual organisms at one trophic level can vary greatly in size. It is not the size of an individual producer organism that matters; it is the total mass of producers in a given area that is important.

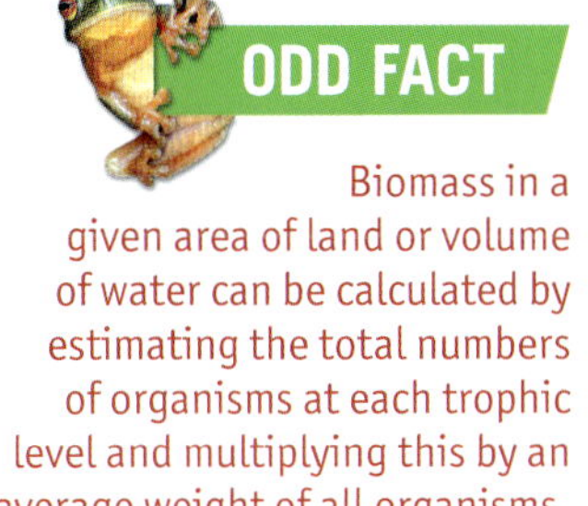

Biomass in a given area of land or volume of water can be calculated by estimating the total numbers of organisms at each trophic level and multiplying this by an average weight of all organisms.

A **pyramid of biomass** records the total dry organic matter of organisms (biomass) at each trophic level in a given area of an ecosystem (see figure 14.22a). This type of pyramid does not take into account the time during which the biomass of organisms at different trophic levels has been built up. For example, long-lived herbivorous fish in an ocean ecosystem build up their biomass over a period of several years. In contrast, short-lived algae in the same ecosystem have only a few weeks to build up their biomass before they die. As a result, if the biomass in an ecosystem is sampled over a short period, a pyramid of biomass can be inverted (see figure 14.22b).

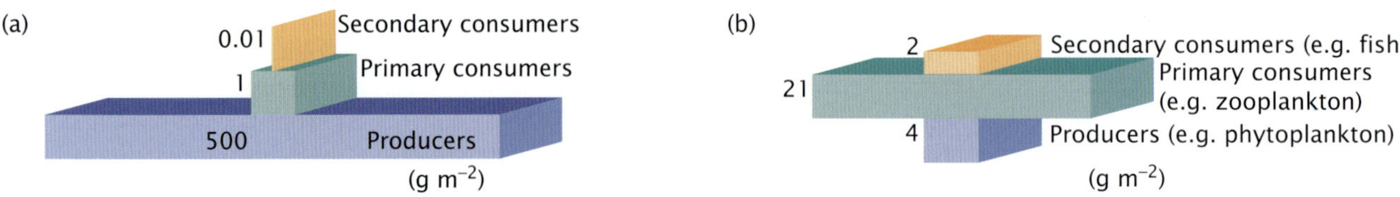

Figure 14.22 **(a)** Pyramid of biomass for an area in a grassland ecosystem. This is a typical pyramid of biomass. Which trophic level has the greatest dry mass of organisms? **(b)** Pyramid of biomass for an ocean ecosystem sampled over a short time period. Why is this described as an 'inverted' pyramid?

Pyramid of energy

A **pyramid of energy** shows the amount of energy input to each trophic level in a given area of an ecosystem over an extended period, often one year (see figure 14.23). Because the flow of energy through all ecosystems is reduced as energy flows from one trophic level to the next, pyramids of energy are never inverted. In a pyramid of energy, the length of the bars denotes the chemical energy at each trophic level. The total length of a bar at a lower trophic level must be greater than the lengths at higher trophic levels. Why?

Figure 14.23 Pyramid of energy for a river ecosystem

KEY IDEAS

- Trophic levels in an ecosystem can be represented by ecological pyramids.
- A pyramid of numbers shows the numbers of organisms at each trophic level.
- A pyramid of biomass shows the total dry organic matter at each trophic level.
- A pyramid of energy shows the amount of chemical energy present at each trophic level.

QUICK-CHECK

7 Identify the following as true or false.
   a A pyramid of numbers denotes the kinds of producers in an ecosystem.
   b In a pyramid of biomass, the length of the bars denotes the biomass at each trophic level.
   c A pyramid of energy cannot be inverted.
8 Consider a mountain ash tree and the consumers that live on it as an ecosystem. What would be distinctive about the pyramid of numbers for this ecosystem?

Ecosystems differ in productivity

As well as looking at ecosystems in terms of food chains and food webs, or in terms of ecological pyramids, we can also look at ecosystems in terms of their **productivity**. Productivity is the rate at which chemical energy is produced in an ecosystem. Productivity is identified through questions such as:

- How fast do the producers in an ecosystem make chemical energy?
- What is the rate of production of organic matter by the producers?

The answer is a rate that is expressed in units such as 'gram of organic matter per square metre per year'.

Production: gross and net

In an ecosystem, the chemical energy produced in the form of organic matter by photosynthesis over a given period of time is termed the **gross primary production (GPP)**. However, some of this organic matter is used by the producers themselves as part of their own cellular respiration for the processes of being alive; we can show this fraction as R. What is left after this is called **net primary production (NPP)**. The NPP is the amount of chemical energy available to the consumers in an ecosystem. GPP and NPP are expressed in units such as 'mass of organic matter per unit area per unit time'.

We can show this by the following equation:

$$NPP = GPP - R.$$

In general, ecosystems with the highest net productivity are those with the greatest accumulated producer biomass:

More producer biomass → more photosynthesis → more chemical energy stored in organic matter

The rate of primary production of organic matter through photosynthesis by producers varies in different ecosystems, as shown in figure 14.24. The rate of 'food' production, or primary productivity, is expressed as chemical energy/square metre/year. Among terrestrial ecosystems, tropical rainforests have very high production rates and, for aquatic ecosystems, coral reefs have very high productivity.

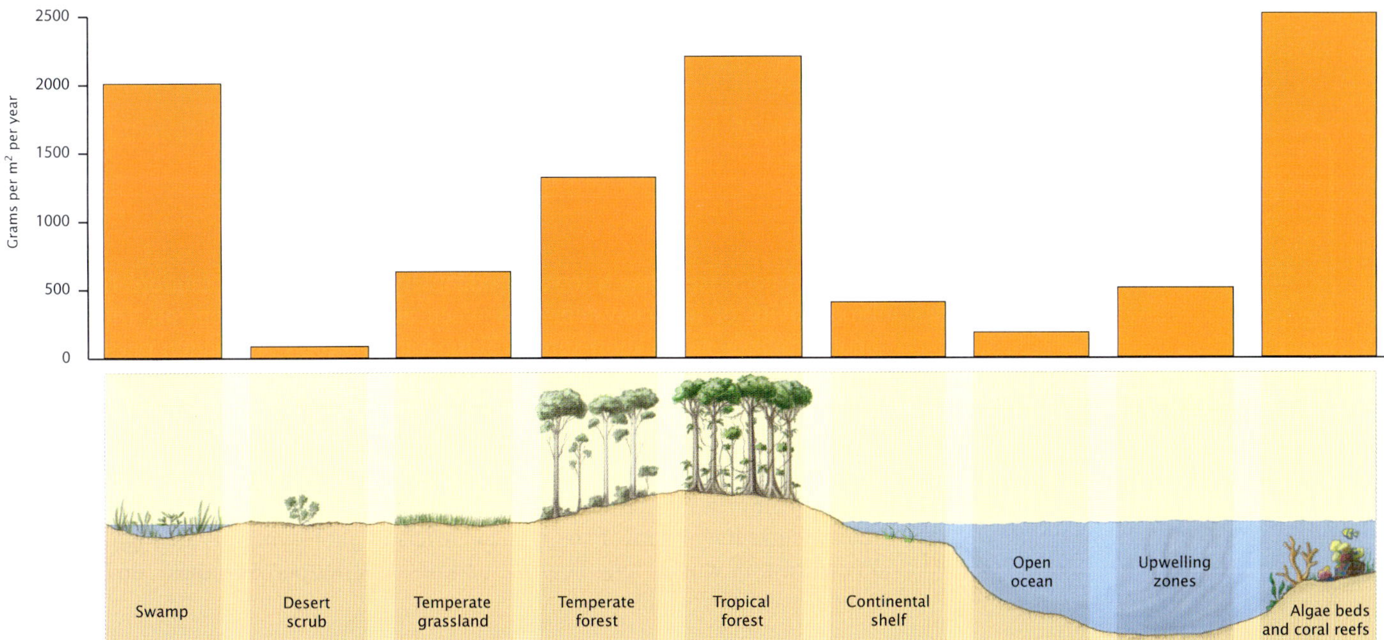

Figure 14.24 Average rates of production of organic matter by producers in various ecosystems (chemical energy (grams) produced per square metre per year)

Upwellings are not present in waters around the Australian continent. As a result, the nutrient levels in ocean waters around Australia are low to moderate. What impact would this have on fisheries in our waters?

The higher the net primary production of an ecosystem, the more food is available for consumer organisms and the greater the biomass of consumers living in that ecosystem. Which ecosystem would be expected to have the smaller mass of consumer organisms per unit area: temperate forest or desert scrub?

Shallow seas above a continental shelf have higher concentrations of producer organisms and a higher primary productivity than open ocean waters. Coastal seas can support larger quantities of consumer organisms, such as fish, than an equal volume of ocean water.

In general, richer fishing grounds are expected in seas over the continental shelf rather than in the open ocean. Among the richest fishing grounds in the world is an area off the west coast of South America where the water has very high mineral nutrient levels because it is a region where nutrient-rich dense cold water from the depths comes to the surface in a large-scale movement known as **upwelling**. Because of this upwelling, this region of the Pacific Ocean supports rich growths of phytoplankton, has a high production rate and the net primary productivity supports the world's richest anchovy fisheries. Not surprisingly, these waters are also the feeding grounds of large numbers of birds that feed on these fish (see figure 14.25).

Figure 14.25 A region of coastline with rich bird life close to an ocean upwelling creates high mineral nutrient levels in the water.

Why do production rates differ?

A desert ecosystem has a lower productivity than a tropical rainforest. Productivity of an ecosystem falls when any of the requirements for photosynthesis are in short supply. Temperature affects production rates or productivity; for example, a forest ecosystem in a colder climate has a lower productivity than a forest ecosystem at warmer temperatures (see figure 14.26). Productivity in ecosystems at high latitudes is also influenced by the season, such as a tundra ecosystem in the northern hemisphere or a marine Antarctic ecosystem. In summer, these ecosystems receive long hours of sunlight and levels of photosynthesis are high. In contrast, in winter, long hours of darkness restrict photosynthesis and hence productivity falls.

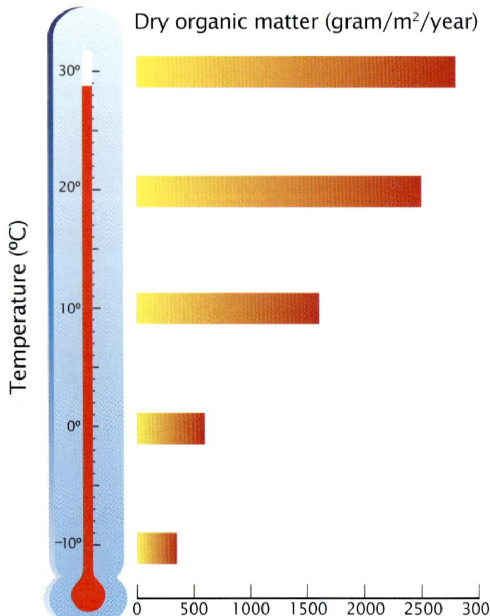

Figure 14.26 Productivity for forests at different air temperatures. What happens to forest ecosystem productivity as temperature rises?

The rate of photosynthesis is controlled by the availability of the raw materials needed for photosynthesis — carbon dioxide and water — and by the intensity of light and the availability of chlorophyll. If one of these factors is limited in its availability relative to the others, that factor is a **limiting factor**. Table 14.2 identifies factors that restrict the rate of photosynthesis in several ecosystems.

Table 14.2 Some limiting factors for photosynthesis in various ecosystems

Ecosystem	*Limiting factor*
tropical rainforest	availability of light
desert scrub	availability of water

There is plenty of water and dissolved carbon dioxide in open ocean ecosystems. So, what limits photosynthesis and reduces productivity in the ocean as compared with coastal waters? There are fewer phytoplankton in the open ocean than in coastal waters. Why?

In 1996, an experiment was carried out in which a 72 square kilometre patch of ocean with a low concentration of phytoplankton was seeded with iron. The result was a 20-fold increase in the concentration of phytoplankton. Would this increased growth rate have been expected to continue over time?

Mineral nutrients are needed for growth of phytoplankton. In coastal seas, mineral nutrients are supplied constantly by rivers draining from the land. In open ocean waters far from land, however, mineral nutrients are not available from this source. Another source of mineral nutrients is from the decomposition of dead organisms. Again, coastal waters are provided with mineral nutrients from this source. In an open ocean ecosystem, however, dead organisms drift down to the ocean floor many kilometres below the surface and are lost from the surface ecosystem. At the ocean floor, dead organisms are consumed by bottom-dwelling scavengers and decomposed by bacteria, and the mineral nutrients that are released from this decomposition remain trapped for long periods in the cold, mineral-rich waters at the ocean depths. Can you understand why regions of upwelling (refer back to page 456) are rich in phytoplankton?

Because Australia is an arid country, with ancient and weathered soils, the water run-off from the land into the sea is not high in mineral nutrients, such as nitrates (see figure 14.27). This lower mineral nutrient level means that the concentration of phytoplankton in Australian coastal waters is not as high as in seas surrounding parts of other continents. As a result, Australia does not have rich fishing grounds near the coast as do some other countries.

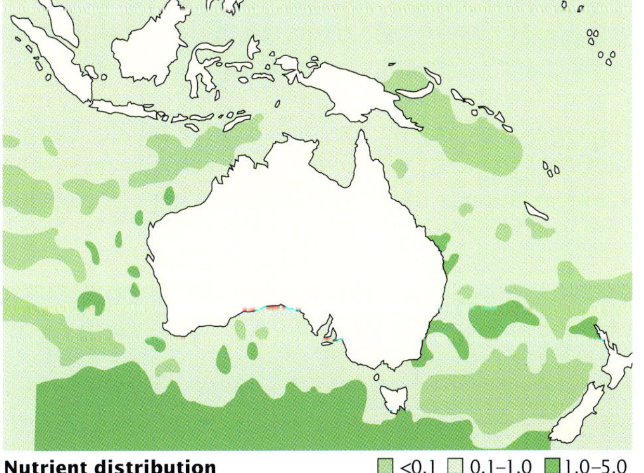

Nutrient distribution
nitrate concentration (µg-atom per litre)
<0.1 0.1–1.0 1.0–5.0

Figure 14.27 Nitrate concentrations in seas and oceans surrounding Australia. Nitrate is needed for the growth of phytoplankton, algae and plants. Do major upwellings occur in Australian waters?

KEY IDEAS

- Different ecosystems vary in their productivity (production rates) because of factors that limit photosynthesis.
- Factors that limit primary productivity in ecosystems include temperature, sunlight, nutrients and, in terrestrial ecosystems, water.
- The total chemical energy produced by producers in an ecosystem is the gross primary productivity.
- Chemical energy in organic matter stored by producers is the net primary production of an ecosystem.

QUICK-CHECK

9 Identify the following as true or false.
   a Net primary production in an ecosystem is greater than gross production.
   b Primary productivity refers to the rate of production of chemical energy by producers.
   c NPP is greater in forests than in desert scrubland.

10 Which of GPP and NPP supports the consumers in an ecosystem?

11 Give an example of a limiting factor for primary productivity in:
   a desert ecosystems
   b tropical rainforest ecosystems.

12 Explain the equation: NPP = GPP – R.

The euphotic or 'lit' waters of the ocean extend from the surface to about 200 metres. Below that depth, no sunlight penetrates and from there to about 1500 metres is a twilight zone. From there to the ocean floor is the abyssal zone of total darkness (see below).

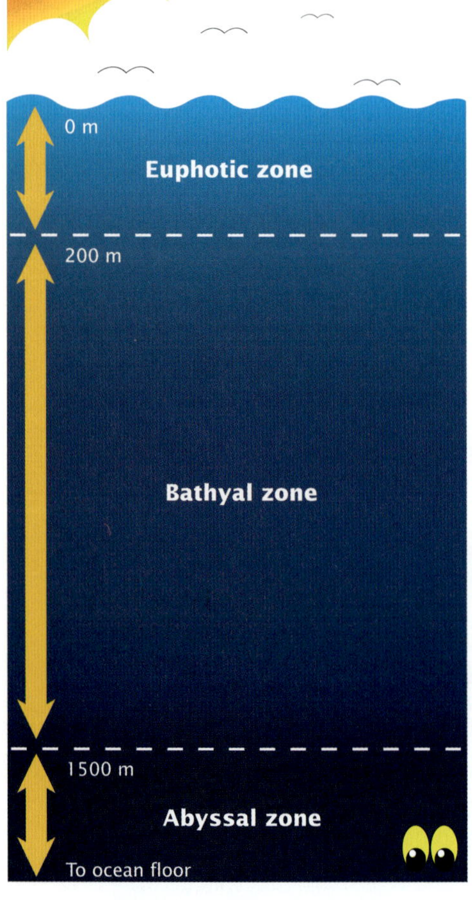

Figure 14.28

Ecosystems in darkness

In the dark water column of the deep ocean, dead organisms and detritus drift down from more shallow waters that receive sunlight. These 'dark' waters depend on the radiant energy of sunlight since the dead organisms and detritus originate directly or indirectly from sunlight. There are no producer organisms in these dark waters.

Not surprisingly, fish and other consumers in these dark waters are few and far between. This poses a problem for sexual reproduction in dioecious fish since the chance of the separate sexes meeting by chance is very low. However, when females of these deep-sea fish species are ready to mate, they release pheromones that create a chemical pathway along which males can travel to them. The triplewart sea devil (*Cryptopsaras couesi*) (see figure 14.29), one of the deep sea anglerfish, shows an even more interesting reproductive adaptation. In this species, the male is very much smaller than the female. Males consist of little more than a pair of mobile testes plus an olfactory organ and a mouth. Once a male reaches a female, it attaches to her by his mouth and eventually fuses with her, becoming an ectoparasite. At mating time, the female and her ectoparasitic male release eggs and sperm synchronously.

Figure 14.29 A female triplewart sea devil. A tiny male attaches to her side and is dependent on the female for nutrition. What function does the male serve? Are anglerfish herbivorous or carnivorous? Explain.

Refer back to page 407 to revise hydrothermal vents.

Producers at the ocean depths

For almost all ecosystems, sunlight is the external energy source that is captured by producers and transformed by photosynthesis to chemical energy stored in organic matter. However, hydrothermal vents — ecosystems on the ocean floor — receive no sunlight but support a dense community of consumer organisms. What are the producers in thriving hydrothermal vent ecosystems in complete darkness on the ocean floor?

The living community in a hydrothermal vent ecosystem is too large to depend on detritus drifting down from surface waters. Hydrothermal vent ecosystems have no green plants, no algae and no photosynthetic bacteria to produce 'food' by photosynthesis. There must be producer organisms to capture energy, bring it into the ecosystem and make it available to the hydrothermal vent community. Who are the producers in this ecosystem of perpetual darkness? Special autotrophic bacteria — let's meet them!

Chemosynthetic bacteria capture chemical energy

The essence of photosynthesis is the use of the sunlight energy to produce organic molecules, such as glucose, that store chemical energy. Organisms that do this are said to be **photosynthetic**. In contrast, some bacteria can also produce organic molecules, such as glucose, but do *not* use the energy of sunlight. In the case of the hydrothermal vent, the energy source for these autotrophic bacteria is the chemical energy present in molecules such as hydrogen sulfide (H_2S). Hydrogen sulfide occurs in high concentrations in the waters around a hydrothermal vent and chemical energy is released when this compound reacts with oxygen.

Bacteria that can produce organic matter using the energy released from certain chemical reactions are said to be **chemosynthetic**. The producers in the hydrothermal vent ecosystem are chemosynthetic bacteria such as sulfur bacteria, *Thiobacillus* spp. In the hydrothermal vent ecosystem, all consumer organisms feed directly or indirectly on organic matter produced by the chemosynthetic bacteria and obtain their energy from the chemical energy stored in this organic matter. Table 14.3 compares the processes of photosynthesis and chemosynthesis.

Figure 14.30 Some bacteria use inorganic matter as their source of energy. Because of their unusual 'food', these kinds of bacteria are able to survive in places where other kinds of living things cannot survive. They are said to be *chemo*synthetic, meaning that they use energy from chemical reactions to *synthesise* organic matter from inorganic substances.

Table 14.3 Comparison of the processes of photosynthesis and chemosynthesis

	PHOTOsynthesis	*CHEMOsynthesis*
Energy source	sunlight	chemical energy
Source of carbon atoms	carbon dioxide	carbon dioxide
Source of hydrogen	water (H_2O)	hydrogen sulfide (H_2S)
End product	glucose	glucose
Waste product	oxygen	sulfur

KEY IDEAS

- Some deep ocean ecosystems survive on organic matter that drifts down from the organisms living in surface waters.
- Sustainable ecosystems in permanent total darkness have producer organisms that are chemosynthetic rather than photosynthetic.
- Chemosynthetic producers include certain bacteria.
- Chemosynthetic bacteria use the energy from chemical reactions with hydrogen sulfide to build organic matter from carbon dioxide.

QUICK-CHECK

13 Identify the following as true or false.
   a Every community in a sustainable ecosystem must include some producer organisms.
   b Chemosynthetic bacteria can carry out the process of photosynthesis.
   c Hydrothermal vent ecosystems rely on chemosynthetic bacteria to bring energy into the ecosystem in the form of organic matter.

14 From where do the hydrothermal bacteria obtain energy to make organic matter?

15 What is the source of carbon atoms in both photosynthesis and chemosynthesis?

Matter cycles within ecosystems

In any ecosystem, the various organisms depend for survival on other organisms and on their surroundings. Organisms obtain the matter that they need to build their organic substance from other organisms and from their surroundings. Unlike energy, matter cycles within any ecosystem and is reused.

The less visible living things in ecosystems, including detritivores and decomposers, such as fungi and some bacteria, play an important role in the recycling of matter and in converting organic matter to simple inorganic matter that can be taken up by producers and rebuilt into new organic substances.

Matter, such as carbon, nitrogen and phosphorus, continually cycles within an ecosystem. As these elements cycle, they are sometimes found in the living component of an ecosystem, such as producers or consumers. At other times, these elements are in non-living components, such as in rocks, in the soil or dissolved in water. We can show this cycling of matter using a simple diagram (see figure 14.31).

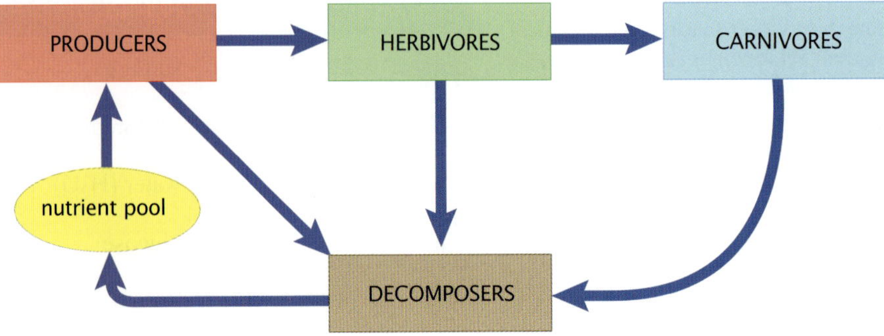

Figure 14.31 Cycling of matter within an ecosystem. What group of organisms are essential to ensure recycling of matter from dead organisms or from parts of organisms that are shed, for example, leaves?

As well as cycling within an ecosystem, matter can also enter or leave an ecosystem. Matter enters an ecosystem, for example, when animals migrate into a region or when leaves or soil are blown into a region. Matter leaves an ecosystem when seeds or larvae and young disperse, when animals emigrate from a region and when soil is removed by water or wind erosion.

Cycling of elements means that the same atoms are reused and, over time, are incorporated into different molecules. In each of these cycles, the elements in question undergo chemical change during their cycling as the element becomes part of different chemical compounds. This cycling also means that, over time, elements are exchanged between living and non-living parts of an ecosystem. For example:

- A carbon atom in a glucose molecule in your blood can later appear in the carbon dioxide that you breathe out into the atmosphere.
- This carbon dioxide molecule could stay in the atmosphere for some time, drift across the globe, be taken up by a plant growing in New Zealand and incorporated into glucose through photosynthesis.
- Later, this carbon atom may become part of a starch molecule in the root of this plant. A hedgehog that digs up this plant may eat the starchy root and, following digestion, this carbon atom could become part of a protein molecule in that herbivore.
- Later, if that herbivore is killed and eaten by a carnivorous stoat, the carbon atom then becomes part of the protein structure of this carnivore.
- When the stoat dies and decomposers act on the organic matter in its carcass, the carbon atom returns to the atmosphere as part of a molecule of carbon dioxide produced by a decomposer.
- The carbon atom at some stage may be incorporated into the calcium carbonate of the shell of a marine invertebrate. When that animal dies, the shell (and many others) may become compressed and form limestone rock.

Biogeochemical cycles

Matter cycles through ecosystems. In this section, we explore the cycling of three elements: carbon, nitrogen and phosphorus. These cycles are termed **biogeochemical cycles**. Why?

- The item that cycles is a chemical element; hence *chemical*.
- For some of the cycle, the chemical is in organisms or biological material; hence *bio*.
- For some of the cycle, the chemical is present in non-living parts of an ecosystem, for example, geological structures such as soil or rocks; hence *geo*.

Biogeochemical cycles are of two types. In one type of cycle, the element enters the atmosphere at some point, with the atmosphere acting as the major reservoir or store of the element in question. As a result, these cycles are not confined to one ecosystem but can spread more globally. Examples of this type of cycle are the **carbon cycle** and the **nitrogen cycle**. As a result, carbon dioxide emitted in the exhaust of cars or in the emissions from coal-burning factories in one region can spread to other regions via the atmosphere.

In a second type of cycle, the element does not enter the atmosphere at any stage and, as a result, these cycles are more localised. Soil or rocks act as the major store of the chemical. An example of this type of cycle is the phosphorus cycle.

Cycling of carbon atoms

Carbon atoms are recycled through the living and non-living parts of an ecosystem. Producers take up carbon dioxide and build this into organic molecules (sugars and starches).

This organic matter is eaten by consumer organisms and some carbon atoms are released as inorganic carbon dioxide when consumers break down sugars during cellular respiration. Carbon dioxide is also released when detritivores and decomposers consume and use the organic carbon in dead organic matter.

The carbon cycle in a terrestrial ecosystem is shown in figure 14.32.

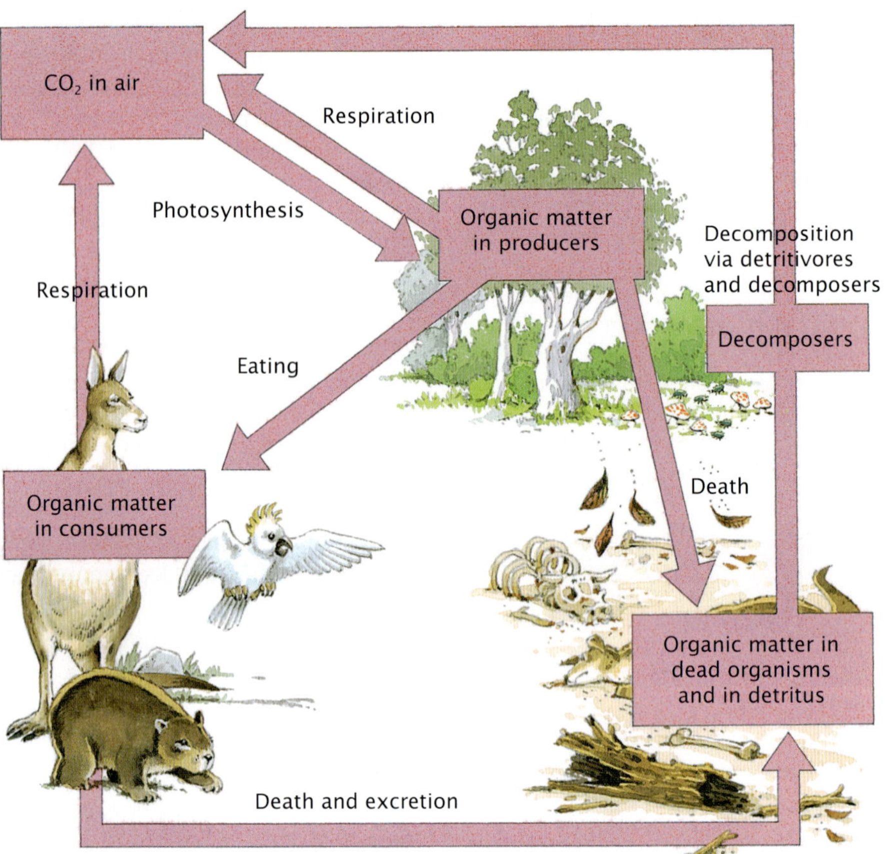

Figure 14.32 Carbon atoms cycle within an ecosystem. Identify one movement of carbon atoms from the inorganic to the organic pool.

Important decomposers that play a role in this and other cycles are fungi. The box on page 467 tells about a botanist who works with fungi.

For part of the cycle, the carbon is present in organic matter in living members of the ecosystem, such as the plants and the animals. The carbon moves to the atmosphere as inorganic carbon dioxide as a result of cellular respiration. The action of decomposers on the organic matter from dead organisms also releases carbon back to the atmosphere as carbon dioxide. Carbon is taken up from the atmosphere by plants by photosynthesis. Because carbon enters the atmosphere, it can move through the atmosphere on a global scale.

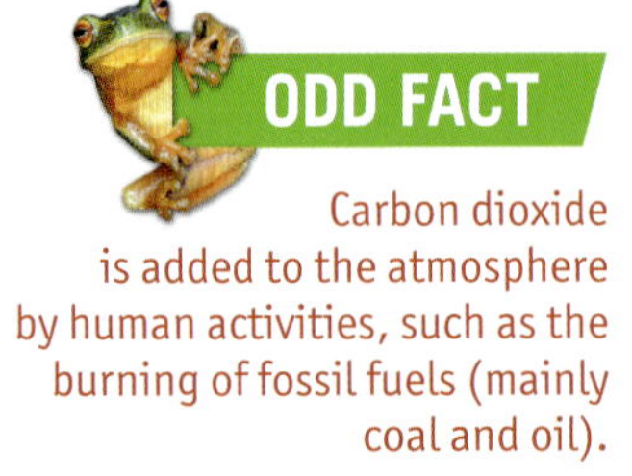

Carbon dioxide is added to the atmosphere by human activities, such as the burning of fossil fuels (mainly coal and oil).

Cycling of nitrogen atoms

Figure 14.33 shows a simplified version of the nitrogen cycle in a terrestrial ecosystem. Nitrogen atoms can be present in inorganic matter such as nitrogen gas in the atmosphere and as nitrate, nitrite and ammonium ions in the soil and water. Nitrogen atoms are also present in organic matter, mainly in the proteins and amino acids of living things and in waste products, such as urea.

Only a few kinds of organism can make use of the nitrogen gas present in the atmosphere, and these organisms include so-called nitrogen-fixing bacteria, such as *Azotobacter* which lives in the soil; *Rhizobium*, which is present in nodules on the roots of plants such as clovers, peas and beans, as well as native species including wattles; and *Frankia*, which lives in nodules on the roots of native she-oaks of the genus *Casuarina*. Nitrogen-fixing bacteria can shift nitrogen from the atmosphere to the soil by converting nitrogen gas to ammonium ions.

Other kinds of bacteria also contribute to the cycling of nitrogen in an ecosystem (see table 14.4).

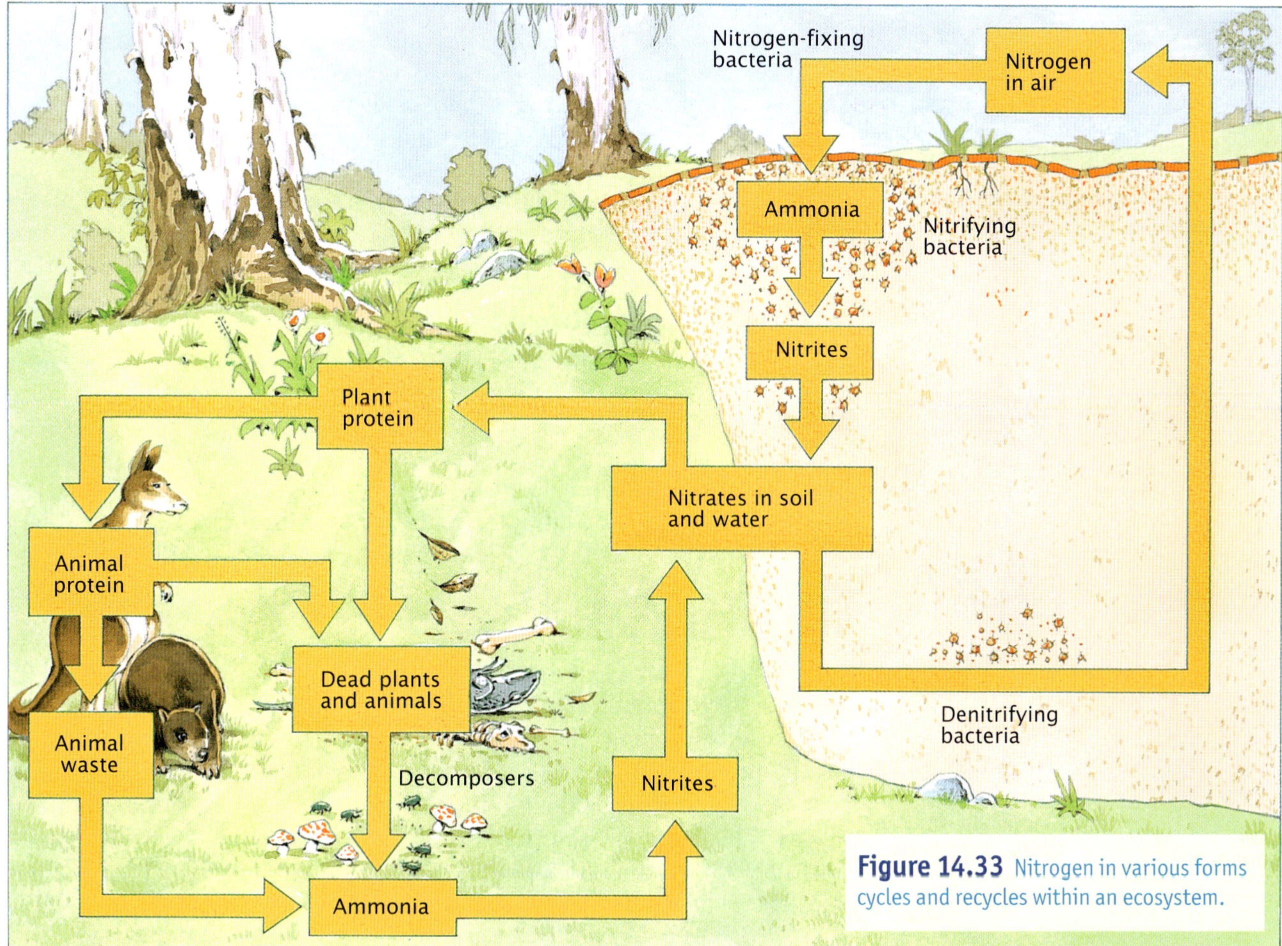

Figure 14.33 Nitrogen in various forms cycles and recycles within an ecosystem.

Table 14.4 Bacteria involved in nitrogen cycling

Bacteria	*Role in nitrogen cycling*
nitrogen-fixing bacteria	convert atmospheric nitrogen (N_2) to ammonium ions (NH_4^+)
nitrifying bacteria	oxidise ammonium ions to produce nitrates (NO_3^-) and nitrites (NO_2^-)
denitrifying bacteria	convert nitrates and nitrites to gaseous nitrogen
decomposer bacteria	convert organic nitrogen to inorganic form

Through the action of nitrogen-fixing bacteria, nitrifying bacteria and decomposer bacteria, nitrogen is made available in a usable form for plants. Nitrifying bacteria flourish best in well-aerated soils. Why? The decomposer bacteria break down the organic nitrogen in dead organisms to humus in the soil and then to ammonium ions that can be taken up by plants. The action of denitrifying bacteria, which thrive in anaerobic conditions (where oxygen is absent), as in waterlogged soils, causes a loss of inorganic nitrogen from the soil and a loss of soil fertility.

Plants take up inorganic nitrogen in the form of ammonium or nitrate ions from soil or water and build it into organic nitrogen present in the amino acids of their proteins. The role of bacteria in producing these forms of nitrogen is critical for the producers in an ecosystem. Plant proteins are eaten in turn by consumers who use the nitrogen-containing amino acids from the plants to build their own proteins. When living things die, their organic nitrogen is broken down by decomposers and returns to the soil. Nitrogen is also returned to the non-living part of the ecosystem when animals excrete nitrogen wastes in urine.

ODD FACT

In many countries, rice crops receive no artificial nitrogen fertilisers. Nitrogen in usable form is made available to the growing rice plants by the action of nitrogen-fixing cyanobacteria that grow in the rice paddies.

In natural ecosystems, recycling of nitrogen atoms from organic to inorganic form occurs over time and, as organisms die, their stores of organic nitrogen are returned to the soil as inorganic nitrogen through the action of decomposers.

The flow of nitrogen is altered by human intervention, such as the harvesting of plant crops. Plants take up nitrates from the soil and incorporate them into organic nitrogen present in the proteins in the plant tissues. When plants are harvested, this organic nitrogen is removed from the area and so is not available for recycling. Over a period of time, constant cropping causes soil to become impoverished in mineral nutrients, such as nitrates. The addition of fertiliser can replace this loss.

Other mineral nutrients are also removed from an ecosystem by harvesting. For example, one study showed that the logging of trees produced a nutrient loss from each hectare of soil, including:

- phosphorus (P) 1 kilogram
- magnesium (Mg) 20 kilograms
- calcium (Ca) 90 kilograms.

These elements were taken up from the soil by the growing trees and built into their organic matter.

ODD FACT

Phosphorus is a limiting factor for plant growth in many ecosystems because phosphorus-containing compounds have a low solubility in water. In agriculture, use is made of phosphate fertilisers in order to overcome this limitation.

Cycling of phosphorus atoms

Phosphorus (P) is found in the living components of an ecosystem; it is an essential nutrient for plants and animals and is found in many biological compounds in all living cells, including ATP, nucleic acids (DNA and RNA). Phosphorus also occurs in the plasma membranes of plant and animal cells and is an important component of the teeth and bones of animals. In the non-living components of ecosystems, phosphorus occurs as minerals in rocks, and as dissolved phosphate in soils and rivers. In contrast to carbon and nitrogen, the **phosphorus cycle** does *not* include cycling through the atmosphere in gaseous form. Figure 14.34 shows the phosphorus cycle.

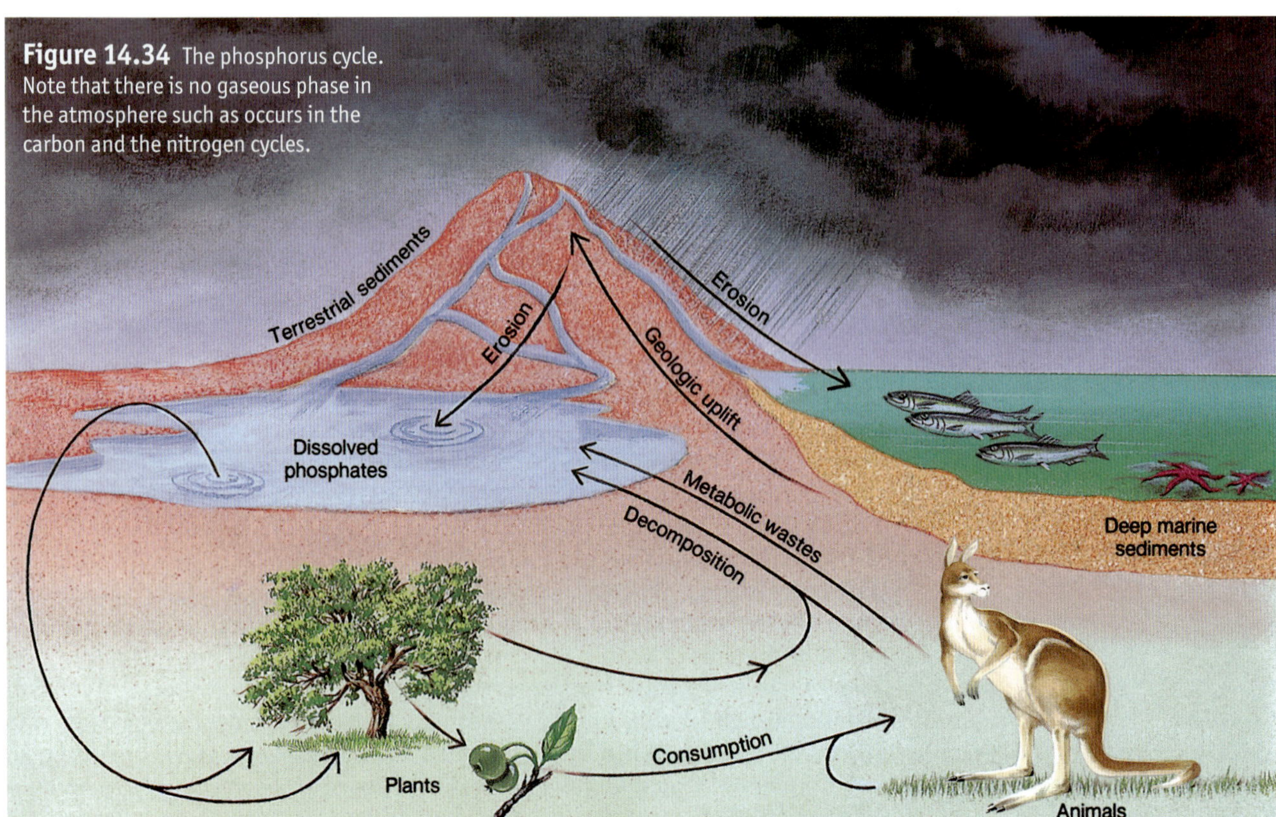

Figure 14.34 The phosphorus cycle. Note that there is no gaseous phase in the atmosphere such as occurs in the carbon and the nitrogen cycles.

The major store of phosphorus is P-containing rocks and sediments and P is released from these rocks by erosion. Producer organisms, such as plants, take up phosphorus in the form of dissolved phosphates from the soil in the case of terrestrial ecosystems or from the waters of lakes and seas in the case of aquatic ecosystems.

Phosphorus moves from plants to animals when consumers eat plants or when animals eat herbivores. Phosphorus is returned to the non-living component of an ecosystem when dead plants and animals decay through the action of decomposers. Phosphorus is also returned to the non-living component of an ecosystem through the release of animal wastes. The phosphorus may then be taken up again by plants or it may be washed away into rivers and seas and settle on the ocean floor to form new phosphorus-containing sediments.

ODD FACT

Guano was mined for use as a phosphate fertiliser from Pacific islands such as Nauru.

Waste products of fish-eating birds are rich in phosphates. Many generations of sea birds nesting on the same sites on oceanic islands produced these wastes that accumulated into a hard substance known as **guano**.

The water cycle

As well as the cycling of elements, water also cycles within ecosystems. In contrast to the carbon and nitrogen cycles, water does not change its chemical form during its cycling. It changes, however, in its physical state during this cycling — it can be in liquid or in vapour form and, in some ecosystems, can be in solid form (glaciers, ice, snow).

ODD FACT

Water vapour in the atmosphere remains there on average for about 10 days, while water in glaciers can be held as solid ice for thousands of years.

Processes involved in the **water cycle** include the biological process of transpiration and the physical processes of evaporation, precipitation, surface run-off and seepage or percolation of water into the soil and deeper into the ground water store (see figure 14.35).

The energy source for the cycling of water is the sun. Solar energy evaporates water, mainly from the ocean surface but also from the surface of lakes, streams and the land. Solar energy also removes water vapour from plant surfaces, a process known as transpiration. The major store of water is the oceans.

As well as recycling water within ecosystems, parts of the water cycle play other roles, such as erosion, run-off of nutrients from soils to streams and dissolving mineral nutrients.

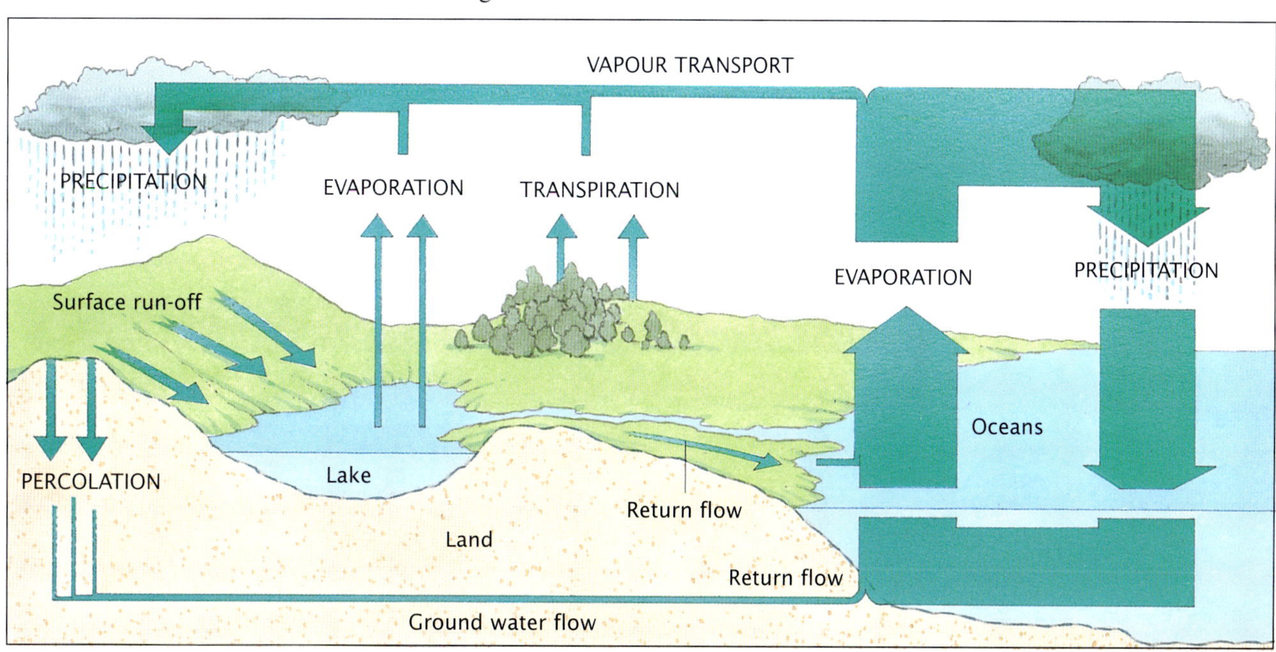

Figure 14.35 Water cycles through ecosystems. The wider the arrow, the greater the amount of water that moves through that part of the cycle. Is the movement of water by transpiration greater or less than the movement by evaporation from the ocean surface?

Bioaccumulation

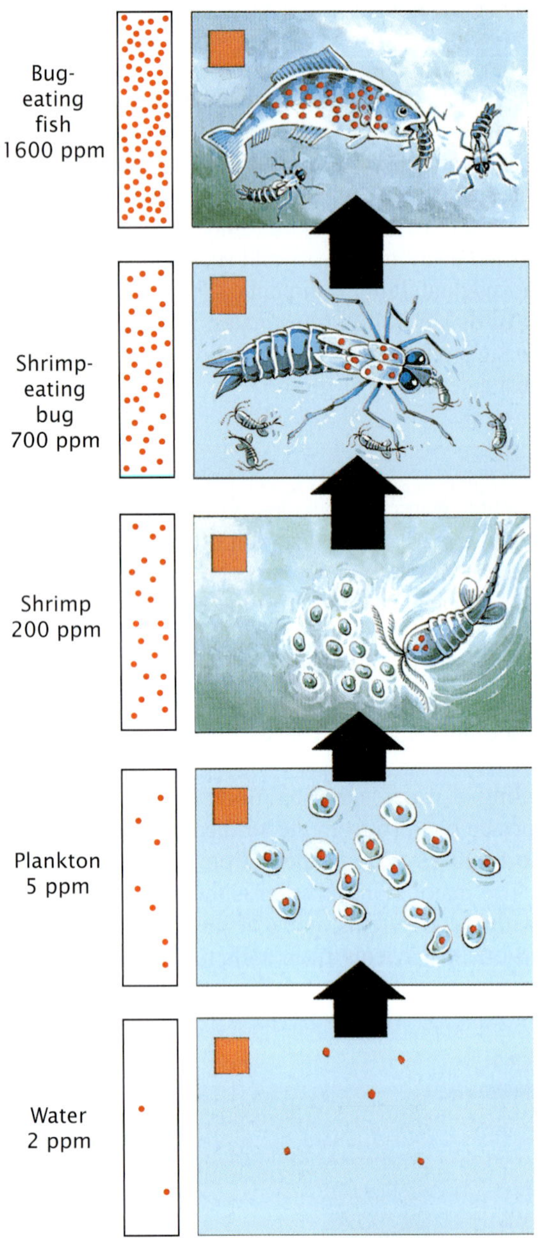

Seal pups were found dead in a breeding colony. An autopsy showed that their tissues contained high levels of chemicals known as polychlorinated biphenyls (PCBs). These chemicals were barely detectable in the sea water, but were present at high concentrations in the pups' tissues. How did this happen?

When a chemical pollutant enters an ecosystem through accidental or deliberate discharge to the soil, water or air, this chemical may enter the food chain. If it is not readily broken down by organisms, it will accumulate and become progressively concentrated as it enters organisms at higher trophic levels. This concentration is termed **bioaccumulation** or **biological magnification** (see figure 14.36). Progressive accumulation involves substances that cannot be excreted or broken down by metabolism. Such substances are said to be **non-biodegradable** or **persistent**. Persistent chemicals include PCBs, DDT and dieldrin.

Recall that production of one unit of body tissue in a primary consumer typically requires the intake of about 10 units of plant material, and that production of one unit of body tissue in a secondary consumer typically requires an intake of about 10 units of herbivore tissue. In the case of the seals, PCBs were present in low concentrations in seas where the adult seals fed. Algae and phytoplankton took up small quantities of this chemical. Small fish consumed large numbers of algae over time and accumulated higher concentrations of the chemical in their tissues than were present in the algae. Larger fish fed on these fish. Tertiary consumers in this ecosystem, such as seals, ate large quantities of fish over time. As a result, the seals accumulated even higher concentrations of the chemical in their tissues. Seal pups fed on their mothers' milk. The lipid part of this milk contained very high concentrations of the pollutant that accumulated to toxic levels in some of the pups.

Figure 14.36 Magnification or bioaccumulation of a persistent pollutant occurs at each trophic level in a food chain. (Note that ppm is a unit of concentration meaning parts per million.)

KEY IDEAS

- Matter cycles within an ecosystem and is re-used.
- Matter that cycles within ecosystems includes water in various physical forms and the elements carbon, nitrogen and phosphorus in various chemical forms.
- As it cycles, matter is sometimes in the living components of an ecosystem and sometimes in non-living components.
- With movement up trophic levels, non-degradable chemicals accumulate.
- Bioaccumulation involves the accumulation of non-degradable substances in organisms at higher trophic levels.

QUICK-CHECK

16 Give an example of the following:
 a an inorganic form in which carbon atoms can be present
 b an organic form in which nitrogen atoms can be present.

17 Identify a cycle that fits the following:
 a one that depends on the action of many kinds of bacteria
 b one that does not involve the atmosphere as a store.

18 Chemical A is water-soluble and can be excreted; chemical B dissolves in lipids and is non-biodegradable. Which would be expected to accumulate?

19 Which group of organisms in an ecosystem would be expected to have the highest accumulation of a persistent chemical? Explain.

BIOLOGIST AT WORK

Dr Pina Milne — researcher/curator and fungi expert

After completing a Bachelor of Education with a major in Biology and Geography, I spent almost 11 years teaching VCE Biology, Science and Geography. After a short break, I became a university tutor and taught practical classes in first-year Biology, and later in human physiology and genetics.

My passion was always plants and, as I was now working in a university, I decided that it was a perfect opportunity to begin a higher degree. I enjoyed field work so the obvious choice was an ecological project. I began a PhD looking at the reproductive biology and ecology of mosses that occur in a rainforest near Mt Donna Buang, northeast of Melbourne. Completing a higher degree not only allowed me to specialise in a particular field of botany and develop skills in research, but also gave me the opportunity to travel to conferences overseas and interstate.

After completing the PhD, I took up a position at the Royal Botanic Gardens Melbourne, in the National Herbarium of Victoria where collections of Australian and overseas fungi, lichens and plants are lodged. Apart from the curation of these collections, research also plays an important part. I began working with a mycologist (fungi expert) and have been involved in the production of a book that lists the scientific names of large fungi found in Australia and references to these fungi. This information is now also available electronically as the *Interactive Catalogue of Australian Fungi*.

While at the Botanic Gardens, I have completed taxonomic work on Australian mosses, looked at soil crusts and the invertebrates that are found among them and fungi that grow on mosses. I found that most information on mosses is written for specialist scientists and is not easily accessible to teachers, students and those

Figure 14.37 Dr Pina Milne with one of her posters

interested in the natural environment. This and our passion for the organisms with which we work inspired the *Forgotten Flora* project. Together with a mycologist and an illustrator, I embarked on the task of producing posters on bryophytes (mosses, liverworts and hornworts), fungi and lichens and interactive CDs in order to increase awareness and encourage an appreciation of these amazing organisms that are often overlooked.

The *Forgotten Flora* project has provided the opportunity to work with Science and Biology teachers who are interested in incorporating fungi, lichens, mosses and liverworts into their teaching.

BIOCHALLENGE

In many countries, the numbers of birds of prey decreased after organochloride pesticides such as DDT came into widespread agricultural use. Biologists now accept that the fall in numbers was due to the thinning of the shells and breakage of eggs as a result of DDT contamination.

a Suggest why birds of prey were more affected by DDT than seed-eating birds.

The diagram in figure 14.38 records the thicknesses of eggshells of peregrine falcons (*Falco peregrinus*) over the period from 1885 through to the mid-1970s. The eggs were taken from museum collections and came from many different areas of Australia.

b Describe what has happened to the average thickness of the eggshells over time.

c Suggest a possible explanation for the range of thickness observed in eggshells from the same year.

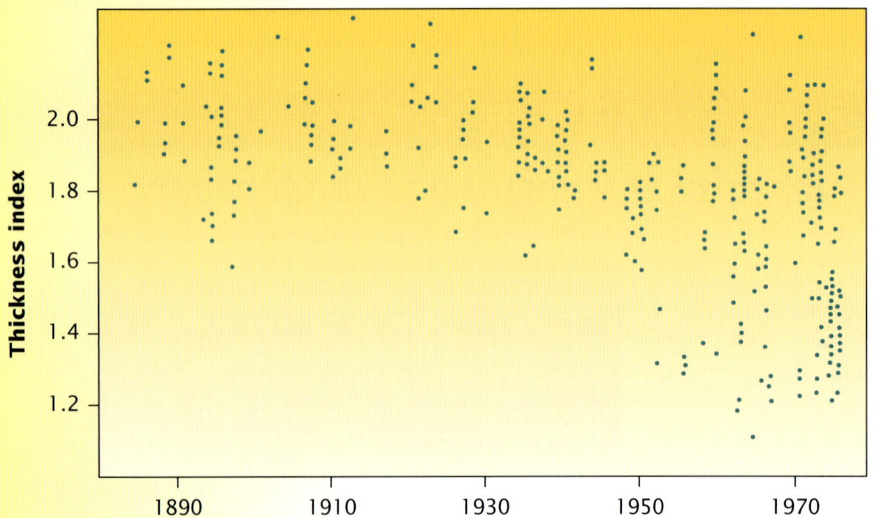

Figure 14.38 Graph showing the changes in thickness of peregrine falcon eggshells

Figure 14.39 A peregrine falcon feeding its young. This was one of the birds of prey affected by the DDT. What trophic level do these birds occupy?

d Based on this graph, suggest the most likely period when DDT began to be used extensively in agriculture in Australia:
- just before 1910
- just before 1930
- just before 1950.

On what evidence have you based your decision?

e Consider the following three observations and identify them as:

S: supports idea that DDT causes thinning of the shells of eggs

N: neither supports nor disproves this idea

C: provides conclusive evidence that DDT does not cause thinning.

Observation 1: Thinning of eggshells is greater in eggs from locations where DDT has been used in crop production than it is in areas where little or no DDT has been used.

Observation 2: Eggshell thickness is unaffected by the length of time the eggs have been incubated on the nest.

Observation 3: Shells of eggs laid during seasons of drought are thinner than those of eggs laid in non-drought seasons.

f The table shows DDT levels in the breast milk of Australian women from urban and rural areas in 1971 and 1979 (based on data from the Natural Resources Defense Council, figures rounded).

i Why did a difference exist in 1971 between the urban and rural samples?

ii Identify a likely cause for the change seen between 1971 and 1979.

	1971		1979	
	Urban	Rural	Urban	Rural
DDT concentration (ng/g lipid)	2300	17 000	1200	1200

CHAPTER REVIEW

CROSSWORD

Key words

bioaccumulation
biogeochemical cycles
biological magnification
biomass
carbon cycle
carnivores
cellular respiration
chemical energy
chemosynthetic
detritivores
ecological pyramids
ecosystems
energy
energy flow
food chains
food webs
gross primary production (GPP)
guano
herbivores
kinetic energy
limiting factor
net primary production (NPP)
nitrogen cycle
non-biodegradable
persistent
phosphorus cycle
photosynthetic
phytoplankton
potential energy
primary consumers
producers
productivity
pyramid of biomass
pyramid of energy
pyramid of numbers
radiant energy
secondary consumers
tertiary consumers
trophic level
upwelling
water cycle

Questions

1 *Making connections* ▸ Prepare a concept map, using key words and phrases from the list above.

2 *Applying your understanding* ▸ Give one example of organisms in an ecosystem that:

a capture and transform radiant energy

b are primary consumers

c occupy the first trophic level

d form part of the biomass of producers of ocean ecosystems.

3 *Analysing information in new contexts* ▸

a What trophic level can be assigned to the following organisms?

i green mosses in a damp gully
ii insects that feed on the mosses
iii birds that feed on the moss-eating insects
iv parasitic mites that live on the birds

b What trophic level(s) is occupied by an Australian 'carnivorous' plant, the alpine sundew (*Drosera arcturi*), that obtains some of its nutrients from insects trapped by its sticky filaments, and that also carries out photosynthesis?

4 *Analysing data and communicating understanding* ▸ Pedra Branca lies off the south coast of Tasmania (see map, figure 14.40). Living on the island and in the surrounding waters is a community that forms part of an ecosystem and includes krill, various species of fish, squid and birds such as the Australasian gannet (*Morus serrator*) and the shy albatross (*Diomedia cauta*).

a Draw a food web showing part of the energy flow in this ecosystem.

b How many trophic levels exist in this ecosystem?

c Draw a possible pyramid of energy for the Pedra Branca ecosystem.

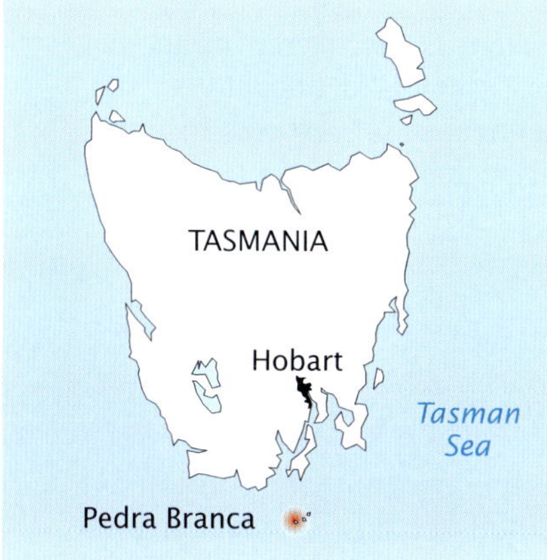

Figure 14.40

5 ***Applying your understanding*** ▸ Refer back to fig 14.5 on page 440. Is this a food chain or a food web?

a In figure 14.5, identify the following:

i an energy flow pathway that involves just two transfers
ii an energy flow pathway that involves three energy transfers

b What is the energy input into this ecosystem?

c Who are the producers in this ecosystem?

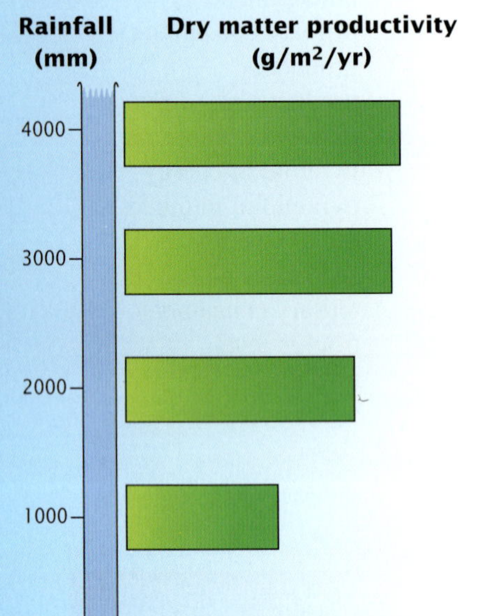

Figure 14.41

6 ***Interpreting data*** ▸ Figure 14.41 shows the relationship between net primary productivity (measured in dry matter) for a forest ecosystem and annual rainfall. Examine this figure and answer the following questions.

a In what units is productivity measured?

b Explain why rainfall affects productivity in a forest.

c As annual rainfall increases, what happens to productivity?

7 ***Applying understanding*** ▸ Explain the following observations in biological terms.

a When fish and mammals can survive on the same food pellets, the cost of producing a given biomass of fish is less than the cost of producing the same biomass of mammals.

b In an ecosystem, the number of large carnivores is typically far less than the number of herbivores.

c More fish and other consumer organisms are found in a volume of coastal sea than in an equivalent volume of open ocean.

d Tropical rainforests have more trophic levels than a desert scrub.

e The cost of one kilogram of steak is greater than the cost of one kilogram of rice.

8 ***Applying biological principles*** ▸ A food chain consists of:

leaves ⟶ caterpillar ⟶ sparrow ⟶ eagle.

Consider that a caterpillar eats 100 grams of leaf organic matter. Based on the 10-per-cent rule, how much of the chemical energy in this organic matter would be available for consumption by (a) a sparrow and (b) an eagle?

9 Consider the lamb shown in figure 14.42.

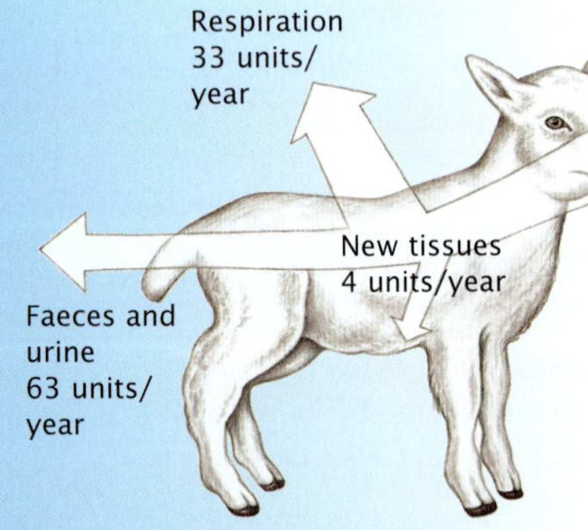

Figure 14.42

a What is the trophic level of this herbivorous lamb?

b What percentage of the energy intake by the lamb is expected to:

i be used in cellular respiration
ii be egested as faeces or excreted as urine
iii appear as new tissue?

10 ***Discussion*** ▸ Imagine that another solar system is discovered in another galaxy. The star at the centre of this new solar system is identical to our sun. One of the planets, named Terranova, is identical in all respects to Earth. Which of the following features would be expected to exist in ecosystems on Terranova?

a use of sunlight as the primary energy source

b recycling of matter

c recycling of energy

d existence of ten trophic levels

11 ***Using the web*** ▸ Go to www.jaconline.com.au/natureofbiology/natbiol1-3e and click on the 'Food web' weblink for this chapter. Click on each organism that appears in the box and drag it into the relevant block that denotes its trophic level.

15 Population dynamics

KEY KNOWLEDGE

This chapter is designed to enable you to:

- extend your knowledge of population dynamics
- develop an understanding of factors affecting distribution and abundance of populations
- identify techniques for estimating various populations.

Figure 15.1 Here we see a population or school of diagonal-striped sweetlips (*Plectorhinchus lineatus*) that forms part of the living community in a Great Barrier Reef ecosystem. Each population is a dynamic biological unit. In this chapter, we will explore features of various populations, including their abundance, distribution, age structure and models of population growth, and we will identify techniques for sampling populations.

A population down south

A journey to the Ross Sea in summer will take us to locations such as Cape Adare that is located 71°S of the equator. As we approach land, a stunning sight will greet us — a rookery of thousands of Adelie penguins (*Pygoscelis adeliae*) (see figure 15.2). As well as this sight, your other senses will register much noise and a strong fishy smell. While large in numbers, all these penguins are members of just one **population**; that is, members of one species living and reproducing in the same region at the same time.

Adelie penguins spend most of their life at sea and come ashore only to breed and to moult. Their breeding season is from October to March. Each penguin pair builds a nest of pebbles in which the female lays one or two eggs. Incubation lasts 35 days and is shared by each parent in turn. After hatching, a chick is guarded by one parent while the other feeds at sea (see figure 15.3) and then the parents reverse duties. After 7 to 10 weeks, Adelie chicks are ready for independent life.

(a)

(b)

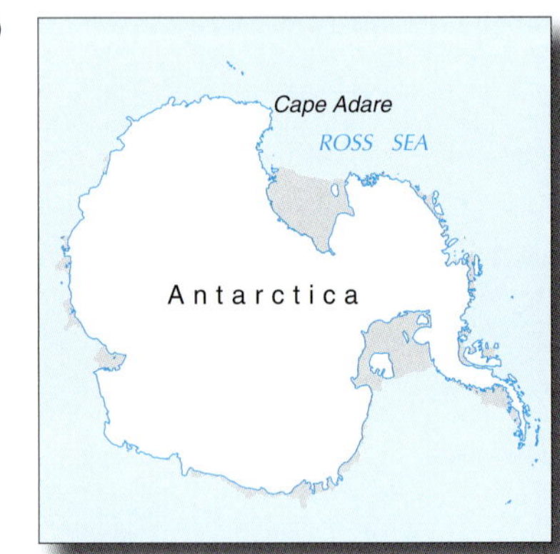

Figure 15.2 **(a)** The Adelie penguin rookery at Cape Adare in Antarctica. Would you find this population here throughout the year?
(b) Location of Cape Adare, Antarctica

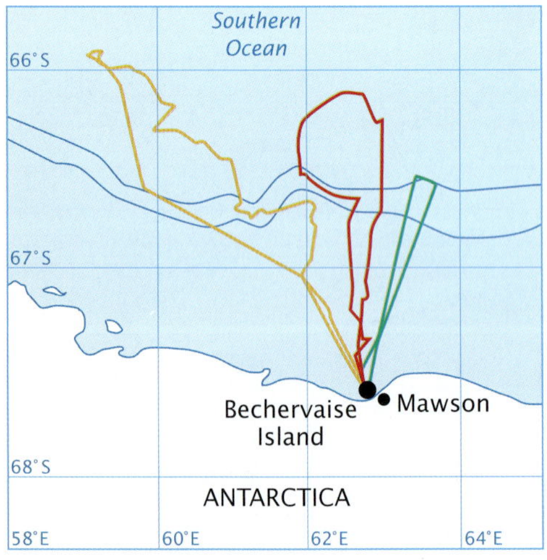

Figure 15.3 Satellite tracking of three individual penguins near Mawson Station showed that Adelie penguins make extensive trips offshore to the continental shelf in search of food, sometimes as far as 300 km.

Populations: how many?

Different communities vary in the number of populations that they contain. The terrestrial **community** at Cape Adare in summer appears to comprise just one population. The situation in a coral reef community is very different with many different populations present. Because each population is one discrete species that lives in the same area at the same time, the number of populations in a community corresponds to the number of different species or the **species richness** of a community. Factors that affect the number of species in an area include:

- physical size of area
- latitude (or distance from the Equator).

Area affects species richness

The number of different populations in terrestrial communities in the same region is related to the physical size of the available area. For example, the Caribbean Sea contains many islands that differ in their areas. Figure 15.4 shows the results of a study into the relationship between the size of the islands and the number of species of reptiles and amphibians (and hence the number of different populations) found on them. In general, if an island has an area 10 times that of another that is in the same region, the larger island can be expected to have about twice the number of different species.

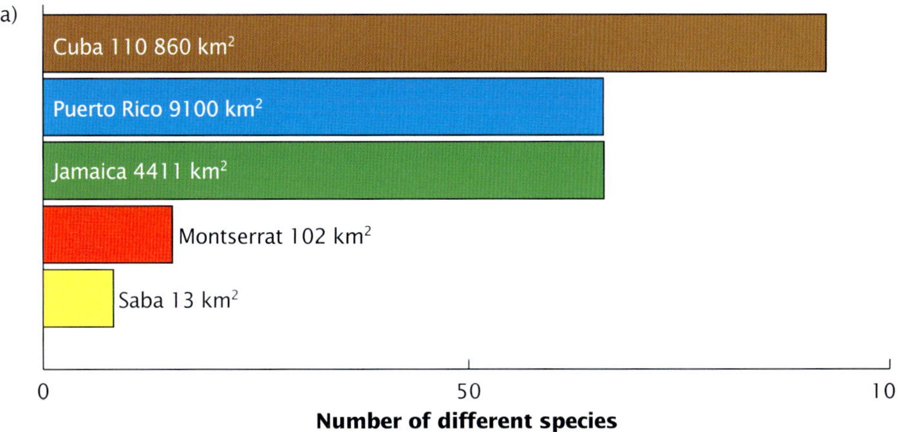

Figure 15.4 **(a)** Relationship between the area of an island and the number of populations of different species that it contains (based on data from R. H. MacArthur and E. O. Wilson, *The Theory of Island Biogeography*, Princeton University Press, 2001) **(b)** Location of the islands in the Caribbean Sea

ODD FACT

Australia has a land area of about 7 600 000 square kilometres and the mainland lies between latitudes 10°S (Cape York) and 39°S (Wilson's Promontory).

Latitude affects species richness

The number of different populations in a terrestrial area is also related to the latitude or distance from the equator. For example, terrestrial Antarctica covers about 14 000 000 square kilometres and the continent lies south of latitude 60°S. Most of the Antarctic continent is an ice desert — dry and ice covered.

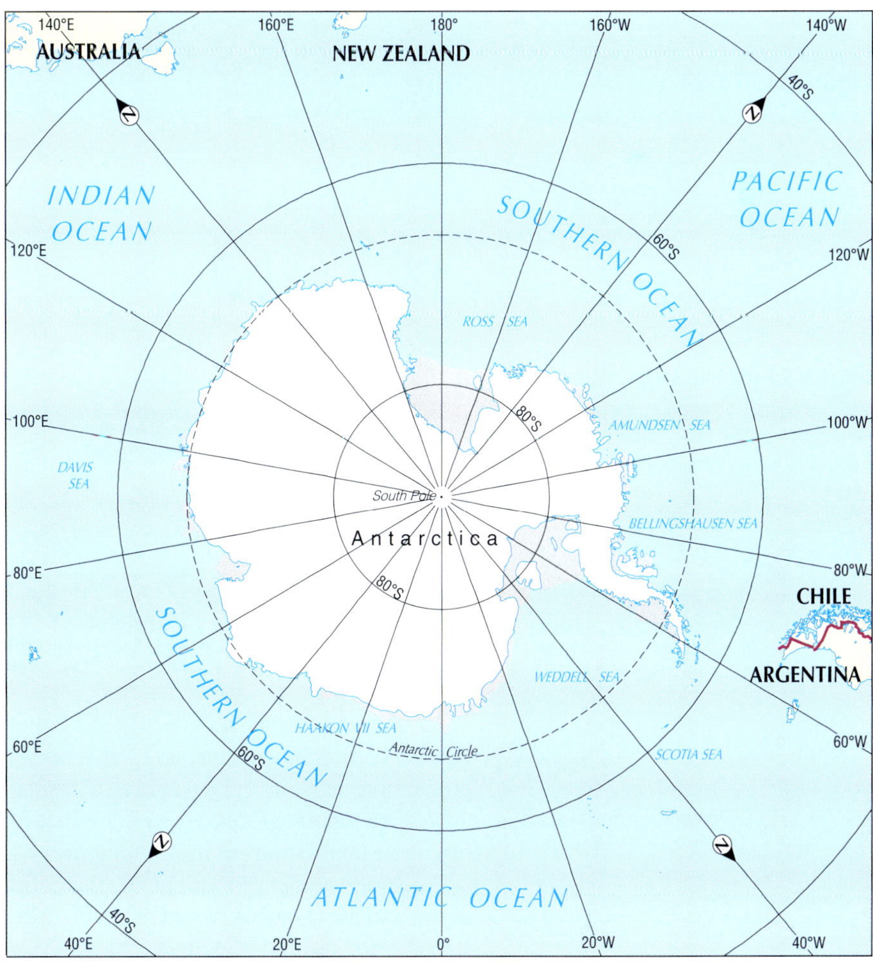

Figure 15.5 The Antarctic circle lies south of latitude 60°S.

The temperature and light levels in Antarctica vary greatly between the extremes of winter and summer. Only three species of flowering plant survive in this habitat — they live along the west coast of the Antarctic Peninsula (see figure 15.6). These are two species of grasses, *Deschampsia parvula* and *D. elegantula*, and a cushion plant, *Colobanthus crassifolius*. Other plants that are found in terrestrial Antarctica are mosses, as well as many species of lichen, which are partnerships of fungi and algae (see figure 15.7). Terrestrial algae, such as *Prasiola* sp. grow on open ground and damp rocks and there is also a pink snow algae (see figure 15.8).

Only a few animal species survive on terrestrial Antarctica, and these include insects, such as species of springtails (*Cryptopygus antarcticus*), wingless flies (*Belgica antarctica*) and midges. There are also other invertebrates such as species of mites, *Alaskozetes antarctica* and *Tydeus tilbrooki*, as well as brine shrimps and nematodes. Why are seals and penguins excluded from this list?

Figure 15.6 Location of the Antarctic Peninsula

Figure 15.7 Vegetation in terrestrial Antarctica in summer. What would a winter picture show?

Figure 15.8 A biological scientist examines a patch of 'pink snow' in an area of melting ice on Antarctica's fringe. The pink effect is caused by algae.

ODD FACT

In contrast to Antarctica, the island of Madagascar lies in the tropical zone about 20°S of the equator and has a land area of about 580 000 square kilometres. It has an estimated 13 000 plant species — more than one thousand times the number in Antarctica.

As we move from the poles to the equator, in general, species richness of terrestrial communities increases. This means that more species and hence more populations exist in a given area of a tropical rainforest ecosystem than in a similar area of a temperate forest ecosystem. In turn, an area of temperate forest ecosystem has more populations than a similar area of a conifer (boreal) forest ecosystem in cold regions of the northern hemisphere. For example, a two-hectare area of forest in tropical Malaysia has more than 200 different tree species while a similar area of forest at 45° north of the equator contains only 15 different tree species.

KEY IDEAS

- The number of populations in a community is a measure of the number of species or species richness.
- In general, species richness and hence number of populations in a terrestrial community decreases from the tropics to the poles.
- For two communities in a similar region, species richness is related to the area available to each community.

QUICK-CHECK

1 Identify the following statements as true or false.
  a All forest communities have a similar number of populations.
  b The species richness of a terrestrial ecosystem is in general related to latitude of the ecosystem.
  c Species richness gives a measure of the number of different populations in a community.
2 Two grassland communities, A and B, cover a similar area, but community A has a greater number of populations than community B.
  a Which ecosystem has the greater species richness?
  b Which ecosystem is expected to be closer to the equator?

A general look at populations

Any population can be characterised in terms of several attributes, including:
- abundance or density
- distribution
- age structure of population
- rate of growth.

In addition to these attributes, the study of a particular population will also involve an examination of other specific features, such as its habitat requirements, breeding season and reproductive strategy.

Abundance of populations

Abundance or **density** is defined as the number of individuals of a given species per unit area. Abundance can be expressed qualitatively, as for example, in order of increasing abundance:
- scarce or rare
- infrequent
- frequent
- abundant
- very abundant (see figure 15.9).

Figure 15.9 Abundance can be expressed in qualitative terms like 'rare' or 'abundant'.

Refer back to figure 15.2 on page 472. In qualitative terms, how would you describe the abundance of the Adelie penguins?

Abundance can also be expressed in quantitative terms as, for example, the number of ants per square metre or the number of insects per tree or the number of plants per hectare. For example, one study identified 54 Tasmanian devils (*Sarcophilus harrisii*) living in an area of 80 square kilometres in Tasmania. This corresponds to an abundance of about 0.7 animals per km^2.

In the case of organisms living in soil or water, the abundance is expressed as the number of organisms per unit volume of soil or water.

When assessing the abundance of a population, the size of the area over which a population is counted may be large, such as an extensive grassland, or it may be small, such as one tree (see figure 15.12 on page 477).

Total count or sample to measure abundance

It is sometimes possible to carry out a **total count** or **true census** of a population by counting every member of a population that lives in a given area. Total counts can be done with populations of animals that are large or conspicuous (such as ground-nesting birds on an island or seals on an isolated beach) or animals that are slow moving or sessile (such as limpets or barnacles on rocks in the inter-tidal zone). Likewise, total counts can be done of populations of large plant species.

Carrying out a total census of a population, however, generally poses difficulties. A true census is not possible for small, shy or very mobile animals because many animals will probably be missed. In any case, the cost of a total census in time and person power may be unacceptably high, especially if a large area of a habitat is involved. Instead, when an entire population cannot be counted, **sampling** techniques are used. Typically one or more samples are taken randomly from a population and the samples are assumed to be representative of the entire population. Sampling from a known area allows biologists to make estimates of both the abundance of a population and the size of the population (see pages 482–6).

Orange-bellied parrots are an endangered native bird species. Their total population is estimated at less than 200 birds.

The abundance of a population cannot usually be based on just one count. This is because of the chance of sampling errors. In order to avoid sampling errors, counts of population are typically repeated several times.

In addition, changes in the abundance of a population can occur over time owing to factors such as migration and breeding patterns; for example, orange-bellied parrots (*Neophema chrysogaster*) migrate annually from south-west Tasmania (where they breed from about October to March) to the southern coast of Victoria and South Australia (where they over-winter from about April to October) (see figure 15.10). In order to measure the abundance of this population in its Victorian breeding grounds, the population must be surveyed at the right time of year when all the members of the population have returned from south-west Tasmania.

Figure 15.10 **(a)** The orange-bellied parrot. **(b)** Annual migratory path of the orange-bellied parrot. Where would you expect to find this parrot in summer? in winter?

Why measure population abundance?

Biologists are interested in the abundance of populations for various reasons:

- Biologists concerned with conservation must measure the abundance of populations of endangered species over time to decide if the populations are stable, increasing or decreasing in abundance. If the abundance of the population of an endangered species falls, the risk of extinction increases. For example, regular counts of the population of the endangered orange-bellied parrot are carried out to see if conservation measures are being successful in rebuilding this population.

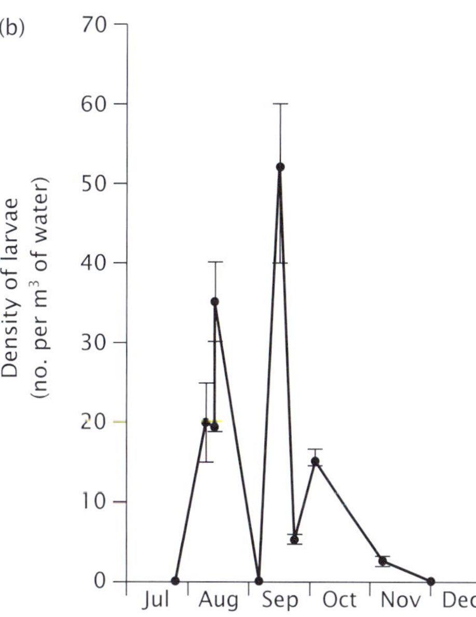

Figure 15.11 **(a)** The northern Pacific seastar, an introduced pest in areas of south-eastern Australian waters. How did this species that is native to the north Pacific reach Australia? **(b)** Abundance of Pacific seastar larvae in samples from the Derwent River during 2001. Abundance is given in number of larvae per cubic metre of water. Note the variation in larval density over time. Based on these data, would ballast water be more likely to contain starfish larvae if taken into a ship in May or in August? (Data from Craig Johnson et al. (2004), in Report to the Department of Sustainability and Environment, Victoria)

- Biologists concerned with the control or elimination of exotic (non-native) pest species need to monitor changes in their abundance and range. For example, the northern Pacific seastar (*Asterias amurensis*) is native to waters of the north Pacific and is now found in Port Phillip Bay in Victoria and in the Derwent estuary in Tasmania (see figure 15.11a). This pest is spread as tiny starfish larvae in the ballast water of ships. In order to identify the risk that this pest could spread further around Australia via this means, biologists measured the population density of Pacific seastar larvae in water samples (see figure 15.11b).
- Biologists interested in measuring the diversity of a community need to count the numbers of various populations in that community. Figure 15.12 shows how biologists measured the invertebrate populations in the leafy crown of one kind of eucalyptus tree. They set up collecting traps high in the trees. The leafy crowns were then sprayed with an insecticide with short-term action. Insects and spiders falling into the traps were collected and identified. From one kind of tree alone, narrow-leaved ironbark (*E. cebra*), more than 700 different kinds of insect and spider were collected — a very diverse community (see figure 15.13).

Figure 15.12 Biologists exploring the diversity of insects and spiders living in the leafy crown of selected eucalyptus trees. The collecting traps are yellow.

Total 722

- Spiders (*Araneae*)
- Mites (*Acarina*)
- Sucking bugs (*Hemiptera*)
- Thrips (*Thysanoptera*)
- Booklice (*Psocoptera*)
- Beetles (*Coleoptera*)
- Flies (*Diptera*)
- Other

Figure 15.13 Different kinds of invertebrate found on one kind of eucalypt. The invertebrates are organised into eight broad groups: spiders, mites, bugs, thrips, booklice, beetles, flies and others.

In 1979, at Green Island in the Great Barrier Reef, the population of crown-of-thorns starfish exploded to reach about 3 million. Each crown-of-thorns starfish consumes and kills between five and six square metres of coral each year.

- Biologists interested in understanding why some populations can 'explode' or sharply increase in numbers must measure population abundance regularly in order to detect patterns and identify possible causes of these explosions. On the Great Barrier Reef, for example, populations of the crown-of-thorns starfish (*Acanthaster planci*) periodically explode. The increased numbers of starfish cause great damage to the corals by eating the coral-producing polyps (see figure 15.14).

In the box on page 479, Ian Miller, a marine biologist at the Australian Institute of Marine Science (AIMS), describes some of his research including sampling populations of marine species using various techniques.

(b)
Number of reefs surveyed
246 140 169 100 82 87 103 74 108 92 98 91 115 98 89
Crown-of-thorns starfish per tow
1.5
1
0.5
0
1986 1988 1990 1992 1994 1996 1998 2000 2002 2004
Survey year

Figure 15.14 (a) A cluster of crown-of-thorns starfish on coral. How many can you count in this small area? The white areas are 'feeding scars' consisting of dead coral that remains after the living coral polyps have been eaten. **(b)** Average crown-of-thorns starfish density across the Great Barrier Reef. Note the changes in population density over time. When was the population density higher: in 1992 or in 1998? (*Source:* Australian Institute of Marine Science)

A — Uniform

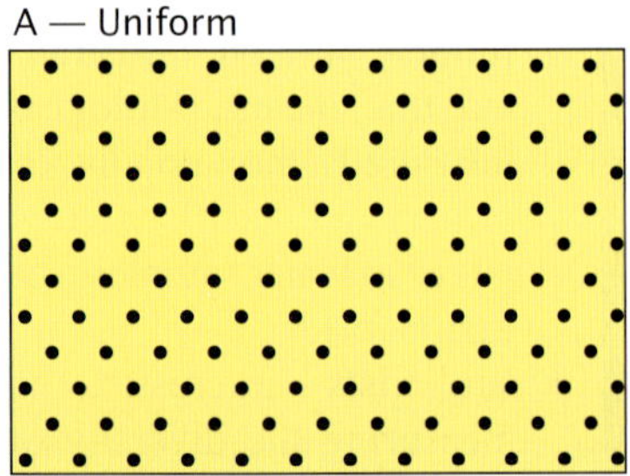

B — Random

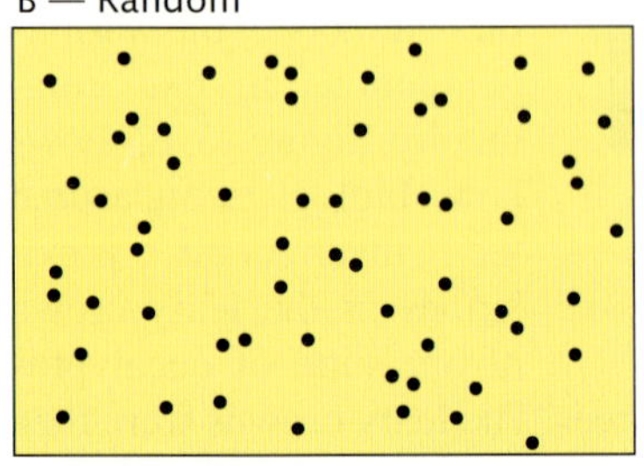

C — Clumped

Distribution of populations

Distribution refers to the spread of members of a population over space. Populations may have identical densities but their distributions can differ. Figure 15.15 shows three populations with identical densities but their horizontal distributions differ, being uniform in A, random in B and clustered or clumped in C. Clumped and uniform distributions are both non-random patterns. The most common pattern observed in populations is a clumped distribution. (What pattern is apparent in the Adelie penguins seen in figure 15.2, page 472?)

Changes in the distribution of populations can occur over time. Animal populations that have a random distribution at one period, such as the non-breeding season, may show a different distribution during the breeding season.

A clumped distribution of a plant population may indicate that some areas only within a sample area are suitable for germination and survival of a plant species and that areas without plants are unsuitable for survival because of the pH of the soil or the lack of water or the ambient temperature.

Figure 15.15 Three populations, A, B and C, with different distributions. What might cause a clumped distribution?

BIOLOGIST AT WORK

Ian Miller — monitoring crown-of-thorns starfish on the Great Barrier Reef

Ian Miller is an experimental scientist employed at the Australian Institute of Marine Science (AIMS). He writes:

I work as a marine biologist. In my role as coordinator of broadscale surveys, I am responsible for the day-to-day management of the crown-of-thorns starfish (COTS) component of a larger Long Term Monitoring Program (LTMP). The LTMP was set up in 1992 and is an extension of a previous monitoring initiative that began in 1985 to describe the pattern and extent of COTS activity on the Great Barrier Reef (GBR). This groundbreaking program was the first to sample the GBR over its entire geographic range on an annual basis. I joined the program in 1989 after obtaining a BSc in Marine Biology from James Cook University. Monitoring the GBR has proved to be an exciting and challenging career.

Figure 15.16 Ian Miller in the 'office'. Observers are towed around the reef perimeter to visually assess the impact of crown-of-thorns starfish outbreaks on the Great Barrier Reef.

COTS outbreaks remain a major management problem on the GBR and are responsible for more coral mortality than any other factor. To determine pattern and extent of COTS activity on the GBR we use the manta tow technique. This is a plotless transect survey method, where the scale of interest is the whole reef or large parts of the reef (i.e. kilometres). Data on COTS counts are collected and visual estimates of live coral, dead coral and soft coral are made. Approximately 100 reefs are surveyed by manta tow annually from Cape Grenville in the north to the Capricorn Bunker Reefs to the south. The method relies on making visual estimates and provides observers with a thrilling ride as they are towed around the reef. By tilting the manta board you can dive to a depth of up to 10 metres and literally fly through the reef environment. Stunning vistas of drop-offs on the reef fronts and bommie fields on the reef backs provide a unique experience. The broad range of reef habitats encountered has given me a greater appreciation of how the GBR changes through time and space.

The manta tow surveys have provided an unprecedented record of change on the GBR and are an invaluable resource for reef managers and scientists alike. The results have led to a greater understanding of the pattern and extent of COTS activity and their effects on coral reefs. The results also provide insight into the dynamic nature of coral reef ecosystems and highlight their vulnerability to large-scale impacts that include not only COTS infestation but also other factors such as cyclones, floods, disease and coral bleaching.

As a team member of the LTMP, I also participate in site-specific surveys of 48 reefs from Cooktown in the north to Gladstone in the south. These surveys involve scuba diving on fixed transects, usually on the north-eastern flanks of reefs, to gather detailed information on corals, algae and fishes. Each survey consists of three fixed sites, which in turn are composed of five 50-metre transects. At each site, visual counts of reef fish (some 200 species) are conducted along the transects as belts, 5 metres wide for large roving demersal species and 1 metre wide for small habitat dependent species. Bottom dwellers are sampled on each transect by video (for later analysis in the laboratory) and factors causing coral mortality are also recorded along 2 metre wide belts using visual counts. These fine-scale surveys allow the monitoring team to define small-scale changes in community structure through time and pinpoint factors that are driving these changes. Results have shown that the reef environment is a far more diverse and dynamic system than was previously imagined and that, following a disturbance, reefs can and do recover to their previous condition depending on the size of the initial disturbance and given enough time before the next disturbance.

As part of the AIMS LTMP, I look forward to being at the forefront of defining the pattern and extent of impacts from new and emerging threats to the GBR, such as coral bleaching and disease. In the immediate future, the LTMP will be a major contributor to defining the role that water quality plays in shaping the fate of inshore reefs on the GBR, which is a current topic of extreme interest for coral reef managers. I continue to find my job an exciting and challenging one where I can make a real contribution to extending our knowledge on coral reefs.

For more information on the AIMS LTMP, visit the website by going to www.jaconline.com.au/natureofbiology/natbiol1-3e and clicking on the 'AIMS Research — Reef Monitoring Index' weblink for this chapter.

Mosses grow in open forests. Their distribution, however, is far from random; they are confined to damp, sheltered areas. A distribution map of mosses in a forest corresponds to the distribution of damp, sheltered areas.

Likewise, some parts of a habitat may be more shaded or more protected or closer to water than other parts. Animal populations will aggregate in the more favourable parts, producing a clumped distribution (see figure 15.17). Clumped distributions are also seen in populations of mammals that form herds or schools as a strategy for reducing predation. Clumped distributions are also seen in populations of plant species that reproduce asexually by runners or rhizomes with new plants appearing very close to the parental plant.

A uniform distribution may indicate a high level of intra-specific competition so that members of a population avoid each other by being equidistant from each other. Uniform spacing is seen in plants when members of a population repel each other by the release of chemicals (see figure 15.18). In animals, uniform distribution occurs when members of a population defend territories.

A random distribution is expected (i) when the environmental conditions within the sample area are equivalent throughout the entire area and (ii) when the presence of one member of a population has no effect on the location of another member of the population. Both of these conditions rarely occur and, as a result, a random distribution pattern of members of a population is rare in nature.

Figure 15.17 A group of feral goats in central Australia — an example of clumped distribution

Figure 15.18 Spinifex covers the level ground and hills of this area of Western Australia — an example of uniform distribution

Age structure of populations

In a population, individual members vary in their ages and life spans. The age structure of a population identifies the proportion of its members that are:

- at pre-reproductive age (too young to reproduce)
- at reproductive age
- at post-reproductive age (no longer able to reproduce).

The age structure of a population is important since it indicates whether or not the population is likely to increase over time. Where the majority of individuals in a population are at reproductive age or younger, that population is expected to increase over time. If a population has most members at post-reproductive stage, then, regardless of its size, this population will decline.

The age structure of populations can be plotted as a series of bars whose lengths indicate the relative numbers in each group or cohort. When the low bars are longest more members of the population are at or below reproductive age and that population will increase. The age structure plot of such a population, with most members at or below reproductive age, is a 'pyramid' shape (see figure 15.19a). In contrast, a population whose age structure is a 'vase' shape is either at zero population growth or is decreasing (see figure 15.19b).

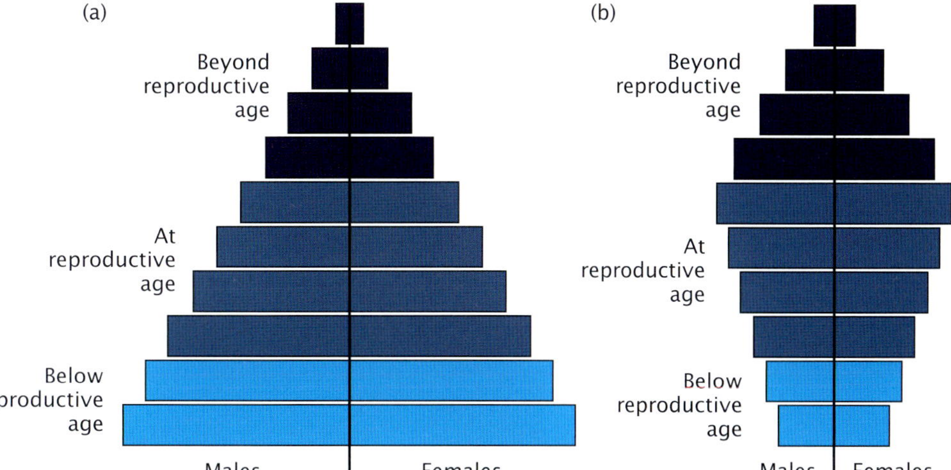

Figure 15.19 Age structures of two populations.
(a) Population with a broad base with most individuals being at reproductive age or younger. Is this population expected to grow?
(b) Population with a narrow base. Is this population expected to grow?

In human populations, where life spans are long, the population structure is generally shown in terms of both age and sex. Figure 15.20 shows the contrasting age–sex structures of the populations of two countries, Australia and Nigeria, in the year 2000.

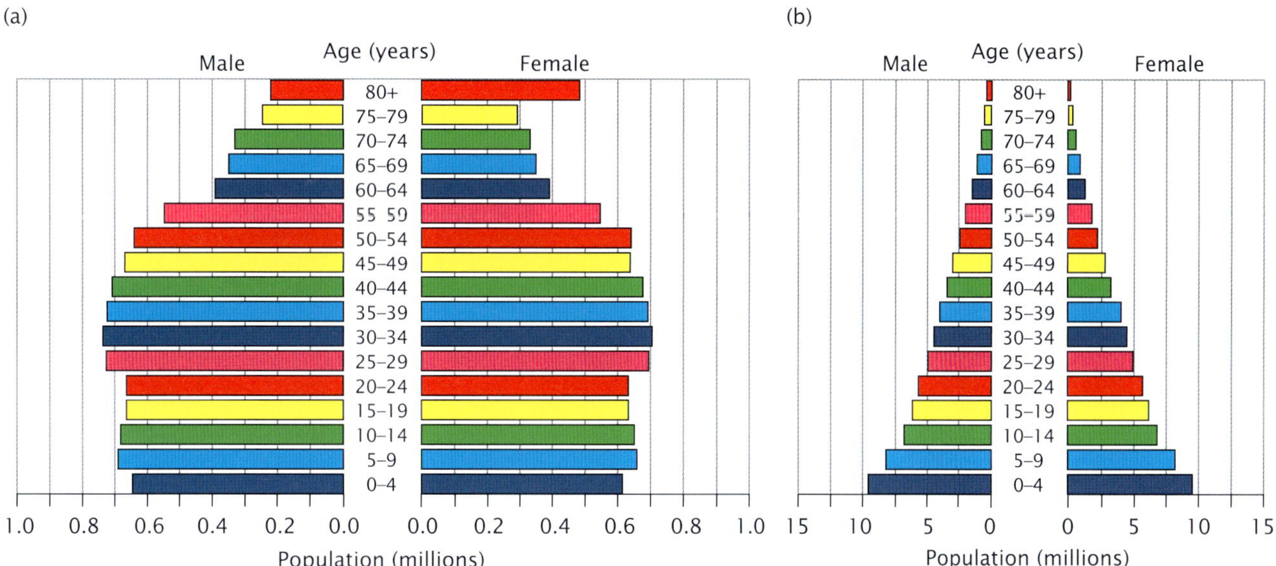

Figure 15.20 Age–sex structures for **(a)** Australia and **(b)** Nigeria in 2000. Note the different shapes of the age–sex structures for the two countries and note the different x-axis scales. Which population will be expected to increase?

KEY IDEAS

- The number of individuals per unit area (or volume) gives the abundance or density of a population.
- Total counts or sampling are used to assess population abundance.
- Distribution of a population identifies how members of a population are spread over space.
- The shape of the plot of the age structure of a population indicates its reproductive capacity.

QUICK-CHECK

3 Identify the following as true or false.
   a An age structure plot with a pyramid shape is indicative of a growing population.
   b Abundance of a population is also known as population density.
   c Total counts are more commonly carried out than the use of sampling techniques.
4 Give one possible explanation for a clumped distribution in a population.
5 Give two reasons a biologist would estimate the population size of one or more species.
6 Identify three problems with total counts of populations.

Sampling populations

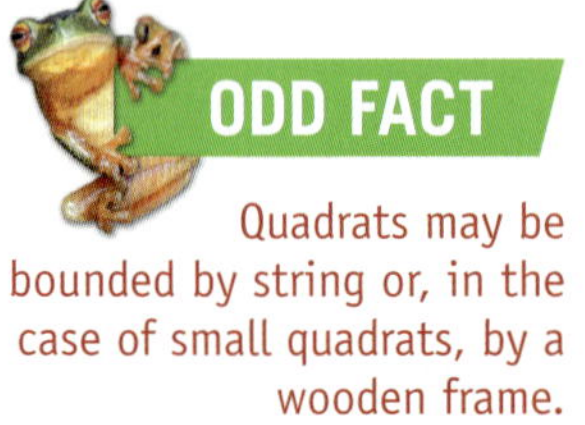

Quadrats may be bounded by string or, in the case of small quadrats, by a wooden frame.

How can we estimate the size of a particular population? It is rarely possible in terms of cost and time or practicable to carry out a total count of a population. Instead, it is common to take samples and from these to estimate the population size. Sampling techniques include:

- use of **quadrats**
- use of **transects**
- **mark-recapture** technique.

Each technique has its particular applications.

Figure 15.21 Sampling a population of sessile animals (oysters) using a quadrat. Quadrats are also used to sample plant populations.

Use of quadrats: within a square

Quadrats are square areas of known size and are often subdivided into smaller units. For example, if a one-metre square quadrat is used and the total number of plants of species A recorded in 10 quadrats is 208, then the abundance of species A = 208 plants per 10 square metres or 20.8 plants per square metre.

In using a quadrat to sample a plant population, certain decisions are critical, as for example:

- What size should the quadrat be? If trees are the objects of study, a large quadrat such as a 20-metre square is used. If lichens on rocks are the objects of study, a one-metre square quadrat might be used.
- How many areas should be sampled? If too few, the result may be subject to sampling error.
- How should sampling sites be chosen? The locations should be determined objectively and randomly in order to prevent sampling bias.

Successful use of quadrats depends on the fact that members of the population being studied will remain in place to be counted. Quadrats can be used to estimate the abundance or population density of plants, of sessile animals like oysters (see figure 15.21), mussels, limpets and anemones, and of slow-moving animals such as chitons and snails. Use of quadrats, however, is useless for fast-moving animals that will not wait around to be sampled!

Use of transects: along a line or within a strip

Another technique is the use of transects. A transect is a line or a strip laid across the area to be studied. Line transects are particularly useful in identifying changes in vegetation with changes in the environment. For example, changes in the plant community growing on a sand dune can be recorded along a transect laid out along the sand dune (see figure 15.22). The line is marked at fixed intervals. As with the use of quadrats, when either line or strip transects are used, repeat counts are carried out in order to minimise sampling errors.

Strip transects (also known as belt transects) are used to estimate animal populations and can be carried out by observers on foot, in cars, in helicopters or fixed-wing aircraft or underwater. The type and the size of the transect (see page 484) will depend on the animal population that is being surveyed.

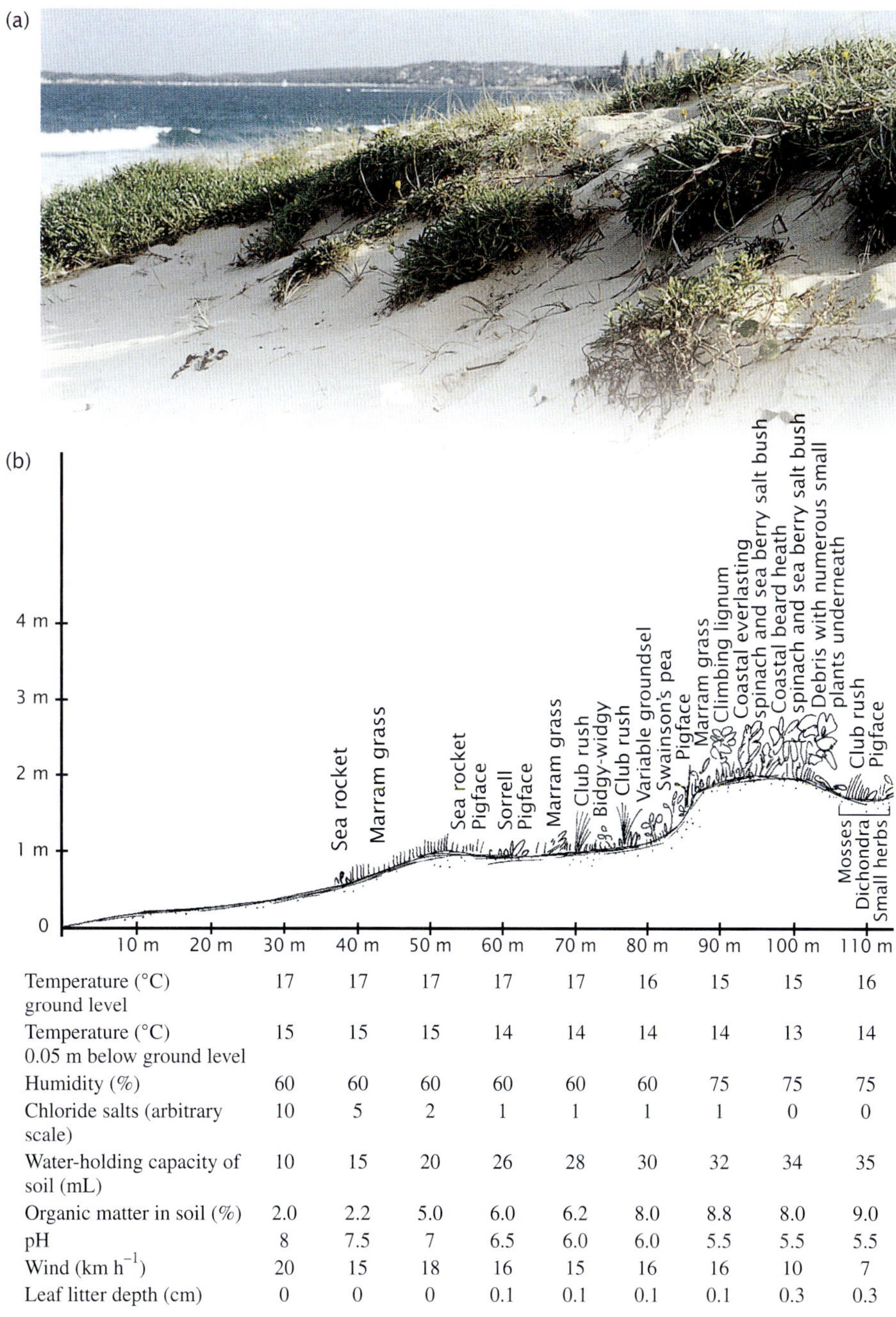

	30 m	40 m	50 m	60 m	70 m	80 m	90 m	100 m	110 m
Temperature (°C) ground level	17	17	17	17	17	16	15	15	16
Temperature (°C) 0.05 m below ground level	15	15	15	14	14	14	14	13	14
Humidity (%)	60	60	60	60	60	60	75	75	75
Chloride salts (arbitrary scale)	10	5	2	1	1	1	1	0	0
Water-holding capacity of soil (mL)	10	15	20	26	28	30	32	34	35
Organic matter in soil (%)	2.0	2.2	5.0	6.0	6.2	8.0	8.8	8.0	9.0
pH	8	7.5	7	6.5	6.0	6.0	5.5	5.5	5.5
Wind (km h^{-1})	20	15	18	16	15	16	16	10	7
Leaf litter depth (cm)	0	0	0	0.1	0.1	0.1	0.1	0.3	0.3

Figure 15.22 **(a)** Part of a sand dune showing some of the vegetation nearest the sea, including marram grass (*Ammophila arenaria*) and sea rocket (*Cakile edentula*) **(b)** Changes in vegetation along a sand dune as recorded along a 110-metre line transect, starting from the waterline and moving inland

Transects from the air

Aerial strip transects are used to estimate the abundance of certain kangaroo species across broad areas, provided the kangaroo species concerned is active by day and lives in an open flat habitat. The procedure involves a trained observer in an aircraft that flies at a speed of 185 km per hour and at an altitude of 76 metres above the ground. Use of global positioning receivers and altimeters enables the aircraft to maintain constant height and speed.

The observer records the numbers of a particular kangaroo species that are seen between two markers on the aircraft that represent 200 metres width on the ground. Over a 97-second period, the plane travels 0.5 kilometre. A belt transect that is 200 metres wide (0.2 kilometre) and 0.5 kilometre long is an area of one square kilometre, so each 97-second period of flight corresponds to one square kilometre. By counting the target species over many 97-second periods, the observer surveys many square kilometres of countryside (see figure 15.23). The accuracy of aerial surveys using strip transects depends on several factors, including the skill of the observer in identifying the species in question.

ODD FACT

From 1978 to 2004, populations of red kangaroos (*Macropus rufus*) have been surveyed in a large area of South Australia using aerial belt transects. Over that time, the population size has varied from a high of 2 175 200 in 1981 to a low of 739 700 in 2003. The major factor influencing abundance was drought.

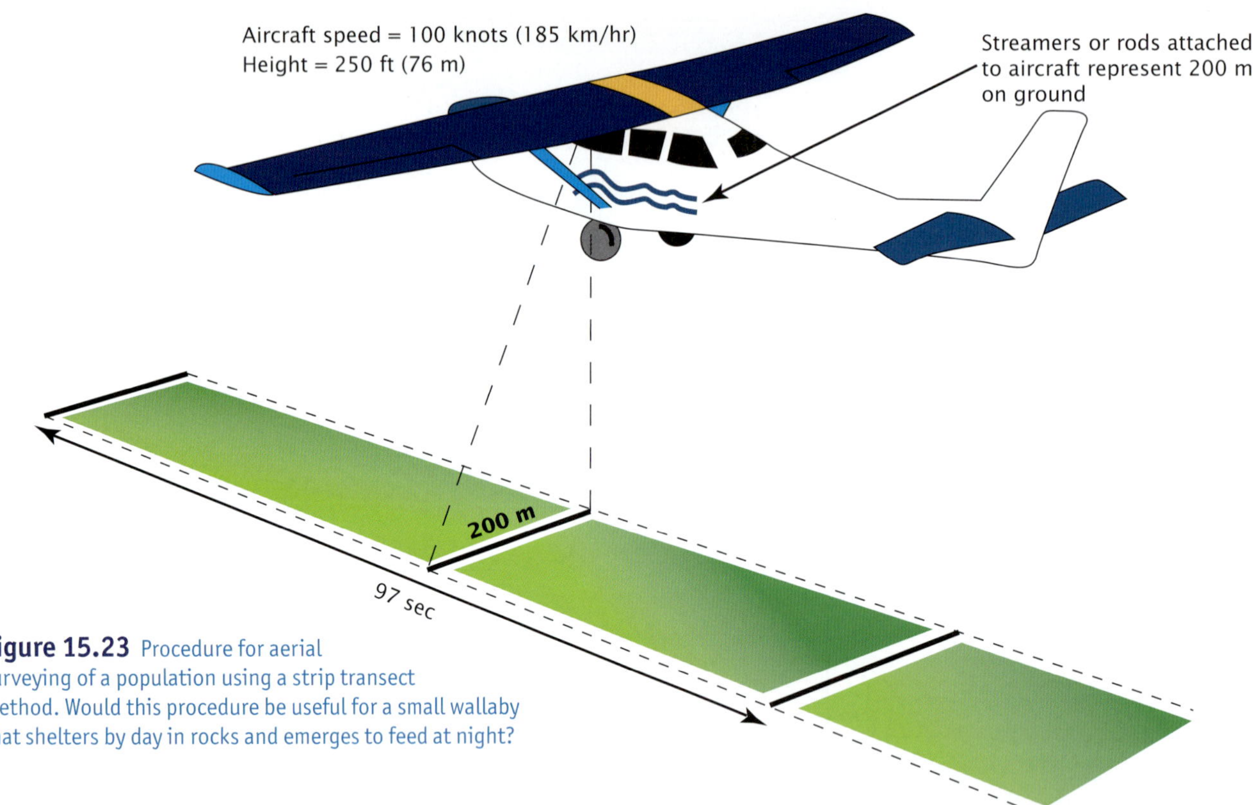

Figure 15.23 Procedure for aerial surveying of a population using a strip transect method. Would this procedure be useful for a small wallaby that shelters by day in rocks and emerges to feed at night?

Transects under the sea

Scientists use strip transects when they study populations of the crown-of-thorns starfish (*Acanthaster planci*) on the edges of reefs in the Great Barrier Reef. To do a strip transect, a snorkel diver holds a so-called manta board (as described by Ian Miller on page 479) that is attached to a boat by a long rope (17 metres) (see figure 15.24a and b). The diver is towed at a constant speed of about 4 km per hour for a two-minute period. The diver counts crown-of-thorns starfish numbers on the reef below and, at the end of that period, the diver records the data (on waterproof paper, of course). Each year about 100 reefs are surveyed using this method. Knowing the population abundance of the crown-of-thorns starfish is important. When the population reaches a density of greater than one starfish per two-minute tow over a particular reef, the situation is identified as an 'active' outbreak or population explosion.

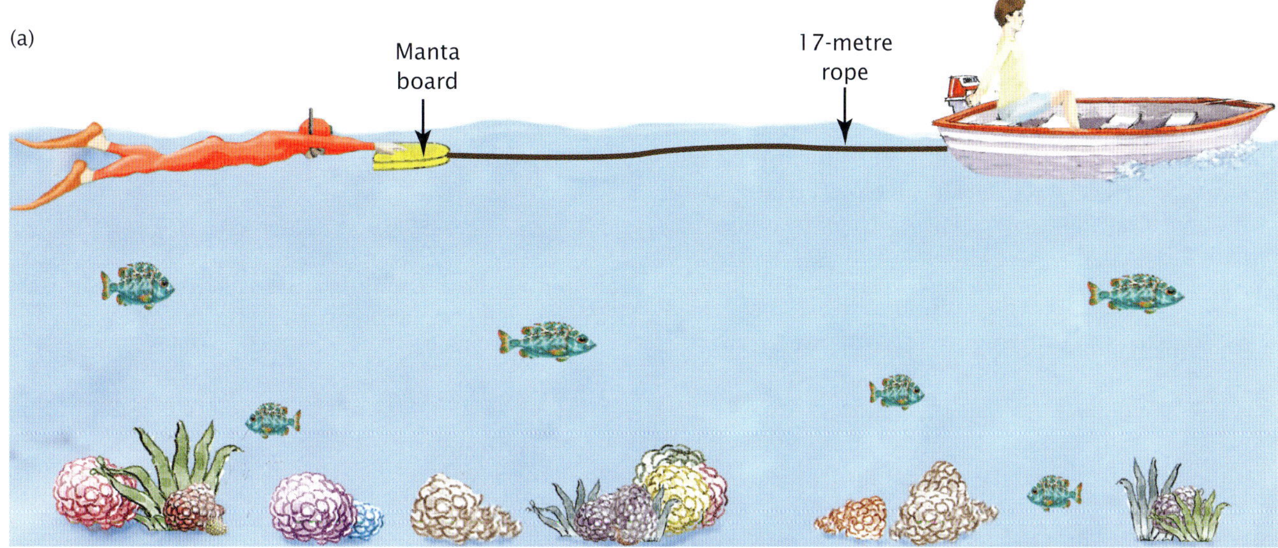

Figure 15.24 (a) Arrangement for an underwater strip transect using a tow. **(b)** A diver under tow. The diver can ascend or descend by changing the angle of the board.

Figure 15.25 A plankton net can be used to sample a population.

Fine-scale surveys of reefs are also performed to calculate the population density of the crown-of-thorns starfish at particular locations. In these surveys, several scuba divers swim down and examine one transect (50 metres long by 5 metres wide) and record all the crown-of-thorns starfish, including juveniles too small to be detected by the strip transect using a manta tow.

'Mark-recapture' technique

Members of many animal populations are small or are difficult to see or are very mobile — here one moment, gone the next! Various sampling techniques have been developed to survey and assess the abundance of animal populations. The technique used will be influenced by factors such as the size of the species and its habitat. For example, microscopic aquatic organisms, such as plankton from a pond, can be sampled using a plankton net (see figure 15.25). Their abundance is then estimated by placing a given volume of the water to a depth of 1 mm in a rectangular cavity chamber measuring 5 × 2 mm and counting the various species.

Sampling of zooplankton in the open ocean uses traditional nets and trawls as well as submersibles and other autonomous vehicles.

The so-called 'mark-recapture' technique can be used with various animal species. The mark-recapture technique involves collecting a sample of the animal population being studied by trapping with mist nets in the case of birds, or using light traps in the case of moths. The trapped animals are marked in some way (leg bands for birds, tiny spots of harmless paint for moths) and are then released (see figure 15.26). Later, another sample of the population is trapped. From these data, it is possible to estimate the size of the population (see the box on page 486).

Figure 15.26 Mark-recapture is a technique used to estimate the size of some animal populations.

MARK-RECAPTURE TO ESTIMATE POPULATION SIZE

A biologist captures a sample of 36 possums in a woodland, marks them all and releases them back into the woodland. One week later, the biologist returns to the same area and captures a second sample of possums. The size of this second sample turns out to be 30 possums and the biologist notes that this sample consists of 12 marked and 18 unmarked possums.

Let the total number of animals marked and released
= M = 36

Let the total number of captured animals
= C = 30

Let the number of captured animals that are marked
= R = 12

We can estimate the size (N) of the possum population in this woodland using these data. The estimate depends on the assumption that the proportion of marked possums in the captured sample ($\frac{R}{C}$) is the same as the proportion of marked possums in the total population ($\frac{M}{N}$).

Using this assumption, we can state that $\frac{M}{N} = \frac{R}{C}$

Solving for N, we can write:

$$N = \frac{MC}{R}$$

Substituting our possum data, we can estimate the size of the possum population as:

$$N = \frac{36 \times 30}{12} = \frac{1080}{12}$$

$$= 90 \text{ possums.}$$

Problems can arise with the mark-recapture procedure if:

- the released animals do not mix at random with the rest of the population after release
- animals migrate into or from the area over the period of study.

Variables affecting population size

The size of the population of a particular species in a given area is not always stable. **Population dynamics** is the study of changes in population size over time.

Variables that influence population size include:

- **birth rate**, that may be expressed numerically (for example, 55 offspring per year)
- **death rate**, that may also be numerically expressed (for example, 10 organisms per week)
- **migration rate**, or the net gain or loss over a given time by movement of individuals either *into* the population from other areas (*immigration*) or *out of* the population to other areas (*emigration*).

The combined action of these variables — birth, death, and migration rates — produce changes in the size of a population over time. This may be represented by the equation:

growth rate = (births + immigration) – (deaths + emigration) per unit time.

The growth rate is positive (e.g. 27 organisms per year) when population size increases. The growth rate is negative when the population size decreases. When gains by births and immigration are matched by losses by deaths and emigration, a population is said to have **zero population growth**.

A population is defined as either *open* or *closed* depending on whether or not migration can occur. Migrations into or out of **closed populations** is nil, unlike **open populations**. Closed populations are isolated from other populations of the same species, as for example, a lizard population on an isolated island. Other closed populations include monkey populations in closed forests on various mountains where the mountains are separated by open grassland and desert that the monkeys cannot cross. Closed populations are less common among bird species. Why?

Information about the abundance and the distribution of a population just tells us about the population at one point in time. The area of population dynamics deals with changes in population size over time. Models relating to changes in population size over time have been developed and we will explore some of these models in the next section.

KEY IDEAS

- To assess population sizes, sampling techniques are more common than total censuses.
- Population sampling can involve the use of quadrats and transects.
- The 'mark-recapture' technique is one method of estimating population size.
- Variables that affect population size are births, deaths, immigration and emigration.
- Population dynamics is the study of changes in population size over time.

QUICK-CHECK

7 Identify the following as true or false.
   a Quadrats can only be used to sample plant populations.
   b 'Mark-recapture' techniques are used for plant populations.
   c An open population has no migration.
   d Population sampling using strip transects can be done from the air.
8 Which is more common: a total count or a sampling of a population?
9 What is the area of one belt transect that is 200 metres wide and 500 metres long?
10 List four variables that affect the size of an open population.
11 In a closed population with zero population growth, how do birth and death rates compare?

Models of population growth

Figure 15.27 Exotic species may undergo exponential growth when first introduced into a new environment with ideal conditions.

Population dynamics deals with changes in population size over time. Is a population increasing, decreasing or remaining unchanged in size? If changing, how fast is the change occurring? Models relating to changes in population size over time have been developed. Models of growth in closed populations include:

- exponential or unlimited growth model
- logistic or density-dependent growth model.

These models are simplifications and use just a few variables. In reality, the dynamics of populations in the real world are very complex and involve many variables. These models, however, give some insight into how populations change in size over time.

Exponential growth: no limits!

Exponential growth is the unlimited growth of a population. This pattern of growth can occur for several generations at least as long as resources are abundant. Exponential growth would be expected to occur when a new species is introduced into a lake or onto an island where there are no predators, no disease and where food and other resources are in plentiful supply (see figure 15.27). The rapid growth in numbers of some introduced pests in Australia, such as cane toads (*Bufo marinus*), is due in part to a period of exponential growth.

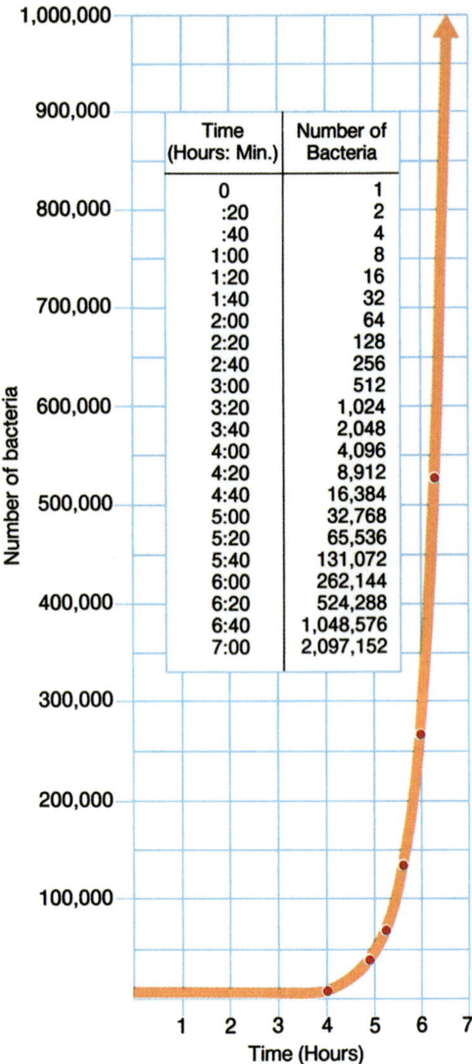

Time (Hours: Min.)	Number of Bacteria
0	1
:20	2
:40	4
1:00	8
1:20	16
1:40	32
2:00	64
2:20	128
2:40	256
3:00	512
3:20	1,024
3:40	2,048
4:00	4,096
4:20	8,912
4:40	16,384
5:00	32,768
5:20	65,536
5:40	131,072
6:00	262,144
6:20	524,288
6:40	1,048,576
7:00	2,097,152

Figure 15.28 Exponential growth of bacteria over a seven-hour period starting with a single cell. What increase in population size occurs over the first three generations (from time = 0 to time = 1 hour)? What increase in population size occurs over the period from four to five hours?

Exponential growth is also seen in the growth of bacteria. For example, consider a bacterial species in which each bacterial cell divides by asexual reproduction to give two cells every 20 minutes. Figure 15.28 shows the theoretical outcome of this pattern of growth starting with a single bacterial cell.

With exponential growth, the increase in population size over each generation is not identical. So, in a population that is growing exponentially, the population size in one generation depends on (i) the population size in the previous generation and (ii) the rate of increase. This means that as the population increases in size, the growth over each generation also becomes larger.

The pattern for exponential growth is always the same, namely an initial slow increase in the size of the population when the population size is small. (This is the **lag phase** that corresponds to the nearly flat section of the graph.) Then, as the population size increases, the numbers increase more sharply in each generation. (This is the **acceleration phase** and begins when the curve turns around the bend.) After this, the size grows more and more rapidly with each generation and the gradient of the graph is almost vertical.) The graph of exponential growth is called a **J-shaped curve** (for obvious reasons). The generation times involved may be minutes, days or weeks depending on the species. What is the generation time in the bacteria as shown in figure 15.28?

Exponential growth in populations with discrete generations can be expressed by the equation below ('discrete generations' means that all the individuals in a population belong to the same **cohort** or age group):

$$P_{next} = r \times P_{previous}$$

where P_{next} = new population size

r = rate of increase

$P_{previous}$ = population size in previous generation.

In the bacterial population shown in figure 15.28, the rate of increase (r) is two because each microbe produces two new microbes.

Table 15.1 Exponential growth in a bushfly population over 8 generations with r = 50

Generation	*Total population*
0	2
1	100
2	5 000
3	250 000
4	12 500 000
5	625 000 000
6	31 250 000 000
7	1 562 500 000 000
8	78 125 000 000 000

Consider the example of the Australian bushfly (*Musca vetustissima*). Let's start a population with just one female bushfly and her mate. Assume that she lays 100 eggs and dies soon after. Of the eggs, assume that 50 develop into females with a generation time of 8 weeks. Table 15.1 shows the growth in the bushfly population that would occur if the population grew exponentially, assuming a rate of increase of 50.

If exponential growth occurred, this single female bushfly and her mate would have 31 billion descendants in just under one year! In reality, however, this number *cannot* materialise because exponential growth of populations cannot occur indefinitely.

The conditions required for exponential growth — unlimited resources such as food and space — can last for only a few generations. Every habitat has limited resources and can support only populations of a limited size. So, let's look at another model of population growth that gives a better fit with reality.

ODD FACT

The concept of unlimited or exponential growth of populations was introduced by Thomas Malthus in his 1798 essay, *The Principle of Population*. Malthus wrote about growth of human populations and linked over-population with human misery.

ODD FACT

In bacterial populations, population growth can slow and stop from the build-up of bacterial waste products that pollute the environment in which the bacteria live.

Logistic growth: S-shapes

A pair of rabbits in a suitable habitat with abundant food and space initially multiplies and, over several generations, the population grows faster and faster; this is a period of exponential growth. However, this rate of growth does not continue. Growth slows and finally stops when the so-called **carrying capacity** of the habitat is reached. The carrying capacity is the maximum population size that a habitat can support in a sustained manner and is denoted by the symbol, K.

To survive, members of a population must have access to particular resources in their habitat. In the case of plants, these resources include space, sunlight, water and mineral nutrients. In the case of animals, necessary resources include food, water and space for shelter and breeding. Resources are limited. Look at figure 15.29 and identify a resource that would be expected to be limited.

Figure 15.29 What resource would be expected to be limited in these habitats for plants? for animals?

As a population increases in size, the pressure on resources increases and population growth slows and then stops. This results from several factors including competition between members of the same population for scarce resources as well as with members of other populations living in the same habitat. This competition can lead to starvation, overcrowding, spread of disease and parasites and increased predation and slows population growth. The impact of these factors on individuals in a population depends on the population size. When a population is small, the impact of these factors is low or absent. When a population becomes large, however, these factors have a major impact on each member of a population. Factors whose impact is related to population size are said to be **density dependent**.

Populations are also affected by factors, such as drought, bushfire and flood, that are **density independent**. The impact of these factors is not influenced by the size of a population; for example, a fire has the same effect on members of a small population as a large population. Collectively, the impact of factors that act as natural checks on growth of populations is termed the **environmental resistance** to growth.

Population growth in the presence of these limiting factors follows a pattern that is termed **logistic growth**, also known as density-dependent growth. When the population size is well below the carrying capacity (K), the growth of the population is rapid, but as the population size approaches carrying capacity, growth slows and stops. The box on page 490 introduces the mathematical basis of this model of growth.

The growth of a population under the density-dependent condition is shown in figure 15.30. The growth is at first like the exponential growth pattern, but as the population grows, the rate of growth slows and finally stabilises at the carrying capacity. This pattern is known as an **S-shaped curve**. The early part of an S-shaped curve has both a lag phase and an acceleration phase. As environmental resistance to growth becomes stronger, population growth slows. This period of slowing is the **deceleration phase**. Then the population size stabilises at the value of the carrying capacity and this is shown in the flattening out of the growth curve. At this stage, the growth rate is zero.

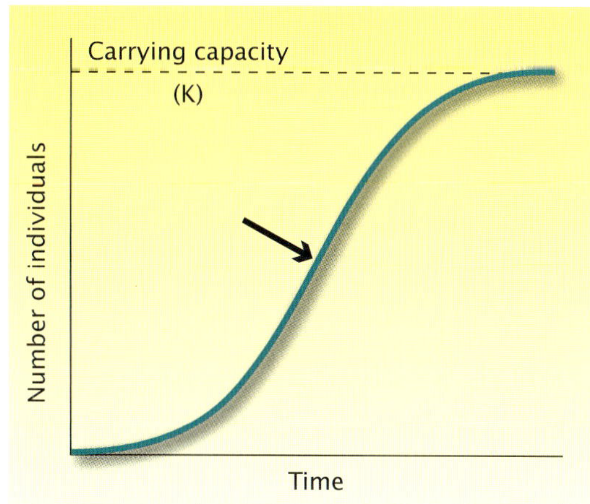

Figure 15.30 An S-shaped curve that is typical of the growth of most populations. The upper limit of this curve is determined by the carrying capacity of the habitat. The arrow marks the point of maximum growth of the population.

ODD FACT

A sad consequence of the use, in long-line fishing, of millions of baited hooks is the death of large numbers of albatrosses (*Diomedea exulans*). When the hooks do not sink immediately, the birds may take the floating baits. This massive destruction of albatrosses has, in turn, endangered this species.

Understanding the S-shaped growth curve can assist decision making about populations. If, for example, a slow-breeding population, such as shark species, is over-exploited by excessive harvesting, then the population size can drop to low levels that push its growth rate back to the lag phase of the S curve. In this case, the recovery time for the population to return to larger size will be very slow, if recovery does occur. This is particularly important for species that have a long period of development before sexual maturity is reached and produce small numbers of offspring (the K-selected species).

The dramatic effects of over-exploitation can be seen in the case of southern bluefin tuna (*Thunnus maccoyii*) and the effects of Japanese long-line fishing in the period 1952–1991 (see figure 15.31). Southern bluefin tuna live for 30-plus years and do not reach sexual maturity until they are 8 to 12 years old.

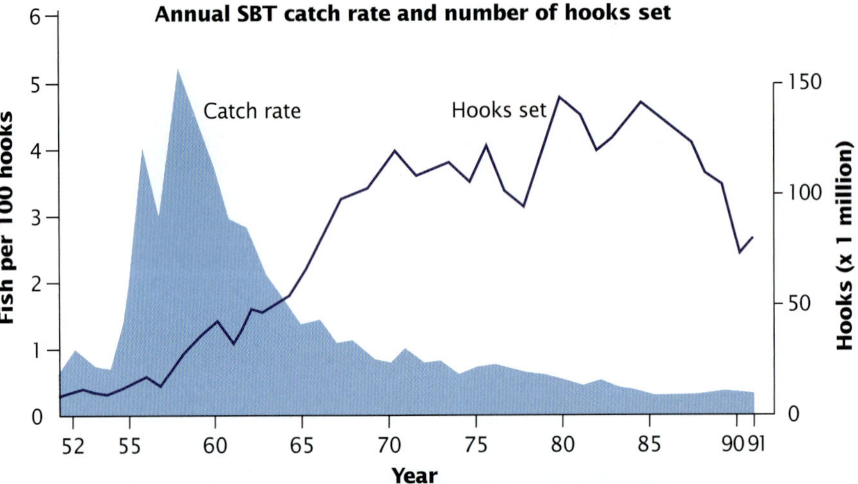

Figure 15.31 Over-exploitation of southern bluefin tuna (*Thunnus maccoyii*) by Japanese long-line fishing in the period 1952 to 1991. Notice how the catch rate has fallen dramatically and how the number of hooks set has increased. When was it first apparent that this rate of harvesting was unsustainable?

LOGISTIC (DENSITY-DEPENDENT) GROWTH: S-SHAPED CURVES

The information in this box is for the mathematically inclined student. Do not worry if that description does not fit you — just skip this box.

Density-dependent or logistic growth is expressed by the following equation:

$$P_{next} = P_{previous} + r\,P_{previous}\,\frac{(K - P_{previous})}{K}$$

where P_{next} = new population size

r = rate of increase

$P_{previous}$ = population size in previous generation

K = carrying capacity.

Part of this equation is like the exponential growth equation ($rP_{previous}$). Note that this value is multiplied by another expression, $\frac{(K - P_{previous})}{K}$.

When the population is low, the value of $\frac{(K - P_{previous})}{K}$ is nearly 1 and the growth is exponential. However, as the population size increases, the value of this expression decreases, and as the carrying capacity is approached, the value of $\frac{(K - P_{previous})}{K}$ approaches zero and growth stops.

Table 15.2 The value of $\frac{(K - P_{previous})}{K}$ varies depending on population size. In this case, the carrying capacity (K) of the habitat is 1000.

$P_{previous}$	$\frac{K-P_{previous}}{K}$
1	(1000–1)/1000 = 0.99
100	(1000–100)/1000 = 0.90
500	(1000–500)/1000 = 0.50
900	(1000–900)/1000 = 0.10
999	(1000–999)/1000 = 0.001
1000	(1000–1000)/1000 = 0.00

ODD FACT

In 1984, Australia's southern bluefin tuna quota was set at 14 500 tonnes but, in 1988, this was cut to 6250 tonnes and a further cut was made in 1989 to the current level of 5265 tonnes. The initial 1984 quota was unsustainable.

Overfishing using surface long-lining put the southern bluefin tuna populations at risk as well as so-called 'bycatch' species, that is, non-target species that are also caught by baits on the long-lines — mainly seabirds, such as petrels and various albatross species. It is estimated that the bycatch from long-line fishing can be up to 20 times the catch of tuna, the target species.

In 1984, Australia, New Zealand and Japan formed the Commission for the Conservation of Southern Bluefin Tuna (CCSBT) in order to ensure sustainable management of the tuna population. There are indications that the tuna population is stabilising but further research is needed to test this conclusion.

Populations affect other populations

Population size of one species can be affected by the size of the population of another species. For example, the size of a plant population will be affected by the size of the populations of herbivores that feed on that plant.

Other density-dependent factors that influence the size of one population include the size of populations of its parasites and its predators. Let's look at how predator and prey populations interact and the impacts on their population sizes.

Predator and prey population numbers

Population size of a prey species can be affected by the size of the population of a predator species that feeds on it. Over time, several outcomes are possible:

- If the predators are absent, the prey population will increase exponentially but will eventually 'crash' when its numbers become too high to be supported by the food resources in the habitat.
- If the prey population is too small, the predator population will starve and die.

In some cases, cycles of 'boom-and-bust' can be seen in both populations, with the peak in the predator population occurring after the peak in the prey population. Why? Figure 15.32 shows the theoretical expectation of these boom-bust cycles while figure 15.33 shows the result obtained in an actual experimental study.

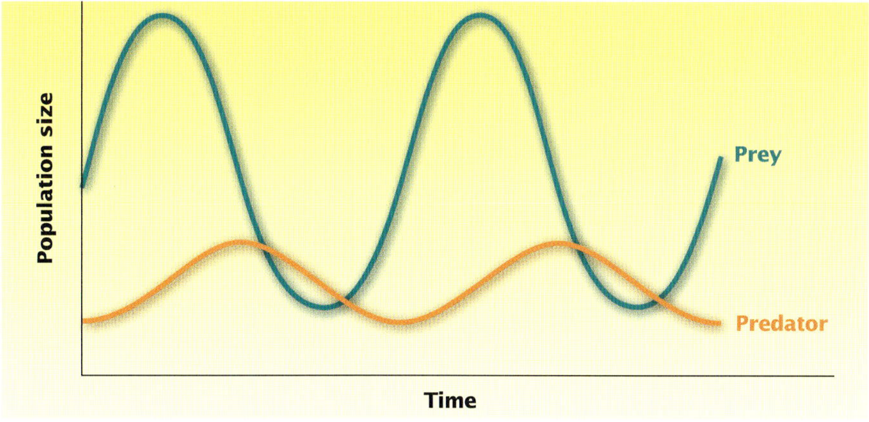

Figure 15.32 Fluctuations in population size in a prey population and in the predator population that feeds on it. Which population peaks first in each cycle — predator or prey? Can you suggest why?

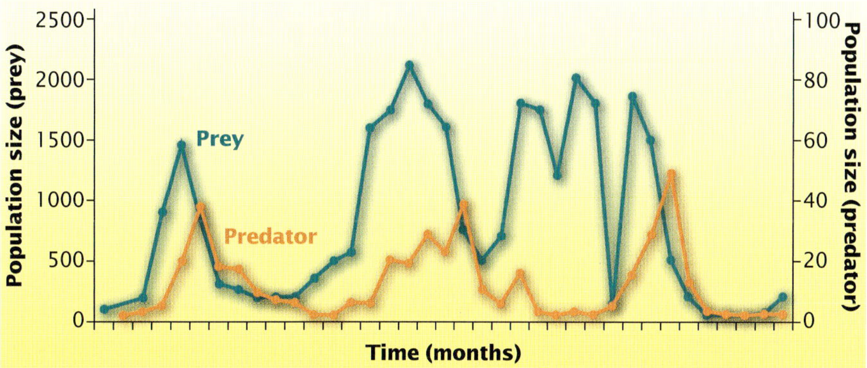

Figure 15.33 Results from an actual study of boom-bust cycles in predator and prey populations

KEY IDEAS

- Population size is the result of birth, death and migration rates.
- Exponential growth follows a J-shaped curve with a lag phase and an acceleration phase.
- Exponential growth cannot continue indefinitely.
- Factors that act to slow and stop population growth are collectively termed environmental resistance to growth.
- Logistic or density-dependent growth follows an S-shaped curve.
- In the logistic growth model, population size typically stabilises at the carrying capacity of the ecosystem concerned.
- Population size of one species is affected by population size of another species in the community.

QUICK-CHECK

12 Identify the following as true or false.
  a In exponential growth, the increase in population size each generation is the same.
  b Floods are an example of a density-independent environmental factor.
  c Environmental resistance to growth produces the deceleration phase in logistic growth.
  d Increases in prey population size are expected to be followed by increases in the predator population size
13 What is the name given to the curve for logistic population growth?
14 In which pattern(s) of growth is a deceleration phase present?
15 Identify a density-dependent factor that would be expected to limit population growth.
16 Give one cause for the 'crash' of a prey population.

Intrinsic growth rates

The letter 'r' comes from the equation for population growth where r = rate of increase per generation.

In the previous section we looked at growth of populations in general. Populations of different species vary in their intrinsic rates of increase, typically denoted by the symbol r.

Populations of some species are short-lived and produce very large numbers of offspring. As we saw in chapter 12 (pages 381–2), species that use this 'quick-and-many' strategy put their energy into reproduction and are said to be **r-selected**. Examples of r-selected species include bacteria, oysters, cane toads, crown-of-thorns starfish, many species of reef fish, clams, coral polyps, many weed species, rabbits and mice. In general, r-selection is directed to quantity of offspring.

The letter 'K' comes from the equation for population growth where K = carrying capacity of a habitat.

At the other end of the spectrum are populations of other species that produce small numbers of offspring at less frequent intervals. Species that use this 'slower-and-fewer' strategy put their energy into development and are said to be **K-selected**. Examples of K-selected species include gorillas, whales, elephants, albatrosses, penguins and many shark species. In general, K-selection is for quality of offspring.

Table 15.3 identifies some of the differences between the extremes of r-selected and K-selected species. Not every species will display all the features of one strategy and many have intermediate strategies.

Table 15.3 Extremes of reproductive strategies compared

Feature	*r-selected strategy*	*K-selected strategy*
general occurrence	commonly seen in oysters, clams, scallops, bony fish, amphibians, some birds and some mammals, such as mice and rabbits	commonly seen in sharks, some birds, such as penguins, and some mammals, such as whales and gorillas
life span	shorter-lived	longer-lived
number of offspring	many	fewer
energy needed to make an organism	smaller	greater
survivorship	very low in young	higher in young
growth rate	faster	slow
age at sexual maturity	early in life	later in life
period that embryo is retained in egg (or in mother's body)	shorter	longer
parental care	little, if any	extensive
adapted for	rapid population expansion	living in densities at or near carrying capacity
relative energy investments	higher investment into numbers of offspring; lower into rearing offspring	lower investment into numbers of offspring; higher into rearing offspring

Figure 15.34 The spawn of cane toads can be readily distinguished from those of native frogs by their appearance as black eggs embedded in long jelly-like strings. How many eggs are produced on average in a cane toad spawning?

r-selected strategy

Populations of species that operate using the 'quick-and-many' strategy can show major fluctuations in population sizes. When resources are plentiful and other environmental conditions are favourable, the numbers of an r-selected species can increase very rapidly because of their short generation times and the large numbers of offspring that they produce. A short generation time means that the lag phase in the growth of an r-selected population is relatively much shorter (weeks) than in a K-selected species where generation times may be several years. When conditions become unfavourable, the very low survival rates of offspring mean that population numbers drop sharply. These r-selected populations, however, can recover quickly because of their high growth rates.

The r-selected populations are adapted for life in 'high risk' and unstable environments, where factors such as flood and drought operate, and in 'new' habitats such as on fresh lava flows.

Some r-selected strategists

The following examples identify species that are r-selected in terms of the numbers of offspring that they produce.

- Cane toads (*Bufo marinus*) are an introduced pest species (see also page 261). Cane toads spawn twice yearly and, at each spawning, one mature female cane toad produces an average of 20 000 eggs that are fertilised externally. Within one to three days, fertilised eggs (see figure 15.34) hatch into tadpoles that metamorphose into small toads very quickly. Within a year, these toads are ready to reproduce and will do so for a period of several years.

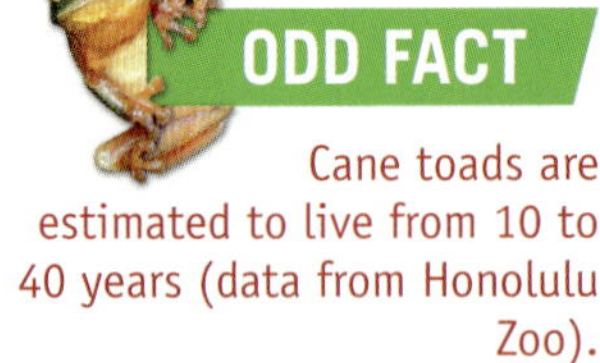

Cane toads are estimated to live from 10 to 40 years (data from Honolulu Zoo).

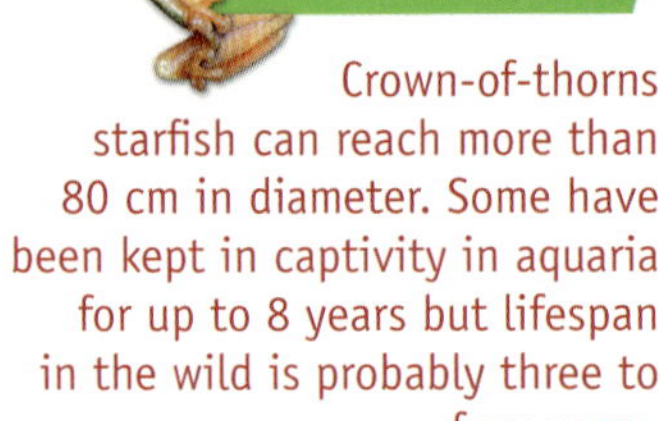

Crown-of-thorns starfish can reach more than 80 cm in diameter. Some have been kept in captivity in aquaria for up to 8 years but lifespan in the wild is probably three to four years.

- The crown-of-thorns starfish (*Acanthaster planci*) (see figure 15.14, page 478) breeds in the waters of the Great Barrier Reef when water temperatures reach about 28°C. Female starfish release eggs into the water where they are fertilised by sperm released by nearby male starfish. In one spawning season, a female crown-of-thorns starfish may release up to 60 million eggs. Water currents disperse the fertilised eggs and larval development is completed over a two-week period. The tiny larvae settle on coral reefs and develop into juvenile starfish about 0.5 mm in size. These juveniles feed on algae for about 6 months and then, when they are about 1 cm in diameter, they begin to feed on coral polyps. Sexual maturity is reached after about two years.

K-selected strategy

Populations of species that operate using the 'slower-and-fewer' strategy live in habitats in stable environments and with their population sizes at or near the carrying capacities of their habitats. K-selected species are adapted to cope with strong competition for resources. If the population size of a K-selected species drops sharply as a result of fire, flood or habitat loss, the population will not recover quickly because of their long generation times and their low rates of increase. K-selected species are at great risk of extinction if their population numbers fall.

Some K-selected strategists

The following examples identify some species that are K-selected in terms of their reproductive strategy.

- In the late autumn and winter each year, humpback whales (*Megaptera novaeangliae*) (see figure 12.14, page 382) migrate up the east coast of Australia to breeding grounds in tropical or semi-tropical waters. It is here that the whales mate and it is also here that pregnant females give birth during the southern hemisphere winter. Humpback whales show many of the features of a K-selected species as follows:
  - — sexual maturity in humpback whales does not occur until whales are about five years old
  - — gestation (pregnancy) in humpback whales lasts about 11.5 months
  - — each female gives birth to just one calf every one or two years
  - — for an average of 10 months after its birth, a mother suckles her calf on milk
  - — the life span of humpback whales is up to 50 years.
- The orange roughy (*Hoplostethus atlanticus*) (see figure 15.35) is a species of fish found in deep waters off south-east Australia. This species is slow-growing and does not reach sexual maturity until it is about 30 years old. At this stage, the fish has a length of about 30 cm.

ODD FACT

How can you tell the age of a fish? It is possible to tell the age of fish from an examination of the growth rings of their scales or from the stony otoliths in their ears. Using this method, some orange roughy have been identified as more than 70 years old.

All species do not dovetail into r-selected or K-selected species. For example, green turtles (*Chelonia mydas*) exhibit r-selection in terms of the numbers of eggs produced and the lack of parental care but they also show some K-selection through features such as their slow growth rates, the period required for sexual maturity (estimated at 40 to 50 years) and their long life span (estimated at 70 years).

A female green turtle lays a clutch of about 100 eggs on a sandy beach at night, covers them with sand and returns to the water (see figure 15.36b). During the breeding season, she returns to the same beach about every two weeks and may lay between three to nine clutches of about 100 eggs per clutch. About two months later, baby turtles hatch from the eggs, dig their way out of the nest and make their way to the sea (see figure 15.36c).

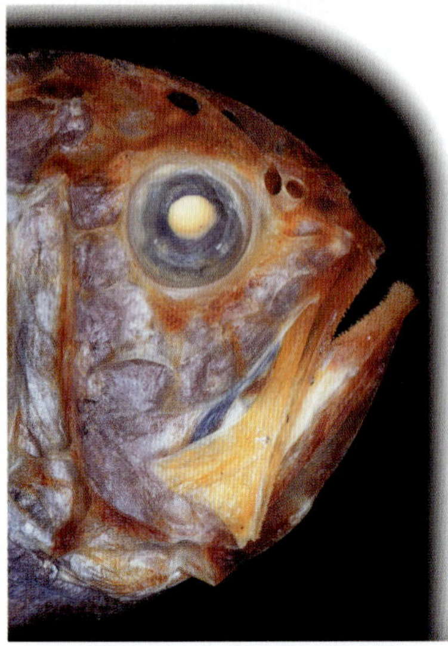

Figure 15.35 The orange roughy (*Hoplostethus atlanticus*)

Figure 15.36 **(a)** Green turtles are common in the waters of the Great Barrier Reef. This species has a worldwide range in tropical and semi-tropical waters. **(b)** Female green turtle laying eggs. The temperature at which the eggs develop determines the sex of the turtles: lower temperatures produce males, while higher temperatures produce females. **(c)** Turtle hatchlings making their way to the sea. If this occurs during the day, predators such as seabirds and crabs will take many of the hatchlings before they reach the sea.

KEY IDEAS

- Populations differ in their intrinsic rates of growth.
- Species can be identified as being r-selected or as K-selected.
- r-selection and K-selection are the extremes of a range.
- r-selected species are adapted for living in newly created and in unstable habitats.
- K-selected species are adapted for living in stable habitats and at densities at or near the carrying capacity of a habitat.

QUICK-CHECK

17 Which species, r-selected or K-selected, would be expected:
   a to recover more quickly after its population was reduced?
   b to be at greater risk of extinction through habitat destruction?
18 Contrast r-selection and K-selection in terms of:
   a number of offspring
   b growth rates
   c age at sexual maturity.
19 Give an example of:
   a a K-selected species
   b an r-selected species.

BIOCHALLENGE

Examine diagrams A to D in turn.

1 For each of diagrams A to D, briefly outline the story that it tells about populations.
2 For diagrams C and D, identify a label for both the horizontal and the vertical axes.
3 Now refer to photograph M. Select one of the diagrams (A to D) that is relevant to photograph M and describe how the two relate.
4 Now refer to photograph N. Select one of the diagrams (A to D) that is relevant to photograph N and describe how the two relate.

A

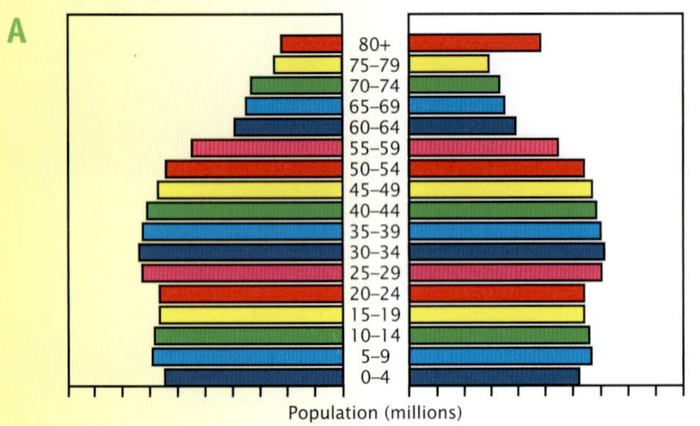

B

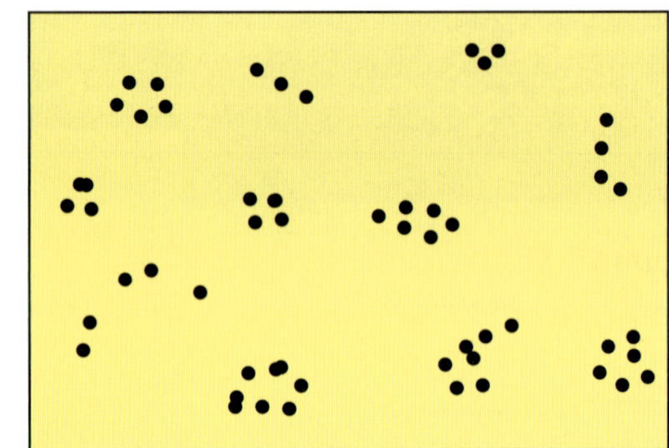

C

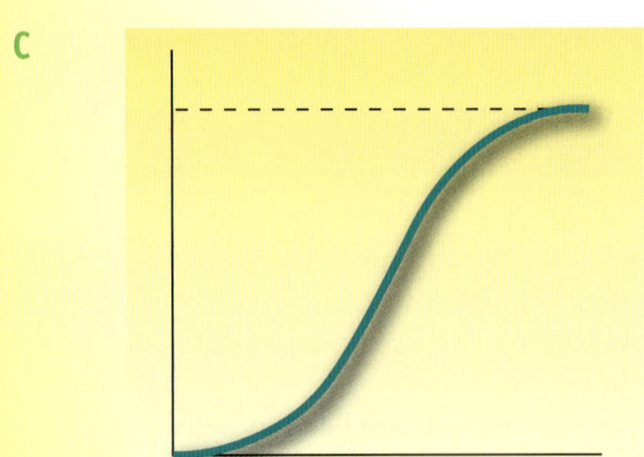

D

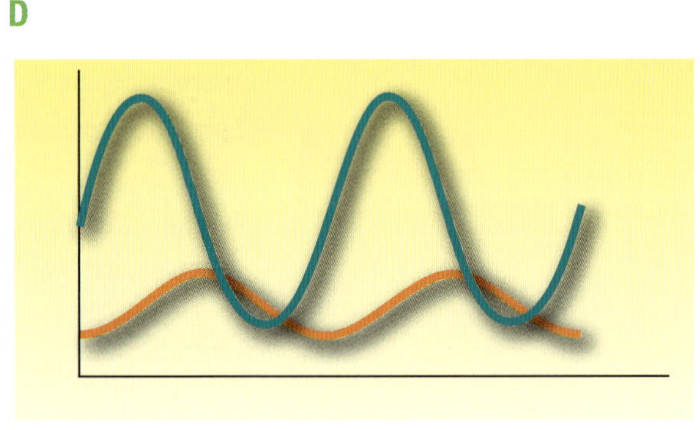

M

N

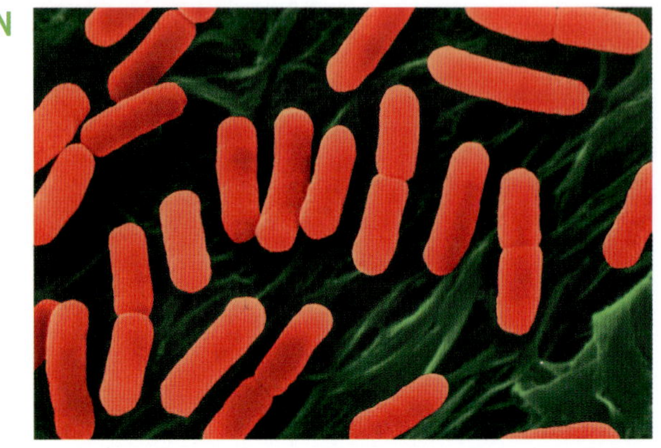

CHAPTER REVIEW

Key words

CROSSWORD

abundance
acceleration phase
birth rate
carrying capacity
closed populations
cohort
community
death rate
deceleration phase
density
density dependent
density independent
distribution
environmental resistance
exponential growth
J-shaped curve
K-selected
lag phase
logistic growth
mark-recapture
migration rate
open populations
population
population dynamics
quadrats
r-selected
S-shaped curve
sampling
species richness
total count
transects
true census
zero population growth

Questions

1 *Making connections* ▸ Using as many as possible of the key words or terms above, prepare a concept map.

2 *Demonstrating your understanding* ▸ Complete the following sentences by choosing the more/most appropriate word.
   a When population size equals K, growth (accelerates/slows/stops).
   b At zero population growth in a closed population, the death rate is (greater than/equal to/less than) the birth rate.
   c r-selected species are adapted for life in (stable/unstable) environments.
   d K-selected species (are/are not) adapted for living in populations at or near the carrying capacity.
   e (Exponential/logistic) growth is typical of a population over the long term.
   f Abundance refers to the (density/distribution) of a population.

3 *Applying your understanding and using data* ▸ Assume a model of exponential population growth in bacteria with a rate of increase (r) equal to 2. Starting with 10 bacteria, how many bacteria will be present after five generations?

4 *Applying your understanding* ▸
   a Identify one method of estimating the population size of:
      i crown-of-thorns starfish on a reef of the Great Barrier Reef
      ii red kangaroos on a large area of open grassland
      iii limpets on a rocky seashore.
   b Of the methods, is any equivalent to a total census? Explain.

5 *Analysing information and drawing conclusions* ▸ Two islands (C and D) in temperate seas differ in their species richness, with island C having twice as many species as D. Of the following four statements, which is the most reasonable?
   a No conclusion is possible.
   b Islands C and D have the same area.
   c Island C has twice the area of island D.
   d Island C has about ten times the area of island D.

6 ***Analysing data*** Table 15.4 contrasts the reproductive strategies in the Pacific black duck (*Anas supercilliosa*) and the wandering albatross (*Diomedea exulans*).

Table 15.4 Attributes of wandering albatross (*Diomedea exulans*) and Pacific black duck (*Anas superciliosa*). Which species breeds earlier? Which breeds once every two years?

Bird species	*Age at first breeding (yr)*	*Clutch size (no. of eggs)*	*Incubation period (days)*	*Parental feeding period*	*Frequency of breeding*
black duck	1–2	11	26–28	n/a	two per year
albatross	8–10	1	80	11 months	one per two years

a Which species would be expected to have the higher offspring mortality?

b Which species is more r-selected than K-selected?

7 ***Applying your understanding*** Identify the following statements as true or false.

a A population that is large must be increasing in size.

b Quadrats can be used to sample populations of fast-moving animals.

c The presence of more than one crown-of-thorns starfish per two-minute tow indicates the start of a population explosion.

d Exponential growth can occur in a population provided resources are not limited.

Figure 15.37 **(a)** The wandering albatross, the most common albatross of Australian oceans, has a wing span of 260–350 cm. What's the span of your outstretched arms? **(b)** Pacific black duck

8 ***Discussion question*** Consider the graph in figure 15.38 that shows the growth of the human population in the period from 10 000 BC to the turn of the 21st century.

a A 'dip' occurred in the curve at about AD 1400. Suggest a possible cause for this dip.

b Does this curve appear to fit an exponential growth model or a logistic growth model?

c What phases can be identified in this graph?

d What time interval was required for the following increases?

i from 1 billion to 3 billion

ii from 3 billion to 5 billion

iii from 5 billion to 6 billion

e Do the data shown in the graph support the assertion that the growth rate of the human population is increasing as the population size grows?

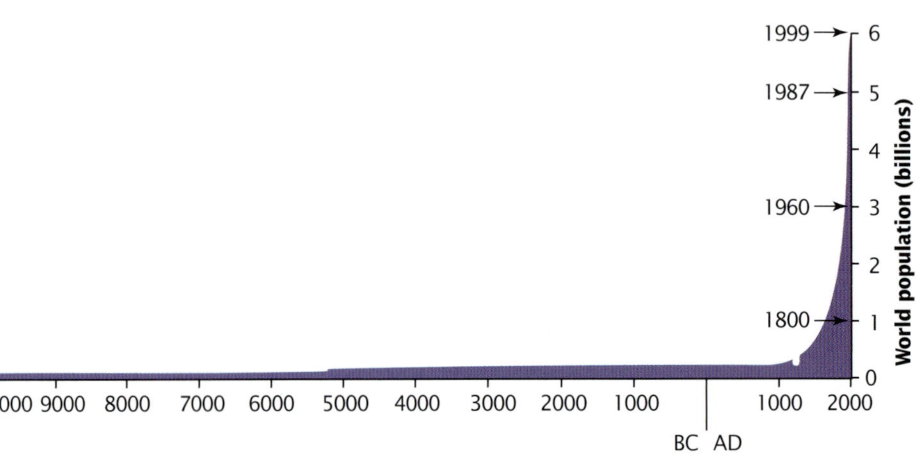

Figure 15.38

9 ***Analysing data*** ▸ The time for a particular human population to double (the 'doubling time') is given by the expression: doubling time = $\frac{70}{\text{annual growth rate}}$. For example, the annual growth rate of the Australian population is 0.93 per cent. So, Australia's doubling time = $\frac{70}{0.93}$ = 75 years.

a What is the theoretical doubling time of Nigeria with an annual growth rate of 2.53 per cent?

b The world's current growth rate is about 1.3 per cent. The world's population in 2005 was 6 billion and assuming this growth rate continues, when would the world's population be expected to reach 12 billion?

10 ***Applying your understanding in new contexts*** ▸ Give an explanation in biological terms for the following observations.

a To assess the diversity of bird species in woodlands in the Mount Lofty Ranges, biologists surveyed bird populations on six different days.

b Bacteria in culture grew exponentially at first but the growth rate then slowed and later stopped.

c An r-selected species can recover more rapidly following an environmental disaster than a K-selected species.

d A population of plants in a particular habitat shows a clumped distribution.

11 ***Discussion question*** ▸

a Biologists used the 'mark-recapture' method to estimate the population size of a particular fish species in a large lake. They captured 240 fish, marked them and returned them to the lake. A second sample of 320 fish was captured some weeks later and 12 of these were found to be marked. Based on these data, what is the estimated size of this population?

b Person MM wished to estimate the fish population in this large lake. He quickly captured and marked a group of fish, returned them to the water and immediately recaptured a second sample of fish. Is the estimate of population size based on this sample likely to be accurate? Explain.

12 ***Web question*** ▸ Go to www.jaconline.com.au/natureofbiology/natbiol1-3e and click on the 'Sharks' weblink for this chapter. This site will allow you to view a simulation that shows changes in the numbers of sharks (predators) shown in red and their fishy prey (shown in green). To start the simulation, click on the right-hand red arrow at the bottom of the screen. Run the simulation for some time and watch as the numbers in the shark and fish populations change.

a Which population is the first to reach maximum numbers?

b Is there any evidence of boom-and-bust cycles?

c Do both predator and prey populations show boom-and-bust cycles?

d Are your findings in agreement with the data shown in figure 15.33 (page 491)? Explain.

16 Changes in ecosystems

KEY KNOWLEDGE

This chapter has been designed to enable you to:

- develop understanding of changes to ecosystems over time
- recognise the scope, intensity and impact of various natural changes
- extend awareness of impacts of human-induced changes on ecosystems
- identify techniques for monitoring and maintaining ecosystems.

Figure 16.1 Here we can see the destructive effects of the spread of salinity in Australia: salt-encrusted land; dead trees; waters with such high levels of salt that native fish can no longer survive; water that is too salty to drink. The change shown here is replicated many times over in Australia and increasing salinity has become an environmental problem of national significance. Presently about five per cent of cultivated land (about 2.5 million hectares) is affected by salinity. Over the next decades, unless remedial action is taken, this could increase to 15 million hectares. In this chapter, we will explore changes in ecosystems arising from both natural causes and from human interventions.

ODD FACT

Gould's petrels form breeding pairs for life and, each year, each pair returns to the same nest site where they produce a single egg.

Cabbage Tree Island

Cabbage Tree Island, located near Port Stephens, New South Wales, is a small rugged area of tropical rainforest that takes its name from the cabbage palms (*Livistona australis*) growing there. The island is home to many bird species, including the endangered Gould's petrel (*Pterodroma leucoptera*) (see figure 16.2), the rarest of Australia's endemic seabirds. Gould's petrels spend most of their life at sea. Each year, after they reach about six years of age, the birds return to shore to breed (October to May).

Figure 16.2 Gould's petrel on Cabbage Tree Island

The major breeding sites of Gould's petrels are two gullies on the western side of the island where they nest in cavities among rocky debris or under fallen palm fronds. In 1953, to protect these breeding sites, Cabbage Tree Island was declared a nature reserve and public access was restricted. However, the number of petrels was declining because of changes resulting from the deliberate introduction of rabbits to the island in 1906.

Once on the island, the rabbits rapidly multiplied and their browsing destroyed the herb and shrub layers. The rabbits also ate seedlings that would eventually have become part of the upper canopy of the closed tropical forest. Over time, the loss of the understorey and the lack of replacement canopy trees changed the closed forest to an open forest in which petrels were exposed to predation by pied currawongs (*Strepera graculina*) and ravens (*Corvus coronoides*).

The loss of the understorey also exposed the petrels to another deadly agent — the fruits of the native bird-lime tree (*Pisonia umbellifera*). The fruits, which are produced during the petrels' breeding season, are covered in a viscous substance that remains sticky for months. When the forest understorey was present, these sticky fruits were caught in foliage above the petrels' nests. However, after the rabbits destroyed the understorey, the bird-lime fruits could fall to the forest floor and petrels moving through the forest became entangled in them. Even a single fruit stuck to its wing is a death sentence for a petrel which, unable to fly, dies either from starvation or by predation.

In 1970, the population of Gould's petrels numbered about 2000. By 1992, it had declined to about 1400 birds because of predation by currawongs and incapacitation by bird-lime fruits. Each year saw less than 50 fledglings successfully reared, a number too low to compensate for adult deaths. The species was declared 'endangered' (see figure 16.3).

Three categories identify the degree of threat of extinction of a species — 'presumed extinct', 'endangered' and 'vulnerable'. Gould's petrels were 'endangered'; that is, they were in danger of extinction unless the causal factors were removed. 'Vulnerable' means likely to become 'endangered'.

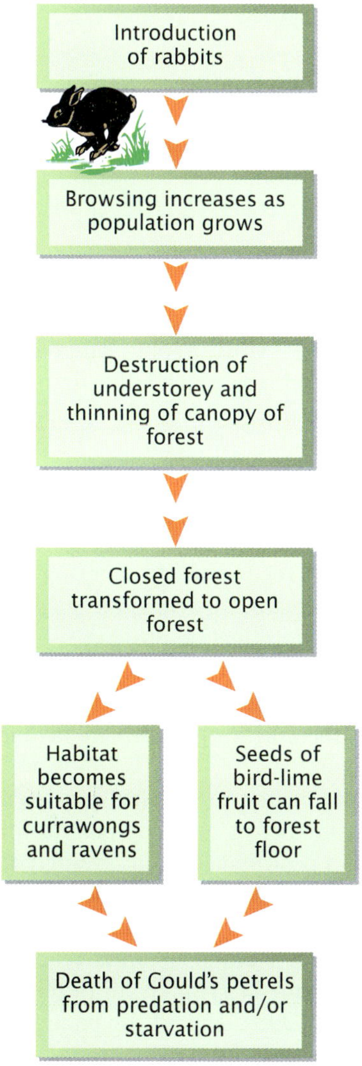

Introduction of rabbits

Browsing increases as population grows

Destruction of understorey and thinning of canopy of forest

Closed forest transformed to open forest

Habitat becomes suitable for currawongs and ravens

Seeds of bird-lime fruit can fall to forest floor

Death of Gould's petrels from predation and/or starvation

Figure 16.3 Changes in the Cabbage Tree Island ecosystem and their consequences. Notice that a single primary change to an ecosystem can trigger a cascade of secondary effects.

ODD FACT

In March 1999, a colony of Gould's petrels was established by translocating 100 nestlings to Boondelbah Island to artificial nesting boxes. In the 2003–2004 breeding season, 10 of these now mature birds returned to this island to breed. They joined another 30 birds attracted to the new breeding site.

ODD FACT

As part of the recovery program for the Gould's petrel population, artificial nest boxes were used to provide nesting sites additional to those naturally available on Cabbage Tree Island.

Measures to restore the balance

Because of the decline of the Gould's petrel population, the New South Wales National Parks and Wildlife Service began a management program in late 1993 that involved:

- culling (destroying) currawongs and ravens and their nestlings
- eradicating bird-lime trees from the petrels' breeding sites
- exterminating rabbits.

Culling the predators

The currawongs and ravens were attracted to Cabbage Tree Island when the closed forest was opened up through the feeding activities of the rabbits. Culling these predatory birds resulted in fewer petrel deaths (see table 16.1).

Table 16.1 Removal of predatory birds and deaths of petrels

	1992–93	1993–94	1994–95	1995–96
currawongs removed	0	22	20	3
ravens removed	0	3	10	0
petrels killed by predatory birds	43	3	3	1
petrel deaths: other causes	8	8	6	6

Removing bird-lime trees

The bird-lime trees in the breeding sites of the Gould's petrels were poisoned with a herbicide in mid-1992. When the petrels arrived to breed in October 1992, the bird-lime trees were dead. No more sticky bird-lime fruits!

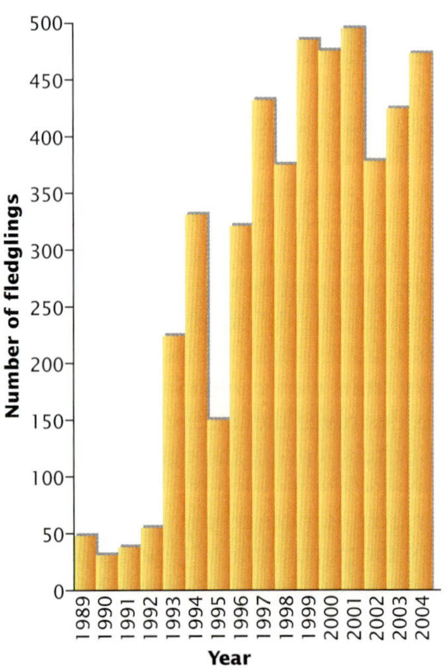

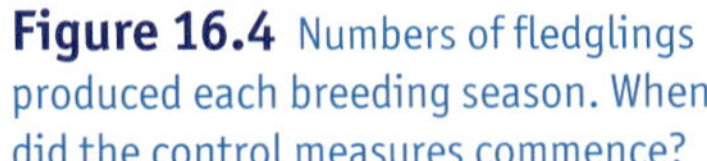

Figure 16.4 Numbers of fledglings produced each breeding season. When did the control measures commence?

At last, the rabbits go!

In March 1997, a program to eradicate rabbits from Cabbage Tree Island commenced. The program used a combination of myxomatosis virus, rabbit calicivirus and poisoning with baits that kill mammals. (This last measure could be used because the island has no native mammals.) Extermination of rabbits was completed in eight weeks. Immediately afterward, the vegetation began to recover and the rainforest began to regenerate. Monitoring for a further six months confirmed the success of the rabbit extermination program.

As the understorey recovered, the forest became closed and the habitat became unsuitable for currawongs, which favour open woodland.

The management program has increased the annual number of breeding pairs from an average of 166 before the program started to a post-management average of almost 800, with a high in 2001 of over 1000 pairs. The number of fledglings increased in the period from 1989 to 2004 (see figure 16.4) and, in 2004, 470 young petrels were successfully reared. As each group of birds reaches maturity, they will return to breed on a Cabbage Tree Island that is far different from the ecologically devastated island of the early 1990s. The box on page 503 identifies the two biologists who saved the Gould's petrel from extinction.

BIOLOGIST AT WORK

Restoring the Cabbage Tree Island ecosystem

'Working as a field biologist can often be extremely arduous, but when it means saving a species from extinction, you easily forget the hardships associated with the uncomfortable hours and physically demanding locations.' These are the comments of David Priddel and Nicholas Carlile, two dedicated conservation biologists with the New South Wales National Parks and Wildlife Service. Their work to save Australia's endangered fauna takes them into remote places, from the dusty arid interior of the continent to the windswept islands that fringe our coastline. By investigating the processes responsible for the decline of a species, these scientists attempt to discover innovative ways to manage and protect Australia's unique biodiversity.

Dr David Priddel is a senior research scientist with over 20 years experience who has worked with a diverse range of species, including mallee fowl, bats, whales and seabirds. His research results are published in scientific journals, magazines and as book chapters. Nicholas Carlile is a research officer who completed tertiary studies in both zoology and botany. With more than 15 years experience in field biology, his research has ranged from devising methods to reduce the overabundance of gulls to recovery actions to save endangered seabirds.

The remarkable success of the Gould's petrel recovery program is a highlight in the careers of both scientists. The increasing number of fledglings departing the island each year vindicates the success of the prescribed management practices and is a great reward for their years of dedicated effort. The most memorable achievement, however, was undoubtedly the eradication of rabbits from Cabbage Tree Island after 90 years of occupation and devastation. This provides continued incentives to apply the newly developed techniques of eradication to other islands where rabbits are a problem.

There can be no better result for field biologists than to know that their work has led to both the recovery of a species and the rehabilitation of an entire ecosystem that was so degraded by alien pests that it was on the verge of collapse. David Priddel and Nicholas Carlile have demonstrated that, provided recovery actions are underpinned by sound scientific research, the recovery of critically endangered species is achievable. For these scientists, however, the challenge lies beyond that of saving and protecting one species. They strive to develop techniques that can be applied by conservation managers elsewhere in Australia and around the globe. Only by sharing knowledge and experience can we be successful in halting the decline of the world's biodiversity.

Figure 16.5 Nicholas Carlile, one of the conservation team, with a Gould's petrel chick on Cabbage Tree Island

KEY IDEAS

- Primary changes in ecosystems may produce a cascade of secondary changes that may not become apparent immediately.
- The Cabbage Tree Island case study provides an example of how damage to an ecosystem can be reversed.

QUICK-CHECK

1 What was the primary change in the Cabbage Tree Island ecosystem that eventually led to the fall in abundance of the Gould's petrels?
2 List two measures that were important in restoring this ecosystem to its original state.
3 Refer back to table 16.1 on page 502.
   a What was the major cause of petrel deaths in the period 1992–93?
   b What was the death rate from the same cause in the period 1993–94?

Changes in ecosystems

Look at figure 16.6. Believe it or not, these photographs are of the same ecosystem — Beatrice Lagoon in the Northern Territory. At left is the lagoon in 1986 when it was a healthy freshwater ecosystem with a diverse community of native birds, fish and other animals. At right is Beatrice Lagoon about ten years later. What happened to this once healthy freshwater ecosystem?

Figure 16.6 Beatrice Lagoon in the Northern Territory at left in 1986. At right is Beatrice Lagoon in 1995.

We can ask the following questions about changes in ecosystems:

1. *What is the primary cause of the change?* The primary cause may be a natural event, such as flood, or the primary cause may be human interventions, such as releasing an exotic (non-native) plant or animal into an ecosystem. Human interventions in ecosystems may be deliberate actions, such as the clearfelling of a forest, or the intervention may be accidental, such as the escape of toxic wastes onto soil or into waterways. In the case of Beatrice Lagoon, the cause of the change in the ecosystem was the introduction to central Queensland of a semi-aquatic exotic grass for cattle food. Unfortunately, this grass, known as olive hymenachne (*Hymenachne amplexicaulis*), spread over wetlands and waterways in northern Queensland and in the Northern Territory, displacing native plants.
2. *What part of the ecosystem is initially changed?* The primary change to an ecosystem may be to the biotic component, that is, the living community, or it may be to the abiotic component, as for example, water, soil or access to light. Primary changes impacting on the biotic part of an ecosystem include actions such as the harvesting of timber in an old-growth forest. Primary changes to the abiotic part of an ecosystem include the release of untreated sewage into waterways. What was the primary cause of the change in the Beatrice Lagoon ecosystem?

Frequency of change

Changes in an ecosystem may be due to:

- regular and predictable events, such as tides and seasons
- sporadic (irregular) events, such as floods
- one-off events, planned or unpredictable, such as a massive oil spill that impacts on a marine habitat.

Changes in some ecosystems due to natural causes may occur over time cycles that are as short as a few hours. For example, ecosystems in the inter-tidal zone along a coastline (see figure 16.7) and ecosystems in the **estuary** of a river (see figure 16.8) are subject to regular change owing to tidal movements.

Figure 16.7 Tidal movements are so predictable that they can be published in advance.

(a)

(b)

Figure 16.8 Regular tidal movements change the environment of mangrove trees whose roots are periodically submerged and exposed. Here we see a mangrove tree **(a)** at high tide and **(b)** at low tide in a river estuary.

Other regular changes that occur in ecosystems include:

- daily (diurnal) changes in light intensity that are part of the day–night cycle
- yearly (annual) changes in climate that are part of the seasons.

Day length (hours of sunlight each day) is an important variable in an ecosystem since it affects the primary productivity of an ecosystem. The seasonal variation in day length depends on latitude (see figure 16.9) with the difference between day length in mid-summer and mid-winter being greater the further an area is from the equator. So, primary productivity of the Antarctic Ocean ecosystem varies greatly throughout the year while, in contrast, the primary productivity of a tropical rainforest ecosystem is fairly constant throughout the year.

Location	Time of year	Daylight (hrs)	Darkness (hrs)
Equator 0°	21 December	12.00	12.00
	21 June	12.00	12.00
Darwin 12.5°S	21 December	12.51	11.9
	21 June	11.24	12.36
Melbourne 38°S	21 December	14.48	9.12
	21 June	9.33	14.27
Heard Island 54°S	21 December	17.11	6.49
	21 June	7.51	16.9
Davis Base 68°S	21 December	24.00	0
	21 June	0	24.00

Figure 16.9 Ecosystems at different latitudes are subject to changing periods of daylight during the year. How would this affect producers in the ecosystem?

Other changes in ecosystems occur less regularly, perhaps on 10- to 20-year cycles, as for example, those due to fires. Figure 16.10 (page 506) shows the pattern of occurrence of **bushfires** in various parts of Australia.

One spectacular example of an irregular change in an ecosystem occurs at Lake Eyre in South Australia. Most of the time, Lake Eyre is not a lake — it is a saltpan composed of layers of salt overlying black mud (see figure 16.11a). About every 30 years, Lake Eyre actually becomes a lake as a result of summer monsoon rains that fall in south-western Queensland, hundreds of kilometres from Lake Eyre. This rain enters dry riverbeds and flows south. Usually, the water soaks into the ground and the rivers dry up long before reaching Lake Eyre. Very rarely, the rainfall is so heavy and persistent that the rivers reach Lake Eyre, filling it with water. When this happens, the saltpan of Lake Eyre is transformed into an inland sea that becomes a breeding ground for many animals — water beetles, fairy shrimp, fish and waterbirds, including pelicans (see figure 16.11b and c).

Lake Eyre is the largest lake in Australia. The first recorded filling of this lake was in 1949.

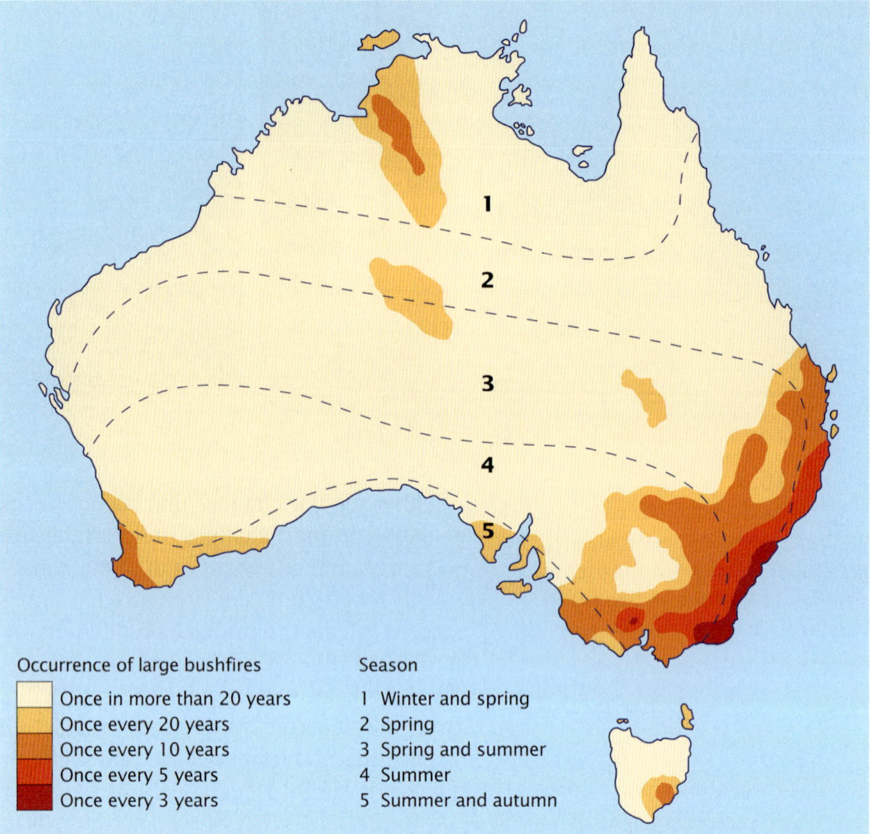

Figure 16.10 Occurrence of bushfires in Australia. The colour code denotes the frequency of fire and the numbers indicate the season of the fires. Can you suggest why northern Queensland is not subject to frequent fires?

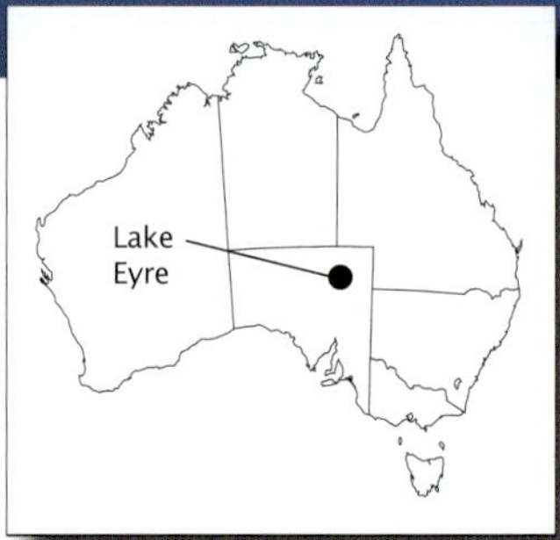

Figure 16.11
(a) Lake Eyre in inland South Australia is usually a dry saltpan. **(b)** Lake Eyre as an inland sea after torrential and persistent rains in south-west Queensland. **(c)** Thousands of pelicans fly great distances to Lake Eyre when it fills, and the pelicans breed on the shores and on mounds in the lake.

Most recently, Tasmania was linked to the mainland by a land bridge until about 13 500 to 8000 years ago when rising sea levels gradually formed Bass Strait.

Ecosystems are also subject to agents that bring about slow, gradual changes. For example, slow rises and falls in sea level occurring over a long period have successively caused the separation and rejoining of Tasmania and the Australian mainland. Even slower changes in ecosystems result from movements of plates of the Earth's crust over millions of years. These movements have, over geological time, changed the patterns of ocean currents and transformed Antarctica from an ecosystem dominated by temperate forests millions of years ago to the present frozen continent.

Changes in ecosystems may be small or large scale in terms of space. For example:

- the floodwaters that occasionally transform arid areas of Australia to an inland sea are large-scale changes in space
- a very small-scale change in space is the collapse of an old tree in a closed tropical forest ecosystem which opens a small area of forest floor to light
- depletion of ozone in the stratosphere is on a global scale.

In some ecosystems, natural changes occur in their communities when the physical or the chemical features of the surroundings change. This natural change in which one community is progressively replaced by another community is known as succession (see page 533).

Small-scale studies of ecosystems over time and space will not reveal changes that are occurring slowly over long time scales. Such gradual changes may be detected only if observations are taken and records are kept over long periods, even over several human lifetimes.

Detection of some changes to ecosystems may require the use of instruments that extend human sensory perception, such as thermographs that measure infrared radiation. Other instruments, such as satellite-mounted sensors, operate over an expanse of space that cannot be replicated by human observers.

Gathering data from a distance, such as from a satellite, is known as **remote sensing**. Remote sensing is in contrast to direct gathering of data by observers on the ground.

Global monitoring of ecosystems

The satellite *Terra* (see figure 16.12) is just one of many orbiting satellites that gathers data about our planet and the ecosystems that it supports. These satellites form the Earth Observing System (EOS). *Terra* carries five sophisticated instruments (see the box on page 508) that gather data on how Earth is changing in response to natural and human-induced changes.

Other satellites carry instruments that can measure the temperature of the sea surface. Figure 16.13 shows a colour-coded image of the surface temperature of the world's oceans on a particular day in 2005. Measurements like this are important in predicting the occurrence of El Niño events that bring drought to many areas of Australia causing changes in ecosystems. During an El Niño event, the surface temperature of the Pacific Ocean rises.

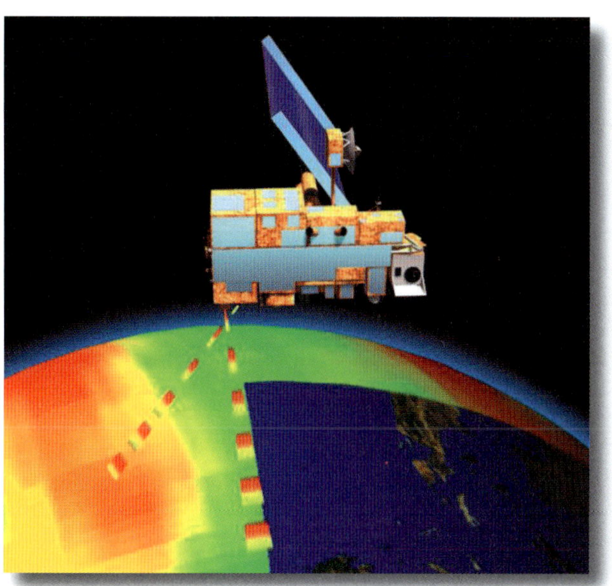

Figure 16.12 *Terra*, the flagship satellite of the Earth Observing System. Specialised instruments carried by *Terra* collect data on the land, oceans and atmosphere of our planet that will provide a record of changes over time.

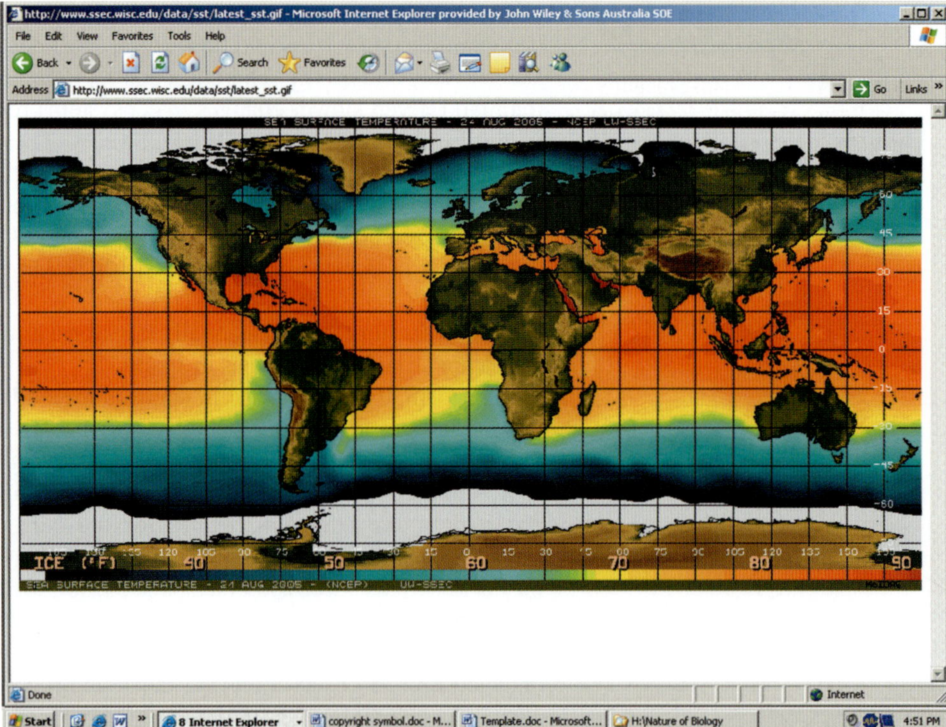

Figure 16.13 Colour-coded image of the sea surface temperature as revealed by an AVHRR (Advanced Very High Resolution Radiometer) carried on a satellite. Red represents the hottest and purple the coolest sea surface temperature.

SATELLITES OF THE EARTH OBSERVING SYSTEM

The satellite *Terra* orbits at an altitude of 705 kilometres above the Earth. Every 99 minutes, it completes one orbit around Earth and each successive orbit is slightly displaced from the previous one. As a result, over 16 days, *Terra* covers the entire surface of Earth and then begins again. A companion satellite, known as *Aqua*, orbits several hours behind *Terra* and its instruments gather data about the water cycle, including clouds, water content of the atmosphere and the soil, snow and ice cover.

Instruments aboard the satellite *Terra* are as follows:

MODIS (**Mod**erate-resolution **I**maging **S**pectroradiometer) records data from 2300 kilometre-wide bands of the Earth's surface. MODIS records various properties of clouds, land, atmosphere and ocean and produces global images of snow and ice cover, cloud cover and type, vegetation cover, ocean temperatures and the chlorophyll content of oceans (see figure 16.14).

MISR (**M**ulti-angle **I**maging **S**pectro**R**adiometer) carries nine cameras pointing at different angles, takes images 360 km wide and can detect objects as small as 275 metres. MISR measures variation in the land surface, clouds and particles in the atmosphere (aerosols).

MOPITT (**M**easurements **O**f **P**ollution **I**n **T**he **T**roposphere) gathers data from a 640 km-wide band of the Earth below. This instrument can record the origin, global distribution and concentration of methane and carbon monoxide in the lower atmosphere.

CERES (**C**loud and **E**arth's **R**adiant **E**nergy **S**ystem) measures the energy emitted from land, ocean and atmosphere and also measures the reflection of sunlight by different surfaces. Data from CERES assists in predicting climate change.

ASTER (**A**dvanced **S**paceborne **T**hermal **E**mission and **R**eflection **R**adiometer) gathers data from a 60 km-wide band of the Earth below. ASTER measures surface temperatures, composition and topography (elevation).

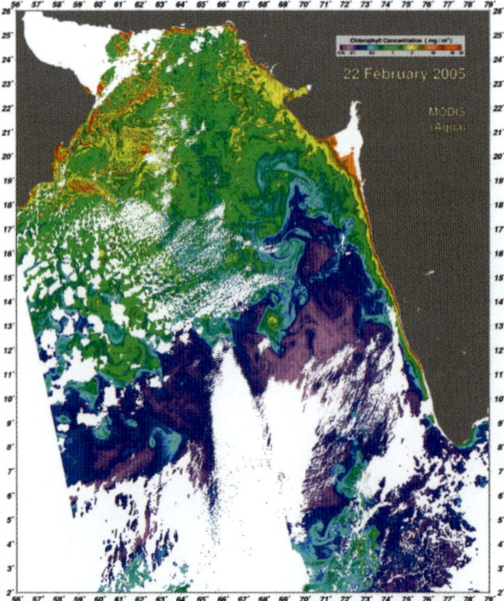

Figure 16.14 Concentration of chlorophyll in Arabian Sea as revealed by MODIS. Chlorophyll signals the abundance of phytoplankton, with highest concentrations shown in red and lowest in blue.

Predicting effects of change in ecosystems

Because ecosystems are complex biological units, it is not easy to predict the short- and long-term effects of a given change. Even when a primary change to an ecosystem is identified, its secondary effects may be less apparent. The scientists who released rabbits on Cabbage Tree Island in 1906 (see page 501) would not have been aware that this action would be responsible, nearly 90 years later, for immobilising Gould's petrels with sticky bird-lime fruits.

Changes that occur naturally and have been repeated many times over geological time periods are not expected to disrupt an ecosystem permanently. For example, bushfires have been acting on **sclerophyll** forest ecosystems in Australia for millions of years, but have not destroyed these ecosystems. Why? The communities in these ecosystems evolved in the presence of this change agent so that they successfully survive the effects of fire.

In contrast, sudden changes in an ecosystem caused by human activities are likely to disrupt the ecosystem for a long time or even permanently. For example, the construction of dams and massive spills of crude oil produce major changes in ecosystems.

KEY IDEAS

- Primary changes in ecosystems may be due to natural agents or to human intervention.
- The primary change in an ecosystem may involve either its living community or its non-living surroundings.
- Changes in ecosystems may occur regularly, sporadically or may be due to one-off events.
- Detecting changes in ecosystems may require long time frames.
- Global monitoring of ecosystems occurs through projects such as the Earth Observation System.

QUICK-CHECK

4 Identify a change agent that acts on an ecosystem as follows:
  a regularly
  b sporadically (irregularly)
  c as a one-off event.

5 Give an example of a change in an ecosystem that is due to:
  a human intervention
  b a naturally occurring agent of change.

6 List one advantage of monitoring of ecosystems by remote sensing as compared with direct observation 'on the ground'.

Human impacts on ecosystems 1

ODD FACT

Accidental introduction, probably in cargo during World War II, of brown tree snakes (*Boiga irregularis*) to the Pacific island of Guam has caused the extinction of 10 of 13 native bird species, many native lizard species and 2 of 3 bat species.

Changes in ecosystems due to human impacts are usually associated with economic development and with meeting the needs of growing human populations. These actions may involve:

- flood control measures, such as damming rivers and building levee banks
- fire prevention measures, such as controlled burns
- agricultural activities, such as land clearing, irrigation, and use of fertilisers
- mining activities
- industrial activities, including processing of raw materials and generation of industrial wastes
- needs of urbanised societies, such as power generation, construction of water storage capacity, road construction and sewage disposal
- introduction of exotic flora and fauna, deliberately or accidentally.

These human actions can change ecosystems. We will examine a number of instances of this in the following section.

ODD FACT

The cost to Australian agriculture of invasive exotic weeds is estimated at more than $4000 million per year. (*Source:* WWF Report 2004). The majority of the exotic plant species that have become major weeds in Australia were originally introduced as ornamental plants for gardens. Would you expect that these invasive exotic weeds and pests would be r-selected or K-selected?

Introduction of exotic species

The introduction to Australia of some **exotic** or non-native or **alien species**, both animal and plant, has caused major changes in many natural ecosystems. Not all exotic species become major ecological pests. Some do not spread naturally in the wild and others have significant economic value, for example, cereal and fruit crops and beef and dairy cattle. Other exotic species, however, are **invasive**; these are the exotic species that spread rapidly in natural ecosystems and produce widespread detrimental changes. We generally give the label 'weed' to invasive plant species and the label 'pest' to invasive animal species.

Impacts of invasive exotic species

When exotic animals and plants are introduced to local ecosystems, negative changes inevitably occur such as displacement and loss of native species.

Displacement and loss of native species from an ecosystem can occur:

1. when introduced species are successful predators of native species. Introduced predators such as foxes and feral cats and dogs that prey on species of native mammals. Native species evolved in the absence of predation by foxes and feral cats and have not evolved defences against these introduced predators.
2. when introduced species bring into a community new diseases to which native species have not previously been exposed and are not resistant. For example, an exotic fungus, *Phytophthora cinnamomi*, was introduced to Australia probably in the 1800s. This fungus is the cause of dieback disease of trees and other plants in native forests, woodlands and heaths.
3. when introduced species use the same limited resources as native species, such as food or shelter, and compete with them more effectively for those resources. For example, redfin perch (*Perca fluviatilis*), an exotic fish, have spread in Tasmanian freshwater lakes and rivers and have caused a decline in the native fish populations with which they compete more efficiently for food.
4. when introduced species change the environment of an ecosystem so that native plants and animals can no longer survive. For example, when swamp buffalo (*Bubalis bubalis*) were introduced to flood plains at Australia's Top End, the buffalo created trails and wallow holes that caused fresh water to drain away and be replaced by salt water (see figure 16.15). This inflow of salt water killed the natural vegetation.

As well as their negative ecological impacts, invasive exotic species also have significant negative economic impacts (see Odd Fact, lower left).

Figure 16.15 Wallow holes and trails made by water buffalo create channels that let fresh water drain away and allow salt water to enter. What effect does this have on the natural vegetation?

ODD FACT

Exotic species arriving in Australia at the time of the First Fleet (1788) or with early European settlers include mice (*Mus musculus*), pigs (*Sus scrofa*), goats (*Capra hircus*) and cats (*Felis catus*).

ODD FACT

Every year, more than 6000 ships come to Australian ports and release about 60 million tonnes of ballast water. By this means, more than 100 exotic marine organisms have reached Australian waters.

Exotics down under

Australia's history since the First Fleet is filled with stories of the introduction — either deliberate or accidental — of exotic species to this country. Deliberate introductions were to make Australia 'more like home' for early European settlers and included animals for sport/hunting, pet animals for companionship, aquarium animals, ornamental plants for gardens, and plants for pastures, windbreaks, shade and preventing soil erosion.

Accidental introductions of exotic species have occurred, for example, as escapees from gardens and aquaria, in the ballast water of ships and in the cargoes of ships and aircraft. Because of the high frequency, volume and speed of intercontinental transport of people and goods, the risk of further introductions of exotic species continues. Figure 16.16 shows some of these introductions.

Let's examine the impacts on Australian ecosystems of:

- three exotic pests — rabbits, cane toads and European carp — that are regarded as among the most damaging pest animals in Australia
- one exotic weed — athel trees.

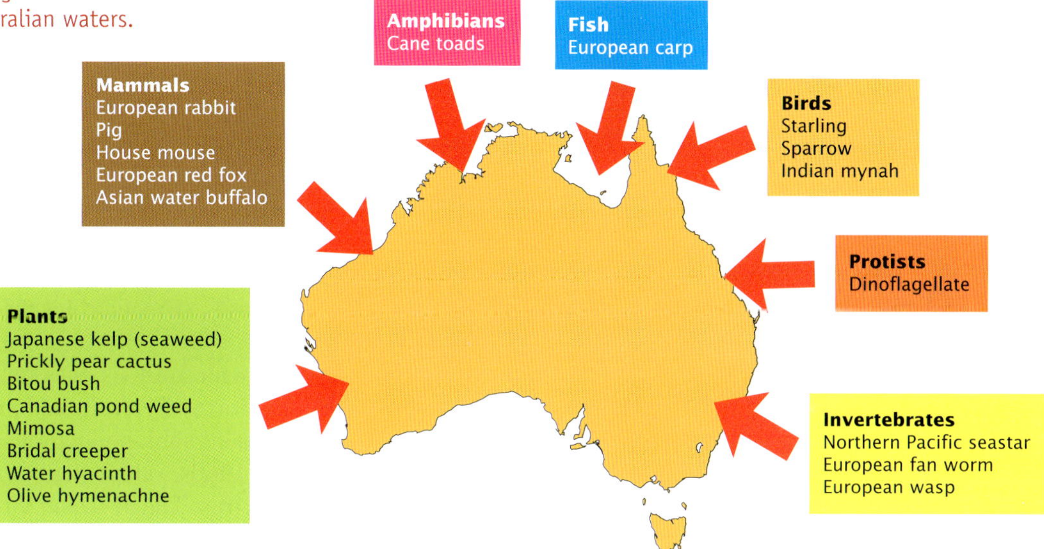

Figure 16.16 Some of the exotic species introduced to Australia since the late 1700s (the arrows do *not* represent the point of introduction)

ODD FACT

In 1882, the New South Wales Government offered a very large cash prize to anyone who could rid the colony of rabbits. No-one was successful in claiming the prize.

Case 1: Rampaging rabbits

The spread of the European rabbit (*Oryctolagus cuniculus*) across large areas of Australia began with several dozen European rabbits brought to the property of Thomas Austin near Winchelsea, Victoria, in 1859. He released rabbits on Christmas Day of that year for hunting purposes and their descendants now number in the hundreds of millions. They rapidly spread north through New South Wales and reached the Queensland border in 1886. The rabbits also spread west, reaching the west coast by the early 1900s (see figure 16.17).

Rabbits have a high reproductive rate, and one female can produce nearly 30 kittens each year. In their natural habitats in Europe, rabbit populations remain more or less constant because of winter food shortages and the presence of predators, parasites and competitors. In Australia, however, these controlling factors are not present to the same extent and rabbit populations exploded, causing great damage to Australian ecosystems.

Rabbits compete more efficiently than native species for food and, over time, have displaced native species from many large areas of Australia.

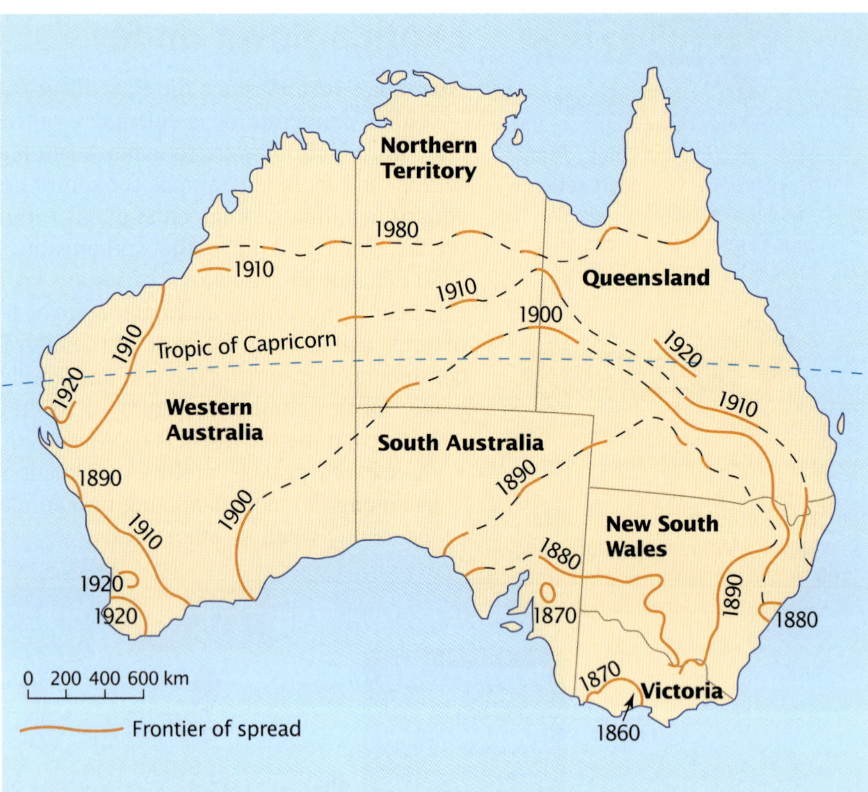

Figure 16.17 Increase in range of feral rabbits in mainland Australia over time. Rabbits are not a pest in their natural habitats in southern Europe where their populations are stable. Why?

Before the release of RCV, rabbit densities in arid areas of Australia could reach 600 rabbits per square kilometre in a good season. After the release of RCV, the same areas had rabbit densities of only one to five rabbits per square kilometre.

Attempts to control rabbit populations have involved **biological control or biocontrol** through the release of viruses. (Biocontrol is discussed in more detail on page 514.)

- *Myxoma* virus, initially carried by mosquitoes, was released in the summer of 1950–51 near Corowa on the Murray River. Infection produced myxomatosis, resulting in widescale deaths in rabbit populations, but not in arid areas of Australia. (Why?) In the mid-1960s, rabbit fleas (*Spilopsyllus cuniculi*) were also used as carriers of the virus. The *Myxoma* virus initially had mortality rates of 99 per cent in some areas; but as rabbit populations became resistant to its effect, mortality rates dropped to 20 per cent.
- Rabbit calicivirus (RCV), which causes rabbit hemorrhagic disease, has been released at hundreds of sites on the Australian mainland since October 1995, dramatically reducing the rabbit population from an estimated 300 million.

Case 2: Cane toads — never in the cane fields

The New South Wales entomologist, W. W. Froggatt, argued strongly against the introduction of cane toads in the 1930s. The release was held up briefly but Froggatt's advice was rejected and the release of cane toads continued.

Cane toads are native to South and Central America. In 1935, about 100 cane toads were introduced to Australia from Hawaii to control greyback beetles (*Dermolepida albohirtum*), which cause serious damage to sugarcane crops. The toads were released near Bundaberg in northern Queensland. They thrived in their new habitat and their numbers increased. Although the toads had voracious appetites, their preferred food did not include greyback beetles! Cane toads have now spread through eastern and northern Australia, and are estimated to be increasing their range by about 35 km per year (see the map in figure 9.11, page 261). Cane toads have few natural enemies in Australia.

Cane toads have many negative ecological impacts. They secrete a creamy venom from glands on their shoulders and this causes the death of many native animals, including goannas, snakes, dingoes and quolls that are known to eat cane toads.

Case 3: Carp, carp, go away!

The European carp (*Cyprinus carpio*) was brought to Australia in the 1870s and escaped from Victorian fish farms in 1960 into the Murray–Darling River system. By 1976, carp had reached Queensland.

Carp feed at the bottom of water. They take up mouthfuls of mud, spit it out so that the material becomes suspended in the water, then swallow particles of organic matter or small organisms. Because of this behaviour, carp destroy water plants and change aquatic habitats so that native fish cannot survive. Sunlight cannot penetrate carp-infested waters to the same depth as in clear waters and, as a result, some aquatic plant life is lost from aquatic ecosystems. Carp excretions increase the levels of dissolved nutrients leading to excessive algal growth on the water surface and this further decreases the penetration of sunlight into the water.

Case 4: Athel pines in the Centre

One of the species of exotic, drought-resistant trees imported as seeds to Australia as shade trees for homesteads in arid areas was the athel pine (*Tamarix aphylla*), also known as the tamarisk. It is both drought- and salt resistant.

Athel pines became established along the Finke River in central Australia after severe floods in 1974 uprooted the river red gums (*E. cameldulensis*) that originally dominated these waterways. The floodwaters also carried in quantities of salt from saltpans and provided conditions suitable for the growth of athel seedlings that now dominate these waterways.

Figure 16.18 (a) Athel pines (*Tamarix aphylla*) growing in the Finke River bed near Alice Springs. Why were they introduced to Australia? Athel pines are one of the 20 Weeds of National Significance, a list announced in 1999. **(b)** Current distribution and **(c)** potential distribution of athel pines (*Source:* www.weeds.org.au)

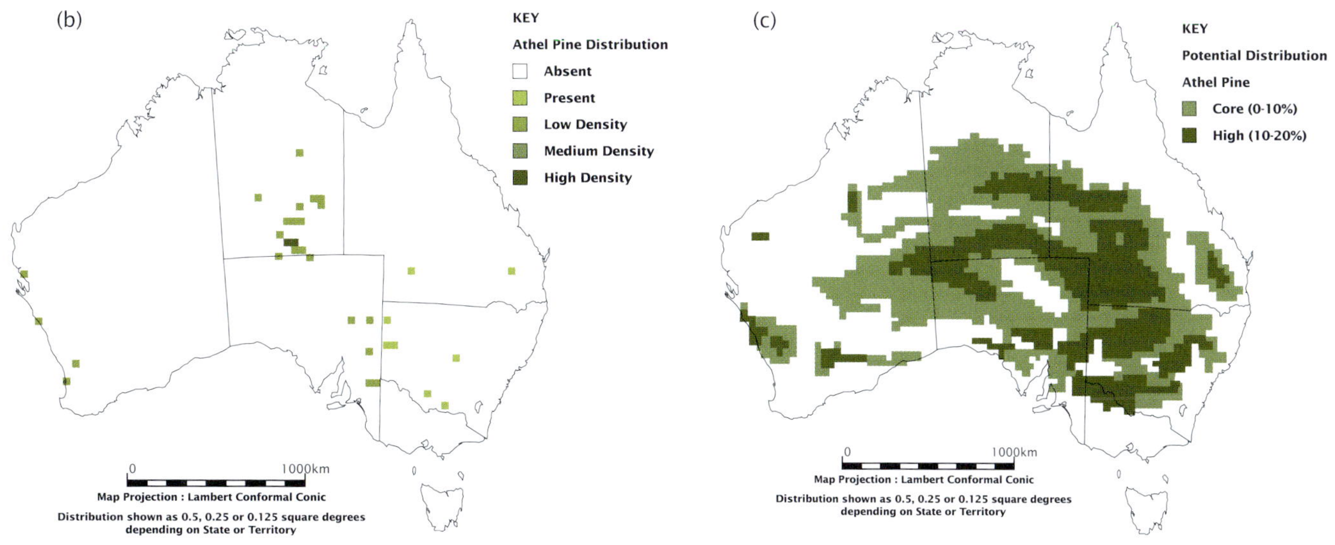

ODD FACT

Serrated tussock (*Nassella trichotoma*), a South American grass, is a noxious weed in south-eastern Australia that makes pastures useless for grazing. A mature plant can produce more than 140 000 seeds each year and these can be carried on light breezes more than 20 kilometres.

Athel pines changed waterway ecosystems because:

- they take up salty water from the ground water supply and lose water vapour from their leaves through transpiration, but the salt is left behind in the leaves. The resulting leaf litter under athel pines is salty.
- leaves of athel pines are eaten by only a few insects, compared with the large numbers of different kinds of leaf-eating insect found on river red gums
- flowers of athel pines are tiny with smaller supplies of nectar compared with those of river red gums.

Table 16.2 compares the features of ecosystems dominated by river red gums and those dominated by athel trees.

Table 16.2 Some differences between inland waterway ecosystems dominated by river red gums and those dominated by athel pines

Feature	*River red gum-dominated ecosystem*	*Athel-dominated ecosystem*
ground cover	many plant species	few plant species
litter layer	loose litter layer	compacted, fine litter
concentration of salt in litter and soil	low	high
fallen logs and branches	many present	few, if any, present
hollows in trees	present	absent
leaf-eating insects	many species present	few species present
biodiversity	higher	lower

ODD FACT

Invasive exotic animals cost Australasia at least $720 million each year (*Source:* AIA CRC 2005).

Responding to exotic invaders

The Australian Quarantine and Inspection Service (AQIS) is responsible for preventing the entry to Australia of new exotic organisms through strict control measures. Quarantine measures assist in keeping Australia free of pests and diseases (see figure 16.19). This is the first line of defence.

Measures to reduce or eliminate populations of exotic pests and weeds and to prevent their spread include:

- declaring a species to be **noxious**; once this occurs, land owners are required to act to exterminate the noxious species from their land and relevant state and federal authorities are responsible for the control of the species on public land
- physical control measures, such as trapping to remove pests from an area, or the building of barriers to prevent their spread from an existing area
- chemical control measures, such as the use of selective herbicides and poisons
- biological control measures, such as the use of another kind of organism that preys on or is a parasite of the exotic pest.

Figure 16.19 The quarantine beagle brigade is an important asset in safeguarding Australia from the import of materials that may place our unique flora and fauna at risk. Banned materials include eggs, meats and plant material.

Biological control measures

Biological control or biocontrol is used to control or eliminate exotic pests. Biocontrol refers to the use of natural enemies (predators, herbivores, parasites or disease-causing organisms) to control exotic pests and weeds after they have invaded a new habitat. Three different types of biocontrol can be identified:

1. classical biocontrol
2. conservation biocontrol
3. use of biopesticides.

In **classical biocontrol**, natural enemies of pests and weeds, mainly insects and mites, are imported from offshore and released into the areas where the exotic species has become established. Populations of these natural enemies may become self-sustaining. By feeding on or causing disease in the exotic species, these natural enemies reduce the abundance of the populations of exotic pests and weeds. However, careful testing must be carried out beforehand to ensure that any introduced biocontrol agent will not itself become an exotic pest (as for cane toads).

Prickly pear cacti (*Opuntia stricta* and *O. inermis*) were introduced to Australia from Central America and rapidly spread across Queensland grasslands during the 1920s. These invasive weeds were brought under control by larvae of the cactus moth (*Cactoblastis cactorum*), a natural enemy of these cacti. The moth was imported into Australia from South America in 1926, large numbers were bred and released and the moth population became established and self-sustaining. Moth larvae feed on the cacti. By the mid-1930s, populations of prickly pear cacti had shrunk and were under control.

Conservation biocontrol involves the use of naturally occurring beneficial agents to control exotic pests and weeds. Beneficial agents commonly used include insects and mites. How does conservation biocontrol differ from classical biocontrol?

Biopesticides are a third form of biocontrol. Biopesticides are naturally occurring agents, such as bacteria, fungi and tiny nematode worms, that kill exotic species. These agents are bred up and applied to the exotic species in greater concentrations than would normally occur. A successful use of biopesticides involved the use of nematodes to control populations of Sirex wasps that were destroying pines.

ODD FACT

The earliest recorded case of classical biocontrol occurred in India in the late 1830s. The cochineal insect (*Dactylopius ceylonicus*) was found by accident to feed on an exotic cactus weed. These beetles were then deliberately spread to other areas of India and to Sri Lanka where this cactus was causing damage.

Biotechnology for control of exotic pests

At the Australasian Invasive Animal Cooperative Research Centre (AIA CRC), scientists are exploring the use of biotechnology to control populations of invasive exotic species. The following are two examples of research projects under development as at 2005.

Blocking conception in rabbits

This project aims to prevent conception in female rabbits by causing them to produce antibodies against a protein present in rabbit eggs and so make them infertile. Researchers intend to achieve this by infecting rabbits with a modified virus that carries a rabbit gene that controls the production of a protein essential for reproduction. When a female rabbit is infected with this virus, it is expected that she will produce this protein and her immune system will respond by producing antibodies against the protein. These antibodies would then attack her eggs (see figure 16.20). Because the immune system is involved, this technique is termed **immunocontraception**.

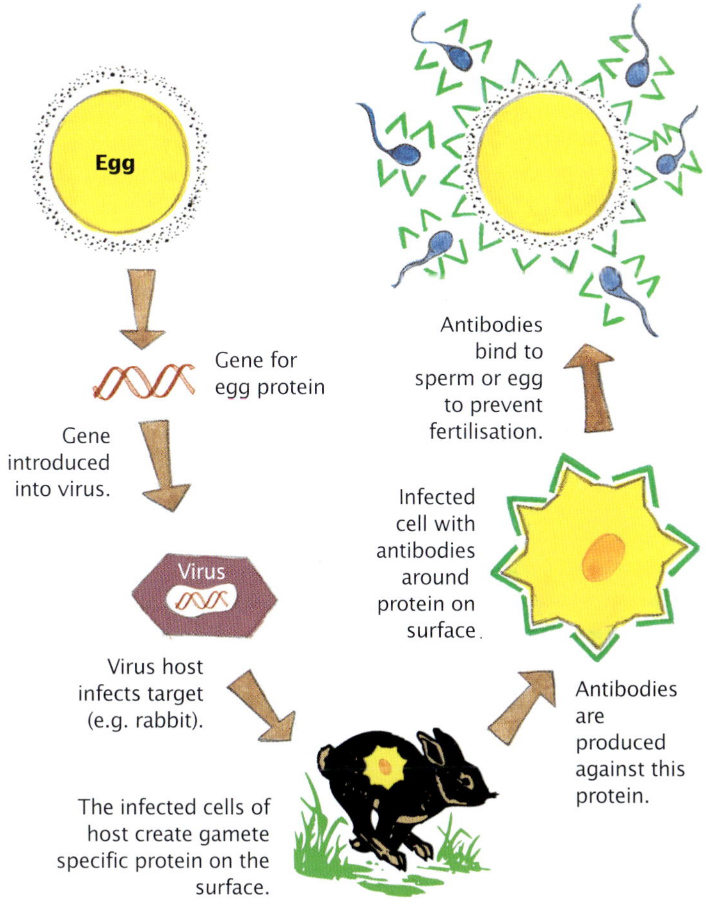

Figure 16.20 The technique involved in immuno-contraception

No daughters, only sons

The Daughterless Carp project began in 2002. The aim of this research is to reduce the abundance of European carp (*Cyprinus carpio*) by creating populations of carp that can produce only male offspring. As existing females in a carp population die off, each successive generation would have fewer and fewer females and, over time, carp population numbers would drop. Researchers intend to achieve this by using gene manipulation. This would be done by 'switching off' one specific gene, namely the gene that controls the production of an enzyme that causes carp embryos to develop as females.

No functional gene → no enzyme → no females!

The following box introduces a scientist working on the project.

Figure 16.21 Switching off a certain gene prevents female development.

BIOLOGIST AT WORK

Dr Jawahar Patil and the European carp

Dr Jawahar Patil is Senior Research Scientist working for CSIRO Marine Research in Hobart, Tasmania. After school, Jawahar obtained a Bachelor of Fisheries Science and a Master of Fisheries Science degree from the University of Agricultural Science in Bangalore, India. He then completed a PhD at the National University of Singapore. He has worked on gene therapy for diabetes and joined CSIRO Marine Research Laboratories in 1998. Jawahar writes:

My work as an aquatic molecular biologist encompasses three major areas of research. My primary focus is on developing molecular mechanisms and strategies for control of feral aquatic animals that are of concern to Australia, particularly European carp. I am a key member of the Daughterless Carp project, which is funded by the Australasian Invasive Animal Control Cooperative Research Centre. My role is to plan, design, build and integrate molecular mechanisms into fish populations to obtain all male or 'daughterless' populations of fish. The research is a genetic technology that aims to reduce carp populations by biasing the sex ratio towards males to reduce the number of breeding females.

The second strand of my research involves developing genetic probes for the detection of several marine pests, including the Pacific seastar and Pacific oyster. These studies will help to investigate the potential to develop technology for detection of key marine pests of concern to Australia, and protection of coasts against invasion by alien species.

The third strand of my research involves DNA or 'genetic' vaccination. The process involves introducing plasmid DNA encoding an antigenic protein that is then expressed within cells of the host organism. This approach has the ability to produce a strong immune response. We hope to use this approach to produce a vaccine for protecting Atlantic salmon against amoebic gill disease (AGD). AGD is estimated to cost the Australian salmon industry $15 million annually in lost revenue.

Apart from project work, I spend a large amount of time preparing and reviewing scientific papers, seeking new research projects and providing supervision to students and staff. I enjoy working as part of a multi-skilled team of biologists and applying the principles of modern molecular biology to solve ecological and industry problems.

Figure 16.22 Dr Jawahar Patil, seen here with Keith Bell of K&C Fisheries, injecting brood carp

KEY IDEAS

- Invasive exotic species cause negative changes in ecosystems.
- Invasive exotic species reduce biodiversity.
- Control of exotic pests and weeds has in some cases been achieved using various biocontrol measures.
- Biotechnology is being explored as a means of controlling invasive pests.

QUICK-CHECK

7 Identify the following statements as true or false.
   a All exotic species are invasive.
   b All exotic species are non-native.
   c Prickly pear cacti infestations were brought under control through classical biocontrol.
8 Name one weed of national significance in Australia.
9 List three different types of biocontrol.
10 Give an example of an invasive exotic species that:
   a was introduced for sporting/hunting purposes
   b failed to control infestations of greyback beetles in sugarcane
   c reached Australia in the ballast water of ships.
11 Give an example of how biotechnology might control pests.

Human impacts on ecosystems 2

In this section, we will explore human interventions that cause changes in ecosystems that involve:

1. over-harvesting of a biological resource, and
2. land degradation producing salinity.

Over-harvesting: lack of sustainability

Chapter 15 (page 490) also refers to over-harvesting.

Over-harvesting is an unsustainable use of a biological resource. Continual excessive harvesting or over-harvesting can push populations to a vulnerable point causing a collapse or 'crash'. Once a population drops below a critical level, it may require many generations to recover, if it recovers at all, especially populations of K-selected species that have low reproduction rates.

Harvesting of any species must be carried out at sustainable levels that allow the species to recover its numbers naturally between harvesting. Important biological knowledge to prevent over-exploitation includes:

- where, when and at what age a species breeds
- its rate of growth
- the time required for sexual maturity.

This information is often not available and many ocean fish stocks are at risk of loss from commercial over-exploitation, including Patagonian toothfish, southern bluefin tuna, Atlantic cod and gemfish. Initial large catches have depleted the breeding stock to levels that mean recruitment of young fish is insufficient to replace the fish being caught.

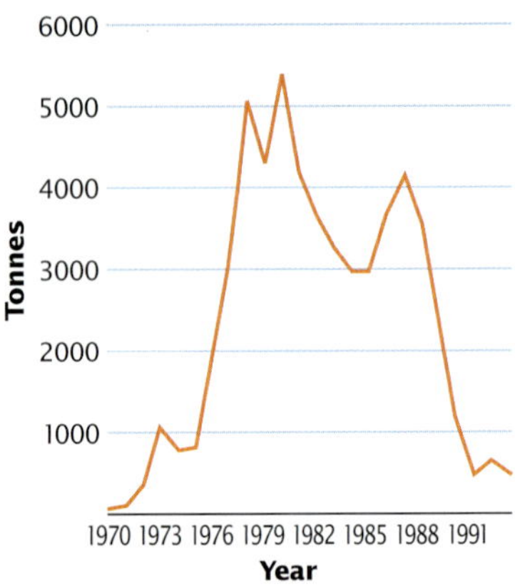

Figure 16.23 Graph showing annual catch of eastern gemfish (*Rexea solandri*) from 1870 to 1991. In which year did the catch peak?

Figure 16.23 shows the decline in the annual catch of gemfish between 1970 and 1991. In order to allow the remaining gemfish stock to recover, the targeted or direct commercial catch was reduced to zero in 1993. The only gemfish allowed to be caught were those obtained from harvesting other species. In 1997, gemfish were again allowed to be harvested, but a strict annual quota of 1200 tonnes was set, with a focus on older fish, aged about seven years.

Collapse of Canadian cod populations

An article in the journal *Nature* (Vol. 423, 15 May 2003) reported that populations of large predatory fish, for example, cod, tuna and marlin, have fallen sharply since industrialised fishing began during the 1950s and 1960s. Globally, the biomass of these large predatory fish in the world's oceans is estimated to have fallen to about 10 per cent of their levels before industrialised fishing began. (Industrialised fishing refers to the use of large ships equipped with sonar and radar and the use of techniques such as large-scale long-line operations.)

For example, populations of Atlantic cod (*Gadus morhua*) in the major fisheries off the Canadian coast were severely overfished from the 1950s to the 1970s and, by 2004, cod populations had collapsed to about 2 per cent of their original levels. As older and bigger cod disappeared as a result of over-harvesting, the population collapse was accelerated when younger and smaller cod that were just reaching sexual maturity were harvested. Instead of being recruited into the breeding stock, these young cod were removed from the population.

In April 2003, the Canadian Minister of Fisheries and Oceans announced that the major cod fisheries off the Canadian coast were closed to both commercial and recreational fishing as the first step in rebuilding the cod populations. Other measures identified were: (1) protecting of cod spawning areas; (2) protection of habitats where juvenile cod aggregate; and (3) reduction in allowable catches of a fish species that is a major prey of cod.

Sustainability and conservation

Use of biological resources in a sustainable manner requires a commitment to **conservation**. The goal of conservation is to maintain living things in their diverse ecological settings and to permit the use of natural resources in a sustainable manner. Conservation may mean keeping a habitat in an almost undisturbed state, free of human impact. In other cases, conservation may be compatible with joint use of habitats to meet human needs, such as recreation and limited harvesting.

Conservation biology is a complex subject that draws on knowledge from areas including ecology, population genetics and evolutionary biology. Ecology provides a framework for making decisions about protecting habitats and ecosystems. Population genetics and evolutionary biology provide a framework for making decisions about protecting species. Conservation is not an isolated biological matter and conservation measures, such as habitat protection or restrictions on industry, have social, political and economic impacts.

The acceptance of conservation measures may involve increasing the knowledge and understanding of the value of biological diversity, changing personal and community attitudes and changing consequent behaviours such as patterns of use of land, energy and other resources, and waste disposal.

ODD FACT

A 10-year-old schoolgirl, Bethany Henderson, from the south coast of New South Wales successfully campaigned to ensure that balloons were not released at the Sydney Olympics. Such balloons can choke to death many marine animals.

Legislating for conservation

Australian federal and state governments have passed legislation to conserve species. The Victorian Government's *Flora and Fauna Guarantee Act 1988* is intended to 'promote the conservation of Victoria's flora and fauna and to provide for a choice of procedures which can be used for the conservation, management or control of flora and fauna and the management of potentially threatening processes'.

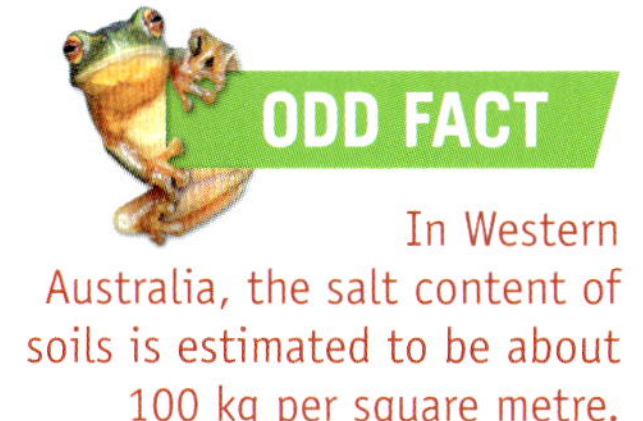

In Western Australia, the salt content of soils is estimated to be about 100 kg per square metre.

Salt in excess

Salinity refers to the salt content of water or soil at a level where the salt content damages soil and degrades water quality. Excessive salt in soil and water places many native plant species and aquatic animal species at risk of extinction. In addition, salinity causes economic loss through loss of productive agricultural land and damage to roads and buildings and puts the supply of acceptable drinking water at risk.

Two kinds of salinity exist in Australia:
- dryland salinity
- irrigation salinity.

Dryland salinity

Soils over much of Australia have naturally high salt levels and this salt is stored below the soil surface in the subsoil. Salt has been carried inland by winds from the sea over many tens of thousands of years. Because of Australia's low rainfall, high evaporation rates and flat terrain, this salt has not been washed away but has been absorbed into the soil.

The natural vegetation in terrestrial ecosystems in Australia consists of deep-rooted trees, shrubs and grasses. When rain falls, native trees stop some of the water reaching the ground by trapping it in their leafy canopies from where the water evaporates. Deep-rooted native plants also take up large amounts of water from the soil and this water is lost by transpiration. Under these conditions, the water table (top of the ground water) in the soil has remained at a constant depth below the soil surface and the ancient salt in the subsoil has remained undisturbed (see figure 16.24).

Since European settlement of Australia, native grasslands and forests have been progressively cleared for mining and urban development, and for agriculture, where deep-rooted trees have been replanted with shallow-rooted cereal crop and pasture plants.

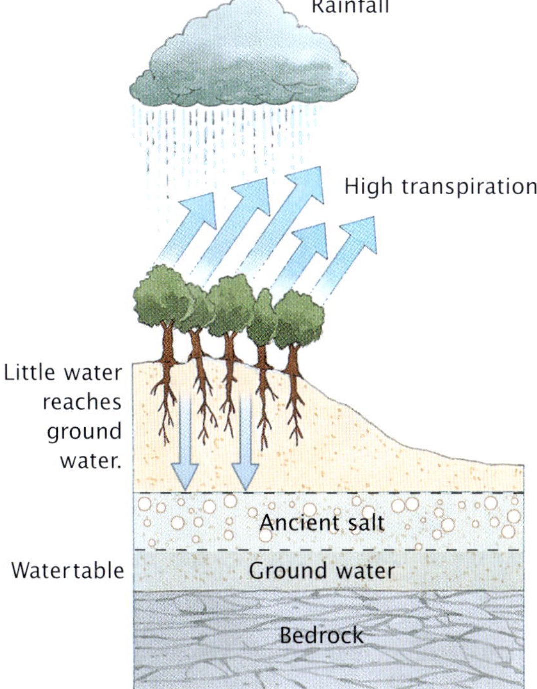

Figure 16.24 Deep-rooted trees take up water and maintain the water table at depth below the soil surface.

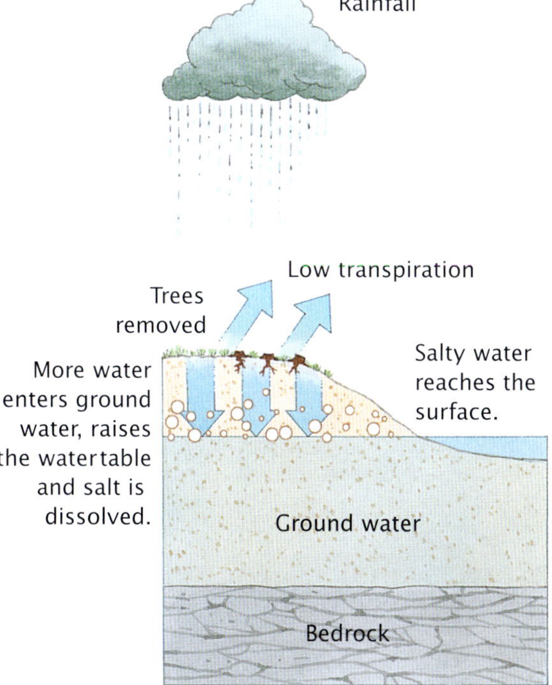

Figure 16.25 Removal of trees results in more water entering the ground water, so that salty water comes close to or reaches the surface.

As land is cleared of trees, more water enters the soil and reaches the ground water, causing the watertable to rise. As the watertable rises, it dissolves salt and brings up salty water that enters streams and rivers. In low-lying areas, salt patches or 'salt scars' may form and plants die (see figure 16.25). This type of salinity, resulting from excessive tree clearing and reduced loss of soil water by transpiration, is termed **dryland salinity**.

Satellite technology can detect soil salinity and monitor changes over time. Figure 16.26 shows changes in salt-affected land in Western Australia as detected by satellite-based sensors. Other techniques for detecting salinity include conductivity profiles that are carried out on the ground or from the air. (The higher the conductivity, the higher the salt concentration.)

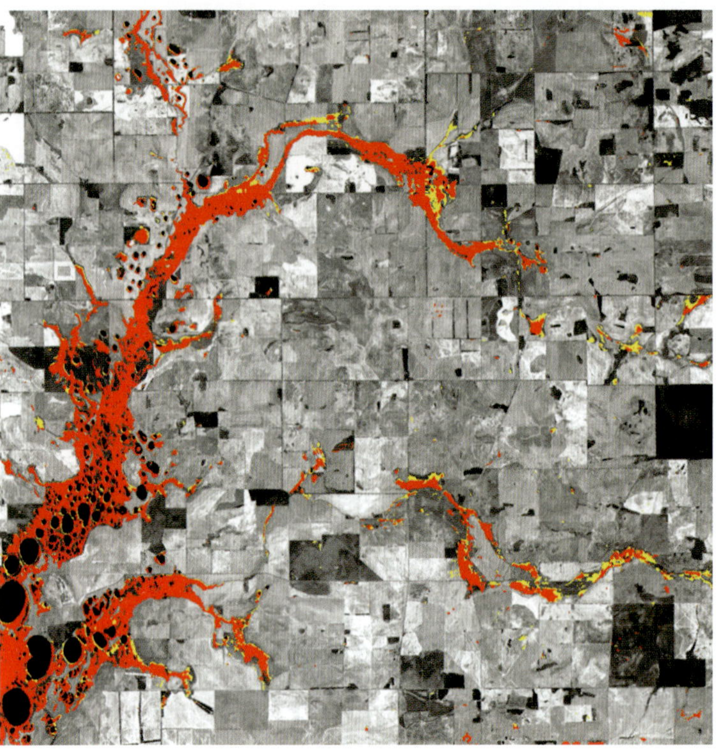

Figure 16.26 Changes in salt-affected land over a 27.5 km^2 area in the northern wheatbelt region of Western Australia. Red represents salt-affected land in 1990; yellow shows additional salt-affected land in 1998. (*Source:* CSIRO Mathematical and Information Sciences)

When the top of a saline watertable reaches to within two metres of the land surface, salinity causes crop damage, structural damage and loss of biodiversity.

Irrigation salinity

Salinity can also occur as a result of excessive irrigation. This type of salinity, termed **irrigation salinity**, occurs when excess irrigation seeps into the soil, causing the watertable to rise. As the watertable rises, it brings dissolved salts with it. Eventually the watertable reaches the root systems of plants and, if they are not salt-tolerant, they are killed. When the irrigation stops and the soil dries out, salt is left at the surface. This process continues each time that irrigation occurs and, as a result, the surface soil becomes more and more salty.

Attacking salinity, the white plague

In the late 1990s, clearing of native vegetation was in excess of one million hectares per year, with Queensland estimated to have the highest rates.

In December 2000, the *National Action Plan for Salinity and Water Quality* (NAPSWQ), supported by both federal and state governments, was announced. The goals of this plan are to:

- prevent, stabilise and reverse trends in dryland salinity affecting the sustainability of production, the conservation of biological diversity and the viability of our infrastructure
- improve water quality and secure reliable allocations for human uses, industry and the environment.

Figure 16.27 Trees and other vegetation in this area of the Murray Valley, Victoria, have been killed due to irrigation salinity.

ODD FACT

The federal government has announced that it will not invest in any NAPSWQ plans with states unless land clearing is prohibited in areas where clearing would lead to increased problems with salinity or degrade water quality.

To achieve these goals, various actions will be taken including:

- using engineering to improve stream water quality through **salt interception** devices and ground water pumping
- maintaining and improving existing native vegetation
- developing agricultural production systems that are tailored for Australian conditions, such as farming brine shrimp and using salt-tolerant crop and pasture species
- buying back of some irrigation allocations.

KEY IDEAS

- Over-harvesting or excessive use of a biological resource can cause populations to crash.
- Salinity is a major environmental problem in Australia.
- Two kinds of salinity exist: dryland and irrigation salinity.
- Satellite-based sensors can be used to monitor areas affected by salinity.
- A National Action Plan for Salinity and Water Quality has been implemented in Australia.

QUICK-CHECK

12 Identify the following statements as true or false.
   a Soils over large areas of Australia have a high salt content.
   b Shallow-rooted plants are responsible for lowering the watertable.
   c Over-harvesting is the sustainable use of a biological resource.
   d Dryland salinity results from excessive clearance of deep-rooted native plants.
13 Explain the effect on the watertable of excessive irrigation.
14 Identify two populations that have 'crashed' from overharvesting.
15 List two methods by which the national problems with salinity will be attacked.

Human impacts on ecosystems 3

In this section, we will explore human interventions that involve changes in the abiotic part of ecosystems, including:

1. damming of rivers and diversion of water from rivers
2. nutrient overload of waterways and waste disposal.

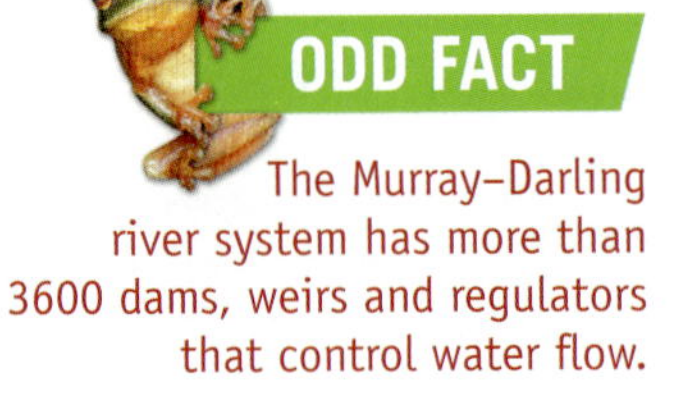

The Murray–Darling river system has more than 3600 dams, weirs and regulators that control water flow.

Let the rivers run

The flow of water in rivers that have no dams or weirs is said to be **unregulated**. Unregulated rivers in southern Australia have seasonal variable flows, with periods of flooding in winter and periods of greatly reduced flow in summer. In contrast, rivers whose natural flow is impeded by dams or weirs or levees are said to be **regulated** with more constant flow over the year (see figure 16.28).

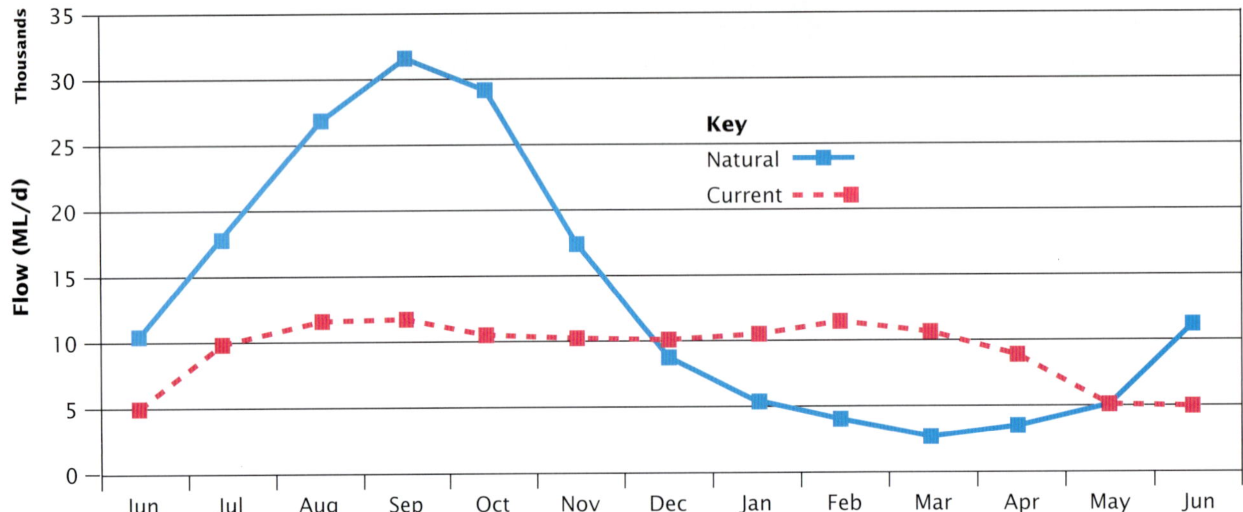

Figure 16.28 Average monthly flow in an unregulated river (shown in green) and in a regulated river (shown in red). Which river has the more variable flow?

In the past, there was no concern for the ecological impacts of damming rivers or diverting water from rivers for irrigation and other purposes. These actions were seen as desirable measures to control floods and enhance agricultural outputs.

The Murray–Darling river system has many important wetlands (see figure 16.29) that are the habitats and breeding grounds of waterbirds, including some migratory waterbirds. These wetland ecosystems have been degraded and destroyed in the past by draining, by clearance and by being permanently flooded as a result of dam construction.

The health of rivers and associated habitats, including wetlands and estuaries, depend on natural water flow to maintain their biodiversity. When natural flow of rivers is stopped, negative ecological impacts follow, as for example:

- release of cold water from the bottom of dams can lower water temperatures to a level where fish can no longer breed or may even be killed; for example, the temperature of the Macquarie River in summer is about 25°C but below the Burrendong Dam, the water temperature in the river is only about 13°C for a long distance
- wetlands that depend on periodic flooding begin to dry out and populations of native fish, waterbirds and invertebrate species decline
- movements of native migrating fish stop or slow, with many species becoming locally extinct in stretches of the river where dams and weirs prevent their migration (this problem can be addressed by the construction of 'fish ladders' beside dams and weirs that allow fish to swim upstream).

More than 1500 fish, including seven native species, used a 'fish ladder' on the Murray River at Lock 8 in its first week of operation in January 2004.

(a)

(b)

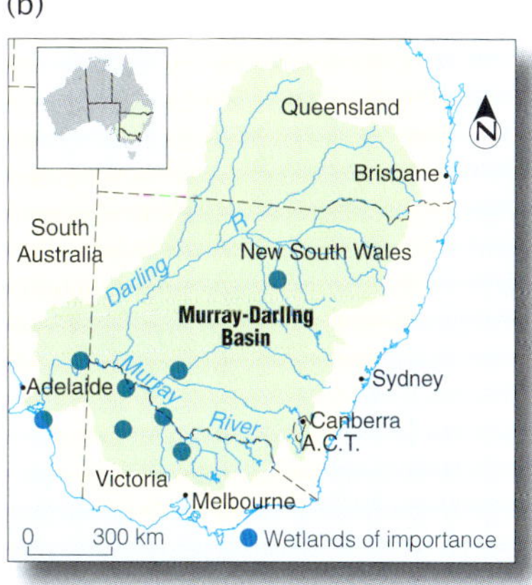

Figure 16.29 **(a)** Wetlands on the Murray River. This is a meandering section of the river between Robinvale and Swan Hill. In their natural state, wetland ecosystems support a diversity of plant and animal life and contribute to the health of rivers. **(b)** The Murray–Darling Basin (MDB) extends from Queensland through New South Wales and Victoria to South Australia and covers an area that includes the catchments for the Darling and the Murray rivers and their tributaries. The MDB is the site of many important wetlands.

Restoring at least some part of the natural flows of rivers is critical for the health of rivers and their surrounding ecosystems. Fortunately, this has begun to happen to a limited extent with the Barmah–Millewa Forests along the Murray and with the Snowy River.

The Barmah-Millewa ecosystems

The flood plain along an undammed river is covered with water each time the river overflows its banks. Organisms occurring naturally in flood plain ecosystems include river red gums (*Eucalyptus cameldulensis*) (see figure 16.30) whose survival depends on regular flooding.

River red gums grow in the Barmah–Millewa ecosystem on the flood plain that covers an area of about 70 000 hectares and includes marshes, swamps, rushlands, red-gum forests, billabongs and creeks.

The Barmah–Millewa ecosystem was significantly changed by the construction, upstream of the forest, of the Hume Weir in 1936 and by the building of levee banks to control flooding. Before these structures were built, the normal pattern for the Barmah Forest was flooding in winter and spring and dry periods in summer and autumn — an environment in which the communities thrived.

After the construction of the weir, the water flow was regulated and the environment became more uniform. Winter flooding became rare and the dry periods in summer were removed. As a result, the Barmah–Millewa ecosystems changed:

- swamps, marshes and rushlands were altered
- the river red gums became unhealthy and tree growth slowed
- breeding sites of birds and fish declined
- populations of many species declined or died out so that species abundance and species diversity were reduced.

ODD FACT

Tasmania's Hydro-Electric Corporation (HEC) decided in late 1998 to release more water into the Mersey River, which was dammed in 1973. In summer, the flow was less than one cubic metre per sec and fish habitats were destroyed. The HEC increased the flow to two cubic metres per sec.

Figure 16.30 River red gums in flood conditions in the Barmah Forest. A water release in 2005 created flood conditions that boosted breeding of native fish and waterbird species.

Trial flooding of the red gum forests along the Murray River commenced in 1998 in order to restore the Barmah–Millawa ecosystem to its former healthy condition. In 1998, the Murray–Darling Basin Commission agreed to an annual allocation of 100 billion litres of water to restore the annual inundation of the flood plain, improve the growth of the river red gums and restore the breeding sites of birds and native fish. Later, a total of 341 gigalitres of water was released to prolong the flood that had occurred in the spring of 2000. As a result, high water levels were sustained in the forests until January 2001 and this replicated conditions of natural flooding.

At present, the Murray–Darling Basin Commission (MDBC), in consultation with the Barmah-Millewa Forum, manages the environmental water allocation to the Barmah-Millewa Forests.

The Snowy River

In 1967, when the Snowy Mountains Hydroelectric Scheme was established, 99 per cent of the natural flow of the Snowy River was diverted at the Jindabyne Dam. Below the dam, the Snowy River was a tiny flow of water, just one per cent of its former flow (see figure 16.31). Habitats downstream of the dam were destroyed and the river course was filled with sediment and became overgrown by willow trees and weeds.

Figure 16.31 The Snowy River below the Jindabyne Dam. This picture, taken in October 1999, shows the metre wide trickle of water from the dam into the river.

On 28 August 2002, the process of rejuvenating the Snowy River began as part of the Snowy Project Agreement between the Commonwealth, Victorian and New South Wales governments. The first step was the release of 38 gigalitres of water back to the Snowy River that increased its average natural flow to six per cent. This was achieved by allowing water from a tributary to flow into the Snowy River. Previously, this water had been diverted via an aqueduct into the Jindabyne Dam.

Over the period from 2002–2012, more water will be returned to the Snowy River to increase its flow to 21 per cent of the original flow. This is one of the most significant river rehabilitation projects in the world. A report published in May 2005 states that signs of recovery are apparent with the river beginning to wash away the sediment that built up along its course when its flow was just one per cent.

Challenges that remain are to restore river ecosystems by removing exotic fish and weeds, by replanting of native trees, shrubs and grasses, and by repopulating the river with native fish species from stocks bred in captivity.

One gigalitre = one thousand million litres

Nutrient overload of waters

A change in the non-living part of an aquatic ecosystem such as an estuary or a lake can occur when the levels of phosphate (a mineral nutrient) in the waters increases, triggering a cascade of secondary changes. The accumulation of dissolved mineral nutrients in a body of water is termed **eutrophication**.

In an estuary, many secondary effects can follow eutrophication of waters. Cyanobacteria (*Nodularia* sp.) can multiply rapidly and form a 'bloom' that covers much of the water surface (see figure 13.15, page 416). As the cyanobacteria bloom spreads, it reduces the light intensity in the water depths so that submerged seagrasses and algae die and the amount of organic matter in the water

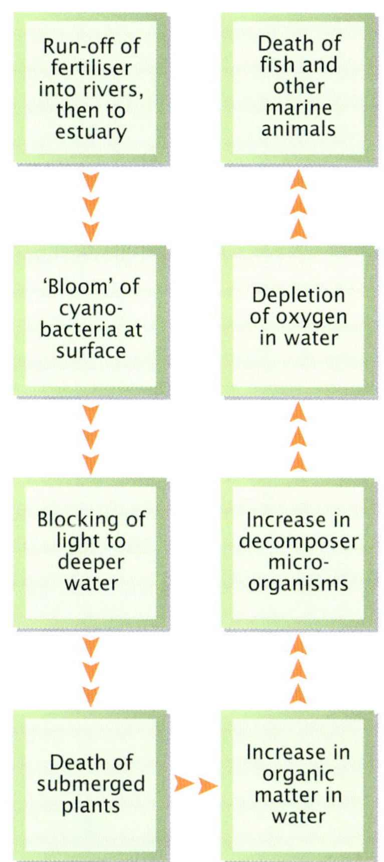

Figure 16.32 Cascade of effects in an estuary ecosystem resulting from eutrophication

increases. Decomposer organisms that feed on organic matter then multiply and use up the dissolved oxygen. As the oxygen levels decline, fish and other animals become starved of oxygen and they also die. Figure 16.32 at left summarises a cascade of events resulting from eutrophication.

The mineral nutrients that accumulate in inland waters and cause 'blooms' of algae and cyanobacteria come from various sources, including agricultural use of fertilisers, discharge from inland sewage treatment plants, run-off from intensive livestock industries and household use of phosphate-rich detergents.

Blooming cyanobacteria

Eutrophication affects many inland waters of Australia — rivers, lakes, dams, rural water supplies and estuaries:

- The Darling River was affected by a toxic 'bloom' of cyanobacteria in 1991 and animals drinking this water were poisoned.
- A 'red tide' that occurred in Lake Macquarie was a toxic 'bloom' of a protist, the dinoflagellate *Noctiluca* sp.

Blooms of cyanobacteria are commonly termed 'algal blooms'. These blooms occur on the surface of water that has high levels of mineral nutrients and are more common in warm stagnant water. Of the many species of cyanobacteria, more than one dozen produce toxins that damage cells.

The cyanobacteria species that produce toxic blooms include species of *Anabaena*, *Microcystis* and *Nodularia* (see figure 16.34). Bodies of water affected by algal blooms include lakes used for recreational purposes, reservoirs that provide public water supplies, dams and billabongs. As well as causing illness or deaths of cattle and sheep, toxic blooms of cyanobacteria pose serious public health problems.

ODD FACT

The toxic bloom of cyanobacteria in the Darling River in New South Wales in 1991 was a world record for the length of river affected — about 1000 kilometres.

Figure 16.33 A scientist measuring the cloudiness or turbidity of water polluted by a cyanobacteria bloom. (This image was taken at Swan River, Perth, in 1992.)

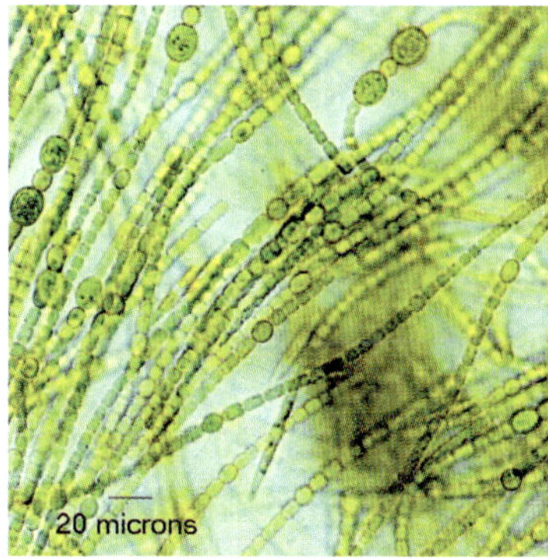

Figure 16.34 Photomicrograph of cyanobacteria that can multiply and form visible green scum on water surfaces.

ODD FACT

The first published scientific report of a toxic algal bloom appeared in 1878. The journal *Nature* reported on a 'Poisonous Australian Lake' in South Australia that caused the death of sheep and cattle that drank water from this lake.

Death of the seagrasses

Seagrasses are flowering plants that grow in tropical and temperate marine habitats. Australia has the world's greatest diversity of seagrasses with 37 known species including eelgrasses (*Zostera* spp.), southern strapweed (*Poisidonia australis*) and paddleweed (*Halophila* spp.).

Seagrass meadows are the major producer organisms in some marine ecosystems along sheltered coasts and in estuaries. Seagrass meadows also provide shelter for species that live on the leaf blades, such as sponges and diatoms, and for organisms that live within the protective shelter of the meadow, such as the juveniles of fish and crustaceans. Tropical seagrass meadows provide food for dugongs (*Dugong dugon*) and green turtles (*Chelonia mydas*). Dead seagrass fragments form detritus that is decomposed by bacteria or eaten by small crustaceans, that are, in turn, eaten by fish.

Significant losses of seagrass meadows have occurred as a result of human actions such as dredging, land reclamation and discharge of particulate industrial wastes in coastal waters that increases the turbidity of the water so that the light levels reaching seagrasses are too low for photosynthesis. Discharge of sewage or fertiliser-rich run-off from agricultural land causes increased growth of phytoplankton in the nutrient-rich water that blocks access to sunlight by the seagrasses, causing them to die (see figure 16.35).

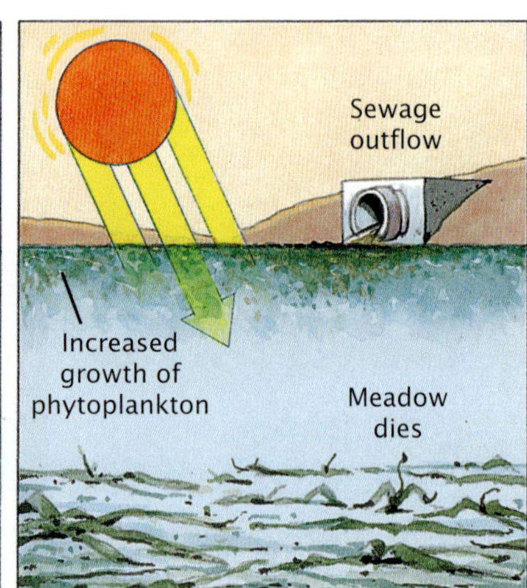

Figure 16.35 High levels of mineral nutrients in the water favour increased growth of phytoplankton that float on the surface. What effect does this have on light intensity in the deeper water?

Waste disposal

While large quantities of paper, bottles, metal and plastic are recycled, large volumes of waste must still be disposed of. Procedures for disposing of waste include:

- disposal in landfills
- incineration
- disposal of effluent into waterways and into the sea.

Decisions about waste disposal can be viewed from many perspectives, such as whether to allow a factory to discharge waste into a river (see figure 16.36).

However, each of these procedures can have negative ecological impacts, such as the release into the atmosphere of toxic chemicals that can be brought back by rain to enter the food chain. Where the wastes are non-biodegradable, they can undergo bio-accumulation (refer back to chapter 14, page 466).

Headlines such as those shown in figure 16.37 remind us that waste that is 'out-of-sight' may have an ecological impact, the effects of which may only become apparent many years after the waste was generated. Sites that were once occupied by factories but have been cleared and used for residential development may later be found to have soils with high levels of persistent chemical contamination.

The box on pages 535–6 tells the story of the restoration of an ecosystem.

Economic

Can't afford alternative waste treatment, factory may have to close, goods will need to be imported.

Biological

Rare species will be endangered if the factory discharges wastes into the river.

Social

Jobs lost from town means reduced spending money — affects shopkeepers.

Consumer

Imported product will cost more.

Political

Worse balance of payments if imports increase, loss of votes from locals at next election.

Figure 16.36 Should a factory be allowed to discharge untreated wastes into a river system where an endangered fish species has been discovered? Many perspectives on the question can exist.

CLEAN-UP OF CONTAMINATED SOIL TO COST MILLIONS

CONTAMINATED SOIL IN BACKYARDS POSES A HEALTH PROBLEM

MOTHERS' BREAST MILK CONTAMINATED BY DIOXINS

BEEF CONTAMINATED WITH PESTICIDE RESIDUE

Figure 16.37 Release of non-biodegradable wastes into the environment can result in bioaccumulation (refer back to chapter 14, page 466) and ecological damage.

KEY IDEAS

- Regulation of river flow produces negative ecological impacts.
- Rehabilitating rivers involves restoring some of their natural flow and the variation in this flow.
- Eutrophication occurs in aquatic ecosystems when high nutrient concentrations stimulate excessive growth of cyanobacteria.
- Waste disposal, unless carefully managed, may produce negative and far-reaching ecological consequences.

QUICK-CHECK

16 What is the difference between regulated and unregulated river flow?
17 Identify two negative consequences of regulated river flow.
18 List an essential condition for an algal bloom to occur.
19 Give a single word for the presence of excessive levels of mineral nutrients, such as phosphates and nitrates, in water.
20 Identify the first step that was taken in the rejuvenation of the Snowy River.

Natural change agents: fire

Fire has been a change agent in Australian ecosystems for many millions of years. Grassland ecosystems and open sclerophyll forest ecosystems, particularly in south-eastern Australia, have been subject to fires often started by lightning strikes during summer storms and fire is a familiar event in south-eastern Australia. After Aboriginal people reached Australia, they learned to make use of fire for land management and for hunting (see the box on page 529). Fire became a familiar and terrifying event for early European settlers to south-eastern Australia (see figure 16.38).

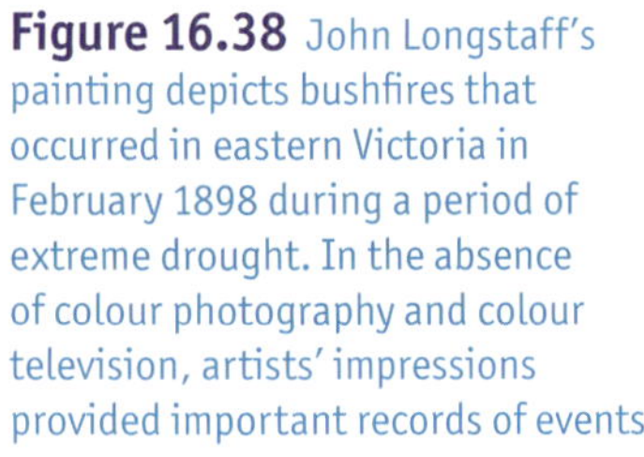

John Longstaff
Australia 1862–1941, lived in Europe 1887–95, 1901–20
Gippsland, Sunday night, February 20th, 1898 1898
oil on canvas
144.8 × 198.7 cm
Purchased, 1898
National Gallery of Victoria, Melbourne

Figure 16.38 John Longstaff's painting depicts bushfires that occurred in eastern Victoria in February 1898 during a period of extreme drought. In the absence of colour photography and colour television, artists' impressions provided important records of events.

Adaptations for living in fire-prone ecosystems

In the following sections, the fires discussed are low-intensity bushfires, not high-intensity wildfires.

Species living in fire-prone ecosystems must be capable of surviving the effects of fires, either as individuals or through their offspring.

Mobile organisms, such as birds or fast-moving mammals, may escape flames and avoid the lethal radiant heat by flying or running away. Slower-moving animals escape by retreating into burrows or sheltering under thick bark where they are insulated from the fire. Soil is a good insulator from the heat of a surface fire. When air temperatures are lethally high (up to 1200°C), the temperature at depth below the soil surface is much lower. Some animals exploit the opportunities provided by fire for feeding. Figure 16.39 shows a native hawk flying in front of a fire front looking for prey that may be flushed from vegetation as it escapes the approaching fire.

Figure 16.39 A native raptor flies ahead of an approaching fire front. What is the benefit of this behaviour for the bird?

ABORIGINAL USE OF FIRE

Aboriginal people used fire for many thousands of years as part of their hunting activities and also as a sophisticated land management tool.

In northern and central Australia, Aboriginal people burned patches of vegetation frequently and early in the fire season when the moisture content was high. This produced low-intensity fires over limited areas that removed the litter that would otherwise build up to fuel a destructive high-intensity **wildfire**. Adjacent patches were left unburned so that a mosaic of burned and unburned areas was created. This patch burning:

- drove animals to areas where they could be trapped and killed
- encouraged new green growth that provided food (see figure 16.40)
- replaced perennial shrubs that are killed by fire with annual grasses that survive frequent burning.

As a result, open grasslands which otherwise would have reverted to woodlands were maintained and biodiversity was preserved. The new growth in burnt areas provided food for grazing animals while the unburnt patches provided them with shelter.

In Arnhem Land in the Northern Territory, areas that continue to be burned by Aboriginal people have greater biodiversity than areas left unburned for many years. Areas that are regularly burned support native plants, such as cypress pines, that are in decline elsewhere. At present, farmers on Australia's rangelands, on advice from CSIRO scientists, are using prescribed burns to revitalise pastures that are being overgrown by woody shrubs. As part of the process of pasture recovery, some areas are burned and are then left free from grazing for two to three years after the burn. Can you suggest why?

Figure 16.40 New growth after a bushfire

Plants in a forest community cannot escape from the flames and heat of bushfires, but plants survive through several strategies:

- vegetative reproducer (VR) strategy
- obligate seeder (OS) strategy.

Vegetative reproducers

After fire, plants that are **vegetative reproducers** regrow through means of buds located under the bark or in underground stems. The VR strategy is seen, for example, in austral bracken and in some species of *Eucalyptus*. The foliage of austral bracken (*Pteridium esculentum*) is destroyed by fire, but the plant regenerates from underground stems (**rhizomes**) (see figure 16.41).

Thick-barked eucalyptus species have **epicormic buds** under their bark. This bark burns poorly and insulates the living tissue within the trunk of the tree. Epicormic buds are normally dormant, but they begin to grow when the leafy crown of the tree is destroyed (see figure 16.42a).

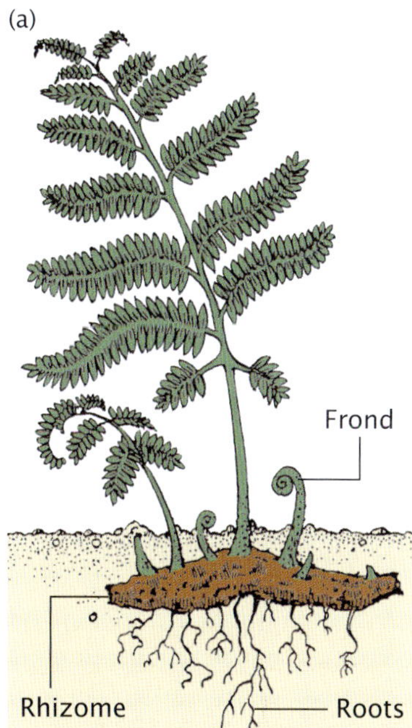

Figure 16.41 **(a)** Austral bracken reproduces asexually from underground stems (rhizomes) when buds from the rhizome develop into new fronds. **(b)** After a bushfire, austral bracken quickly becomes re-established in a burnt area. Can you suggest why?

(b)

Other eucalypt species, commonly known as 'gums', have thin bark and the living tissues above ground are typically killed by fires. These eucalypts have woody underground stems, termed **lignotubers**, that are insulated from the heat of the fire and contain dormant buds from which new growth occurs after fire (see figure 16.42b).

Figure 16.42 **(a)** Epicormic growth from the trunk of a eucalypt tree after a bushfire **(b)** Regrowth of gum from an underground woody lignotuber

Obligate seeders

Obligate seeder plants do not usually survive bushfires. Their seeds, however, do survive and germinate after fires in the mineral-rich soil that is produced by fire. The OS strategy is seen in some plants in the following native genera — *Acacia*, *Banksia*, *Casuarina* and *Hakea*. Mountain ash (*Eucalyptus regnans*) also survives fire through its seeds. Some OS species store seeds in woody fruits on the mature plant, while other species produce seeds that drop to the forest floor.

Most *Hakea* plants are small woody shrubs that, when mature, produce nutlike woody fruits. The plants are killed in bushfires but the heat causes these fruits to split and each releases two seeds (figure 16.43b). *Banksia* seeds are also protected in woody fruits which open when exposed to the heat of a bushfire (figure 16.43a).

Plants of the genus *Acacia* shed seeds with hard seed coats that cannot germinate until water penetrates the outer coat. So, although mature wattle plants are killed by fire, their hard-coated seeds survive and the heat splits the seed coats so that, when the next rains fall, these seeds germinate to produce a new generation of *Acacia* plants.

Figure 16.43 **(a)** A woody *Banksia* fruit opened after a bushfire **(b)** *Hakea* fruit (nut) after a bushfire. How do *Hakea* species survive bushfires?

Fire frequency affects ecosystems

After bushfires, the native plant populations in the community of a woodland or forest ecosystem can either be replaced (OS plant species) or can regenerate (VR plant species). This does not mean, however, that the plant community of an ecosystem will be unchanged after a fire. Why?

The interval between fires can affect the diversity of a plant community. If fires occur frequently, some species may be lost from the plant community and biodiversity is reduced. Assume that the plant community includes:

- tall trees, mainly vegetative reproducers of the genus *Eucalyptus*
- tall shrubs consisting of obligate seeders, such as *Banksia*, *Hakea* and *Casuarina*
- low shrubs and grasses, including vegetative reproducers such as *Goodenia*.

Forest with plants of various ages comprising both OS and VR species

↓ Summer bushfire ↓

Seed bank from OS plants germinates with potential to grow into new tall shrub layer of even-aged plants.

VR plants also recover.

↓ Bushfire next summer ↓

Young OS shrubs killed. Why?

No replacement of OS plants. Why?

VR plants increase, since competition from OS plants for space and other resources is removed.

↓

New community develops with only tall trees and low shrubs and grasses.

Which layer is missing? Why?

Figure 16.44 A possible sequence of events in a forest subject to frequent fires. Which plant types (VR or OS) will become dominant if fires occur frequently? (OS = obligate seeder; VR = vegetative reproducer)

Assume that a summer fire burns the area. After this fire, the VR plants regenerate from buds or lignotubers and new OS plants grow from seed. These new OS plants require time to mature and reproduce to make a new seed bank. A fire-free period of several years may be needed before this OS seed bank is available.

Figure 16.44 shows what happens if another fire occurs the following summer before the new OS plants have matured and formed seeds. After this second fire, the forest community will become dominated by VR plants. The OS species will disappear unless winds carry in seeds from other regions. If the OS plants are not replaced, tall shrub species will be lost. This change in the forest ecosystem will affect the consumers that depend directly on the middle layer of shrubs for food or nesting sites or shelter.

As part of preventing wildfires, controlled burning is carried out before summer to reduce the litter load. If controlled burns are carried out too often, the OS plants will, over time, be lost from forest ecosystems.

Absence of fire affects ecosystems

Figure 16.45 Golden-shouldered parrots rely on grasslands for survival. Fire regenerates grasslands and prevents replacement (succession) by woodlands.

Fire for the parrots

On Cape York Peninsula, golden-shouldered parrots (*Psephotus chrysopterygius*) thrive in open grasslands where they nest in termite mounds, feed on grass seeds growing in the surrounding area and can see approaching predators (see figure 16.45). The grasslands also support the termites that feed on grasses.

The open grassland habitat of the parrots depends on fire. If burning does not occur, shrubs such as the broad-leaved ti-tree invade the grassland and shade out the grasses, causing their decline and death. These shrubs also provide cover for pied butcherbirds that prey on the parrots. As the grasses die, the termites lose their food source and die, the termite mounds crumble and the parrots lose their nesting sites. The Queensland Department of Environment is assisting landholders to develop burning practices that will safeguard the grasslands.

Figure 16.46 Mountain ash, *E. regnans*, occurs in wet sclerophyll forests in south-east Australia, including the Dandenong Ranges, Victoria.

Fire for forest giants

The largest members of the plant community in wet sclerophyll forest ecosystems of south-east Australia are mountain ashes, *Eucalyptus regnans*, that can reach heights of up to 110 metres and have a life span of about 400 years (see figure 16.46). Unlike other *Eucalyptus* species, mountain ash trees are damaged by bushfire and regenerate only from seed. (Are they VR or OS plants?)

Mountain ash trees produce many seeds that fall to the forest floor but, at ground level in an unburnt forest, the low light conditions are unsuitable for survival of any mountain ash seedlings.

Fires that can kill the mature mountain ash trees create favourable conditions for the seedlings by removing the light-blocking canopy and allowing light to reach the forest floor. If the habitat is not burned during the lifetime of mountain ash trees, no seedlings will be available to replace them when they die and this species will disappear from the ecosystem. No fires — no forest giants!

Human interventions

Low-intensity fire is a natural change agent in Australian ecosystems. However, human interventions in fire management have produced changes to the frequency and intensity of bushfires.

Until recently, the practice was to prevent fires. As a result, in fire-free seasons, masses of litter accumulated in forests and grasslands that could later support high-intensity wildfires that burned uncontrollably and often over vast areas.

Monitoring fires on a global level

Fires throughout the world are detected by satellites, such as *Terra*, using an instrument known as MODIS (refer back to page 508). Figure 16.47a shows fires burning in northern Australia in the region of Darwin. These fires are planned burns carried out before the dry season to reduce the risk of more serious fires later in the year. Figure 16.47b shows devastating wildfires that burnt in January 2003 in south-eastern Australia as captured by MODIS.

(a)

(b)

Figure 16.47 **(a)** Fires burning in northern Australia on 17 May 2005 as revealed by satellite-borne sensor, MODIS **(b)** Wildfires burning in Victoria and New South Wales over weeks in January 2003. Smoke from these fires spread eastwards and reached beyond New Zealand.

KEY IDEAS

- Fire is an important agent of change in some Australian ecosystems.
- Various strategies have evolved in Australian flora that enable survival of the species in fire-affected ecosystems.
- The frequency of fire (fire regime) can have a major effect on ecosystems.
- Monitoring of fires at a global level occurs using sensors on satellites.

QUICK-CHECK

21 Identify the following statements as true or false.
   a Mature wattle trees are typically destroyed by fires.
   b Seeds of mature wattle trees are stimulated to germinate by fire.
   c Epicormic growth is a feature of obligate seeder plants.
   d Aboriginal people used fire for many generations for hunting
22 Distinguish between the meanings of the terms:
   a obligate seeder (OS) and vegetative reproducer (VR)
   b bushfire and wildfire
   c fire intensity and fire frequency.

Natural succession in ecosystems

Communities in some ecosystems do not remain stable, and slight changes in the physical or the chemical features of a habitat can cause the community to change without any human intervention. The natural replacement over time of one community by another community with different dominant species is termed **succession**. There are two kinds of succession:

- **primary succession** in which different communities in turn become established on land that has not previously been colonised, such as a lava flow (see figure 16.48), a sand dune or land uplifted from the sea as a result of an earthquake
- **secondary succession** in which different communities in turn become established in an area that was previously colonised but which has been disturbed, such as a ploughed field, mine wastes, an abandoned paddock and a drained dam.

The first species to become established in a 'new' habitat are termed **pioneer species** or members of a pioneer community. These are species that can survive under harsh conditions and are adapted for dispersal and rapid reproduction — pioneer species are typically r-selected species. The process of succession stops when a stable community becomes established, with no further change in the dominant species. This stable community is known as the **climax community**. The climax community will depend on the physical features in the area, including aspect, altitude, temperature, rainfall and soil type.

Refer back to page 531 that described the grasslands on Cape York Peninsula. These grasslands are not the climax community in this area. It is only the occurrence of fires that maintains the grasslands and, in the absence of fires, woodlands would become established as a climax community.

Figure 16.48 With time, this new habitat in Hawaii created by volcanic lava flow will be colonised by various communities. Lichens will appear on the lava surface and slowly cause chemical breakdown. Wind and rain will cause physical breakdown. Soil particles gradually accumulate. Moss and fern spores blown into the area germinate and grow in sheltered crevices among the rocks. Over time, further changes will occur until a climax community is finally established. What type of succession is this?

KEY IDEAS

- Replacement of one community by another is termed succession.
- Succession can be primary or secondary.
- Primary succession occurs in areas that have not been previously colonised.
- Secondary succession occurs in areas that were once colonised but have been disturbed.

QUICK-CHECK

23 Define the following terms:
 a a pioneer species
 b a climax community.

24 Consider the following features. For each feature, identify whether it is more likely to be present in a member of a pioneer community or in a member of a climax community.
 a able to survive on thin or no soil
 b able to survive with little water
 c produces only a few large seeds annually
 d produces very large numbers of small seeds annually
 e regarded as a weed

Restoring the balance

Humans can impact negatively on ecosystems. Human actions at various levels can also protect or restore ecosystems, either directly or indirectly.

Personal level

Actions at a personal level include:

- recycling glass, aluminium, paper and other products
- switching off unneeded lights
- composting biodegradable household waste
- using water thoughtfully
- supporting community organisations such as local conservation groups, Land for Wildlife (see figure 16.49) and the Australian Conservation Foundation.

Individuals do make a difference. For example, four farmers in the Gwydir wetlands in north-west New South Wales voluntarily signed agreements that make part of their farmlands protected wetlands conservation sites.

Figure 16.49 Individuals and groups can set aside land corridors on their property for coexistence with wildlife. This sign is on a golf course in Western Australia.

Local council level

Actions at a local council level include:

- providing facilities for the collection of recyclable products
- organising clean-ups of local sites
- developing urban forests.

State government level

Actions at a state government level include:

- sewage treatment
- passing legislation to protect ecosystems, such as the *Flora and Fauna Guarantee Act 1988* and the *National Parks (Marine National Parks and Marine Sanctuaries) Act 2002*.

The Australian Government Envirofund is the community-based component of the government's $3 billion Natural Heritage Trust. The Trust provides grants of up to $30 000 to target local problems.

Every year, an estimated 6000 million kilograms of litter is added to the seas from merchant shipping and passenger vessels.

Federal government level

Actions at a federal government level include:

- proclaiming legislation such as the *Commonwealth Endangered Species Act 1992* and the *Environment Protection and Biodiversity Conservation Act 1999*
- funding the Quarantine and Inspection Service
- fulfilling responsibilities under international agreements including the Conservation of Migratory Species and the Montreal Protocol
- funding of community-based National Heritage Trust projects including Bushcare, Landcare, Rivercare, Coastcare
- cooperating through the development of international agreements (protocols) on environment protection
- action through the International Union for Conservation of Nature and Natural Resources (IUCN) and through the United Nations Environment Program to assist developing countries to meet the costs associated with conservation measures.

The following box tells the story of how human actions can restore an ecosystem.

CASE STUDY

Restoring the Roxby Downs arid ecosystem

Twenty kilometres north of Roxby Downs in South Australia is a large reserve comprising a variety of habitats including saltbush shrublands, mulga/native pine woodlands and *Acacia* dunes. Arid Recovery is a joint initiative of BHP Billiton (which took over from WMC Resources in 2005), the South Australian Department for Environment and Heritage, the University of Adelaide and a local Friends group. The project provides a case study of community cooperation to restore an ecosystem and to allow sustainable commercial activities.

The vegetation in part of this area had been severely degraded by over-grazing from introduced herbivores. In many areas the mature saltbush plants (*Atriplex* spp.) had disappeared with no recruitment of new seedlings. Other degraded areas were the calcareous plains that support bluebush (*Maireana* spp.) and saltbush and the sand dunes that support mulga (*Acacia aneura*) and horse mulga (*A. ramulosa*). The local native fauna was depleted, with 60 per cent of the mammal species which used to occur at Roxby Downs now locally or completely extinct.

The long-term aim is to restore the arid zone ecosystem and to achieve multiple and sustainable land use including mineral exploration, local mining and pastoral industries. In order to restore the Roxby Downs arid ecosystem, several goals were identified:

- removal of rabbits from the area. *Why?* Rabbits and other introduced herbivores browse on new seedlings of native plants and ringbark young trees. Rabbits also remove grass cover which provides important shelter for ground-dwelling native species. *How?* Through various measures including fumigation, baiting and use of rabbit calicivirus.
- regeneration of perennial plants that are native to the region. *Why?* Native plants are a key part of the communities in arid zone ecosystems because they reduce soil erosion because of their deep root systems and sustain native mammals and birds which depend on these plants for food and shelter. *How?* Through removal of rabbit populations and reduction in cattle numbers.
- reduction in the number of cattle in some areas. *Why?* Reducing stock numbers reduces grazing pressure on native vegetation. *How?* Cooperation with local pastoralists, including demonstration of land management practices to optimise biodiversity.

Figure 16.50 When grazing and browsing pressures are reduced, the arid zone vegetation can regenerate, including the scarlet Sturt desert-pea (*Swainsona formosa*).

(continued)

- control of introduced feral mammals, such as foxes and cats, in the region. *Why?* Feral predators are a major cause of the loss of native mammal species. Feral pest control is an essential step before native species can be successfully reintroduced to an area. *How?* Through various control measures, including baiting and trapping.
- exclusion of feral pests from some areas in the region. *Why?* To establish an area for release of and research into the survival of native mammals when they are free from the pressure of introduced predators and of rabbits that compete with them for limited resources; to examine how vegetation recovers when freed from grazing pressures. *How?* By building a pest-proof fence around a 60 square kilometre reserve initially as a physical barrier (see figure 16.51). The enclosure was completed in 2000 and more than 5000 rabbits and 50 cats and foxes were removed. Approximately 300 monitoring sites for plant regeneration and animals have been established. By January 2001, the reserve was declared free of rabbits, cats and foxes.
- reintroduction of locally extinct native animals into the reserve and conduct of research into their survival. *Why?* This is a first step in identifying whether the biodiversity of the natural community ecosystem can be restored. Selection of species for reintroduction is based on various evidence, including:
  1. sub-fossil evidence that identifies which species formerly lived in the Roxby Downs area
  2. recollections of station owners who, for example, recall seeing bilbies in the Roxby Downs area early this century
  3. identification of native mammals that live close to Roxby Downs; for example, stick-nest rat nests have been found within 20 kilometres of Roxby Downs.

  How? The timing and order of reintroductions was dependent on the availability of animals. The first species, the stick-nest rat or wopilkara (*Leporillus conditor*) was reintroduced in April 1999 (see figures 16.52a, b). Burrowing bettongs were reintroduced from Western Australia in 1999, greater bilbies in 2000 and western barred bandicoots in 2001.

An additional 20 square kilometres of reserve was recently fenced to accommodate the native animal populations. Each species has been reintroduced successfully and, compared with outside the reserve, there are now three times as many of the small mammals within the fenced zone.

Research is continuing to provide a greater understanding of how the ecosystem functioned before the arrival of Europeans and the role of reintroduced species in ecological processes. For example, bilbies dig within the reserve for food and their diggings provide areas for seed collection and germination, thus assisting the regeneration of vegetation. Removing feral animals and restoring locally extinct species is an important step in improving ecosystem health in arid areas.

Figure 16.51 The fence enclosing the pest-free area on Roxby Downs

(a)

(b)

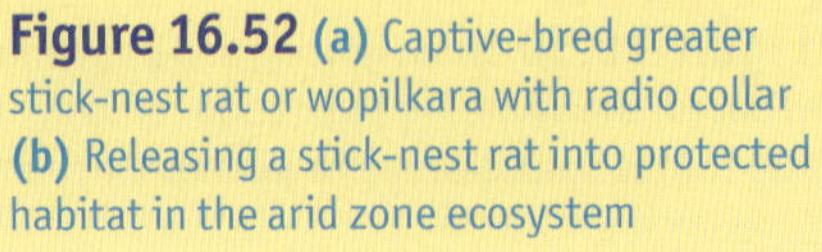

Figure 16.52 **(a)** Captive-bred greater stick-nest rat or wopilkara with radio collar **(b)** Releasing a stick-nest rat into protected habitat in the arid zone ecosystem

ODD FACT

Life doesn't stop at the bottom of the garden bed! The depths of the Earth's crust were once thought to be sterile but in the late 1980s, scientists extended the search for life into the sub-surface domain. Deep below the surface, many types of bacteria and fungi have been found, including micro-organisms that live many kilometres below the Earth's surface in very hot oxygen-free conditions.

The global ecosystem: the biosphere

The **biosphere** is the life-support system of planet Earth and its source of energy is the radiant energy of the sun. The biosphere contains all the ecosystems of planet Earth and the continued existence of all living things depends on a functioning biosphere. The biosphere includes a living part, the **biota**, comprising all living organisms. It also has non-living parts, made up of the gases of the atmosphere, the surface sediments of the **lithosphere**, and the waters of the **hydrosphere**. The biosphere includes the interactions that occur between these parts, as shown in figure 16.53. The interactions within the biosphere involve the cyclic movement of essential elements, such as carbon, that move between the various parts of the biosphere.

The results of actions that affect the atmosphere and the hydrosphere in one region of the planet can have an effect at a distance. So, for example, the radioactive wastes that escaped into the atmosphere from the explosion at the Chernobyl nuclear reactor in 1986 did not remain in that region but were spread by air currents to other parts of the globe. Likewise, the carbon emissions from car exhausts in the United States contribute to the global increase in atmospheric carbon dioxide. Sulfur dioxide emission from volcanoes in Indonesia spread through the atmosphere to other parts of the world.

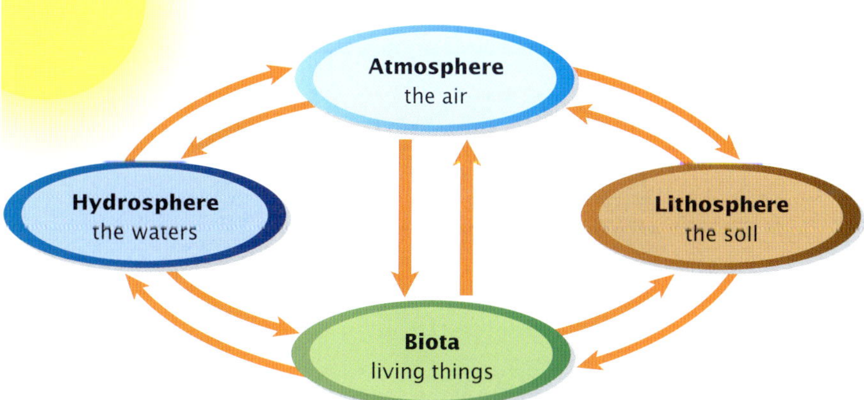

Figure 16.53 The biosphere is made up of the hydrosphere, lithosphere, atmosphere and biota, as well as the interactions between them and an energy input from the sun.

Ozone layer: a protective blanket

Some human actions are truly global in their impact. Chlorofluorocarbons (CFCs) were used for many decades as cooling agents, as industrial solvents and as propellants in aerosols. CFCs and related chemicals have drifted into the stratosphere where they still continue to react, causing the breakdown of ozone (O_3) in the **ozone layer**.

The ozone layer protects Earth from the lethal effects of shortwave ultraviolet (UV) radiation. Since living organisms are harmed by shortwave UV radiation, thinning of this layer (or the so-called 'hole in the ozone layer') has many negative impacts, including damaging crops, harming marine life and causing cancers.

Remote sensing through instruments carried on satellites allows us to see changes in components of the biosphere, including the ozone layer. These satellites include the Earth Probe satellite that carries a Total Ozone Mapping Spectrometer (TOMS). Figure 16.54 shows the mission patch for Earth Probe and the TOMS instrument that it carries. TOMS gathers data on ozone levels over the planet by measuring the total amount of ozone in a 'column' of air from the top of the atmosphere to the Earth's surface. Scientists analyse the data to see if the loss of ozone is continuing or stabilising. Figure 16.55 shows ozone levels as mapped by TOMS for Antarctica and other regions of the southern hemisphere in October 2004 and in February 2005. The ozone hole corresponds to ozone levels of less than 220 Dobson units (DU).

An Ozone Monitoring Instrument (OMI) launched on the EOS-Aura satellite in July 2004 superseded Earth Probe (EP) TOMS in 2005. Visit the website at http://toms.gsfc.nasa.gov.

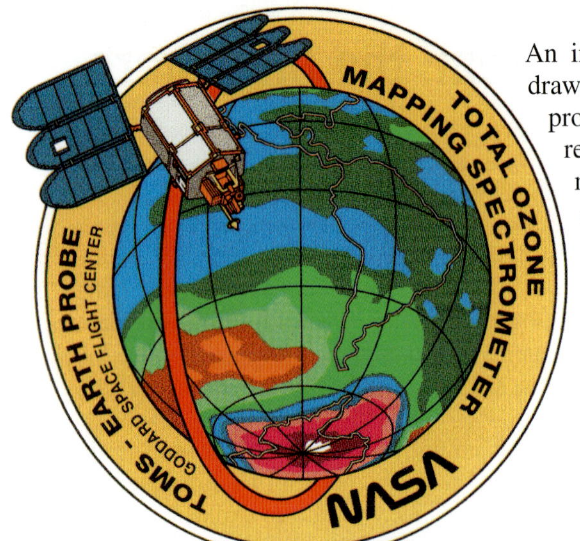

Figure 16.54 Patch for the TOMS Earth Probe

An international and legally binding agreement, the Montreal Protocol, drawn up in 1987, was signed by many nations including Australia. This protocol, most recently strengthened in 1999 in Beijing, obliges nations to reduce to zero their use of ozone-depleting chemicals. (For more information, go to www.jaconline.com.au/natureofbiology/natbiol1-3e and click on the 'Montreal Protocol' weblink for this chapter.)

(a)

(b)

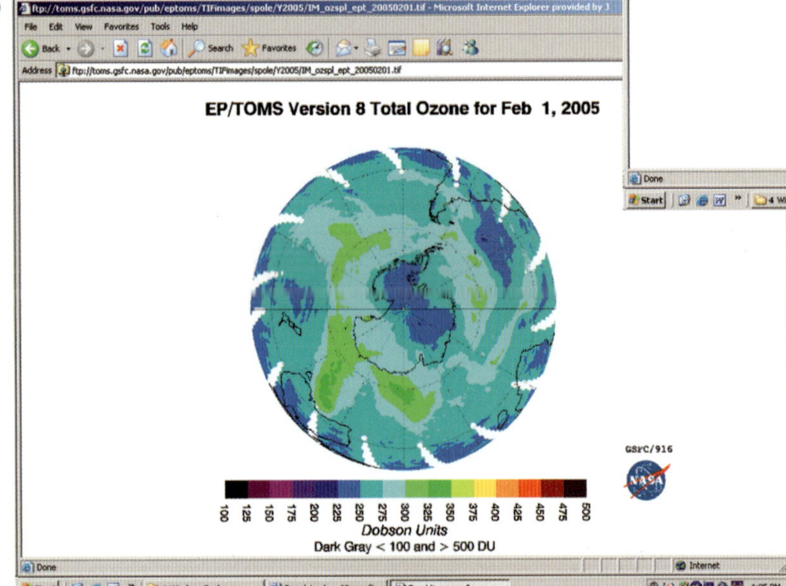

Figure 16.55 Total ozone over Antarctica and southern hemisphere as recorded by EP/TOMS on **(a)** 1 October 2004 and **(b)** 1 February 2005. What was the lowest level of ozone recorded on 1 October 2004 and where did this occur?

KEY IDEAS

- Actions to protect or restore ecosystems must occur at various levels.
- The Roxby Downs case study illustrates ecosystem restoration.
- The biosphere is part of the global ecosystem.
- The ozone layer that protects Earth from short wave UV radiation has been depleted by ozone-destroying chemicals.

QUICK-CHECK

25 List two actions at a personal level that could assist in protecting ecosystems.

26 Briefly explain the significance to the global ecosystem of:
a TOMS b Montreal Protocol.

BIOCHALLENGE

The National Action Plan for Salinity and Water Quality (NAPSWQ), supported by both federal and state governments, identifies several strategies for attacking the salinity problem in Australia (refer back to pages 520–1). These strategies include an engineering process known as salt interception. Salt interception is the removal by pumping of saline ground water out of the ground and storing it.

The first salt interception scheme on the Murray River was established at Buronga in 1981. This scheme has prevented more than 200 tonnes of salt a day from entering the Murray River. Figure 16.56 shows the details of a salt interception scheme.

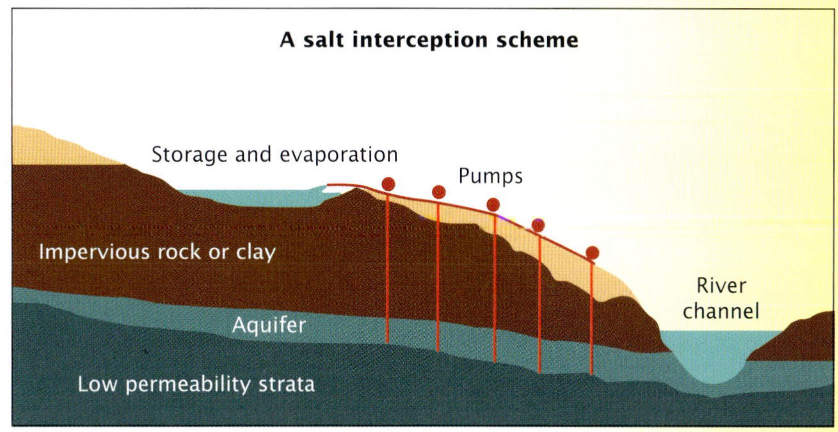

Figure 16.56 A salt interception scheme (Access the following organisations' websites for more details: www.murrayusers.sa.gov.au; www.murraycare.com.au)

1. There are several ways of lowering the watertable. One is to reduce the recharge by surface water into the ground water, while another is to remove water from the ground water system. What happens with a salt interception scheme?
2. An important part of a salt interception scheme relates to the disposal of the saline water that is pumped from the ground water. From the following alternatives, identify how the saline water is treated after pumping.
   a The saline water from the ground water is allowed to drain into the river.
   b The saline water from the ground water is stored in a basin that is impervious to water.

   On farmland, lowering the watertable to more than two metres below the land surface reduces the stress on crops and pastures. On a property at Pyramid Hill, Victoria, about half the land was affected by dryland salinity. The owners pumped saline water from the underlying watertable and transferred it to evaporation ponds lined with polythene. Pumping lowered the watertable from just half a metre below the land surface to about six metres, and soil that was previously highly saline was reclaimed for vegetable crops. As an added bonus, salt obtained by evaporation is now purified and sold through a company, Pyramid Salt Pty Ltd, set up by the owners.
3. Why were the evaporation ponds lined with polythene?
4. Why did it become possible to grow vegetables again on land that had previously been too salty?

Figure 16.57 shows the average salinity levels of the Murray River at Morgan in South Australia from 1980 to 2003, with a blue curve indicating the effects of salinity management and the red curve showing the salinity that would exist without any interventions.

5. What trend was apparent in the period from 1980 to 1983?
6. What was the highest level of salinity recorded?
7. In 2003, what was the approximate reduction in salinity as a result of salinity management?

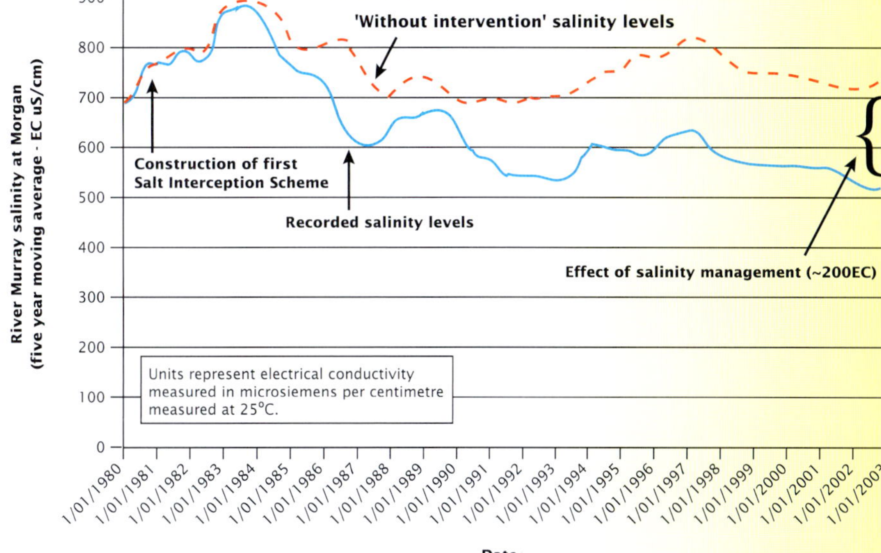

Figure 16.57 Average salinity levels of the Murray River (*Source:* Murray-Darling Basin Commission)

CHAPTER REVIEW

Key words

CROSSWORD

alien species
biological control or biocontrol
biopesticides
biosphere
biota
bushfires
classical biocontrol
climax community
conservation
conservation biocontrol
dryland salinity
epicormic buds
estuary
eutrophication
exotic species
hydrosphere
immuno-contraception
invasive
irrigation salinity
lignotubers
lithosphere
noxious
obligate seeder
over-harvesting
ozone layer
pioneer species
primary succession
regulated flow
remote sensing
rhizomes
salinity
salt interception
sclerophyll
secondary succession
succession
unregulated flow
vegetative reproducers
wildfire

Questions

1 ***Making connections*** ▸ Use as many as possible of the key words above to create a concept map.

2 ***Developing explanations*** ▸ Suggest a possible explanation for each of the following observations:

a Revegetation of cleared land with deep-rooted trees can result in a drop in the watertable.

b After a bushfire, clumps of new leaves appear on the trunks of some trees.

c Austral bracken is one of the first plants to reappear in a forest area after a bushfire.

d The rate of growth of algae in a lake increases after large areas of adjacent land are cleared of forests and used for farming purposes.

e Eutrophication is observed in lakes but not in open oceans.

3 ***Making predictions and evaluating alternatives*** ▸ An exotic floating weed pest, water hyacinth (*Eichhornia crassipes*), becomes established in a lake ecosystem.

a Describe possible effects on the community of this lake ecosystem as the water hyacinth spreads and covers the lake surface.

b Identify three different kinds of control measure that might be used against the water hyacinth.

c Three imported insect species were assessed for possible use in the biocontrol of the water hyacinth. The target preferences of the insects are shown below.

	Insect species A	*Insect species B and C*
Target preference	Insects eat many plant species	Insects eat the exotic weed

Which species should be rejected as a possible biocontrol agent?

d Further studies showed that, when the intended target weed disappeared, insect species B ate native plants, but insect species C died of starvation. Which insect species appears to be a suitable biocontrol agent?

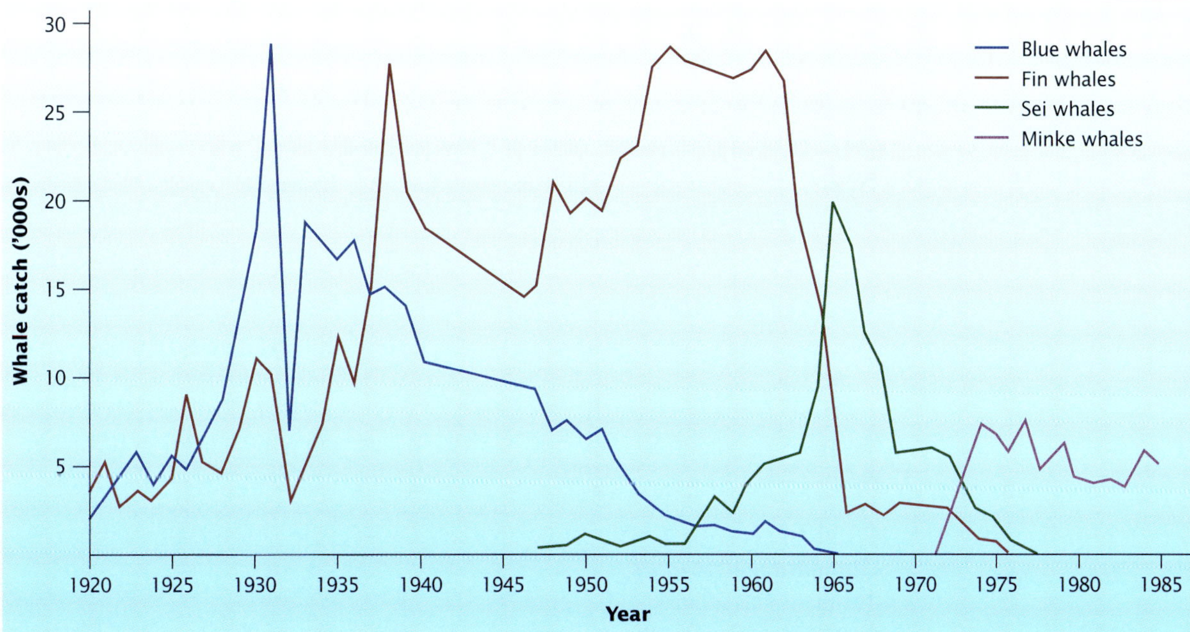

Figure 16.58

4 ***Interpreting data*** ▸ Figure 16.58 shows the annual harvest of whales before the International Whaling Commission imposed bans.

a What species provided the bulk of the initial harvest?

b Was this harvesting sustainable? Explain.

c How did whaling nations respond when one species declined?

5 ***Applying understanding in new contexts*** ▸ Consider two imaginary fish: *Pisces rapid* and *Pisces slow*. *P. rapid* grows quickly, reaching 20 centimetres after just one year and breeds when it is two years old. *P. slow* grows very slowly, reaching a length of 20 centimetres after 10 years and does not breed until it is 10 years old. Which is at greater risk of being harvested at unsustainable levels? Explain.

6 ***Developing hypotheses*** ▸ In an experiment, it was observed that coral-producing animals die if placed in water contaminated by suspended particulate matter. Suggest a hypothesis to explain why small particulate matter might kill coral. How would you test this hypothesis?

7 ***Demonstrating your understanding*** ▸ Describe how the plant community in a tall open forest ecosystem might change under the following fire regimes:

a fires occurring in successive years

b no fires for centuries.

8 ***Using the web*** ▸ The Australian National Botanic Gardens has published information on major invasive weeds in Australia. Go to www.jaconline.com.au/natureofbiology/natbiol1-3e and click on the 'Invasive weeds' weblink for this chapter, then answer the following questions.

a What percentage of the total flora of Australia today consists of introduced plants?

b How many of these are or have the potential to become serious weeds?

c What is the most common source of our serious environmental weeds?

d List three invasive plants that are identified as being 'of major concern'.

e Go to the entry on the rubber vine (*Cryptostegia grandiflora*) and click on the camera icon to see the impact of this weed. Briefly describe how this weed produces changes in ecosystems.

APPENDIX A Scientific measurement

Prefixes used with metric units of measurement

Name	Symbol	Multiplication factor	Example
Greater than one:			
mega	M	10^{6}	Mg = megagram = 10^{6} gram
kilo	k	10^{3}	km = kilometre = 10^{3} metre
Less than one:			
deci	d	10^{-1}	dl = decilitre = 1/10 litre
centi	c	10^{-2}	cm = centimetre = 1/100 metre
milli	m	10^{-3}	mV = millivolt = 1/1000 volt
micro	μ	10^{-6}	μm = micrometre = 10^{-6} metre
nano	n	10^{-9}	ng = nanogram = 10^{-9} gram
pico	p	10^{-10}	ps = picosecond = 10^{-12} second

Standard metric units

Standard unit of mass	gram	g
Standard unit of length	metre	m
Standard unit of volume	litre	L

Some standard units of mass

Unit	Abbreviation	Equivalent
kilogram	kg	10^{3} g
gram	g	
milligram	mg	10^{-3} g
microgram	μg	10^{-6} g
nanogram	ng	10^{-9} g
picogram	pg	10^{-12} g

Some standard units of length

Unit	Abbreviation	Equivalent
metre	m	
centimetre	cm	10^{-2} m
millimetre	mm	10^{-3} m
micrometre	μm	10^{-6} m
nanometre	nm	10^{-9} m
picometre	pm	10^{-12} m
angstrom	Å	10^{-10} m (a non-SI unit of measurement)

Some standard units of volume

Unit	Abbreviation	Equivalent
litre	L	
millilitre	mL	10^{-3} L
microlitre	μL	10^{-6} L

Some standard units of area

Unit	Abbreviation	Equivalent
square metre	m^2	Area enclosed by a square, each side 1 m in length
hectare	ha	1 ha = 10 000m^2 = 10^{4} m^2
square centimetre	cm^2	1 cm^2 = 1/10 000 m^2 = 10^{-4} m^2

Relationship between mass and volume of water (at 20 °C)

1 g = 1 cm^3 = 1 mL = 1 cc (cubic centimetre)

APPENDIX B Amino acids

Chemical structure	Amino acid
$HO-C(=O)-CH_2-C(H)(NH_2)-C(=O)OH$	Aspartic acid *asp*
$HO-C(=O)-CH_2-CH_2-C(H)(NH_2)-C(=O)OH$	Glutamic acid *glu*
$NH_2-CH_2-CH_2-CH_2-CH_2-C(H)(NH_2)-C(=O)OH$	Lysine *lys*
$NH_2-C(=NH)-NH-CH_2-CH_2-CH_2-C(H)(NH_2)-C(=O)OH$	Arginine *arg*
$HC{=}C(-N{=}CH-NH-)-CH_2-C(H)(NH_2)-C(=O)OH$	Histidine *his*
$HS-CH_2-C(H)(NH_2)-C(=O)OH$	Cysteine *cys*
$CH-S-CH_2-CH_2-C(H)(NH_2)-C(=O)OH$	Methionine *met*
$NH_2-C(=O)-CH_2-C(H)(NH_2)-C(=O)OH$	Asparagine ($AspNH_2$) *asn*
$NH_2-C(=O)-CH_2-CH_2-C(H)(NH_2)-C(=O)OH$	Glutamine ($GluNH_2$) *gln*
$CH_2-CH_2-CH_2-N(H)-CH-C(=O)OH$ (ring: CH_2—CH_2, CH_2—N, N—CH)	Proline *pro*
$H-C(H)(NH_2)-C(=O)OH$	Glycine *gly*
$CH_3-C(H)(NH_2)-C(=O)OH$	Alanine *ala*
$(CH_3)_2CH-C(H)(NH_2)-C(=O)OH$	Valine *val*
$(CH_3)_2CH-CH_2-C(H)(NH_2)-C(=O)OH$	Leucine *leu*
$CH_3-CH_2-CH(CH_3)-C(H)(NH_2)-C(=O)OH$	Isoleucine *ile*
$HO-CH_2-C(H)(NH_2)-C(=O)OH$	Serine *ser*
$CH_3-C(H)(OH)-C(H)(NH_2)-C(=O)OH$	Threonine *thr*
$HO-C_6H_4-CH_2-C(H)(NH_2)-C(=O)OH$	Tyrosine *tyr*
$C_6H_5-CH_2-C(H)(NH_2)-C(=O)OH$	Phenylalanine *phe*
$C_8H_6N-CH_2-C(H)(NH_2)-C(=O)OH$ (indole ring: HC, HC, C–H, C–H, C, C, CH, N–H)	Tryptophan *trp*

APPENDIX C Classification of living things

This appendix uses the five-kingdom scheme for classifying living organisms, as outlined in chapter 8, pages 245–6. This differs from the three domains scheme described on pages 246–9. The scheme used here is incomplete because only some phyla, classes and orders are listed. Approximate worldwide numbers of species are given for many of the taxa.

KINGDOM MONERA

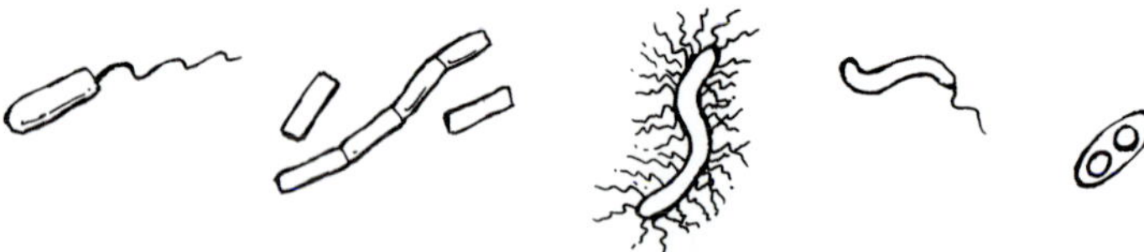

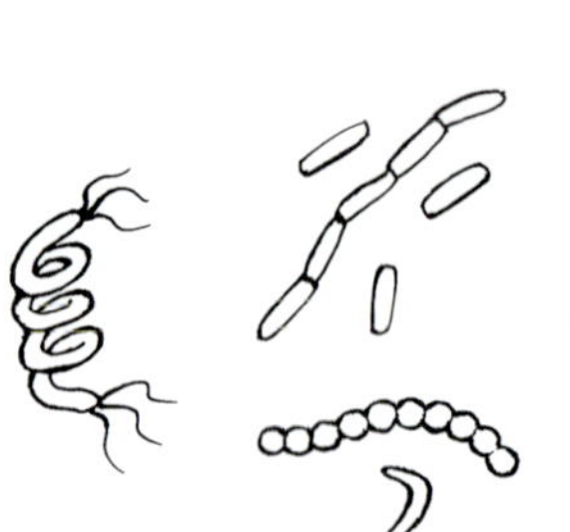

Prokaryotes, usually unicellular, but sometimes occur in long filaments formed by the joining end-to-end of cells. Members include photosynthetic and chemosynthetic autotrophs, as well as heterotrophs that feed by absorption.

Subkingdom ARCHAEBACTERIA: Bacteria that thrive in oxygen-free (anaerobic) conditions, including bacteria that live in salt lakes and hot acidic conditions and methane-producing bacteria that use chemicals such as methanol and acetate as their source of energy and which are found in swamps, marshes and in the gut of cattle.

Subkingdom EUBACTERIA: Includes photosynthetic, chemosynthetic and heterotrophic bacteria.

Phylum SCHIZOPHYTA: Heterotrophic bacteria living in both oxygen-rich and oxygen-free regions; some of these bacteria cause disease.

Examples: *Escherischia coli, Chlamydia, Streptococcus*

Phylum CYANOPHYTA: Photosynthetic blue-green bacteria.

Examples: *Nostoc, Nodularia*

KINGDOM PROTISTA

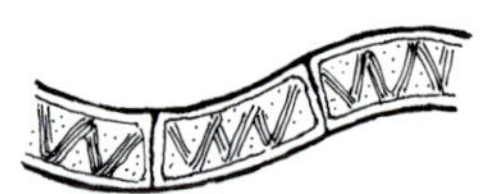

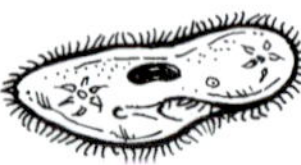

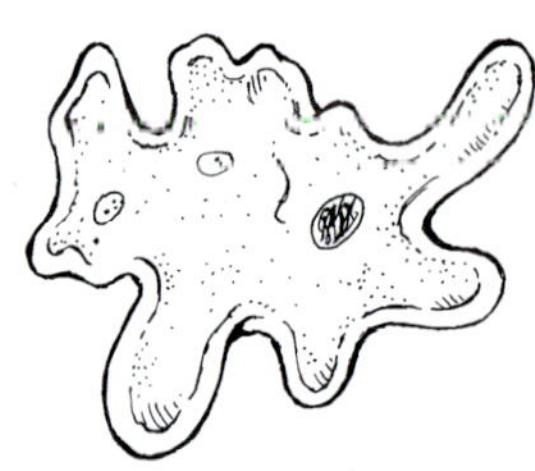

Includes unicellular and multicellular photosynthetic algae and unicellular and multicellular heterotrophs that are not fungi, plants or animals.

Phylum SARCODINA: Unicellular organisms that capture prey by engulfing it. Found in fresh water, salt water and in the bodies of some animals. Approx. 11 500 spp.

Examples: *Amoeba*, a free-living organism, *Entamoeba* found in the gut and the cause of one form of dysentery.

Phylum CILIOPHORA: Unicellular organisms moving by means of beating cilia on their surfaces and feeding by engulfing food. Approx. 7200 spp.

Example: *Paramecium*

Phylum SPOROZOA: Unicellular organisms living as parasites; move by means of flagella. Approx. 5000 spp.

Example: *Plasmodium*, cause of malaria

Phylum PYRROPHYTA: Unicellular photosynthetic organisms; form part of the phytoplankton that floats at the ocean surface. Approx. 2000 spp.

Examples: dinoflagellates

Phylum CHRYSOPHYTA: Unicellular photosynthetic organisms; in some the cell wall is thickened with silica. Approx. 13 000 spp.

Examples: diatoms, golden-brown algae

Phylum EUGLENOPHYTA: Unicellular photosynthetic organisms, found mainly in fresh water; move by means of a single flagellum. Approx. 1000 spp.
Example: *Euglena*
Phylum CHLOROPHYTA: Green algae found in both fresh water and salt water; includes unicellular and multicellular members. Approx 9000 spp.
Examples: unicellular *Chlamydomonas*, multicellular Ulva, the sea lettuce, and *Spirogyra*, a freshwater green alga
Phylum PHAEOPHYTA: Multicellular brown algae living in salt water, and varying in size from small tufts to massive kelps. Contain brown pigment, fucoxanthin, in addition to chlorophylls. Approx. 1500 spp.
Examples: *Hormosira* (Neptune's necklace), *Durvillaea potatorum* (bull kelp) found on rocky shores along southern Australia
Phylum RHODOPHYTA: Mainly multicellular red algae living in salt water. Contain red pigments, phycobilins, in addition to chlorophylls. Approx. 4000 spp.
Examples: *Amphiroa*, a red coralline algae, and *Gracilaria*, common in rock pools

KINGDOM FUNGI

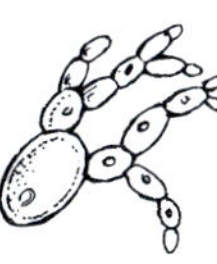

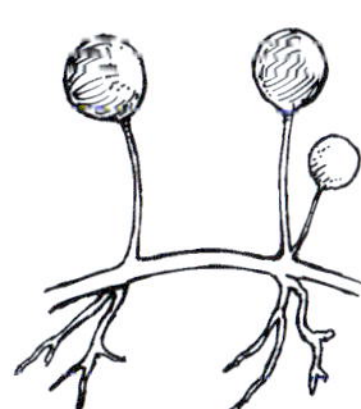

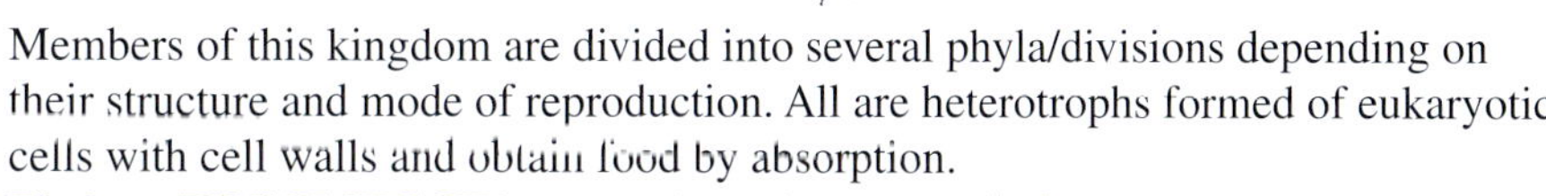

Members of this kingdom are divided into several phyla/divisions depending on their structure and mode of reproduction. All are heterotrophs formed of eukaryotic cells with cell walls and obtain food by absorption.
Phylum ZYGOMYCOTA: Reproduce either sexually by conjugation or asexually.
Examples: black bread mold (*Rhizopus*) and fungi that grow in association with roots of flowering plants. Approx. 600 spp.
Phylum ASCOMYCOTA: Reproduce either sexually by spores that develop in a sac-like structure known as an ascus or asexually.
Examples: multicellular fungi such as powdery mildews, truffles and unicellular yeasts. Approx. 30 000 spp.
Phylum BASIDOMYCOTA: Reproduce sexually by spores that form in a structure known as a basidium and also reproduce asexually.

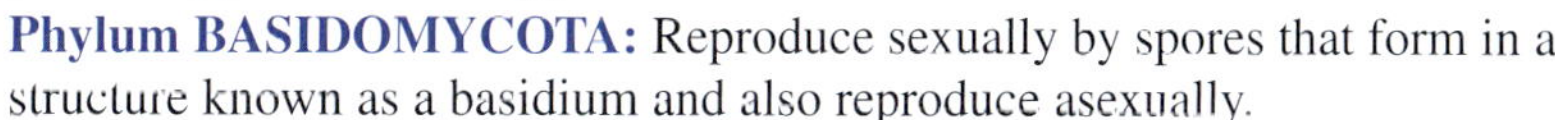

Examples: mushrooms, toadstools, rusts, bracket fungi and puffballs. Approx. 25 000 spp.
Phylum DEUTEROMYCOTA

Examples: cheese molds, *Penicillium*, and fungi that cause skin diseases, such as athlete's foot. Approx. 25 000 spp.

KINGDOM ANIMALIA

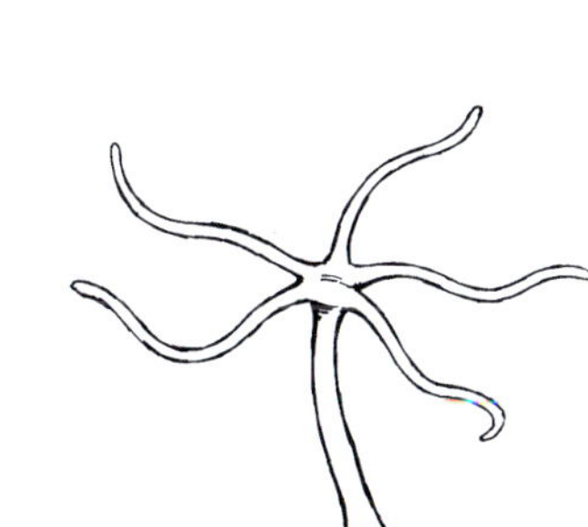

Members of this kingdom are multicellular heterotrophs formed of eukaryotic cells with no cell walls. Only 12 of more than 30 phyla are listed.
Phylum PORIFERA: Mainly marine animals known as sponges; bodies with many holes that let in water carrying food particles. Approx. 5000 spp.
Phylum CNIDARIA: Radially symmetrical animals with a single opening for ingestion and egestion; possess tentacles with stinging cells. Approx. 9000 spp.
Examples: *Hydra*, jellyfish such as *Physalia* (Portuguese man-of-war), corals, sea anemones

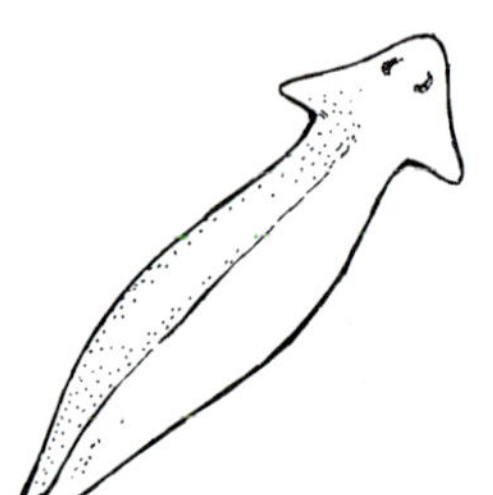

Phylum PLATYHELMINTHES: Animals with bilateral symmetry, flattened bodies and a single gut opening. Approx. 12 200 spp.

Class Turbellaria: free-living flatworms, such as *Planaria*

Class Trematoda: parasitic flukes with suckers, such as *Schistosoma*

Class Cestoda: parasitic flatworms with no digestive systems, such as tapeworms, *Taenia*

Phylum NEMATODA: Roundworms with digestive system with mouth and anus; mainly parasites. Approx. 12 000 spp.

Example: hookworms such as *Ascaris*, free-living vinegar eels

Phylum ANNELIDA: Worms consisting of many segments, possess digestive system with mouth and anus. Approx. 12 000 spp.

Class Oligochaeta: earthworms and marine worms with indistinct head

Class Polychaeta: marine worms with distinct heads

Class Hirudinea: parasitic worms with suckers at one or both ends of the body, such as leeches

Phylum POGONOPHORA: A phylum containing giant tubeworms found in mid-ocean ridge ecosystems. One species only.

Phylum MOLLUSCA: Unsegmented soft-bodied animals with head and muscular 'foot'; often with protective shell(s); mainly aquatic. Approx. 200 000 spp.

Class Gastropoda: usually with single spiral shell, such as snails and whelks, also slugs. Approx. 35 000 spp.

Class Bivalvia: molluscs with two shells, such as clams, mussels, scallops and oysters

Class Cephalopoda: molluscs with eight or ten tentacles and well-developed eyes; octopuses, squid, *Nautilus* and extinct ammonites. Approx. 1000 spp.

Class Polyplacophora: molluscs with eight shell plates on their backs, such as chitons

(Note: *The following three phyla are sometimes grouped together as Phylum Arthropoda.)

***Phylum CHELICERATA:** Segmented animals with body divided into two distinct regions — a cephalothorax and an abdomen; typically possess four pairs of legs. About 60 000 spp.

Class Arachnida: spiders, harvestmen (daddy-long-legs), scorpions, mites and ticks

***Phylum CRUSTACEA:** Segmented animals with jointed limbs and protective exoskeleton made of chitin; mainly marine; body consists of head, thorax and abdomen; possess two pairs of antennae. About 25 000 spp.

Class Malacostraca: lobsters, crayfish, shrimps, crabs, yabbies, prawns and slaters; usually aquatic; possess gills

Class Cirripedia: barnacles

Class Copepoda: copepods

***Phylum UNIRAMIA:** Segmented animals with jointed limbs and protective exoskeleton made of chitin. Mainly terrestrial. More than 1 million spp. Subdivided into eight classes including:

Class Chilopoda: centipedes; one pair of legs per body segment; body comprising 15 to more than 170 segments

Class Diplopoda: millipedes; two pairs of legs per body segment

Class Insecta: ants, bees, butterflies, beetles, fleas, wasps, flies; terrestrial and air breathing; body consists of head, thorax and abdomen, with three pairs of legs and often two pairs of wings on thorax. This class is subdivided into 29 orders including: Order Lepidoptera: butterflies, moths. Approx. 112 000 spp.

Order Diptera: flies. Approx. 98 500 spp.
Order Hymenoptera: wasps, ants. Approx. 103 000 spp.
Order Coleoptera: beetles. Approx. 290 000 spp.
Order Thysanura: silverfish
Order Ephemeroptera: mayflies
Order Siphonaptera: fleas
Order Hemiptera: bugs, leafhoppers
Order Orthoptera: grasshoppers, crickets and locusts

Phylum ECHINODERMATA: Radially symmetrical animals with internal skeleton; possess a water vascular system and often tube feet that are used in movement; marine. Approx. 6100 spp.

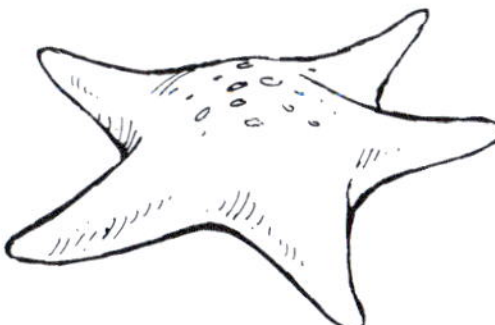

Class Asteroidea: starfish, brittle stars
Class Echinoidea: sea urchins
Class Holothuroidea: sea cucumbers

Phylum CHORDATA: Animals that possess at some stage of their development a dorsal notochord, gill slits, and a hollow dorsal nerve cord. Approx. 43 000 spp.

Class Agnatha: lampreys; aquatic chordates with jawless mouth. 45 spp.
Class Chondrichthyes: sharks, skates and rays; marine; skeleton made of cartilage. More than 700 spp.
Order Selachii: sharks. Approx. 200 spp.
Order Batoidea: skates and rays. Approx. 350 spp.

Class Osteichthyes: fish with bony skeleton, such as flathead, trout, eels, tuna, lungfish. More than 21 000 spp.
Class Amphibia: frogs, newts, toads; aquatic larval stage (tadpoles) with gills; adult stage with lungs; body covered by moist skin. Approx. 4200 spp.
Order Urodela: newts and salamanders
Order Aneura: frogs and toads
Class Reptilia: turtles, crocodiles, snakes; possess lungs; body covered with scales; ectothermic; produce eggs with leathery shells. Approx. 6300 spp.

Order Chelonia: turtles, tortoises, terrapins
Order Squamata: geckos, lizards, goannas, snakes
Order Crocodylia: crocodiles, alligators, gharial (gavial)
Order Rhynchocephalia: one living member only, the tuatara, occurs in New Zealand
Class Aves: birds; body covered with feathers; endothermic; produce eggs with hard calcareous shells. This class has approx. 9000 spp and is divided into 27 orders. It includes:
Order Anseriformes: ducks, geese, swans
Order Pelecaniformes: gannets, boobies, cormorants, pelicans
Order Procellariiformes: albatrosses, petrels
Order Falconiformes: hawks, eagles and falcons
Order Struthioniformes: large flightless birds, such as emus, cassowaries, ostriches, rheas, kiwis
Order Sphenisciformes: penguins
Order Psittaciformes: cockatoos, parrots, lorikeets
Order Coraciiformes: kingfishers, kookaburras
Order Strigiformes: owls
Order Passeriformes: perching songbirds, such as larks, swallows, flycatchers, wagtails; possess foot with three unwebbed front toes and one hind toe

Order Galliformes: mallee fowls, brush turkeys, pheasants, quail

Class Mammalia: mammals, body typically covered with hair or fur; endothermic; young feed on milk from mother. Approx. 4600 spp.

Subclass Prototheria: egg-laying monotremes, such as echidna and platypus

Subclass Marsupialia: marsupial mammals, often pouched; produce tiny undeveloped young that attach to the mother's nipple and undergo further development

Order Dasyuromorphia: carnivorous marsupials, such as quolls, dunnarts, Tasmanian devils, antechinuses and phascogales; usually with four or five pairs of front teeth

Order Peramelemorphia: bandicoots and bilbies

Order Diprotodonta: herbivorous marsupials, such as kangaroos, wombats, possums, marsupial mice; typically with one pair of front teeth

Order Notoryctemorphia: marsupial mole

Subclass Eutheria: placental mammals whose young develop in the mother's uterus. This group is subdivided into 17 orders, including:

Order Chiroptera: bats; flying mammals with forelimbs modified as wings

Order Sirenia: aquatic mammals such as dugongs, manatees

Order Edentata: armadillos, sloths

Order Perissodactyla: odd-toed, hoofed mammals, such as horses, zebras, donkeys

Order Rodentia: rats, mice, squirrels

Order Lagomorpha: hares, rabbits

Order Artiodactyla: hoofed mammals with even number of 'toes', such as rhinoceros, pigs, camels, deer, goats, buffalo

Order Carnivora: carnivorous hunters, such as cats, wolves, dogs, bears, weasels, seals, sea lions

Order Proboscidea: elephants, extinct mammoths

Order Cetacea: aquatic mammals, such as whales, dolphins, porpoises

Order Cetacea: whales, dolphins, porpoises

Order Insectivora: moles, hedgehogs, shrews

Order Pholidota: pangolins

Order Primates: monkeys, orang-outans, chimpanzees, gorillas, human beings

Order Tubulidentata: aardvarks

KINGDOM PLANTAE

Multicellular organisms made of eukaryotic cells with cell walls, photosynthetic, mainly terrestrial; sexual reproduction.

Division BRYOPHYTA: Small multicellular plants without vascular tissue; lack true roots, leaves and stems; gamete-producing stage dominant.

Class Hepaticae: liverworts. Approx. 6000 spp.

Class Musci: mosses. Approx. 9500 spp.

Division PTERIDOPHYTA: Plants with vascular tissue specialised for transport of water and organic material; possess true roots; spore-producing stage dominant.

Class Lycopsida: club mosses, small plants with roots, stems and leaves. Approx. 1000 spp.

Class Sphenopsida: horsetails. Approx. 15 spp.

Class Filicopsida: ferns, such as soft tree fern *Dicksonia antarctica*, rough tree fern, *Cyathea australis*, bracken *Pteridium esculentum*, filmy ferns *Hymenophyllum*, and floating ferns *Azolla*. Approx. 12 000 spp.

Division SPERMATOPHYTA: seed-producing plants

Class Cycadopsida: cycads, small tropical shrubs

Class Ginkgopsida: plant with separate sexes with one plant producing pollen on cones and a second producing seeds; the only living member of this class is *Ginkgo biloba*, a deciduous tree with fan-shaped leaves, which is native to China

Class Coniferopsida (Gymnospermae): cone-producing plants, such as pines, cypresses, cedars, firs, sequoias; needle-like leaves; produce cones with seeds; mainly wind-pollinated. Australian members of this group include the bunya pine and hoop pine (*Araucaria* spp.). About 550 spp.

Class Magnoliopsida (Angiospermae): flowering plants with seeds developing inside ovaries; pollination by animals, such as insects, birds and bats, or by wind. Approx. 240 000 spp. in about 470 families.

Subclass Liliidae (monocotyledons): embryo with a single seed leaf (cotyledon); leaves mainly with parallel veins; flower parts in threes or multiples, vascular bundles are scattered in the stem

Family Poaceae: grasses, such as *Poa* and *Triodia*, cereals, such as wheat, corn and rice

Family Orchidaceae: orchids

Family Xanthorrhoeaceae: grass trees *Xanthorrhea* and kangaroo paw *Anigozanthos*

Family Juncaceae: sedges and rushes

Family Liliaceae: lilies

Family Iridaceae: irises

Subclass Magnoliidae (dicotyledons): embryo with two cotyledons; leaves mainly with net-like patterns of veins; flower parts in fours or fives or multiples; vascular bundles arranged in rings in the stem

Family Proteaceae: includes *Banksia, Hakea, Grevillea*

Family Myrtaceae: includes eucalypts *Eucalyptus*, tea trees *Leptospermum*, bottle brushes *Callistemon*

Family Loranthaceae: mistletoes

Family Mimosacea: wattles Acacia

Family Casuarinaceae: she-oaks *Casuarina*

Family Fagaceae: southern beeches *Nothofagus*

Family Rosaceae: roses, peaches, apples, almonds, strawberries

Family Brassicaceae: cabbages, broccoli, cauliflower, radishes

Family Fabaceae: peas, beans

Family Rutaceae: includes *Boronia*, *Eriostemon*

Family Epacridaceae: southern heaths, such as *Epacris*

Family Chenopodiaceae: salt bushes

Family Asteraceae: daisies

Family Cactaceae: cacti

GLOSSARY

A

abiotic: non-living (p. 270)

absorption: movement of dissolved substances such as digested food and oxygen across a plasma membrane into the interior of a cell (p. 103)

abundance: number of individuals of a given species per unit area; also termed abundance (p. 475)

acacia shrublands: major vegetation type dominated by mulga, a species of *Acacia*, occurring in arid inland Australia (p. 284)

acceleration phase: stage of population growth denoted by that part of a J-curve (or an S-curve) that starts at the upward-turning section of the curve and continues sharply upwards; stage of growth of a population when its size is sufficient to result in much higher population growth in each generation as compared with the lag phase (p. 488)

accessory pigments: pigments, other than chlorophyll, that trap light energy and transfer it to chlorophyll (p. 67)

active site: that part of an enzyme molecule that binds with a substrate (p. 65)

active transport: movement of dissolved substance across a plasma membrane in an energy-requiring process that results in a net movement of that substance against a concentration gradient from a region of lower concentration to a region of higher concentration (p. 28)

adaptations: features that appear to equip an organism for survival in a particular habitat (pp. 283, 317)

adenine: one of the bases (A) found in the nucleotides that are the building blocks of DNA (and RNA) (p. 59)

adenosine triphosphate (ATP): compound that is the common source of chemical energy for cells and whose structure comprises one adenosine molecule and three phosphate molecules (p. 29)

aerobic cellular respiration: (also called aerobic respiration) the breakdown of glucose to simple inorganic compounds in the presence of oxygen and with release of energy that is transferred to ATP (p. 70)

aerobic respiration: (also referred to as aerobic cellular respiration) the breakdown of glucose to simple inorganic compounds in the presence of oxygen and with release of energy that is transferred to ATP (p. 70)

affector neurons: also called sensory neurons; type of nerve cell that transmits information about changes in external or internal conditions to the central nervous system (p. 302)

aggression: any action designed for self defence, to establish position in a hierarchy, or for defence of territory (p. 355)

aldosterone: hormone produced by the adrenal cortex; acts on kidney tubules to stimulate sodium ion reabsorption (p. 326)

alien species: a species that does not occur naturally in a region, but which has been introduced into the region by an external agent; also known as an exotic species (p. 510)

alimentary canal: muscular tube through which food passes; in humans it extends from the mouth to the anus and includes the pharynx, oesophagus, stomach, duodenum, and small and large intestines; also known as the digestive tract or gut (p. 108)

allelochemicals: substances produced by plants that prevent or limit damage by herbivores (p. 421)

allelopathy: a form of competition involving the release of a chemical by one plant species that inhibits the growth of other species (p. 421)

alternation of generations: type of life cycle occurring in all plants (and a few animals only) in which an asexually reproducing haploid generation alternates with a sexually reproducing diploid generation (p. 189)

altricial: birds that are blind and nearly naked when hatched and are dependent on their parents for warmth, food and protection for several weeks (p. 377)

alveolar duct: entry to an alveolus, one of the microscopic air sacs of lungs (p. 145)

alveoli: (singular = alveolus) microscopic air sacs whose moist surfaces are the sites of gas exchange in the lung (p. 145)

amino acids: nitrogen-containing compounds that are the building blocks of proteins (p. 57)

ammonia: highly soluble form of nitrogen-containing waste produced by some aquatic organisms belonging to class Cephalopoda (p. 150)

amnion: in mammals, the fluid-filled sac enclosing the embryo (later a fetus) (p. 384)

amniote eggs: eggs produced by reptiles and birds with features including an outer shell, a series of internal membranes that enabled these eggs to be laid and for embryonic development to occur in terrestrial environments (p. 384)

amylase: group of enzymes that break down the polysaccharides starch and glycogen into smaller units (p. 109)

anabolism: energy-requiring reactions that result in the production of complex substances from more simple ones (p. 53)

anaerobic respiration: respiration that occurs without the involvement of oxygen; the end products of anaerobic respiration in human muscle are lactic and carbon dioxide (p. 70)

anaphase: stage of mitosis during which single-stranded chromosomes move to opposite poles of the spindle fibre within a cell (p. 83)

angioplasty: technique in which a fine catheter is inserted into arteries to remove a blockage caused by a build-up of fatty deposits (p. 139)

annuals: plants that germinate, mature and produce seed and then die in one growing season (p. 287)

antidiuretic hormone (ADH): hormone produced by neurosecretory cells in the hypothalamus; increases reabsorption of water into the blood from distal tubules and collecting ducts of nephrons in the kidney (p. 325)

antifreeze: compound that when added to a fluid lowers the freezing point of the fluid (p. 320)

anus: posterior or terminal opening of the alimentary canal through which faeces are expelled from the body (p. 116)

aorta: major artery that carries oxygenated blood away from the heart to body tissues (p.136)

apoptosis: the natural death of cells, also called programmed cell death (pp. 36, 88)

Archaea: one of the three domains in the Woese system of classification that includes those prokaryotic organisms known as archaeans (p. 246)

archaeans: prokaryotic organisms that are members of the Archaea domain, one of the three domains in the Woese system of classification (p. 24)

asexual reproduction: method of producing offspring that does not involve the fusion of different gametes; for example, binary fission (p. 176)

ATP: see adenosine triphosphate

atrium: chamber of vertebrate heart that receives blood from veins; also called auricle (p. 137)

autosomes: refers to any one of a pair of homologous chromosomes that are identical in appearance in males and females of a species (p. 83)

autotrophic: describes an organism that, given a source of energy, can produce its own food from simple inorganic substances; also known as a producer (pp. 66, 96)

autotrophs: organisms that, given a source of energy, can produce their own food from simple inorganic substances; also known as producers (pp. 95, 415)

axon: extension of a nerve cell along which impulses are transmitted away from that cell (p. 301)

B

bacteria: (singular = bacterium) microscopic, usually unicellular organism, and member of Kingdom Monera (p. 24)

Bacteria: one of the three domains in the Woese system of classification that includes those prokaryotic organisms known as bacteria (p. 246)

behaviour: activities that animals perform in response to internal and external stimuli (p. 342)

behavioural features: activities that an animal performs in response to internal and external stimuli (p. 317)

biennials: plants that live for two growing seasons, germinating and growing in the first and producing seed and dying in the second (p. 287)

bile: secretion produced by the liver and stored in the gall bladder that contains bile salts that emulsify lipids (p. 112)

bile duct: tube that transports bile from the gall bladder (p. 112)

binary fission: process of cell multiplication in bacteria and other unicellular organisms in which there is no formation of spindle fibres and no chromosomal condensation (p. 178)

binomial system of naming: system of naming organisms, introduced by Linnaeus, in which all species are given a two-part scientific name, the first part being the generic name and the second being the trivial (specific) name (p. 225)

bioaccumulation: also known as biological magnification; refers to the increasing accumulation of a non-biodegradable substance in organisms at higher trophic levels in an ecosystem (p. 466)

biodiversity: total variety of life forms, their genes and the ecosystems of which they are a part; biodiversity can relate to planet Earth or to a given region (p. 229)

biogenesis: accepted view that new cells are produced from existing cells (p. 8)

biogeochemical cycles: cycling of a chemical substance in an ecosystem in which the chemical spends some part of the cycle in the biological matter of organisms and some part of the cycle in geological matter such as rocks (p. 461)

bioinformatics: use of computer technology in handling large bodies of biological data (p. 234)

biological clock: internal timing mechanism in plants and animals (p. 365)

biological control or biocontrol: use of one kind of organism that is a predator or parasite of a pest species in order to reduce or eliminate populations of the pest (p. 512)

biological magnification: progressive accumulation of a non-biodegradable substance in the tissues of consumers at higher trophic levels in a food chain (p. 466)

biomass: total mass of organic matter present in all living things in a given space within an ecosystem; biomass will vary depending on environmental conditions prevailing in the ecosystem (p. 453)

biopesticides: one form of biocontrol that uses naturally occurring agents, such as bacteria, fungi and nematode worms, to attack an exotic pest species (p. 515)

biosphere: life-support system of Earth that consists of the atmosphere, upper levels of the lithosphere, the hydrosphere, the biota and the interactions between them, and for which the sun is the primary energy source (p. 537)

biota: the sum total of all living things on Earth (p. 537)

biotic: living (p. 270)

bird-pollinated: refers to flowering plants for which the transfer of pollen to other flowers of the same species occurs through the agency of birds (p. 398)

birth rate: number of organisms born over a given period in a population (p. 486)

bladder: muscular organ that stores urine (p. 151)

blood circulatory system: mechanism that delivers nutrients and oxygen to all cells of a multicellular organism (p. 131)

bolus: rounded mass of food, formed during chewing and ready for swallowing (p. 109)

bonds: in biology, forces of attraction between atoms that bring them together to form molecules (p. 54)

bower: structure constructed by native Australian bowerbirds that forms an essential part of the display of male bowerbirds in attracting females (p. 379)

Bowman's capsule: part of a nephron surrounding a glomerulus in the kidney (p. 153)

breeding season: the period when mature members of an animal population have sperm and eggs ready for release and fertilisation (p. 393)

broadcast spawning: release of eggs and sperm into the external environment by groups of fish (p. 374)

bronchi: (singular = bronchus) airways in the lung, branching from the trachea (p. 144)

bushfires: uncontrolled burning of forest or woodland (p. 505)

C

camouflage: structural method of protection against predators in which the appearance of an animal (p. 424)

carbohydrates: class of organic molecules that includes sugars, starch and cellulose (p. 56)

carbon cycle: cyclic process of exchange of carbon-containing matter between the various parts of the biosphere (p. 461)

carbon dioxide: inorganic form of carbon present in the atmosphere as a gas; form of carbon taken up by plants and converted to carbohydrate by the process of photosynthesis (p. 145)

carnivores: organisms that kill and eat animals (pp. 108, 417, 445)

carpels: reproductive organs in plants consisting of a single ovary, style and stigma (p. 187)

carrying capacity: stabilised size reached by a given population in a given habitat (p. 489)

caste system: formation of specialised groups to carry out different functions in an animal society (p. 353)

catabolism: energy releasing reactions in an organism that result in the breakdown of complex molecules to more simple ones (p. 53)

cell body: that part of a neuron that contains the nucleus (p. 301)

cell cycle: time taken for a newly formed cell to mature and give rise to two new cells (p. 81)

cell membrane: partially permeable boundary of a cell that separates it from its physical surroundings; also known as plasma membrane (p. 24)

Cell Theory: unifying theory of biology that identifies all living things as being made of cells (p. 8)

cell wall: semi-rigid structure located outside the plasma membrane in cells of plants, algae, fungi and bacteria (p. 31)

cells: basic structural and functional units of all living organisms (p. 3)

cellular respiration: process of converting chemical energy of food into a form usable by cells, typically ATP (pp. 32, 69, 123, 447)

cellulose: complex carbohydrate composed of chains of glucose molecules; the main component of plant cell walls (pp. 23, 56)

central nervous system (CNS): in vertebrates, the brain and spinal cord (p. 301)

centromere: position where the chromatids are held together in a chromosome (p. 79)

cephalopods: group of animals that includes octopi, squid and nautaloids (p. 388)

cervix: in mammals, a ring of muscular tissue at the junction of the uterus and vagina (p. 207)

chemical energy: potential energy stored in substances that becomes available when certain types of chemical reactions occur (pp. 66, 163, 438)

chemosynthetic: refers to autotrophic bacteria that use energy from chemical reactions to synthesise organic matter from inorganic substances (p. 459)

chenopod shrublands: major vegetation type dominated by saltbushes and bluebushes occurring in arid and semi-arid regions of Australia (p. 284)

chitin: carbohydrate material present in the exoskeletons of arthropods and also in the cell walls of some fungi (p. 57)

chlorophylls: green pigment required for photosynthesis that traps the radiant energy of sunlight (pp. 37, 67)

chloroplasts: chlorophyll-containing organelles that occur in the cytosol of cells of specific plant tissues (pp. 36, 67, 96)

chromatid: one of the strands in a double-stranded chromosome (p. 79)

chromatin: stained material in the nucleus of a eukaryotic cell (p. 32)

chromosomes: thread-like structures composed of DNA and protein, and visible in cells during mitosis and meiosis (p. 77)

chyme: soupy mixture of food, saliva and gastric juice found in the stomach (p. 110)

cilia: (singular = cilium) in eukaryote cells, whip-like structures formed by extensions of the plasma membrane involved in synchronised movement (p. 38)

circadian rhythm: refers to a pattern of activity following a 24-hour cycle (p. 365)

cladistics: in reference to classification, the grouping of organisms according to the number of derived characters that they share (p. 243)

cladodes: fine green branches that carry out photosynthesis in plant species where leaves are very reduced (p. 414)

cladogram: diagram, based on cladistic study, that shows the inferred relationship between different groups of organisms (p. 243)

class: taxonomic grouping consisting of members of related orders (p. 240)

classical biocontrol: a method of biocontrol that uses natural predators of a pest species imported from the home country of the pest in order to reduce or eliminate it (p. 514)

classical conditioning: type of learned behaviour in which an animal forms an association between two stimuli, one significant and one insignificant, so that the response normally associated with the significant stimulus occurs when only the insignificant stimulus is presented (p. 358)

classification: artificial process of organising things into groups according to one or more criteria (p. 236)

climax community: final stable community resulting from succession in a particular area (p. 533)

clones: groups of cells, organisms or genes with identical genetic make-up (p. 176)

closed circulatory system: type of circulatory system in which blood remains in vessels throughout its circulation (p. 140)

closed forests: type of forest in which the foliage of the upper storey covers 70 per cent or more of the sky (when viewed from below) (p. 283)

closed populations: populations that are not subject either to immigration into or emigration from the populations (p. 486)

cochlea: fluid-filled coiled tube in the inner ear containing fine hairs that can detect sound vibrations and which leads to the auditory nerve (p. 306)

coenzyme: organic compound that acts with an enzyme to alter the rate of a reaction (p. 65)

cohesion: the tendency of molecules of the same kind to stick together (p. 160)

cohesive: refers to the tendency of molecules of the same kind to stick together (p. 54)

cohort: members of a particular age range in a population, as for example, the 19–25 year cohort (p. 488)

colon: portion of the large intestine leading from the small intestine to the rectum (p. 115)

colonial: refers to a few to many single-celled organisms living as a single colony (p. 374)

commensalism: association between two different species in a community in which one benefits and the second apparently neither gains nor is harmed (p. 431)

communication: transmission of information between organisms (p. 349)

community: biological unit consisting of all the populations living in a specific area at a specific time (pp. 267, 409, 472)

competition: interaction between individuals of the same or different species that use one or more of the same resources in the same ecosystem (pp. 262, 355, 421)

complex carbohydrates: carbohydrates made up of many sugar molecules linked together; examples include starch, cellulose and glycogen (p. 56)

compound light microscopes: light microscopes with at least two sets of lenses (p. 10)

conception: production of a zygote from fertilisation of an egg by a sperm (p. 207)

conduction: transfer of heat between two materials that are in direct contact with each other (p. 310)

connecting neurons: also called interneurons; type of nerve cell, located in the central nervous system, that links sensory and effector neurons (p. 302)

conservation: preservation of living things in their diverse ecological settings (p. 518)

conservation biocontrol: a method of biocontrol that uses insect species that occur naturally in the region where a pest species has become established in order to reduce or eliminate that pest (p. 515)

consumers: organisms that obtain their energy and organic matter by eating or ingesting the organic matter of other organisms; also termed heterotrophs (p. 415)

contraception: any technique that prevents production of a zygote or its implantation into the uterus wall (p. 176)

convection: transfer of heat by air or water currents (p. 310)

cooperate: refers to combined action of a group of organisms to achieve an outcome different from that achieved by an individual (p. 355)

core temperature: temperature of internal cells of the body; in humans, core temperature is around 37°C (p. 300)

corpus luteum: tissue that forms in the follicle in the ovary of mammals after the egg is released and which produces progesterone (p. 207)

countercurrent: situation in which two fluid systems flowing adjacent to each other, but in opposite directions, enables the transfer of heat or compounds from one system to the other by diffusion (p. 148)

countercurrent exchange: anti-parallel arrangement of vascular tissue that enables heat to be transferred from a vessel carrying fluid from the core to an extremity, to a vessel carrying fluid from an extremity to the core (p. 321)

covalent bond: type of bond between an oxygen atom and each of the two hydrogen atoms in a water molecule; electrons are shared between the atoms involved in the bonding (p. 53)

criterion: basis for separating a large group into smaller groups (p. 237)

cross-pollination: process in which pollen from one plant is transferred to the stigma of another plant of the same species (p. 196)

cuticle: waxy layer on the outer side of epidermal cells; waxy outer layer on leaves (p. 331)

cyanobacteria: one group of photosynthetic bacteria (p. 247)

cycads: an ancient group of palm-like land plants that produce toxic compounds (p. 216)

cystic fibrosis: a condition, inherited as an autosomal recessive, in which an affected person produces abnormal mucus secretions (p. 111)

cytokinesis: division of cytoplasm occurring after mitosis (p. 77)

cytosine: one of the pyrimidine bases (C) found in the nucleotides that are the building blocks of DNA (and RNA) (p. 59)

cytoskeleton: network of filaments within a cell (p. 24)

cytosol: fluid contents only of a eukaryotic cell (p. 24)

D

day-neutral plants: plants that flower regardless of the length of day (p. 366)

death rate: number of organisms dying in a given period in a population (p. 486)

deceleration phase: stage of population growth denoted by the final flat section of an S-curve during which the growth rate progressively declines to zero as the carrying capacity of the habitat is approached (p. 489)

decomposers: organisms, such as fungi, that can break down and absorb organic matter of dead organisms or their products (p. 415)

demersal: describes eggs that are laid in rock crevices (in contrast to free floating or pelagic eggs); also refers to spawners, that is, fish that produce eggs and sperm and guard the eggs after fertilisation (p. 375)

dendrites: branched extensions of nerve cell that transmit impulses to that cell (p. 301)

density: the number of individuals of a given species per unit area; also termed abundance (p. 475)

density dependent: refers to factors whose impact on members of a population is dependent on the size of the population (p. 489)

density independent: refers to factors whose impact on members of a population is not affected by the size of the population (p. 489)

deoxyribonucleic acid (DNA): nucleic acid that contains the genetic information and which is organised into chromosomes in eukaryotic cells (pp. 32, 59, 77)

derived characters: modified features that were not present in the common ancestor of a group but evolved later in some members only of the group; features useful in cladistic analysis (p. 243)

desiccation: drying out (pp. 273, 385, 412)

detritivores: organisms that eat particles of organic matter found in soil or water (pp. 417, 445)

detritus: fragments of organic material present in soil and water (p. 417)

diapause: a state of inactivity characterised by low metabolism (p. 263)

diatoms: unicellular eukaryotic organisms, members of the Kingdom Protista, that form part of the phytoplankton that float on the ocean surface (p. 23)

dichotomous key: means of identifying organisms based on a series of questions, each able to be answered by selecting one of two alternatives (p. 222)

diffusion: net movement of a substance from an area of high concentration of the substance to an area of lower concentration by a process that does not require energy (pp. 27, 166)

digestion: chemical breakdown of large organic molecules into smaller units that can pass across plasma membranes (p. 103)

digestive system: alimentary canal and associated organs that secrete chemicals into the canal (p. 109)

dioecious: describes flowering plants in which individual plants have either flowers with functional carpels or flowers with functional stamens, but not both; also refers to some conifer species in which one plant has either female cones or male cones (p. 191)

diploid: refers to organisms or cells having two copies of each specific chromosome, that is, having a paired set of chromosomes (pp. 83, 186)

distribution: pattern of spread of members of a population over space; may be described as random, clumped or uniform (p. 478)

diversity: measure of 'species richness' or the number of different species in a community (p. 409)

DNA: see deoxyribonucleic acid

DNA barcoding: a new and rapid means of differentiating species based on the DNA sequence of a particular gene (p. 235)

DNA sequencing: process of determining the order of bases along one chain of a DNA segment (p. 244)

domains: three major groups — Bacteria, Archaea and Eukarya — each containing several kingdoms (p. 246)

dormancy: condition of inactivity resulting from extreme lowering of metabolic rate in an organism (p. 290)

double fertilisation: process of fertilisation in flowering plants in which one sperm nucleus fertilises an egg to form a zygote and a second sperm nucleus fuses with two nuclei to form the endosperm of the seed (p. 202)

drought: periods of deficiency in rainfall compared with the average (p. 280)

dryland salinity: increased salt levels in soil caused when the water table rises; typically due to excessive tree clearing (p. 520)

duodenum: portion of the small intestine leading from the stomach and into which secretions from the pancreas and the liver are released (p. 111)

E

echolocation: technique in which animals emit sounds and detect the presence of objects by the echoes produced as the sound is reflected (p. 352)

ecological pyramids: diagrammatic representation of aspects of trophic levels in an ecosystem, such as a pyramid of numbers, a pyramid of biomass and a pyramid of energy (p. 453)

ecology: study of communities in their habitats and the interactions between them and their environment (p. 410)

ecosystems: biological units comprising the community living in a discrete region, the non-living surroundings and the interactions occurring within the community and between the community and its surroundings (pp. 409, 441)

ectothermic: organism whose body temperature is governed by external sources of heat (p. 316)

effector neurons: also called motor neurons; type of nerve cell that transmits information from the central nervous system to muscle cells or glands (p. 302)

egg: female gamete produced in an ovary (p. 186)

El Niño: global weather event, occurring every two to seven years, that produces drought conditions in eastern Australia (p. 281)

electron tomography: technique using an electron microscope in which many images of 'slices' of cells at different angles are reconstructed to form a three-dimensional image (p. 82)

embryo: early stage of a developing organism; in humans, includes the first eight weeks of development (p. 207)

emulsification: breakdown of large fat droplets into smaller fat droplets (p. 112)

endocrine glands: also called ductless glands; glands that produce one or more hormones, and secrete them directly into the bloodstream where they circulate to particular targets (p. 307)

endocrine system: system of ductless glands that produce hormones and release them directly into the bloodstream (p. 307)

endocytosis: bulk movement of solids or liquids into a cell by engulfment (p. 29)

endoparasites: organisms that live inside another organism (the host) and obtain food from it, such as a tapeworm (p. 425)

endoplasmic reticulum (ER): cell organelle consisting of a system of membrane-bound channels that transport substances within the cell (p. 34)

endosomes: membrane-bound organelles found in animal cells; when material enters a cell by endocytosis, endosomes pass on the newly ingested material to lysosomes for digestion (p. 37)

endosperm: food store in seed plant ovules for the developing embryo (p. 202)

endothermic: organism whose body heat is generated from internal metabolic sources (p. 316)

energy: any non-material agent that can cause matter to change, such as chemical energy and mechanical (kinetic) energy (p. 441)

energy flow: flow of chemical energy through an ecosystem; can be shown by food chains and food webs (p. 450)

environmental resistance: collection of factors that act as natural checks to limit growth of a population (p. 489)

enzymes: typically, specific proteins that act as catalysts to increase the rate of a particular chemical reaction in living organisms (pp. 58, 109)

ephemeral: refers to plant species that germinate, grow and produce seed within a short period of weeks (p. 280)

epicormic buds: dormant buds present under the bark of some trees, for example, *Eucalyptus* species, which begin to grow when the leafy crown of the tree is destroyed (p. 529)

epicormic shoots: growth occurring from dormant buds under the bark after crown foliage is destroyed (p. 86)

epididymis: coiled tube leading from the testis where sperm are stored (p. 205)

ergastic substances: the excretory products of plants which include stored materials such as starch grains and some insoluble compounds, some of which are waste products (p. 168)

erythrocytes: red blood cells (p. 132)

estuary: wide area at the mouth of a river subject to tidal movements (p. 504)

ethology: study of animal behaviour (p. 342)

eucalypt woodlands: major vegetation type dominated by widely dispersed eucalypt trees (p. 284)

Eukarya: one of the three domains in the Woese system of classification that includes all eukaryotic organisms (p. 246)

eukaryotes: refers to cells or organisms with a membrane-bound nucleus (p. 24)

eukaryotic cells: describes cells that have a membrane-bound nucleus (p. 24)

eutrophication: accumulation of dissolved mineral nutrients in a body of water (p. 524)

evaporation: loss of water in vapour form from the surface of a body of water (p. 310)

evaporative cooling: cooling of the body achieved when heat energy is used to evaporate liquid sweat from the skin and heat energy is thus lost from the body (p. 313)

excretion: elimination of waste products produced within and by the cell(s) of an organism (p. 150)

exocytosis: movement of material out of cells via vesicles in the cytoplasm (p. 30)

exoparasites: organisms, such as fleas, that live on the exterior of another organism (the host) and obtain food from it (p. 425)

exotic species: a species that does not occur naturally in a region, but which has been introduced into the region, either deliberately or accidentally; also known as a non-native species (p. 510)

exponential growth: in population biology, a type of growth in which the rate of growth increases as the population size increases (pp. 178, 487)

external environment: environment outside an organism (p. 299)

external fertilisation: type of fertilisation in which eggs and sperm fuse in the external environment, outside the body of either parent (p. 191)

extreme environments: describes environmental conditions in so-called hostile habitats (from a human perspective) (p. 259)

extremophiles: microbes that live in extreme environmental conditions, such as high temperature and low pH (p. 248)

eyepiece (ocular) lenses: in a microscope, the eyepiece lens is closest to the eye of the observer; it receives an inverted image from the objective lenses and enlarges it further to be visible to the observer (p. 10)

F

facilitated diffusion: form of diffusion involving a specific carrier molecule for the substance that diffuses (p. 28)

faeces: undigested material eliminated from the gastrointestinal tract (p. 116)

family: taxonomic grouping consisting of members of related genera (p. 240)

fatty acids: sub-units of the structure of fats, oils and waxes (p. 59)

fecundity: refers to the number of eggs produced by an animal species in one breeding season; fecundity is said to be high when the number of eggs is high (p. 381)

fertilisation: union of egg and sperm to form a zygote (p. 191)

field guides: identification tools containing verbal descriptions and/or images of organisms found in a particular region (p. 220)

filter feeders: aquatic organisms that obtain their food by filtering small particles of organic matter or small organisms from water that passes across specialised structures that can trap this food (p. 273)

flaccid: state of limpness of a cell because of water loss (p. 164)

flagellum: (plural = flagella) whip-like cell organelle involved in movement; structure of a bacterial flagellum differs from that occurring in eukaryotes (p. 38)

fluorescence microscope: type of microscope that uses ultraviolet light to reveal particular compounds that have been stained with fluorescent dyes (p. 11)

follicle: structure in an ovary where an egg develops (p. 207)

follicle stimulating hormone (FSH): hormone, produced by the pituitary gland, that stimulates the growth of ovarian follicles and the ripening of an egg (p. 207)

food chains: one kind of representation to show chemical energy flow within an ecosystem beginning with producers (p. 450)

food webs: one kind of representation of energy flow in an ecosystem; a food web comprises interrelated food chains (p. 450)

foregut fermenters: herbivorous mammals in which bacterial digestion of plant material occurs in a greatly enlarged stomach or oesophagus (p. 119)

fossils: evidence or remains of an organism that lived long ago (p. 229)

free water: water available for an animal to use, including to drink (p. 298)

freeze fracture: technique in which tissue is rapidly frozen in liquid nitrogen then fractured for further treatment before microscopic examination (p. 17)

fruit: mature structure formed from the ovary of a plant and which commonly encloses the seeds (p. 202)

fungi: group of heterotrophic organisms (such as mushrooms, moulds and yeasts) that play a role in decomposition of organic matter; fungi release to the exterior enzymes that break down organic matter, then absorb the products of this digestion (fungi differ from animals that engulf or digest organic matter into their bodies prior to digestion) (p. 24)

G

gall bladder: organ in which bile is stored (p. 111)

gametes: eggs or sperm cells (p. 177)

gametophyte: haploid stage of a plant life cycle that produces gametes (p. 189)

gaseous exchange: the movement of oxygen into the blood and the exit of carbon dioxide from the blood (p. 143)

gastric juice: fluid secreted by various cell types in the stomach wall; contains enzymes, water, mucus and hydrochloric acid (p. 110)

generic name: name of the genus to which an organism belongs; first part of a scientific name (p. 225)

genus: (plural = genera) taxonomic group consisting of members of related species (p. 240)

geotropism: response of a shoot or root to gravity (p. 364)

gills: out-folded structures for gas exchange present in many aquatic animals including fish, sharks and lampreys (p. 147)

global positioning system (GPS) tracking: method of tracking animals in their habitats through use of a tag fitted to an animal that receives a signal transmitted from a special satellite (p. 265)

glomerulus: cluster of capillaries inside the Bowman's capsule of a nephron (p. 153)

glucose: a six-carbon sugar; a common monosaccharide (pp. 56, 299)

glycerol: molecule to which three fatty acids are linked to form a lipid (triglyceride) (p. 59)

glycogen: polysaccharide that is the storage carbohydrate in liver and muscle tissue (pp. 56, 142)

Golgi complex: organelle that packages material into vesicles for export from a cell (also known as Golgi apparatus or Golgi body) (p. 34)

gram-positive bacteria: group of microbes within the Bacteria domain that includes members responsible for many human and animal infectious diseases (p. 248)

grana: (singular = granum) stacks of membranes on which chlorophyll is located in chloroplasts (pp. 37, 67)

gross primary production (GPP): chemical energy produced in the form of organic matter by producer organisms over a given period of time (p. 455)

guanine: one of the purine bases (G) present in the nucleotides that are the building blocks of DNA (and RNA) (p. 59)

guano: consolidated excreta of fish-eating birds (p. 465)

guard cells: specialised cells that control the opening of pores (stomata), typically present in the epidermis of leaves (p. 164)

guilds: groups of different species that use similar food resources in a similar way in an ecosystem, such as seed-eating birds (p. 418)

H

habitat: part of an ecosystem in which an organism lives, feeds and reproduces (p. 258)

habituation: type of learned behaviour in which the response to a repeated stimulus gradually decreases (p. 358)

halophilic: describes those organisms that thrive in habitats with high salt concentration (p. 248)

haploid: having one copy of each specific chromosome, that is, having an unpaired set of chromosomes (p. 186)

harem polygyny: situation in which one dominant male in a group mates with each mature female in that group during one breeding season (p. 378)

haustoria: (singular = haustorium) thin strand of tissue through which a plant parasite makes connection with its host (p. 429)

heart: muscular organ that pumps blood (p. 136)

heliotropism: also called solar tracking; ability of some plants to move so that they remain either perpendicular or parallel to the sun's rays throughout the day (p. 365)

hemi-parasitism: form of parasitism in which a plant parasite obtains some nutrients and water from its host plant but also makes some of its own food through photosynthesis (p. 427)

herbarium: (plural = herbaria) institution where scientific collections of plant specimens are held (p. 222)

herbivore–plant relationship: one of many relationships existing between different species in a community, this one being the interaction between plants and the animals that eat them (p. 425)

herbivores: organism that eats living plants or parts of them (pp. 108, 417, 445)

hermaphrodites: organisms with functional male and female reproductive organs (p. 189)

heterotrophic: describes an organism that cannot make its own food and must ingest or absorb organic material from its environment (pp. 66, 96)

heterotrophs: organisms that ingest or absorb food in the form of organic material from their environment; also known as consumers (pp. 95, 415)

hindgut fermenters: herbivorous mammals in which bacterial digestion of plant material occurs in modified regions of the colon or the caecum (p. 119)

holo-parasitism: form of parasitism in which a plant parasite depends completely on its host for nutrients and water (p. 427)

homeostasis: condition of a relatively stable internal environment maintained within narrow limits (p. 300)

homeothermic: refers to an organism that is able to maintain an internal body temperature within a narrow range (p. 316)

homologous: refers to members of a matching pair of chromosomes (p. 83)

hormones: in animals, chemicals produced in endocrine glands, that are released into and transported via the bloodstream to other parts of the body where they act (p. 307)

host: organism on or in which a specific parasite lives (p. 425)

humidity: measure of the amount of water vapour in the atmosphere (p. 298)

hummock grassland: major vegetation type dominated by spinifex grasses and occurring over one-quarter of Australia (p. 283)

hydrogen bonds: weak non-covalent bonds that form between complementary nucleotides in different DNA strands; hydrogen bonds are responsible for stabilising the structure of the DNA double helix (p. 54)

hydrophilic: refers to substances that dissolve easily in water; also called polar (pp. 27, 54)

hydrophobic: refers to substances that tend to be insoluble in water; also called non-polar (p. 54)

hydrosphere: waters of Earth in both liquid and solid form (p. 537)

hydrothermal vents: regions at the ocean depths where mineral-laden superheated water escapes from the Earth's crust, typically through a 'chimney' composed of precipitated mineral deposits (p. 407)

hypertonic: refers to a solution having a higher concentration of dissolved substances than the solution to which it is compared (pp. 28, 154)

hypothalamus: tiny region of brain below the thalamus that controls various essential functions, including those associated with the autonomic nervous system (p. 309)

hypotonic: refers to a solution having a lower concentration of dissolved substances than the solution to which it is compared (pp. 28, 154)

I

ileum: part of the small intestine leading from the jejunum to the large intestine (p. 111)

immuno-contraception: experimental procedure for blocking contraception in an exotic pest species by causing females to produce an immune response against her own eggs (p. 515)

imprinting: form of rapid and irreversible learning occurring during the early stage of an animal's life (p. 360)

induced ovulation: ovulation occurring as a result of mating; induced ovulation contrasts with spontaneous ovulation (p. 394)

innate or inborn behaviour: behaviour that is essentially the same in all members of a species and which can occur without an individual having had prior experience of the behaviour; previously known as instinctive (p. 342)

insect-pollinated: refers to flowering plants for which the transfer of pollen to other flowers of the same species occurs through the agency of insects (p. 396)

insight learning: type of learned behaviour in which an animal applies previous experience to the solution of a new problem (p. 359)

insulating layers: typically thick layers of fat, hair or feathers of an animal that assist the animal to retain heat and withstand extremely low temperatures (p. 320)

inter-specific competition: competition for resources in an ecosystem involving members of different species (p. 421)

internal environment: the fluid surrounding living cells within a multicellular organism (p. 300)

internal fertilisation: union of sperm and egg occurring inside the body of the female parent (p. 191)

interphase: in the mitotic cell cycle, period of cell growth and DNA synthesis (p. 81)

intra-specific competition: competition for resources in an ecosystem involving members of the same species (p. 421)

invasive: describes an exotic species that has become a pest in a country to which it has been introduced (p. 510)

invertebrate: any animal that lacks a backbone (pp. 176, 374)

irrigation salinity: increased salt levels in soil resulting from excessive irrigation (p. 520)

isotonic: refers to a solution having the same concentration of dissolved substances as the solution to which it is compared (pp. 28, 155)

J

J-shaped curve: describes the plot of population size over time under conditions of exponential growth (p. 488)

jejunum: short section of the small intestine leading from the duodenum to the ileum (p. 111)

K

K-selected: refers to a species that, in comparison with an r-selected species, breeds less frequently and has fewer offspring in which great parental care is invested (p. 492)

K-selection: a situation in which a species, in comparison with an r-selected species, breeds less frequently and has fewer offspring in which great parental care is invested (p. 381)

keys: means for identification of organisms based on a series of questions (p. 222)

kidneys: excretory organs that filter wastes, mainly nitrogenous, from the blood and form urine (pp. 151, 325)

kinetic energy: the energy of motion, including heat energy and energy of moving objects (p. 442)

L

La Niña: global weather event, occurring every two to seven years, that produces conditions of higher than average rainfall in eastern Australia (p. 281)

lacteals: lymph vessels in the villi of the small intestine (p. 113)

lag phase: stage of population growth, denoted by the initial flat section of both a J-curve and an S-curve, that covers the period when the population size is small and growth rates are low (p. 488)

lamellae: thin plate-like structures that are the gas-exchange surfaces of gill filaments in fish (p. 147)

large intestine: part of the digestive system between the small intestine and anus (p. 115)

larynx: chamber containing vocal cords located between the pharynx and the trachea; also known as the voice box (p. 143)

leaching: gradual loss of mineral nutrients from soil when they are dissolved and washed away (p. 281)

learned behaviours: behaviours that develop or change as a result of experience (p. 343)

lek: area in which a number of males gather for the purpose of displaying to attract mates (p. 379)

lekking: the gathering of males in one area for the purpose of displaying to attract mates (p. 379)

lenticels: areas of loosely packed parenchyma cells in the outer woody layer of some stems, allowing the diffusion of oxygen to living tissue below (p. 166)

leucocytes: white blood cells (p. 132)

light energy: radiant energy, for example, sunlight (pp. 66, 163)

light microscope: instrument for viewing cells that uses visible light to illuminate and pass through a specimen and magnify the object by way of one or more glass lenses (p. 10)

lignotubers: woody underground stems of some eucalypts that contain dormant buds from which new growth can occur after a fire (p. 530)

limiting factor: environmental condition that restricts the types of organism that can survive in a given habitat (pp. 277, 457)

lipids: general term for fats, oils and waxes (p. 56)

lipophilic: refers to a substance that will dissolve in or mix uniformly with lipids (p. 27)

lithosphere: rocky crust of Earth, many kilometres thick, which is covered by sediments known as soil (p. 537)

littoral (inter-tidal) zone: inter-tidal zone along a coast between high- and low-water marks (p. 411)

liver: in mammals, large organ located in upper abdomen and which has many functions including removal of some toxic materials from the blood, storage of glycogen and production of bile (p. 111)

logistic growth: a model of population growth in which growth eventually slows and the population size stabilises at the carrying capacity of the habitat concerned (p. 489)

long-day plants: plants that flower when day-length increases (night-length shortens) to a particular critical point, generally in late spring or early summer (p. 366)

loop of Henle: section of tubule in a nephron where urine becomes concentrated by removal of water (p. 153)

lungs: organs consisting of infolded sacs that are present in land vertebrates and which function in gas exchange (p. 143)

luteinising hormone: pituitary hormone that causes a follicle to rupture and then become the corpus luteum (p. 205)

lymphatic system: network of lymph vessels and lymph nodes (p. 141)

lysosome: membrane-bound vesicle containing digestive enzymes (p. 29)

M

macroscopic: of a size that is visible to the unaided eye (p. 219)

magnification: enlargement of view of a specimen, generally with instrument (lens or microscope), to facilitate examination of the specimen (p. 10)

Malpighian tubules: excretory organs of insects and spiders; blind ending tubules that empty into the gut (p. 155)

mariculture: commercial growing of organisms under artificial conditions in sea water (p. 197)

mark-recapture: technique used to estimate size of an animal population that entails capturing a sample, marking and then releasing each captured animal and later, recapturing a second sample of the same population (p. 482)

marsupials: type of mammal that gives birth to very undeveloped embryo-like young; generally have a pouch; member of subclass Marsupialia of class Mammalia (p. 391)

mass spawning: refers to the spawning when very large numbers of fish gather in one area and synchronously release their eggs and sperm into the external environment (p. 374)

mating season: breeding season for animals with internal fertilisation (p. 394)

meiosis: process of cell division that results in the production of new cells, each containing half the number of chromosomes of the original cell (p. 177)

menstrual cycle: monthly cycle under hormonal control that occurs in sexually mature human females and involves egg formation and release, and preparation of the uterus for a possible pregnancy (pp. 207, 394)

meristem: plant tissue found in tips of roots and shoots and made of unspecialised cells that can reproduce by mitosis (p. 86)

metabolism: total of all chemical reactions occurring in a living thing (p. 53)

metaphase: stage of mitosis during which chromosomes align around the equator of a spindle (p. 83)

methanogens: microbes within the Archaea domain that gain their energy for living from chemical reactions that produce methane (p. 248)

micro-environments: conditions in a small region of a habitat (p. 275)

microhabitat: small region within a habitat that may have environmental conditions that differ from those prevailing in the larger habitat (p. 259)

micropyle: the opening at the end of the female gametophyte close to the egg cell in a flowering plant (p. 201)

microscopes: instruments used for viewing cells and other small objects to give an increase in both magnification and resolution (p. 4)

microscopic: of a size that can be seen only through a light microscope (p. 219)

microtubules: part of the supporting structure or cytoskeleton of a cell, made of subunits of the protein tubulin (p. 24)

mid-ocean ridges: series of ridges and mounts located on the ocean floor (p. 407)

migration: refers to the predictable movements of organisms over large distances, which may occur once in the lifetime of an organism (e.g. eels) or yearly (e.g. muttonbirds) (p. 344)

migration rate: number of organisms immigrating to or emigrating from a population over a given period; one of the factors affecting population growth (p. 486)

migratory: refers to a population that moves to a new habitat at a predictable time, either on an annual basis or once in a lifetime (p. 262)

mimicry: a situation in which one species has an appearance similar to that of a different but distasteful species where that similarity apparently gives protection against predators (p. 424)

minerals: inorganic substance required as a nutrient (p. 56)

mitochondria: (singular = mitochondrion) in eukaryotic cells, organelles that are the major site of ATP production (p. 32)

mitosis: process involved in the production of new cells genetically identical with the original cell; an essential process in asexual reproduction (pp. 77, 178)

monoecious: refers to flowering plants in which an individual plant has flowers with both functional carpels and functional stamens (p. 190)

monogamy: situation in which animals form a pair bond and one male mates with one female for one or more breeding seasons or for life (p. 377)

monomers: small sub-units of the larger units of cells (polymers); also known as the building blocks of cells and include sugars, fatty acids, amino acids and nucleotides (p. 56)

monosaccharide: any kind of simple sugar molecule, for example, glucose (p. 56)

moults: replacements of skin and other structures during larval stage of some insects (p. 88)

mucus: lubricating substance secreted by mucous glands particularly in the alimentary canal and the bronchii (p. 109)

multiple fission: process of division in which multiple cells are produced from a single starting cell (p. 179)

mutualism: an association between two different species in a community in which both gain some benefit (p. 430)

mycorrhiza: fine threads formed by a fungus that form a large surface area for uptake of nutrients (p. 430)

N

nastic movement: plant movement in which direction of movement is independent of direction of stimulus causing the movement (p. 365)

negative feedback: a regulatory mechanism that maintains a relatively steady value for body variables, such as blood glucose concentration, temperature and water concentration (p. 308)

nephrons: functional units of a kidney (p. 152)

nerve cells: neurons, each generally consisting of a cell body, dendrites and an axon and specialised to initiate, receive and transmit electrical signals (p. 301)

nervous control system: brain, spinal cord and all nerves of the body (p. 301)

net primary production (NPP): amount of chemical energy in organic matter stored by producers and available as 'food' to consumers (p. 455)

neurons: nerve cells (p. 301)

niche overlap: situation in an ecosystem in which different species are in competition for the same energy and space resources; in reality, niche overlap in natural ecosystems is typically zero or minimal (pp. 269, 433)

niche separation: situation in which two species in the community use different resources such that the two species have no degree of niche overlap (p. 433)

niches: ways of life of organisms in an ecosystem; roles of species in a community (p. 268)

nitrogen cycle: cyclic process of exchange of nitrogen-containing matter between various parts of the biosphere (p. 461)

nitrogen-fixing bacteria: bacteria able to convert nitrogen from the atmosphere into ammonium ions (pp. 248, 430)

nitrogenous wastes: nitrogen-containing waste compounds such as urea and uric acid (p. 150)

non-biodegradable: refers to substances that cannot be excreted or broken down by living organisms (p. 466)

non-polar: refers to substances that tend to be insoluble in water; also called hydrophobic (p. 54)

noxious: harmful in some way; for example, exotic organisms that cause damage to Australian ecosystems (p. 514)

nuclear envelope: membrane surrounding the nucleus of a eukaryote cell (p. 24)

nucleic acids: compound, such as DNA or RNA, built from nucleotide sub-units (p. 56)

nucleoli: (singular = nucleolus) structure present in the nucleus and which is a store of ribosomal RNA (rRNA) (p. 32)

nucleus: in eukaryotic cells, membrane-bound organelle containing the genetic material DNA (p. 8)

O

objective lenses: in a microscope, the lower lens or lenses closest to the object that receive light rays then invert and magnify the initial image and bring it into focus (p. 10)

obligate seeder: type of plant that is killed in a bushfire but whose seeds survive and germinate after the fire (p. 530)

oesophagus: muscular tube that transports food from the pharynx at the back of the mouth to the stomach (p. 109)

oestrogen: female sex hormone secreted by follicle cells in the ovaries (p. 207)

oestrous cycle: refers to the cycle of egg production in female mammals and, depending on the species, there may be one or more cycles during a year; in humans, the oestrus cycle is termed the menstrual cycle (p. 394)

oestrus: period of sexual receptivity in female mammals, sometime termed 'being in heat' (p. 394)

oil immersion objective lens: a type of lens that uses oil of the same refractive index as glass, which is placed between the glass slide and the objective lens, reducing the loss of light due to refraction and enabling higher magnifications (p. 10)

omnivores: organisms that eat both plants and animals (pp. 105, 417)

open circulatory system: circulatory system in which blood is contained in vessels for only part of its circuit (p. 140)

open forests: type of forest in which the foliage of the upper storey covers less than 70 per cent of the sky (when viewed from below) (p. 283)

open populations: populations whose numbers are affected by immigration and emigration (p. 486)

operant conditioning: form of trial-and-error learning (p. 358)

operculum: in fish, the flaps covering the gills (p. 147); in molluscs, a hard impermeable lid that closes the shell, making a watertight compartment for the animal inside (p. 290)

order: taxonomic grouping consisting of members of related families (p. 240)

organ: structure consisting of several tissues that is specialised to carry out a particular biological function, for example, liver, kidney (p. 44)

organ system: group of organs that combine to perform a particular function within an organism, for example, the digestive system (p. 45)

organelles: structures present in a eukaryotic cell, for example, mitochondrion, chloroplast (p. 24)

organic compounds: carbon-containing compounds present in living matter (p. 52)

osmoreceptors: specialised nerve cells in the hypothalamus of the brain that detect changes in blood concentration and release hormones in response (p. 326)

osmoregulation: maintenance of constant internal salt and water concentrations in internal fluids (homeostasis) in spite of different concentrations in the external environment (p. 325)

osmosis: net movement of water across a partially permeable membrane without an input of energy and down a concentration gradient (pp. 27, 327)

ovary: egg-producing organ (p. 186)

over-harvesting: commercial harvesting of a species that removes too many breeding individuals from a population such that the rate of removal exceeds the rate of natural replacement (p. 517)

oviparity: mode of reproduction in which embryonic development occurs outside the mother's body but within an egg that provides nutrition from the egg yolk (p. 383)

oviparous: describes animals that lay eggs from which young hatch after laying (p. 383)

ovulation: release of an egg from the ovary (p. 207)

ovules: parts of the ovary of a flower that later become a seed (p. 201)

ovum: female gamete or egg (p. 207)

ozone layer: layer in upper atmosphere containing high concentrations of ozone (p. 537)

P

palynology: the branch of biology concerned with the study and identification of pollen (p. 217)

pancreas: organ that secretes digestive enzymes into the duodenum and hormones into the bloodstream (p. 111)

parasite: organism that lives on or in another organism and feeds from it usually without killing it (p. 425)

parasite–host relationships: form of interaction within a community that involves one species, the parasite, living on or in another species, the host, typically without killing the host (p. 425)

parasitoids: adult females of some wasp and fly species that are like parasites but that slowly kill their hosts (p. [illegible])

parthenocarpy: in the absence of fertilisation, production by a plant of seedless fruit (p. 203)

parthenogenesis: one form of asexual reproduction in which new individuals are produced from unfertilised eggs (p. 181)

partially permeable: describes a boundary that allows only some materials to pass through it; sometimes termed semi-permeable (p. 25)

pecking order: hierarchy within an animal group that reflects different ranks (p. 354)

pelagic: refers to eggs that are free floating in water (p. 374)

pellets: mass of indigestible fur, bones and other material from captured prey that is regurgitated by a bird of prey (p. 219)

penis: part of the reproductive organs in male vertebrates that deposits sperm in the reproductive tract of a female (p. 205)

pepsin: protein-digesting enzyme produced by cells in the stomach (p. 110)

perennials: plants that live for several years and produce seeds in two or more different years (p. 287)

peripheral nervous system (PNS): in vertebrates, all nerve cells that in whole or part lie outside the brain and spinal cord (p. 301)

peristalsis: movement of food along the gut by means of contraction and relaxation of muscle in the gut wall (p. 108)

peroxisomes: small membrane-bound organelles rich in the enzymes that detoxify various toxic materials that enter the bloodstream (p. 37)

persistent: refers to substances that cannot be excreted or broken down by living organisms (p. 466)

phagocytic: describes the bulk movement of solid material into cells (p. 134)

phagocytosis: bulk movement of solid material into cells (p. 29)

pharynx: area of the throat that leads to both the oesophagus and the larynx (p. 143)

phase contrast microscope: type of microscope that enables observation of unstained, intact living cells (p. 11)

pheromones: chemicals secreted by some animals to communicate with other members of the species; chemical messengers (pp. 352, 422)

phloem: vascular tissue in plants that transports sugar (mainly sucrose) and other organic compounds (p. 158)

phospholipids: major type of lipids found in plasma membranes (p. 59)

phosphorus cycle: movement of phosphorus within an ecosystem involving the presence of the element in the organic matter of living organisms and in soil and rocks (p. 464)

photic zone: region of ocean from the surface to the greatest depth to which light can penetrate, typically about 200 metres (p. 273)

photoperiod: refers to the relative lengths of day and night in a 24-hour period (p. 365)

photoperiodism: response of plants to particular periods of light and dark in terms of flowering or germination of seeds (p. 365)

photosynthesis: process by which plants trap the radiant energy of sunlight to produce carbohydrates from carbon dioxide and water (pp. 36, 66, 96, 416)

photosynthetic: refers to autotrophic organisms, typically plants, algae and some bacteria, that use the energy of sunlight to make organic compounds from inorganic substances such as carbon dioxide and water (p. 459)

phototropism: movement of a plant in response to light (p. 362)

phyllodes: leaf-like structures derived from a petiole (stem of a leaf) (p. 334)

phylum: (plural = phyla) taxonomic grouping consisting of members of related classes (p. 240)

physiological features: particular aspect of the function of an organism or any its parts (p. 317)

phytoplankton: tiny photosynthetic organisms that float on or near the surface of a body of water (pp. 38, 443)

pinocytosis: bulk movement of liquids into cells (p. 30)

pioneer species: first species to become established in a previously uncolonised habitat (p. 533)

pistil: female reproductive structure of a flower (p. 187)

pituitary: endocrine gland attached to the hypothalamus, influences the production of thyroxin by the thyroid (p. 312)

placenta: organ that permits exchange of materials between the blood of a mother and her fetus (p. 386)

plasma membrane: partially permeable boundary of a cell separating it from its physical surroundings; boundary controlling entry to and exit of substances from a cell (p. 24)

play behaviour: patterns of behaviour in young animals that appear to provide practice of behaviours that are seen in adults (p. 358)

poikilothermic: refers to an organism having a body temperature that can vary greatly with changes in the external environment (p. 316)

polar: refers to substances that dissolve easily in water; also called hydrophilic (p. 54)

polar body: small cell resulting from unequal cytoplasmic division during meiosis in female mammals (p. 206)

polar nuclei: two of the nuclei in the female gametophyte of a flowering plant that, when fused with one of the sperm nuclei, form the triploid endosperm of a seed (p. 201)

pollen: grains that contain the male reproductive cells of a seed-producing plant (p. 187)

pollination: transfer of pollen from anther to stigma (p. 395)

polyandry: one kind of polygamy in which one female has multiple male partners during a breeding season (p. 378)

polygamy: situation in which one male or one female has multiple partners during a breeding season (p. 378)

polygyny: one kind of polygamy in which one male has multiple female partners during a breeding season (p. 378)

polymers: large molecule made of identical or similar single units, for example, starch which is made of many glucose units (p. 56)

polynomial system: early system of naming organisms using long Latin descriptions; now superseded by the binomial system (p. 225)

polysaccharides: carbohydrate made of many monosaccharide units (p. 56)

pop-off archival tag (PAT): tag fitted to a marine animal that collects data on environmental conditions and, after a given period, detaches and pops to the ocean surface from where it transmits its stored data to a satellite network (p. 266)

population: members of one species living in a specific habitat at a particular time (pp. 409, 472)

population dynamics: study of changes in populations over time (p. 486)

potential energy: stored energy including the chemical energy of compounds (p. 442)

precocial: birds that are covered with down, are able to stand and can fend for themselves and feed when they hatch, although they may not be able to fly (p. 378)

predator: an animal that actively seeks out other animals as its source of food (p. 422)

predator–prey relationship: a form of interaction within a community that involves the eating of one species, the prey, by another species, the predator (p. 422)

prey: living animal that is captured and eaten by a predator (pp. 95, 422)

primary cell wall: the first layer of cellulose and other polysaccharides forming the cell wall outside a newly formed plant cell (p. 31)

primary consumers: herbivores that obtain chemical energy by consuming organic matter made by producers (p. 445)

primary succession: ecological succession occurring in an area that was not previously colonised (p. 533)

primitive characters: features that were present in the common ancestor of a group of organisms (p. 243)

producers: photosynthetic organisms and chemosynthetic bacteria that, given a source of energy, can build organic matter from simple inorganic substances (pp. 163, 415, 443)

productivity: rate of production of organic matter by producers, typically as a result of photosynthesis, in a given area over a given period of time (p. 455)

progesterone: female sex hormone secreted by the corpus luteum that prepares the uterus lining for pregnancy (p. 207)

prokaryotes: any cells or organisms without a membrane-bound nucleus (p. 24)

prokaryotic cells: cells that do not have their genetic material contained within a discrete nucleus and which lack cell organelles such as mitochondria (p. 24)

prophase: stage of mitosis in which the chromosomes contract and become visible, the nuclear membrane begins to disintegrate and the spindle forms (p. 83)

protein filaments: part of the cytoskeleton or internal framework of a cell that supply strength and support for the cell; made of a variety of proteins, depending on the particular cell, and are very tough (p. 24)

protein sequencing: process of identification of the order of the various amino acids in a protein (p. 244)

proteins: large molecules made of many amino acids joined by peptide bonds (pp. 33, 56)

protists: one of the groups of organisms within the five kingdom scheme; also identified in some texts as protoctists (p. 24)

puberty: onset of processes leading to sexual maturity including development of secondary sex characteristics (p. 205)

pulmonary artery: major artery that carries deoxygenated blood from the heart to the lungs (p. 137)

pulmonary veins: major veins that return oxygenated blood from the lungs to the heart (p. 137)

pupa: developmental stage between larva and adult of some insects (p. 88)

pyloric sphincter: a muscular opening through which food is forced from the stomach into the small intestine (p. 111)

pyramid of biomass: diagram that records the total dry organic matter of organisms at each trophic level in a given area of an ecosystem (p. 454)

pyramid of energy: diagram that shows the amount of energy input to each trophic level in a given area of an ecosystem over an extended period (p. 454)

pyramid of numbers: diagram that shows the numbers of organisms at each trophic level per unit area of an ecosystem (p. 453)

Q

quadrats: areas of known size, often one-metre square, used to outline an area for sampling a population of plants or sessile animals (p. 482)

qualitative: refers to a description of an environment in imprecise terms, such as 'warm', 'shady' (p. 274)

quantitative: refers to a description of an environment using numerical values, such as '10°C', '75% cover' (p. 274)

R

r-selected: refers to a species which, in comparison with a K-selected species, breeds more often and produces larger numbers of offspring in which little or no parental care is invested (p. 492)

r-selection: a situation in which a species, in comparison with a K-selected species, breeds more often and produces larger numbers of offspring in which little or no parental care is invested (p. 381)

radiant energy: kind of wave energy which includes X-rays, ultraviolet rays, and most importantly, visible sunlight (p. 438)

radiation: heat energy, in the form of infrared heat rays, lost from the surface of a warm body (p. 310)

radio tracking: general term to describe methods of tracking the movements of animals in their habitats where the animals do not need to be within sight of an observer (p. 265)

range: geographic area enclosing all the habitats where a given species lives; sometimes referred to as distribution (p. 260)

raptors: birds of prey, such as eagles, hawks, owls (p. 219)

rectum: final section of the large intestine that ends at the anus (p. 115)

reference collections: collections of plant or animal species held in herbaria and museums respectively (p. 222)

regeneration: replacement of damaged or lost parts of an organism by asexual reproduction of cells (p. 77)

regulated flow: refers to the flow of water in rivers where that flow is impeded by dams and weirs (p. 522)

remote sensing: data-gathering about an object from a distance, such as with LandSat satellite (pp. 264, 507)

renin: hormone released by the kidney that acts to increase water reabsorption from nephron tubules (p. 325)

reproductive behaviours: behaviour patterns associated with courtship, mating and care of the young (p. 353)

reproductive strategies: contrasting modes of reproduction that commit energy either to producing very large numbers of offspring with little or no parental care (r-selection) or smaller numbers but with greater parental care (K-selection) (p. 376)

resolution: measure of the ability of a microscope to distinguish fine detail in a specimen; higher resolution means finer detail can be seen (p. 11)

resource use: refers to how different species use resources in a habitat (p. 268)

respiratory system: organ system that carries out gas exchange; includes the nose, pharynx, larynx, trachea, bronchi and lungs (p. 143)

rhizoids: fine root-like structures present in some plants such as mosses (p. 87)

rhizomes: horizontal underground stems (p. 529)

rhythmic behaviours: behaviours that are repeated at regular intervals, such as feeding or sleeping behaviours (p. 344)

ribonucleic acid (RNA): type of nucleic acid consisting of a single chain of nucleotide sub-units which contain the sugar, ribose and the bases A, U, C and G; RNA includes messenger RNA (mRNA), transfer RNA (tRNA) and ribosomal RNA (rRNA) (pp. 32, 59)

ribosomes: organelles containing RNA that is major site of protein production in cells (p. 33)

rough endoplasmic reticulum: endoplasmic reticulum with ribosomes attached (p. 34)

ruminants: herbivorous mammals with a stomach comprising several compartments, one of which is known as the rumen where fermentation of food occurs (p. 119)

S

S-shaped curve: describes the plot of population size over time under conditions where resource availability limits population size to the carrying capacity of the habitat concerned (p. 489)

salinity: refers to the amount of salt concentration in soils or water (p. 519)

salt interception: scheme for removal of salt through pumping saline groundwater from below ground and storing it (p. 521)

sampling: technique by which part of a population is examined and, from this sample, estimates are made about the size of the total population (p. 476)

satellite tracking: method of tracking animals in their habitats through use of a tag fitted to an animal that transmits a unique signal to a ground station via a satellite (p. 265)

SA:V ratio: (surface-area-to-volume ratio) a measure that identifies the number of units of surface area available to 'serve' each unit of internal volume of a cell, tissues or organism (pp. 26, 103)

scanning confocal microscope: type of microscope that uses laser light to produce in-focus images of successive layers of a specimen (p. 12)

scanning electron microscope: type of microscope that enables observation of cell and tissue surfaces (p. 14)

scats: animal faeces or droppings (p. 219)

sclerophyll: describes Australian forests that are dominated by trees and other plants that have hard, often small leaves (p. 509)

scrotum: loose pouch of skin that contains the testes of a mammal (p. 204)

secondary cell walls: walls of lignin and cellulose deposited on the primary cell wall of some plant cells after cell growth has ceased (p. 31)

secondary consumers: organisms that obtain chemical energy by eating primary consumers (p. 445)

secondary succession: establishment over time of replacement communities in an area following an event that removes the original community (p. 533)

sedentary: refers to a population whose members feed and breed within a small area (p. 382)

seed: matured ovule; contains a plant embryo and a food supply enclosed within a seed coat (p. 202)

self-incompatibility: condition in which a chemical signal prevents the growth of pollen from one plant that lands on the stigma of one of its flowers so preventing self-pollination (p. 197)

self-pollination: process that occurs when pollen from one flower settles on the stigma of the same flowers and grows through the style to the egg where it can fertilise it (p. 197)

semen: nutrient-rich fluid containing sperm (p. 205)

sequential hermaphrodites: condition in which an individual animal during its lifetime changes over time from producing gametes of one sex then producing gametes of the other sex (p. 190)

serial polygyny: situation in which one male displays and attracts passing females in turn for mating (p. 378)

sex chromosomes: the pair of chromosomes that differ in males and females of a species (p. 83)

sexual reproduction: method of producing offspring in which an egg and a sperm fuse to form a zygote (p. 177)

shadowing: technique in which fractured pieces of specimen within a vacuum chamber are exposed to, and partly covered by, heavy metal that is evaporated; the metal film reflects the contours of the parts that are covered and the parts are then covered with a layer of carbon atoms that is transparent to electrons. The specimen fragments are dissolved away leaving metal replicas that can be examined with an electron microscope. (p. 17)

short-day plants: plant that flowers when day-length shortens (night lengthens) to a particular critical point, generally in late summer, autumn or winter (p. 366)

simple light microscopes: light microscopes with only one lens (p. 10)

simultaneous (synchronous) hermaphrodism: condition in an individual animal in which it is producing both sperm and eggs at the same time (p. 190)

small intestine: part of the alimentary canal that comprises the duodenum, jejunum and ileum, and is located between the stomach and the large intestine (p. 111)

social hierarchy: the different ranks of importance or dominance within an animal group (p. 354)

social interactions: interactions that involve two or more individuals and may involve cooperation, as in mating, or conflict as in defending a territory or competing for a mate (p. 353)

solar tracking: also called heliotropism; ability of some plants to move so that they remain either perpendicular or parallel to the sun's rays throughout the day (p. 365)

spawning: simultaneous release of large numbers of gametes into water by fish or other aquatic organisms that have external fertilisation (pp. 95, 192)

spawning season: period of year during which fish or other aquatic organisms release their eggs and sperm into the external environment (p. 394)

species: taxonomic unit consisting of organisms capable of mating and producing viable and fertile offspring (p. 218)

species richness: the number of different species, that is, different populations, in a community (p. 472)

specific name: second part of the scientific name of a species, such as sapiens (in *Homo sapiens*) (p. 225)

sperm: male gametes produced in the testes (p. 186)

sphincters: rings of smooth muscle able to relax and contract and so permit or restrict flow of material into or out of a cavity (p. 134)

spinal cord: dorsal nerve cord within the vertebral column that together with the brain and associated nerves makes up the central nervous system (p. 301)

spindle: fine protein fibres that form between the poles of a cell during mitosis and to which chromosomes become attached (p. 79)

spiracles: in an insect, openings on the body that lead into a series of tubes (called trachea) that transport oxygen to cells (p. 148)

spontaneous generation: early nineteenth century belief that living things could arise from non-living matter or dead matter (p. 8)

spores: reproductive cells produced by the sporophyte generation of a plant (p. 188)

sporophyte: diploid stage of the life cycle of a plant that produces spores (p. 189)

staining: the addition of a colour to stain compounds in cells that are normally colourless, for identifying them under magnification (p. 11)

stamens: male reproductive parts of a flower consisting of anther and filament (p. 187)

starch: complex carbohydrate (polysaccharide), polymer of glucose units, main storage carbohydrate in many plant cells (p. 56)

stem cells: cells with the capacity to reproduce themselves and then differentiate into either one or different kinds of cells; in bone marrow, a type of cell that reproduces and differentiates into the different kinds of blood cells (p. 132)

stigma: tip of the pistil; part of the female reproductive organ in a flower (p. 196)

stoma: (plural = stomata) an opening, typically on a leaf surface, through which water vapour and carbon dioxide can move (p. 164)

stomach: expanded part of the gut into which the oesophagus opens (p. 110)

stomata: (singular = stoma) openings, typically on a leaf surface, through which water vapour and carbon dioxide can move (pp. 286, 331)

stroma: in chloroplasts, the semi-fluid substance between the grana and which contains enzymes for some of the reactions of photosynthesis (pp. 37, 67)

structural features: particular aspects of the structure of an organism or any of its parts (p. 317)

substrate: compound on which an enzyme acts (p. 65)

succession: natural process by which the community living in an area changes over time (p. 533)

sulfur bacteria: one group of bacteria that gain their energy by oxidising sulfur compounds (p. 408)

sulfur-oxidising bacteria: chemosynthetic microbes within the Bacteria domain that gain their energy for living from chemical reactions involving the oxidation of sulfur (p. 248)

surface-area-to-volume ratio (SA:V ratio): measure that identifies the number of units of surface area available to 'serve' each unit of internal volume of a cell, tissues or organism (pp. 26, 103)

symbiosis: prolonged association between different species in a community in which at least one partner benefits; includes parasitism, mutualism and commensalism (p. 432)

systematics: area of study concerned with identifying relationships between organisms and making inferences about their evolutionary history (p. 229)

T

taxon: (plural = taxa) any taxonomic group, for example, class, family, genus (p. 239)

taxonomists: biologists who specialise in the study of the classification of living things (p. 229)

taxonomy: area of study concerned with the describing, naming and classification of organisms (p. 229)

telemetry: method of monitoring from a distance (p. 265)

telophase: stage of mitosis in which new nuclear membranes form around the separated groups of chromosomes (p. 83)

territorial behaviour: behaviour associated with defending an area against other individuals, usually of the same species (p. 354)

tertiary consumers: organisms that consume the organic matter of secondary consumers (p. 445)

testis: (plural = testes) male gonad that produces both sperm and male sex hormones (p. 186)

testosterone: major sex hormone produced by the testes that causes the development of male secondary sex characteristics (p. 205)

thematic mapper (TM): data-gathering sensors on satellites that detect and record radiation emitted or reflected from the land or sea surface (p. 264)

thermophilic: describes those organisms that thrive in habitats with high temperatures (p. 248)

thigmotropism: a plant response to contact with a solid object (p. 364)

thymine: one of the pyrimidine bases (T) found in the nucleotides that are the building blocks of DNA (p. 59)

tissue: group of many similar cells organised to perform a particular function (p. 42)

tissue fluid: fluid that bathes the surface of cells within a multicellular organism (p. 134)

tolerance range: extent of variation in an environmental factor within which a particular species can survive (pp. 276, 299)

total count: result from counting all members of a species in a given area; also known as a true census (p. 476)

trachea: in terrestrial vertebrates, tube leading from the larynx to the bronchi, and also called the windpipe; in insects, air-filled tubes leading from spiracle openings into the body (p. 111)

tracheoles: in insects, narrow tubes branching from trachea and making direct contact with cells to facilitate diffusion of oxygen (p. 148)

transects: techniques for sampling the species present in a habitat; may be a line transect or a belt transect (also known as a strip transect); with a line transect, various species present at regular intervals along the line crossing the area of study are recorded; with a belt transect, the abundance of one or more species within a band of known width and length is recorded (p. 482)

translocation: transport of organic material, including sugars, through the phloem of a vascular plant (p. 161)

transmission electron microscope: type of microscope that enables observation of very highly magnified images of cell sections; often abbreviated to TEM (p. 14)

transpiration: loss of water from the surfaces of a plant (pp. 286, 322)

transpiration stream: flow of water through a plant from roots to leaves (p. 331)

triglycerides: organic molecules made of one molecule of glycerol combined with three fatty acid molecules (p. 59)

trophic level: refers to the feeding level of organisms in an ecosystem (p. 445)

tropism: a directional growth response of a plant to an environmental stimulus (p. 362)

true census: result from counting all members of a species in a given area; also known as a true count (p. 476)

tubule: minute tubular structure in the kidneys (p. 153)

turgid: state of a cell that is firm because of water uptake (p. 164)

U

unregulated flow: describes the flow of water in rivers with no artificial structures present, such as dams or weirs (p. 522)

upwelling: large-scale movement of cold nutrient-rich ocean water from the ocean depths to the surface (p. 456)

urea: nitrogen-containing waste product typically produced by mammals (p. 150)

ureters: tubes that transport urine from kidney to bladder (p. 151)

urethra: tube that transports urine from the bladder to the exterior and, in males, also transports sperm (pp. 151, 205)

uric acid: a nitrogen-containing waste product typically produced by birds (p. 151)

urinary system: system primarily responsible for water, salts and acid-base balance and removal of nitrogenous wastes; includes kidneys, ureters, bladder and urethra (p. 151)

urine: liquid waste filtered from the blood by the kidneys (p. 151)

uterus: muscular organ typical of female mammals in which the embryonic and fetal development occurs (p. 207)

V

vacuoles: structures within plant cells that are filled with fluid containing materials in solution, including plant pigments (p. 37)

vagina: in female mammals, muscular tube, leading from the uterus to the exterior; also termed the birth canal (p. 207)

vas deferens: tube leading from the testes to the urethra through which sperm move (p. 205)

vascular plants: plants that have xylem and phloem tissue (p. 216)

vascular tissue: plant tissue specialised for transport of nutrients, water and minerals, and which provides a plant with support (p. 158)

vasopressin: the antidiuretic hormone (ADH) produced by neurosecretory cells in the hypothalamus; increases reabsorption of water into the blood from distal tubules and collecting ducts of nephrons in the kidney (p. 325)

vectors: refers to insects or other animals that carry pollen from one flower to another flower of the same species; can also refer to a carrier of a disease from one host to another (p. 396)

vegetative reproducers: refers to plants that, after bushfire, regenerate through structures such as underground stems and epicormic buds, such as seen in many eucalypts (p. 529)

vegetative reproduction: asexual reproduction in plants (p. 181)

veins: relatively thin-walled blood vessels that transport blood back to the heart from tissues; vascular tissue in a leaf (p. 135)

ventricle: heart chamber that pumps blood out of the heart (p. 137)

venules: small blood vessels joining capillaries to a vein (p. 135)

vertebrate: animal with a backbone (p. 374)

very high frequency (VHF) radio tracking: method of tracking animals in their habitats through use of a radio transmitter fitted to the animal and a signal detector carried by the person monitoring the animal's movement (p. 265)

vesicle: membrane-bound sac found within a cell, typically fluid-filled; for example, lysosome (p. 29)

villi: (singular = villus) small projections of tissue that function to increase surface area (p. 113)

vitamins: in mammalian nutrition, refers to an organic compound required in minute amounts (p. 56)

viviparity: mode of reproduction in which embryonic development occurs within the mother's body, with young being born as miniature adults (p. 383)

viviparous: describes animals that give birth to live young (in contrast to laying eggs) (p. 383)

volatile: refers to a chemical substance that vaporises at ambient temperatures and diffuses through the air (p. 422)

W

warning colouration: bright pattern of colour markings on a species that apparently signals to predators that the members of the species contains distasteful chemicals (p. 425)

water cycle: cyclic process by which water moves between the parts of the biosphere (p. 465)

water tappers: refers to trees that have a single main root extending to depths near the water table before forming lateral branches (p. 286)

wildfire: destructive and uncontrollable bushfire occurring after a long fire-free period and which burns at high intensity (p. 529)

wind-pollinated: refers to flowering plants for which the transfer of pollen to other flowers of the same species occurs through the agency of wind (p. 396)

X

xerophytes: plants with characteristics that enable resistance to drought and life in very dry areas (p. 332)

xylem: vascular tissue that transports water and minerals throughout a plant and which provides a plant with support (p. 158)

Z

zero population growth: refers to a stable population with no net growth in which increases from births and immigration are matched by decreases from deaths and emigration (p. 486)

zooplankton: small free-floating organisms living in ocean that include the larvae of fish and other aquatic species (p. 375)

zygote: fertilised egg that results from the fusion of haploid gametes (p. 191)

INDEX

C

F

O

P

Q

R

S